GLOBAL CHANGE AND MOUNTAIN REGIONS

mountain research initiative

Affiliated Institutions

GLOBAL IGBP CHANGE

IHDP

GTOS

UNESCO

MaB

sc | nat

Swiss Academy of Sciences
Akademie der Naturwissenschaften
Accademia di scienze naturali
Académie des sciences naturelles

ADVANCES IN GLOBAL CHANGE RESEARCH

VOLUME 23

The titles published in this series are listed at the end of this volume.

GLOBAL CHANGE AND MOUNTAIN REGIONS
An Overview of Current Knowledge

Edited by

Uli M. Huber

University of Bern, Switzerland

Harald K. M. Bugmann

Swiss Federal Institute of Technology, Zurich, Switzerland

and

Mel A. Reasoner

The Mountain Research Initiative, Bern, Switzerland

Springer

A C.I.P. Catalogue record for this book is available from the Library of Congress.

ISBN-10 1-4020-3507-1 (PB)
ISBN-13 978-1-4020-3507-4 (PB)
ISBN-10 1-4020-3506-3 (HB)
ISBN-13 978-1-4020-3506-7 (HB)
ISBN-10 1-4020-3508-X (e-book)
ISBN-13 978-1-4020-3508-1 (e-book)

Published by Springer,
P.O. Box 17, 3300 AA Dordrecht, The Netherlands.

www.springeronline.com

Printed on acid-free paper

Printed in the Netherlands.

TABLE OF CONTENTS

INTRODUCTION

PART I: PALEOENVIRONMENTAL CHANGES

PART II: CRYOSPHERIC CHANGES

PART IV: ECOLOGICAL CHANGES

Introduction: The International Year of Mountains Challenge and Opportunity for Mountain Research

Thomas Hofer
Sustainable Mountain Development, UN Food and Agriculture Organization, Viale delle Terme di Caracalla, 00100 Rome, Italy
phone +39 06 5705-3191, fax +39 06 5705-5137, e-mail Thomas.Hofer@fao.org

1. Mountain ecosystems: Vulnerable to global changes

Mountains are complex and fragile ecosystems characterised by verticality, highly differentiated climatic conditions and often by an abundance of water and rich biodiversity. Mountains are high-risk environments: avalanches, glacial lake outbursts, landslides and earthquakes threaten life in mountain areas. Remoteness and difficult access hamper development in mountain regions. Therefore, mountain areas are often marginalized. Despite these constraints, mountains offer significant opportunities. Mountain dwellers have adapted to life in steep and harsh conditions and have developed sophisticated techniques for farming, water use, forestry and communication. The agro-biodiversity as a function of altitude, exposition and farmers' crop selection is huge. Mountain inhabitants have also developed a rich cultural diversity. Therefore, people living in lowland areas or in big cities increasingly prefer mountains for recreation.

Global change may increasingly threaten, or at least alter, the capacity of mountain ecosystems to provide goods and services for both highland and lowland people. On-going glacier recession, for example, poses a significant threat to water supply and public safety, especially in tropical and subtropical regions where a large proportion of the most vulnerable global population resides (e.g. Thompson 2000; Mark and Seltzer 2003). Permafrost degradation is likely to be associated with an increased magnitude and frequency of mountain slope instability (Harris et al. 2001). Further, climate change may lead to more frequent and severe fires in mountain ecosystems,

U. M. Huber et al. (eds.), Global Change and Mountain Regions, 1-8.

such as the temperate forests of the Rocky Mountains (e.g. Keane et al. 1996). Tropical mountain areas, such as Mount Kenya, are undergoing rapid population growth, land use intensification and change. This results in increased competition for water resources in regions where water is already scarce and conflicts over water are growing (e.g. Gichuki et al. 1998). From a scientific point of view, the high sensitivity of mountain regions provides unique opportunities to detect, model and analyse global change processes and their effects on the socio-economic conditions of mountain areas.

The awareness about the global importance of mountain areas, the fragility of their resources in the context of global change and the difficult living conditions of many mountain people has increased significantly over the last decades. Along with this, the need for a better understanding of the functioning of mountain ecosystems and of the global change impacts on these ecosystems has grown. The International Year of Mountains (IYM) 2002 gave mountain research a new impetus. The Mountain Research Initiative (MRI) with its integrated interdisciplinary approach to global change in mountain regions responded to this growing need for better information with the timely and highly relevant commissioning of this State of Knowledge Overview. Part 2 (Paleoenvironmental changes) of this book sets the long-term perspective against which recent environmental changes can be assessed. Part 3 to 4 discuss the biophysical (cryospheric, hydrological and ecological) responses of mountain regions to climate and land use change, whereas part 5 addresses the human dimension of global change and provides ideas for sustainable mountain development.

This introduction puts global change research in mountains into the context of the IYM. It discusses its importance, highlights challenges and expectations for future programmes, and elaborates on emerging partnership opportunities.

2. The International Year of Mountains and Chapter 13

Based on an initiative of the Republic of Kyrgyzstan, the United Nations General Assembly declared 2002 as the International Year of Mountains in November 1998. The Year reinforced the implementation of Chapter 13 of Agenda 21 placing mountains on an equal footing with climate change, tropical deforestation and desertification. The mission statement of the IYM was as follows: "The IYM promotes the conservation and sustainable development of mountain regions, thereby ensuring the well-being of mountain and lowland communities" (FAO 2000). This statement put the main emphasis of the Year on people and highlighted the close linkages between highlands and lowlands. Further, the statement emphasized that the Year is not just about conservation and protection of mountain ecosystems, but that mountain areas deserve investment to prevent out-migration of mountain people in search for better opportunities. The United Nations General Assembly invited FAO to coordinate the IYM because of the agency's role as task manager for Chapter 13 of Agenda 21 and its extensive experience in key mountain issues (e.g. forestry, agriculture, sustainable livelihoods, food security, watershed management and biodiversity). In spite of the obvious diversity of conditions and priorities in the different mountain areas of the world, it was possible to identify and formulate a few overarching principles, which were to be followed in the implementation of the IYM. The Year was supposed to:

- focus on mountain people and contribute to the improvement of their livelihoods;
- further increase awareness of the global importance of mountain regions and their fragile resources;
- trigger long-term action reaching far beyond 2002;
- encourage inter-disciplinary and multi-sectoral approaches and stimulate new ways of collaboration among stakeholders from different disciplines;
- trigger attention and action on the national level and contribute to the formulation of mountain-specific policies;
- initiate new mountain research programmes; and
- contribute to peace-making in the world.

3. Achievements of the IYM

The IYM was a major success thanks to the strong involvement of a large number of stakeholders and the contributions from various disciplines. The mountain research community, and the Mountain Research Initiative in particular, made a major contribution to the success of the IYM that can be summarised as follows:

3.1 Awareness-raising

In 2002, a number of major global, regional, national and sub-national mountain-related events were organised. Countless newspaper articles, television programmes, and educational materials were produced and research programmes developed, all stimulating a new way of thinking about mountains and their fragile resources. Today, many people know that

- mountains are crucial to all life on earth;
- more than half of humanity depends on freshwater resources from the mountains;
- the genetic and biological diversity of mountain ecosystems is crucial for our future;
- mountain areas are particularly sensitive to global change;
- the cultural diversity and gender issues merit particular attention;
- the future of mountain areas and their inhabitants is threatened by social, ethnic and religious tensions and armed conflicts, by climate change and natural hazards, and by unsustainable mining, forestry and agricultural practices.

3.2 Attention on national level

Worldwide, 78 countries have established a national committee for the IYM. Many of these are currently in the process of institutional transformation to be able to guide long-term follow up to the IYM on the national level. Together with the Ville de Chambéry in France, FAO has convened a meeting of IYM national focal points in June 2003 to share experiences and to discuss the follow up. The well-attended meeting created a lot of enthusiasm to work on sustainable mountain development at national level far beyond 2002. Important elements in this follow up will be the establishment of sustainable mountain development strategies as well as the formulation of mountain-specific policies and laws. Many countries have already started initiatives in these directions. Global change and its impact on mountain ecosystems are important elements in these discussions and are instrumental in the formulation of national strategies and policies.

3.3 Partnerships

Initiated by the Swiss Government, UNEP and FAO, a formal alliance, the International Partnership for Sustainable Development in Mountain Regions, known as the "Mountain Partnership," was launched at the World Summit on Sustainable Development in 2002 in Johannesburg. The partnership was further strengthened at the Bishkek Global Mountain Summit, the closing event of the IYM. The objective of the partnership is to promote and strengthen the cooperation between donor organisations, governmental and non-governmental organisations, the private sector, mountain communities, science and other stakeholders. Under the umbrella of this partnership, a number of thematic and regional initiatives are being established, one of them focusing specifically on mountain research. The Mountain Partnership is considered an evolving voluntary alliance, which will allow the assessment of the complexity, variety and scale of mountain-specific themes and problems with the highest possible flexibility. The director general of FAO has offered to host a secretariat for this partnership. By spring 2004, 39 countries, 14 Intergovernmental Organizations and 46 Major Groups (e.g. civil society, private sector) had already joined the Mountain Partnership. A milestone in the development of this partnership was a conference in Merano (Italy) in 2003, in which high-level representatives of those Governments and institutions, which had so far signed up to the partnership, participated and gave their political blessing. The partnership got the clear mandate to establish the membership criteria, governance structure and the secretariat at FAO, as well as to promote the formation of partnership initiatives. In the context of this publication it is particularly interesting to note that the Secretariat of the United Nations Framework Convention on Climate Change (UNFCCC) has joined the Mountain Partnership.

4. Mountain research: A key to sustainable mountain development

4.1 Importance of mountain research

A solid knowledge base about mountain ecosystems and their responses to global change is a pre-condition for the successful follow up to the IYM, the implementation of Chapter 13, the development of national strategies for sustainable mountain development, and the formulation of mountain-specific policies. Accordingly, mountain research scholars and initiatives have a major responsibility and a key role to play in the follow up efforts to the IYM. The importance of mountain research is spelled out in three key documents:

- Programme area A (Chapter 13 of Agenda 21) asks for "generating and strengthening knowledge about the ecology and sustainable development of mountain ecosystems" (UN 1992).
- One main objective of the IYM was to "increase awareness of, and knowledge on, mountain ecosystems, their dynamics and functioning, and their overriding importance in providing a number of strategic goods and services essential to the well being of both rural and urban, highland and lowland people, particularly water supply and food security" (FAO 2000).
- The United Nations General Assembly Resolution A/RES/58/216 on Sustainable Development in Mountain Regions (UNGA 2004) "encourages Member States to collect and produce information and to establish databases devoted to mountains so as to capitalize on knowledge to support interdisciplinary research, programmes and projects and to improve decision-making and planning."

Research in the context of global change in mountain regions is particularly important and requires high priority in future research programmes.

4.2 Research initiatives in the framework of the IYM

There are a number of international research programmes, which were initiated in the framework of the IYM. One of the most prominent ones is the Mountain Research Initiative (MRI), which looks at environmental change in mountain areas, at processes along altitudinal gradients and at sustainable land use and natural resource management. The EU-funded project on Global Change in Mountain Regions (GLOCHAMORE) is a particularly promising research project, which the MRI and UNESCO are jointly implementing in mountain biosphere reserves. The Global Mountain Biodiversity Assessment (GMBA) and the Global Observation Research Initiative in Alpine Environments (GLORIA) are also key international mountain research programmes. The Millennium Ecosystem Assessment, another major global research effort, is charged with providing decision makers within governments, civil society, and the private sector with the latest scientific information about the relationships between ecosystem change and human well-being. Chapter 27 within the Conditions Working Group of the Millennium Ecosystem Assessment focuses on mountain ecosystems. Early in 2003, IUCN established a Mountain Task Force with the objective to assist the organization to streamline and coordinate its mountain-related activities, but more broadly to be a platform for discussion about the links between mountain research and policy development. The Consultative Group on International Agricultural Research (CGIAR) renewed its system-wide Global Mountain Programme. A number of UN Agencies have contributed to knowledge creation on mountain ecosystems: The United Nations Environment Programme (UNEP) has published "Mountain Watch," which documents environmental change and sustainable development in mountains (UNEP/WCMC 2002). Jointly with the Centre for Development and Environment

(CDE) of the University of Berne, the United Nations University (UNU) embarked on a Global Mountain Partnership Programme. In its normative work programme, FAO has carried out research on vulnerability in mountain areas (Huddleston et al. 2003), on mountain legislation (Villeneuve et al. 2002), on household food security and nutrition (Jenny and Egal 2002) and on mountain fisheries (Petr 2003). In addition, FAO launched the initiative "Preparing the next generation of watershed management programmes" with the overall objective (i) to promote worldwide dissemination and exchange of information regarding achievements and gaps in watershed management and (ii) to develop guidelines for future watershed management projects and programmes, considering various scenarios of global change.

A number of mountain research conferences have been held during the IYM 2002, such as the global research meeting of the United Nations University on "Conservation of Mountain Ecosystems" held in Tokyo, Japan, a regional mountain research conference organised by the Royal Academy of Sciences in Kathmandu, and the Forum Alpin in Alpbach, Austria. In a scientific workshop, convened by the Swedish Academy of Sciences already prior to the IYM, the progress of the implementation of Chapter 13 in the ten years since Rio was reviewed from the research perspective and the priorities for mountain research for the next years were defined (AMBIO 2002). Finally, mountains were the core theme of the 54th Congress of German-speaking geographers (Deutscher Geographentag), which was held in late 2003 in Berne (Jeanneret et al. 2003).

4.3 Mountain research in the follow up to the IYM: Challenges and expectations

Mountain research needs to orient itself on global priorities and, at the same time, place itself within the framework of national programmes for sustainable mountain development. Although specific research priorities vary in the different mountain areas, issues such as global change, water, biodiversity, tourism/recreation, economy, cultural heritage, gender issues and highland-lowland linkages require research attention and efforts all over the world (AMBIO 2002). The information and knowledge needs in the follow up to the IYM call for inter-disciplinary and applied mountain research. However, disciplinary and basic research is crucial to fill specific knowledge gaps. Researchers engaged in disciplinary and basic mountain research should never lose sight of the larger, inter-disciplinary context.

Today, nobody can afford to carry out isolated research. Although this is a well-known fact, in practice, there are still too many gaps and not enough dialogue between the research world and the world of application and decision making. The December 2002 issue of the SDC journal "Eine Welt" discusses this challenge under the heading "research and development – a difficult dialogue between academicians and practitioners" (SDC 2002). There is a clear need to intensify the links between researchers and practitioners engaged in sustainable mountain development. Scientists need to make practitioners aware of the potential threat that global change poses for mountain people. Mountain researchers have to be proactive in the identification of emerging issues, anticipate future information needs and make politicians and decision makers aware of mountain research issues and priorities. On the other hand, politicians, NGOs, decision makers and other stakeholders need to be receptive and supportive and assist researchers to work on relevant issues. In this process, it is essential that researchers listen to concerns and information needs expressed by practitioners and try to respond to such requests. This does not reduce the research freedom but guarantees mutual benefits for scientists and the users of research results. Ideally, practitioners and future users should be involved in the conceptual and formulation phase of a research project and maintain a constant dialogue during the project implementation to ensure that research activities remain on track and user-friendly.

The results of successful mountain research projects need to be documented, publicised and communicated. Traditionally this is done through scientific journals with a scientific readership. However, equally important is the transfer of research results to users. Human resources, time, and money need to be invested for the synthesis of research results and for their translation into user-friendly, operational and practical products. In mountain research projects, this is an often under-valued or neglected effort. Therefore, future mountain research programmes should earmark the last project phase specifically for the preparation of policy-relevant documentation and for the transfer of the research results (e.g. through training materials and curricula for education and capacity building on sustainable mountain development for school children, students, technicians, and decision makers). The journal "Mountain Research and Development," edited at the Centre for Development and Environment in Berne, as well as the Mountain Forum with its regional nodes offer excellent communication platforms.

4.4 Partnership opportunities

Apart from offering significant opportunities for mountain research, the increased demand for sound information, created through the IYM, poses new challenges. Increased coordination, information exchange, and use of synergie are needed to effectively fill knowledge gaps and to avoid duplication. All this calls for the establishment of stronger research partnerships across the continents and across mountain regions. The Mountain Partnership, which was launched at the World Summit on Sustainable Development (see 3.3), is an umbrella alliance under which all partners can join specific initiatives according to their interest, competence and priorities. Such initiatives can be organized around thematic areas or by geographic regions. There is considerable scope for existing mountain research institutions and programmes to create a partnership initiative under the umbrella of this global Mountain Partnership. Initial steps towards such a mountain research partnership have already been taken. Such an arrangement will strengthen existing research programmes by creating synergies and providing a common framework. It is obvious that global change in mountain regions will be a priority theme to be researched and discussed under this partnership initiative.

5. Conclusion

Each mountain researcher contributes to the implementation of the global mountain agenda and towards real action in sustainable mountain development. Applied and basic research, disciplinary and inter-disciplinary research, general and location-specific research are equally important and useful as long as the research activities are driven by the burning issues and needs in mountain areas and root in true enthusiasm to contribute to the sustainable development of mountain regions. In this context, the publication of this State of Knowledge Overview by the Mountain Research Initiative is very timely and is highly appreciated. Contributions by excellent scholars from all over the world and from a large variety of expertise make this book a highly relevant and strategic document. We hope that this volume will help orient future mountain research initiatives. The IYM 2002 and the International Year of Freshwater 2003 have provided an ideal platform for reviewing the state of knowledge on Global Change and mountain ecosystems, for initiating new mountain research programmes and for strengthening research partnerships around the world.

6. References

AMBIO (2002). The Abisko Agenda: Research for mountain area development. A contribution to the United Nations Year of Mountains 2002. The Royal Swedish Academy of Sciences, Stockholm.

FAO (2000). International Year of Mountains: Concept paper. Food and Agriculture Organization of the United Nations. Rome, 30pp.

Gichuki, F. N., Liniger, H. P., MacMillan, L., Schwilch, G., and Gikonyo, G. (1998). Scarce water: Exploring resource availability, use and improved management. *In* "Resources, actors and policies – towards sustainable regional development." *Eastern and Southern Africa Journal* **8**, 15-28.

Harris, C., Haeberli, W., Vonder Mühll, D., King, L. (2001). Permafrost monitoring in the high mountains of Europe: The PACE Project in its global context. *Permafrost and Periglacial Processes* **12**, 3-12.

Huddleston, B., Ataman, E., and Fe d'Ostiani, L. (2003). "Towards a GIS-based analysis of mountain environments and populations." Environment and Natural Resources Working Paper 10. FAO, Rome.

Jeanneret, F., Wastl-Walter, D., and Wiesmann, U. (2003). Welt der Alpen - Gebirge der Welt: Ressourcen, Akteure, Perspektiven. *Jahrbuch der Geografischen Gesellschaft Bern* **61**.

Jenny, A. L., and Egal, F. (2002). Household food security and nutrition in mountain areas: An often forgotten story. Nutrition Programmes Service, FAO-ESNP.

Keane, R. E., Morgan, P., and Running, S. W. (1996). "FIRE-BGC: A mechanistic ecological process model for simulating fire succession on coniferous forest landscapes of the northern Rocky Mountains." USDA Forest Service Research Paper INT-RP-484, 122 pp.

Mark, B. G., and Seltzer, G. O. (2003). Tropical glacial meltwater contribution to stream discharge: A case study in the Cordillera Blanca, Perú. *Journal of Glaciology*.

Petr, T. (2003). "Mountain fisheries in developing countries." FAO, Rome.

Swiss Agency for Development and Cooperation (2002). *Eine Welt* **4**, December 2002.

Thompson, L. G. (2000). Ice core evidence for climate change in the Tropics: Implications for our future. *Quaternary Science Reviews* **19**, 19-35.

UNEP/WCMC (2002). Mountain Watch. Environmental change and sustainable development in mountains. UNEP World Conservation Monitoring Centre, Cambridge.

UNGA (2004). Resolution A/RES/58/216 on Sustainable Development in Mountain Regions, adopted by the General Assembly. United Nations Headquarters, New York.

UNITED NATIONS (1992). Earth Summit: Agenda 21. The United Nations Programme of Action from Rio. 3-14 June 1992, Rio de Janeiro, Brazil, 294 pp.

Villeneuve, A., Castelein, A., and Mekouar, M.A. (2002). Mountains and the law: Emerging trends. *FAO Legislative Study* **75**. Rome.

Part I : Paleoenvironmental changes

The world's climate had always been changing and those changes have had considerable impact on aspects of the environment that are now important resources, among them human habitat, forests, and water. Paleoenvironmental change shows us what the planet can do, and gives us insights regarding how climate change and its impacts occur.

Discerning paleoenvironmental change is a detective story, in which researchers unearth clues and attempt to infer exactly what happened. A theme that arises from many of these papers is the great difficulty of discerning the past. The proxy data are themselves the result of complex physical processes and therefore their development needs to be understood before one can extract data, much less information, from the record. The temporal resolution and length of proxies vary greatly, with these papers covering four orders of temporal magnitude (from centuries to millions of years). Thus one researcher's conclusions on a trend may refer to phenomena that another paleo researcher would consider noise.

Nevertheless, in the first paper *Alverson et al.* encourage us to aim high in paleoreconstructions, specifically to abandon the story telling mode and to move toward quantitatively calibrated records based on multiple lines of evidence. They offer two examples of how to improve inference from annually resolved proxy records at the century scale.

Shafer et al. focus on key physical questions: how mesoscale factors such as topography modified the larger scale forcing of climate in the past, and from that, to infer how future forcing might change the climate. They examine changes in an E-W gradient across the northern Rocky Mountains at the scale of tens of thousands of years.

Thompson et al. discuss ice cores from tropical mountain glaciers as unique archvies of past climate and environmental change on the scale of decades to millennia.

Gosse returns to methods, in this case to the use of nuclides in rocks exposed to cosmic rays to provide data at scales up to millions of years. He provides excellent examples of how an understanding of proxy development can improve inference.

Nesje et al. consider glaciers as sources of information. They use estimates of summer temperature to estimate precipitation in a variety of sites in Norway over the past ten thousand years

Solomina summarizes multiple lines of evidence (tree rings, lichen, pollen, lake sediments) for environmental change over the last thousand years gathered at widely dispersed sites in Russia. While she tentatively identifies a general pattern across this immense area, she emphasizes the need for accurate and intercomparable data sets to support accurate assessments.

Kerschner focuses on the use of glaciers for the estimation of precipitation in the Alps over the Holocene. He raises the interesting image of an extremely arid central Alpine region but as with other authors, pleads for much for accurate temperature reconstructions in order to improve estimation of precipitation.

Overpeck et al. take us to the Himalaya and Tibet and note evidence for past abrupt changes in the monsoon that today would have huge impacts on the population. They see the need for much more extensive data principally from lakes on the Tibetan Plateau to understand the full range of potential climate behavior.

Grosjean and *Veit* study a quite the opposite environment, the arid Atacama of South America. They point out the roles of previously humid phases during the late glacial and the Holocene and rare extreme events in the more recent past in creating the water resources of today.

The next three papers focus on lakes as recorders of environmental information. *Lotter* describes some of the limitations of the use of biological data from lakes, such as the lack of sensitivity to winter conditions and the great sensitivity to thermal stratification. He notes the need to disentangle human from climate impacts on the biology of lakes.

Battarbee et al. focus on pollutants in European lakes with data gathered from a wide network of study sites. They conclude that climate change can affect multiple biological and chemical properties of lakes. They note a particularly interesting possibility of cold trapping serving to concentrate certain volatile organic compounds.

Rose et al. go further to emphasize that even remote lakes, particularly those in cold locations cannot be considered as pristine. Concentrations of certain pollutants reach their highest levels in remote, cold lakes.

Tinner and *Ammann* summarize information on forest composition in the Alps as a measure of environmental change over the last several thousand years. They emphasize that while some plant communities have remained relatively intact, others have exhibited large scale species replacement. They note several issues related to inferences from pollen data.

The last two papers focus on climate variability in the central and southern Andes in both Chile and Argentina. *Lara et al.* present a picture of climate changes along the length of the Cordillera over the past 1000 years from tree ring data. *Villalba et al.* look at both tree ring and glacial data. They note that data from trees must be interpreted with an understanding of tree demography: that mortality can be triggered by severe droughts on a short annual scale, while recruitment requires more favorable conditions over the course of many years.

A Dynamical Perspective on High Altitude Paleoclimate Proxy Timeseries

Keith Alverson[1*], Christoph Kull[1], G. W. K. Moore[2], and Patrick Ginot[3,4]
[1]*PAGES International Project Office, Bärenplatz 2, CH-3011 Bern, Switzerland*
[2]*University of Toronto, Toronto, Canada*
[3]*University of Bern, Bern, Switzerland*
[4]*Paul Scherrer Institute, Villigen PSI, Switzerland*
**phone +41-31-312 3133, fax +41-31-312 3168, email alverson@pages.unibe.ch*

Keywords: Cerro Tapado, Ice core, Mount Logan, Paleoclimate, Reanalysis.

1. Introduction

Mountain paleoarchives, including glaciers, laminated lake sediments, and trees near the limits of their habitable range, provide much information relevant to the study of past climatic changes (Alverson and Kull 2002). Properties recorded in these archives offer quantitative climate-related information at annual or higher temporal resolution. In addition, by nature of their occurrence at high elevation, they provide information about climate variability in the free atmosphere, not just its surface expression. However, interpreting these proxy records in terms of large-scale climatic change is a difficult task. Mountains are generally regions of strong climatic gradients and inherently high natural variability, making interpretation of local records difficult. Additional difficulties exist due to the fact that the proxies do not respond to climate alone, but are influenced by myriad additional factors. In this chapter, we highlight two methods which use dynamical constraints, either from the climate system or the underlying archives themselves, to help tease out the climatic information contained in point-based proxy timeseries. Although the examples that we present are applied in conjunction with ice core records, the techniques are relevant to the interpretation of annually resolved climate proxy timeseries in high altitude regions. Past climatic

U. M. Huber et al. (eds.), Global Change and Mountain Regions, 11-20.

changes are often either reconstructed using paleoproxy data or modeled using a numerical representation of the underlying dynamics of either the climate system or paleoarchive development. A more holistic approach is to combine these underlying dynamical equations with proxy reconstructed variables as complementary constraints within the context of a single reconstruction.

2. Dynamical interpretation using reanalysis data

Climatic interpretation of ice cores, as with any archive, relies first and foremost on an accurate chronology. Visual identification of the annual accumulation and ablation layers is possible only in the uppermost portion of the ice, due to increasing compaction with depth. Independent age determination can be done with annual oscillations of $\delta^{18}O$ or other chemical species. Time markers, such as datable organic material, chemical signatures from known volcanic eruptions and atmospheric nuclear weapon tests, serve as additional, independent dating checks and help demonstrate that a given ice core does not contain time hiatuses due to ice motion or negative mass balance.

If a reliable, annually resolved chronology can be established and it extends into the modern period, there exists the possibility of using instrumental data to put the ice core record in a spatial and dynamical perspective and to address the drivers behind the observed climatic changes. In this regard, the recent reanalysis projects, carried out in both Europe and the USA, are of particular interest. The key innovation employed in these reanalyses is the use of a single state-of-the-art data assimilation system for the entire period from the late 1940s onward. In this way, all available climate data are incorporated to produce an estimate of the climatic evolution that is consistent with all available data and a modern atmospheric dynamical model and, importantly, is not corrupted by changes in analysis or modeling techniques that have occurred during the period under consideration. The Mt. Logan ice core record from 5340 m altitude in the Saint Elias mountains of the Yukon, Canada, has been analyzed in detail with the help of reanalysis data (Moore et al. 2001; 2002; 2003). The three hundred year long ice core derived snow accumulation timeseries shows a clear secular trend towards increasing snow accumulation, a trend that has accelerated in recent years. However, climatic interpretation of this trend, if it were the only relevant information at hand, would be highly speculative. Figure 1 puts this trend in a dynamical perspective by regressing the Mt. Logan accumulation record against the NCEP reanalysis climate reconstruction, thus indicating the climatic pattern consistent with both the ice core data and the model equations governing atmospheric motions. The regression of mean winter geopotential height, moisture transport and atmospheric temperature from the NCEP reanalysis data against the Mt. Logan snow accumulation time series is shown during their overlap period from 1948-2000. Fields are only shown where the regression is significant at the 95% level. As is evident in Figure 1, the accumulation record at Mt. Logan is linked to moisture transport associated with the strength of the Pacific North American (PNA) pattern. This dipole pattern of warmth over the Northwest American continent and cool conditions over the North Pacific Ocean has a major impact on the spatial distribution of precipitation in North America, with higher precipitation in the northwestern region as the PNA pattern strengthens. The connection between the snow accumulation time series and this large-scale climatic pattern is statistically significant and, more importantly, the reanalysis data provide a plausible dynamical mechanism behind this statistical correlation.

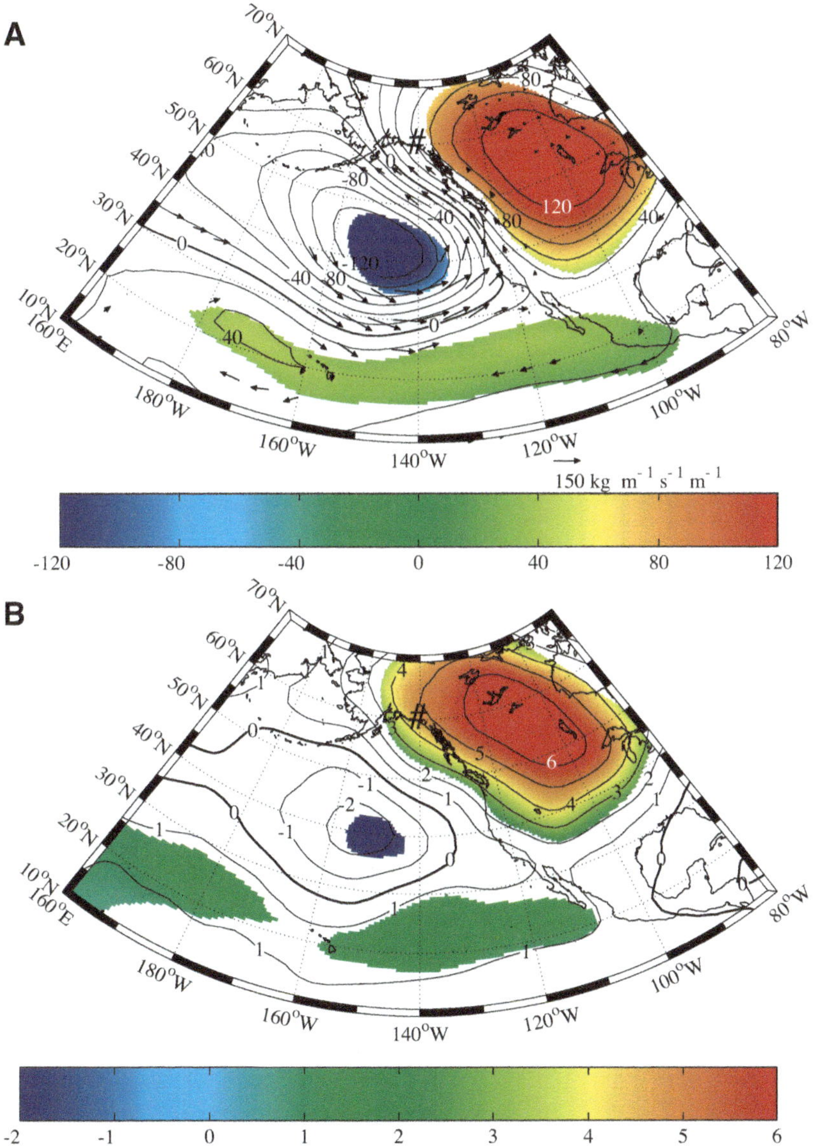

Figure 1: Regression of climate parameters derived from the NCEP reanalysis against the annual snow accumulation at Mount Logan: (A) 500mb geopotential height field (m m^{-1}) and the vertically integrated moisture transport vector field (kg m^{-1} s^{-1} m^{-1}) and (B) mean temperature in the 1000-500mb layer (°C m^{-1}) during the winter (JFM).

Figure 2 shows the regional trends in winter mean surface temperature, in °C per decade, from the HADCRUTv dataset over the period 1870-1999 and the NCEP reanalysis dataset from 1948-2000. In both cases, shading indicates that the trend is significant at the 95% level. Since the ice core accumulation record is correlated with surface temperature fields associated with the PNA, this figure puts the secular trend seen in the ice core data in a larger spatial perspective. As is evident in Figure 2, the trend is part of a large-scale regional amplification of the PNA pattern that has

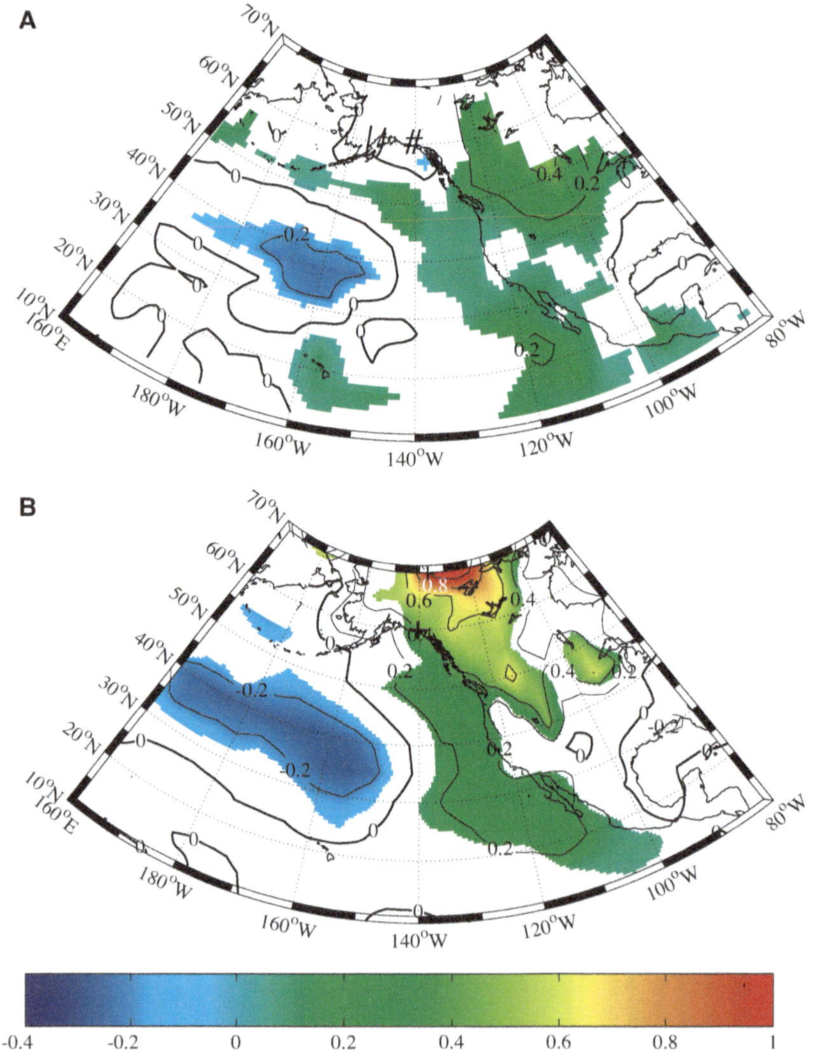

Figure 2: Trend in the surface winter (JFM) temperature field (°C decade^{-1}) from (A) the HADCRUTv dataset from 1870-1999 and (B) the NCEP Reanalysis from 1948-2000. In each instance, shading indicates where the trend is significant at the 95% level in the presence of temporally autocorrelated noise. The "#" indicates the location of Mount Logan.

been occurring for the last ~150 years and has been accelerating in recent decades. Assuming that the correlation between the ice core record and this climatic pattern also held in the pre-reanalysis period, the ice core record indicates that over the period from 1700 to 1850, there was no statistically significant trend in this pattern. The implication, given this evidence for a recent and accelerating trend in the PNA pattern, is that this may be an example of anthropogenic climate change manifesting itself as an amplification of a naturally occurring climatic mode. This plausible interpretation could not be made with the ice core data alone. Rather, it depends on finding a dynamically consistent large-scale climatic pattern correlated with the ice core record during the recent NCEP period, and the assumption that this correlation also held in earlier times.

3. Modeling archive-specific processes

3.1 Problem

High-altitude ice cores provide unique information about past climate. As noted above, extracting quantitative information depends first and foremost on accurate dating of the ice layers and on understanding the mechanisms, which produced the recovered records preserved in the ice (e.g. stable isotopes, major ion concentrations, dust). Water accumulation, and isotopic and chemical species concentrations are a product of numerous processes, which must first be understood in order to be able to back out climatic variability at the site. Recorded ice accumulation is influenced by local precipitation, avalanche feeding, snow drift, melt and sublimation.

Although $\delta^{18}O$ is often used for temperature reconstruction, this parameter depends also on a range of site-specific climatic conditions, on changes in moisture source properties (e.g. ocean temperature, oceanic vs. continental air masses) and on moisture transport history. This points to the importance of understanding synoptic atmospheric circulation in order to interpret an ice core based $\delta^{18}O$ history. After deposition, processes, such as percolation of melt water or surface sublimation, are able to alter the isotopic record. Chemical species captured in the ice (e.g. Cl^-, SO_4^{2-}, Ca^{2+}, Na^+) are accumulated by dry and humid deposition and their concentrations are influenced by both changes in their respective sources and climatic conditions. In drier areas, interpretation of chemical concentration records is often difficult due to the strong influence of sublimation. Thus, ice core records arise as a result of a complicated accumulation history, modified by a myriad of different processes. In the past, climate modes may have been markedly different, including periods of sufficiently low accumulation that were associated with negative mass balance, and therefore a hiatus in the core. Similar gaps in a supposedly "continuous" core can also arise in shear zones. Failure to diagnose such a hiatus in the record leads to an erroneous chronology, an underestimation of past accumulation and an overestimation of the deposition of chemical species - in other words, a completely flawed climatic interpretation. Such effects are often difficult to detect but may play an important role when interpreting "continuous" ice core records back to the Last Glacial Maximum, especially from susceptible high-altitude glaciers in the tropics and subtropics.

3.2 Solution

One way to provide a more rigorous ice core based climatic reconstruction is to model the site- and archive-specific accumulation history, the $\delta^{18}O$ and chemical species record and the related climate conditions. In such a study in the North Chilean Andes (Fig. 3, Ginot et al. 2001; Stichler et al. 2001; Kull et al. 2002; Ginot et al. submitted; Schotterer et al. 2003) we carried out field experiments on Cerro Tapado Glacier (5550 m asl, 30°S/69°W) to quantitatively estimate the climatic controls on local mass balance (sublimation, melt, accumulation) and to assess the post-depositional effects on environmental tracers stored in the firn (Fig. 3b). These experiments confirmed that post-deposition processes, mainly sublimation, have a substantial influence on the ice and snow surface. The loss of water by sublimation (around 2 mm per day) during fair weather resulted in an enrichment of conservative chemical species and a reduction in accumulation. This process may even lead to a negative mass balance in particularly dry years and has wide implications for interpreting ice core records. In a second step, the mass balance (sublimation, accumulation, melt) at the coring site and the related changes in the concentration of conservative chemical species was modeled (Ginot et al. 2001; Kull et al. 2002). These models were based on local climatic data and field measurements of sublimation and enrichment of some chemical species (Fig. 3b). Such models allow us to identify dynamical processes that affect the concentration of chemical constituents in ice and thus lead to an improved interpretation of the ice core record.

Mass loss and modification of chemical constituents have significant consequences for the interpretation of the paleo-record from the long Cerro Tapado ice core. Bedrock was reached at 36 m, a depth corresponding to 28 m of water equivalent (Fig. 3c). Dating was performed by a combination of annual layer counting (assuming that regular wet and dry periods lead to low and high concentrations of chemical constituents), ^{210}Pb, tritium fallout from nuclear weapons tests, and a firn densification model.

Our results show that the upper half of the accumulated ice is younger than 50 years, as indicated by the pre-bomb tritium level of 1952, and that more than 80% of the ice has accumulated during the 20th century (Schotterer et al. 2003; Fig. 3c). However, the lowermost ice must have been formed under very different climatic conditions in the more distant past. Below 23 m water equivalent (weq), a distinct change is apparent in both the ice core stratigraphy and the concentration profiles of isotopes and chemical constituents compared to the upper part of the ice core (Fig. 3c). The reconstructed accumulation history in this part of the core must have been driven by massive sublimation losses during the buildup of the glacier (Schotterer et al. 2003; Ginot et al. submitted; Fig. 3c). The climatic interpretation, consistent with both the ice core data and the climate-mass-balance model (Kull and Grosjean 2000, Kull et al. 2002), points to lower temperatures, higher precipitation and increased seasonality in the moisture supply to the glacier (Fig. 3d). High precipitation in the humid winter season is responsible for the necessary accumulation while the extended dry season relates to the pronounced sublimation. The timing of this climatic regime remains unclear, because of problems with age determination for the lower section of

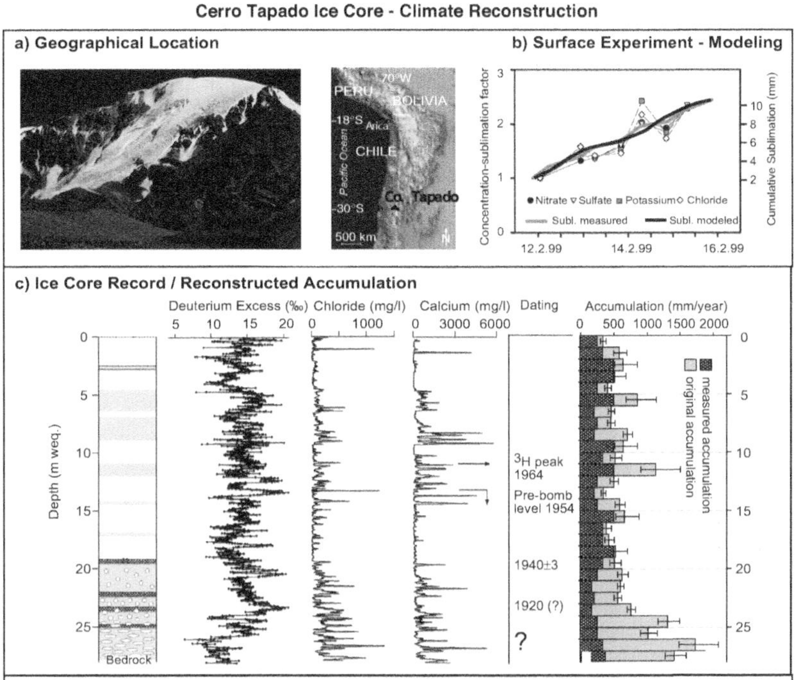

d) Climatic Interpretation

Parameter	1998 / 1999 measured /modeled	1920 / 1999 modern average	23-28 m depth paleo-conditions	Difference paleo-modern
Mean annual Temp. (°C)	-11.5	-12.4 ± 0.2	- 15 ± 1.5	- 3 (± 1.5)
Total accumulation (mm/y)	750	540 ± 45	830 ± 50	+ 290 ± 150
Winter accumulation (mm)	500	310 ± 45	780 ± 50	+ 470 ± 100
Summer accumulation (mm)	250	230 ± 45	50 ± 50	- 180 ± 100
Sublimation (mm/y)	490	240	620	+380
Net accumulation (mm/y)	260	300	210	-80

Figure 3: (A) Location of Cerro Tapado (5550m asl / 30°S) in the North Chilean Andes where an ice core was drilled to bedrock in 1999; (B) Results of the surface experiment, showing the measured and modeled daily sublimation amounts as well as the related concentrations of chemical species. There is a clear linear relationship between the measured concentrations of chemical species and measured sublimation. This relationship can then be applied to model past sublimation from measured chemical species concentrations in ice cores. A mass balance model (Kull and Grosjean 2000; Kull et al. 2002) and a model for the concentration of chemical species (Ginot et al. 2001) can be used to calculate specific mass balance and climate conditions from the ice core record; (C) Ice core record from Cerro Tapado: The ice core stratigraphy and ^{2}H excess both show clear differences between the core sections below and above 23 m water equivalent (weq.). The variation in the concentration of the chemical species, which is directly linked to sublimation changes, results from climate variability (Schotterer et al. 2003; Ginot et al. submitted). Reconstruction of the original ice accumulation as corrected for sublimation losses, using a model that calculates the concentration of chemical tracers as a function of postdepositional processes (Ginot et al. 2001). A clear change in accumulation is visible at ca. 23m weq (Ginot et al. submitted); (D) Climate reconstruction from the accumulation history in the core using the mass balance model from Kull and Grosjean (2000) and Kull et al. (2002). A significant change in climate is found at ca. 23m weq. This points towards a core hiatus of unknown age and duration, in agreement with the stratigraphy and ^{2}H history. Below 23m weq. the reconstruction suggests more humid and colder conditions with a very pronounced precipitation seasonality, in marked contrast to 20th century conditions (Schotterer et al. 2003).

this ice core. However, the presence of a time and accumulation hiatus of unknown age and duration is clearly indicated and raises concerns.

3.3 Conclusion

An accumulation and time hiatus is recorded in the Cerro Tapado ice core. A viable climatic interpretation of the Cerro Tapado ice core record rests upon archive-specific modeling work and field experiments. Traditional methods of ice core interpretation would fail in this case due to the highly variable and complex accumulation history. There is no reason to believe that the Cerro Tapado case is unique, rather the possibility of such effects should be considered (or reconsidered) in all cores. Site-specific mass balance–climate models can provide more robust interpretations of ice core records, and their use is particularly vital at high-altitude, low-latitude sites where complex climatic conditions produce paleo-proxy records that can often not be explained by a simple set of climate variables.

4. Future research directions

Mountain regions are characterized by climatic variations on all timescales. Dramatic changes in these environments, due to a combination of external forcings and internal system dynamics, are known to have occurred on timescales from decades to hundreds of thousands of years. Furthermore, their history is replete with examples of nonlinear dynamics, wherein various system components demonstrate large, abrupt or irreversible changes in response to forcings that are both small and smooth as for example the infinitely nonlinear (step function) phase transition from water to ice. Instrumental climate and ecosystem related data, on the other hand, are characterized by a short and, at least in relative terms, uneventful history. Meteorological data from a global network of stations have been available for about a century, but few of these stations are at high altitudes (see Diaz this volume). Satellite measurements date back only about a decade.

Paleoclimate research in mountain regions seeks to bridge the vast gulf that looms between these two extremes - the long and rich record of the past on the one hand, and the short and comparatively uneventful instrumental record on the other. Closing this gap is the only way to make the past record relevant, for example, to resource sustainability. A quantitatively calibrated, chronologically well-constrained record based on multiple lines of empirical evidence alongside a hierarchy of dynamical models is the foundation for such understanding.

Over the past decade, paleoscience in mountain regions has come a long way. The climatic and environmental history recorded in paleoarchives is now widely recognized as being of great relevance to societal concerns (Alverson and Oldfield 2000; Alverson et al. 2002). Based solely on such proxy records of the past, we can make some fairly strong statements about the present. Greenhouse gas levels are higher than they have been for hundreds of thousands of years. Global average temperatures are warmer now than they have been for the past millennium. Rapid, large amplitude environmental change can occur in response to smooth, small amplitude forcing. Remote and seemingly pristine mountain lake ecosystems, when compared to the records of the past millennium or longer, stored in their sediments, are acidified and contaminated by both heavy metals and a wide range of persistent organic pollutants (see Rose et al. and Batterbee et al. this volume).

These are all messages that the past decade of paleoresearch has brought to the public attention. As paleodata comes into the limelight however, so will it come under increasing outside scrutiny. In order to be a genuine partner in the broader earth system science community, the paleocommunity must dispense with qualitative methods. In some cases, quantitative calibrations are not yet available – this is especially true in mountains where, for example, long-term climate stations are scarce. In these situations, proxy records are of course still useful, but every effort needs to be made to calibrate them with modern data. We need to publish our uncertainties. We need to be clear about our chronological accuracy. We need to make the data behind published results freely available (e.g. at the World Data Center for Paleoclimatology).

Mountain climate and environmental systems are enormous and complex, while our data remain sparse and our models crude. The system can therefore only be underdetermined. As every first year college mathematics student knows, there are infinitely many solutions to an underdetermined problem. Thus, stories, which appear to fit our data, though often plausible and exciting, are not necessarily conclusive or significant. The paleocommunity needs to abandon the storytelling mode of research. Rather than collecting data and then interpreting them, we need to adopt a hypothesis-driven approach, which explicitly seeks out certain data in order to rule out specific scenarios. Moreover, a single core, a single proxy, a single model or a single researcher cannot possibly answer questions about the evolution of mountain climates or ecosystems with the degree of sophistication that we wish to address them. More than ever before, our research must be grounded on quantitative calibration against instrumental datasets, detailed understanding of the behavior of paleoarchives and how the proxy records therein are created, along with inverse modeling for robust, quantitative, dynamically consistent, past climate state estimation.

4.1 Relevance to societal concerns

Since the industrial revolution, mountains have become increasingly affected by human activities. Some might suggest that anthropogenic change has been so dramatic as to render dynamics of the past irrelevant to current concerns. This is not the case. Natural processes are now woven together with human induced changes in a complex tapestry of forcings, responses, feedbacks, and consequences. However, the past record remains of great significance for the future. For example, there is much evidence to suggest that global anthropogenic climate change may be expressed by the strengthening or weakening of naturally occurring climatic modes (e.g. the PNA pattern and its effect on mountain precipitation, discussed above). Another example is mountain biodiversity. The degree and range of modern biodiversity is not explainable based on current climatic conditions alone. Rather, it has arisen in response to the integrated history of conditions in the past. Understanding the basis for the persistence of high mountain biodiversity in the face of past disturbances is the key to ensuring its future survival in the face of modern change (see Tinner and Ammann this volume). A globally inclusive, coordinated effort to decipher the complexity of natural climatic variability and ecosystem change in mountain regions, whenever possible concentrating on those aspects most relevant to modern concerns, must remain the primary goal of paleoresearch in mountain regions.

5. References

Alverson, K., Bradley, R., and Pedersen, T., Eds. (2002). "Paleoclimate, global change and the future." IGBP Book Series, Springer Verlag, Heidelberg.

Alverson, K., and Kull, C. (2002). Understanding future climate change using paleorecords. *In* "Global climate: Current research and uncertainties in the climate system." (X. Rodó, and F. A. Comín, Eds.), pp. 153-185. Springer Verlag, Heidelberg.

Alverson, K., and Oldfield, F. (2000). PAGES - Past global changes and their significance for the future: An introduction. *Quaternary Science Reviews* **19**, 3-7.

Ginot, P., Kull, C., Schwikowski, M., Schotterer, U., and Gäggeler, H. W. (2001). Effects of postdepositional processes on snow composition of a subtropical glacier (Cerro Tapado, Chilean Andes). *Journal of Geophysical Research* **106**, 32375.

Kull, C., and Grosjean, M. (2000). Late Pleistocene climate conditions in the North Chilean Andes drawn from a climate-glacier model. *Journal of Glaciology* **46**, 622-632.

Kull, C., Grosjean, M., and Veit, H. (2002). Modeling Modern and Late Pleistocene glacio-climatological conditions in the North Chilean Andes (29°S - 30°S). *Climatic Change* **52**, 359-381.

Moore, G. W. K., Holdsworth, G., and Alverson, K. (2001). Extra-tropical response to ENSO 1736-1985 as expressed in an ice core from the Saint Elias mountain range in northwestern North America. *Geophysical Research Letters* **28**, 3457-3461.

Moore, G. W. K., Holdsworth, G., and Alverson, K. (2002). Climate change in the North Pacific region over the last three centuries. *Nature* **420**, 401-403.

Moore, G. W. K., Alverson, K., and Holdsworth, G. (2003). On the effect that elevation has on the ENSO related climate signal contained in precipitation records from northwestern North America. *Climatic Change* (in press).

Schotterer, U., Grosjean, M., Stichler, W., Ginot, P., Kull, C., Francou, B., Gäggeler, H., Gallaire, R., Hoffmann, G., Pouyaud, B., and Schwikowski, M. (2003). Glaciers and climate in the Andes between the Equator and 30°S: What is recorded under extreme environmental conditions? *Climatic Change* (submitted).

Stichler, W., Schotterer, U., Fröhlich, K., Ginot, P., Kull, C., Gäggeler, H., and Pouyaud, B. (2001). Influence of sublimation on stable isotope records recovered from high-altitude glaciers in the tropical Andes. *Journal of Geophysical Research* **106**, 22613.

Understanding the Spatial Heterogeneity of Global Environmental Change in Mountain Regions

Sarah L. Shafer[1*], Patrick J. Bartlein[2], and Cathy Whitlock[2]
[1]*U.S. Geological Survey, 200 SW 35th Street, Corvallis, OR 97333, USA*
[2]*Department of Geography, University of Oregon, Eugene, OR 97403, USA*
**phone +1-541-754-4498, fax +1-541-754-4799, e-mail sshafer@usgs.gov*

Keywords: Future climate change, Holocene, Paleoenvironmental change, Rocky Mountains, Spatial heterogeneity.

1. Introduction

One of the challenges for global environmental change research is to understand how future climate changes will be expressed in mountain regions. The physiographic complexity of mountains creates environments that can be highly variable over relatively short distances. This spatial heterogeneity reflects a hierarchy of environmental controls. At regional scales, insolation and atmospheric circulation features determine the dominant regional climate patterns that affect mountain regions. At finer spatial scales, substrate, aspect, elevation, and a number of other environmental factors influence ecosystem dynamics. Vegetation, for example, is affected by all levels of this hierarchy, from regional-scale climate regimes down to site-specific features, such as substrate type (cf. Körner, this volume).

The spatial heterogeneity of mountain environments will significantly influence the ways in which mountain ecosystems respond to future climate change, yet the potential spatial complexity of ecosystem responses continues to be underestimated. For example, vegetation response to future climate change in the northern hemisphere is often described in terms of "moving northward and upward." Although this type of movement may approximate the response of plant species in regions with relatively little topographic variability, such as the eastern United States (US), the response is

U. M. Huber et al. (eds.), Global Change and Mountain Regions, 21-30.

unlikely to be as simple in topographically complex mountain regions, such as the Rocky Mountains of the western US. In fact, preliminary results from vegetation modeling indicate that plant taxa in mountain regions may move not just northward in response to future climate change, but in all directions, including southward in some instances (e.g. Shafer et al. 2001). This simulated complexity of future vegetation response to climate change is supported by the paleoecological record, which indicates that species have responded individualistically to climate changes in the past (Huntley 1995; Webb 1995). Moreover, vegetation change is just one of the potential complex responses of mountain ecosystems to future climate change.

We have been using a combination of numerical models and paleoenvironmental data to investigate the spatial heterogeneity of environmental response to climate change in mountain regions. Of particular interest to us is how large-scale forcings, such as global warming, are mediated by mesoscale physiographic features, such as mountain ranges. We are examining this mediation process in the Yellowstone region of the northern Rocky Mountains (US) where large-scale atmospheric circulation features interact with the region's topography to create contrasting precipitation regimes. These mesoscale interactions between hemispheric-scale circulation features and regional-scale physiographic features have received relatively little attention in discussions of the potential effects of future climate change, and yet they may greatly increase the variability of ecosystem responses to changing climate. Moreover, an understanding of these mesoscale dynamics is necessary to determine the potential range of mountain ecosystem responses to climate change, and to develop appropriate policy and management strategies to address future global change impacts in mountain regions.

2. Spatial heterogeneity in mountain regions: An example from the northern Rocky Mountains of the western US

In the Yellowstone National Park region of the northern Rocky Mountains there are two contrasting precipitation regimes. As illustrated by the precipitation bar graphs in Figure 1 (middle and right), to the north and east of Yellowstone are areas that receive much of their annual precipitation during the summer (summer-wet regions), while to the south and west of Yellowstone are areas that receive much of their annual precipitation during the winter (summer-dry regions). These two precipitation regimes are part of a broader-scale pattern of precipitation seasonality in the western US (Fig. 1, left) controlled by different large-scale atmospheric circulation features (Whitlock and Bartlein 1993; Whitlock et al. 1995). The areas with a summer-wet precipitation regime receive summer moisture from the Gulf of Mexico and the Gulf of California via monsoonal circulation. In contrast, the areas with a summer-dry precipitation regime are dominated in summer by the eastern Pacific subtropical high-pressure system that suppresses precipitation in the region. These summer-dry regions receive winter moisture from Pacific Ocean air masses that are funneled to the interior of the continent through physiographic low areas in the northwestern US, such as the Columbia Gorge (Bryson and Hare 1974; Whitlock and Bartlein 1993). Although

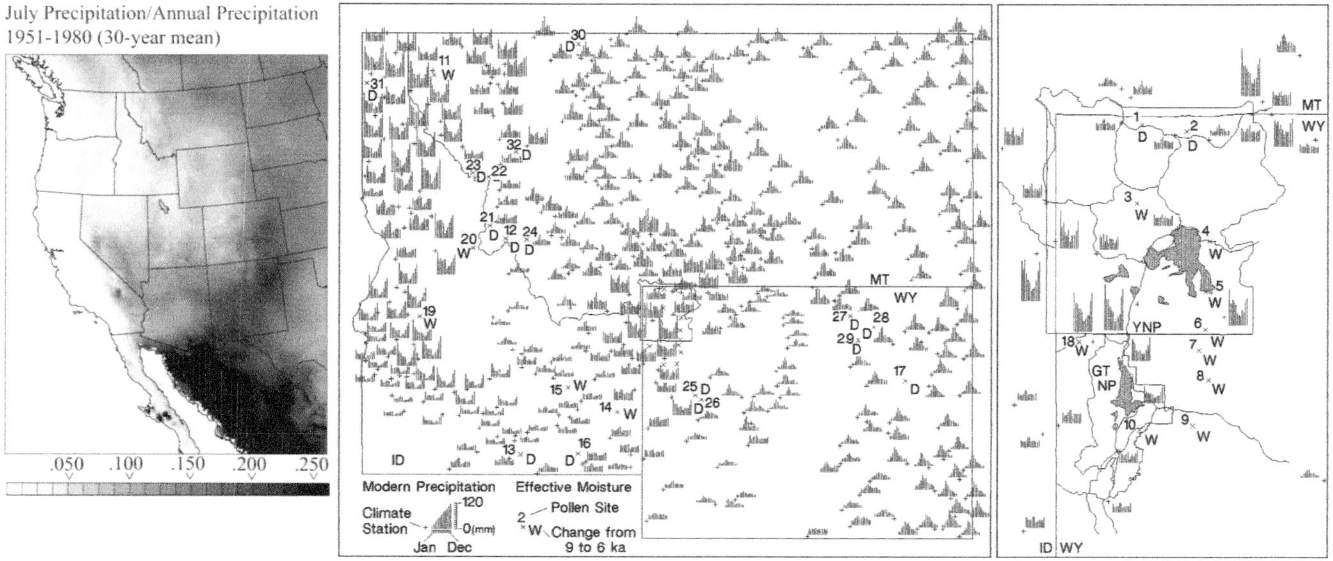

Figure 1: (Left) Ratios of July-to-annual precipitation for 1951-1980 (30-year mean; Thompson et al. 1999). (Middle) Annual distribution of precipitation at climate stations in Idaho, Montana, and Wyoming (World WeatherDisc Associates n.d.), and inferred effective moisture changes from 9000 to 6000 cal yr BP at paleoecological sites. "W" indicates an inferred effective moisture change to wetter conditions, "D" an inferred effective moisture change to drier conditions, and "-" no inferred direction of change. (Right) The same information for the region of Yellowstone National Park (YNP) and Grand Teton National Park (GTNP). Sources: 1. Gennett & Baker (1986) Palynology 10:61-71; 2. Whitlock & Bartlein (1993) Quaternary Research 39:231-38; Millspaugh & Whitlock (2003) *In* After the Fires: The Ecology of Change in Yellowstone National Park, Wallace (Ed.); 3. Millspaugh et al. (2000) Geology 28:211-14; 4. Waddington & Wright (1974) Quaternary Res. 4:175-84; 5. Baker (1976) USGS Professional Paper 729-E:E1-E48; 6.-10. Whitlock (1993) Ecol. Monogr. 63:173-98; 11. Mack et al. (1983) Quaternary Res. 20:177-93; 12. Mehringer et al. (1977) Arctic & Alpine Res. 9:345-68; 13. Davis et al. (1986) Quaternary Res. 26:321-39; 14. Beiswenger (1991) Ecol. Monogr. 61:165-82; 15. Davis et al. (1986) Quaternary Res. 26:321-39; 16. Bright (1966) Tebiwa 9:1-47; 17. Markgraf & Lennon (1986) Plains Anthropologist 31:1-12; 18. Whitlock et al. (1995) Quaternary Res. 43:433-36; 19. Doerner & Carrara (1999) Arctic, Antarctic, & Alpine Res. 31:303-11; 20.-22. Brunelle-Daines (2002) Dissertation, Univ. of Oregon; 23. Karsian (1995) Master's Thesis, Univ. of Montana; 24. Brunelle-Daines (2002) Dissertation, Univ. of Oregon; 25. Fall et al. (1995) Quaternary Res. 43:393-404; 26. Lynch (1998) Ecology 79: 1320-38; 27, 28, 29. Burkart (1976) Dissertation, Univ. of Iowa; 30. Barnosky (1989) Quaternary Res. 31:57-73; 31. Mack et al. (1995) Quaternary Res. 10:241-55; 32. Mehringer (1985) *In* Pollen records of Late-Quaternary North American sediments, Bryant & Holloway (Eds.).

the amplitude of the seasonal cycle of precipitation in the Yellowstone region is controlled by these large-scale circulation patterns, the spatial pattern of the summer-wet versus summer-dry areas is controlled by the region's topography (Whitlock and Bartlein 1993).

The juxtaposition of these two precipitation regimes creates a mesoscale level of environmental complexity across the Yellowstone region that affects many aspects of the region's environment. To explore the influence of this mesoscale complexity on the region's response to future climate change, we examined paleoenvironmental data to evaluate how this region responded to climate changes in the past. From approximately 12,000 to 6000 cal yr BP (i.e. calendar years before present), summer insolation in the northern hemisphere was greater than it is at present. Although the direct effect of higher-than-present summer insolation was likely to have increased evapotranspiration, creating drier conditions in the Yellowstone region, an important indirect effect may have been to increase the strength of both the summer monsoon and the eastern Pacific subtropical high-pressure system (Whitlock and Bartlein 1993; Mock and Bartlein 1995; Mock and Brunelle-Daines 1999). The strengthening of these two circulation features had opposing consequences on the precipitation regimes of the Yellowstone region. The indirect strengthening of the summer monsoon likely overwhelmed the direct insolation effect, creating wetter-than-present summer conditions in the summer-wet areas in the early Holocene. In contrast, the indirect strengthening of the eastern Pacific subtropical high-pressure system likely augmented evapotranspiration, creating drier-than-present summer conditions in the summer-dry areas at this time (Fig. 1, middle).

These opposing responses to increasing insolation had significant impacts on the region's ecosystems, particularly its vegetation and fire regimes. The paleoecological record indicates that vegetation in the Yellowstone region responded to these seasonal precipitation changes, with vegetation communities in the summer-dry region supporting an increased number of taxa that tolerate drier summer conditions, and vegetation communities in the summer-wet regions supporting an increased number of taxa that require more mesic summer conditions (Whitlock and Bartlein 1993; Whitlock et al. 1995; Millspaugh and Whitlock 2003). Fire regimes in the region, which can be reconstructed for centennial-to-millennial time periods from charcoal analysis of lake-sediment cores, were also affected by the changing strength of the region's precipitation regimes. Charcoal records indicate that enhanced dryness in summer-dry regions resulted in increased fire frequencies (Millspaugh et al. 2000; Millspaugh and Whitlock 2003). As insolation decreased from 9000 to 6000 cal yr BP, the strength of both the eastern Pacific subtropical high-pressure system and the southwestern monsoonal circulation likely decreased as well. As a result, summer-wet areas became drier in summer and summer-dry areas became wetter than they were during the early Holocene, and this change is recorded in the paleoenvironmental record (Fig. 1; middle and right).

2.1 Simulating past and future environmental changes in mountain regions

The evidence that the precipitation regimes in the Yellowstone region responded in opposite ways to insolation forcing in the Holocene raises important questions about how the region's climate and ecosystems may respond to future climate change. We are currently investigating these questions using a variety of physically-based process models to simulate the potential response of vegetation and soil moisture (as a proxy for fire regimes) to both past and future climate changes in the Yellowstone region.

General circulation model (GCM) simulations of precipitation for 6000 cal yr BP resolve the broad-scale pattern of summer-wet and summer-dry regions in the western US (Fig. 2). This pattern includes summer-wet conditions in the southwestern US attributed to strengthened monsoonal circulation, and summer-dry conditions in the

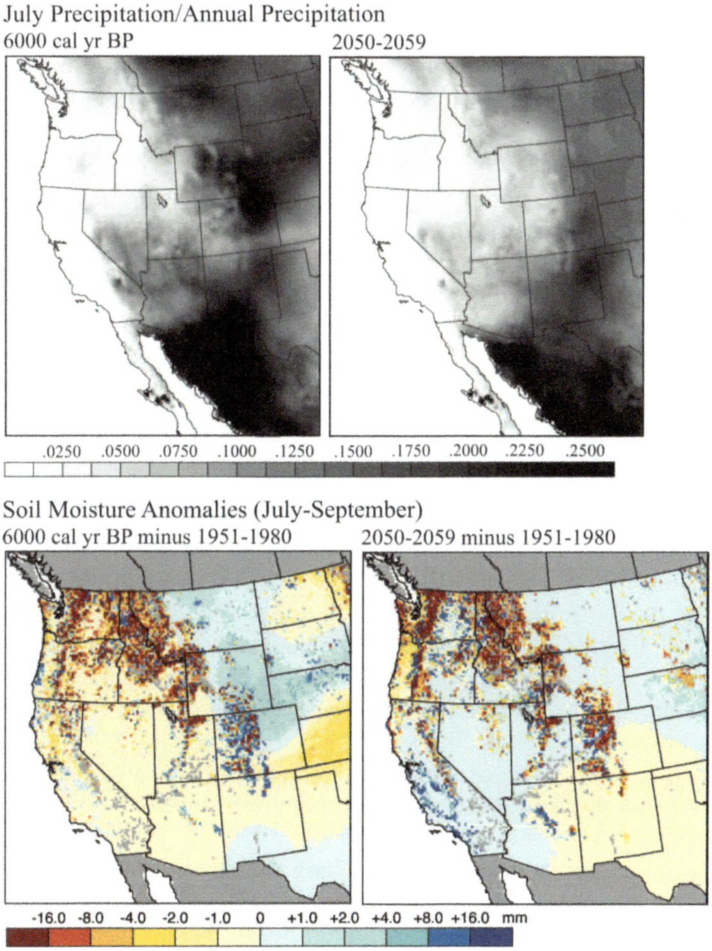

Figure 2: Ratios of July-to-annual precipitation for 6000 cal yr BP (top left) and 2050-2059 (10-year mean; top right), and mean summer (July-September) soil-moisture anomalies (mm) for 6000 cal yr BP (bottom left) and 2050-2059 (bottom right) compared to present (1951-1980, 30-year mean) in the western US. Climate data: 1951-1980 (Thompson et al. 1999); 6000 cal yr BP, CCM3 (Bonan 1996); 2050-2059, HADCM2 HCGSa (Mitchell and Johns 1997). Soil data: CONUS-Soil (Miller and White 1998).

Pacific Northwest and parts of the northern Rocky Mountains arising from a stronger-than-present eastern Pacific subtropical high-pressure system. Soil moisture simulated for 6000 cal yr BP, using a simple water-balance model, reflects the broad-scale summer-wet and summer-dry precipitation patterns of the western US (Fig. 2), with drier-than-present summer soil conditions simulated for much of the northwestern US, and wetter-than-present conditions simulated for the southwestern mountains and the areas north and east of the Yellowstone region (Fig. 2). The soil moisture simulations also capture the local-scale effects of different soil types and variations in the amount and duration of snow cover across the region. These local-scale variations create the mosaic of both drier and wetter soil moisture conditions simulated for the mountain areas in the western US. The largely drier-than-present soil moisture patterns simulated for 6000 cal yr BP in the Yellowstone region are consistent with the paleoecological data, which includes evidence of increased fire frequency in the region at 6000 cal yr BP compared to present (Millspaugh et al. 2000; Millspaugh and Whitlock 2003).

The summer-wet versus summer-dry pattern is maintained in the simulated future July-to-annual precipitation ratios for 2050-2059, although the summer-wet areas are simulated to be slightly drier in the future, relative to annual precipitation, than they are at 6000 cal yr BP (Fig. 2). Potential future summer soil moisture simulated for 2050-2059 contains drier-than-present patterns similar to those simulated for 6000 cal yr BP at high elevations. At low elevations there is slightly more moisture simulated for 2050-2059 than for 6000 cal yr BP, particularly in the summer-dry regions (Fig. 2). This increased soil moisture is a result of simulated future increases in summer precipitation compared to present in the western US (Mitchell and Johns 1997).

These potential future soil moisture changes could have important impacts on vegetation and fire regimes of the Yellowstone region. Decreased summer soil moisture simulated for high elevations in the Yellowstone region relative to present could lead to increased drought stress on vegetation and create conditions suitable for increased future fire frequencies (Clark 1989). Warmer and drier conditions would also decrease the moisture content of fuels (e.g. downed wood, duff), making them more flammable. Under the simulated future climate conditions, increased summer soil moisture at low elevations could also lead to increased fire frequencies if there is sufficient moisture to increase the productivity and connectivity of fine fuels, particularly grasses (Gitay et al. 2001). This situation could occur if effective moisture increased during the growing season, for example, allowing increased grass growth, which was then followed by summer climate conditions that dried out the grass during the fire season.

There are a number of caveats that go along with these simulations of potential future climate, soil moisture, and vegetation. That we find similarities in simulated patterns of certain variables, such as soil moisture, for past and future time periods does not mean that the physical processes creating those patterns necessarily will be the same in the future as they were in the past. Additional model experiments will allow us to refine our understanding of these processes and determine, for example, how robust the atmospheric circulation features implicated in past climate changes in the region may be under future climate scenarios. There are also uncertainties associated with model simulations of both past and future time periods. For example, the atmosphere-ocean general circulation models (AOGCMs) used to simulate potential future climate change have difficulty simulating precipitation patterns and fine-scale climate features, which limits the confidence we can have in the model simulations of increased soil moisture for particular regions (Giorgi et al. 2001). These uncertainties continue to be reduced as the models simulating climate, soil moisture, and vegetation are improved. In the meantime, model simulations can be used to indicate the potential magnitude and direction of changes that may occur in the future.

It is also important to remember when interpreting model simulations of future climate change that climate change is transient, with the current global warming trend predicted to continue for centuries (IPCC 2001). As a result, the mesoscale summer-wet versus summer-dry pattern that exists in the Yellowstone region may not persist in the future as climate changes exceed particular environmental thresholds beyond which other climatic processes may become dominant. For example, the increased summer moisture in the summer-wet area during the early Holocene was likely the result of increases in summer precipitation that were large enough to exceed concomitant increases in evapotranspiration. If future global temperatures continue to increase, the opposite result may ensue: future summer evapotranspiration rates may at some point exceed precipitation in the summer-wet regions of Yellowstone. Although the region might still have a summer-wet precipitation regime, drier summer soils could lead to significant changes in the region's vegetation and fire regimes.

Finally, a strength of our modeling efforts is that they often explicitly rely on paleoecological data (e.g. Whitlock et al. 2003). Although the paleoenvironmental record cannot be used as an analogue for future change in the Yellowstone region, it provides important evidence of how mountain environments have responded to large environmental changes in the past - changes similar in magnitude to potential future climate changes simulated to occur over the next few centuries. Paleoecological data also can be used in conjunction with models to reconstruct paleoenvironments and thus to validate model simulations of past time periods. Models that can accurately simulate environmental changes in the past give us more confidence in their ability to simulate future environmental conditions.

3. Future priorities for investigating the spatial heterogeneity of environmental response to climate change

An important priority for spatial heterogeneity research in mountain regions is to identify other areas where topographically controlled circulation features exhibited complex responses to past climate changes or are anticipated to have complex responses to future climate change (e.g. see Fall et al.'s (1995) study of the Wind River Range in Wyoming and Brunelle-Daines' (2002) study of the Bitterroot Range in Montana). Interactions of mountain ranges with large-scale circulation features are unique, with different ranges having different thresholds for change. The topographic control of climate by smaller mountain ranges may be overwhelmed as circulation features increase or decrease in strength with changing climate. Additional studies are needed to understand these dynamics in other regions of the globe.

3.1 Improving model simulations of environmental change

A variety of physically-based process models are used by researchers to simulate environmental dynamics in mountain regions. Refinement of these models will lead to improvements in our ability to simulate both the past and potential future spatial complexity of environmental responses to climate change in mountain regions. Of particular importance is improving the spatial resolution of climate simulations. AOGCMs are run at too coarse a spatial resolution to adequately resolve mountain regions, although they provide important information about large-scale atmospheric circulation patterns. Regional circulation models (RegCMs) have a finer spatial resolution and, although they are too complex to run at global scales, they can be nested within AOGCM simulations for hypothesis testing of regional-scale questions (e.g. Hostetler et al. 2000; Leung, this volume). However, even finer-scale resolution models are needed to resolve the topographic complexity of mountain regions that creates the spatial heterogeneity observed at regional scales.

3.2 Increasing data-model comparisons

Paleoecological research in mountain regions is improving our understanding of the spatial variability of environmental responses to past climate change. In some mountain regions, such as the western US, there is now sufficient paleoenvironmental data coverage (e.g. NOAA/NGDC paleoclimatology data: www.ngdc.noaa.gov/paleo/data.html/) for regional syntheses of the dynamics of past environmental changes. A wide variety of paleoenvironmental data (e.g. pollen, charcoal, diatoms, isotopes) are available and multi-proxy studies are beginning to reveal patterns of environmental change in mountain regions during the Holocene. These studies include information on past patterns of species migration (MacDonald and Cwynar 1985; Anderson 1996; Whitlock and Millspaugh 2001), disturbance regimes, such as fire and drought (Meyer et al. 1995; Long et al. 1998; Millspaugh et al. 2000; Mohr et al. 2000), and how these processes varied with past climate change. Designing studies that utilize paleoenvironmental and modern environmental data in conjunction with climate models and other physically-based process models will help us move from identifying the spatial patterns associated with environmental change in mountain regions to identifying the processes controlling those patterns.

3.3 Developing model simulations of climate change impacts for policy and management

Finally, in order to develop policy and management strategies for responding to the impacts of future climate change in mountain regions, we must improve our ability to predict the magnitude, variability, and spatial complexity of potential future environmental changes. Policy-makers and managers would like to have predictions of the exact timing, location, and type of climate change impacts on environmental systems that will occur in the future, but this level of accuracy is not now possible. However, we do have increasing confidence in the ability of models to simulate the magnitude and direction of potential future environmental changes, and efforts should be made to translate these results and their uncertainties into language that is accessible to policy-makers and managers.

4. Acknowledgements

Support for this research was provided by NSF grants ATM-9910638 and ATM-0117160, and the U.S. Geological Survey Earth Surface Dynamics Program. We thank Andrea Brunelle-Daines for assistance with the figures, and Allen Solomon, Peter Van de Water, Harald Bugmann and Ulli Huber for their comments on the text.

5. References

Anderson, R. S. (1996). Postglacial biogeography of Sierra lodgepole pine (*Pinus contorta* var. *murrayana*) in California. *Ecoscience* **3**, 343-351.

Bonan, G. B. (1996). "A land surface model (LSM version 1.0) for ecological, hydrological, and atmospheric studies: Technical description and user's guide." NCAR Technical Note NCAR/TN-417+STR, NCAR, Boulder.

Brunelle-Daines, A. R. (2002). "Holocene changes in fire, climate, and vegetation in the northern Rocky Mountains of Idaho and western Montana." Unpublished Ph.D. thesis, University of Oregon, Eugene.

Bryson, R. A., and Hare, F. K. (1974). The climate of North America. *In* "Climates of North America. World survey of climatology. Volume 11." (R. A. Bryson, and F. K. Hare, Eds.), pp. 1-47. Elsevier, Amsterdam.

Clark, J. S. (1989). Effects of long-term water balances on fire regime, north-western Minnesota. *Journal of Ecology* **77**, 989-1004.

Fall, P. L., Davis, P. T., and Zielinski, G. A. (1995). Late Quaternary vegetation and climate of the Wind River Range, Wyoming. *Quaternary Research* **43**, 393-404.

Giorgi, F., Whetton, P. H., Jones, R. G., Christensen, J. H., Mearns, L. O., Hewitson, B., vonStorch, H., Francisco, R., and Jack, C. (2001). Emerging patterns of simulated regional climatic changes for the 21st century due to anthropogenic forcings. *Geophysical Research Letters* **28**, 3317-3320.

Gitay, H., Brown, S., Easterling, W., and Jallow, B. (2001). Ecosystems and their goods and services. *In* "Climate change 2001: Impacts, adaptation, and vulnerability." (J. J. McCarthy, O. F. Canziani, N. A. Leary, D. J. Dokken, and K. S. White, Eds.), pp. 235-342. Cambridge University Press, Cambridge.

Hostetler, S. W., Bartlein, P. J., Clark, P. U., Small, E. E., and Solomon, A. M. (2000). Simulated influences of Lake Agassiz on the climate of central North America 11,000 years ago. *Nature* **405**, 334-337.

Huntley, B. (1995). How vegetation responds to climate change: Evidence from palaeovegetation studies. *In* "Impacts of climate change on ecosystems and species: Environmental context." (J. C. Pernetta, R. Leemans, D. Elder, and S. Humphrey, Eds.), pp. 43-63. IUCN, Gland.

Intergovernmental Panel on Climate Change (IPPC) (2001). "Climate change 2001: The scientific basis." (J. T. Houghton, Y. Ding, D. J. Griggs, M. Noguer, P. J. van der Linden, X. Dai, K. Maskell, and C. A. Johnson, Eds.). Cambridge University Press, Cambridge.

Long, C. J., Whitlock, C., Bartlein, P. J., and Millspaugh, S. H. (1998). A 9000-year fire history from the Oregon Coast Range, based on a high-resolution charcoal study. *Canadian Journal of Forest Research* **28**, 774-787.

MacDonald, G. M., and Cwynar, L. C. (1985). A fossil pollen based reconstruction of the late Quaternary history of lodgepole pine (*Pinus contorta* ssp. *latifolia*) in the western interior of Canada. *Canadian Journal of Forest Research* **15**, 1039-1044.

Meyer, G. M., Wells, S. G., and Jull, A. J. T. (1995). Fire and alluvial chronology in Yellowstone National

Park: Climatic and intrinsic controls on Holocene geomorphic process. *Geological Society of America Bulletin* **107**, 1211-1230.

Miller, D. A., and White, R. A. (1998). A conterminous United States multilayer soil characteristics dataset for regional climate and hydrology modeling. *Earth Interactions* **2**, 1-26.

Millspaugh, S. H., and Whitlock, C. (2003). Postglacial fire, vegetation, and climate history of the Yellowstone-Lamar and Central Plateau provinces, Yellowstone National Park. *In* "After the fires: The ecology of change in Yellowstone National Park." (L. Wallace, Ed.). Yale University Press, New Haven (in press).

Millspaugh, S. H., Whitlock, C., and Bartlein, P. J. (2000). A 17,000-year history of fire for the Central Plateau of Yellowstone National Park. *Geology* **28**, 211-214.

Mitchell, J. F. B., and Johns, T. C. (1997). On modification of global warming by sulfate aerosols. *Journal of Climate* **10**, 245-267.

Mock, C. J., and Bartlein, P. J. (1995). Spatial variability of late-Quaternary paleoclimates in the western United States. *Quaternary Research* **44**, 425-433.

Mock, C. J., and Brunelle-Daines, A. R. (1999). A modern analogue of western United States summer palaeoclimate at 6000 years before present. *The Holocene* **9**, 541-545.

Mohr, J. A., Whitlock, C., and Skinner, C. N. (2000). Postglacial vegetation and fire history, eastern Klamath Mountains, California. *The Holocene* **10**, 587-601.

Shafer, S. L., Bartlein, P. J., and Thompson, R. S. (2001). Potential changes in the distributions of western North America tree and shrub taxa under future climate scenarios. *Ecosystems* **4**, 200-215.

Thompson, R. S., Anderson, K. H., and Bartlein, P. J. (1999). "Atlas of relations between climatic parameters and distributions of important trees and shrubs in North America." *U.S. Geological Survey Professional Paper* **1650** A & B.

Webb, T., III (1995). Pollen records of late Quaternary vegetation change: Plant community rearrangements and evolutionary implications. *In* "Effects of past global change on life." (National Research Council Commission on Geosciences, Environment, and Resources), pp. 221-232. National Academy Press, Washington, DC.

Whitlock, C., and Bartlein, P. J. (1993). Spatial variations of Holocene climatic change in the Yellowstone region. *Quaternary Research* **39**, 231-238.

Whitlock, C., Bartlein, P. J., and Van Norman, K. J. (1995). Stability of Holocene climate regimes in the Yellowstone region. *Quaternary Research* **43**, 433-436.

Whitlock, C., and Millspaugh, S. H. (2001). A paleoecological perspective on past plant invasions in Yellowstone. *Western North American Naturalist* **61**, 316-327.

Whitlock, C., Shafer, S. L., and Marlon, J. (2003). The role of climate and vegetation change in shaping past and future fire regimes in the northwestern U.S. and the implications for ecosystem management. *Forest Ecology and Management* **178**, 5-21.

World WeatherDisc Association (no date). "World WeatherDisc CD-Rom." World WeatherDisc Associates, Inc., Seattle.

Ice Cores from Tropical Mountain Glaciers as Archives of Climate Change

Lonnie G. Thompson[1,2*], Mary E. Davis[1], Ping-Nan Lin[1], Ellen Mosley-Thompson[1,3], and Henry H. Brecher[1]
[1]Byrd Polar Research Center, The Ohio State University, Columbus OH 43210, USA
[2]Department of Geological Sciences, The Ohio State University, Columbus OH 43210, USA
[3]Department of Geography, The Ohio State University, Columbus OH 43210, USA
**phone 614-292-6652, fax 614-292-4697, e-mail thompson.3@osu.edu*

Keywords: Abrupt change, Climate, Ice cores, Isotopes, Tropics, Warming.

1. Introduction

The 20th century has seen the acceleration of unprecedented global and regional-scale climatic and environmental changes to which humans are vulnerable, and by which we will become increasingly more affected in the coming centuries. One-half of the Earth's surface area lies in the tropics between 30°N and 30°S, and this area supports almost 70% of the global population. Thus, temporal and spatial variations in the occurrence and intensity of coupled ocean-atmosphere phenomena such as El Niño and the Monsoons, which are most strongly expressed in the tropics and subtropics, are of worldwide significance. Unfortunately, meteorological observations in these regions are scarce and of short duration. However, ice core records are available from low-latitude, high-altitude glaciers, and when they are combined with high-resolution proxy histories such as those from tree rings, lacustrine and marine cores, corals, etc., they provide an unprecedented view of the Earth's climatic history over several millennia. This paper provides an overview of these unique glacier archives of past climate and environmental changes on millennial to decadal time scales. Also included is a review of the recent, global-scale retreat of these alpine glaciers under present climate conditions, and a discussion of the significance of this

U. M. Huber et al. (eds.), Global Change and Mountain Regions, 31-38.

retreat with respect to the longer-term perspective, which can only be provided by the paleoclimate records.

Over the last 25 years the principal objective of the Ice Core Paleoclimatology Research Group (ICPRG) at the Byrd Polar Research Center has been the acquisition and analysis of a global array of ice cores that can provide high-resolution climatic and environmental histories, which contribute to our understanding of the complex interactions within the Earth's coupled climate system. With the help of new light-weight drilling equipment, we have achieved one of our main scientific objectives by expanding our research from the polar regions to remote ice fields on some of the highest tropical and subtropical mountains. Ice core records from mountains in Africa, South America, and China make it possible to study processes in the subtropical and tropical latitudes where human activities are concentrated. We utilize an ever-expanding ice core database of multiple proxy information (i.e. stable isotopes of oxygen and hydrogen, or $\delta^{18}O$ and δD, respectively, insoluble dust, major and minor ion chemistry, precipitation reconstruction) that spans the globe in spatial coverage and is of the highest possible temporal resolution.

The records contained within the Earth's alpine ice caps and glaciers provide a wealth of data that contribute to a spectrum of critical scientific questions. These range from the reconstruction of high-resolution climate histories to help explore the oscillatory nature of the climate system, to the timing, duration, and severity of abrupt climate events, to the relative magnitude of 20th century global climate change and its impact on the cryosphere. The information from these ice core studies complements other proxy records that compose the Earth's climate history, which is the ultimate yardstick by which the significance of present and projected anthropogenic effects will be assessed.

2. Recent results

The sites from where the ICPRG has retrieved high-altitude ice cores are shown in Figure 1. The first program to drill a low-latitude mountain core to bedrock was carried out on the Quelccaya ice cap in southern Peru (14°S, 71°W) in 1983, and the most recent was accomplished in 2000 on the Puruogangri ice cap (34°N, 89°E) in the center of the Tibetan Plateau. In between, we have recovered cores (Dunde, Guliya, Dasuopu) from other regions of the Tibetan Plateau, from the Andes (Huascarán and Sajama) and from Kilimanjaro in East Africa. With the exception of Puruogangri, all the cores have been analyzed and their overall climate records have been published.

Low-latitude, high-altitude ice core records have revealed the nature of climate variability over both glacial and interglacial time scales, specifically from the Last Glacial Maximum (LGM) 18 to 20 thousand years ago, to the present. Two records from the South American Andes (Huascarán in northern Peru at 9°S, 78°W and Sajama in Bolivia at 18°S, 69°W) and one from the western Tibetan Plateau (Guliya at 35°N, 81°E) extend to or past the LGM and confirm, along with other climate proxy records (e.g. Guilderson et al. 1994; Stute et al. 1995; Colinvaux et al. 1996; Weyhenmeyer et al. 2000), that the LGM was much colder in the tropics and subtropics than previously believed (Thompson et al. 1995; 1997; 1998). Although this period was consistently

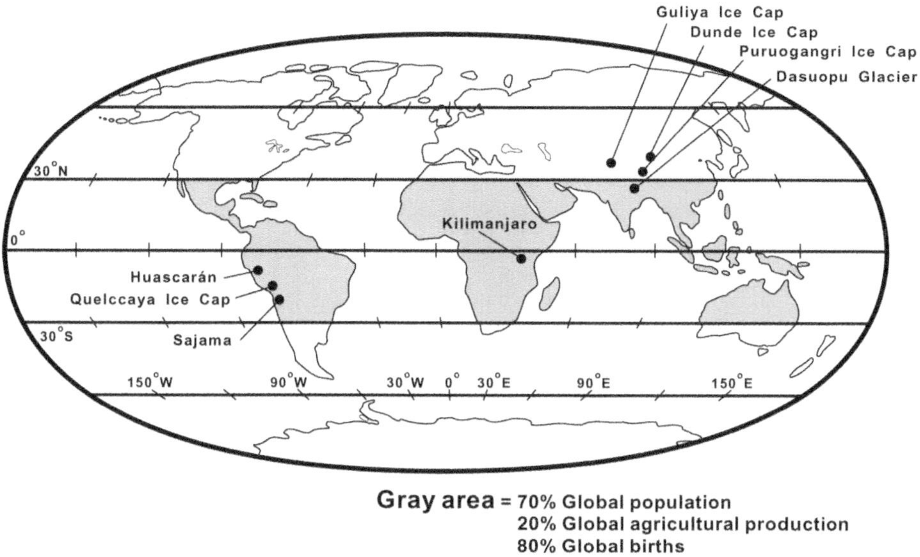

Figure 1: Locations of sites from where ice cores have been taken by the Ice Core Paleoclimate Research Group. The shading depicts the extent of the tropics over land from 30°N to 30°S.

colder, it was not consistently drier through the lower latitudes, unlike in the polar regions. For example, the effective moisture along the axis of the Andes Mountains during the end of the last glacial stage was variable, being much drier in the north than in the Altiplano region in the central part of the range (Thompson et al. 1995; 1998; Davis 2002). In another example, the Guliya ice cap is partly affected by the variability and strength of the Southwest Indian Monsoon system, which was much weaker during the last glacial stage than during the Holocene. However, this region of the Tibetan Plateau also receives (and received) moisture generated from the cyclonic activity carried over Eurasia by the prevailing wintertime westerlies. Not only were many lake levels in the western Tibetan Plateau higher than tropical lakes during the LGM (Li and Shi 1992), but the dust concentrations in the Guliya ice core record were comparable with those of the Early Holocene when the summer Asian Monsoons became stronger, suggesting that local sources of aerosols were inhibited during this cold period by higher precipitation and soil moisture levels (Davis 2002).

Tropical and subtropical ice core records during the Holocene show evidence of major climatic disruptions, specifically droughts. Major dust events, beginning between 4.2 and 4.5 ka and lasting several hundred years, are observed in the Huascarán and Kilimanjaro ice cores (Thompson 2000; Thompson et al. 2002, respectively), and the timing and character of the dust spike is similar to one seen in a marine core record from the Gulf of Oman (Cullen et al. 2000) and a speleothem $\delta^{13}C$ record from a cave in Israel (Bar-Matthews et al. 1999). This dry period is also documented in several other proxy climate records throughout Asia and Northern Africa (see contributions in Dalfes et al. 1994). Two other periods of abrupt, intense climate change in East Africa are observed in the Kilimanjaro ice core at ~8.3 ka and

5.2 ka (Thompson et al. 2002). The latter event is associated with a sharp decrease in $\delta^{18}O$, indicative of a dramatic but short-term cooling.

More recently, a historically documented drought in India in the 1790s, which was associated with monsoon failures and a succession of severe El Niños, was recorded in the insoluble and soluble aerosol concentration records in the Dasuopu ice core (Thompson et al. 2000). Another recorded Asian Monsoon failure in the late 1870s (Lamb 1982; Charles et al. 1997) is noticeable in the Dasuopu dust flux record (Davis 2002), which is a parameter that incorporates both the dust concentration and the annual accumulation rate of ice on the glacier surface.

High-resolution records of Late Holocene variations in temperature are available from low-latitude alpine ice cores. Composites of the $\delta^{18}O$ profiles of the South American cores (Huascarán, Quelccaya, and Sajama) and three of the Tibetan Plateau cores (Dunde, Guliya, and Dasuopu) show similar trends in decadal averages over the last millennium (Thompson et al. 2003) (Fig. 2). When all six of the records from these mountain glaciers are combined, the resulting composite is similar to the Northern Hemisphere temperature records of Mann et al. (1998) and Jones et al. (1998) covering the last 1000 years. As in polar ice cores, the dominant factor controlling mean $\delta^{18}O$ values in Andean snowfall on decadal, centennial, and millennial timescales must be temperature, while on seasonal to annual time scales both temperature and precipitation influence the local $\delta^{18}O$ signal (Vuille et al. 2003). $\delta^{18}O$ variations in ice cores from Bolivia and Peru are highly correlated with sea surface temperatures (SSTs) across the equatorial Pacific Ocean, which are closely linked to ENSO variability (Bradley et al. 2003). Likewise, $\delta^{18}O$ variations in the

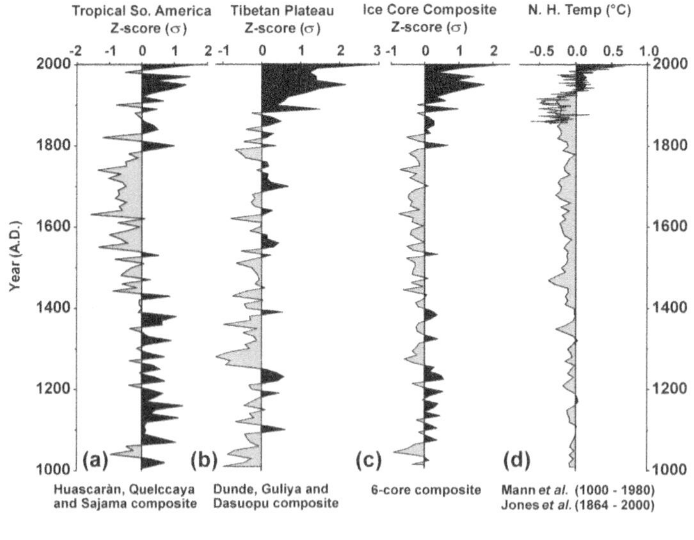

Figure 2: Composite records of decadal averages of $\delta^{18}O$ from ice cores from (a) the South American Andes (Huascarán, Quelccaya, Sajama) and (b) the Tibetan Plateau (Dunde, Guliya and Dasuopu) from A.D. 1000 to the present. All six ice-core records are combined (c) to give a total view of variations in $\delta^{18}O$ over the last millennium in the tropics, which is compared with the Northern Hemisphere reconstructed temperature record (d).

Dasuopu ice core from the Himalayas also reflect SST variations in the equatorial Pacific Ocean. The controlling factor on $\delta^{18}O$ is a matter of debate; however, not only do these comparisons argue for the important role of temperature in the composition of oxygen isotopic ratios in glacier ice, but they also demonstrate abrupt warming from the late 19th century through the 20th century. Indeed, they suggest that the 20th century was the warmest period in the last 1000 years in the tropics, which also encompasses the time of the "Medieval Warming".

The recent warming is recorded in tropical alpine glaciers in other ways, both within the ice core records and by the rapid retreat of many of the ice fields. In the Andes, on the Tibetan Plateau and in the East Africa Rift Valley region this climate change has left its mark. On the Tibetan Plateau, the trend is amplified and is accelerating with increasing elevation (Thompson et al. 2000). The lower elevation ice caps in the Andes are experiencing damage to their seasonal $\delta^{18}O$ signals from the lifting of the 0°C isotherm (Davis et al. 1995). For example, not only is the seasonal isotope signal on the Quelccaya ice cap at 14°S in southern Peru being smoothed out as meltwater percolates through the upper layers of the snow (Thompson et al. 1993), but the ice margins are undergoing rapid and accelerating retreat. The rate of this retreat from 1983 to 1991 (14 m/yr) was almost three times that between 1963 and 1983 (5 m/yr), and in the 2000/2001 year reached 205 m/yr. The many ice fields on Kilimanjaro covered an area of 12.1 km^2 in 1912, but today only 2.6 km^2 remains. If the current rate of retreat continues, the perennial ice on this mountain will likely disappear within the next 20 years (Thompson et al. 2002).

3. Future Priorities

Meteorological data from around the world suggest that the Earth's globally averaged temperature has increased 0.6°C since 1950. The El Niño year of 1998 saw the highest globally averaged temperatures on record, while 2002 (a non-El Niño year) was the second warmest. The marked warmth of the last two decades has contributed to the widespread melting of low-latitude, high-altitude glaciers. During this time, the ICPRG has been monitoring the accelerating retreat of this tropical ice in conjunction with its global ice core drilling and climate reconstruction program.

Seasonal and annual resolution of chemical and physical parameters in ice core records from the Andes Mountains have allowed reconstruction of the variability of the ENSO phenomenon over several hundred years (Thompson et al. 1984; 1992; Henderson 1996; Henderson et al. 1999). Because the effects of El Niño and La Niña events are spatially variable, ice core records from the northernmost (Colombia) and southernmost (Patagonia) reaches of the Andes Mountains will help further resolve the frequency and intensity of ENSO along with temperature variations long before human documentation. This will aid in placing the modern climate changes and the modern ENSO into a more comprehensive perspective.

Variability of the South Asian Monsoon is also of vital importance for a large percentage of the world's population that lives in the affected areas. The ICPRG has drilled four cores on the Tibetan Plateau that have yielded millennial-scale histories of monsoon variability across this large region and information on the interaction between the monsoon system and the prevailing westerlies that are traced back to the Atlantic Ocean. Although marine cores from the Arabian Sea show that the intensity of the South Asian Monsoon has increased over the last four centuries (Anderson et al. 2002), the Dasuopu record from the Himalayas demonstrates that since the early 19th century the amount of precipitation falling on this region has decreased (Thompson et al. 2000). However, the Dunde record from the north side of the Plateau shows an accumulation history that is opposite to that in the Himalayas (Davis and Thompson submitted). Like ENSO, therefore, the South Asian Monsoon systems have varying geographical effects. Retrieval of ice core records from the west side of the Himalayas, which is more directly affected by the SW Indian Monsoon than is the east side where Dasuopu is located, will provide a more comprehensive overview of the precipitation and temperature histories of the Himalayas as a whole. The glaciers on these mountains are vital sources of stream water for the populations of Nepal and India during the dry seasons, and their recent disappearance should be a source of great concern for these countries.

Compelling evidence for major climate warming underway today comes from the tropical glaciers, recorded in both the ice core records and in the drastic retreats of both total area and total volume. The rapid retreat causes concern for two reasons. First, these glaciers are the world's "water towers", and their loss threatens water resources necessary for hydroelectric production, crop irrigation and municipal water supplies for many nations. The ice fields constitute a "bank account" that is drawn upon during dry periods to supply populations downstream. The current melting is cashing in on that account, which was built over thousands of years but is not currently being replenished. As Figure 3 illustrates, all the mountain glaciers in the tropical latitudes are currently retreating (see Kaser and Noggler 1991; Francou and Ribstein 1995; Mark and Seltzer, this volume), as are glaciers in middle and subpolar latitudes (Dyurgerov, Haeberli, both this volume). Although the land between 30°N and 30°S is home to most of the world's population and 80% of the world's births, the average gross domestic product per capita of the 72 tropical nations is only about one-third that of the 78 extratropical nations. Only 20% of the global agricultural production takes place in these climatically sensitive regions and the dwindling water resources for dry-season irrigation will further threaten this production.

The second concern that is brought about by the disappearance of these ice fields is that they contain paleoclimatic histories that are unattainable elsewhere and, as they melt, the records preserved therein are forever lost. These records are needed to discern how climate has changed in the past in these regions and to assist in predicting future changes.

The manifestations of the current global warming remain a topic of much debate, but the scientific evidence verifies that the Earth's globally averaged surface temperature is indeed increasing. At the same time, global water resources are at risk, and mountain glaciers and their unique climate histories are disappearing at an ever-increasing rate. In order to preserve these records that are essential for examining how climate has changed in the past and to predict future changes, we must accelerate the rate at which ice cores are being recovered and focus on those ice fields that are at the greatest risk.

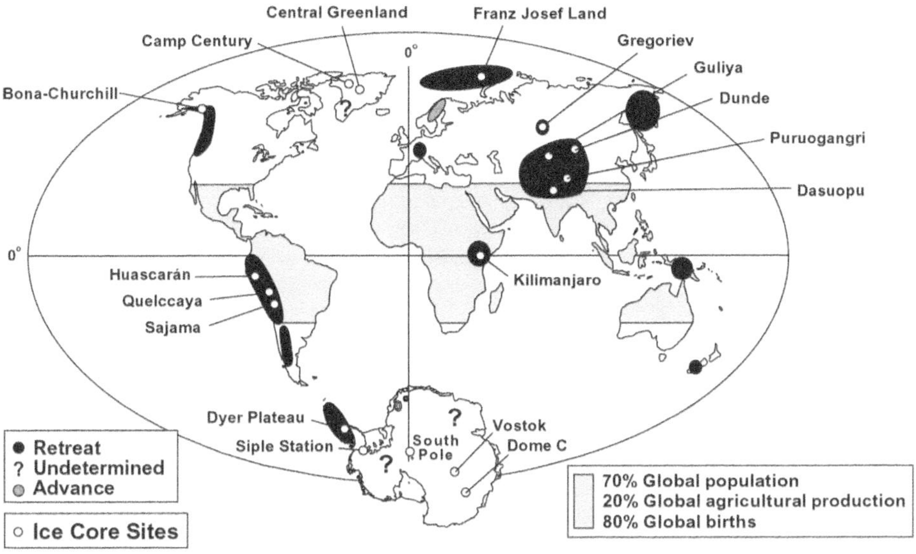

Figure 3: Map demonstrating the current condition of the Earth's cryosphere. Dark shading depicts regions where glacier retreat is underway, while lighter shading depicts where glacier advance is occurring. Shading over land between 30°N and 30°S indicates the tropical regions where much of human activity is currently concentrated.

4. References

Anderson, D. M., Overpeck, J. T., and Gupta, A. K. (2002). Increase in the Asian Southwest Monsoon during the past four centuries. *Science* **297**, 596-599.

Bar-Matthews, M., Ayalon, A., Kaufman, A., and Wasserburg, G. J. (1999). The Eastern Mediterranean paleoclimate as a reflection of regional events: Soreq Cave, Israel. *Earth and Planetary Letters* **166**, 85-95.

Bradley, R. S., Vuille, M., Hardy, D., and Thompson, L. G. (2003). Low latitude ice cores from the Andes record Pacific sea surface temperatures. *Geophysical Research Letters* **30**, 1174 (doi: 10. 1029/2002GL 016546).

Charles, C. D., Hunter, D. E., and Fairbanks, R. G. (1997). Interaction between the ENSO and the Asian Monsoon in a coral record of tropical climate. *Science* **277**, 925-928.

Colinvaux, P. A., DeOliveira, P. E., Moreno, J. E., Miller, M. C., and Bush, M. B. (1996). A long pollen record from lowland Amazonia forest and cooling in glacial times. *Science* **274**, 85-88.

Cullen, H. M., deMenocal, P. B., Hemming, S., Hemming, G., Brown, F. H., Guilderson, T., and Sirocko, F. (2000). Climate change and the collapse of the Akkadian empire: Evidence from the deep sea. *Geology* **28**, 379-382.

Dalfes, H. N., Kukla, G., and Weiss, H. (1994). "Third millennium BC climate change and Old World collapse." Springer, Berlin.

Davis, M. E., Thompson, L. G., Mosley-Thompson, E., Lin, P.-N., Mikhalenko, V. N., and Dai, J. (1995). Recent ice-core climate records from the Cordillera Blanca, Peru. *Annals of Glaciology* **21**, 225-230.

Davis, M. E. (2002). "Climatic interpretations of eolian dust records from low-latitude, high-altitude ice cores." Unpublished Ph.D. thesis, The Ohio State University, Columbus.

Davis, M. E., and Thompson, L. G. (2003). Four centuries of climatic variation across the Tibetan Plateau from ice-core accumulation and $\delta^{18}O$ records. Submitted to "Earth paleoenvironments: Records preserved in mid and low latitude glaciers." Kluwer, New York (submitted).

Francou, B., and Ribstein, P. (1995). Glaciers et évolution climatique dans les Andes Boliviennes: glacier de Zongo et glacier de Chacaltaya Cordillère Royale, 16°S. *Bulletin de l'Institut Français d'Études Andines* **24**, 23-36.

Guilderson, T. P., Fairbanks, R. G., and Rubenstone, J. L. (1994). Tropical temperature variations since 22,000 years ago: Modulating inter-hemispheric climate change. *Science* **263**, 663-665.

Henderson, K. A. (1996). "The El Niño-Southern Oscillation and other modes of interannual tropical climate variability as recorded in ice cores from the Nevado Huascarán col, Peru." Unpublished M.S. thesis, The Ohio State University, Columbus.

Henderson, K. A., Thompson, L. G., and Lin, P.-N. (1999). "Recording of El Niño in ice core $\delta^{18}O$ records from Nevado Huascarán, Peru." *Journal of Geophysical Research* **D104**, 31,053-31,065.

Jones, P. D., Briffa, K. R., Barnett, T. P., and Tett, S. F. B. (1998). High-resolution palaeoclimatic records for the last millennium: Interpretation, integration and comparison with general circulation model control-run temperatures. *The Holocene* **8**, 455-471.

Kaser, G., and Noggler, B. (1991). Observations on Speke Glacier, Ruwenzori Range, Uganda. *Journal of Glaciology* **37**, 315-318.

Lamb H. H. (1982). "Climate history and the modern world." Methuen, London.

Li, S., and Shi, Y. (1992). Glacial and lake fluctuations in the area of the west Kunlun Mountains during the last 45,000 years. *Annals of Glaciology* **16,** 79-84.

Mann, M. E., Bradley, R. S., and Hughes, M. K. (1998). Global-scale temperature patterns and climate forcing over the past six centuries. *Nature* **392**, 779-787.

Stute, M. et al. (1995). Cooling of tropical Brazil (5°C) during the last glacial maximum. *Science* **269**, 379-383.

Thompson, L. G., Mosley-Thompson, E., and Arnao, B. M. (1984). El Niño-Southern Oscillation events recorded in the stratigraphy of the tropical Quelccaya ice cap, Peru. *Science* **226**, 50-52.

Thompson, L. G., Mosley-Thompson, E., and Thompson, P. A. (1992). Reconstructing interannual climate variability from tropical and subtropical ice-core records. *In* "El Niño: Historical and paleoclimatic aspects of the southern oscillation." (H. F. Diaz, and V. Markgraf, Eds.), pp. 295-322. Cambridge University Press, Cambridge.

Thompson, L. G., Mosley-Thompson, E., Davis, M. E., Lin, P.-N., Yao, T., Dyurgerov, M., and Dai, J. (1993). Recent warming: Ice core evidence from tropical ice cores with emphasis upon Central Asia. *Global and Planetary Change* **7**, 145-146.

Thompson, L. G. et al. (1995). Late Glacial Stage and Holocene tropical ice core records from Huascarán, Peru. *Science* **269**, 47-50.

Thompson, L. G. et al. (1997). Tropical climate instability: The last glacial cycle from a Qinghai-Tibetan ice core. *Science* **276**, 1821-1825.

Thompson, L. G. et al. (1998). A 25,000 year tropical climate history from Bolivian ice cores. *Science* **282**, 1858-1864.

Thompson, L. G. (2000). Ice-core evidence for climate change in the Tropics: Implications for our future. *Quaternary Science Reviews* **19**, 19-36.

Thompson, L. G., Yao, T., Mosley-Thompson, E., Davis. M. E., Henderson, K. A., and Lin, P.-N. (2000). A high-resolution millennial record of the South Asian Monsoon from Himalayan ice cores. *Science* **289**, 1916-1919.

Thompson, L. G. et al. (2002). Kilimanjaro ice core records: Evidence of Holocene climate change in tropical Africa. *Science* **298**, 589-593.

Thompson, L. G., Mosley-Thompson, E., Davis, M. E., Lin, P.-N., Henderson, K., and Mashiotta, T. A. (2003). Tropical glacier and ice core evidence of climate change on annual to millennial time scales. *Climatic Change* (in press).

Vuille, M., Bradley, R. S., Healy, R.,Werner, M., Hardy, D. R., Thompson, L. G., and Keimig, F. (2003). Modeling $\delta^{18}O$ in precipitation over the tropical Americas, Part II: Simulation of the stable isotope signal in Andean ice cores. *Journal of Geophysical Research* **108**, 10.1029/ 2001 JD 002039.

Weyhenmeyer, C. E., Burns, S. J., Waber, H. N., Aeschbach-Hertig, W., Kipfer, R., Loosli, H. H., and Matter, A. (2000). Cool glacial temperatures and changes in moisture source recorded in Oman groundwaters. *Science* **287**, 842-845.

The Contribution of Cosmogenic Nuclides to Unraveling Alpine Paleoclimate Histories

John C. Gosse
Department of Earth Sciences, Dalhousie University, Halifax, Nova Scotia, Canada
phone +00-1-902-494-6632, fax +00-1-902-494-6889, e-mail john.gosse@dal.ca

Keywords: ^{10}Be, Exposure dating, Glaciers, Moraine, Paleoclimate

1. Introduction

Moraines are non-continuous short-term records of ice marginal positions. Moraines help provide important paleo-glaciological mass balance information (e.g. glacier surface area, ice volume, terminus elevation, snowline altitudes, longitudinal ice surface gradient below the paleo-snowline) which in part controls the geometry of the glacier and the rate of advance and retreat of an ice margin. Therefore, chronologies on these ancient glacial landforms can be directly tied to local paleo-temperature and paleo-precipitation estimates for specific times during and after a glaciation. In the past two decades, the terrestrial cosmogenic nuclide (TCN) exposure dating method has made a revolutionary contribution to the study of alpine paleo-glacial histories and paleoclimatology. (i) Exposure dating of boulders on moraines provides the time since a boulder was deposited from an ice margin. It directly determines when the glacier reached a measurable mass-balance condition, whereas other chronometers, such as radiocarbon, U-series, and luminescence dating, typically provide only minimum or maximum limiting ages on ice margin positions. (ii) The method can provide a precise estimate of the timing of initial ice retreat. Timing of when an alpine glacier reaches its maximum position is not only a function of local climate but also of numerous glaciological and hydrological conditions. Initial retreat is the most discrete short-lived climate-response event in a moraine record. Unlike the timing of initial retreat, initial advance is not recorded in moraine records because glaciers override

U. M. Huber et al. (eds.), Global Change and Mountain Regions, 39-49.

their moraines during advance (Gibbons et al. 1984). The ability to precisely date the timing of initial retreat may be the technique's most important yet under-utilized contribution to global change research. (iii) In the alpine environment, TCN's have been used to exposure date outwash terraces, bedrock and tor surfaces, and perched erratics to establish histories of ice marginal positions and study glacier processes. (iv) TCN methods are applicable from 10^1 to 10^6 years and are therefore useful in linking alpine events (especially those that occur before ~60 ka - the radiocarbon limit) with long-term continuous records of climate change.

Paleoclimatology has benefited from the increased reliability of TCN dating. Moraines can be dated directly and affiliated with targeted Quaternary events, such as cooling intervals associated with the Younger Dryas chron (Gosse et al. 1995a) or Heinrich Events (Phillips et al. 1996). This helps to indicate if a climate signal was of local, regional, hemispheric, or global extent and therefore provides constraints on the climate change triggers and the roles of atmospheric and oceanic circulation that must have controlled glacier extent. Regional climate changes, inferred from moraine chronologies, must be reproduced by paleoclimate simulations before the same models can be used for climate predictions. As the precision of the TCN dating method continues to improve, a new level of questions can be tackled: Are there subtle spatial trends in the timing of initial deglaciation across large regions (e.g. Eurasia) that have been assumed synchronous? What atmospheric conditions could explain such trends? Are there any alpine moraine records of the 2000-year phase lag in climate change indicators between the high latitude Southern Ocean and the North Atlantic ice core stratigraphy? What was the natural rate of retreat of alpine glaciers from the last major stadial (Younger Dryas) and how does this compare to current rates of anthropogenically-induced shrinkage?

2. Principles of the TCN method

For a recent review of the theory and applications of the TCN method the interested reader is referred to Gosse and Phillips (2001), as only a cursory review can be provided here. Galactic cosmic radiation, primarily high-energy protons, penetrates Earth's magnetic field and interacts with nuclei of atoms in the atmosphere. The high-energy interactions produce a cascade of secondary particles, including protons, neutrons, and muons. As the secondary radiation penetrates the atmosphere and Earth's surface, the average flux attenuates. Fast neutrons account for the majority of the production of useful TCN at Earth's surface, and the mean attenuation length of fast neutrons in granite near the surface is roughly 50 cm (distance over which the flux is decreased by a factor of $1/e$). The concentration **C** (atoms·g^{-1}) of TCN in a mineral is proportional to their rate of production **P** (atoms·g^{-1}·y^{-1}, varies spatially and is different for each isotope-mineral system) and to the time **T** (yr) that the rock was exposed. Hence, measured concentrations of these nuclides can be used to determine the exposure time of a rock surface. For stable isotopes on an ideal rock surface, **T = C/P**. Most surfaces are not ideal for dating. The dependence of the TCN dating method on careful sampling approaches cannot be overstated as the technique is extremely sensitive to the exposure history of the rock surface. Rock surfaces that erode will have a lower concentration for the same exposure duration than a more resistant surface. Rocks buried by ice, snow, ash, dense forests, loess or other sediments are partially or completely shielded from cosmic radiation. Spatial and temporal changes in production rates due to geomagnetic, atmospheric, and local geometry effects are becoming better established. These factors, along with loss of radioisotopes due to decay and a correction for non-cosmogenic concentrations of the isotopes, need to be considered when interpreting exposure ages.

Although a wide spectrum of TCN is produced from these interactions, only six are routinely used. Cosmogenic ^{14}C, ^{36}Cl, ^{26}Al, and ^{10}Be are radionuclides with half-lives of 5.7 kyr, 0.3 Myr, 0.7 Myr, and 1.5 Myr respectively, and cosmogenic ^{3}He and ^{21}Ne are stable noble gases. The TCN are sufficiently stable to be useful for dating on geological timescales ranging from 10^1 to 10^6 yr. There is sufficient production of the TCN in common minerals to allow measurement above natural, non-cosmogenic, or chemical background abundances of the isotope or isobars in the mineral or rock type. All six TCN have provided alpine paleo-glacier chronologies. The stable isotopes and ratios of two radioisotopes have been used to date surfaces exposed for > 1 Myr. Isotopes with relatively high production rates (e.g. ^{3}He) are more useful for determining the age of shorter exposures, but even ^{10}Be, with the lowest sea level surface production rate of the six at 5 atoms·g^{-1}·yr^{-1}, can date high altitude surfaces that have been exposed during the past 1000 years. Essentially any rock type can be dated. The most commonly utilized minerals are quartz (^{10}Be, ^{14}C, ^{21}Ne, ^{26}Al, ^{36}Cl), and olivine and pyroxene (^{3}He). Cosmogenic ^{36}Cl is the only TCN measured routinely in a "whole rock" sample. Other isotope-mineral systems continue to be evaluated for their usefulness for geological applications.

3. Applications and key contributions of TCN to alpine global change research

This section describes some noteworthy contributions of TCN to global change research. Alpine glaciers are useful in monitoring climate change on scales of 10^1 to 10^5 years because their volume responds quickly to changes in temperature and precipitation (Oerlemans and Fortuin 1992; Oerlemans 1994). Moraine maps permit the reconstruction of paleoequilibrium line altitudes, which are approximately equivalent to annual snowline altitudes and rough indications of the average elevation of the annual zero-degree isotherm. In general, an alpine glacier responds sensitively to an increase in winter precipitation or decrease in summer temperature because its equilibrium line altitude (ELA, separating zones of net annual ablation and accumulation on the glacier surface) will decrease and glacier volume will expand to maintain a mass balance above and below the ELA. In the case of drying or warming, the ELA rises in accordance to local adiabatic lapse rates and the glacier will shrink. Paleo-mass balance calculations can constrain ranges of past precipitation and temperature changes reflected by paleo-ice volume change. However, precisely dating ice margin positions in the Quaternary has proven difficult. Radiocarbon ages reflect the timing of death of an organism (cessation of gas exchange with the atmosphere) in the glacial sediment and rarely tie a precise age to a particular paleo-ice volume (see Reasoner et al. (1994) and Madole (1986) for exceptions). Other dating methods, such as U-series dating, have similar disadvantages in that they typically date some event (e.g. travertine deposition or petrocalcic horizon in till), indirectly related to when a glacier margin was at a specific position (see Sturchio et al. (1994) for an exception). The lack of sufficiently precise chronologies was one of the chief factors identified (Clark and Bartlein 1995) for the apparent lack of global synchronicity of glacier maximum advances, according to Gillespie and Molnar (1995).

It is important to point out that the accuracy of an exposure age derived from a single boulder with a single isotope will probably never be better than 10% (2σ). The majority of this uncertainty is a reflection of the uncertainty in the various production rate algorithms. Additionally, there is a real probability that environmental conditions that influence the concentration of the cosmogenic isotopes may not be well established for the entire duration of exposure. We recognize that a thin cover of till may have persisted on a large boulder for a few thousand years following deposition, snow cover thickness, density, and duration have likely varied, and small amounts of inheritance may exist in even the most deeply plucked boulders. For this reason, it is critical to rely only upon ages derived from multiple measurements from a moraine. Where the random errors contribute negligibly, the precision of an average age may be sufficiently precise to distinguish ages at the 4% (2σ) level or better. This means it is possible to establish the relative timing of a certain short-lived event, such as the Younger Dryas cooling (12.8 to 11.5 ka) or the initial retreat from the last major glaciation (approximately 20 ka). Documenting synchronicity and spatial trends in the responses of alpine glaciers is critical to our understanding of the climate change mechanisms and telekinesis among energy sources (i.e. the Sun, ocean, and atmosphere). A compilation of glacial boulder exposure ages measured in the past 3 years (in an attempt to use similarly scaled age calculations) suggests that the distribution of age occurrences may begin to reflect the ages of moraines worldwide (Fig. 1). Although the magnitudes of the occurrences are influenced by many non-climate factors, such as funding, availability of suitable boulders, and various glaciological conditions, the distribution shows that there appears to be a Younger Dryas age signal throughout the world, whereas a 16-kyr event that may be related to Heinrich event 1 seems to be more prevalent in the Northern Hemisphere. Clearly, more consideration of site attributes and more data are needed to make any conclusions based on this distribution. However, if Younger Dryas-age moraines are demonstrably synchronous throughout the world, then the dramatic cooling could not have been communicated solely by ocean circulation because of the long (several kyr) period for complete circulation. Similarly, if few or no moraine ages in the Southern Hemisphere are found concordant with the North Atlantic Heinrich events, but many moraines in the Northern Hemisphere are, this indicates that any associated climate changes may have only weakly influenced the Southern Hemisphere and suggests that the source of the climate change was in the Northern Hemisphere. A primary objective of exposure dating of moraines is to help direct climate simulations by providing certain boundary conditions and to test hypotheses of ocean-atmosphere dynamics used in simulations for predicting future climate trends.

3.1 Dating ice marginal positions

An important breakthrough in moraine chronology was realized with the development of accelerator and noble gas mass spectrometry. Utilizing the prescient theoretical foundation set in the preceding decades, Phillips et al. (1986) used AMS to measure cosmogenic ^{36}Cl in boulders on moraine ridge crests to determine the age of Sierra Nevadan glaciations. Phillips et al. determined that the technique would yield exposure durations that were similar to the published ages of the moraines based on geomorphometric indices and a scant number of radiocarbon dates. For the younger (Tioga) moraine sequences, the technique was shown to yield reproducible ages that were consistent with the independently attained ages of the moraines. This was a significant result for the TCN technique as well as the field of glacial geology. However, the exposure ages Phillips et al. obtained for the older moraine sequences were not in stratigraphic sequence. This result brought attention to the influences of moraine crest denudation (Hallet and Putkonen 1994) and boulder erosion (Zimmerman et al. 1994) on exposure dating. Since then, more than 1000 boulders on alpine moraines have been dated with cosmogenic nuclides on all seven continents (Fig. 1). Subsequently, the Sierra Nevadan Tioga glacial chronology (Phillips et al. 1986) was correlated directly via outwash sedimentation to well-dated glacial-fed lake sediment records of the Owens Valley (Benson et al. 1996). These results led

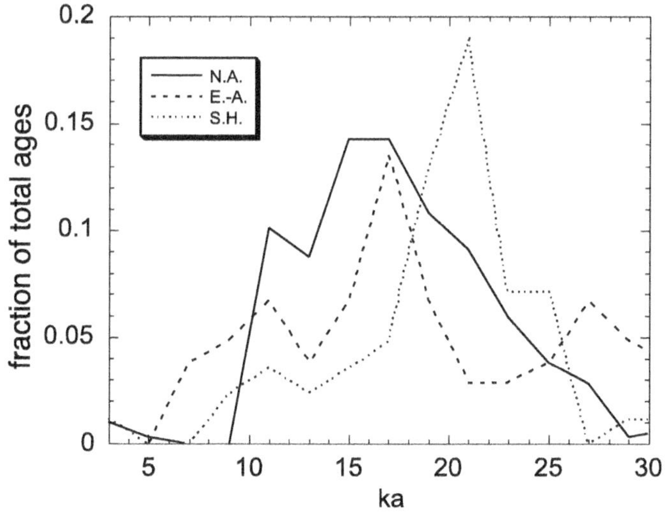

Figure 1: Recurrence of TCN exposure ages on glacial erratics worldwide (*n*=441), showing only the last 30 ka (2-kyr bins). N.A. North America; E.-A. Europe and Asia; S.H. Southern Hemisphere. Data compiled from publications using recently revised production rates and altitude scaling for ^{10}Be, ^{26}Al, ^{3}He, or ^{21}Ne. Ages remain unaltered from the original publications: Marsella et al. 2000; Phillips et al. 2000; Barrows et al. 2002; Licciardi et al. 2001; Jackofsky et al. 2001; Owen et al. 2002; Schaefer et al. 2002; Easterbrook et al. 2003; Macdonald 2003; Marquette et al. in press.

to the conclusion that Heinrich Events and other rapid north-Atlantic type climate change signals were evident in the southwestern USA glacial moraine and lake records. A more recent development in alpine and piedmont moraine chronology has revealed that the oxygen isotope stage 4 glaciation (approximately 75 to 60 ka) was much more significant in the southern hemisphere than the northern (Barrows et al. 2002; Jackofsky 2001). Additional chronologies are needed to confirm this, and future climate simulations will be used to help fully understand the asymmetry.

3.2 Multi-nuclide approaches to ice histories

By the end of the 1980's, a group of chemists and physicists, including K. Nishiizumi, D. Lal, J. Klein, R. Middleton, and J. Arnold, devised a means to overcome the limiting influence of surface erosion and burial on TCN exposure dating. They showed that when two isotopes with different half lives are measured in an exposed sample, their ratio could be used to evaluate the steady state erosion rate of the rock or provide a minimum estimate of how long exposure of a rock may have been interrupted by a burial event. Although earlier isotopic ratio results were published, their article (Nishiizumi et al. 1991) described the first use of $^{26}Al/^{10}Be$ to document ice history and low erosion rates in the Antarctic Dry Valleys region and Transantarctic Mountains. This multi-nuclide approach takes advantage of the influence of nuclear decay rate and erosion rate on TCN concentration. Secular equilibrium (when the number of isotopes lost to decay equals the number of isotopes produced) is attained sooner by a shorter-lived isotope. Erosion causes a radionuclide concentration to equilibrate even earlier. The isotopic ratio between the shorter- and the longer-lived radionuclide can be used to simultaneously determine the exposure duration and erosion rate. In the most common way to depict the ratio plot, the exposure duration increases from left to right along the upper solid curve (Fig. 2).

The multi-nuclide approach also helps to identify complications to exposure histories. For instance, the technique can be used to show if ice had ever covered a landscape without eroding it. This non-erosive ice cover is referred to as "frozen-based" or "cold-based". Cold-based ice cover has been particularly important in polar or alpine regions where thick permafrost may have existed prior to glaciation. However, it may have been more expansive than the glaciology community realizes because its geomorphic effects are minor relative to the effects of warm-based (erosive) ice, which, over just a short time, has such a profound effect on the landscape. In the case of the $^{26}Al/^{10}Be$ ratio plot, ratios plotting below the erosional equilibrium line must have experienced shielding during exposure (Fig. 2). The technique therefore provides a minimum estimate of the burial duration (assuming no erosion, one burial event, and no post-burial exposure). Of course it is possible that multiple burial (glacial) and re-exposure (interglacial) events have been recorded by a surface but the total duration of these cannot be calculated.

In theory, the multi-nuclide approach is invaluable and should be considered for many applications of TCN in order to ensure simple exposure histories and adjust for erosion. However, the precision of ratio measurements limits the use of $^{26}Al/^{10}Be$, for example, to surfaces with longer exposure histories and inhibits unique solutions of erosion-exposure history (Gosse and Phillips 2001). Ratios involving shorter-lived radionuclides (e.g. ^{14}C) are more appropriate for shorter burial histories, and ratios including thermal neutron capture nuclides (^{36}Cl) are more useful for characterizing erosion rates (Phillips et al. 1997).

The multi-nuclide approach, coupled with other evidence (marine multi-bathymetry data, more detailed glacial geology mapping, and soil mineralogy and geochemistry), has helped to develop a significantly different conceptual history of ice cover in eastern Canadian highlands. Many summits along the coastal mountains have scant evidence of glaciation. As a result, the summit areas had been misinterpreted as ice-free during at least the last glaciation. This meant that paleo-ice volume was interpreted to be lower than actual. The proposal of ice-free enclaves on these summits was also used to explain the observation of disjunct populations of bryophytes and beetles in coastal highlands in eastern and western Canada, Greenland, Scandinavia, and Antarctica. As other areas with mostly cold-based ice cover are being recognized, we may see a substantial change in ice volumes and paleoclimate parameters for much of the northern hemisphere. Using ^{10}Be to date perched erratics (very rare on the highest summits) and the ratio of two nuclides, instead of one, to examine the exposure history of tor-like outcrops of bedrock, Gosse and others (Marquette et al. in press) have shown in eastern Canada that the summit erratics post-date the last glaciation (many average 12 ka, Younger Dryas-age), yet the minimum exposure history of adjacent bedrock is 0.5 Ma. The erratics indicate that the last ice cover

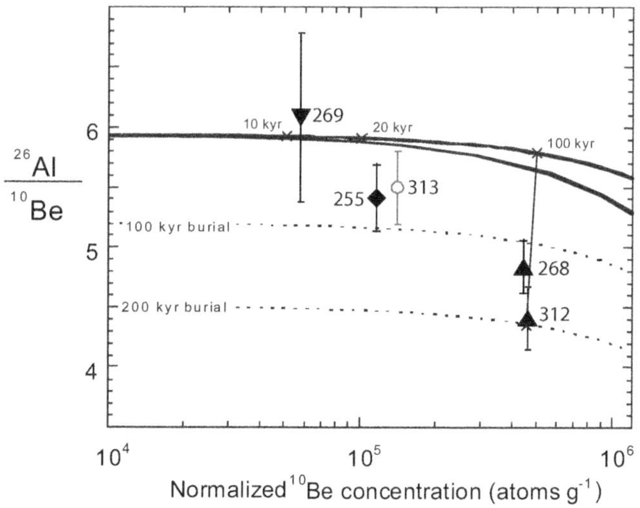

Figure 2: $^{26}Al/^{10}Be$ ratio plot for bedrock samples in eastern Canada (modified from Marquette et al. in press). ▲: Bedrock tors on high summits; ▼ and ◆: Valley bedrock; **O**: Erratic on highest summit; error bars 1s analytical precision. Samples below the solid thick curves indicate that the exposure of the surface was interrupted. Many combinations of multiple exposure, erosion, and partial or complete burial periods can explain the data below the thick curves. In the case of sample 312, the minimum burial duration is 200 kyr after an initial exposure of 100 kyr, although the actual total history could be much longer than the implied 300 kyr.

persisted until the end of the Younger Dryas, but the old exposure age on the summit bedrock surfaces reveal that there was little or no glacial erosion (i.e. the summits were covered by cold-based ice). The existence of non-erosive ice cover also explains the stark contrast between summit soils, with higher abundances of gibbsite, kaolinite, and chlorites, and the valley soils, which are derived from the same parent material but were clearly rejuvenated by glacierization.

3.3 High-precision time control on high-amplitude regionally correlable climate events

In this example of the application of TCN for mountain climate change study, a distinction is made between dating and correlation. The reliability of the former is controlled by the accuracy of the method, whereas correlation requires a high degree of precision. In other words, if the same systematic errors apply everywhere to the TCN methods used, then the systematic errors may be ignored when correlating moraine records dated by this method over long distances. In order to use ice marginal records to test hypotheses of regional or global synchronicities of climate events, the events must have been significant enough to have affected the glacier's mass balance, but short enough so that the synchronicity can be adequately evaluated. Individual glaciers respond to a variety of forcings and environmental conditions. For instance, under the same climate change, the time it takes individual glaciers to advance and re-equilibrate will differ because factors, such as basal hydrology and glacier size, influence flow and response time. Similarly, the length of time a glacier occupied its terminal position during a glaciation will vary for each glacier, and may be as much as 5 kyr for some large alpine systems (Pinedale Wyoming, Gosse et al. 1995b; 2004). The only events that are useful to attempt to correlate among regional and global glacier records are the initial responses to rapid high amplitude climate changes. A record of initial retreat is preserved in all moraine sequences.

By the mid-1990s it was shown by two independent groups concurrently working on Younger Dryas-age moraines in the Titcomb Basin of the Wind River Range, Wyoming (Gosse et al. 1995a) and the famous Julier Pass in the Swiss Alps (Ivy-Ochs et al. 1996) that a single glacial landform could be exposure dated with a coefficient of variation about the mean age of the moraine (s/m) of greater than 5%. This is better than the estimated total random uncertainty (1σ) in an individual exposure age even with today's technology. If additional adjustments were made to compensate for neutron loss due to the size and geometry of the boulders (Masarik and Wieler 2003) the standard deviation would approach 2%. The mean age of the boulders at both sites fell within the 1300-year duration of the Younger Dryas Chron. This result gave further credence to the accuracy of the TCN method, at least for short exposures where erosion was not an influence. The precision demonstrated that inheritance was probably not a significant factor in dating glacial boulders at those sites. The precision (about 600-years at Younger Dryas times) is equivalent or better than the 1σ precision of calibrated radiocarbon dates for the same period. The moraines could be reliably correlated around the world. In the case of the Wind River Range experiment, lakes dissected by the moraine were later cored to confirm with fossil pollen, sedimentology, and 30 AMS radiocarbon dates on terrestrial macrofossils and bulk sediment that the moraine was indeed Younger Dryas in age (Gosse et al. 2004). This high precision was reproduced when dating multiple late-Pleistocene recessional moraines at the Pinedale type locality for the last glaciation (four tightly nested recessional moraines, 9 boulder dates (Gosse et al. 1995b). The results showed that ice marginal retreat from the terminal moraines must have proceeded at a rate faster than the resolution of the method (e.g. > 2 km/kyr).

The reproducibility of multiple boulder ages on single moraines or a group of recessional moraines led to an important application of exposure dating to climate change. It is possible to date the timing of initial retreat by exposure dating large boulders near the back (up-ice direction) edge of the terminal crest and the foremost boulders on the top of the first recessional moraine. The moraine chronology at Pinedale, Wyoming, (Gosse et al. 1995b) demonstrates this approach. Samples (n=14) collected on the crest of the Pinedale (Last Glacial Maximum) terminal moraine show a decreasing age distribution (24.7 to 18.2 ka) from front to back. The mean age of the first recessional moraine in a set of tightly nested recessional moraines of the adjacent piedmont lobe is 18.2 ka, and this is also the mean age of all four recessional moraines. The exact agreement of the two boulders on the back rim of the terminal and boulders on top of the first recessional moraines is a coincidence. Nevertheless, if suitable boulders exist on the targeted moraines and sufficient samples are measured, the timing of initial retreat after the last glaciation (i.e. local warming) may be resolved to within 1-kyr worldwide. By establishing precise chronologies for glacial events, we are becoming better prepared to address questions of how past ocean and atmospheric circulation impacted regional, hemispheric and global climate change, using local proxies.

4. Future alpine global change studies involving TCN

In many high-latitude mountainous regions, new results from geomorphology, soils, and TCN dating (e.g. Marquette et al. 2003) are challenging previous estimates of ice cap and ice sheet volume. More analyses of TCN ratios are needed to demonstrate that ice cover may have been significantly more extensive during the Younger Dryas and late Pleistocene glaciations. Coupling soils, geomorphology, and TCN data to provide constraints on thermomechanical ice cover models will help direct the new research.

As the systematic and random errors for TCN dating continue to be reduced and new isotope-mineral systems become established, new questions may be addressed. For example, what was the average annual snow cover ($g \cdot cm^{-2}$) in a given region during the Little Ice Age? In the next decade, measurements of ^{14}C in quartz will become routine. This will open up some new questions regarding cold-based ice cover, and will undoubtedly provide important insights regarding ice dynamics and glaciological processes.

Where possible, I urge the glacial community to stop acquiring low precision ages on broad events such as the last glacial maximum. Terminal moraine records reflect multiple processes and provide little direct data to increase our understanding of climate change. In order to maximize benefit of paleo-glacier moraine records, the timing of initial retreat from the terminal moraine must be dated precisely using TCN from as many boulders on the terminal and recessional moraines as needed to attain the required precision. This will provide a robust and sufficiently precise age constraint on single events that is useful for climate modelers and glaciologists.

5. Acknowledgements

A. Murphy and J. Willenbring provided comments on the original manuscript. Data for eastern Canada (Fig. 2) were generated from NSF grant (OPP-9906280) and ACOA grant AIF-1005052.

6. References

Barrows, T. T., Stone, J. O., Fifield, L. K., and Cresswell, R. G. (2002). The timing of the last glacial maxima in Australia. *Quaternary Science Reviews* **21**, 159-173.

Benson, L. V., Burdett, J. W., Kashgarian, M., Lund, S. P., Phillips, F. M., and Rye, R. O. (1996). High-resolution records of climatic and hydrologic change in the Owens Lake Basin and adjacent Sierra Nevada. *Science* **274**, 746-748.

Clark, P. U., and Bartlein, P. J. (1995). Correlation of late Pleistocene glaciation in the western U.S. with North Atlantic Heinrich events. *Geology* **23**, 483-486.

Easterbrook, D. J., Pierce, K., Gosse, J. C., Gillespie, A., Evenson, E. B., and Hamblin, K. (2003). Quaternary geology of the western United States. In "Quaternary Geology of the United States." (D. J. Easterbrook, Ed.), pp. 19-80. Denver, Geological Society of America.

Gibbons, A. B., Megeath, J. D., and Pierce, K. L. (1984). Probability of moraine survival in a succession of glacial advances. *Geology* **12**, 327-330.

Gillespie, A., and Molnar, P. (1995). Asynchronous maximum advances of mountain and continental glaciers. *Reviews of Geophysics* **33**, 311-364.

Gosse, J. C., Evenson, E. B., Klein, J., Lawn, B., and Middleton, R. (1995a). Precise cosmogenic ^{10}Be measurements in western North America: Support for a global Younger Dryas cooling event. *Geology* **23**, 877-880.

Gosse, J. C., Klein, J., Evenson, E. B., Lawn, B., and Middleton, R. (1995b). Beryllium-10 dating of the duration and retreat of the last Pinedale glacial sequence. *Science* **268**, 1329-1333.

Gosse, J. C., Klein, J., Davis, P. T., Evenson, E. B., Jull, A. J. T., and Burr, G. (2004). Cosmogenic ^{10}Be and ^{26}Al production rates at mid-latitude high altitude sites for exposures of 10 to 15 kyr. *Earth and Planetary Science Letters* (submitted).

Gosse, J. C., and Phillips, F. M. (2001). Terrestrial *in situ* cosmogenic nuclides: Theory and applications. *Quaternary Science Reviews* **20**, 1475-1560.

Hallet, B., and Putkonen, J. (1994). Surface dating of dynamic landforms: Young boulders on aging moraines. *Science* **265**, 937-940.

Ivy-Ochs, S., Schlüchter, C., Kubik, P., Synal, H.-A., Beer, J., and Kerschner, H. (1996). The exposure age of an Egesen moraine at Julier Pass, Switzerland, measured with the cosmogenic radionuclides ^{10}Be, ^{26}Al and ^{36}Cl. *Eclogae Geologicae Helvetiae* **89**, 1049-1063.

Jackofsky, D. S. (2001). "Quaternary glacial chronology and climate dynamics in Tierra del Fuego, Chile and at Lago Nahuel Huapi, Argentina." Masters thesis, University of Kansas, Lawrence.

Licciardi, J. M., Clark, P. U., Brook, E. J., Pierce, K. L., Kurz, M. D., Elmore, D., and Sharma, P. (2001). Cosmogenic ^{3}He and ^{10}Be chronologies of the late Pinedale northern Yellowstone ice cap, Montana, USA. *Geology* **29**, 1095-1098.

MacDonald, F. (2003). "Glacial geology and geochronology of the Peggy's Cove region." Honours Thesis, Dalhousie University, Halifax, CA.

Madole, R. F. (1986). Lake Devlin and Pinedale glacial history, Front Range, Colorado. *Quaternary*

Research **25**, 43-54.

Marquette, G. C., Gray, J. T., Gosse, J. C., Courchesne, F., Stockli, L., Macpherson, G., and Finkel, R. (in press). Felsenmeer persistence through glacial periods in the Torngat and Kaumajet Mountains, Quebec-Labrador, as determined by soil weathering and cosmogenic nuclide exposure dating. *Canadian Journal of Earth Sciences*.

Marsella, K. A., Bierman, P. R., Davis, P. T., and Caffee, M. W. (2000). Cosmogenic ^{10}Be and ^{26}Al ages for the last glacial maximum, eastern Baffin Island, Arctic Canada. *Geological Society of America Bulletin* **112**, 1296-1312.

Masarik, J. and Wieler, R. (2003). Production rates of cosmogenic nuclides in boulders. *Earth and Planetary Science Letters* **216**, 201-208.

Nishiizumi, K., Kohl, C. P., Arnold, J. R., Klein, J., Fink, D., and Middleton, R. (1991). Cosmic ray produced ^{10}Be and ^{26}Al in Antarctic rocks: Exposure and erosion history. *Earth and Planetary Science Letters* **104**, 440-454.

Oerlemans, J. (1994). Quantifying global warming from the retreat of glaciers. *Science* **264**, 243-245.

Oerlemans, J., and Fortuin, J. P. F. (1992). Sensitivity of glaciers and small ice caps to greenhouse warming. *Science* **258**, 115-117.

Owen, L. A., Spencer, J. Q., Haizhou, M., Barnard, P. L., Derbyshire, E., Finkel, R. C., Caffee, M. W., and Zeng Yong, N. (2002). Timing of late Quaternary glaciation along the southwestern slopes of the Qilian Shan, Tibet. *Boreas* **32**, 281-291.

Phillips, F. M., Leavy, B. D., Jannik, N. O., Elmore, D., and Kubik, P. W. (1986). The accumulation of cosmogenic Chlorine-36 in rocks: A method for surface exposure dating. *Science* **231**, 41-43.

Phillips, F. M., Zreda, M. G., Benson, L. V., Plummer, M. A., Elmore, D., and Sharma, P. (1996). Chronology for fluctuations in Late Pleistocene Sierra Nevada glaciers and lakes. *Science* **274**, 749-751.

Phillips, F. M., Zreda, M. G., Gosse, J. C., Klein, J., Evenson, E. B., Hall, R. D., Chadwick, O. A., and Sharma, P. (1997). Cosmogenic ^{36}Cl and ^{10}Be ages of Quaternary glacial and fluvial deposits of the Wind River Range, Wyoming. *Geological Society of America Bulletin* **109**, 1453-1463.

Phillips, W. M., Sloan, V. F., Shroder, J. F., Jr., Sharma, P., Clarke, M. L., and Rendell, H. M. (2000). Asynchronous glaciation at Nanga Parbat, northwestern Himalaya Mountains, Pakistan. *Geology* **28**, 431-434.

Reasoner, M. A., Osborn, G., and Rutter, N. W. (1994). Age of the Crowfoot advance in the Canadian Rocky Mountains: A glacial event coeval with the Younger Dryas oscillation. *Geology* **22**, 439-442.

Schaefer, J. M., Tschudi, S., Zao, Z., Wu, X., Ivy-Ochs, S., Wieler, R., Baur, H., Kubik, P. W., and Schluechter, C. (2002). The limited influence of glaciations in Tibet on global climate over the past 170 000 yr. *Earth and Planetary Science Letters* **194**, 287-297.

Sturchio, N. C., Pierce, K. L., Murrell, M. T., and Sorey, M. L. (1994). Uranium-series ages of travertines and timing of the last glaciation in the northern Yellowstone area, Wyoming-Montana, USA. *Quaternary Research* **41**, 265-277.

Zimmerman, S. G., Evenson, E. B., Gosse, J. C., and Erskine, C. P. (1994). Extensive boulder erosion resulting from a range fire on the type–Pinedale moraines, Fremont Lake, Wyoming. *Quaternary Research* **42**, 255-265.

Holocene Glacier Fluctuations and Winter Precipitation Variations in Southern Norway

Atle Nesje[1,3*], Svein Olaf Dahl[2,3], Øyvind Lie,[3] and Jostein Bakke[2,3]
[1]Department of Earth Science, University of Bergen, Allégt. 41, N-5007 Bergen, Norway
[2]Department of Geography, University of Bergen, Breiviksveien 40, N-5045 Bergen, Norway
[3]Bjerknes Centre for Climate Research, Allégt. 55, N-5007 Bergen, Norway
**phone +47-55583502, fax +47-55584330, e-mail atle.nesje@geo.uib.no*

Keywords: Glaciers, Holocene, Southern Norway, Winter precipitation.

1. Introduction

Glacier fluctuations provide important information on climate variations as a result of changes in the mass and energy balance at the Earth's surface. Variations in glacier mass balance (e.g. Paterson 1994) are the direct reaction of a glacier to climatic variations. Fluctuations in the length of valley and cirque glaciers, on the other hand, are the indirect, filtered, and commonly enhanced response. Available mass balance records are, however, relatively short compared to the longer records of glacier length variations.

The main aims of the study of Holocene climate variations in southern Norway have been to: (a) reconstruct continuous, high-resolution records of Holocene glacier variations at a regional and local scale; (b) develop improved methods for the reconstruction of past glacier and climate variations from glacio-lacustrine and glacio-fluvial sedimentary sequences; (c) investigate whether the Norwegian glaciers melted completely during the early Holocene thermal optimum from ca. 8000-6000 cal. yr BP (calendar years before present). In case they did, our study will focus on what caused their disappearance, when did they form again, whether they formed simultaneously, and which climatic factors caused their regrowth; (d) investigate whether the maritime and continental glaciers fluctuated synchronously throughout

U. M. Huber et al. (eds.), Global Change and Mountain Regions, 51-60.

the Holocene, and (e) combine glacier equilibrium-line altitudes (ELAs) with independent biological proxy data for summer temperature to reconstruct variations in Holocene winter precipitation.

Research on glacier variations is important for current global change issues in mountain regions, because it provides a better understanding on which climatic factors (accumulation-season precipitation, ablation-season temperature, wind direction and wind strength during the accumulation season) drive glacial changes on decadal to multi-millennial timescales. It also helps to improve our understanding of the potential impact of climatic gradients on glacier variability, and of natural climate variability against which present-day changes can be assessed.

2. Methodology

A variety of approaches and research techniques have been applied to reconstruct Holocene glacier and climate variability. These have included mapping of terminal moraines and sedimentary analyses (loss-on-ignition (LOI), water content, dry weight, bulk density, magnetic susceptibility, grain size) of Holocene lake sediments and peat sections in both pro-glacial and non-glacial settings. We have studied the relation between modern glacier mass-balance and climate, in particular the North Atlantic Oscillation (NAO), and have modeled present-day glacial equilibrium-line altitudes (ELA). A solid understanding of modern relationships is an important prerequisite for extracting palaeoclimatic evidence from Holocene glacial fluctuations.

3. Holocene glacial history in southern Norway

The Holocene glacial history of several glaciers in southern Norway, ranging from the maritime (Nordre Folgefonna) to continental (Snøhetta, Dovre) climate regime has been reconstructed (Fig. 1). The early part of the Holocene was characterised by retreating glaciers, however, this retreat was punctuated by some significant glacier readvances. The studied glaciers had completely melted at least once during the mid Holocene. Millennial-scale glacier variability during the Holocene was most pronounced during the last 4000-5000 calendar years. New morphological and stratigraphical evidence has been found about the age and nature of the earliest Holocene glacial advance in Norway, the so-called Erdalen event, first described from the Jostedalsbreen area (e.g. Nesje et al. 1991). In front of present outlet glaciers from Jostedalsbreen, Hardangerjøkulen and Folgefonna, and glaciers in Jotunheimen, sets of pre-"Little Ice Age" (LIA) terminal moraines can be morphostratigraphically correlated with the Erdalen event. Lithostratigraphy in the basal part of a bedrock basin in Jostedalen provides evidence for a two-phase Erdalen event associated with two glacier readvances of Nigardsbreen (Dahl et al. 2002). Radiocarbon dates suggest that the first event took place at approximately 10,000 cal. yr BP, while the second occurred close to 9700 cal. yr BP.

High-resolution (0.5-1-cm intervals) LOI analyses of cores from pro-glacial and non-glacial lakes revealed a specific pattern, characterised by a two-peaked reduction

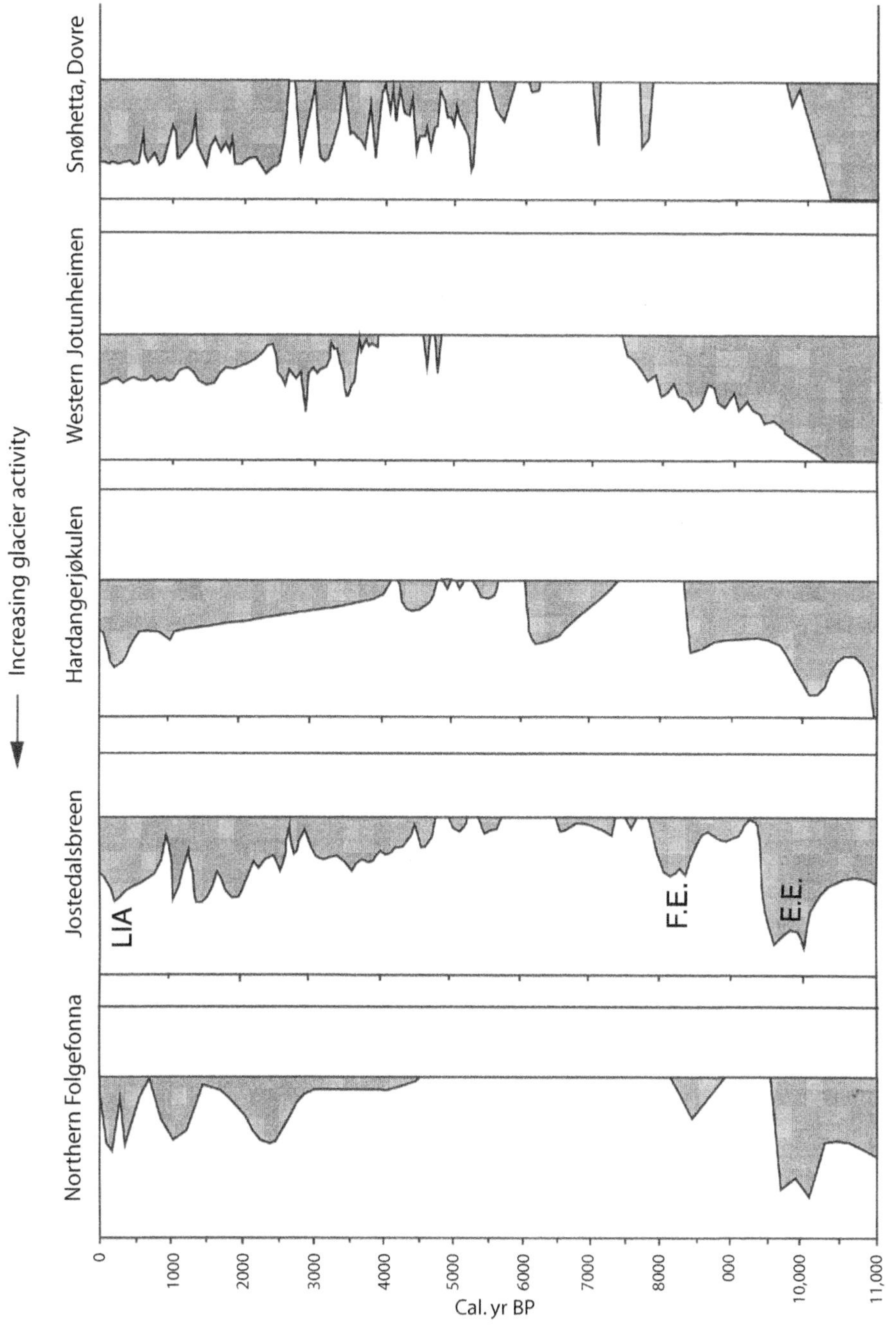

Figure 1: Comparison of Holocene glacier fluctuations along a climatic transect from the maritime Nordre Folgefonna to the continental Snøhetta at Dovre. E.E.=Erdalen event, F.E.=Finse event, LIA="Little Ice Age".

of LOI in the lower part of the cores (Nesje and Dahl 2001). In non-glacier-fed lakes, LOI reductions may reflect a decline in lake productivity and hence a decrease in summer temperature, whereas reductions in the organic content in proglacial lakes mainly reflect increased glacier activity in the catchment. AMS ^{14}C dates on bulk fine detritus gyttja and terrestrial plant macrofossils in lake sediments and peat sections suggest that this event, termed the Finse event (Dahl and Nesje 1994; 1996), occurred between ~8500 and 8000 cal. yr BP. This event correlates with the "8200 BP event" recorded in the Greenland ice cores (e.g. Alley et al. 1997), in European lake sediments (Karlén 1976; 1988; Karlén et al. 1995; Grafenstein et al. 1998; Snowball et al. 1999; Willemse and Törnquist 1999; Matthews et al. 2000; Nesje et al. 2000a; 2001) and in North Atlantic marine records (Bond et al. 1997; Klitgaard-Kristensen et al. 1998). The Finse event or "8200 BP event" has been attributed to the breaching of Laurentide ice sheet dams impounding glacial lakes Agassiz and Ojibway, which led to perturbations of North Atlantic thermohaline circulation (Barber et al. 1999).

Our investigations indicate that most, if not all, glaciers in southern Norway may have melted completely at least once during the early to mid Holocene (ca. 8000-4000 cal. yr BP). Most glaciers seem to have started to reform/advance between 6000 and 4000 cal. yr BP. Most glaciers experienced their maximum post-Erdalen event position during the LIA. However, the timing of the maximum LIA advance in different parts of southern Norway varies considerably, ranging from the early 18th century to the 1940s. Reconstructed millennial-scale Holocene glacier variations and decadal-scale LIA glacier variations in southern Norway do not show a consistent pattern. Therefore, regional differences between the glacier response are affected by differences in the relative importance of summer temperature and winter precipitation operating on different timescales (Winkler et al. 2003). Net mass balance for the maritime glaciers of southern Norway is more influenced by the winter balance, whereas the opposite holds for the more continental glaciers (e.g. Nesje et al. 1995; 2000).

4. Norwegian glaciers, modern climate, and the North Atlantic Oscillation

The North Atlantic Oscillation (NAO) is one of the major modes of climate variability in the North Atlantic region (e.g. Hurrell 1995; Hurrell et al. 2003). Atmospheric circulation during winter commonly displays a strong meridional (north-south) pressure contrast, with low pressure (cyclone) centred close to Iceland and high pressure (anticyclone) near the Azores. This pressure gradient drives mean surface winds and mid-latitude winter storms from west to east across the North Atlantic, bringing mild moist air to NW Europe. Interannual atmospheric climate variability in NW Europe, especially over Great Britain and western Scandinavia has mainly been attributed to the NAO, causing variations in winter weather over the NE North Atlantic and the adjacent land areas. A comparison between NAO and winter precipitation between AD 1864 and 1995 in western Norway shows that these are strongly linked.

Glacier variations are related to both changes in summer temperature and winter precipitation. In Europe, a high correlation between decadal variations in the North Atlantic Oscillation (NAO) and glacier mass balance has been demonstrated (Nesje et al. 2000b; Reichert et al. 2001; Six et al. 2001), the dominant factor being the strong relationship between winter precipitation and the NAO. The NAO winter index is highly correlated with the winter mass balance from maritime glaciers in southern Norway (e.g. Ålfotbreen $r = 0.71$).

A positive NAO phase means enhanced winter precipitation for maritime glaciers in Scandinavia and reduced winter precipitation for glaciers in the European Alps (Reichert et al. 2001; Six et al. 2001). This internal climate system mechanism explains the observed strong positive mass balance on maritime glaciers in western Scandinavia, associated with NAO-related changes in precipitation patterns since the 1970s, and partly explains the strong negative mass balance of glaciers in the European Alps. However, in the European Alps, negative mass balance is also related to increased summer temperatures. Some authors suggest that the recent trend in NAO may potentially be related to global warming (e.g. Hurrell et al. 2003).

5. Reconstruction of Holocene variations in winter precipitation

The equilibrium-line altitude (ELA) on a glacier is mainly controlled by precipitation as snow during accumulation-season and summer temperature during the ablation season. It has been demonstrated that there is a close exponential relationship between mean ablation-season temperature t (1 May-30 September) and winter accumulation A (1 October-30 April) at the ELA of modern Norwegian glaciers (Liestøl in Sissons 1979; Sutherland 1984), which is expressed by the regression equation (Ballantyne 1989; for further details, see also Dahl and Nesje 1996; Nesje and Dahl 2000):

$$A=0.915\ e^{0.339t}\ (r^2=0.989,\ P<0.0001)$$

where A is in meters water equivalent and t is in °C.

Based on the regression equation above, present and past mean winter precipitation (A) at the ELA of a steady state glacier can be quantified when mean ablation-season temperature (t) is known (Dahl and Nesje 1996; Nesje and Dahl 2000). Variations in winter precipitation at other elevations can be calculated by using a precipitation-elevation gradient. In southern Norway this gradient has been calculated to ca. 8% precipitation change per 100 m (Haakensen 1989; Dahl and Nesje 1992; Laumann and Reeh 1993).

To reconstruct the variations in Holocene winter precipitation at Hardangerjøkulen and Jostedalsbreen, we used Holocene ELA fluctuations (adjusted for land uplift) inferred from terrestrial and lacustrine sites, and Holocene summer (July)-temperature variations at Finse, central southern Norway, reconstructed from chironomids in lake sediments (Velle 1998). The Holocene winter precipitation variations (percentage variations) (Fig. 2) are calculated on the basis of the modern relationship between ablation-season temperature and accumulation-season precipitation normals (1961-90) at adjacent meteorological stations. The reconstructed winter precipitation curves for the Hardangerjøkulen and Jostedalsbreen areas show significant, millennial-scale variations. The Holocene winter-precipitation curves may reflect fluctuations between periods with prevailing mild and wet winter conditions ("positive NAO index weather mode") and periods with prevailing cold and dry winters ("negative NAO index weather mode"), and thus indicate large-scale Holocene variability in the atmospheric winter circulation over NW Europe.

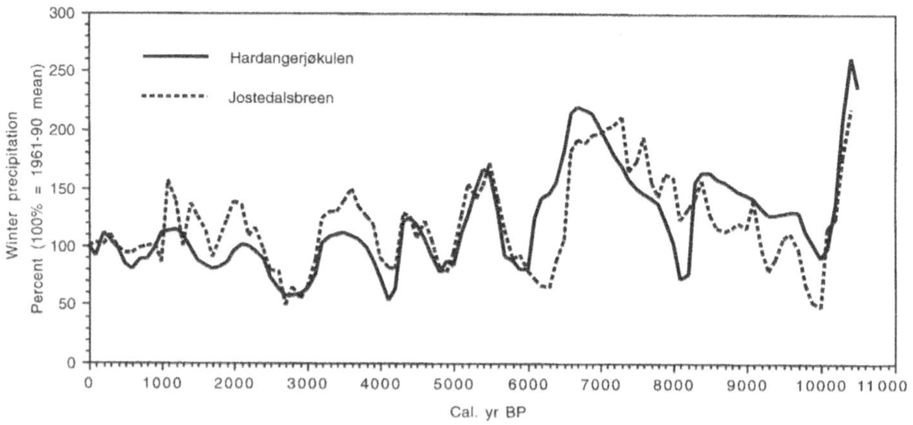

Figure 2: Reconstructed Holocene variations in winter precipitation at Hardangerjøkulen and Jostedalsbreen. The curve from Hardangerjøkulen is modified from Dahl and Nesje (1996), whereas the curve from Jostedalsbreen is adapted from Nesje et al. (2001).

6. Mapping of the modern theoretical equilibrium-line altitude in areas not presently covered by glaciers

In an attempt to improve the palaeoclimatic inferences from reconstructions of glacial ELA, we carried out a detailed study, which involved modeling, Geographical Information Systems (GIS), and a digital elevation model. In contrast with previous studies, ELA was modeled not only in relation to climate, but also topography and landscape features. Based on the close exponential relationship between mean ablation-season temperature and winter precipitation at the ELA of ten Norwegian glaciers, three equations have been derived (Lie et al. 2003a,b). The first enables the calculation of the minimum altitude of areas that are climatically suited for glacier formation today, and is termed the altitude of instantaneous glacierization (AIG) (Fig. 3). The second is based on the "principle of terrain adaptation", enabling the quantification of glacial build-up sensitivity (GBS) in an area. In other words, GBS quantitatively estimates how susceptible an area is for glacier build-up taking into consideration the local topographical features. The theoretical climatic temperature-precipitation ELA (^{c}TP-ELA) in presently non-glaciated areas is calculated in the third equation by combining GBS with terrain altitude. The ^{c}TP-ELA is thus used to calculate the theoretical ELA in areas not presently covered by glaciers (Fig. 3).

Mass-balance records from four glaciers in maritime to continental climate regimes were used to test the equations. The correlation between AIG and net balance measurements (b_n) is high (r = -0.80 to -0.84). Calculated AIGs correspond well with observed ELAs at three glaciers but they deviate somewhat from observed ELAs on Storbreen, perhaps due to leeward accumulation of windblown snow.

Lie et al. (2003a) demonstrated how regionally representative climatic ELAs can be calculated for non-glaciated areas where instrumental records of ablation-season temperature and winter precipitation are available. The three equations have been implemented in a GIS. Using meteorological stations for the normal period 1961-1990, the GIS is based on 122 temperature stations and 197 precipitation stations recalculated to sea level. Interpolation of these data was carried out in the GIS using an "inverse square interpolation routine" to fill each cell of a Digital Elevation Model (resolution 5x5 minutes) with calculated climate data. The present glacier distribution can be modeled in great detail, and a map showing the modern theoretical ^{c}TP-ELA in both glacierized and non-glacierized areas has been constructed. Maps of the modern theoretical ELA may, for example, be used to quantify how much the ELA must be lowered in a region to start build-up of glaciers.

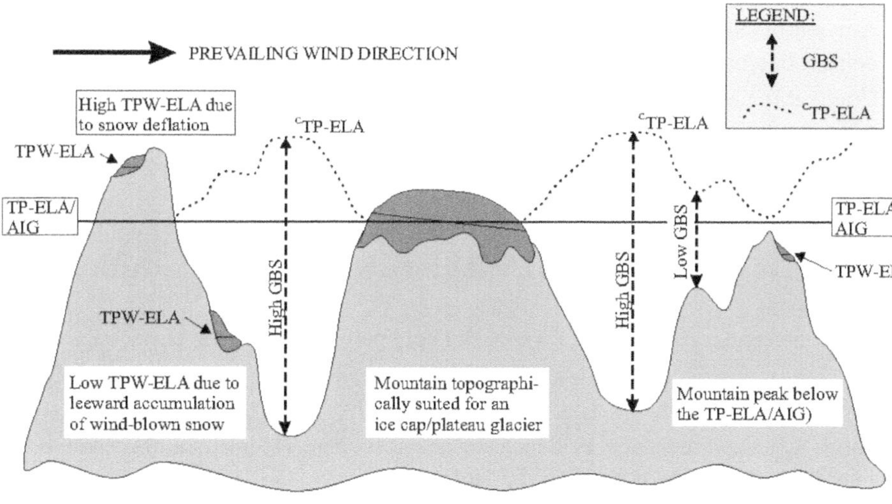

Figure 3: Schematic examples showing the difference between the temperature-precipitation equilibrium-line altitude (TP-ELA) on plateau glaciers and temperature-precipitation-wind equilibrium-line altitude (TPW-ELA) on cirque glaciers. The altitude of instantaneous glacierization (AIG) is the climatically calculated value of the observed TP-ELA. The glacial buildup sensitivity (GBS) is defined as the difference between the altitude of a certain terrain and the altitude where conditions are favourable for glacier formation, taking into account the "principle of terrain adaptation". The climatic temperature-precipitation equilibrium-line altitude (^{c}TP-ELA) defines the altitude where conditions are favourable for glacier formation. Note how the GBS and ^{c}TP-ELA are conditioned by the terrain according to the "principle of terrain adaptation" (Dahl et al. 1997; Lie et al. 2003a, modified from Dahl and Nesje 1992).

7. Future visions for mountain research in Norway

Our research on modern and past glacier variations on different time scales - decadal to multi-millennial - in southern Norway has demonstrated that glacier variations cannot be interpreted in the context of summer (ablation-season) temperature alone. In maritime regions like western Norway, winter precipitation changes may be the dominant climatic factor for observed glacier mass balance variations and changes in frontal position. These results indicate that under future warming scenarios, increased accumulation-season precipitation may, at least initially, compensate for higher summer (ablation-season) temperatures and cause glacier readvance, in contrast to other glaciated regions where glacier net mass balance is primarily controlled by ablation-season temperature.

Future palaeoenvironmental studies in Norwegian mountain regions should specifically address the following scientific questions:

1. Has the Holocene climatic variability in the Norwegian mountains stayed within the range predicted by model scenarios of future climatic development? Is there evidence to suggest that recent climate change is unprecedented in the context of the Holocene?
2. Has the Holocene been subject to rapid, high-amplitude climatic and environmental changes? Are there specific diagnostic features for these abrupt events? What are their causes?
3. Are warm and cold periods over decadal to millennial time scales spatially coherent across the S-N climate gradient in Norway?
4. Could climate variability have affected human activity and society in Norwegian mountain regions during the Holocene?

The main objectives of future palaeoenvironmental studies in Norwegian mountains are to:

- reconstruct modes of Holocene environmental variability from different archives;
- reconstruct periodicity, amplitude and environmental effects of observed climate variations;
- study the spatial coherency of warm and cold periods during the Holocene;
- study the nature and possible causes of periods of abrupt climate change identified in the different proxy records;
- study whether there is evidence to suggest that recent climate change is unprecedented in the context of the Holocene;
- study possible effects of human impact;
- produce reliable, high-resolution reconstructions of environmental change for the validation of coupled atmosphere-ice-ocean climate models.

8. References

Alley, R. B., Mayewski, P. A., Sowers, T., Stuiver, M., Taylor, K. C., and Clark, P. U. (1997). Holocene climate instability: A prominent, widespread event 8200 yr ago. *Geology* **25**, 483-486.

Ballantyne, C. K. (1989). The Loch Lomond readvance on the Island of Skye, Scotland: Glacier reconstruction and palaeoclimatic implications. *Journal of Quaternary Science* **4**, 95-108.
Barber, D. C., Dyke, A., Hillaire-Marcel, C., Jennings, A. E., Andrews, J. T., Kerwin, M. W., Bilodeau, G., McNeely, R., Southon, J., Morehead, M. D., and Gagnon, J.-M. (1999). Forcing of the cold event of 8,200 years ago by catastrophic drainage of Laurentide lakes. *Nature* **400**, 344-348.
Bond, G., Showers, W., Chesby, M, Lotti, R., Almasi, P., deMenocal, P., Priore, P., Cullen, H., Hajadas, I., and Bonani, G. (1997). A pervasive millennial-scale cycle in North Atlantic Holocene and glacial climates. *Science* **278**, 1257-1266.
Dahl, S. O., and Nesje, A. (1992). Paleoclimatic implications based on equilibrium-line altitude depressions of reconstructed Younger Dryas and Holocene cirque glaciers in inner Nordfjord, western Norway. *Palaeogeography, Palaeoclimatology, Palaeoecology* **94**, 87-97.
Dahl, S. O., and Nesje, A. (1994). Holocene glacier fluctuations at Hardangerjøkulen, central-southern Norway: A high resolution composite chronology from lacustrine and terrestrial deposits. *The Holocene* **4**, 269-277.
Dahl, S. O., and Nesje, A. (1996). A new approach to calculating Holocene winter precipitation by combining glacier equilibrium-line altitudes and pine-tree limits: A case study from Hardangerjøkulen, central southern Norway. *The Holocene* **6**, 381-398.
Dahl, S. O., Nesje, A., Lie, Ø., Fjordheim, K., and Matthews, J. A. (2002). Timing, equilibrium-line altitudes and climatic implications of two early-Holocene glacier readvances during the Erdalen Event at Jostedalsbreen, western Norway. *The Holocene* **12**, 17-25.
Dahl, S. O., Nesje, A., and Øvstedal, J. (1997). Cirque glaciers as morphological evidence for a thin Younger Dryas ice sheet in east-central southern Norway. *Boreas* **26**, 161-180.
Grafenstein, U. von, Erlenkeuser, H., Müller, J., Jouzel, J., and Johnsen, S. (1998). The cold event 8200 years ago documented in oxygen isotope records of precipitation in Europe and Greenland. *Climate Dynamics* **14**, 73-81.
Haakensen, N. (1989). Akkumulasjon på breene i Sør-Norge vinteren 1988-89. *Været* **13**, 91-94.
Hurrell, J. W. (1995). Decadal trends in the North Atlantic Oscillation: Regional temperatures and precipitation. *Science* **269**, 676-679.
Hurrell, J. W., Kushnir, Y., Ottersen, G., and Visbeck, M. (2003). An overview of the North Atlantic Oscillation. *In* "The North Atlantic Oscillation: Climatic significance and environmental impact." (J. W. Hurrell, Y. Kushnir, G. Ottersen, and M. Visbeck, Eds.), *Geophysical Monograph* **134**, 1-35.
Karlén, W. (1976). Lacustrine sediments and tree-limit variations as indicators of Holocene climatic fluctuations in Lapland: Northern Sweden. *Geografiska Annaler* **58A**, 1-34.
Karlén, W. (1988). Scandinavian glacier and climatic fluctuations during the Holocene. *Quaternary Science Reviews* **7**, 199-209.
Karlén, W., Bodin, A., Kuylenstierna, J., and Näslund, J.-O. (1995). Climate of northern Sweden during the Holocene. , Holocene Cycles: Climate, Sea Levels, and Sedimentation. *Journal of Coastal Research Special Issue* **17**, 49-54.
Klitgaard-Kristensen, D., Sejrup, H.-P., Haflidason, H., Johnsen, S., and Spurk, M. (1998). A regional 8200 cal. yr BP cooling event in northwest Europe, induced by final stages of the Laurentide ice-sheet deglaciation? *Journal of Quaternary Science* **13**, 165-169.
Laumann, T., and Reeh, N. (1993). Sensitivity to climate change of the mass balance of glaciers in southern Norway. *Journal of Glaciology* **39**, 656-665.
Lie, Ø., Dahl, S. O., and Nesje, A. (2003a). A theoretical approach to glacier equilibrium-line altitudes using meteorological data and glacier mass-balance records from southern Norway. *The Holocene* **13**, 365-372.
Lie, Ø., Dahl, S. O., and Nesje, A. (2003b). Theoretical equilibrium-line altitudes and glacier buildup sensitivity in southern Norway based on meteorological data in a geographical information system. *The Holocene* **13**, 373-380.
Matthews, J. A., Dahl, S. O., Nesje, A., Berrisford, M., and Andersson, C. (2000). Holocene glacier variations in central Jotunheimen, southern Norway based on distal glaciolacustrine sediment cores. *Quaternary Science Reviews* **19**, 1625-1647.
Nesje, A., and Dahl, S. O. (2000). "Glaciers and environmental change". Arnold, London.
Nesje, A., and Dahl, S. O. (2001). The Greenland 8200 cal. yr BP event detected in loss-on-ignition profiles in Norwegian lacustrine sediment sequences. *Journal of Quaternary Science* **16**, 155-166.
Nesje, A., Johannessen, T., and Birks, H. J. B. (1995). Briksdalsbreen, western Norway: Climatic effects on

the terminal response of a temperate glacier between AD 1901 and 1994. *The Holocene* **5**, 343-347.

Nesje, A., Kvamme, M., Rye, N., and Løvlie, R. (1991). Holocene glacial and climate history of the Jostedalsbreen region, western Norway; evidence from lake sediments and terrestrial deposits. *Quaternary Science Reviews* **10**, 87-114.

Nesje, A., Dahl, S. O., Andersson, C., and Matthews, J. A. (2000a). The lacustrine sedimentary sequence in Sygneskardvatnet, western Norway: A continuous, high-resolution record of the Jostedalsbreen ice cap during the Holocene. *Quaternary Science Rewiews* **19**, 1047-1065.

Nesje, A., Lie, Ø., and Dahl, S. O. (2000b). Is the North Atlantic Oscillation reflected in Scandinavian glacier mass balance records? *Journal of Quaternary Science* **15**, 587-601.

Nesje, A., Matthews, J. A., Dahl, S. O., Berrisford, M. S., and Andersson, C. (2001). Holocene glacier fluctuations of Flatebreen and winter precipitation changes in the Jostedalsbreen region, western Norway, based on glaciolacustrine records. *The Holocene* **11**, 267-280.

Paterson, W. S. B. (1994). "The physics of glaciers." Elsevier Science, London.

Reichert, B. K., Bengtsson, L., and Oerlemans, J. (2001). Mid latitude forcing mechanisms for glacier mass balance investigated using general circulation models. *Journal of Climate* **14**, 3767-3784.

Sissons, J. B. (1979). Palaeoclimatic inferences from former glaciers in Scotland and the Lake District. *Nature* **278**, 518-521.

Six, D., Reynaud, L., and Letréguilly, A. (2001). Bilans de masse des glaciers alpins et scandinaves, leurs relations avec l'oscillation du climat de l'Atlantique nord. *Earth and Planetary Sciences* **333**, 693-698.

Snowball, I., Sandgren, P., and Petterson, G. (1999). The mineral magnetic properties of an annually laminated Holocene lake sediment sequence in northern Sweden. *The Holocene* **9**, 353-362.

Sutherland, D. G. (1984). Modern glacier characteristics as a basis for inferring former climates with particular reference to the Loch Lomond Stadial. *Quaternary Science Reviews* **3**, 291-309.

Velle, G. (1998). "A palaeoecological study of chironomids (Insecta: Diptera) with special reference to climate." Unpublishd thesis, Museum of Zoology, University of Bergen.

Willemse, N. W., and Törnquist, T. E. (1999). Holocene century-scale temperature variability from West Greenland lake records. *Geology* **27**, 580-584.

Winkler, S., Matthews, J. A., Shakesby, R. A., and Dresser, P. Q. (2003). Glacier variations in Breheimen, southern Norway: Dating Little Ice Age moraine sequences at seven low-altitude glaciers. *Journal of Quaternary Science* (in press).

Glacier and Climate Variability in the Mountains of the Former Soviet Union during the last 1000 Years

Olga N. Solomina
Institute of Geography, Russian Academy of Sciences, 109017 Staromonetny-29, Moscow, Russia
phone +7-095-939-01-21, fax +7-095959-00-33, e-mail solomina@gol.ru

Keywords: Climate Change, Former Soviet Union, Glacier variations, Lichenometry, Little Ice Age Tree-rings

1. Introduction

Pollen analysis, ^{14}C and lichenometric dating of moraines, former elevations of the upper tree limit, and dendroclimatological and limnological data are some of the most relevant proxies for the reconstruction of climate variability and glacier behavior during the last millennium. A considerable number of paleoclimate reconstructions exist for the mountains of the Former Soviet Union. In this paper, we provide a regional overview of these datasets. Only regions with chronologically controlled and, preferably, high-resolution reconstructions will be considered here, namely, the Khibiny, the Urals, the Cherskogo Range, the Putorana Plateau, the Birranga Mountains, the Suntar-Khayata, the Kamchatka, the Caucasus, the Pamir-Alay, the Tien Shan, and the Altay Mountains (Fig. 1). This paper is a brief summary of the glacier and climate history of the last millennium and identifies achievements as well as gaps in our knowledge of paleoclimate in these regions. Ultimately, the identification of regional patterns of past climate changes will allow us to gain a better understanding of the causes behind climate variability on inter-annual to centennial timescales.

U. M. Huber et al. (eds.), Global Change and Mountain Regions, 61-72.

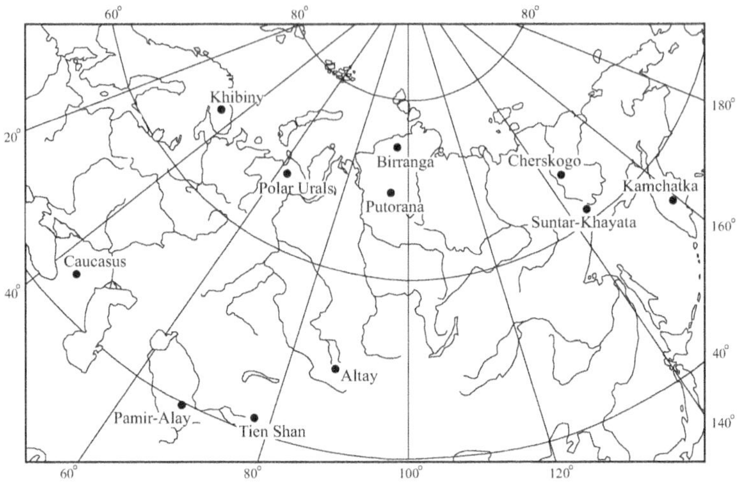

Figure1. Former Soviet Union. Regions mentioned in the text.

2. The Khibiny Mountains

Radiocarbon-dated macrofossils of *Pinus sylvestris* indicate that forests grew at least 100-140 m above the modern pine treeline in the Khibiny Mountains from ca. AD 1000 to 1300 (Fig. 2a; Hiller et al. 2001). Based on pollen analyses, winter, summer and mean annual temperatures at the turn of the first millennium were all approximately 1°C higher than today, and annual precipitation exceeded modern values by 50 mm (Klimanov 1989). Based on modern analogues, the warming must have been substantial enough for glaciers to completely disappear. Buried soil horizons that developed in snow avalanche debris between 1130±70 and 660±50 ^{14}C years BP (twelve ^{14}C dates; Vaschalova 1987) indicate that avalanche activity was low between ca. AD 800 and 1350 (see Fig. 2a). Pollen results indicate that, during the subsequent Little Ice Age (LIA) cooling (13th-19th century; Grove 1988), winter temperatures were as much as 3°C cooler than today, and precipitation was a little less than today. Glaciers re-appeared: fresh-looking presumably Little Ice Age moraines (undated), are found in many cirques. Tree-ring chronologies, sensitive to May-July temperatures, show that, during the LIA, conditions for pine growth were least favorable in the 1590s and 1640s (Grippa 1999). Temperatures recorded at the Arkhangel'sk meteorological station since AD 1813 correlate well with the shorter Apatity and Kola meteorological observations in the Khibiny Mountains. Summer coolings in the region were observed from the beginning of the 1830s to the middle of the 1840s and from the 1870s to the 1900s. The timing of the earlier cooling agrees well with historical evidence that some summits of the Khibiny Mountains were permanently snow-covered during the 1830s. The positive trend in annual temperature that began in the second decade of the 20th century is mostly related to warmer winters. As response to this warming, the glaciers retreated and most of them disappeared. Only four glaciers still existed in the Khibiny Mountains in the middle of the 20th century.

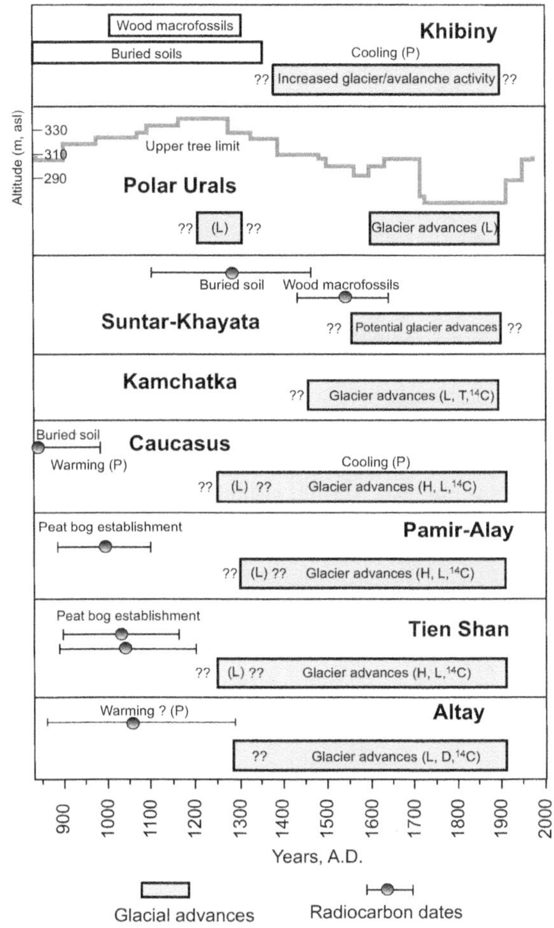

Figure 2a. Warm and cold intervals based on radiocarbon dating of wood macrofossils, buried soils, moraine dating, pollen analyses and peat bog establishment near the glaciers (see text for references). (H) - historical data, (L) – lichenometric dates, (^{14}C) – radiocarbon dates, (P) – pollen data, (T) – tephrochronological dates, (D) – dendrochronological dates.

3. The Polar Urals

Radiocarbon and dendrochronological dating of subfossil wood allowed detailed reconstructions of fluctuations in the upper tree limit over the last millennium (Fig. 2a; Shiyatov 2003). From AD 850 to 1280, treeline increased to a position higher than at any time during the last 1000 years. From AD 1280 to 1580, upper treeline retreated from 340 m asl to 295 m asl. Between AD 1580 and 1710, treeline variations were minor. The most significant treeline retreat (from 305 m asl to 270 m asl) occurred around AD 1710 and treeline remained low until AD 1910. Since then treeline rose by 20-40 m, whereas the June-July isotherm rose by 120-130 m in altitude. Shiyatov

(2003) explains this discrepancy by the deficiency of seeds above the uppermost locations of trees. In other words, the rise in treeline lags the recent increase in temperature. Hence, reconstructed temperatures based on treeline variations in the Polar Urals should be multiplied by a factor of 4. Although the temperature-sensitive ring-width chronologies of larch (Fig. 2b; Briffa et al. 1995) show some agreement with the treeline dynamics described above (Fig. 2a), some substantial details are different. For instance, the warmest period prior to the 20th century, reconstructed from larch ring widths and density, is substantially shorter, only lasting from the 13th to the 15th century. Also, the summer temperature decrease between the end of the 10th and the end of the 12th century, recorded by ring widths, is not evident from the treeline dynamics. The disagreement between the two proxies may reflect the different nature of the climatic signal recorded: in general, tree line dynamics better reflect long-term changes, whereas ring width and density chronologies are more useful for the reconstruction of decadal variations. Tree-ring records from the Polar Urals show a LIA cooling between approximately the beginning of the 16th and the end of the 19th century, interrupted by somewhat warmer conditions in the 18th century.

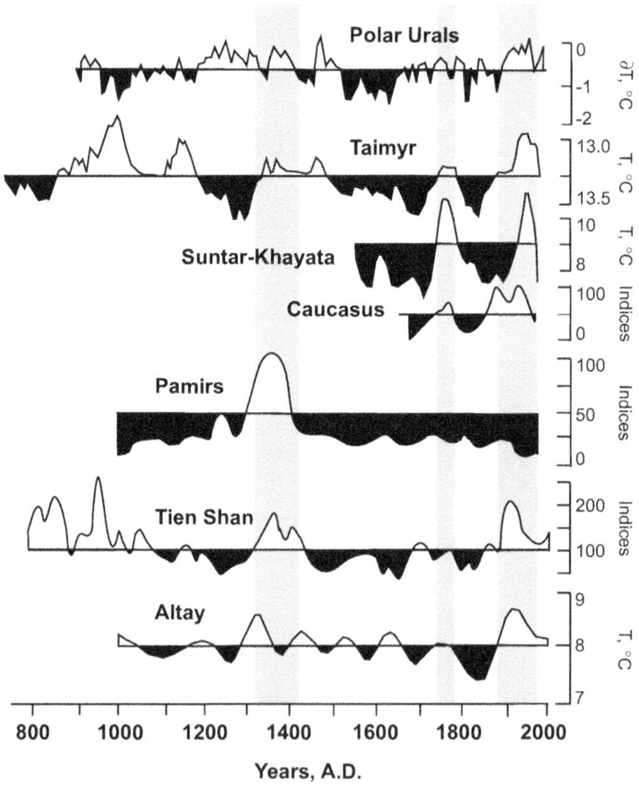

Figure 2b. General trends of (early) summer temperatures based on tree-ring reconstructions (see text for references). Where no decadal variations are indicated in the original publications the chronologies were smoothed manually.

The Holocene moraines in the Polar Urals are poorly dated. One to three sets of moraines are found up to 200 m below the present-day glacier fronts. Periods of moraine stabilization have been dated by lichenometry to approximately the 13th, 17th, 18th and late 19th centuries, but these estimates are still tentative. The mass balance of glaciers in the Polar Urals was reconstructed for the period 1818-1998 by correlating measured mass balance of the IGAN Glacier with climate observations of the Bol'shaya Khadata, Vorkuta, Ust'-Vorkuta and Syktyvkar meteorological stations (Troitsky et al. 1966; Lebedeva, personal communication). In the 19th century, mass balance was close to zero. By the end of the 19th century, it became negative, with the exception of the 1880s, when glacier volumes increased. During the 20th century, climate conditions were generally unfavorable for glaciers, especially during the warming of the first half of the century. Troitsky (1966) estimated that the glaciers of the Polar Urals have thinned by about 20-30 m in their accumulation areas and by 40-50 m in the ablation areas from the 1880s to the 1960s. In many cirques, the glaciers have disappeared completely. Data for more recent decades are sparse. Based on the extent of modern warming, glacier degradation should have continued.

4. The Birranga Mountains (Taimyr Peninsula) and the Putorana Plateau

This area is very poorly studied and moraines have not been dated. The decadal temperature variations in Taymir (Fig. 2b; Naurzbaev et al. 2002) agree rather well with those of the Polar Urals for the period from the second half of the 14th century to the end of the chronology in 1996, but show opposing trends between the 10th and the early 14th century. In general, the trees in this region were very sensitive to early summer temperature, although this response has weakened since the 1960s. Future research needs to address, which climatic mechanisms might lead to a shift from synchronous to opposing temperature trends between these regions.

5. The Cherskogo Range

No moraines are dated in this region. A *Larix gmelinii* chronology, spanning from AD 1545-1989 and based on 25 living trees, was established for the eastern periphery of the Cherskogo Range (Earle et al. 1994). According to this reconstruction, the coldest periods during the Little Ice Age interval were observed from 1571-1587, 1663-1681, 1721-1748, and 1902-1927, and warmer intervals date to 1612-1639, 1698-1720, 1790-1809, and 1866-1882. The record also shows a strong temperature increase between 1942-1968.

6. The Suntar-Khayata

Wood macrofossils from 200-250 m above the modern treeline (420±50 ^{14}C yr BP, i.e. AD 1420-1640) and a buried soil horizon (660±90 ^{14}C yr BP, i.e. AD 1100-1470) are evidence for former warmer conditions (Fig. 2a; Nekrasov et al. 1973). One larch chronology at the upper tree limit in the Suntar valley reveals a period of relatively cold summers between at least AD 1550 and ca. 1900, which is interrupted by a short interval of warmer summers centered around AD 1750 (Fig. 2b; Nekrasov et al. 1973). Young unvegetated moraines near the modern glaciers most probably originate from glacier advances triggered by LIA coolings. The tree-ring record also shows a strong temperature increase in the first half of the 20th century.

7. Kamchatka

Tephrochronology, lichenometry and tree-ring analyses have been used to date moraines of the Kamchatka glaciers. The most prominent advances of the last millennium occurred in the late 15th to mid-16th, the early to mid-17th and the late 19th centuries (Solomina 1999). The ages of moraines in Kamchatka, although sparse, are similar to those of the Brooks Range and the Gulf of Alaska.

Several larch ring-width chronologies, which reflect mostly early summer temperature, were established for this region. According to these data, particularly cold early summers during the LIA occurred in the 1640s-1650s, 1710s-1730s, 1760s-1790s, 1810s, 1860s-1890s, and 1950s. Ice core records from the Ushkovsky Ice Cap (Shiraiwa et al. 1999) provide very important paleoclimatic information concerning conditions during the last few centuries. There is a good agreement between the percentage of melt features in the Ushkovsky ice core and ring widths, especially in the 18th-19th century.

The longest meteorological records in Kamchatka (Petropavlovsk-Kamchatsky, measurements since 1891) show a slight positive trend in annual precipitation during the 20th century, whereas there is no clear trend in summer temperature. However, several summer warmings of decadal duration, combined with decreased solid precipitation, resulted in a general regression of glaciers since the late 19th century.

8. Caucasus

Archeological data, a buried soil horizon (1280±100 ^{14}C yr BP: AD 610-980), pollen evidence, and absence of moraines (Fig. 2a) all indicate a relatively warm climate at the end of the first and the beginning of the second millennium AD. Both pollen data and lichenometrically dated moraines indicate a period of significant cooling thereafter. A number of glacier advances occurred both on the northern and southern slopes of the Central Caucasus in the 13th, 14th, 16th, late 17th, early and late 18th, 19th, and early 20th century (Serebryanny et al. 1984). Most of the preserved moraines were deposited within the last two centuries. Historical data, though rather sparse, yield evidence of much larger glaciers than at present from the mid-19th to the early 20th century, in agreement with morainal evidence. The timing of glacier fluctuations in the Caucasus, revealed by moraine dating, is similar to the Alps. Panov (1993) demonstrated that summer temperature anomalies in the Caucasus were negative from 1890-1920 and positive thereafter. The mass balance of Dzhankuat glacier was positive from 1877 to the 1920s, but mostly negative since then. On average, glaciers in the Caucasus have retreated by approximately 1000 m in length since the end of the 19th century, and the altitudes of glacier fronts have risen by 200 m. Despite the great potential for tree-ring research in this area, no comprehensive dendroclimatic reconstructions exist so far. The longest available pine chronology, which reflects both temperature and precipitation signals, shows a distinct growth depression between the end of the 18th and the mid-19th century and an increase thereafter (Fig. 2b; Turmanina 1971).

9. The Pamir-Alay

Phenological and historical data point to reduced continentality centered on the 10th century. The longest temperature-sensitive juniper chronology in the Zeravshanky range (Mukhamedshin 1977) shows cooler conditions between at least the beginning of the 11th and the end of the 13th century; a very prominent warming is registered in the 14th century (Fig. 2b). This is in contrast to pollen data, which suggest a humidity peak between the 13th and 14th century, accompanied by a summer temperature decrease (Abramova 1994). However, these differences between the records may be related to the poor chronological control of the pollen data. According to tree-ring data, conditions once again became cooler after ca. AD 1400. Several shorter juniper ring-width chronologies, which correlate with June-July temperature, have been developed in the Pamir-Alay Mountains in the 1970s and 1980s, but these must be checked for potential missing rings. According to dendrochronological data, the coldest periods in the Fanskye Gory occurred in the 12th to 13th, the 17th, the 19th, and the mid-20th centuries. In the Zeravshansky and Alaisky Ridges, climate was coldest in the second half of the 16th, the late 17th to early 18th, the late 18th to early 19th and the mid-19th centuries.

Glacial advances in the Alaisky Range have been tentatively dated by lichenometry on the forefields of four glaciers to the 14th to 15th, the mid-17th to the late 18th, the mid- and early 19th, and the early 20th centuries (Solomina 1999). There is no significant warming trend during the last century in this region. An exception is the interval between the 1930s and 1960s, which showed a warming for both summer and winter. During the 1891-1900 interval, which was the coldest decade of the instrumental period (measurements since the 1880s), glacier mass balance was positive and many glaciers advanced. By the middle of the 20th century, glacier length had decreased on average by 500-1000 m and glacier front altitudes had risen by 100-200 m. During the second half of the 20th century, this retreat continued.

10. Tien-Shan

Extensive and comprehensive archeological evidence and radiocarbon data suggest that the level of Lake Issyk-Kul fell to about 5-7 m below its present level from approximately the 9th to the 16th century. Pollen spectra and charcoal occurrence also indicate greater warmth and aridity in the early second millennium. Glaciers were as small as now or even smaller. Small peat bogs were forming on the outwash plains near the glacier fronts or between end moraines. Two radiocarbon dates, 1020±70

^{14}C yr BP (AD 880-1200) and 1030±35 BP ^{14}C yr (AD 890-1160) (Fig. 2a), from the bottom of two such peat bogs in the Western Tien Shan suggest that peat establishment was approximately synchronous (Solomina 1999).

The majority of moraines in the Tien Shan date from the 17th-19th centuries, based on lichenometry. Moraines deposited between the 13th and 17th centuries, though not so numerous, occur in all areas of Tien Shan, but they are represented only by ridge fragments preserved in lateral moraines. Thus, from the 13th-16th century the glaciers were smaller than from the 17th-19th century, but larger than in the 20th century. The glacier advances in the 1850s, 1880s and 1900s-1910s are also documented historically. The general temperature trend since 1882 is positive, and the temperature increase is particularly pronounced after the 1970s. Also, the warming is much more distinct in the foothills than in the high mountains. Glacier front altitudes are now on average 150-200 m higher than during the LIA maximum, marked by prominent terminal moraines.

Since the 1970s, a number of spruce and juniper chronologies have been established for the Tien Shan. The ring widths of *Picea schrenkiana*, even at the upper tree limit, depend mainly on the sum of October-May precipitation, reflecting the importance of snow cover as a water source for spruce in this arid region. Juniper ring widths at the upper tree limit correlate with summer or annual temperature. A comprehensive temperature reconstruction for the Tien Shan and Karakorum Mountains, based on juniper ring widths, was published recently (Fig. 2b; Esper et al. 2002) and reveals clear centennial temperature trends during the last ca. 1300 years. Temperatures are generally warmer between at least the end of the 8th to the end of the 11th century. Cooler conditions occur between approximately the end of the 11th to the end of the 19th century, with the exception of a warm interval from the early 14th to early 15th century. Also, the tree-ring record shows a distinct warming in the 20th century. These conclusions fit with reconstructions of glacier variations, based on moraine dating.

11. Altay

Between the end of the first and the beginning of the second millennium AD, warmer conditions are indicated by vegetation changes, decreased eolian activity, lake regression and the decrease of river runoff (Ivanovsky et al. 1982; Butvilovsky 1993). A new millennium-long temperature reconstruction, based on larch tree-rings (Ovchinnikov 2002), shows similar century-scale trends as in the Tien Shan (Fig. 2b). The 14th and 20th centuries were the warmest of the whole period. The longest and most pronounced June-July cooling occurred in the mid 19th century.

Glacial evidence indicates the onset of cooler conditions after the early 15th century. Similar to many other areas considered here, the LIA moraine complexes in the Altay Mountains often comprise the traces of many glacier advances. The LIA moraine complex of Malii and Bol'shoy Aktru Glaciers contains wood remnants dated by ^{14}C to the mid-15th to early 17th century. The LIA moraine of the Bol'shoy Berelsky Glacier consists of at least three fragments of stadial moraines deposited in the 1840s and between 1915 and 1925, according to dendrochronology, and at approximately 520±80 ^{14}C yr BP (AD 1290-1520 or AD 1570-1630), based on a ^{14}C date of buried wood (Mikhailov 1987). At Katunskiy Glacier, one of the moraines is dated to about the 1830s, based on historical descriptions, and another one was deposited in the 17th century or earlier, based on tree-ring counts (Okishev 1982). Most of the moraines dated by lichenometry were deposited during the 17th-early 20th century. Moraines of the 15th-18th century are partially crosscut by those of later advances. The mass balance of Malii Aktru glacier was reconstructed for 1838-1985 with the help of meteorological observations from the Barnaul and Aktru meteorological stations (Narozhniy 1986). The lowest ablation rates were estimated for 1838-1889; the highest rates occurred between 1890 and 1936. The maximum mass accumulation between 1901 and 1914 led to a glacier advance from 1909-1914. Since the 20th century, temperatures have been rising and glaciers have been retreating. From the mid-19th century to the 1960s-1980s their length decreased on average by 500 m.

12. Conclusions and research needs

Tree-ring based reconstructions with annual resolution, covering the whole of the last millennium, have been established for the Tien Shan, Altay, Taymir and Polar Urals and provide the possibility of very precise inter-regional comparisons of annual to decadal (early) summer temperature variations. These chronologies might also be suitable for defining centennial trends, but the ring-width for all series must be detrended using a uniform approach that preserves long-term fluctuations (e.g. Esper et al. 2002). A millennium-long temperature sensitive reconstruction is being developed in the Khibiny Mountains, but in many regions of the Former Soviet Union, such as the Caucasus, the Pamir-Alay, most parts of East Siberia and the Far East, this is a task for the future.

Unfortunately, the comparison between individual chronologies is rather complicated due to different approaches to index calculation and averaging, differences in the definition of "anomalies," variable reference periods etc. used by different authors. Also, the chronologies have different degrees of replication, length, and sensitivity. In many cases, data were not submitted to international tree-ring databases and are not available in numerical format. Considerable effort is still necessary to build well-replicated millennium-long regional chronologies and make these accessible to the scientific community. This process will be critical for establishing regional patterns of interannual to centennial-scale climate change and for linking these patterns to different climate forcings.

Despite of all the problems mentioned, there are several climatic events of decadal to centennial length that can be detected using available chronologies. Several tree-ring records indicate cooler conditions from some time after AD 1400-1500 until the 19th century. Also, most chronologies register pronounced warming in the first half of the 20th, the mid 18th and the 14th century.

The recent development of tree-ring studies in the mountains of the Former Soviet Union (FSU) is far ahead of other proxies, such as moraine chronologies, ice-cores, lake sediments or speleothems. Great potential exists to develop these methods and improve chronologies in most of the areas mentioned, which will provide more detailed paleoclimatological information.

Lichenometric age determinations on moraines in the mountains of the FSU are often of quite low accuracy. However, in the regions where they are available (Tien Shan, Altay) they agree with decadal-scale summer temperature minima, which demonstrates a certain potential of glacier variations as a climatic proxy. The coincidence between glacier advances and a series of cold summers might be explained by a combination of reduced ablation and increased proportion of solid summer precipitation during cool episodes.

The data considered here (pollen, macrofossils, moraines, buried soils) can contribute to discussions of the century-scale climate trends of the last millennium. Based on these proxies, the period of approximately the 10th-12th century seems to be in general warm in the Khibiny, Polar Urals, Tien Shan, Altai and Caucasus (Fig. 2a). However, this assumption is mostly based on low-resolution data. Several tree-ring chronologies do not show this warming, but instead show a cooling during at least part of this interval. Moreover, in some regions of Central Asia and the Altay the most prominent peak in ring width prior to the 20th century only occurred in the 14th century. It is probably too early to discuss the potential climatic reasons for this dissimilarity, which might be a simple result of the low number of tree ring chronologies that extend to the beginning of the millennium, or the poor chronological control of "event" chronologies, such as buried soils, and moraines. It is clear that further investigations, based on glacial chronology and high-resolution paleo-climate proxies, are necessary in order to resolve the disagreement between proxies and to assess whether apparent regional differences are real.

Glaciers probably advanced in the 13th century in the Caucasus and Tien Shan, but the timing of this advance is not well constrained. The cooling in the 13thcentury (Fig. 2b) is most probably the climatic reason for this advance. The 14th century clearly deserves more attention from paleoclimatologists in the future. According to tree-ring reconstructions, climate at that time seems to be very warm in the Altay and in some regions of Central Asia.

Glacier advances are recorded in the 15th-16th centuries in all regions for which lichenometrically dated moraine chronologies are available. During the last millennium, very prominent advances occurred in the 13th century in the Caucasus, in the 17th and 19th centuries in the Caucasus, Tien Shan and Altay, and in the 19th century in the Kamchatka. In the Tien Shan, Altay and Kamchatka, the periods of main glacier expansion roughly coincide with the peaks of cooling recorded by tree rings. However, the accuracy of moraine dating is often not high enough for a precise comparison. Since the second half of the 19th century, glacier sizes have been diminishing everywhere in the mountain ranges mentioned, indicating a general warming over the entire region.

13. Acknowledgment

I thank the unnamed reviewers for valuable comments and criticism on earlier drafts of this paper. I am most grateful to Ulli Huber, whose help and patience was crucial for the preparation of this manuscript.

14. References

Abramova, T. A. (1994). Environmental changes of some arid regions of Eurasia during the last two millennia (Izmeneniya prirodnoi sredi nekotorikh aridnikh raionov Evrazii za dva poslednikh tisiacheletiya). Vestnik MGU, Series 5. *Geography* **5**, 59-66 (in Russian).
Butvilovsky, V. V. (1993). "Paleogeography of the last glaciation and Holocene in the Altay" (Paleogeografiya poslednego oledeneniya i Golotsena na Altaye). Tomsk State University Press, Tomsk (in Russian).
Briffa, K. R., Jones, P. D., Schweingruber, F. H., Shiyatov, S. G., and Cook, E. R. (1995). Unusual twentieth-century summer warmth in a 1000-year temperature record from Siberia. *Nature* **376**, 156-158.
Earle, C. J., Brubaker, L. B., Lozhkin, A. V., and Anderson, P. M. (1994). Summer temperature since 1600 for the Upper Kolyma Region, Northeastern Russia, reconstructed from tree rings. *Arctic and Alpine Research* **26**, 60-65.
Esper, J., Schweingruber, F. H., and Winiger, M. (2002). 1300 years of climatic history for Western Central Asia. *The Holocene* **12**, 267-277.
Grippa, S. P. (1999). "Dendro-indication of natural and anthropogenic environmental changes in Fennoscandia" (Dendroindikatsiya prirodnikh i antropogennikh izmenenii prirodnikh uslovii Fennoskandii). Unpublished PhD thesis, Russian State Pedagogic University, Sankt-Petersbourg (in Russian).
Grove, J. M. (1988). "The Little Ice Age." Methuen, London.
Hiller, A., Boettger, T., and Kremenetski, C. (2001). Medieval climatic warming recorded by radiocarbon dated alpine tree-line shift on the Kola Peninsula, Russia. *The Holocene* **11**, 491-497.
Ivanovsky, L. N., Panichev, V. A., and Orlova, L. A. (1982). The age of the moraine stages 'Aktru' and 'Historical' in the Altay Mts. (Vozrast konechnikh moren stadii 'Aktru' i 'Historicheskoy' na Altaye. *In* "Late Pleistocene and Holocene of the South of East Siberia" (Pozdnii Pleistotcen i Golotsen yuga Vostochoi Sibiri). pp. 57-64. Nauka, Novosibirsk (in Russian).
Klimanov, V. A. (1989). Cyclicity and quasi-periodicity of climatic fluctuations in the Holocene. (Tsiklichnost' i kvaziperiodichnost' klimaticheskikh kolebanii v golotsene). *In* "Paleoclimates of the Late Pleistocene and Holocene" (Paleoklimati pozdnelednikov'ia i golotsena). (N. A. Khotinsky, Ed.), 29-33. Nauka, Moscow.
Mikhailov, N. N. (1987). Dynamics of the Belukha glaciers (Altay) in historical time (Dinamika lednikov Belukhi (Altay) v istoricheskoye vremia). Vestnik LGU. *Geology and Geography* **3**, 100-103 (in Russian).
Mukhamedshin, K. D. (1977). "Juniper forests of the Tien Shan and their economic significance" ("Archevniki Tian'-Shania i ikh lesokhoziaistvennoye znacheniye"). Frunze, Ilim (in Russian).
Narozhniy, Yu. K. (1986). Mass balance reconstruction and ice formation conditions of Malii Aktru glacier during the last 150 years (Rekonstruktsiya balansa massi i uslovii l'doobrazovaniya lednika Malii Aktru za 150 let). *Glaciology of Siberia* (*Gliatsiologiya Sibiry*) **3**, 85-104.
Naurzbaev, M. M., Vaganov, E. A., Sidorova, O. V., and Schweinguber, F. H. (2002). Summer temperatures in eastern Taimyr inferred from a 2427-year late-Holocene tree-ring chronology and earlier floating series. *The Holocene* **12**, 727-736.
Nekrasov, I. A., Maksimov, E. V., and Klimovsky, I. V. (1973). "Last glaciation and cryolithic zone of the southern Verkhoyaniye" (Poslednee oledeneniye i kriolitozona yuzhnogo Verkhoyania). Yakutsk Yakutskoye Knizhnoye Izdatel'stvo (in Russian).
Okishev, P. A. (1982). "Dynamics of glaciers in the Altay mountains in the Late Pleistocene and Holocene" (Dinamika oledeneniya Altaya v pozdnem pleistotcene i golotsene). Tomsk State University Press, Tomsk (in Russian).
Ovchinnikov, D. B. (2002). "Reconstruction of the climate variations in the Altay Mountains by dendrochronological methods" (Rekonstruktsiya izmenenii klimata gor Altaya dendrokhronologicheskimi metodami). Unpublished PhD thesis, Institute of Geography, Siberian Branch of Russian Academy of Sciences, Irkutsk (in Russian).
Panov, V. D. (1993). "Evolution of the modern glaciation in the Caucasus" ("Evolutsiya sovremennogo oledeneniya na Kavkaze"). Gidrometeoizdat, Sankt-Petersburg (in Russian).
Serebryanny, L. R., Golodkovskaya, N. A., Orlov, E. V., and Maliasova, E. S. (1984). "Glacier variations and moraine accumulation processes in the Central Caucasus" (Kolebaniya lednikov I protsessi morenonakopleniya na Kavkaze). Nauka, Moscow (in Russian).

Shiraiwa, T., Nishio, F, Kameda, T., Takahashi, A., Toyama, Y., Muraviev, Y., and Ovsyannikov, A. (1999). Ice core drilling at Ushkovsky ice cap, Kamchatka, Russia. *Seppyo* **61**, 25-40.

Shiyatov, S. G. (2003). Rates of change in the upper treeline ecotone in the Polar Ural Mountains. *PAGES News* **11**, 8-10.

Solomina, O. N. (1999). "Mountain glaciation of Northern Eurasia in the Holocene" (Gornoye oledeneniye Severnoy Evrazii v Golotsene). Moscow, Nauchniy Mir (in Russian).

Troitsky, L. S., Khodakov, L. S., and Mikhalev, V. I. (1966). "Urals Glaciation" (Oledeneneye Urala). AN SSSR Press. Moscow (in Russian).

Turmanina, V. I. (1971). Prospects of the use of phytoindication in glaciology (Perspektivy primeneniya fitoindikatsionnikh metodov v gliatsiologii). *In* "Methods of phytoindication in glaciology" (Fitoindikatsionniye metody v gliatsiologii). (G. K. Tushinsky, Ed.), pp. 5-19. Moscow State University Press, Moscow (in Russian).

Vaschalova, T. (1987). "Paleogeographical approach to reconstruct avalanche activity for long-term forecasting (Khibiny example)". (Paleogeograficheskiy podkhod k rekonstruktsii lavinnoy aktivnosti v tseliakh dolgosrochnogo prognoza (na primere Khibiny). Unpublished PhD thesis, Moscow State University, Moscow (in Russian).

Glacier-Climate Models as Palaeoclimatic Information Sources: Examples from the Alpine Younger Dryas Period

Hanns Kerschner
Institut für Geographie, Universität Innsbruck, Innrain 52, A-6020 Innsbruck, Austria
phone +43-512-507 5409, fax +43-512-507 2895, e-mail Hanns.Kerschner@uibk.ac.at

Keywords: Alps, Glaciers, Glacier-climate models, Precipitation, Younger Dryas.

1. Introduction

The regional distribution of precipitation in a mountain range like the European Alps is a good indicator for continental-scale atmospheric circulation patterns. This is particularly true when precipitation is primarily caused by the advection of air masses to the Alps from the North Atlantic or the Mediterranean Sea, as is the case under cold conditions. Alpine precipitation patterns during the Lateglacial period can hence be interpreted in terms of past atmospheric circulation patterns in continental Europe. In this paper, glacier-climate models are used for the reconstruction of Younger Dryas precipitation patterns based on changes in equilibrium line altitudes of Alpine glaciers. This type of research provides important information concerning the range of past precipitation variability against which present climatic changes in the Alps can be assessed. Also, unravelling the spatial patterns of Alpine precipitation allows us to gain a better understanding of forcing mechanisms behind precipitation changes.

The „Alpine Lateglacial" (ca. 19,000 – 11,500 cal BP) was characterised by a rapid downwasting of the large piedmont and valley glacier systems, which was interrupted by several successively smaller readvances associated with short cold and/or more humid phases („Stadials"). The morainic systems associated with these advances form the basis for the definition of a glacial event sequence (Maisch 1982).

U. M. Huber et al. (eds.), Global Change and Mountain Regions, 73-81.

The Younger Dryas (YD) cold phase (Greenland Stadial 1, 12,650 – 11,500 cal. BP; Björck et al. 1998) in the Alps is represented by the „Egesen Stadial". In many valleys, three distinct sets of YD moraines can be found (Egesen-I to Egesen-III), jointly representing the last major cold event prior to the Holocene. Widespread rock glacier development after the Egesen-I phase documents the existence of permafrost down to altitudes of 2000 m asl and below (Sailer and Kerschner 2000). Surface exposure ages of an Egesen-I moraine at Julier pass (Switzerland) show that the moraine stabilised a few centuries after the onset of the Younger Dryas (Ivy-Ochs et al. 1996).

Lateglacial glaciers play an important role as a palaeoclimatic data source, because glacier behaviour is directly controlled by climate on a time-scale on the order of several decades to a few centuries. The spatial coverage of glacier data for selected time-slices, such as the early Younger Dryas, is very dense in the Alps. The data set for this study comprises 160 data points.

2. Theoretical framework

The accumulation area of a glacier is separated from the ablation area by the equilibrium line (EL), where accumulation exactly equals ablation. The equilibrium line altitude (ELA) of former glaciers can be determined from glacier maps, if past glacier extents are sufficiently well documented. Under alpine conditions, the accumulation area is usually twice as large as the ablation area, if the glacier is in equilibrium (Gross et al. 1977). The ELA and the vertical shift of the ELA (ΔELA) are used as principal input parameters for the reconstruction of paleoclimate with glacier-climate models.

2.1 The glacial-meteorological approach

A vertical shift of the ELA is entirely caused by climatic factors, be it a change in accumulation (c) or in ablation (a), or both. If one parameter (c or a) can be estimated reasonably well, the other can be, at least in principle, calculated from ΔELA. The perturbation analysis of the mass and energy balance equation by Kuhn (1981, 1989) allows us to analyse the physical mechanisms controlling ELA fluctuations. This approach is referred to as the „glacial-meteorological model" (GMM) below.

A change in accumulation (Δc) and/or ablation (Δa) causes a shift of the ELA along the vertical gradients of accumulation ($\partial c/\partial z$) and ablation ($\partial a/\partial z$) until balanced conditions are achieved again (1).

$$\Delta c + \frac{\partial c}{\partial z}\Delta ELA = \Delta a + \frac{\partial a}{\partial z}\Delta ELA \qquad (1)$$

Expressing ablation as the consequence of the energy supplied to the glacier surface, equation (1) can be written as

$$\Delta c + \frac{\partial c}{\partial z}\Delta ELA = \frac{\tau}{Lm}[-G\partial r + \partial A + \frac{\partial A}{\partial z}\Delta ELA + \alpha(\partial Ts + \frac{\partial Ts}{\partial z}\Delta ELA)] - \tau\partial S\,(1 - \frac{Ls}{Lm}) \qquad (2)$$

with τ as the duration of the ablation period, Lm and Ls as the latent heat of fusion and of sublimation, respectively, G as the global radiation, r as the albedo, A as the atmospheric long wave radiation, α as the turbulent heat exchange coefficient, Ts as the air temperature of the ablation period and S as the sublimation. All changes of variables which are not incorporated in equation (2) are considered to be zero. As a first approximation it is assumed that all vertical lapse rates are linear. Finally, the change in precipitation can be calculated as $\Delta p = \Delta c / 1.5$ under the assumption that accumulation on a glacier surface is 50% larger than precipitation due to wind drift and avalanching (for details see Kuhn 1981, 1989; Kaser and Osmaston 2002).

2.2 Statistical models

Statistical models relate climatic parameters governing ablation with parameters governing accumulation using regression techniques. This approach was used for the first time by Ahlmann (1924) for glacierized basins in Norway. It shows that precipitation at the ELA increases exponentially with increasing summer temperatures. Since then, various authors derived equations with summer temperature as a parameter for ablation and with precipitation-related data as a parameter for accumulation. This works surprisingly well, since summer temperature seems to be a good parameter for all processes causing ablation (Ohmura 2001).

Statistical models should use physically meaningful variables, but they are not physical models. Therefore, they may only be used within the boundary conditions of the dataset from which they were derived and require a careful climatological calibration and adaptation of the input data. Input data, which are not compatible with the model parameters, may cause considerable errors. Table 1 lists various statistical models and the parameters used for their calculation.

In order to calculate precipitation changes for a time interval in the past (e.g. early Younger Dryas), a reliable estimate of present-day precipitation in the currently glacierized areas is necessary. Measuring precipitation in glacierized high mountain regions is, however, extremely difficult and the spatial resolution of gridded precipitation data (Frei and Schär 1998) is too coarse to be used as a reference. Therefore, it is appropriate to determine present-day precipitation indirectly from present-day ELAs using a glacier-climate model. Under Alpine conditions, the normal period 1931-60 seems to be most useful as a reference period because it is well documented and provides the climatic background for the glacier inventories in the European Alps. Modern precipitation at the ELA can then be determined with the glacier-climate models by Ohmura et al. (1992), Krenke (1975) or Khodakov (1975). Wherever these models have been tested against measured Alpine data, the results fit quite well. Results obtained from the Norwegian model and the relation by Shi et al. (1992) are systematically too high in warmer areas of the Alps.

For the reconstruction of past precipitation, all models require external information on temperature change from non-glacial proxy data. Glacial and non-glacial data sources have to be synchronised, and temperature proxy data have to be available on a time scale, which is roughly equivalent to the response time of glaciers (i.e. several decades to a few centuries).

Table 1: Statistical glacier-climate models (temperatures are in °C, accumulation / precipitation in kg $m^{-2}a^{-1}$).

$$c = 75\ e^{(0.9\ Ts\ -\ 0.45)} + 1173$$ (Ahlmann 1924)

c: Runoff from glacierized basins in Norway, Ts: standard mean summer temperature June-September, vertical temperature lapse rate -0.7 K/100 m.

$$c = a = 0.96\ (Ts + 10)^{2.93}$$ (Khodakov 1975)

a: annual ablation, Ts: standard mean summer temperature (June-August), data from glacierized areas in the Soviet Union.

$$c = a = (Ts + 9.5)^3$$ (Krenke 1975)

Same glacier data as Khodakov, Ts: mean temperatures from long temperature series (1881 – 1960) in the Soviet Union, extrapolated to the ELA and reduced by 1 K to account for temperatures near the glacier surface.

$$P(ELA) = 645 + 296\ Ts + 9\ Ts^2$$ (Ohmura et al. 1992)

P(ELA): Precipitation at the ELA (winter mass balance + summer precipitation), Ts: mean summer temperature (June-August) of the free atmosphere at the ELA, world-wide data set.

$$Pw = 915\ e^{(0.339\ T)}$$ (Norwegian Model; "Liestøl-equation", e.g. Dahl and Nesje 1996)

Pw: winter precipitation at the ELA, T: summer temperature at the ELA (May-September), temperature lapse rate -0.65 K /100 m, data from Norwegian glaciers by O. Liestøl.

$$P(ELA) = 441\ e^{(0.4225\ T)}$$ (Shi et al. 1992)

P(ELA): annual precipitation at the ELA, T: summer temperature (not specified), data from glacierized areas in West China (mainly Tibet).

3. Application: Alpine precipitation patterns during the Younger Dryas

The early Younger Dryas in the Alps provides a good testing ground for the determination of palaeoprecipitation with either statistical glacier-climate models or the GMM. Glacier advances are well documented in the field (maximum advance of the "Egesen Stadial") and glacier topographies can be reliably plotted on topographic maps. The modeling approach is outlined in Figure 1; more details can be found in Kerschner et al. (2000).

The crucial first step in palaeoprecipitation reconstructions is the determination of summer temperature depression (ΔTs) at decadal to century-scale resolution. In the Alps, estimates for Younger Dryas ΔTs vary between about -1.5 and -7.5 K (Lotter et al. 2000), depending on the proxy record used.

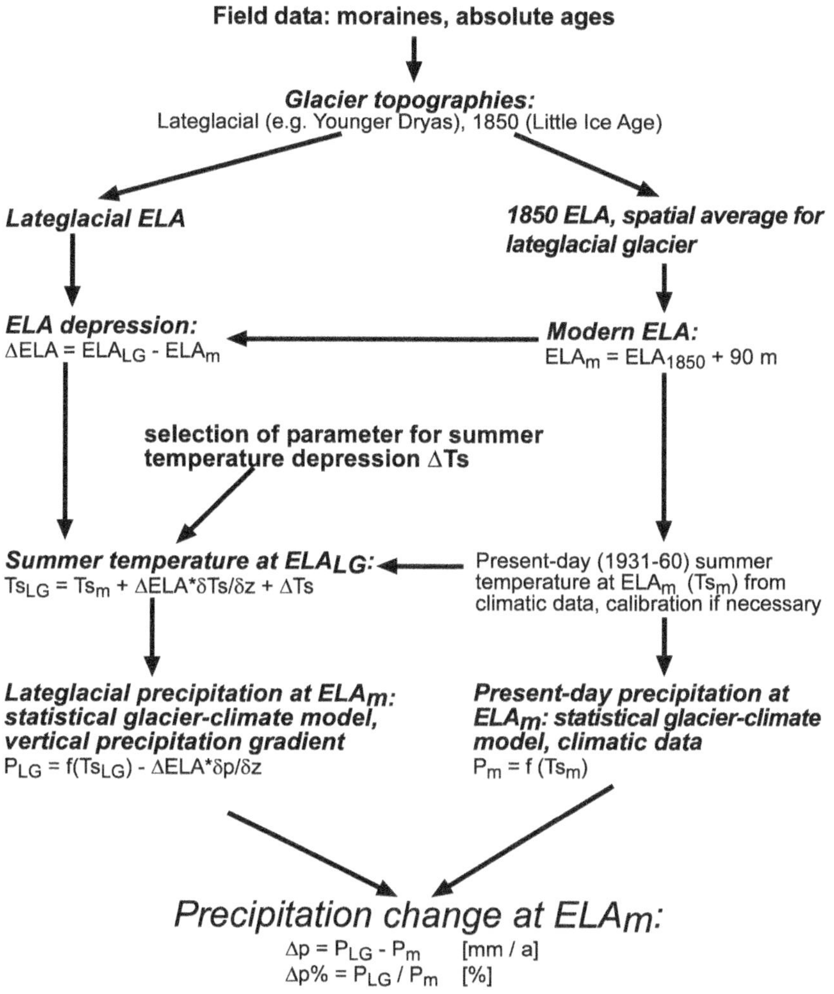

Figure 1: Flow chart for the determination of precipitation change with glacier-climate models.

As upper timberline is largely dependent on summer temperature, timberline fluctuations can be used to calculate ΔTs. However, timberline reconstructions for the YD vary substantially across the Alps. If an average timberline depression on the order of -500 m is assumed (Bortenschlager 1984; Burga and Perret 1998), ΔTs during the early Younger Dryas was about -3.5 K, which is comparable to results from Schwalb et al. (1994). Recent studies suggest a timberline depression on the order of -800 m to -1000 m, corresponding to a ΔTs of -5.6 to -7 K (Tobolski and Ammann 2000; Wick 2000). Future research will have to resolve whether these differences are associated with methodological differences in timberline reconstructions or regional differences in the YD temperature decline across the Alps.

ELA depressions of Egesen-maximum glaciers close to the northern fringe of the Austrian Alps are on the order of -500 m, whereas in the central Alps they are on the order of -200 m relative to the ELA of 1850. Assuming a spatially constant temperature depression and glaciers in equilibrium, the observed spatial variability of ELA depression can be interpreted in terms of spatial differences in precipitation change. With a ΔTs of -3.5 K, precipitation during the early Younger Dryas should have been somewhat higher than today along the northern fringe of the Alps (+15%) and considerably lower than today in the well-sheltered valleys of the interior (up to -30%; Fig. 2). With an assumed ΔTs of -4.5 K, the respective values are ±0% along the northern fringe of the Alps and -50 to -60% in the interior. In any case, such a pattern is typical for a more zonal westerly circulation under cold conditions, as simulated for the YD interval by Atmospheric General Circulation Models (Renssen et al. 2001; Renssen and Isarin 1998).

The Norwegian model indicates that winter precipitation was more reduced than annual precipitation. This is reasonable, as winter precipitation under cold conditions is generally very low. Assuming a NW – SE gradient of temperature change during the YD, with a stronger temperature depression in the NW part of the Alps (Isarin and Bohncke 1998), the contrast in precipitation change between the northern fringe of the Alps and the more continental interior would have been less pronounced.

A summer temperature depression of -7.5 K (Lotter et al. 2000) requires full arid conditions in the central Alps. Depending on the model used, precipitation would have been reduced by 40 to 60% along the northern fringe of the Alps and by 75 to almost 100 % in the valleys of the interior. In more than 75% of the cases, summer temperatures at the ELA would have been negative, which is typical for glaciers in a cold and arid environment. As other glaciological parameters like basal shear stresses of early YD glaciers were rather similar to those of modern, well nourished glaciers (Maisch and Haeberli 1982) in more humid climates, the assumption of a spatially constant severe summer temperature depression requires further study.

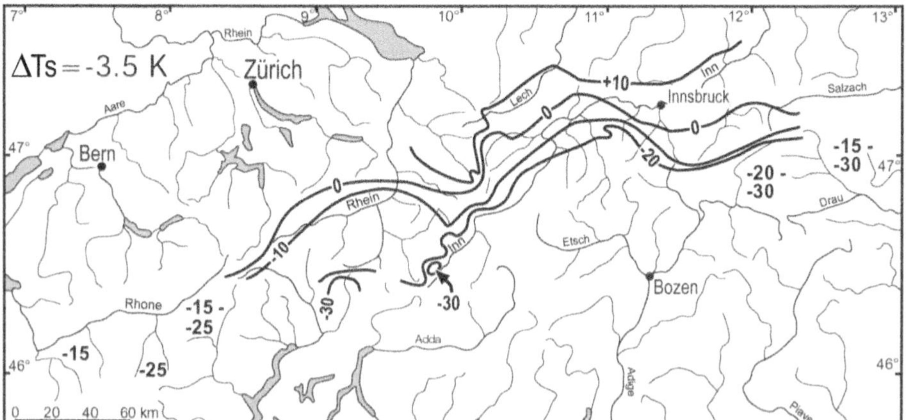

Figure 2: Early Younger Dryas precipitation change in the Alps under the assumption of ΔTs = -3.5 K; data from 160 Egesen-I glaciers.

4. Some perspectives for future research

The reconstruction of precipitation change based on past ELA fluctuations can give at least "semi-quantitative" information about an important factor of the hydrological cycle during the Lateglacial. The spatial patterns of precipitation change can serve as a test for the reliability of Atmospheric General Circulation Models. For future research, a few points should be mentioned, which may serve as guidelines.

- The simulation of mass balance gradients with the glacial-meteorological model, which has shown promising results for modelling tropical glaciers (Kaser and Osmaston 2002), should also be applicable to glaciers of the Alpine Lateglacial. It could be used to test the influence of parameters such as a) shorter ablation periods than today, b) the effects of more pronounced seasonal contrasts in temperature and precipitation, c) a possible increase in sublimation from the glacier surface and d) albedo changes.
- The glacier bed topography of many Lateglacial glaciers is well known, which allows a three-dimensional reconstruction of the glacier topography. This is an important prerequisite for the application of more sophisticated glaciological models, which are largely independent of climatic information from non-glacial proxy-data. Ultimately, these approaches combine climate models, mass and energy balance models and glacier flow models (e.g. Kull and Grosjean 2000). A simplified approach was used by Kerschner et al. (1999) to reconstruct the climatic conditions during the Gschnitz Stadial (ca. 16,000 cal B.P.).
- There is still a substantial need for reliable data on summer temperature change in the Alps during the Lateglacial on a timescale of about 100 to 300 years. A greater spatial coverage of quantitative temperature reconstructions would also help to address whether or not Lateglacial temperature changes were uniform across the Alps. If reliable data on temperature change are available, the resulting maps of precipitation change are probably of similar quality as present-day climatological maps of the Alps.
- Lateglacial glacier data, such as past ELAs, are presently available for wide areas of the Alps in Western Austria and Switzerland. In these areas, we have a fairly good overview of the distribution of Younger Dryas glaciers and also a reasonably good idea about Pre-Bølling Stadials (older than 14,700 cal BP). There are, however, still significant gaps in our knowledge both in space and time. Data from the Northern Alps and from areas south of the Alpine Main Ridge are still scarce or more or less completely missing. These regions should be the focus of future studies.
- There is also an urgent need for more reliable absolute ages of the deposits left behind by glacier advances. Modern dating methods (e.g. surface exposure dating with cosmogenic radionuclides, analysis of proglacial lake sediments and U/Th dating of carbonate sinters) should provide sufficient data to at least bracket several periods of glacier advances. This should permit a more reliable correlation between the glacial record from the Alps, proxy data from the circum-alpine region and the large-scale climatic development, as recorded in ice-cores and ocean sediment cores. There still remains much work to be done.

5. Acknowledgements

This study was supported by the Austrian Science Foundation (FWF) under grants P-12600 GEO and P-15108.

6. References

Ahlmann, H.W:son (1924). Le niveau de glaciation comme fonction de l'accumulation d'humidité sous forme solide. *Geografiska Annaler* **6**, 221-272.

Björck, S., Walker, M. C., Cwynar, L. C., Johnsen, S., Knudsen, K.-L., Lowe, J. J., Wohlfahrt, B., and INTIMATE members (1998). An event stratigraphy for the Last Termination in the North Atlantic region based on the Greenland ice-core record: A proposal by the INTIMATE group. *Journal of Quaternary Science* **13**, 283-292.

Bortenschlager, S. (1984). Beiträge zur Vegetationsgeschichte Tirols I. Inneres Ötztal und unteres Inntal. *Berichte des Naturwissenschaftlich-Medizinischen Vereins in Innsbruck* **71**, 19-56.

Burga, C., and Perret, R. (1998). "Vegetation und Klima der Schweiz seit dem jüngeren Eiszeitalter." Ott, Thun.

Dahl, S. O., and Nesje, A. (1996). A new approach to calculating Holocene winter precipitation by combining glacier equilibrium-line altitudes and pine-tree limits: A case study from Hardangerjøkulen, central southern Norway. *The Holocene* **6**, 381-398.

Frei, Ch., and Schär, Ch. (1998). A precipitation climatology of the Alps from high-resolution rain-gauge observations. *International Journal of Climatology* **18**, 873-900.

Gross, G., Kerschner, H., and Patzelt, G. (1977). Methodische Untersuchungen über die Schneegrenze in alpinen Gletschergebieten. *Zeitschrift für Gletscherkunde und Glazialgeologie* **12**, 223-251.

Isarin, R. F. B., and Bohncke, S. J. P. (1998). Mean July temperatures during the Younger Dryas in northwestern and central Europe inferred from climate indicator plant species. *Quaternary Research* **51**, 158-173.

Ivy-Ochs, S., Schlüchter, Ch., Kubik, P., Synal, H.-A., Beer, J., and Kerschner, H. (1996). The exposure age of an Egesen moraine at Julier Pass, Switzerland, measured with the cosmogenic radionuclides ^{10}Be, ^{26}Al and ^{36}Cl. *Eclogae Geologicae Helvetiae* **89**, 1049-1063.

Kaser, G., and Osmaston, H. (2002). "Tropical Glaciers." Cambridge University Press, Cambridge.

Kerschner, H., Ivy-Ochs, S., and Schlüchter, Ch. (1999). Paleoclimatic interpretation of the early late-glacial glacier in the Gschnitz valley, Central Alps, Austria. *Annals of Glaciology* **28**, 135-140.

Kerschner, H., Kaser, G., and Sailer, R. (2000). Alpine Younger Dryas glaciers as paleo-precipitation gauges. *Annals of Glaciology* **31**, 80-84.

Khodakov, V. G. (1975). Glaciers as water resource indicators of the glacial areas of the USSR. *International Association of Hydrological Sciences Publication* **104**, 22-29.

Krenke, A. N. (1975). Climatic conditions of present-day glaciation in Soviet Central Asia. *International Association of Hydrological Sciences Publication* **104**, 30-41.

Kuhn, M. (1981). Climate and glaciers. *International Association of Hydrological Sciences Publication* **131**, 3-20.

Kuhn, M. (1989). The response of the equilibrium line altitude to climatic fluctuations: Theory and observations. *In* "Glacier fluctuations and climatic change," (J. Oerlemans, Ed.), pp. 407-417. Kluwer, Dordrecht.

Kull, C., and Grosjean, M. (2000). Late Pleistocene climate conditions in the north Chilean Andes drawn from a climate-glacier model. *Journal of Glaciology* **46**, 622-632.

Lotter, A. F., Birks, H. J. B., Eicher, U., Hofmann, W., Schwander, J., and Wick, L. (2000). Younger Dryas and Alleröd summer temperatures at Gerzensee (Switzerland) inferred from fossil pollen and cladoceran assemblages. *Palaeogeography, Palaeoclimatology, Palaeoecology* **159**, 349-361.

Maisch, M. (1982). Zur Gletscher- und Klimageschichte des alpinen Spätglazials. *Geographica Helvetica* **37**, 93-104.

Maisch, M., and Haeberli, W. (1982). Interpretation geometrischer Parameter von Spätglazialgletschern im Gebiet Mittelbünden, Schweizer Alpen. *In* „Beiträge zur Quartärforschung in der Schweiz," (Physische Geographie **1**, M. Gamper, Ed.), pp. 111-126. Geographisches Institut der Universität, Zürich.

Ohmura, A. (2001). Physical basis for the temperature-based melt-index method. *Journal of Applied Meteorology* **40**, 753-761.

Ohmura, A., Kasser, P., and Funk, M. (1992). Climate at the equilibrium line of glaciers. *Journal of Glaciology* **38**, 397-411.

Renssen, H., and Isarin, R. F. B. (1998). Surface temperature in NW Europe during the Younger Dryas: AGCM simulation compared with temperature reconstructions. *Climate Dynamics* **14**, 33-44.

Renssen, H., Isarin, R. F. B., Jacob, D., Podzun, R., and Vandenberghe, J. (2001). Simulation of the Younger Dryas climate in Europe using a regional climate model nested in an AGCM: Preliminary results. *Global and Planetary Change* **30**, 41-57.

Sailer, R., and Kerschner, H. (2000). Equilibrium line altitudes and rock glaciers in the Ferwall-Group (Western Tyrol, Austria) during the Younger Dryas cooling event. *Annals of Glaciology* **28**, 141-145.

Schwalb, A., Lister, G. S., and Kelts, K. (1994). Ostracode carbonate $d^{18}O$- and $d^{13}C$-signatures of hydrological and climatic changes affecting Lake Neuchâtel, Switzerland, since the late Pleistocene. *Journal of Paleolimnology* **11**, 3-17.

Shi, Y., Zheng, B., and Li, S. (1992). Last glaciation and maximum glaciation in the Qinghai-Xizang (Tibet) Plateau: A controversy to M. Kuhle's ice sheet hypothesis. *Zeitschrift für Geomorphologie N.F.,* Supplementband **84**, 19-35.

Tobolski, K., and Ammann, B. (2000). Macrofossils as records of plant responses to rapid Late Glacial climatic changes at three sites in the Swiss Alps. *Palaeogeography, Palaeoclimatology, Palaeoecology* **159**, 251-259.

Wick, L. 2000. Vegetational response to climatic changes recorded in Swiss Late Glacial lake sediments. *Palaeogeography, Palaeoclimatology, Palaeoecology* **159**, 231-250.

Holocene Environmental Change in the Himalayan-Tibetan Plateau Region: Lake Sediments and the Future

J. Overpeck[1*], K. B. Liu[2], C. Morrill[3], J. Cole[4], C. Shen[5], D. Anderson[6], and L. Tang[7]

[1]*Institute for the Study of Planet Earth and Department of Geosciences, University of Arizona, Tucson, AZ 88721, USA*

[2]*Department of Geography and Anthropology, Louisiana State University, Baton Rouge, LA 70803, USA*

[3]*National Center for Atmospheric Research, Boulder, Colorado 80307, USA*

[4]*Department of Geosciences, University of Arizona, Tucson, AZ 88721, USA*

[5]*Department of Geography and Anthropology, Louisiana State University, Baton Rouge, LA 70803, USA*

[6]*National Geophysical Data Center, 325 Broadway, Boulder Co 80303, USA*

[7]*Nanjing Institute of Geology and Paleontology, Academia Sinica, Nanjing 210008, China*

[*]*phone 1-520-622-9065, fax 1-520-792-8795, email jto@u.arizona.edu*

Keywords: Geochemistry, Lakes, Monsoon, Paleoclimate, Pollen, Tibet.

1. Introduction

The South Asian Monsoon system is one of the most important and influential of the Earth's major climate systems. The people of the most heavily populated Asian countries have adapted many aspects of their society to the subtleties of the monsoon rains, and are thus highly susceptible to small changes in the timing and intensity of monsoon precipitation. A monsoon failure can have disastrous effects, and flooding related to extreme monsoon rains has proven to be one of the most deadly of natural catastrophes (e.g. in Bangladesh, China, India and Nepal). These vulnerabilities are likely to increase in the future with continued population growth, intensified land-use and sea-level rise. Although there is a growing effort to improve seasonal to

U. M. Huber et al. (eds.), Global Change and Mountain Regions, 83-92.

interannual monsoon prediction skills via new research, the largest threats to human health and livelihood could come from unanticipated decade- and longer-scale extremes in monsoon. A major goal of this paper is to summarize the state-of-the-art regarding century to millennium-scales of monsoon variability, and to identify the paleoenvironmental research that is most urgently needed in the Himalayan-Tibetan Plateau if society is to be served effectively in the 21st century.

The Southwest (SW) Indian or Asian Monsoon is the dominant monsoon system impacting the Himalayan-Tibetan Plateau region, as well as the heavily populated countries that are downstream of this region and its headwaters. All of the monsoon systems of Asia are powered to some extent by the largest high-elevation region of the world (Fig. 1). The strength of the monsoons is a reflection of the pressure gradient between the south Asian landmass and surrounding oceans. Because the land, and in particular the kilometers-high Himalayan-Tibetan Plateau, warms much more than the surrounding oceans during late spring and summer, low pressure dominates the warm seasons of the Himalayan-Tibetan Plateau region, with higher pressure off-shore. This gradient drives the monsoon flow of moisture-laden air off the oceans and up over land. Monsoon rainfall is concentrated by the orographic uplift of air masses passing over the mountains of South Asia, but monsoon rainfall, particularly from the SW Monsoon flowing north from the Indian Ocean, still extends its dominance deep into the Tibetan Plateau. The kilometers-high Himalayan-Tibetan Plateau produces an immense heat-low that serves to link the dynamics of the SW Monsoon to the other

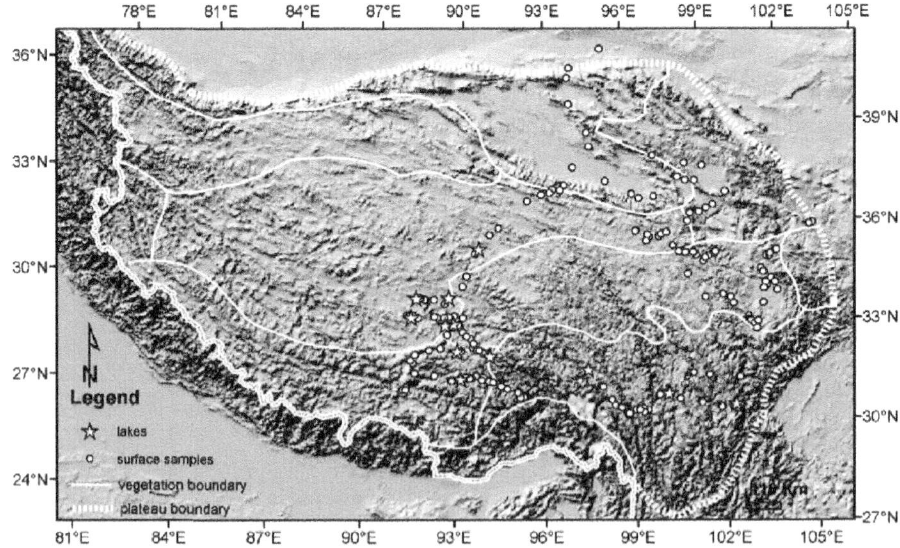

Figure 1: Shaded relief map of the Himalayan-Tibetan Plateau region showing boundaries of major vegetation regions and the Tibetan Plateau. Also shown are both lakes (stars) and surface sediments (circles) that have been sampled by our team over the last several years as part of a program to create a network of lake sediment records that span the entire gradient of climate (e.g. temperature and rainfall) and vegetation change across the region. Results from the samples shown on the map will be published soon, and the next phase of fieldwork will extend westward from ca. 90°E, as well as northward from 30°N.

monsoon systems of south and east Asia, as well as to impact climate variability far from the monsoon regions themselves. For these reasons, the paleoclimatic study of the Himalayan-Tibetan Plateau region, and the SW Monsoon in particular, is of great importance in unraveling the full range of natural climate variability. Knowing the past patterns and causes of climate variability can, in turn, be used to provide a baseline against which present climate changes can be assessed, and future changes anticipated.

2. The paleoenvironmental record

The SW Indian or Asian Monsoon may be the best studied in terms of paleoclimate, but this wealth of research has also uncovered many important unknowns. The purpose of this paper is to focus on the aspects of monsoon research that are relevant to the people of the Himalayan-Tibetan Plateau region. Thus, although there has been important work done on pre-Quaternary and glacial-age monsoon variations, this paper focuses primarily on the Holocene period, which likely provides the most relevant information for the prediction of future changes in monsoon dynamics.

Orbital-scale paleoclimatology. Unraveling the role of astronomical (Milankovitch) variations in pacing the SW Asian Monsoon on glacial-interglacial time scales is a major success story of late-20th century research on past climate dynamics (COHMAP 1988). The glacial to interglacial increase in summertime insolation, caused by variations in the earth's orbit, was the primary driver of the much stronger monsoons in the early to mid-Holocene (ca. 11500 to 4500 yr BP) across most of eastern Africa, the Arabian Sea and southern Asia, as well as the observed waning of the monsoon to the present day (Overpeck et al. 1996). This change is reflected very well in the paleomonsoon records of the Himalayan-Tibetan Plateau region (Gasse and van Campo 1994; Morrill et al. 2003). It is quite comforting that we can now even simulate many aspects of the orbital-scale monsoonal response to insolation (Kutzbach and Street-Perrott 1985; COHMAP 1988). Being able to simulate aspects of climate system behavior means that we understand the underlying processes and causes of this behavior.

The response of the SW Monsoon and rainfall over the Himalayan-Tibetan Plateau region to Milankovitch forcing has not been entirely linear. Overpeck et al. (1996) highlighted how the response of the SW Monsoon lagged insolation forcing over the last deglaciation, and how this lag was likely a result of glacial boundary conditions (i.e. low sea-surface temperatures and glacial ice sheets as far away as the North Atlantic and Europe) retarding the ability of the Tibetan Plateau to warm in synch with the gradual increase in summertime insolation resulting from subtle changes in the earth's orbit. Summer insolation increased steadily from the onset of the last deglaciation to a peak around 12,000 years ago, but the monsoon stayed relatively weak until after 11,500 years ago (all dates in this paper are in calendar time rather than radiocarbon). Only after the demise of glacial-age conditions in and upwind of the Himalayan-Tibetan Plateau was the monsoon able to take full advantage of the insolation forcing that was significantly stronger than today (Overpeck et al. 1996;

Morrill et al. 2003). The result was peak monsoon rainfall in the early Holocene.

Century-scale paleoclimatology. The story gets more complicated when century-scale monsoon dynamics are considered. Given the dominant role of Milankovitch forcing, particularly in the absence of glacial boundary conditions, it would be expected that the SW Monsoon simply weakened gradually over the last 9500-8000 years as the summertime insolation gradually decreased by about 12% (Overpeck et al. 1996). Although there has been a general weakening of the SW Monsoon over this time, each year of research brings new hints that the change has not been gradual or monotonic. Clearly, there is more going on with the SW Monsoon that needs to be understood. It appears that the SW Monsoon also varies significantly on century time scales, and that this scale of variability could be the most important to all those that depend on SW Monsoon rains in one way or another.

The first systematic look at abrupt century-scale variability of the Asian Monsoon since the last glacial period was carried out by Morrill et al. (2003; Fig. 2). In this review of existing literature, it was discovered that the simple hypothesis of Overpeck et al. – of a monsoon that weakened gradually and in synch with insolation over the

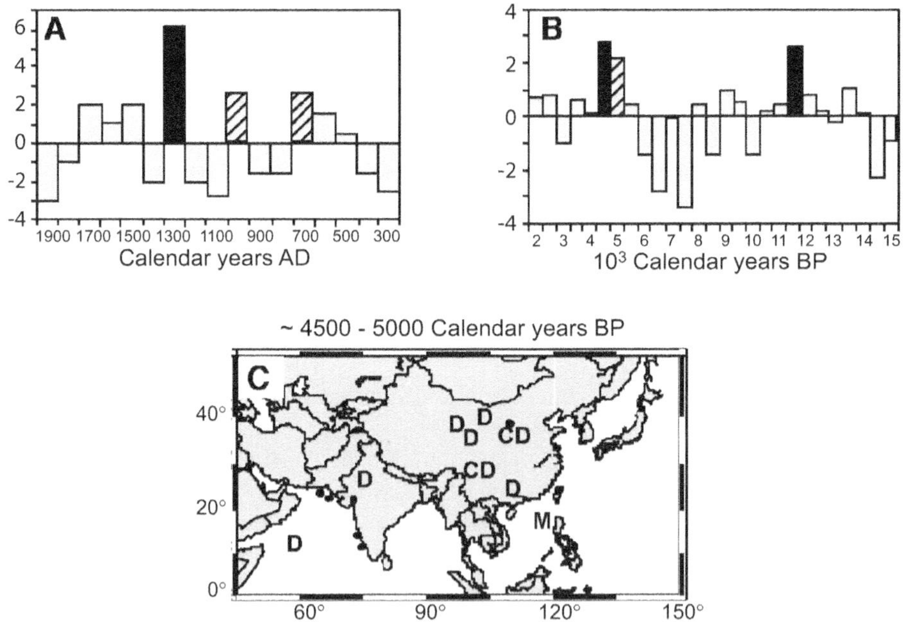

Figure 2: Histograms (A and B) showing the number of abrupt changes in monsoon strength observed in the Asian monsoon region for historical (A) and pre-historical (B) time periods (from Morrill et al. 2003). Frequency-of-events for each bin is expressed as anomalies from expected number of events assuming a random distribution of events through time as described in Morrill et al. (2003). Black (striped) shading indicates bins with a positive anomaly that is statistically different from zero at the 95% (90%) confidence level. Also shown (C) is a map of the entire Asian monsoon region indicating where abrupt events are observed at ca. 4500 to 5000 years B.P. ("D" signifies abrupt drying, "M" moistening, and "C" cooling). Note that the biggest hole in the map is the Himalayan-Tibetan Plateau region where a new network of lake sediment sites is being established.

last ca. 8000 years – must be modified to accommodate additional sources of monsoon variability. In particular, there is evidence that the SW Monsoon weakened abruptly at two times since the early Holocene peak monsoon period. A monsoon weakening at about 4500 years ago appears to coincide with abrupt climate change in the Middle East, North Africa, the North Atlantic and even as far away as the tropical Pacific (Morrill et al. 2003). The cause of these changes is difficult to nail down, but Morrill et al. favor the hypothesis that non-linear feedbacks in the North Atlantic and/or tropical Pacific El Niño-Southern Oscillation (ENSO) system may have played an important role. A shift to more, or stronger, El Niños could have shifted Pacific convection more eastward, and in doing so, reduced the moisture available to the Asian Monsoons. The presence of the ca. 4500 yr BP climate change in the Himalayan-Tibetan Plateau monsoon region, a region generally outside the influence of North African climate change, also casts some doubt on the dominance of non-linear land-atmosphere interactions in North Africa (e.g. Claussen et al. 1999) as a major factor in causing this increasingly well-known abrupt climate shift over Africa, Asia and beyond.

Morrill et al. (2003; Fig. 2) also reveal hints of an abrupt SW Monsoon weakening at about A.D.1300, more or less coincident with the onset of the "Little Ice Age" in the North Atlantic and European region, as well as an apparent change in climate variability over parts of North America (e.g. Overpeck et al. 1997). Once again, this abrupt shift in monsoon dynamics appears to coincide with abrupt climate change in the North Atlantic and North Africa, leading Morrill et al. to speculate that the ultimate cause of the sudden monsoon weakening was a shift in North Atlantic climate that resulted in the advection (by the westerlies) of colder air downstream over Eurasia, and also perhaps a longer snow season over the Himalayan-Tibetan Plateau region. Colder, snowier conditions would delay the ability of the Plateau to warm in the spring and summer, and thus delay and weaken the summer monsoon. This snow-monsoon linkage has been hypothesized for some time, and the paleoclimatic record of the Himalayan-Tibetan Plateau region suggests that it is important on all time scales and for both glacial and non-glacial boundary conditions (Overpeck et al. 1996; Morrill et al. 2003).

Two new studies of SW Monsoon variation over the Arabian Sea have heightened interest in century-scale variability of the Asian Monsoon, as well as increased the importance of paleoclimatic work in the Himalayan-Tibetan Plateau region. In a study of oceanographic change in the monsoon-sensitive Arabian Sea over the last 1000 years, Anderson et al. (2002) reveal that the SW Monsoon apparently intensified dramatically since the middle of the "Little Ice Age" (ca. A.D. 1600), and in particular over the last 150 years. Moreover, present-day monsoon strength appears to be the strongest of the last 1000 years, an observation that led the authors to hypothesize that the unprecedented 20th century warming of the Himalayan-Tibetan Plateau region (Thompson et al. 2000; IPCC 2001) has led to an increasingly invigorated SW Monsoon. However, as pointed out by Anderson et al. (2002), the 20th century rainfall history of India does not reflect this increase, nor does a 1000-year ice-core record from the 7200 m crest of the Himalaya (Thompson et al. 2000). This heightens the possibility that continued global warming will instead increase rainfall mostly over the Tibetan Plateau itself. This hypothesis needs testing, as does the possibility that warming-induced monsoon intensification could be partially countered by the effects of abrupt future cooling in the North Atlantic (Gupta et al. 2003) on monsoon dynamics.

The possible influence of the North Atlantic on century-scale monsoon variability is the topic of a new paper by Gupta et al. (2003). In this work, the high-resolution Arabian Sea record of monsoon variability is extended back through the Holocene and reveals a clear pattern of abrupt centennial monsoon events superimposed on a longer-term and more gradual orbitally-forced pattern of change. The most interesting aspect of this work is that the abrupt monsoon events appear to correlate with a similar series of abrupt cool episodes in the North Atlantic (Bond et al. 2001). This finding needs further testing, but strengthens support for a hypothesized Atlantic-monsoon linkage, and possibly for a hypothesized influence of the sun as the ultimate trigger for the observed series of abrupt climate events as well (Black et al. 1999; Bond et al. 2001). Either way, further work in the Himalayan-Tibetan Plateau region is needed to test these hypotheses, and to work out where the impacts are greatest when the monsoon strengthens or weakens abruptly. In particular, we need to examine how well the Bond/Gupta variations manifest themselves in the terrestrial heartland of the SW Monsoon, and also how these hypothesized variations related to the widely observed events at 4500 yr BP and AD 1300.

The link between the monsoon and North Atlantic variability highlights the question of future monsoon behavior in the Himalayan/Tibetan Plateau region. Simple radiative arguments suggest that the monsoon should strengthen as climate warms, because the land should heat up faster than the surrounding ocean, increasing the land-ocean contrast. Additional work on radiative aspects of this question have included the influence of anthropogenic aerosols (dust and pollutive), which are increasing and projected to continue increasing in this rapidly developing region. Aerosols may weaken the monsoon due to radiative reflection, absorption, and indirect effects on clouds, although their impact is not well simulated by existing models. If the North Atlantic variability is indeed a significant influence on the monsoon, then an additional factor - the behavior of the global thermohaline circulation - must be added to the list of climate variables that need to be well simulated to predict future monsoon variability. A projected weakening of the thermohaline circulation could counteract some or all monsoon enhancement originating from greenhouse warming.

3. Future directions and interdisciplinary linkages

The case has never been stronger for paleoclimatic research in the Himalayan-Tibetan Plateau region. Recent paleoclimatic work has found evidence of monsoon behavior not seen in the instrumental record. Paleoclimatic work also lends some support to solar forcing of the monsoon, as well as connections with climate variability in the Atlantic and Pacific regions. More than ever, the Himalayan-Tibetan Plateau region has been cast as an important focus of climate system research. Moreover, since climate change in this region has, via the monsoons of Asia, the potential to impact billions of people, there is an even greater imperative to study past changes in climate and related environmental systems in the Himalayan-Tibetan Plateau region. Lastly, the most recent published work highlights that the SW Monsoon is currently undergoing rapid change, perhaps in response to global warming (see previous section). There is thus a special urgency to the study of this region.

Paleoclimate priorities. The primary goal has to be focused on understanding the full range of possible monsoon behavior and how the monsoon of the Himalayan-Tibetan Plateau region will likely change in the future. Ongoing work is focused on replacing the extremely sparse array of poorly-dated records with a network of well-dated high-resolution paleoclimatic records for the Himalayan-Tibetan Plateau region (Fig. 2). Experience to date indicates that the most challenging aspect of this work is accurate time control, particularly in the face of widespread, large "old carbon" problems in the lake sediments of the region. This can be overcome by AMS radiocarbon dating of terrestrial plant and animal macrofossils. The second challenge is to build a "multi-proxy" context for reconstructing past climate variability and change. Efforts underway focus on using a variety of approaches, including palynological, sedimentological, geochemical (e.g. stable isotopes and trace element analysis) and paleolimnological methods. The goal has to be to build on the groundbreaking Tibetan Plateau work of Gasse and van Campo (1994) to produce paleoenvironmental inferences based on multiple lines of evidence.

The key to success hinges on our ability to reconstruct the space-time patterns of past change, answer specific questions and test specific hypotheses. Key questions include:

- What is the pattern of "warm climate" abrupt climate events in the Himalayan-Tibetan Plateau region, and to what extent do paleoenvironmental records confirm the existence of major events at ca. 4500 and 700 years ago (Morrill et al., 2003), and/or a series of events that are coincident with those of the Arabian Sea and/or the Atlantic (e.g. Gupta et al. 2003)? Could the major events at 4500 and 700 years ago just be particularly pronounced events in a series of weaker centennial scale events?
- Does the strengthening of the monsoon, observed in the Arabian Sea (Anderson et al. 2002; Gupta et al. 2003), have its greatest rainfall effect inland of the Himalayan crest, or is wind stress in the Arabian Sea decoupled from monsoon rainfall over land on centennial time scales?
- How sensitive are surface air temperature and rainfall on the Tibetan Plateau to changes in radiative forcing, such as those that occurred in the past (e.g. due to Milankovitch forcing, possible solar forcing, and recent anthropogenic forcing (Anderson et al. 2002; Gupta et al. 2003)), and what does this mean for the future (i.e. with continued increases in anthropogenic trace-gases and aerosols)?
- To what extent might future abrupt cooling in the North Atlantic region, related to a global warming-induced weakening of the Gulf Stream (IPCC 2001), reduce monsoon rainfall in the Himalayan-Tibetan Plateau region?

We will address these questions using new, quantitative reconstructions of specific climate variables that we derive through calibrated multi-proxy records from lakes in the region. Through a focus on specific climate variables (e.g. summer surface air temperature, effective moisture) we expect to develop a more nuanced view of past monsoon variability that will help us link reconstructed changes to specific forcings and mechanisms. Efforts are already underway to use modern pollen samples from hundreds of sites across the region (Fig. 1) to enable quantitative pollen-inferred climate reconstructions. We are also working to generate independent paleoclimate estimates using the geochemistry (e.g. stable isotopic and trace-elemental analyses of fossil ostracod valves). The hope is that these efforts will enable independent estimates of Tibetan Plateau climate sensitivity, both in terms of temperature and rainfall.

It should be emphasized that the most important interval of time for answering the above questions is the Holocene. The main reasons for this is that the Holocene includes a period of warmer summers, and also because it should be possible to generate a network of well-dated Holocene-length records for the Himalayan-Tibetan Plateau region. Well-dated glacial age records, to the extent they can be generated, will help understand the dynamics of the glacial world, but will have less relevance to monsoon dynamics during the present warm (generally ice free) period, and hence to the people who live in the Himalayan-Tibetan Plateau region. It will also be worthwhile to have one or more well-situated records that extend back at least through the last interglacial (the last 130,000). Such records would be valuable for estimating the sensitivity of the Himalayan-Tibetan Plateau region and monsoon system to altered forcing. Since the Himalayan-Tibetan Plateau region was likely warmer in summer during the last interglacial than today (e.g. Montoya et al. 2000), it would be helpful to know exactly how much warmer, and also how much wetter. Although this is not a precise analog for future climate, we can use the case of the last interglacial monsoon to identify important processes and feedbacks that may operate in a warmer-than-present scenario.

Interdisciplinary Linkages. The paleoclimatic study of the Himalayan-Tibetan Plateau region also affords rich opportunities for investigating how climate variability and change of the future may influence important aspects of life in the region. Our team is interested in interactions between climate and vegetation, wildfire, hydrology, surficial processes, natural hazards, limnology, and society. Moreover, the examination of climate variability and change on the Himalayan-Tibetan Plateau will also facilitate the testing of hypotheses that suggest this region has had significant impacts on the global climate system. For example, hypothesized influences of Himalayan-Tibetan Plateau climate on ENSO, North American precipitation, and the Australian Monsoon (Morrill et al. 2003) should soon be readily testable.

The multi-proxy approach will allow the study of climate-induced vegetation change, along with the role of vegetation disturbance by fire. For example, how sensitive are the dominant Himalayan-Tibetan Plateau vegetation communities (Fig. 2) to shifts in both temperature and rainfall, and how fast can the vegetation of the region respond to climate change? Simulating past vegetation responses to climate change will be key to assessing models that can, in turn, be used to estimate how the region's vegetation, aquatic systems and ecosystems may change in the future. Are there communities or species that may become extinct as the result of global warming? Although paleoecological data are only part of the puzzle, they are a key part because they facilitate the assessment of past vegetation responses to abrupt climate changes.

In particular, paleoclimatic data can help to address what rates of climate change and types of abrupt climate "surprises" may prove difficult for biodiversity conservation efforts (Overpeck et al. 2002).

Climate variability (i.e. extremes) already poses one of the most costly natural hazard risks in the Himalayan-Tibetan Plateau region, both in terms of lives and property. Paleoclimatic work will make it easier to estimate the probability of drought and flood extremes in the region, the latter of which can have their biggest societal impacts hundreds of kilometers downstream from the Himalayan-Tibetan Plateau. With the advent of anthropogenic global warming, glaciers throughout the world are in steady retreat (Dyurgerov and Meier 2000; Haeberli, this volume), leaving behind a growing number of moraine-dammed lakes. These lakes are susceptible to catastrophic discharge events that can prove especially devastating to people downstream. These events can be exacerbated by floods and mass-movement of surficial material associated with intense convective rainfall. A valuable spin-off of high-resolution paleomonsoon research is thus an improved understanding of the climatic conditions (e.g. extreme runoff and flood events) that can trigger catastrophic lake discharge. Moreover, comparison of moraine-dammed and nearby non-moraine-dammed lakes will provide a better understanding of the sediment and environmental variability associated with the two types of lakes. It should be possible to reconstruct the histories of sediment mass movement into lakes due to a variety of processes such as earthquakes, convective storms, and land use change.

In order for paleoenvironmental studies to be most effective, they need to be coupled with research focused on environment-society interactions. In the Himalayan-Tibetan Plateau region, this kind of work has taken two related forms. There is good evidence that ancient societies such as the Indus Civilization thrived in part due to the stronger monsoons of the middle Holocene, and there is good reason to suspect that the collapse of this civilization downstream of the Himalayan-Tibetan Plateau region was, at least in part, due to the weakening of the SW Monsoon, and associated rainfall reductions, near 4500 years ago. The Himalayan-Tibetan Plateau region itself is full of archeological evidence that needs to be put into a climate-society perspective. If this can be accomplished, then it may be possible to gain new insights into how the current people of the Himalayan-Tibetan Plateau region may be affected by future changes in monsoon rainfall.

Of course, societies of today are uniquely different from those of the past, and it will also be important to apply what we learn about SW Monsoon variability, as well as likely future change, to reducing vulnerability of human systems to climate extremes and change. This is a task that needs to be undertaken in a broad interdisciplinary context, and it need not await complete understanding of the climate dynamics. Indeed, it may be helpful to have "stakeholders" in the Himalayan-Tibetan Plateau region help identify key research questions. The way to begin would be to start breaking down interdisciplinary boundaries in much the same way similar efforts began in other regions (e.g. the Climate Assessment for the Southwest program in the United States, CLIMAS, 2002).

4. Acknowledgements

We thank Jon Pelletier and Jim Morrison for discussions of how lake sediments could help constrain the possible future dynamics of moraine-dammed lakes. We also thank Ulli Huber and Mel Reasoner for their review and comments, and the U.S. National Science Foundation for funding.

5. References

Anderson, D. M., Overpeck, J. T., and Gupta, A. K. (2002). Increase in the Asian Southwest Monsoon during the past four centuries. *Science* **297**, 596-599.

Black, D., Peterson, L., Overpeck, J., Kaplan, A., Evans, M., and Kashgarian, M. (1999). Eight centuries of North Atlantic Ocean atmosphere variability. *Science* **286**, 1709-1713.

Bond, G., Kromer, B., Beer, J., Muscheler, R., Evans, M. N., Showers, W., Hoffmann, S., Lotti-Bond, R., Hajdas, I., and Bonani, G. (2001). Persistent solar influence on north Atlantic climate during the Holocene. *Science* **294**, 2130-2136.

Claussen, M., Kubatzki, C., Brovkin, V., and Ganopolski, A. (1999). Simulation of an abrupt change in Saharan vegetation in the mid-Holocene. *Geophysical Research Letters* **26**, 2037-2040.

CLIMAS (2002). http://www.ispe.arizona.edu/climas/index.html

COHMAP (1988). Climatic changes of the last 18,000 years: Observations and model simulations. *Science* **241**, 1043-1052.

Dyurgerov, M. B., and Meier, M. F. (2000). Twentieth century climate change: Evidence from small glaciers. *Proceedings of the National Academy of Sciences of the United States of America* **97**, 1406-1411.

Gasse, F., and van Campo, E. (1994). Abrupt postglacial climate events in West Asia and North-Africa Monsoon domains. *Earth and Planetary Science Letters* **126**, 435-456.

Gupta, A. K., Anderson, D., and Overpeck, J. (2003). Abrupt changes in the Holocene Asian Southwest Monsoon and their links to the North Atlantic Ocean. *Nature* **421**, 354-357.

IPCC (2001). Climate change 2001: The scientific basis. Contribution of Working Group I to the Third Assessment Report of the Intergovernmental Panel on Climate Change. *In* J. T. Houghton, Y. Ding, D. J. Griggs, M. Noguer, P. J. van der Linden, X. Dai, K. Maskell, and C. A. Johnson (Eds.), pp. 881. Cambridge University Press, Cambridge.

Kutzbach, J. E., and Street-Perrott, F. A. (1985). Milankovitch forcing of fluctuations in the level of tropical lakes from 18 to zero kyr BP. *Nature* **317**, 130-134.

Montoya, M., von Storch, H., and Crowley, T. J. (2000). Climate simulation for 125 kyr BP with a coupled ocean- atmosphere general circulation model. *Journal of Climate* **13**, 1057-1072.

Morrill, C., Overpeck, J. T., and Cole, J. E. (2003). A synthesis of abrupt changes in the Asian summer monsoon since the last deglaciation. *The Holocene* **13**, 465-476.

Overpeck, J., Anderson, D., Trumbore, S., and Prell, W. (1996). The southwest Indian Monsoon over the last 18000 years. *Climate Dynamics* **12**, 213-225.

Overpeck, J., Hughen K., Hardy, D., Bradley, R., Case, R., Douglas, M., Finney, B., Gajewski, K., Jacoby G., Jennings A., Lamoreux, S., Lasca, A., MacDonald, G., Moore, J., Retelle, M., Smith, S., Wolfe, A., and Zielinski, G. (1997). Arctic environmental change of the last four centuries. *Science* **278**, 1251-1256.

Overpeck, J., Whitlock, C., and Huntley, B. (2002). "Terrestrial biosphere dynamics in the climate system: Past and future." *In* Paleoclimate, global change and the future. (K. Alverson, R. Bradley, and T. Pedersen, Eds.), pp. 81-111. Springer, Berlin.

Thompson, L. G., Yao, T., Mosley-Thompson, E., Davis, M. E., Henderson, K. A., and Lin, P. N. (2000). A high-resolution millennial record of the South Asian Monsoon from Himalayan ice cores. *Science* **289**, 1916-1919.

Water Resources in the Arid Mountains of the Atacama Desert (Northern Chile): Past Climate Changes and Modern Conflicts

Martin Grosjean[1*] and Heinz Veit[2]
[1]*NCCR Climate, University of Bern, Erlachstrasse 9a, CH-3012 Bern, Switzerland*
[2]*Department of Geography, University of Bern, Hallerstrasse 12, CH-3012 Bern, Switzerland*
**phone +41-31-631 31 47, fax +41-31-631 43 38, e-mail grosjean@giub.unibe.ch*

Keywords: Andes, Arid zone, Chile, Climate change, Holocene, Quaternary.

1. Introduction

The Atacama Desert of the Central Andes (18°S to 28°S) has become a focal point of environmental research in recent years. Indeed, this area is a key site in several respects. It is located between the tropical and extratropical precipitation belts; the vertical gradients of ecozones range from sea level at the Pacific Coast up to high mountains that reach into the mid-troposphere at 6000 m elevation. The prominent mountain chain of the Andes stretches N-S, perpendicular to the zonal westerly airflow of the mid-latitudes, which creates distinct environmental gradients at meso- and micro-scales. Due to their sensitive location at the juncture between tropical and extratropical climate zones, paleoclimate records from this area may potentially provide important insights into the dynamics of the large-scale atmospheric circulation in the Central Andes in the past. This region therefore provides an ideal natural laboratory for paleoclimatologists.

One of the geoecological key features of this high-mountain desert is the extremely high sensitivity to changes in effective moisture (precipitation minus evaporation). Even smallest changes in the moisture budget result in significant and amplified responses of the mostly saline and shallow lakes, in modifications

U. M. Huber et al. (eds.), Global Change and Mountain Regions, 93-104.

of geomorphological forms and processes (particularly in glacier variability), in vegetation changes and in other variations in the bio-geochemical systems. Different paleoclimate archives show evidence of abrupt, dramatic and high-amplitude moisture changes in the Atacama Desert during the late Quaternary epoch, especially during late-glacial and Holocene times (Betancourt et al. 2000; Baker et al. 2001; Grosjean et al. 2001; 2003; Latorre et al. 2002; and references therein). It is worth noting that such paleoenvironmental information from tropical-subtropical areas balances the view of the "climatically stable Holocene" as is inferred from temperature-sensitive proxy-data in high-latitude ice records (e.g. Blunier et al. 1995).

The highest mountains of this extreme desert range up to 6700 m and show a unique geoecological feature, the lack of glaciers even in the continuous permafrost belt above 5600 m. Glaciers in this extremely arid climate would not form and grow even with colder temperatures. Instead, glacier inception and advances in this area are almost exclusively triggered by increasing moisture (Kull and Grosjean 2000; Kull et al. 2003). Thus, it is moisture and not primarily temperature that plays the key role regarding environmental changes. This observation is critically important when the paleoclimatic history of the Atacama Altiplano is compared with other areas in South America or along the meridional Pole-Equator-Pole transect through the Americas (PEP-1) (Bradbury et al. 2001; Clapperton and Seltzer 2001; Schotterer et al. 2003).

In recent years, scientific evidence has increased that modern groundwater recharge in the extremely arid Atacama Desert is very limited and restricted to high elevation areas above 3500 m (> 100 mm precipitation yr^{-1}), whereas no modern water component was observed in the lower-elevation aquifers (< 20 mm precipitation yr^{-1}) (e.g. Fritz et al. 1979; Aravena 1995; Grosjean et al. 1995; Pourrut and Covarrubias 1995). There is increasing concern that large proportions of the current water resources might have formed during past periods of humid conditions, when the climate was very different from today, and that the water resources are renewed slowly or are non-renewable today and thus might be the limiting factor for economic growth in this region. Furthermore, the centers of water consumption (large cities and mines) are mainly found in the coastal and mid-elevation areas below 2500 m, whereas water resources are located in the high-elevation areas - a "classic" highland-lowland interaction problem.

The aim of this article is to review late Quaternary climatic conditions in the Atacama Desert and to discuss pluvial phases with regard to current water resources. We focus on the area along the western Altiplano (above 2500 m) from 18°S to 28°S (Fig. 1). Beside glacial deposits, pollen profiles from wetlands, paleosols and archaeological sites (Messerli et al. 1993; Núñez et al. 2002; Kull et al. 2003), plant macrofossils in rodent middens (Betancourt et al. 2000; Latorre et al. 2002) and sediments from closed-basin Altiplano lakes (Grosjean 1994; Grosjean et al. 1995; 2001; Geyh et al. 1999; Bobst et al. 2001) proved to be the most suitable paleoclimatic archives to investigate short- and long-term moisture changes and climate variability at decadal to millennial scale. We summarize the geological evidence for lake level changes in the Atacama Altiplano during the pre-LGM humid phase between >35,000 and 23,000 ^{14}C yr B.P. (radiocarbon years before present) and the humid late-glacial/early Holocene phase between ca. 13,000 and 8500 ^{14}C yr B.P. These intervals may have been critical for regional groundwater formation in the Atacama Desert. The mid-Holocene fully arid period between ca. 8500 and 3600/3000 ^{14}C yr B.P. deserves special attention, because short-term (several years to a few decades?) more humid climates and low-frequency but intense storms led to occasional flooding of the plains and this may provide an explanation for groundwater recharge during arid periods. Finally, we place the long-term dynamics of water resources and vegetation within the context of the current economic development, and conclude with some thoughts about pressing research needs for the future.

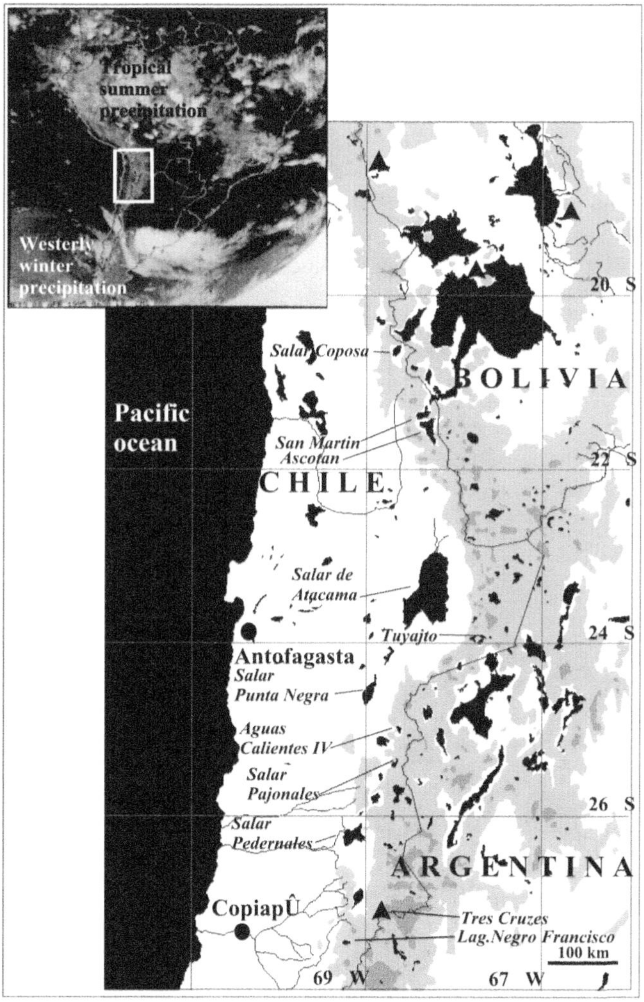

Figure 1: Research area in the Central Andes with sites discussed in the text. The MET3 image of 08 APR 1995 shows the cloud-free (dark) buffer zone between the tropical and extratropical cloud and precipitation belts. Black areas in the map represent lakes and salt lake surfaces; areas above 4000 m a.s.l. are shaded in gray.

2. The humid pre-LGM phase (c.f. "Minchin Phase")

^{14}C and U/Th (Uranium/Thorium) dates in five lake basins between 21°S and 26°S provide evidence of a humid pre-LGM paleolake phase in the Chilean part of the Altiplano (Fig. 1). Salar Coposa (26,165 ± 230 ^{14}C yr B.P., carbonate fraction), Salar Atacama (U/Th dated lake phases between 76,000 to 61,000 yr B.P., and 53,000 to 17,000 yr B.P.; Bobst et al. 2001), Salar Punta Negra (33,580 ± 745 ^{14}C yr B.P., carbonate fraction), Salar Pajonales (30,620 ± 455 ^{14}C yr B.P., carbonate fraction) and Salar Pedernales (29,730 ± 1440 ^{14}C yr B.P., carbonate fraction) show the presence of extended bioherms and stromatolite deposits (near-shore algal reefs) along the shorelines of high paleolakes or along extended springs that show elevated groundwater bodies in alluvial cones. This humid phase is also documented in other parts of the Central Andes and South America and is roughly dated between >35,000 (possibly as old as >70,000 yr B.P.) and ca. 23,000 ^{14}C yr B.P. (Clapperton 1993). The detailed internal structure of this humid phase is poorly known. First data from Bolivia and NW Argentina suggest that it consisted of several sub-phases (Baker et al. 2001; Kunz 2001). Also the chrono-stratigraphic position is poorly constrained: nothing is known about possible ^{14}C reservoir effects, and the finite ^{14}C ages (around 35,000 ^{14}C yr B.P.) might well be due to post-sedimentary re-crystallization of carbonates and uptake of modern atmospheric C (<2 pmC). For details see Geyh et al. (1999). At least in the Salar de Atacama, the contribution of tectonic subsidence to the local lake transgression remains open, thus the purely paleoclimatic interpretation of the lake level changes remains speculative to some extent.

Table 1 shows changes in lake levels and surface areas of the large and deep paleolakes in the Atacama Altiplano and in the Atacama Graben to the west, where river runoff drained large areas of the Andes. The Minchin paleolakes in the Chilean Atacama Desert were most extended, suggesting that annual precipitation rates exceeded the ca. 500 mm calculated for the late-glacial/early Holocene paleolakes (Grosjean 1994). Precipitation in this area is <200 mm today. The thickness of the open-water sedimentary facies in the Lago Pozuelos sediments (Kunz 2001) and the thickness of the shoreline and paleo-spring deposits suggest that the lakes existed

Table 1: Surface area and lake level changes of selected modern lakes and "Tauca" and "Minchin"paleolakes and wetlands in the Atacama Altiplano.

Site	**Basin size**	**Modern lakes**	**Late-glacial/early Holocene lakes/wetlands**		**Pre-LGM lakes and wetlands**	
	[km^2]	Surface area [km^2]	Surface area [km^2]	Lake level [m]	Surface area [km^2]	Lake level [m]
Salar Coposa	1115	0.4	124	+18	150	+33
Salar Ascotan	1536	13	301	30	n.d.	n.d.
San Martin*	480	5	260	70	n.d.	n.d.
Salar Tuyajto	246	3	11	+20	16	+40
S. Pta Negra	4264	2	350	+30	473	+55
Aguas Cal. IV	726	1	31	+25	?	?
S. Pajonales	2003	3	108 (?)	ca. 3 (?)	205	+50
S. Pedernales	2543	9	n.d.	n.d.	>>426	+30

* open system

for a long period of time (centuries to millennia), which suggests a significant and long-term climate change. Although direct data are still missing for the Atacama groundwater bodies, data from the eastern slope of the Andes (Chaco of Paraguay) show that much of today's groundwater resources formed during the humid Pre-LGM phase (Geyh et al. 1996).

At the current stage of knowledge, nothing is known about possible reasons for the changes in large-scale atmospheric circulation patterns that may be responsible for the observed moisture changes.

3. The late-glacial/early Holocene humid phase (c.f. "Tauca Phase" and "Coipasa Phase")

The late-glacial/early Holocene humid phase, dated in the Chilean part of the Atacama Altiplano between ca. 13,000 and 8500 ^{14}C yr B.P. (Geyh et al. 1999; Betancourt et al. 2000; Grosjean et al. 2001) is well documented. Plant diversity and primary production (Latorre et al. 2002) as well as animal diversity (Núñez et al. 2002) were highest during this period: today extinct deer and Pleistocene horse were present at that time. Along the currently hyperarid desert margin between 2400 and 3100 m, Latorre et al. (2002) estimated the late-glacial primary production at 10-20 times higher than today. Vegetation cover was possibly between 50% and 80% in areas where modern vegetation cover is <5%. The high-elevation lakes were much larger than today and covered most of the salt flat surfaces that are currently exposed to the atmosphere (Grosjean et al. 1995; 2001). Water balance and energy budget models of lakes (Grosjean 1994), vegetation–climate models (Latorre et al. 2002) and glacier-climate models (Kull and Grosjean 2000) suggest that annual precipitation rates (mainly summer precipitation) increased by a factor of 3 to 5 compared to today. This increase in effective humidity led to substantial changes of the land surface cover, the surface albedo and the regional radiation budget. Open water surfaces accounted for ~8% of the total area during the Tauca phase as opposed to 0.9% in modern times. At that time, approximately 5% of the land surface were covered by glaciers, whereas glaciers do not exist today. Snow cover above 4000 m was 10% as opposed to 3% in modern times (Kull and Grosjean 1998). Evidently, the late-glacial/early Holocene water cycle was very different from today. Although the methodology is fraught with many difficulties, isotope hydrological investigations in the Atacama area suggest that this humid climate also accounted for much of the regional groundwater formation in the Atacama Desert (Fritz et al. 1979; Aravena 1995; Grosjean et al. 1995).

Núñez et al. (2002) pointed out the close relationship between the fossil shorelines of the paleolakes and the Early Archaic human settlement patterns and showed that the presence of paleolakes was crucial during the initial phase of human colonization in this area between ca. 11,000 and 8000 ^{14}C yr B.P. (Fig. 2).

The increase in humidity is attributed to enhanced tropical summer precipitation with continental, Atlantic moisture sources. Summer flowering plants dominated the species spectrum as recorded in rodent middens (Latorre et al. 2002). Also the mass balance gradients of the late-glacial glaciers in this area are typical for a summer rainfall regime (Kull and Grosjean 2000; contributions of winter June-September precipitation <15% of the total annual precipitation between 18-22°S). These findings are in line with the spatial pattern of equilibrium line elevations that clearly suggests a late-glacial/early Holocene summer rainfall regime for the area north of 25°S. What were the mechanisms and forcings? All attempts to explain the pluvial phases with "Milankovich forcing" are highly inconsistent. Summer insolation was at a minimum during late-glacial/early Holocene times (Latorre et al. 2002). Instead, our results suggest that late Quaternary humidity changes on the Altiplano reflect a collective response to i) environmental changes in the source area of the moisture (e.g. re-expansion of the rain forest and increased release of latent heat over Amazonia and the Chaco, warm sea surface temperatures in the E Pacific), and ii) large-scale circulation patterns and wave structures in the upper troposphere (strength and position of the Bolivian High, divergent flow stimulating convection over the Altiplano), or that they even reflect a response to iii) interhemispherical teleconnections (Kull and Grosjean 1998).

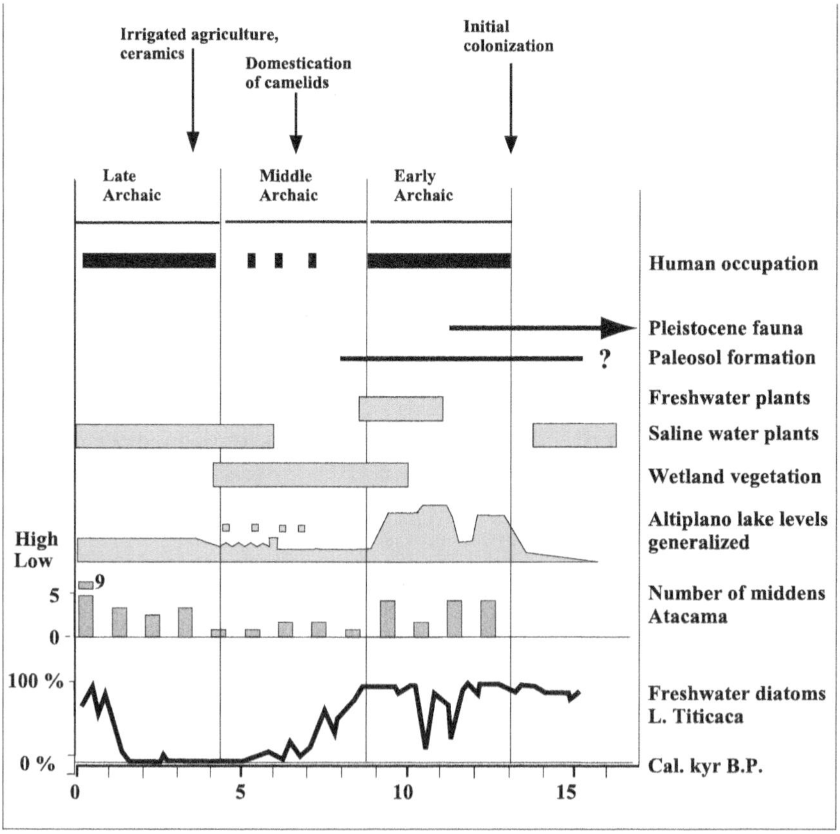

Figure 2: Late-glacial, Holocene paleoclimatic synthesis for the Central Andes and major cultural epochs (data from Baker et al. 2001; Grosjean et al. 2001; Latorre et al. 2002; Nunez et al. 2002).

4. The importance of storms during the arid Mid-Holocene

Paleoclimate archives that record low-frequency climate variability at millennial scale (e.g. larger lakes, lakes fed by groundwater, soils, groundwater bodies) show that climate conditions on the Altiplano were even drier than today between 8500 and ca. 3600/3000 ^{14}C yr B.P. (Abbott et al. 1997; Baker et al. 2001; Grosjean 2001; Grosjean et al. 2001; Nùñez et al. 2002). An exception is a century-scale humid phase centered around 4200 ^{14}C yr B.P. (ca. 5000 to 5500 cal yr B.P., Baker et al. 2001; Bobst et al. 2001; Grosjean et al. 2001; Latorre et al. 2002). This "Mid-Holocene aridity" was also found in the south-central Andes (e.g. Veit 1996; Jenny et al. 2002; Maldonado and Villagran 2002). Our interpretation of millennial-scale mid-Holocene aridity in the Atacama Desert appears to be in conflict with paleovegetation reconstructions based on rodent middens (Latorre et al. 2002) that indicate the invasion of vegetation into absolute desert during this interval. However, the two datasets can be accommodated when the timescales they represent are taken into consideration. Middens can only provide discontinuous records of vegetation history, and plant remains from these archives may correspond to decadal or sub-decadal scale wet intervals in the generally arid but highly variable climate of the mid-Holocene (Grosjean et al. 2003).

In light of this general prolonged arid period, Holocene groundwater recharge during this interval, as reported by Aravena (1995), is difficult to explain. Evidence from lake sediments in the Central and Southern Altiplano, and deposits of individual debris flows (Grosjean et al. 1997) suggest that arid conditions were interrupted by low-frequency storms and (sub)decadal-scale humid spells. This resulted in torrential river run-off and floods, and may have been associated with discrete events of groundwater formation in the adjacent depressions. In the Salar de Atacama, such floods are recorded as fine-grained deposits embedded in windblown sand. Although at a much smaller scale, a similar event was observed in March 1997.

The type of large-scale atmospheric circulation patterns that is responsible for the observed aridity during the mid-Holocene is poorly understood. Both winter and summer precipitation should be considered. Jenny et al. (2002) and Maldonado et al. (2002), and references therein, observed a significant weakening of mid-Holocene westerly winter precipitation in Central Chile. On the other hand, data from the eastern Cordillera in Southern Bolivia and NW Argentina show that summer precipitation with continental moisture sources also decreased significantly. Vuille (1999) and Garreaud (1999) point to two very important modern mechanisms that govern deep convection and thus precipitation over the Altiplano: the mixing ratio of the advected air masses east of the Andes (threshold of 7 g m^{-3}) and the Southern Oscillation index. Both may serve as a model to explain mid-Holocene aridity in the central Andes, and underscore the importance of the source area of moisture (continental lowlands) and the dynamics of the large-scale circulation pattern in the circum-Pacific area. Comparing Australasian and South American paleodata, Shulmeister (1999) concluded a stronger mid Holocene Walker Circulation (cold phase of the Southern Oscillation) that would result in widespread aridity over the Western South-Central Andes at large.

5. Water resources and economic development: Potential for conflicts

What is the role of past climate changes and groundwater formation with regard to the current economic development in this area? The Region de Antofagasta in the core of this hyperarid area hosts unique mineral ore deposits. Exploitation of copper, gold and lithium led to unprecedented economic development of this region during the last decade. Between 1994 and 2000, 25% of all foreign investments in Chile concentrated in this region. Currently, more than 7 billion US$ are invested in new mining, transportation and energy infrastructure. In 1994, the Atacama Desert generated 17% of the world's copper production (43% of Chile's, 4.5 mio t yr^{-1}) and accounted for 30% of total Chilean exports (Romero 2002). This development went hand in hand with the political opening and the economic integration of Chile with Argentina, South America and global markets. The "Peace and Friendship Treaty" of 1984, restoration of democracy in Chile and Argentina, and the globalization of markets led to an ambitious and visionary development plan, the "Corredores Bioceánicos" that connects the countries of southern South America with harbors on the Pacific and Atlantic coasts, making their economies accessible to the Asian and European markets. High capacity infrastructure (roads, rail tracks, gas and water pipelines, power lines) was built to facilitate the exchange of goods and energy, and to supply the urban and industrial centers, located in absolute desert, with food and water (Romero 2002). Last but not least, tourism increased exponentially during the last decade. The pristine spectacular desert environment, the archaic volcanic landscape, the few wetlands, salt flats and lakes with amazing bird life and the blue sky create impressive contrasts with the hostile desert and attract a growing number of domestic and foreign tourists.

Thus, it is not surprising that the increasing water demand for industrial and urban purposes led to severe conflicts with the protection of natural ecosystems and traditional indigenous culture. The currently known surface and groundwater resources of this area (126,000 km^2, ca. three times the area of Switzerland) amount to 12,000 $l\ s^{-1}$. In 1990, 7130 $l\ s^{-1}$ were diverted for domestic use (19%), mining (39%) and agriculture (42%). By 1996, the water consumption for mining had doubled, and the total claims for water extraction rights amounted to 16,000 $l\ s^{-1}$, significantly exceeding the available water resources (Mercurio 1997). Given the importance of mining for the regional and the national economy, greatest pressure is put both on the traditional irrigation agriculture that consumes a large proportion of the available water with a very low economic revenue and on the ecosystems of the remaining wetlands, salt pans and lakes. However, traditional agriculture is an integral part of the several thousand year old cultural heritage and a key attraction for tourism. Wetlands and lakes, which are focal points of any wild life present in this desert, are key areas regarding biodiversity and endemism (Messerli et al. 1997; Arroyo et al. 1998), and symbolize the unique natural heritage of this peculiar area, are also important tourist attractions.

Romero (2002) shows an illustrative example for this conflict: The Aldabarán gold mine near Copiapo (28°S), run by the Canadian Cerro Casale Company, invested 1.4 billion US$ and created 4000 jobs for an expected period of 18 years. The required freshwater is captured in the high Andean area near Piedra Pomez, which is the hydrological recharge area for two National Parks (Nevado Tres Cruces and Laguna del Negro Francisco) and the 11,700 ha of downstream grasslands that are traditionally used by the indigenous communities of the "Kollas". Jointly with other inhabitants of the valley and the city of Copiapo and the industrial farmers who produce for the US and European markets, they are concerned about water shortage and water pollution that is likely to arise from upstream mining activities. We emphasize that paleo-research has shown that most of the groundwater resources come from humid phases (pre-LGM, late-glacial and early Holocene) in the past. These non-renewable resources are maybe already brought close to their limits and the region is therefore extremely vulnerable.

6. Outlook and research needs

In light of the conflict that results from limited water resources and increased water needs in the Atacama region we identify two major research needs that are of fundamental importance for sustainable development in this area. This research requires contributions from bio-geophysical, technical, social, economic and political sciences.

6.1 Water resources research

Knowledge about the long-term (late Quaternary) development, quantity, quality, flowpaths and recharge rates of aquifers in the Atacama is very limited. Traditional hydro-geological methods (e.g. pumping experiments, geological properties of the aquifers) must be combined with isotope hydro-geochemical investigations and research on Quaternary climate changes that provide insight into the origin, the age, the recharge rate and thus the long-term dynamics of the water body. Unfortunately, this type of research is very expensive. It is however indispensable if the available water resources are to be assessed on a sound scientific basis. Given the ca. 20-year investment horizon of a company, this knowledge is of strategic importance and thus most likely available. However, it is kept strictly confidential and is not accessible to regional governments, political decision makers, planners and scientists.

Alternative approaches for the use of renewable water resources in the Atacama Desert have recently been tested. The potential of collecting water from coastal fog in northern Chile has been demonstrated (e.g. Larrain et al. 2002). The site of Alto Patache (21.5°S), for instance, collected 5400 mm of water m^{-2} in 1997. However, the authors point out the large seasonal, inter-annual and spatial variability of fog and comment that better knowledge of the interplay between sea surface temperatures (SSTs), atmospheric circulation, topography of the coastal range and fog formation would substantially improve the site selection and make water harvesting in this region more efficient.

A sound scientific basis and a spatially explicit knowledge about available renewable and non-renewable water resources stands at the beginning of sustainable water use and a long-term management plan.

6.2 Water management, long-term strategy for development and institutional control

Assessment and management of water resources requires a catchment approach that explicitly takes into account all the stakeholders with their respective requirements both in terms of water quantity and quality. Special emphasis must be put on downstream users. Such a catchment approach is of particular importance for the remaining undisturbed wetlands and lakes, many of which are in protected areas. However, the buffer zones ("water protection areas") around these wetlands must be large enough, variable in size and account for the hydro-geological conditions in the upstream catchment. An efficient and independent monitoring system must ensure and enforce that environmental impacts due to water extractions or diversions remain within the accepted range outlined in the environmental impact study of a given project.

Governments have to set up the legal frameworks to force or encourage (with "market based instruments" or "command and control") water users to implement water saving strategies of highest technological standards (within the industrial process, waste water treatment, recycling). These technologies exist in the mining industry and have to be further developed to reduce substantially the freshwater consumption. Reduction of water loss due to leaking pipelines (estimated to about 30%) and recycling of waste water (e.g. in agriculture) have a great potential to reduce the environmental pressure on pristine water bodies. Research and development of technological solutions to water wastage is a task both for governments and the private sector.

Likely the most difficult task for governments and society is to develop a long-term management and development plan that is socially, politically and economically acceptable and balances the long-term needs (cultural and natural heritage) with the short-term needs (private sector, mining), thus ensuring the viability of humans, animals and plants in this desert environment. Developing a scientific basis and appropriate decision making tools and proposing consistent and visionary solutions is a truly interdisciplinary challenge for scientists.

7. References

Abbott, M. B., Seltzer, G. O., Kelts, K., and Southon, J. (1997). Holocene paleohydrology of the tropical Andes from lake records. *Quaternary Research* **47**, 70-80.

Aravena, R. (1995). Isotope hydrology and geochemistry of Northern Chile groundwaters. *Bulletin de l' Institut Français d'Etudes Andines* **24**, 495-503.

Arroyo, M. T. K., Castor, C., Marticorena, C., Muñoz, M., Cavieres, L., Matthei, O., Squeo, F. A., Grosjean, M., and Rodriguez, R. (1998). The flora of Llullaillaco National Park in the transitional winter-summer rainfall area of the northern Chilean Andes. *Gayana Botánica* **55**, 93-110.

Baker, P. A., Rigsby, C. A, Seltzer, G. O., Fritz, S. C., Lowenstein, T. K., Bacher, N. P., and Veliz, C. (2001). Tropical climate changes at millennial and orbital timescales in the Bolivian Altiplano. *Nature* **409**, 698-701.

Betancourt, J. L., Latorre, C., Rech, J. A., Quade, J., and Rylander, K. A. (2000). A 22,000-year record of monsoonal precipitation from northern Chile's Atacama Desert. *Science* **289**, 1542-1546.

Bobst, A. L., Lowenstein, T. K., Jordan, T. E., Godfrey, L. V., Ku, T.-L., and Luo, S. (2001). A 106 ka paleoclimate record from drill core of the Salar de Atacama, northern Chile. *Palaeogeography, Palaeoclimatology, Palaeoecology* **173**, 21-42.

Blunier, T. Chapellaz, J., Schwander, J., Stauffer, B., and Raynaud, D. (1995). Variations in atmospheric methane concentrations during the Holocene epoch. *Nature* **374**, 46-49.

Bradbury, J. P., Grosjean, M., Stine, S., and Sylvestre, F. (2001). Full- and late-glacial lake records along PEP-1 transect: Their role in developing inter-hemispheric paleoclimate interactions. *In* "Interhemispheric climate linkages." (V. Markgraf, Ed.), pp. 265-291. Academic Press, San Diego.

Clapperton, C. M. (1993). "Quaternary geology and geomorphology of South America." Elsevier, Amsterdam.

Clapperton, C. M., and Seltzer, G. O. (2001). Glaciation during Marine Isotope Stage 2 in the American Cordillera. *In* "Interhemispheric climate linkages." (V. Markgraf, Ed.), pp. 173-181. Academic Press, San Diego.

Fritz, P., Silva, C., Suzuki, O., and Salati, E. (1979). Isotope hydrology in northern Chile. *IAEA-SM* **228**, 525-543.

Garreaud, R. (1999). Multiscale analysis of summertime precipitation over the central Andes. *Monthly Weather Review* **127**, 901-921.

Geyh, M., Grosjean, M., Núñez, L. A., and Schotterer, U. (1999). Radiocarbon reservoir effect and the timing of the late-glacial/early Holocene humid phase in the Atacama Desert, northern Chile. *Quaternary Research* **52**, 143-153.

Geyh, M. A., Grosjean, M., Kruck, W., and Schotterer, U. (1996). Sincronopsis del desarrollo morfológico y climatológico del Chaco Boreal y de Atacama en los ultimos 35.000 anos AP. *In* "Memorias del XII Congreso Geológico de Bolivia, Tomo III." pp. 1267-1276. Sociedad Geologica de Bolivia.

Grosjean, M. (1994). Paleohydrology of the Laguna Lejía (Northchilean Altiplano) and climatic implications for lateglacial times. *Palaeogeography, Palaeoclimatology, Palaeoecology* **109**, 89-100.

Grosjean, M., Geyh, M., Messerli, B., and Schotterer, U. (1995). Late-glacial and early Holocene lake sediments, groundwater formation and climate in the Atacama Altiplano. *Journal of Paleolimnology* **14**, 241-252.

Grosjean, M., Núñez, L. A., Cartajena, I., and Messerli, B. (1997). Mid-Holocene climate and culture change in the Atacama Desert, northern Chile. *Quaternary Research* **48**, 239-246.

Grosjean, M., van Leeuwen, J., van der Knaap, W. O., Geyh, M., Ammann, B., Tanner, W., Messerli, B., and Veit, H. (2001). A 22,000 ^{14}C yr BP sediment and pollen record of climate change from Laguna Miscanti 23°S, northern Chile. *Global and Planetary Change* **28**, 35-51.

Grosjean, M. (2001). Mid-Holocene climate in the south-central Andes: Humid or dry? *Science* **292**, 2391-2392.

Grosjean, M., Cartajena, I., Geyh, M. A., and Nuñez, L. A. (2003). From proxy-data to paleoclimate interpretation: The mid-Holocene paradox of the Atacama Desert, northern Chile. *Palaeogeography, Palaeoclimatology, Palaeoecology* (in press).

Jenny, B., Valero-Garcés, B. L., Villa-Martínez, R., Urrutia, R., Geyh, M., and Veit, H. (2002). Evidence of early to mid-Holocene aridity in Central Chile related to the Southern Westerlies: The Laguna Aculeo record (34°S). *Quaternary Research* **58**, 160-170.

Kull, C., and Grosjean, M. (2000). Late Pleistocene climate conditions in the north Chilean Andes drawn from a climate-glacier model. *Journal of Glaciology* **46**, 622–632.

Kull, C., Hänni, F., Grosjean, M., and Veit, H. (2003). Evidence of massive LGM cooling in NW-Argentina

(22°S) derived from a glacier climate model. *Quaternary International* (in press).

Kunz, A. (2001). "Limnologische Analyse von drei Seen in N-Chile und NW Argentinien und ihre paläoklimatische Interpretation." Unpublished Diploma thesis, University of Bern, Bern.

Larrain, H., Velásquez, F., Cereceda, P., Espejo, R., Pinto, R., Osses, P., and Schemenauer, R. S. (2002). Fog measurements at the site "Falda Verde" north of Chanaral compared with other fog stations of Chile. *Atmospheric Research* **64**, 273-284.

Latorre, C., Betancourt, J. L., Rylander, K. A., and Quade, J. (2002). Vegetation invasion into absolute desert: A 45 k.y. rodent midden record from the Calama-Salar de Atacama basins, northern Chile (lat 22-24°S). *Geological Society of America Bulletin* **114**, 349-366.

Maldonado, A., and Villagran, C. (2002). Paleoenvironmental changes in the semiarid coast of Chile (~32°S) during the last 6200 cal years inferred from a swamp-forest pollen record. *Quaternary Research* **58**, 130-138.

Messerli, B., Grosjean, M., Bonani, G., Bürgi, A., Geyh, M., Graf, K., Ramseyer, K., Romero, H., Schotterer, U., Schreier, H., and Vuille, M. (1993). Climate change and natural resource dynamics of the Atacama Altiplano during the last 18,000 years: A preliminary synthesis. *Mountain Research and Development* **13**, 117-127.

Messerli, B., Grosjean, M., and Vuille, M. (1997). Water availability, protected areas, and natural resources in the Andean Desert Altiplano. *Mountain Research and Development* **17**, 229-238.

Mercurio de Santiago (1997). Universidades de II Region: Expertos Temen por Reserva Hidrica en Salar de Atacama. 28 July 1997.

Nùñez, L., Grosjean, M., and Cartajena, I. (2002). Human occupations and climate change in the Puna de Atacama, Chile. *Science* **298**, 821-824.

Pourrut, P., and Covarrubias, A. (1995). Existencia de agua en la II Región de Chile: Interrogantes e hipothesis. *Bulletin de l'Institut Français d'Etudes Andines* **24**, 505-515.

Romero, H. (2002). The Andes of Chile: Clash between economic and sustainable development. *IHDP Newsletter* **1**, 7-9.

Schotterer, U., Grosjean, M., Stichler, W., Kull, C., Ginot, P., Francou, B., Gäggeler, H., Gallaire, R., Hoffmann, G., Pouyaud, B., and Schwikowski, M. (2003). Glaciers and climate in the Andes between the Equator and 30°S: What is recorded under extreme environmental conditions? *Climatic Change* (in press).

Shulmeister, J. (1999). Australasian evidence for mid-Holocene climate change implies precessional control of Walker Circulation in the Pacific. *Quaternary International* **57/58**, 81-91.

Veit, H. (1996). Southern Westerlies during the Holocene deduced from geomorphological and pedological studies in the Norte Chico, Northern Chile (27-33°S). *Palaeogeography, Palaeoclimatology, Palaeoecology* **123**, 107-119.

Vuille, M. (1999). Atmospheric circulation over the Bolivian Altiplano during dry and wet periods and extreme phases of the Southern Oscillation. *International Journal of Climatology* **19**, 1579-1600.

Palaeolimnological Investigations in the Alps: The Long-Term Development of Mountain Lakes

André F. Lotter
University of Utrecht, Botanical Palaeoecology, Laboratory of Palaeobotany and Palynology, Budapestlaan 4, 3584 CD Utrecht, The Netherlands
phone +31-30-2532653, fax +31-30-2535096, e-mail a.f.lotter@bio.uu.nl

Keywords: Alps, Aquatic ecosystems, Climate change, Human impact, Mountain lakes, Palaeolimnology.

1. Introduction

Most mountain lakes and their catchments are, due to their remoteness, less impacted by human actions than lakes in lowland regions. They are, therefore, often considered pristine systems. Nevertheless, even remote, uninhabited areas are polluted via atmospheric deposition of aerosols that transport acid rain, heavy metals, organic compounds, and nutrients.

In mountain lake ecosystems, the ice-cover determines the duration of the open-water season which in turn influences productivity, circulation regimes and hence also the oxygen budget (e.g. Ohlendorf et al. 2000). Due to their special location in sensitive climatic space, these mountain lakes provide not only means of assessing the response rates of different biota to past global change, but also the possibility for estimating leads and lags between different abiotic and biotic systems. Moreover, proxies archived in the sedimentary record of mountain lakes are useful indicators for reconstructing past environmental change and can provide an important baseline against which the magnitude and rates of current changes can be assessed.

Currently, coordinated palaeolimnological investigations in the Alps are carried out in several European research projects (e.g. CHILL-10'000, MOLAR, EMERGE; see e.g. Korhola et al. 2000; Battarbee et al. 2002b) and are focussing on two major

U. M. Huber et al. (eds.), Global Change and Mountain Regions, 105-112.

topics:

i. reconstruction of natural climate variability and its effects on mountain ecosystems (see e.g. Leemann and Niessen 1994; Sommaruga-Wögrath et al. 1997; Battarbee et al. 2002a; Catalan et al. 2002);
ii. study of human impact on mountain lakes to assess pre-industrial background values for nutrients and pollutants (e.g. MOLAR Group 1999; Fernandez et al. 2000) as well as catchment-lake interactions (e.g. Dapples et al. 2002; Lotter and Birks 2003a).

In the following sections, aspects of these topics are illustrated using a selection of examples that is biased by my own research.

2. Lake sediments as records of past environmental change

2.1 Multi-proxy climate reconstructions

Several organism-specific inference models (transfer functions) have been developed recently. They allow quantitative reconstructions of temperature (Wunsam et al. 1995; Lotter et al. 1997) or lake nutrient status (Wunsam and Schmidt 1995; Lotter et al. 1998). Based on a close relationship between air and water temperature (Livingstone and Lotter 1998; Livingstone et al. 1999), these transfer functions model the empirical relationship between the abundance of aquatic organisms and the prevailing air temperature during the open-water season (i.e. summer). Yet, inferences of winter temperatures that would be important to trace changes in seasonality are not possible. There are, however, attempts to assess the length of the winter season for Alpine lakes by taking into account the duration of ice-cover (e.g. Livingstone 1997; Lotter and Bigler 2000).

Currently, only a few studies have attempted quantitative Holocene climate reconstructions in the Alps. Lotter et al. (2000a) inferred summer temperature changes of 2-3°C at the transition from the late-glacial period to the Holocene, using oxygen isotopes, pollen, and Cladocera at Gerzensee. Studying the chironomid assemblages of Hinterburgsee, a small lake in the northern Swiss Prealps, Heiri et al. (2003) observed six Holocene chironomid-inferred July air temperature fluctuations (Fig. 1) on the order of 0.5-1°C. Heiri (2001) correlated them with fluctuations of tree-pollen percentages at a site in the Central Alps, as well as with ice-rafted debris events in the North Atlantic Ocean, thus showing that millennium-scale Holocene climatic oscillations are recorded by independent biota in the Alps.

2.2 Human impact during the Holocene

The study of a 9000 year long, well radiocarbon dated record of environmental dynamics in Sägistalsee (Lotter and Birks 2003b), a small lake at the present day tree line in the northern Swiss Alps, showed major shifts in chironomid assemblages (Heiri and Lotter 2003). These shifts are related to changes in the accumulation rates of the three major chironomid taxa (Fig. 2) and are attributed to natural effects such as shallowing of the lake by infilling of the lake basin throughout the millennia. During

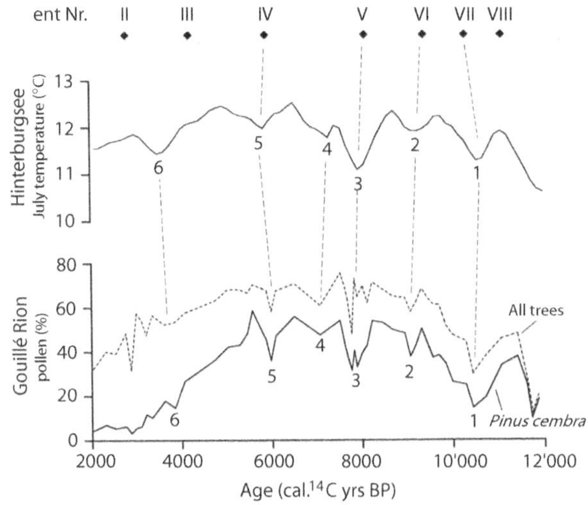

Figure 1: Holocene chironomid-inferred July air temperature reconstruction from Hinterburgsee (northern Swiss Alps, Heiri et al. 2003) correlated with percentages of total tree pollen, *Pinus cembra* pollen from Gouillé Rion (central Swiss Alps, Tinner and Ammann 2001), and ice rafted debris events in the North Atlantic (Bond et al. 1997). After Heiri (2001).

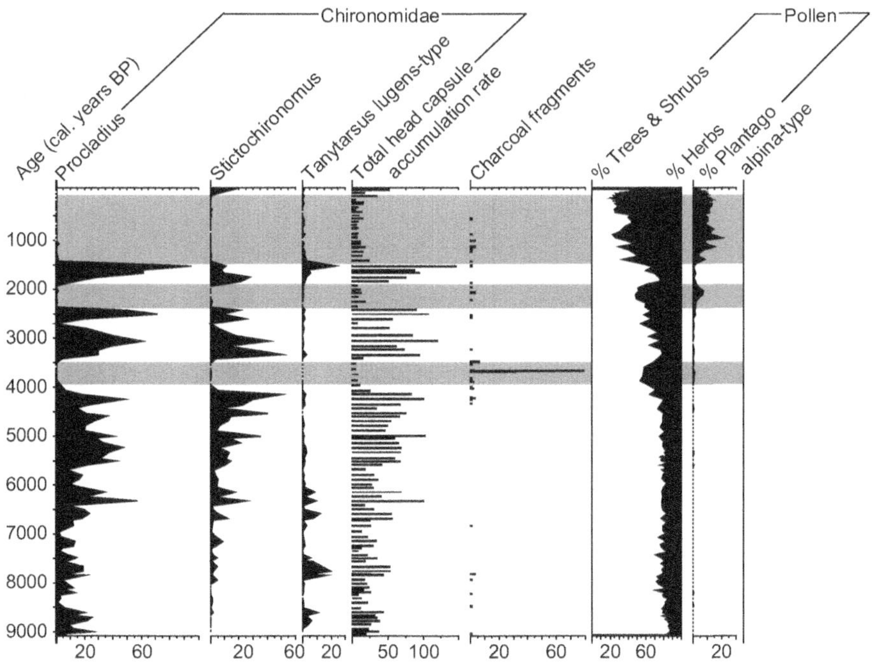

Figure 2: Head capsule accumulation rates (hc/100 cm^2/yr) of selected chironomid taxa from the sediments of Sägistalsee compared with palaeobotanical proxies. Shaded areas are interpreted as phases of anthropogenically induced anoxia that led to low chironomid accumulation. After Heiri and Lotter (2003).

three distinct phases, head capsules of *Procladius*, *Stictochironomus* and *Tanytarsus lugens* are scarce or even absent from the sediments (shaded areas in Fig. 2). All three episodes coincide with phases of increased human activity in the catchment of Sägistalsee (Wick et al. 2003) as evidenced by pollen of plants indicating pasturing as well as by macroscopic charcoal remains (Fig. 2). The absence of chironomids during these phases points to anoxic conditions in the bottom water of the lake through enhanced nutrient loading of the lake due to the presence of humans and their livestock in the catchment. The chironomid fauna reacted in the same way to the intensive pasturing of the past 1500 years as it did to the first clear-cutting during Bronze Age (ca. 3700 cal. yr BP: charcoal peak, see Fig. 2) and more moderate pasturing during the Bronze, Iron, and Roman Ages. This suggests that Alpine lake ecosystems are extremely sensitive to human activity in their catchment. Nonetheless, the chironomid assemblages also show a considerable amount of resilience to human disturbance, as the fauna reverted to the pre-impact stage after the first two periods of human activity. However, even though pasturing has decreased in recent years, the fauna has only partly recovered, which is most likely related to stocking of the lake with fish.

2.3 Disentangling the influence of climate and man during the Little Ice Age

The most pronounced climatic oscillation in historical times in the Alps was most probably the Little Ice Age. Instrumental records as well as documentary data allow the assessment of leads and lags of ecosystem reaction to this climatic oscillation (Lotter et al. 2000b). However, the climatic change associated with the Little Ice Age also had a fundamental impact on human society in the Alps and on socio-economic systems, leading, among other things, to changes in land-use. The interpretation of changes observed in abiotic factors and biotic assemblages from lake sediments is, therefore, often equivocal. Disentangling direct climate effects from indirect effects, due to changes in catchment land-use, is of paramount importance when interpreting palaeolimnological results (e.g. Lotter and Birks 1997).

In a high-resolution, multi-proxy study Hausmann et al. (2002) give an example of how climatic oscillations influenced tree line ecosystems through changes in land-use and pasturing patterns in the Alps during the Little Ice Age. They studied the annually laminated sediments of Seebergsee, a small lake located at the present-day tree line, in the northern Swiss Alps. The occurrence of high amounts of small centric diatoms of the genus *Stephanodiscus* between the middle of the 14th and the late 17th century point to a distinct period of hypertrophy (i.e. strong nutrient enrichment). Older and younger diatom assemblages are dominated by *Cyclotella* species (Fig. 3), typical for meso- to oligotrophic (i.e. intermediate to low nutrient levels) lakes at this elevation in the Alps.

All biological proxies suggest increased pasturing since ca. AD 1345 (Fig. 3): the beginning of the *Stephanodiscus*-dominated assemblages coincides with a threefold increase of grazing-indicating pollen types, which is a reflection of a regional rather than a local, catchment-related signal. However, the amount of coprophilous fungal spores (fungi that live on faeces), as a proxy for local pasturing, also showed a

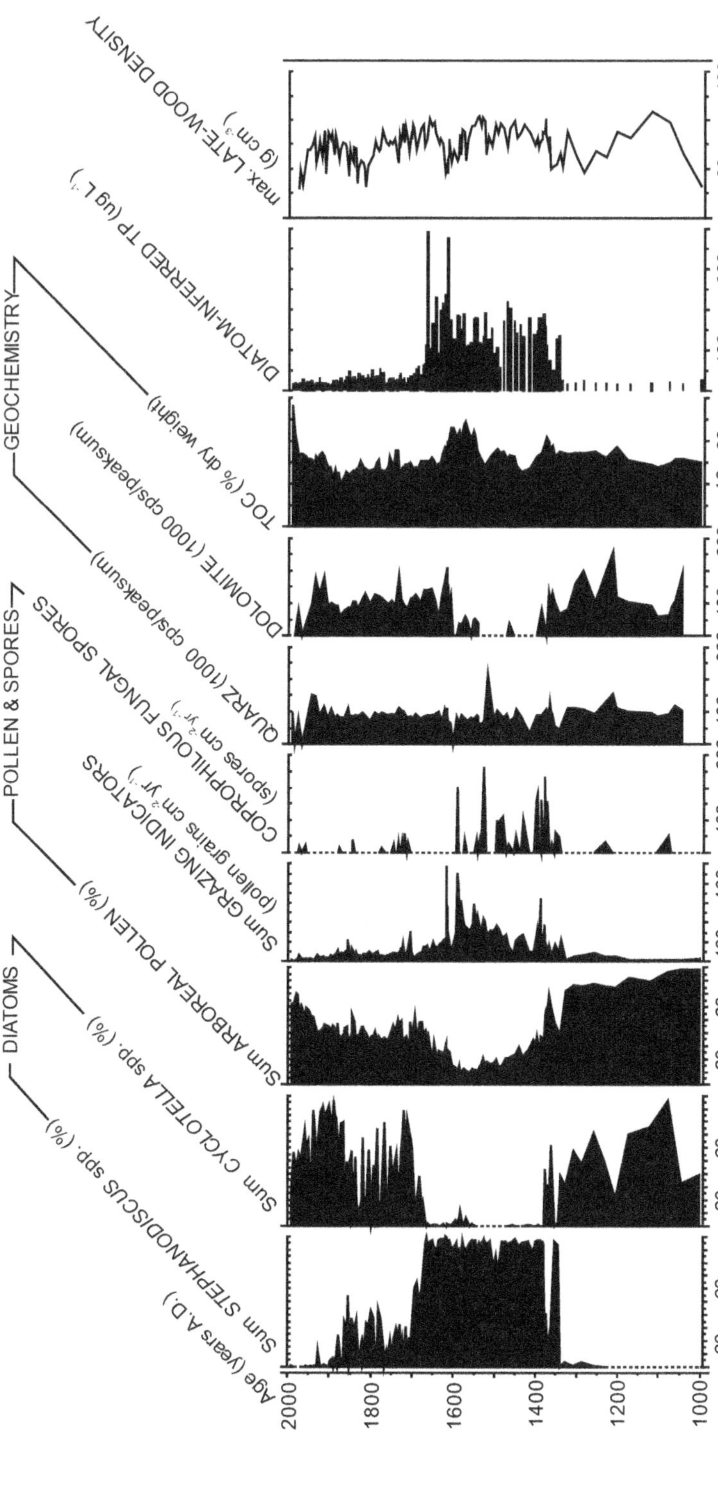

Figure 3: Results of diatom, pollen and spore, geochemical and mineralogical analyses compared to diatom-inferred total phosphorus (TP) reconstructions from Seebergsee (Hausmann et al. 2002) and late-wood densities from the Swiss Alps (Schweingruber et al. 1988).

synchronous threefold increase. The eutrophication (i.e. nutrient enrichment) of the lake around AD 1345 occurred remarkably fast. The diatom-inferred total phosphorus concentration (DI-TP) increased in only eight years from 25 to 140 μgL^{-1}. The absence of dolomite during this hypertrophic phase is interpreted as anoxia-induced lowering of the bottom-water pH and subsequent dissolution of carbonates in the hypolimnion (Ohlendorf et al. 2000). According to the coprophilous fungal spore data, grazing tentatively ended at ca. AD 1595. Interestingly, there seems to be a lag of 35 years before the grazing indicator pollen and a lag of almost 90 years before the DI-TP reached pre-eutrophication levels. Apparently, the lake needed about a century to recover from its phosphorus load. In contrast, the delayed drop of DI-TP during the second half of the 17th century took place very rapidly. Synchronously, late-wood densities indicate a warming during the second half of the 17th century (Schweingruber et al. 1988), which is likely to have triggered strong stratification of the water column and the establishment of meromixis in Seebergsee. As a consequence, nutrients of the stratified bottom water (monimolimnion) were eventually no longer available in the trophogenic zone where planktonic algae live.

To decide whether the diatom-inferred hypertrophy was induced by climate or by land-use changes, the variance in the diatom assemblage data was analysed. Pollen and fungal spore occurrences of plants indicative of pasturing were used to model past land-use, whereas late-wood density data was used as a proxy for climate. Partitioning of the variance of the diatom data showed that grazing indicator pollen and DI-TP concentrations explain ten times more variance than the dendroclimatological data. Nutrient enrichment in Seebergsee played thus an overriding role, whereas climate had an indirect effect by allowing pasturing at higher elevations (Hausmann et al. 2002).

3. Future research

Several transfer functions that numerically model the relationship between different groups of aquatic organisms and important physical and chemical environmental factors, such as pH, temperature, or nutrients have been developed for the Alps. However, these relationships are purely empirical and future research should also focus on the study of the autecology of these aquatic organisms to better understand and predict their response in connection with global change and their potential as indicators of changing environments. Moreover, studies including different independent biotic and abiotic proxies may also help to better reconstruct past global change. When using biotic proxies one of the major challenges is to detect environmental threshold values that lead to major changes in the assemblages. Commonly, such major biotic compositional changes are observed in ecotonal situations where small changes in the physical or chemical environment may already trigger strong biotic responses. Yet, these ecotonal situations and the environmental thresholds do not necessarily need to coincide for different organisms.

In the pre-alpine lowland, man has been an important factor in changing the environment over the past seven millennia, whereas human impact has been delayed at higher elevations. There is evidence for strong human impact in the Alps since at least the early Bronze Age, i.e. for the past 4000 years (e.g. Wick et al. 2003). This has to be taken into consideration when interpreting sediment-derived proxies in terms of climate change. Given the close interaction between vegetation, hydrology, topography, soil, climate, and human activity and the link between catchment processes and physical, chemical, and biological lake responses, the catchment area of a mountain lake is an important and critical unit for study in both palaeoecological and palaeolimnological investigations. Future palaeolimnological studies will on the one hand focus on the reconstruction of regional or global signals. On the other hand, the study of catchment-lake interactions will also be an important research topic. PAGES initiatives such as LIMPACS (Human Impact on Lake Ecosystems; www.geog.ucl.ac.uk/ecrc/limpacs) will most certainly have a strong impact in setting the future research agenda for palaeolimnological studies of mountain lakes.

4. References

Battarbee, R. W., Grytnes, J. A., Thompson, R., Appleby, P., Catalan, J., Korhola, A., Birks, H. J. B., Heegaard, E., and Lami, A. (2002a). Comparing palaeolimnological and instrumental evidence of climate change for remote mountain lakes over the last 200 years. *Journal of Paleolimnology* **28**, 161-179.

Battarbee, R. W., Thompson, R., Catalan, J., Grytnes, J. A., and Birks, H. J. B. (2002b). Climate variability and ecosystem dynamics of remote alpine and arctic lakes: The MOLAR project. *Journal of Paleolimnology* **28**, 1-6.

Bond, G., Showers, W., Cheseby, M., Lotti, R., Almasi, P., De Menocal, P., Priore, P., Cullen, H., Hajdas, I., and Bonani, G. (1997). A pervasive millennial-scale cycle in North Atlantic Holocene and glacial climates. *Science* **278**, 1257-1266.

Catalan, J., Vetura, M., Brancelj, A., Granados, I., Thies, H., Nikus, U., Korhola, A., Lotter, A. F., Barbieri, A., Stuchlik, E., Lien, L., Bitusik, P., Buchaca, T., Camarero, L., Goudsmit, G. H., Kopacek, J., Lemcke, G., Livingstone, D. M., Müller, B., Rautio, M., Sisko, M., Sorvari, S., Sporka, F., Strunecky, O., and Toro, M. (2002). Seasonal ecosystem variability in remote mountain lakes: Implications for detecting climatic signals in sediment records. *Journal of Paleolimnology* **28**, 25-46.

Dapples, F., Lotter, A. F., van Leeuwen, J. F. N., van der Knaap, W. O., Dimitriadis, S., and Oswald, D. (2002). Paleolimnological evidence for increased landslide activity due to forest clearing and land-use since 3600 cal BP in the western Swiss Alps. *Journal of Paleolimnology* **27**, 239-248.

Fernandez, P., Vilanova, R. M., Martinez, C., Appleby, P., and Grimalt, J. O. (2000). The historical record of atmospheric pyrolytic pollution over Europe registered in the sedimentary PAH from remote mountain lakes. *Environmental Science and Technology* **34**, 1906-1913.

Hausmann, S., Lotter, A. F., van Leeuwen, J. F. N., Ohlendorf, C., Lemcke, G., Grönlund, E., and Sturm, M. (2002). Interactions of climate and land use documented in the varved sediments of Seebergsee in the Swiss Alps. *The Holocene* **12**, 279-289.

Heiri, O. (2001). "Holocene palaeolimnology of Swiss mountain lakes reconstructed using subfossil chironomid remains: Past climate and prehistoric human impact on lake ecosystems." Unpublished PhD thesis, Bern University.

Heiri, O., and Lotter, A. F. (2003). 9000 years of chironomid assemblage dynamics in an Alpine lake: long term faunistic trends, sensitivity of the community to disturbance, and resilience of the ecosystem. *Journal of Paleolimnology* **30** (in press).

Heiri, O., Lotter, A. F., Hausmann, S., and Kienast, F. (2003). A chironomid-based Holocene summer air temperature reconstruction from the Swiss Alps. *The Holocene* **13** (in press).

Korhola, A., Lotter, A. F., Birks, H. J. B., and Cameron, N. G. (2000). Climate history as recorded by ecologically sensitive Artic and Alpine lakes in Europe during the last 10,000 years: A multi-proxy approach (CHILL-10,000). *In* "European Climate Science Conference Vienna 1998, CD-ROM." European Commission.

Leemann, A., and Niessen, F. (1994). Varve formation and the climatic record in an Alpine proglacial lake:

Calibrating annually-laminated sediments against hydrological and meteorological data. *The Holocene* **4**, 1-8.

Livingstone, D. M. (1997). Break-up dates of Alpine lakes as proxy data for local and regional mean surface air temperatures. *Climatic Change* **37**, 407-439.

Livingstone, D. M., and Lotter, A. F. (1998). The relationship between air and water temperatures in lakes of the Swiss Plateau: A case study with palaeolimnological implications. *Journal of Paleolimnology* **19**, 181-198.

Livingstone, D. M., Lotter, A. F., and Walker, I. R. (1999). The decrease in summer surface water temperature with altitude in Swiss Alpine lakes: A comparison with air temperature lapse rates. *Arctic, Antarctic, and Alpine Research* **31**, 341-352.

Lotter, A. F., and Bigler, C. (2000). Do diatoms in the Swiss Alps reflect the length of ice-cover? *Aquatic Sciences* **62,** 125-141.

Lotter, A. F., and Birks, H. J. B. (1997). The separation of the influence of nutrients and climate on the varve time-series of Baldeggersee, Switzerland. *Aquatic Sciences* **59,** 362-375.

Lotter, A. F., and Birks, H. J. B. (2003a). The Holocene palaeolimnology of Sägistalsee (1935 m asl) and its environmental history: A synthesis. *Journal of Paleolimnology* **30** (in press).

Lotter, A. F., and Birks, H. J. B. (2003b). Holocene sediments of Sägistalsee, a small lake at the present-day tree-line in the Swiss Alps. *Journal of Paleolimnology* **30** (in press).

Lotter, A. F., Birks, H. J. B., Hofmann, W., and Marchetto, A. (1997). Modern diatom, cladocera, chironomid, and chrysophyte cyst assemblages as quantitative indicators for the reconstruction of past environmental conditions in the Alps. I. Climate. *Journal of Paleolimnology* **18,** 395-420.

Lotter, A. F., Birks, H. J. B., Hofmann, W., and Marchetto, A. (1998). Modern diatom, cladocera, chironomid, and chrysophyte cyst assemblages as quantitative indicators for the reconstruction of past environmental conditions in the Alps. II. Nutrients. *Journal of Paleolimnology* **19,** 443-463.

Lotter, A. F., Birks, H. J. B., Eicher, U., Hofmann, W., Schwander, J., and Wick, L. (2000a). Younger Dryas and Alleröd summer temperatures at Gerzensee (Switzerland) inferred from fossil pollen and cladoceran assemblages. *Palaeogeography, Palaeoclimatology, Palaeoecology* **159**, 349-361.

Lotter, A. F., Hofmann, W., Kamenik, C., Lami, A., Ohlendorf, C., Sturm, M., van der Knaap, W. O., and van Leeuwen, J. F. N. (2000b). Sedimentological and biostratigraphical analyses of short sediment cores from Hagelseewli (2339 m a.s.l.) in the Swiss Alps. *Journal of Limnology* **59**, 53-64.

MOLAR Water Chemistry Group (1999). The MOLAR Project: Atmospheric deposition and lake water chemistry. *Journal of Limnology* **58**, 88-106.

Ohlendorf, C., Bigler, C., Goudsmit, G. H., Lemcke, G., Livingstone, D. M., Lotter, A. F., Müller, B., and Sturm, M. (2000). Causes and effects of long periods of ice cover on a remote high Alpine lake. *Journal of Limnology* **59**, 65-80.

Schweingruber, F. H., Bartholin, T., Schär, E., and Briffa, K. R. (1988). Radiodensitometric dendrochronological conifer chronologies from Lapland (Scandinavia) and the Alps (Switzerland). *Boreas* **17**, 559-566.

Sommaruga-Wögrath, S., Koining, K. A., Sommaruga, R., Tessadri, R., and Psenner, R. (1997). Temperature effects on the acidity of remote alpine lakes. *Nature* **387**, 64-67.

Tinner, W., and Ammann, B. (2001). Timberline paleoecology in the Alps. *PAGES News* **9**, 9-11.

Wick, L., van Leeuwen, J. F. N., van der Knaap, W. O., and Lotter, A. F. (2003). Holocene vegetation development in the catchment of Sägistalsee (1935 m asl), a small lake in the Swiss Alps. *Journal of Paleolimnology* **30** (in press).

Wunsam, S., and Schmidt, R. (1995). A diatom-phosphorus transfer function for alpine and pre-alpine lakes. *Memorie dell'Istituto Italiano di Idrobiologia* **53**, 85-99.

Wunsam, S., Schmidt, R., and Klee, R. (1995). *Cyclotella*-taxa (Bacillariophyceae) in lakes of the Alpine region and their relationship to environmental variables. *Aquatic Sciences* **57**, 360-386.

High Mountain Lakes and Atmospherically Transported Pollutants

Richard W. Battarbee[1*], Simon Patrick[1], Martin Kernan[1], Roland Psenner[2], Hansjoerg Thies[2], Joan Grimalt[3], Bjoern O. Rosseland[4], Bente Wathne[4], Jordi Catalan[5], Rosario Mosello[6], Andrea Lami[6], David Livingstone[7], Evzen Stuchlik[8], Vera Straskrabova[9], and Gunnar Raddum[10]

[1]*Environmental Change Research Centre, University College London, United Kingdom*

[2]*Institute of Zoology and Limnology, University of Innsbruck, Austria*

[3]*Department of Environmental Chemistry, Consejo Superior de Investigaciones Científicas (CSIC), Barcelona, Spain*

[4]*Norsk Institutt for Vannforskning (NIVA), Oslo, Norway*

[5]*Centro de Estudios Avanzados de Blanes (CEAB-CSIC), Blanes, Spain*

[6]*Consiglio Nazionale delle Ricerche (CNR), Istituto Italiano di Idrobiologia, Verbania-Pallanza, Italy*

[7]*Eidgenössische Anstalt für Abwasserreinigung und Gewässerschutz (EAWAG), Dubendorf, Switzerland*

[8]*Department of Hydrobiology, Charles University, Prague, Czech Republic*

[9]*Hydrobiology Institute, Czech Academy of Sciences, Ceske Budejovice, Czech Republic*

[10]*Institute of Zoology, University of Bergen, Norway*

**phone +44 (0) 2076797582, fax +44 (0) 2076797565, e-mail r.battarbee@ucl.ac.uk*

Keywords: Acid deposition, Biology, Chemistry, Climate change, Mountain lakes, Toxic substances.

1. Introduction

Remote mountain lakes, whether found at high altitudes or high latitudes, usually appear to be in pristine condition. In particular, those lakes that are situated above or beyond the tree-line are rarely disturbed by agricultural or forestry practices and few if any people inhabit their catchments. However, recent research indicates that even the

U. M. Huber et al. (eds.), Global Change and Mountain Regions, 113-121.

most remote lakes are impacted by atmospherically transported pollutants, and that greenhouse-gas forced climate change is beginning to have a significant influence on ecosystem functioning. UV-B radiation is also increasing and, in interaction with global warming, may already be changing biogeochemical cycles in many mountain lakes (Vinebrooke and Leavitt, this volume). All sites are subject to multiple stresses, and studies of the ecological response of mountain lakes to such combined stress need to consider interactions between all factors, both natural and anthropogenic. In this chapter, we consider acid deposition, toxic substances and climate change as the three main drivers of ecosystem change in high mountain lakes.

The data presented here are derived mainly from research conducted by a consortium of research groups under the auspices of the EU-funded project "Measuring and modelling the dynamic response of remote mountain lake ecosystems to environmental change" (MOLAR) (Battarbee et al. 2001). To identify the impacts of atmospheric pollution and climate change independently of other environmental stresses, mountain lakes situated above the timberline were selected so that they could be arranged along environmental gradients across Europe (Fig. 1). These gradients extend from high latitude (Arctic), but relatively low altitude sites in Finland, to the Pyrenees, and from Atlantic Seaboard locations in Scotland and Norway, through high sites in the Alps, to the Tatra Mountains of central-eastern Europe. Although all sites are relatively small, oligotrophic (total phosphorus < 10 $\mu g\ l^{-1}$) headwater lakes produced by glacier erosion, they differ considerably in their depth (2-73 m), in their pH (4.5 – 7.3) and in the length of winter ice-cover (up to 8 months). Almost all have good, relatively organic (5-20%) sediment records that allow the history of pollution and climate change to be recorded *in situ*.

At the key sites in the network, physical, chemical, biological and palaeolimnological data have been collected since 1991, using standard techniques. Physical data include meteorological data (based on automatic weather station measurements) and water column temperature (from thermistor chains). Chemical data include precipitation and lake water measurements, especially of major ions and pH (MOLAR Chemistry Group 1999). Biological data include microbial populations (Straskrabova et al. 1999), epilithic diatoms (Cameron et al. 1999), macro-invertebrates (e.g. Raddum and Fjellheim 2002) and fish (Rognerud et al. 2002). Palaeolimnological data principally include diatoms, cladocera, chironomids, fly ash particles, radionuclides, trace metals and persistent organic pollutants (Lami et al. 2000; Battarbee 2002). Full details of the specific methods used can be obtained from the MOLAR project manual (Wathne and Hansen 1997).

2. Acid deposition in mountain lakes

The most severe stress on the ecology of mountain lakes is the deposition of acidic sulphur and nitrogen compounds derived from emissions generated by fossil fuel combustion. Sulphur emissions are mainly derived from coal and oil combustion in power stations. Nitrogen emissions include both oxidised species (NO_x), derived from both power stations and vehicles, and reduced species (NH_3), derived principally from agricultural sources.

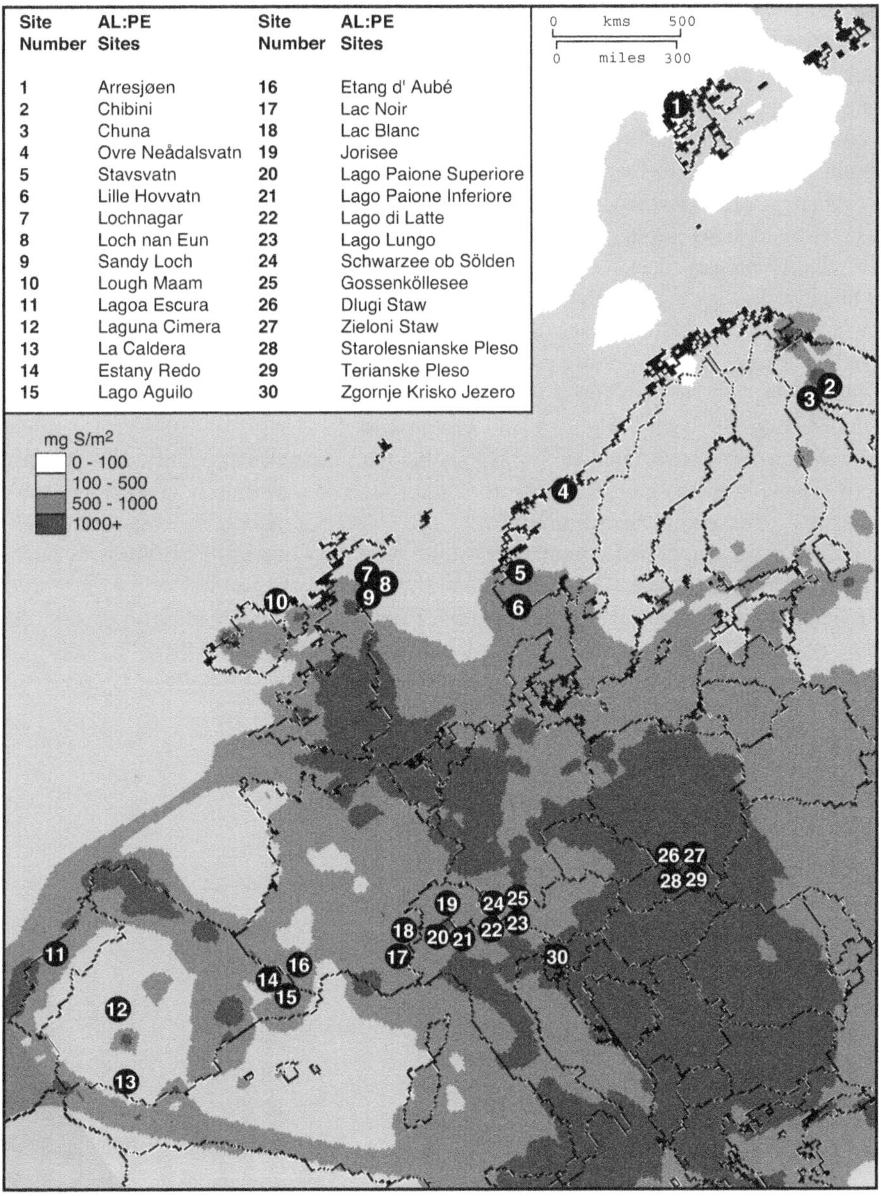

Figure 1: Location of MOLAR sites in the context of European sulphur deposition (1999) (S deposition = mg S m^{-2}, from the EMEP Eulerian acid deposition model).

Evidence for the contamination of mountain lakes by these products can be demonstrated from the high concentration of non-marine sulphate and nitrate in the water column (Fig. 2) and the presence of fly ash particles, especially spheroidal carbonaceous particles (SCP), in lake sediments. The spatial variation in the concentration of these substances (Fig. 1) is in good agreement with the distribution of industrial regions within Europe, and the temporal variation in the concentration of SCPs measured in sediment cores reflects the progressive industrialisation of Europe from the beginning of the 19th century (Rose et al. 2003).

Whilst many mountain lakes in Europe remain un-acidified, either because they have adequate natural alkalinity to neutralise the acidity or because they occur in regions of low acid deposition, lakes with low natural alkalinity in high acid deposition regions have been acidified. This is indicated most clearly by changes in the composition of diatom assemblages preserved in recent lake sediments (Jones et al. 1993). The most severe acidification has taken place in Central and Western Europe. In these regions, some lakes lack zooplankton, and macro-invertebrate populations are impoverished. Some lakes have lost species that are sensitive to acidification, such as *Baetis rhodani* (Raddum and Fjellheim 2002). Also, fish populations, mainly brown trout (*Salmo trutta*) or arctic char (*Salvelinus alpinus*), show signs of acid stress (Rosseland et al. 1999). Whereas sulphur and nitrogen deposition together are responsible for surface water acidification, nitrogen deposition additionally causes problems of eutrophication, especially in situations where vegetation types are nitrogen-limited.

Following international agreements on the reduction of acidifying gases in Europe (UNECE 1998), acid deposition in mountain regions has begun to decline and the pH and alkalinity of some, but not all sites, have begun to recover (Mosello et al. 2002). However, little biological recovery has so far been recognised. This is probably due to time-lags in water chemistry responses (e.g. Kopacek et al. 2002) and problems of species dispersal, re-colonisation and persistence.

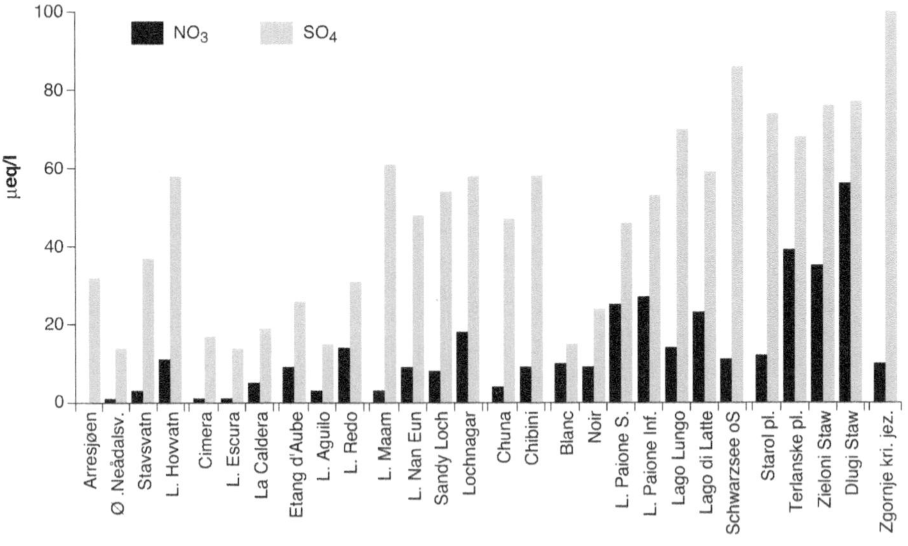

Figure 2: Lake water nitrate and sulphate concentrations (μeq l^{-1}) at selected MOLAR sites. Values are annual mean from 1993.

3. Toxic substances in mountain lakes

In addition to acidity and fly ash contamination, mountain lakes are also contaminated by toxic metals and organic compounds (see Rose et al., this volume). The increases that have taken place over the last century or so can be clearly observed from analyses of sediment cores (e.g. Yang et al. 2002). Changes in the concentration of trace metals, such as Pb, Hg and Cd, and in polycyclic aromatic hydrocarbons (PAHs) correspond closely to the patterns observed for SCPs, suggesting a common fossil-fuel combustion source (Fernandez et al. 2002).

The trends for some of the pesticide residues, on the other hand, are independent, and mainly post-date World War II. For some persistent organic pollutants (POPs), there is evidence that concentrations increase with altitude as these substances become progressively redistributed to colder and more remote regions by volatilisation and cold trapping processes (Grimalt et al. 2001) and contaminate lakes that are distant from production and use of the compounds. Some of the highest concentrations in both metals and POPs are found in Svalbard, one of the most remote regions of the world, where levels of Hg (Fig. 3), and a number of PCB congeners are double those at other sites as a result of food chain biomagnification. Rognerud et al. (2002) have shown that these higher values are the result of a progressive shift to cannibalism in the diet of arctic char between the ages of 11 and 20 years (Fig. 3).

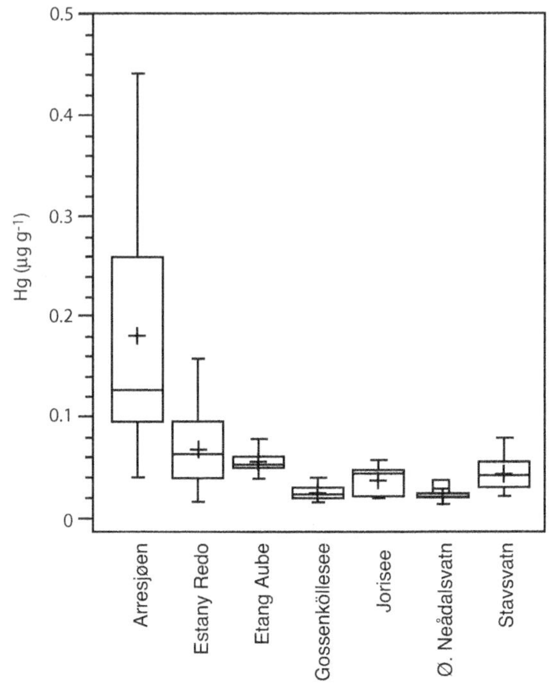

Figure 3: Mercury concentration in fish tissues ($\mu g\ g^{-1}$) at selected MOLAR sites.

4. Climate change impacts on mountain lakes

Most early studies of environmental change in mountain regions assumed that climate, although variable, imposed a relatively constant influence on aquatic ecology. It is now becoming clear, however, that this is not the case, and that climate change is exerting an additional stress on lakes. The climatic impact on mountain lakes is driven mainly through changes in temperature, precipitation and wind regimes that affect snow and ice cover, catchment hydrology, and water column stratification and mixing. These, in turn, control many chemical and biological processes, such as primary production, nutrient cycling, hypolimnetic (bottom-water) O_2 consumption, alkalinity generation and water column pH, and have a direct and strong influence on habitat characteristics and distribution, and on biological life-cycles.

Instrumental temperature reconstructions for the last 200 years by Agustí-Panareda and Thompson (2002) show that decadal-scale fluctuations in mean annual air temperature with an amplitude of up to 2°C have taken place at mountain lake sites and that the most intense warming has taken place over the last few decades. These climate changes are sufficient to cause ecologically important changes in lake-catchment behaviour, and palaeolimnological studies covering this time period provide strong evidence for such changes, specifically through the controlling influence of temperature on lake acidity and lake productivity. For acidity, Psenner and Schmidt (1992) have shown from diatom analysis of recent sediments in the Tyrolean Alps that inter-annual and decadal fluctuations in pH are closely related to changes in mean annual air temperature (Fig. 4). Temperature plays an important role in driving the generation of alkalinity in lakes and lake catchments, and for lakes of naturally low alkalinity relatively small changes in alkalinity can lead to significant shifts in lake water pH that, in turn, have a marked impact on the composition of diatom communities.

At other sites, recent warming appears to be having mainly an effect on primary productivity. This can be seen from both changes in diatom plankton at sites in the Pyrenees (Catalan et al. 2002a), Finland (Sorvari and Korhola, 1998), and Austria (Koinig et al. 2002) and from recent increases in the amount of organic matter accumulating in sediments (Battarbee et al. 2002). These changes can best be explained as a result of changes in nutrient, especially phosphorus, loading through water temperature and water column mixing that drives internal nutrient recycling. Alternatively, they may be the result of a catchment change where an increase in the delivery of nutrients to the lake may be caused by reduced snow cover, enhanced carbon turnover in soils, and increased soil erosion related to changes in catchment hydrology (Catalan et al. 2002b). The specific diatom responses observed may be a combination of increased nutrient availability coupled to lengthening of the growing season and changes in water column stratification and mixing that favour late summer and autumn blooming taxa.

5. Conclusions - Future work

Despite the success of recent studies, mountain lake ecosystems are still poorly understood. High quality datasets (e.g. diatoms, zooplankton, water chemistry) are available for only a few sites, as the physical nature of the high mountain environment presents difficulties for field sampling, surveys and experimental work. Nevertheless, recent research has shown that few if any mountain lakes are pristine, almost all are contaminated in some way by atmospherically transported pollutants, and in some cases the level of contamination is sufficiently high to have caused significant ecological change. Whilst a degree of recovery from acidification might be expected in the future, there are remaining threats from toxic metals and organic compounds, and some of these threats may become greater if climate change causes an acceleration in the transfer of organochlorine compounds to cold regions (see Rose et al., this volume).

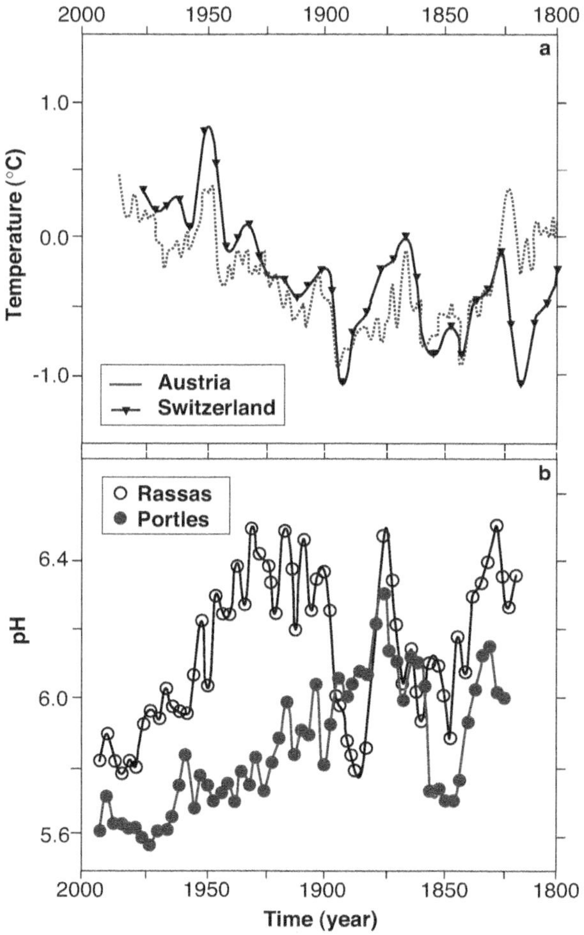

Figure 4: Relationship between Swiss and Austrian temperature curves (a) and pH inferred from diatom assemblages (b) in the sediments of Rassas and Portles See (Tyrolean Alps). Swiss temperatures are smoothed averages (n = 11) of deviation from the mean in Basel (1901-1960). Austrian temperatures (20 stations) are the running averages (n = 9) of the deviation from the mean of the years 1951-1980. pH is reconstructed by weighted averaging (adapted from Psenner and Schmidt 1992).

Understanding how climate change influences mountain lakes both directly and indirectly, by modifying catchment processes and the behaviour of pollutants, is central to future research. Emphasis needs to be placed on the interaction between acidity and climate change in different climate systems, on the transport of metals and organic compounds to lake catchments and their uptake in the lake food chain, and on the impact of climate change on biogeochemical processes that control alkalinity generation and nutrient dynamics in the lake-catchment system. Models that simulate both hydrochemical processes and biological responses need to be developed further, and, given the complexity of the potential interactions between processes, it is essential to maintain a high quality monitoring network to enable models to be tested and to serve as an early warning system for future change.

Many of these ideas are currently being developed under the auspices of EU research programmes such as EMERGE (http://www.mountain-lakes.org) and will also be the focus of future national and international projects. Understanding the global distribution of pollutants and their impacts on mountain environments needs to be the long-term objective of such research.

6. Acknowledgements

We acknowledge all colleagues participating in, and the EU for funding, the following research projects (1991-2003) from which this paper has been drawn: **AL:PE 1**: 1991-1993 Acidification of Mountain Lakes: Palaeolimnology and Ecology. EU contract STEP-CT90-0079; **AL:PE 2**: 1993-1995 Acidification of Mountain Lakes: Palaeolimnology and Ecology. EU contract EV5V-CT92-0205; **MOLAR**: 1996-1999 Measuring and modelling the dynamic response of remote mountain lake ecosystems to environmental change, a programme of mountain lake research (http://www.natur.cuni.cz/hydrobiology/molar/). EU contract ENV4-CT95-0007; **EMERGE**: 2000-2003 European mountain lake ecosystems: regionalisation diagnostics & socio-economic evaluation (http://www.mountain-lakes.org/). EU contract EVK1-CT-1999-0003.

7. References

Agustí-Panareda, A., and Thompson, R. (2002). Reconstructing air temperature at eleven remote alpine and arctic lakes in Europe from 1781 to 1997 AD. *Journal of Paleolimnology* **28**, 7-23.

Battarbee, R. W., Ed. (2002). MOLAR: Mountain lake research. *Journal of Paleolimnology* **28**, 1-179.

Battarbee, R. W. et al. (2001). Measuring and modelling the dynamic response of remote mountain lake ecosystems to environmental change (the MOLAR project). *Verhandlungen der Internationalen Vereinigung für Theoretische und Angewandte Limnologie* **27**, 3774-3779.

Battarbee, R. W. et al. (2002). Comparing palaeolimnological and instrumental evidence of climate change for remote mountain lakes over the last 200 years. *Journal of Paleolimnology* **28**, 161-179.

Cameron, N. G. et al. (1999). Surface-sediment and epilithic diatom pH calibration sets for remote European mountain lakes (AL:PE Project) and their comparison with the Surface Waters Acidification Programme (SWAP) calibration set. *Journal of Paleolimnology* **22**, 291-317.

Catalan, J. et al. (2002a). Lake Redó ecosystem response to an increasing warming the Pyrenees during the twentieth century. *Journal of Paleolimnology* **28**, 129-145.

Catalan, J. et al. (2002b). Seasonal ecosystem variability in remote mountain lakes: Implications for

detecting climatic signals in sediment records. *Journal of Paleolimnology* **28**, 25-46.
Fernández, P., Rose, N. L., Vilanova, R. M., and Grimalt, J. O. (2002). Spatial and temporal comparison of polycyclic aromatic hydrocarbons and spheroidal carbonaceous particles in remote European lakes. *Water, Air and Soil Pollution, Focus 2*, 261-274.
Grimalt, J. O. et al. (2001). Selective trapping of organochlorine compounds in mountain lakes of temperate areas. *Environmental Science and Technology* **35**, 2690-2697.
Jones, V. J. et al. (1993). Palaeolimnologiccal evidence for the acidification and atmospheric contamination of lochs in the Cairngorm and Lochnagar areas of Scotland. *Journal of Ecology* **81**, 3-24.
Koinig, K. A. et al. (2002). Environmental changes in an alpine lake (Gossenköllesee, Austria) over the last two centuries: The influence of air temperature on biological parameters. *Journal of Paleolimnology* **28**, 147-160.
Kopacek, J. et al. (2002). Hysteresis in reversal of Central European mountain lakes from atmospheric acidification. *Water, Air and Soil Pollution Focus* **2**, 92-114.
Lami, A. et al. (2000). Paleolimnology and ecosystem dynamics of remote European Alpine lakes (Mountain Lakes Research programme, MOLAR). *Journal of Limnology* **59**, 1-119.
MOLAR Chemistry Group (1999). The MOLAR Project: Atmospheric deposition and lake water chemistry. *Journal of Limnology* **58**, 88-106.
Mosello, R. et al. (2002). Trends in the chemical composition of high altitude lakes in Europe. *Water, Air and Soil Pollution Focus* **2**, 75-89.
Psenner, R., and Schmidt, R. (1992). Climate-driven pH control of remote alpine lakes and effects of acid deposition. *Nature* **356**, 781-783.
Raddum, G. G., and Fjellheim, A. (2002). Species composition of freshwater invertebrates in relation to chemical and physical factors in high mountains in southwestern Norway. *Water, Air, and Soil Pollution Focus* **2**, 311-328.
Rognerud, S. et al. (2002). Mercury and organochlorine contamination in brown trout *(Salmo trutta)* and arctic charr *(Salvelinus alpinus)* from high mountain lakes in Europe and the Svalbard archipelago. *Water, Air, and Soil Pollution Focus* **2**, 209-232.
Rosseland, B.-O. et al. (1999). The ecophysiology and ecotoxicology of fishes as a tool for monitoring and management strategy of high mountain lakes and lakes and rivers in acidified areas. *Zoology* **190**, 90-100.
Sovari, S., and Korhola, A. (1998). Recent diatom assemblage changes in subarctic Lake Saanajärvi, NW Finnish Lapland, and their palaeoenvironmental implications. *Journal of Paleolimnology* **20**, 205-215.
Straskrabova, V. et al. (1999). Pelagic food web in mountain lakes (Mountain Lakes Research Program). *Journal of Limnology* **58**, 1-222.
UNECE (1998). Convention on long-range transboundary air pollution, protocol on further reduction of sulphur emissions (http://www.unece.org/env/lrtap/fsulf_hl.htm).
Wathne, B. M., and Hansen, H. E. (1997). "MOLAR: Measuring and modelling the dynamic response of remote mountain lake ecosystems to environmental change: A program of mountain lake research." MOLAR Project manual. NIVA Report 0-96061, Oslo.
Yang, H. et al. (2002). Mercury and lead budgets for Lochnagar, a Scottish mountain lake and its catchment. *Environmental Science and Technology* **36**, 1383-1388.

Trace Metals, Fly-ash Particles and Persistent Organic Pollutants in European Remote Mountain Lakes

Neil L. Rose[1*], Handong Yang[1], Pilar Fernández[2], and Joan O. Grimalt[2]
[1]*Environmental Change Research Centre, University College London, 26 Bedford Way, London WC1H 0AP, United Kingdom*
[2]*Department of Environmental Chemistry, Institute of Chemical and Environmental Research (CSIC), Jordi Girona, 18, 08034-Barcelona, Spain*
**phone +44 207 679 5543, fax +44 207 679 7565, e-mail nrose@geog.ucl.ac.uk*

Keywords: Atmsopheric deposition, Biota, Europe, Mountain lakes, Pollution, Sediments.

1. Introduction

Many anthropogenic pollutants emitted to the atmosphere can be transported over large distances and affect ecosystems and human health thousands of kilometres from their source. In recent years, concern has grown over the increased contamination of remote areas, particularly the Arctic and mountain regions, and the unprecedented levels of pollutants observed in areas previously considered to be pristine. Atmospheric transport is one of the most efficient and rapid means by which toxic pollutants, including trace metals and persistent organic pollutants (POPs), can be transferred to remote areas. Understanding the pathways and mechanisms from source to sink is thus vitally important. Atmospheric transport models predict that sources of pollutants to remote areas are widespread and diverse, such that there are contributions from "local" and regional sources, as well as transboundary and even global inputs (e.g. Hanisch 1998).

Remote mountain lakes are known to be excellent indicators of atmospheric pollution and its environmental effects. Sensitive geologies, sparse soils and extreme meteorology conspire to produce fragile ecosystems, frequently isolated from

U. M. Huber et al. (eds.), Global Change and Mountain Regions, 123-132.

direct contamination. The additional stress of atmospheric pollutant inputs to these systems often results in detectable physico-chemical and/or biological change and mountain lakes can therefore provide "early warning" indicators for impacts in less sensitive areas. As the sediments of these lakes accumulate, they store a record of environmental conditions within the lake ecosystem and of contaminants deposited from the atmosphere, whilst biota within the lake store chemical compounds with a high affinity for organic matter and those with a capacity to bioaccumulate through the food chain. Consequently, in the absence of reliable direct measurements in remote areas, these two environmental compartments can be used to provide a natural record of atmospherically deposited pollutants and, in the case of sediments, provide one of the few ways by which temporal trends in this deposition can be determined.

For many pollutant species, the affinity for organic matter and accumulation within lake sediments raises concentrations to more easily measurable levels, whilst fine slicing techniques and the use of radionuclide chronologies (e.g. ^{210}Pb, ^{137}Cs, ^{241}Am) frequently allow a sediment resolution on an annual to sub-decadal scale. The sediment record therefore provides an historical context for contemporary measurements by allowing identification of the direction and rate of change in both pollutant input and any resulting biological response.

2. The historical record

Over the last decade, sediments of remote mountain lakes have been used to study both the spatial and temporal distributions of a range of pollutants, for example POPs (e.g. Fernández et al. 2000; 2002; Grimalt et al. 2001), metals (Rognerud et al. 1998; Yang et al. 2002) and spheroidal carbonaceous particles (SCPs) (Rose et al. 1999; 2002). SCPs are a component of fly-ash, only produced by the high-temperature combustion of fossil fuels, and whilst not directly harmful themselves, do provide unambiguous evidence for contamination from these sources (Rose 2001). In sediments, they have the potential to act as surrogates for other, more toxic pollutants.

The historical record of many trace metals, as evidenced by the sediments of remote mountain lakes, shows that contamination to levels above a "pre-industrial" background, as a result of atmospheric deposition, begins at the start of the 19th century, although elevated levels have also been associated with Roman and Medieval periods (Renberg et al. 2001). Having no natural background, the record of SCPs began in the mid-19th century with the onset of industrial combustion of coal. The simultaneous start of this record with a decrease in reconstructed pH of lake water was used as evidence for atmospheric deposition being the source of surface water acidification in the debate of the 1980s (e.g. Battarbee 1990). The SCP sediment record has been shown to be reliable and replicable across Europe and hence, characteristic features, once dated with an independent technique (e.g. varves, radionuclides), can be readily used as time markers in the lake sediment record. Further, the SCP record is well correlated with those of pollutants from similar sources, e.g. polycyclic aromatic hydrocarbons (PAH) (Fernández et al. 2002), anthropogenic Hg and Pb (Yang et al.

2002) and other trace metals, such as Zn, Cu and Cd (Fig. 1). PAHs produced from the natural combustion of organic material (e.g. from forest fires) show a full record in the sediment, whilst those produced from industrial combustion show elevated levels above a pre-industrial "background" since the beginning of the 20th century. In contrast, other POPs, such as the organochlorine compounds, exhibit different temporal patterns since they were first used at a later date, e.g. the 1940s or 1950s in the case of DDTs and polychlorobiphenyls (PCBs), respectively.

After a period of increasing pollutant concentrations in sediments, reflecting the increase in emissions from extensive industrial growth and energy demand across Europe from the mid-19th century to the mid to late 20th century, the emissions of particulates and some metals have in recent decades become subject to legislation and control. Furthermore, in the 1980s, the production of PCBs was banned and the use of DDT severely restricted. As a consequence, the sediment records of these pollutants show a decline in the uppermost levels of many remote mountain lakes.

However, a more detailed study of trace metals at Lochnagar, a remote lake in Scotland, shows that when full sediment basin fluxes are considered the expected decline in trace metal inputs are not observed. In common with many sites, the sediment trace metal record shows elevated levels starting in the 1860s but despite reductions of more than 70% in the atmospheric deposition of both Hg and Pb across

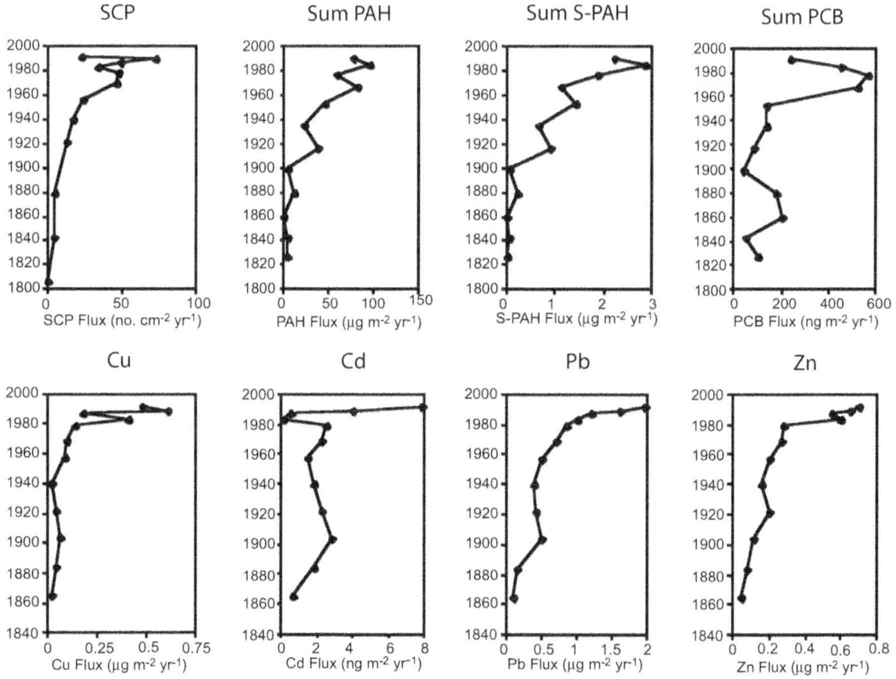

Figure 1: Sediment core profiles for a range of pollutant fluxes from Estany Redó, a Pyrenean mountain lake (42° 38'N; 0°46'E; 2240 m a.s.l.). S-PAH refers to polycyclic aromatic sulphur heterocycles. Like PAHs, these are produced from combustion sources, but principally those with a significant sulphur content.

the UK since the 1970s (Baker 2001), full basin sediment fluxes for these metals show no decline in recent years. Yang et al. (2002) estimated that approximately 90% of total anthropogenic Hg and 86% of total anthropogenic Pb, stored in the system, are retained in the upper levels of the catchment peats (a similar proportion (71%) is observed for SCPs; Rose et al. 2002), and it is the increased input from these catchment sources that may be responsible for negating any improvement resulting from reductions in atmospheric input at this site. Several hypotheses have been put forward to explain the recent lack of decline in sediment metal flux. First, this may be due to a time lag between the retention and release of metals from catchment soils (Schindler et al. 1995). Second, increasing rainfall over recent decades may be causing enhanced erosion of contaminated soils. Third, warmer summers may be increasing decomposition of soil organic matter that is subsequently leached as dissolved organic carbon (DOC), with bound metals. Fourth, climate warming has resulted in longer ice-free periods, allowing more time for algae to scavenge metals from the open-water column, hence increasing the metal flux to the sediment.

Further work is required to determine the relative roles of these various hypotheses and how widespread this problem could be in remote lakes. Certainly, the DOC levels in Lochnagar waters have, in common with many UK upland waters, increased in recent years (Freeman et al. 2001), and positive correlations for both Hg and methyl Hg (a highly toxic form which is efficiently bioaccumulated) with DOC have been observed in high altitude lakes in the United States (Krabbenhoft et al. 2002). If these climate-related hypotheses are correct, then there are significant implications for continued, or even increased, trace metal inputs to mountain lakes as a result of predicted climate change, despite ongoing emissions reductions. Although there have been no direct measurements of this effect for POPs, their affinity for DOC (Gao et al. 1998; Winch et al. 2002) would suggest that similar increases in catchment inputs are likely to have occurred.

3. The spatial dimension

Whilst the sediment record of a single lake provides considerable information on the sources of contamination and their temporal trends, comparisons of contamination levels within or between regions provide data on pathways and transport of pollutants to remote areas. For most trace metals, concentrations, fluxes or inventories (a measure of the full anthropogenic impact through time) tend to decline away from source regions so that, in remote mountain areas, identification of contamination over and above natural levels becomes increasingly difficult (e.g. Rognerud et al. 1998; Boyle et al. 2003). The same distribution trend is true for SCPs, although the lack of natural background allows an easier identification of contamination in remote areas. The spatial distribution of SCP inventories from remote mountain lakes in Europe shows a latitudinal pattern (EU funded AL:PE project; Rose et al. 1999) with highest levels at 50 - 55°N. SCP inventories decline to both north and south, thereby showing a similar spatial distribution to the major European industrial centres. However, at no sites were SCPs absent, and hence a measurable level of atmospheric contamination was detectable even in the most remote areas.

The patterns for POPs and the volatile trace metal, Hg, are more complicated. Atmospheric transport via global distillation (Wania and Mackay 1996) provides a mechanism for the preferential movement and deposition of more volatile organic compounds to higher latitudes. In this process, volatile compounds are preferentially released to the atmosphere in warmer, lower latitude regions and transported northwards where, in cooler climates, they condense again. Repetition of this process results in the distillation of organic compounds by latitude with the more volatile compounds elevated in higher latitude, cooler regions whilst less volatile compounds preferentially remain in warmer, lower latitude areas. A similar phenomenon has also been reported for the transport of volatile compounds to higher altitude sites in temperate areas of Europe (EU funded MOLAR project; Grimalt et al. 2001). The relationship between altitude and concentrations of selected organochlorine compounds in fish in European mountain lakes is shown in Figure 2. These fish concentration data suggest that the less volatile compounds (e.g. 4,4'-DDE; PCB-101; PCB-153) are retained to a major extent in higher altitude sites, reflecting a lower annual average air temperature, whereas the more volatile compounds (e.g. PCB-28) appear to show no trend with altitude. Sedimentary inventories show a similar pattern. However, measurements of organochlorines in atmospheric deposition indicate that all compounds are introduced to higher altitude areas independent of their volatility. Therefore, the lack of correlation between the concentrations of the more volatile compounds in fish and sediment and altitude is most probably due to the loss of these compounds to the atmosphere during warm periods reducing the fraction retained in lakes. This difference between lake retention of low- and high-volatility organochlorine compounds has been termed "selective trapping" (Grimalt et al. 2001).

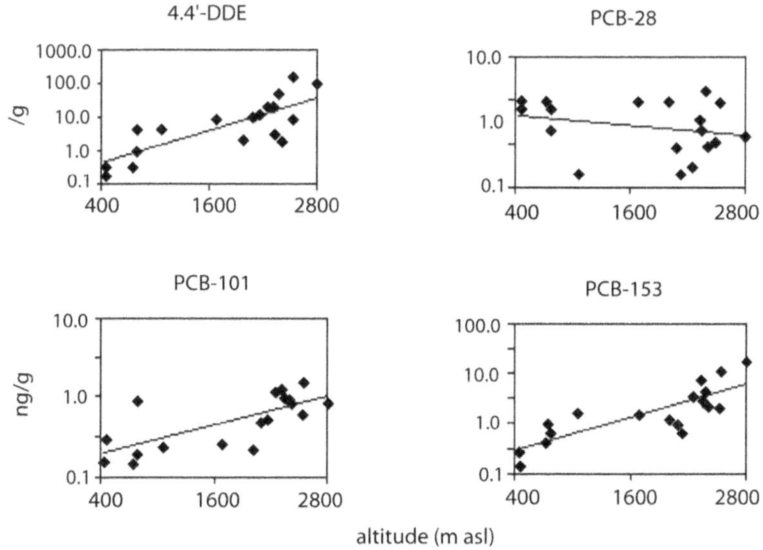

Figure 2: Concentrations of selected organochlorine compounds in fish (ng g^{-1} wet weight) vs. lake altitude.

Comparison of the relative composition of PCB congeners in atmospheric deposition, lake water, sediment and fish from three lakes situated in different mountain areas of Europe, Estany Redó (Spanish Pyrenees), Gossenköllesee (Austrian Alps) and Øvre Neådalsvatn (mid-Norway), confirms this effect. Thus, sediments and fish are significantly enriched in the less volatile congeners than atmospheric deposition and water. The difference is most pronounced for Gossenköllesee, the lake with the lowest annual average air temperature.

Evidence for inter-continental or trans-oceanic transport of POPs has also been recorded in European mountain lake sediments. Rose et al. (2001) showed that the concentration profile of toxaphene in Lochnagar comprised a bi-modal distribution with maxima in the mid-1970s and early 1990s. The earlier toxaphene peak showed agreement with the U.S. source curve and, therefore, was thought to correspond to modelled global patterns, whilst the later peak was thought to correspond to long-range transport from eastern and southern Europe, or from still lower latitudes. Although no toxaphene was ever produced or used in the UK, the concentrations determined for this remote mountain lake equate to those observed in the Great Lakes where there have been additional riverine inputs. These high toxaphene concentrations, in a mountain lake far removed from any source areas, raise concerns over the effects of this pesticide at European sites closer to sources, for example in southern and eastern Europe.

4. Pollutants in biota: Concentrations and role in monitoring

The sediments of mountain lakes provide a valuable tool in detecting spatial and temporal changes across a region, but pollutant measurements must be related to biological impact and many mountain lake species, already living in cold, often nutrient-poor waters are sensitive to the further stress caused by atmospheric pollutant deposition. Monitoring changes in mountain lake biota, either directly by sampling and analysis of the organisms, or indirectly by observing changes in, for example, species composition, can also provide evidence for the levels of impact in these systems.

Contamination of fish with PAHs, organochlorine compounds and Hg has been determined in remote European mountain lakes (EU funded MOLAR and EMERGE projects; e.g. Grimalt et al. 2001; Rognerud et al. 2002). Organochlorine compound concentrations in fish are normally in the order of 10-100 ng g^{-1} and are dependent on the age of the fish and the altitude of the lake. In general, both increasing age of the fish and lower annual average air temperatures are reflected in higher organochlorine concentrations. For Hg, low sediment Hg fluxes, low net production of methyl-Hg and short food chains explain low Hg concentrations in invertebrate-feeding fish (Rognerud et al. 2002), whilst the highest Hg, PCB and DDE concentrations are observed at sites where the fish are piscivorous. Thus, in the oldest fish on Svalbard, Hg exceeds guideline concentrations for human consumption. The high levels in this remote Arctic lake are therefore due to biomagnification, emphasising the need to examine not only the contaminant concentration, but also the ecological context, in this case the trophic level of the measured fish. The trophic level of biota can be determined using $\delta^{15}N$ measurements (Rognerud et al. 2002). Similar biomagnification would be expected for piscivorous fish in temperate mountain lakes.

Studies of the impacts of atmospheric contaminants on other terrestrial and aquatic flora and fauna in catchments and water bodies of remote mountain lakes are few. One study of trace metal levels in a number of ecosystem compartments has been undertaken at Lochnagar, Scotland, where Hg, Cd, Pb, Cu, Ni and Zn have been measured in annual samples of catchment plants and mosses, aquatic macrophytes, epilithic algae, zooplankton and macro-invertebrates. In addition, fortnightly measurements of trace metals in atmospheric deposition and lake water and annual sediment trap samples (Rose and Yang 2002) have been determined, and recently a one-off fish survey (EU funded EMERGE project; B. O. Rosseland, pers. comm.) has been carried out. Annual trends in terrestrial moss species (*Pleurozium schreberi* and *Hylocomium splendens*) have been found to agree well with annual atmospheric deposition fluxes and similar trends have been observed for Pb.

Observed Hg concentrations for various compartments in Lochnagar are given in Table 1. These clearly show the enhanced accumulation in biotic components with respect to lake water and emphasise the role that biota can play in the monitoring of trace pollutants in remote mountain lakes. Monitoring of contaminants in remote lakes using biota has significant advantages over "direct" measurements of lake waters or atmospheric deposition. First, as mentioned above, the affinity of organic matter for trace metals and POPs, via bioaccumulation, elevates concentrations to more easily measurable levels. Second, short-term fluctuations are "smoothed" over months or years, whereas a single water sample may be representative of only hours or days. Third, there is the practical consideration of obtaining representative data from a single visit to sites which are remote and, especially in winter, inaccessible.

Sediment traps have been used as an additional tool to monitor trends in trace metals and SCPs in upland lakes in the UK since 1990. Monitoring using sediment traps in remote lakes has the same advantages as those of monitoring using biota, but they also have three additional advantages. First, they can be deployed in any lake as their use is not dependent on the presence of a particular species at study sites. Second, deployment and retrieval dates, and hence the period covered by the sample, are precisely known, and third, sediments are not subject to biomagnification and hence measurements of trophic level are not required. At the 2002 UNECE ICP workshop "Heavy metals (Pb, Cd and Hg) in surface waters: Monitoring and biological impact" held in Lillehammer, a range of biotic monitoring tools from fish and invertebrates to mosses were recommended to cover different time periods. Sediment traps were also included in these recommendations (Skjelkvåle and Ulstein 2002), as were sediment cores to provide a historical perspective to contemporary monitoring.

Table 1: Typical Hg values in ecological compartments of Lochnagar, Scotland 2000-2001.

Lake water	Epilithic algae	Aquatic macrophytes	Macro-invertebrates	Zooplankton	Brown trout
10 – 20 ng/l	15 – 20 ng/g	30 – 60 ng/g	50 – 60 ng/g	150 – 250 ng/g	50 – 220 ng/g

5. Future challenges

In many industrial areas the emission of trace pollutants to the atmosphere has declined over recent decades as a result of implemented legislation. At the start of the 21st century, the greatest threat to mountain lake biota from atmospherically deposited pollutants results from the current and future effects of climate change. We have already described how climate change may enhance the input of previously deposited trace metals and POPs to remote lakes, but climate warming may further exacerbate inputs of POPs to high-latitude and high-altitude lakes as a consequence of an increase in global distillation of volatile species from temperate and tropical regions. Higher temperatures will result in wider distributions of both volatile and semi-volatile species, resulting in contamination of remote areas from broader source regions. In short, more pollutants will be able to travel over longer distances, and protection of the most remote and sensitive ecosystems will therefore require international, if not global, implementation of policy.

Most studies involving atmospherically-derived pollutants in remote mountain lakes have tended to focus on the distribution of single pollutant species or classes, and it is only recently that direct comparisons between multiple pollutants have been undertaken (e.g. Fernández et al. 2002). However, whilst it is most important that contamination levels are related to impacts on biota, critical loads and thresholds for many pollutants are only now being developed and many have still not been determined for aquatic species. Further, these thresholds are being developed for single pollutants with little attention paid to the synergistic effects of multiple pollutants or indeed, other environmental changes such as climate, elevated UV-radiation or eutrophication (e.g. Vinebrooke and Leavitt this volume). One of the main challenges for pollutant studies in remote mountain lakes is therefore to more fully examine the impact that long-term exposure to ambient levels of multiple pollutants have on a range of aquatic organisms at different trophic levels. Preliminary data from studies in the UK have shown that sediments in many areas, including upland and mountain sites, may already be toxic to sediment- and water-dwelling invertebrate organisms (B. Rippey pers. comm.), and this is likely due to a combination of pollutants at "low" concentrations. Directly affected species are near the bottom of the aquatic food chain and, whilst it is known that there is a toxic effect to these organisms, it is currently uncertain how these effects may biomagnify to top predators.

One certainty for the 21st century is that new organic compounds will be developed for industrial or agricultural processes or to replace those that are no longer effective or banned. These new compounds will, either deliberately or accidentally, be emitted to the environment and many will have a detrimental impact. Vigilance for the impacts of new compounds, or the enhanced role of established ones, is therefore required and the role of mountain lake ecosystems as "early warning" indicators will be important in their detection. To this end, there is a need for ongoing monitoring and assessment of all contaminant classes within a range of ecosystem compartments in both remote mountain lakes and their catchments.

6. Acknowledgements

We would like to thank all our partners in four EU funded mountain lakes projects: AL:PE (Acidification of mountain lakes: Palaeolimnology and Ecology, 1991 - 1993) (EU contract STEP-CT90-0079); AL:PE 2 (Remote mountain lakes as indicators of air pollution and climate change, 1993 - 1995) (EU contract EV 5V-CT92-0205); MOLAR (Measuring and modelling the dynamic response of remote mountain lake ecosystems to environmental change. A programme of mountain lake research, 1996 - 1999) (EU contract ENV4-CT95-0007) and EMERGE (European mountain lake ecosystems: Regionalisation, diagnostics and socio-economic evaluation, 2000 - 2003) (EU contract EVK1-CT-1999-00032) and the EU for funding.

7. References

Baker, S. J. (2001). Trace and major elements in the atmosphere at rural locations in the UK: Summary of data for 1999. National Environment Technology Centre Report AEAT/R/ENV/0264.

Battarbee, R. W. (1990). The causes of lake acidification, with special reference to the role of acid deposition. *Philosophical Transactions of the Royal Society of London* **B327**, 339–347.

Boyle, J. F., Rose, N. L., Appleby, P. G., and Birks, H. J. B. (2003). Recent environmental change and human impact on Svalbard: The lake-sediment geochemical record. *Journal of Paleolimnology* (in press).

Fernández, P., Vilanova, R. M., Martínez, C., Appleby, P. G., and Grimalt, J. O. (2000). The historical record of atmospheric pyrolytic pollution over Europe registered in the sedimentary PAH from remote mountain lakes. *Environmental Science and Technology* **34**, 1906-1913.

Fernández, P., Rose, N. L., Vilanova, R. M., and Grimalt, J. O. (2002). Spatial and temporal comparison of polycyclic aromatic hydrocarbons and spheroidal carbonaceous particles in remote European lakes. *Water, Air and Soil Pollution: Focus* **2**, 261-274.

Freeman, C., Evans, C. D., Monteith, D. T., Reynolds, B., and Fenner, N. (2001). Export of organic carbon from peat soils. *Nature* **412**, 785.

Gao, J. P., Maguhn, J., Spitzauer, P., and Kettrup, A. (1998). Distribution of polycyclic aromatic hydrocarbons (PAHs) in pore water and sediment of a small aquatic ecosystem. *International Journal of Environmental Analytical Chemistry* **69**, 227-242.

Grimalt, J. O, Fernández, P., Berdie, L., Vilanova, R. M, Catalan, J., Psenner, R., Hofer, R., Appleby, P. G., Rosseland, B. O., Lien, L., Massabuau, J. C., and Battarbee, R. W. (2001). Selective trapping of organochlorine compounds in mountain lakes of temperate areas. *Environmental Science and Technology* **35**, 2690-2697.

Hanisch, C. (1998). Where is the mercury coming from? *Environmental Science and Technology* **32**, 176A-179A.

Krabbenhoft, D. P., Olson, M. L., Dewild, J. F., Clow, D. W., Striegl, R. G., Dornblaser, M. M., and Vanmetre, P. (2002). Mercury loading and methylmercury production and cycling in high-altitude lakes from the western United States. *Water, Air and Soil Pollution: Focus* **2**, 233–249.

Renberg, I., Bindler, R., and Brännvall, M.-L. (2001). Using the historical atmospheric lead-deposition record as a chronological marker in sediment deposits in Europe. *The Holocene* **11**, 511–516.

Rognerud, S., Skotvold, T., Fjeld, E., Norton, S. A., and Hobaek, A. (1998). Concentrations of trace elements in recent and pre-industrial sediments from Norwegian and Russian Arctic lakes. *Canadian Journal of Fisheries and Aquatic Sciences* **55**, 1512-1523.

Rognerud, S., Grimalt, J. O., Rosseland, B. O., Fernández, P., Hofer, R., Lackner, B., Lauritzen, B., Lien, L., Massabuau, J. C., and Ribes, A. (2002). Mercury and organochlorine contamination in Brown Trout (*Salmo Trutta*) and Arctic Charr (*Salvelinus alpinus*) from high mountain lakes in Europe and the Svalbard Archipelago. *Water, Air and Soil Pollution: Focus* **2**, 209–232.

Rose, N. L. (2001). Fly-ash particles. *In* "Tracking environmental change using lake sediments: Volume 2. Physical and Chemical Techniques." (W. M. Last, and J. P. Smol, Eds.), pp. 319–349. Kluwer

Academic Publishers, Dordrecht.

Rose, N. L., and Yang, H. (2002). Trace metal monitoring in the Lochnagar ecosystem. *In* "Proceedings from the Workshop on Heavy Metals (Pb, Cd and Hg) in surface waters." (B. L. Skjelkvåle, and M. Ulstein, Eds.) Lillehammer, March 2002. ICP Waters Report 67/2002.

Rose, N. L., Harlock, S., and Appleby, P. G. (1999). The spatial and temporal distributions of spheroidal carbonaceous fly-ash particles (SCP) in the sediment records of European mountain lakes. *Water, Air and Soil Pollution* **113**, 1-32.

Rose, N. L., Backus, S., Karlsson, H., and Muir, D. C. G. (2001). An historical record of toxaphene and its congeners in a remote lake in western Europe. *Environmental Science and Technology* **35**, 1312-1319.

Rose, N. L., Shilland, E., Yang, H., Berg, T., Camerero, L., Harriman, R., Koinig, K., Lien, L., Nickus, U., Stuchlik, E., Thies, H., and Ventura, M. (2002). Deposition and storage of spheroidal carbonaceous fly-ash particles in European mountain lake sediments and catchment soils. *Water, Air and Soil Pollution: Focus* **2**, 251–260.

Schindler, D. W., Kidd, K. A., Muir, D. C. G., and Lockhart, W. L. (1995). The effects of ecosystem characteristics on contaminant distribution in northern freshwater lakes. *Science of the Total Environment* **160/161**, 1-17.

Skjelkvåle, B.-L., and Ulstein, M., Eds. (2002). "Proceedings from the Workshop on Heavy Metals (Pb, Cd and Hg) in Surface Waters." Lillehammer, March 2002. ICP Waters Report 67/2002.

Wania, F., and Mackay, D. (1996). Tracking the distribution of persistent organic pollutants. *Environmental Science and Technology* **30**, 390A-396A.

Winch, S., Ridal, J., and Lean, D. (2002). Increased metal bioavailability following alteration of freshwater dissolved organic carbon by ultraviolet B radiation exposure. *Environmental Toxicology* **17**, 267–274.

Yang, H., Rose, N. L., Battarbee, R. W., and Boyle J. F. (2002). Mercury and lead budgets for Lochnagar, a Scottish mountain lake and its catchment. *Environmental Science and Technology* **36**, 1383-1388.

Long-term Responses of Mountain Ecosystems to Environmental Changes: Resilience, Adjustment, and Vulnerability

Willy Tinner* and Brigitta Ammann
Institute of Plant Sciences, Section Paleoecology, University of Bern, Altenbergrain 21, CH-3013 Bern, Switzerland
**phone + 41 31 631 49 32, fax +41 31 332 20 59, e-mail willy.tinner@ips.unibe.ch*

Keywords: Alps, Climate change, Human impact, Fire history, Paleoecology, Vegetation history.

1. Introduction

The steep environmental gradients of mountain ecosystems over short distances reflect large gradients of several climatic parameters and hence provide excellent possibilities for ecological research on the effects of environmental change. To gain a better understanding of the dynamics of abiotic and biotic parameters of mountain ecosystems, long-term records are required since permanent plots in mountain regions cover in the best case about 50 – 70 years. In order to extend investigations of ecological dynamics beyond these temporal limitations of permanent plots, paleoecological approaches can be used if the sampling resolution can be adapted to ecological research questions, e.g. a sample every 10 years. Paleoecological studies in mountain ecosystems can provide new ecological insights through the combination of different spatial and temporal scales. If we thus improve our understanding of processes across both steep environmental gradients and different time scales, we may be able to better estimate ecosystem responses to current and future environmental change (Ammann et al. 1993; Lotter et al. 1997).

The complexity of ecological interactions in mountain regions forces us to concentrate on a number of sub-systems – without losing sight of the wider context.

U. M. Huber et al. (eds.), Global Change and Mountain Regions, 133-143.

Here, we summarize a few case studies on the effects of Holocene climate change and disturbance on the vegetation of the Western Alps. To categorize the main response modes of vegetation to climatic change and disturbance in the Alps we use three classes of ecological behaviour: "resilience", "adjustment", and "vulnerability". We assume a resilient (or elastic) behaviour if vegetation is able to recover to its former state, regaining important ecosystem characteristics, such as floristic composition, biodiversity, species abundances, and biomass (e.g. Küttel 1990; Aber and Melillo 1991). Conversely, vegetation displacements may occur in response to climatic change and/or disturbance. In some cases, this may culminate in irreversible large-scale processes such as species and/or community extinctions. Such drastic developments indicate high ecosystem vulnerability (or inelasticity or instability, for detailed definitions see Küttel 1990; Aber and Melillo 1991) to climatic change and/or disturbance. In this sense, the "vulnerability" (or instability) of an ecosystem is expressed by the degree of failure to recover to the original state before disturbance and/or climatic change. Between these two extremes (resilience vs. vulnerability), ecosystem adjustments to climatic change and/or disturbance may occur, including the appearance of new and/or the disappearance of old species. The term "adjustment" is hence used to indicate the response of vegetational communities, which adapted to new environmental conditions without loosing their main character. For forest ecosystems, we assume vegetational adjustments (rather than vulnerability) if the dominant (or co-dominant) tree species are not outnumbered or replaced by formerly unimportant plant species or new invaders. Adaptation as a genetic process is not discussed here and will require additional phylogeographical studies (that incorporate the analysis of ancient DNA) in order to fully understand the distributions of ecotypes.

2. Recent results

2.1 Bioclimatic limits of species distributions

Environmental requirements of species (e.g. as input for ecological models) are usually derived from today's presence or absence in e.g. climatic space, rather than from experiments. However, this "biogeographical" approach may be distorted by prehistoric and historic human impact, which may result in a biased view of the potential range of a species. In the past, some species may have benefited from natural environments that did not persist until today. These no-analogue environmental conditions may be used to provide additional information about the behaviour of species under changing and extreme climatic conditions. According to recent quantitative studies on chironomid assemblages (Heiri 2001), the Holocene climatic optimum in the Alps was reached at around 6500 cal. yr BP (calendar years before present, i.e. 1950 AD), when temperatures were about 1.3-1.5°C warmer than today. These results are supported by alpine timberline studies (e.g. Tinner et al. 1996; Tinner and Theurillat, in press), and it is assumed that overall climatic conditions in the Alps at this time were moderately more continental than today (Tinner and Ammann 2001). If paleoecological sites are arranged according to their altitude, it is possible to reconstruct past forest-vegetation belts in order to assess the ecological potential of today's species under warmer (and moderately more continental) conditions, when human impact was still small. The reconstructed vegetation distribution in the Alps at ca. 6500 cal. yr BP (Fig. 1a) is strikingly different from today and challenges some of the ecological "beliefs" that have been derived from present distributions. For example, recent paleoecological studies show that *Abies alba* (white fir, the tallest tree of Europe at ~70 m height) was able to co-dominate stands situated between 200 m in the southern Alps (e.g. Wick Olatunbosi 1996; Tinner et al. 1999) and 1900 m in the northern Alps (Wick et al., in press) under reconstructed July temperatures between ~24° and 10° C, respectively. In contrast, reliance on today's observations implies that the species is competitive only at sites with July means of ~18°-13° C (e.g. Ellenberg 1996). The paleoecological results suggest that today's constricted geographical range is largely determined by disturbance (especially by fire, Tinner et al. 1999; Wick et al. in press). Dynamic vegetation models (Keller et al. 2002) provide the first quantitative evidence in support of the paleoecological conclusions.

2.2 Response modes of Alpine ecosystems to environmental changes

Plant species interact with other organisms and with abiotic factors, so that the response of vegetation to climatic change and/or disturbances depends not only on the environmental requirements of the species, but also on interspecific competition and interference. Depending on the species involved and the agents forcing the environmental change or disturbance (magnitude, severity, frequency), forest ecosystems may respond with "resilience", "adjustment", or "vulnerability". New paleoecological results suggest that many vegetation types in the Alps responded with displacement (vulnerability) or adjustment during the past 6500 cal. yr (Fig. 1), whereas only a few showed resilient behaviour on such long time scales.

2.2.1 Resilient behaviour of ecosystems

Detailed macrofossil analyses near timberline (Lang and Tobolski 1985; Tinner et al. 1996; Kaltenrieder 1999) confirmed the pollen-based view that the subalpine *Larix decidua-Pinus cembra* belt in the central Alps is one of the most resilient ecosystem types in the Alps (Welten 1982). This forest type first appeared more than 10,000 years ago, and, despite repeated, strong climatic changes and anthropogenic disturbances, the original vegetational character has been preserved until today. This is striking, because most of today's *Larix decidua-Pinus cembra* communities are in pronounced ecotonal situations, which should result in higher sensitivities to disturbances and climatic change. However, modelling studies (Bugmann and Pfister 2000) generated vegetational patterns largely comparable to those observed during the Holocene. In the models, *Larix decidua* and *Pinus cembra* stands were able to regain their former timberline positions within 200 yr after climatic cooling during the Little Ice Age. In the simulations, *Larix decidua* and *Pinus cembra* did not respond linearly to temperature changes. Instead, forest disruptions and subsequent recoveries were determined by rare but extreme climatic events.

In the valley bottoms of the Central Alps, ~1000-1500 m below the *Larix-Pinus*

a) Vegetation belts at ca. 6500 cal. yr BP (4550 BC), 1.2-1.6 ° C warmer than today

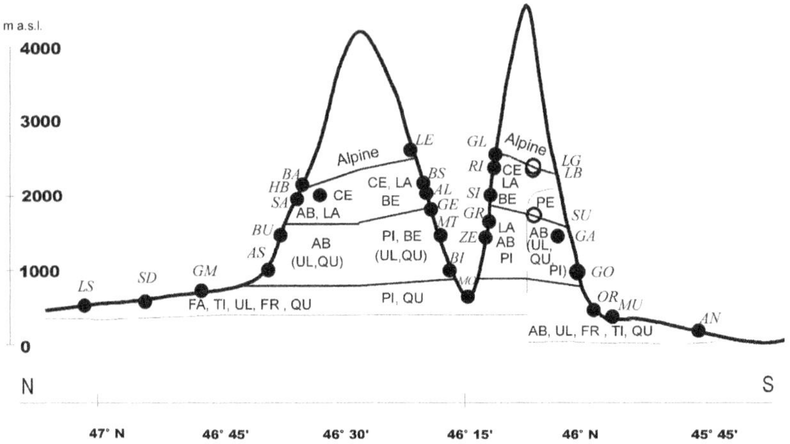

b) Resilience, adjustment, and vulnerability of forest vegetation during the past 6500 yr

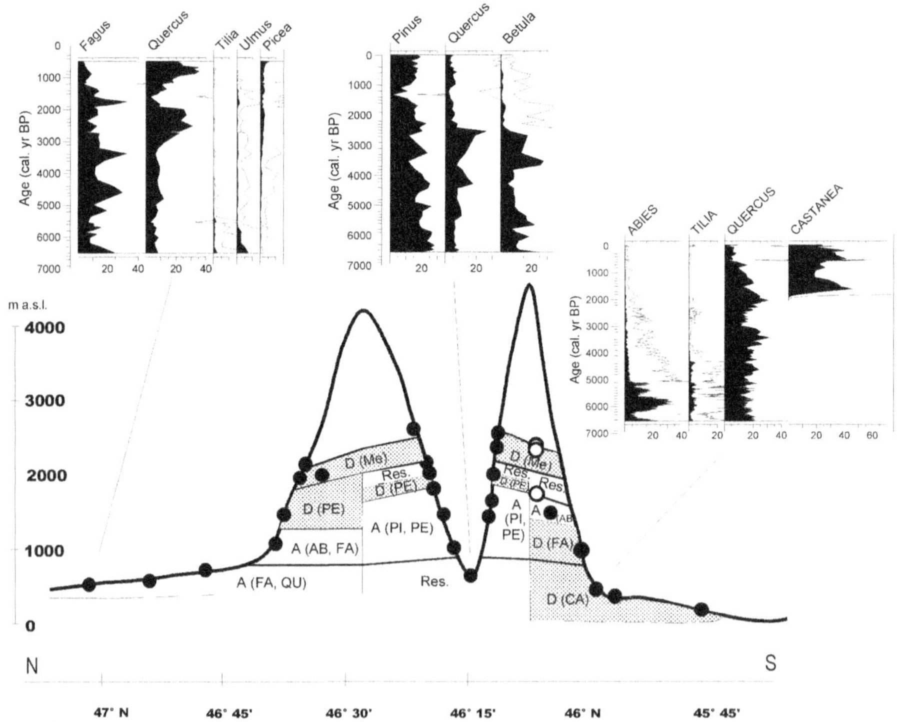

Figure 1: Altitudinal transect through the western Alps: a) Important forest trees at around 6500 cal. yr BP, when summer climate was about 1.5 ° C warmer than today. Less important tree species are in brackets; b) Modern forest distribution in regard to forest changes during the past 6500 yr. The summary pollen diagrams illustrate three modes of forest response to climate change and fire disturbance during the past 6500 yr: vegetational adjustments (Lobsigensee), resilience (Mont d'Orge), and vulnerability/displacement (Origlio). A = Adjustments, D = community displacements as a result of high vulnerability (gray shading), Res. = Resilient behaviour. The distinct change at ca. 2500 cal. yr BP at Mont d'Orge is explainable by a decrease in forested area, however the composition and structure of remaining forests hardly changed. Dominant present-day tree species are in brackets. Taxa and vegetation abbreviations: AB = *Abies alba*, BE = *Betula*, CA = *Castanea sativa*, CE = *Pinus cembra*, FR = *Fraxinus excelsior*, LA = *Larix*, Me = Meadows, PE = *Picea abies*, PI = *Pinus sylvestris*, QU = *Quercus*, TI = *Tilia*, UL = *Ulmus*. Abbreviations of study sites: AL = Aletschwald (Welten 1982), AN = Annone (Wick Olatunbosi 1996), AS = Aegelsee (Wegmüller and Lotter 1990), BA = Bachalpsee (Wick et al. unpubl.), BI = Bitsch (Welten 1982), BS = Böhnigsee (Markgraf 1969), BU = Untere Bunschleralp (Welten 1982), GA = Gondo-Alpjen (Welten 1982), GE = Greicheralp (Welten 1982), GL = Gouillé Loéré (Tinner and Theurillat, in press), GM =Gänsemoos (Welten 1982), GO = Gola di Lago (Zoller and Kleiber 1971), GR = Grächensee (Welten 1982), MT = Montana (Welten 1982), HB = Höhenbiel (Küttel 1990), LB = Lago Basso (Wick 1994), LE = Lengi Egga (Tinner and Theurillat, in press), LG = Lago Grande (Wick 1994), LS = Lobsigensee (Ammann et al. 1986), MO = Mont d'Orge (Welten 1982), MU = Muzzano (Gobet et al. 2000), OR = Origlio (Tinner et al. 1999), RI = Gouillé Rion (Tinner et al. 1996), SA = Sägistalsee (Wick et al. in press), SD= Lac de Seedorf (Richoz 1998), SI = Simplon (Lang and Tobolski 1985), SU = Suossa (Zoller and Kleiber 1971), ZE = Zeneggen (Welten 1982). The latitudinal position is only approximate. Three sites in the southern Alps (open circles) are not part of the north-south transect. These locations are shown according to their position within the different eco-regions.

cembra belt, *Quercus-Pinus sylvestris* forests also resisted combined climatic variation and human disturbance over thousands of years (Fig. 1). If compared to the rest of the Alps, these inner-Alpine longitudinal valleys encompass only minor areas, but their position and environmental history is special: Due to orographic effects, the modern and early Holocene climatic conditions are less different than those observed in regions north and south of the Alps (Welten 1982; Tinner and Ammann 2001), and no indications for any substantial Holocene shifts to more oceanic conditions are apparent (e.g. precipitation increase), which contrasts with reconstructions for the northern and southern Alps (Tinner et al. 1999; Tinner and Lotter 2001). Relatively constant climatic regimes probably endorsed vegetation resilience, but other factors such as species-related human assistance (e.g. differential anthropogenic use of tree species) and adjustment to disturbance must be taken into account. Similarly, in the southern Alps the resilient behaviour of *Picea abies* after 6500 cal. yr BP (the species expanded at around 9000 cal. yr BP) may have been in part favoured by a persistent trend to more oceanic conditions during the past 9100 cal. yr. Ravazzi's (2002) recent summary of the history of *Picea abies* in southern Europe shows that in the millennia before 6000 cal. yr BP, the species expanded rapidly in the upper forest belts of the Alps (~ 1500-2300 m a.s.l.), where it was not subject to competition with other trees (e.g. *Abies alba*).

2.2.2 Invasions and ecosystem adjustments to climatic change and disturbance

The topic of species invasion is attracting much attention and debate in modern ecological studies. However, for long-lived taxa, such as trees, the behaviour of invasive species is especially difficult to assess. Paleoecology offers some tools to study two processes involved in invasion, i.e. migration and population expansion (Bennett 1983). In addition, it also allows the study of the responses of the invaded communities at the level of individual taxa. For instance, in the Alps, many tree species expanded rapidly in response to abrupt warming at 14700-14500 cal. yr BP (e.g. *Pinus sylvestris, Picea abies,* Ammann et al. 1994; Ravazzi 2002) or at 11500 cal. yr BP (e.g. *Quercus, Ulmus, Fraxinus excelsior,* and *Acer* in the southern Alps; Tinner et al. 1999). Other species (e.g. *Abies alba* and *Fagus sylvatica* in the northern Alps) expanded in response to climate cooling and probably precipitation increase around 8200 cal. yr BP (Tinner and Lotter 2001). The invasion by *Fagus* started immediately after climatic change (0-20 yr, Tinner and Lotter 2001). This implies that the species was already locally present before climatic change, despite very few and discontinuous pollen findings. The presence of such isolated individuals or small stands is very difficult to detect in pollen records. For example, Welten (1944) showed that the presence of ~ 70 trees of a particular species in a forested lake catchment with a radius of 2300 m may result in the detection of one grain for every 10,000 pollen grains counted (i.e. one grain in every 10th sample in pollen sums of 1000, approximately corresponding to the situation for *Fagus* before 8200 cal. yr BP). Consequently, he criticised the use of threshold pollen values (e.g. 2 % isopollen lines) for determining the local presence. More recent studies show that this old controversy is still topical. Based on a threshold value of 5 %, the immigration of *Fagus* has been dated to 7000 to 6000 cal. yr BP for the Swiss Plateau, showing a delay of > 500 yr from east to west (e.g. Burga and Perret 1998) and thus a delayed response to climate change due to migrational lags. However, on the basis of different criteria, such as the beginning of the *Fagus* pollen increase in high-resolution studies, the expansion is dated to 8175±45 cal. yr BP, i.e. 1000-2000 yr earlier and immediately following the climate change at 8200 cal yr BP (Tinner and Lotter 2001). Furthermore, according to Tinner and Lotter (2001), the population expansions on the Swiss Plateau occurred without any major migrational delay. The natural population expansion process of *Fagus* into the mixed deciduous forests of the northern Alpine forelands 8200 yr ago may illustrate how rapidly forested ecosystems can change in response to new climatic conditions. However, none of the tree taxa that were present before the invasion were permanently displaced, although the abundance of previously important tree species (*Tilia, Ulmus, Quercus, Corylus*) was reduced considerably. 6500 years ago, the expansion of *Fagus sylvatica* culminated in the dominance of the species, suggesting that the adjustment process gradually evolved into a displacement of the original communities (before 6500 cal. yr BP, *Tilia* and *Ulmus* were co-dominant for millennia; Ammann 1986; 1988). With increasing anthropogenic disturbance (especially fire; Clark et al. 1989; Haas 1996; Richoz 1998), important tree species were completely suppressed in the deciduous forests (e.g. *Tilia, Ulmus*, see Fig. 1b around 5000 cal. yr BP), whereas *Fagus* remained stable or even expanded in phases of reduced human impact (Ammann 1986; 1988). Similarly, *Quercus* was able to hold its ground.

Taken together, the paleoecological results suggest that climatic changes during the Holocene were the main determinants for the establishment of new species in the forelands of the northern Alps (e.g. *Fagus* and *Abies*). Conversely, human impact (fire and wood cutting) possibly caused forest-diversity reductions, but a strong overall increase in biodiversity through the expansion of native and introduced crops and weeds in new habitats (e.g. arable farmlands, heathlands, shrublands, pastures). In the southern Alps, the situation was even more extreme: climatic changes at 11500 and 9200 cal. yr BP allowed the establishment of species-rich forest types (Tinner et al. 1999). However, anthropogenic influence after 4000 cal yr BP resulted in strong forest impoverishment and enduring displacement.

2.2.3 Vulnerability and irreversible processes

Paleobotanical data suggest that in the Alps several tree species as well as forest communities were displaced during the past 6500 yr (compare Fig. 1a and b). Studies from the southern Alps (Tinner et al. 1999) and the northern Alps (Wick et al., in press) provide new insights into the mechanisms of local and regional extinctions. As shown by time-series analyses of charcoal and pollen results, fire was responsible for the regional extinction of *Abies alba* in the lower montane belt of the southern Alps, where today the species is missing at altitudes between 200 and 800m a.s.l. Together with *Abies alba,* other tree taxa were heavily reduced or suffered local extinction (e.g. *Tilia, Ulmus, Fraxinus excelsior, Acer, Hedera*). The area impacted by these displacements encompasses about 17,000 km^2 (Tinner et al. 2000). However, more important than the area losses is the fact that these mixed thermophilous communities were unique for Europe. Today, related vegetation types occur only in the climatically similar Caucasus region, where some of the component species are different (e.g. *Abies nordmanniana*; see Tinner et al. 1999). Such forest impoverishment processes after fire have been widely observed for other regions of the world (Caldararo 2002). In the Alps, responsible factors were inherent in the species' traits. For example, in the case of *Abies alba*, the vulnerability to fire is primarily caused by the absence of an effective resprouting capacity and the short-lived seed bank. The resulting vegetation of the southern Alps was profoundly different, since the only tree taxa to survive increased fire pressure were *Quercus* and *Alnus*. It is likely that this fire-adapted but impoverished forest vegetation was vulnerable to the anthropogenically-favoured invasion by *Castanea sativa*, as suggested by its amazingly rapid and successful expansion 1800 years ago (Fig. 1b). On the basis of the paleoecological results, it has also been hypothesised (Carraro et al. 1999) that the current situation, with vastly reduced abundances of evergreen species (mainly *Abies alba*, but also *Ilex aquifolium* and *Hedera helix*) facilitates the continuing invasion by exotic and native evergreen species (e.g. palms, evergreen oaks, spice laurel; see also Klötzli et al. 1996).

In the northern Alps, the tree species forming the timberline is *Picea abies*. Above these spruce forests, a few relict stands of *Pinus cembra* have survived only in remote areas. A recent paleoecological study shows that *Pinus cembra* and *Abies alba* were displaced from timberline by human deforestation activities (through fire) and possibly by a climatic change to more oceanic conditions at around 3700 cal yr BP (Wick et al., in press). Küttel (1990) attributed the reduction in *Pinus cembra* and *Larix* forest stands north of the Alps to climatic cooling at around 5200 and 3500 cal. yr BP. However, isolated patches of *Pinus cembra* forests in the northern Alps (e.g. near Frutigen, Swiss Alps, at 2000 m a.s.l.) occur on landslide boulder fields and these show that the species is still able to form considerable stands above the *Picea abies* belt, particularly where accessibility is poor. The presence of these stands favours the interpretation that human impact was also important for the enduring reduction in *Pinus cembra* forests.

2.2.4 Summary

On the basis of these few examples, it is difficult to conclusively answer the question whether some communities are more vulnerable than others to climatic change and human impact. In the Alps, most disturbances since ca. 6000 cal. yr BP were of human origin, and there are clues that fire was the decisive tool. In this context, it is intriguing that the most vulnerable species was *Abies alba*. Among all tree species, it had the largest area losses in the western Alps during the past 6500 years. The species is very fire-sensitive (Tinner et al. 2000) and subject to intensive browsing during winter. It was also the most highly prized timber for construction (Küster 1994). However, other fire-sensitive taxa (e.g. *Tilia, Ulmus, Fraxinus*) were also heavily reduced, causing the displacement of entire communities. Conversely, the most resilient communities were situated in the longitudinal inner-Alpine valleys (*Pinus sylvestris-Quercus-Betula* forests, *Pinus cembra-Larix-Betula* forests). Because of the dry climatic regime, these continental communities are naturally more fire-prone than their oceanic counterparts north and south of the Alps. Some of the dominant taxa (e.g. *Pinus sylvestris* and *Quercus*) have evolved strategies for surviving fire injuries. Thus, in addition to the persistence of a continental climate in the inner valleys of the Alps (Tinner and Ammann 2001), the resilience to burning could have contributed to maintain these communities for thousands of years.

3. Priorities for the future

At 6500 yr ago the climate in the European Alps was probably 1.5°C warmer than today, and at most sites, the vegetation was rather different from modern plant communities. Most of the differences are better explained by anthropogenic influence during various intervals than by climatic change. Human activities generated low-competition niches in vegetation. Due to such niches, it seems that the modern vegetation in the Alps is partly in disequilibrium with climate. At the same time, marked societal and land-use changes occurred during the last 50 years, so that forest vegetation is no longer in equilibrium with human activities (e.g. Conedera et al. 1996). Thus, even under constant climatic conditions, large-scale replacement processes would probably occur during the next decades.

The late-glacial and Holocene records (e.g. Ammann et al. 2000; Tinner and Lotter 2001) suggest that large-scale vegetation changes in the Alps occurred when temperature variations exceeded 1.5-2°C. Temperature changes were often accompanied by precipitation variations and by strong and abrupt responses of vegetation. It is hence likely that, if the forecasted temperature increase should exceed 2-3°C, vegetation will respond rapidly, i.e. within a few decades. The modern situation, with vegetation neither in equilibrium with climate nor with human land-use, may amplify and complicate the compositional and structural adjustment processes to future climate change. To better address this issue, we need additional efforts in paleoecology and related fields, including the following:

1) To avoid circular arguments, we need more independent (if possible non-biotic) records of climatic change (e.g. oxygen isotopes). Such investigations should feature a high resolution (5-20 yr per sample) in order to be used for comparison with vegetational and ecological proxies. Some of the new results discussed in our contribution are based on such a procedure (e.g. invasion and expansion of *Fagus* into the forelands of the Northern Alps).
2) The existing pollen information should be complemented by other biotic proxies (e.g. macrofossils, chironomids, diatoms, and ostracods). Such refinements are indispensable for understanding different aspects of long-term ecological processes. The timberline studies discussed above are an example for such a multi-proxy approach (pollen, macrofossils, chironomids).
3) The prominent role of fire in vegetation history and ecology in the Alps has long been neglected by paleoecologists and ecologists. The omission of this important disturbance agent could lead to serious errors in the assessment of vegetation records. For our understanding of ecological processes, but also for conservation and management of endangered vegetation types and species we need more studies of fire ecology and the role of fire in paleoecology.
4) Currently, there are just a few high-resolution studies (5-20 yr/sample) for the Holocene and the late-glacial period in the Alps. Such records are indispensable for addressing ecological questions such as succession and vegetation dynamics on longer time scales.
5) Generalization should be used with caution in paleoecology and ecology. For instance, pollen threshold values may be useful to trace population expansions, but not necessarily the arrival or presence of the species. In ecology, the neglect of long-term forest-dynamics may lead to wrong assumptions, such as the identification of highly anthropogenically transformed communities as natural stands (see discussion in Gobet et al. 2000).
6) More comparative studies that involve both paleoecology and vegetation modelling could provide new insights into the mechanisms of complex ecological processes, such as invasions and displacements. Considering the high relevance of human impact for explaining today's vegetation distribution in the Alps, the models should include important (anthropogenic) disturbance factors such as fire. It is most likely that factors related to land-use (browsing, wildfire, wood cutting for energy and construction purposes) will largely determine the development of future forests in the Alps.
7) New national parks are planned for the Alps. Conservation in these parks should be based on paleoecological and ecological research. In anthropogenically altered ecosystems, such as the Alps, one major contribution of paleoecology is the assessment of human impact on vegetation since prehistoric times. Paleoecology can help to assess the potential natural vegetation of a region and help to identify vegetation types that are dependent on land use. In protected habitats such as national parks, the exclusion of fire, wood cutting and browsing would probably allow, at least at some sites in the Alps, the reestablishment of vegetation types

similar to those at around 6500 cal yr BP. But as shown by the example of the inner-Alpine valleys, an exclusion of fire may not be advisable in every case. For instance in the fire-prone *Pinus* belts, the suppression of fire could be problematic (increased fuel loading associated with higher risk of intense fires) and possibly induce a change in forest type (e.g. through an expansion of more fire-sensitive species).

4. Acknowledgements

We are very grateful to H. Bugmann, U. Huber, and M. Reasoner for many valuable suggestions on the manuscript.

5. References

Aber, J. D., and Melillo, J. M. (1991). "Terrestrial Ecosystems." Saunders College Publishing, Philadelphia.

Ammann, B. (1988). Palynological evidence of prehistoric anthropogenic forest changes on the Swiss Plateau. *In* "The cultural landscape: Past, present and future." (H. H. Birks, H. J. B. Birks, P. E. Kaland, and D. Moe, Eds.), pp. 289-299. Cambridge University Press, Cambridge.

Ammann, B. (1989). Late-Quaternary palynology at Lobsigensee. Regional vegetation history and local lake development. *Dissertationes Botanicae* **137**, 1-157.

Ammann, B., Birks, H. J. B., Drescher-Schneider, R., Juggins, S., Lang, G., and Lotter, A. F. (1993). Patterns of variation in late-glacial pollen stratigraphy along a North-West - South-East transect through Switzerland: A numerical analysis. *Quaternary Science Reviews* **12**, 277-286.

Ammann, B., Eicher, U., Gaillard, M.-J., Haeberli, W., Lister, G., Lotter, A. F., Maisch, M., Niessen, F., Schlüchter, C., and Wohlfarth, B. (1994). The Würmian Late-glacial in lowland Switzerland. *Journal of Quaternary Science* **9**, 119-125.

Ammann, B., Birks, H. J. B., Brooks, S. J., Eicher, U., von Grafenstein, U., Hofmann, W., Lemdahl, G., Schwander, J., Tobolski, K., and Wick, L. (2000). Quantification of biotic responses to rapid climatic changes around the Younger Dryas: A synthesis. *Palaeogeography, Palaeoclimatology, Palaeoecology* **159**, 313-347.

Bennett, K. D. (1983). Postglacial population expansion of forest trees in Norfolk, UK. *Nature* **303**, 164-167.

Bugmann, H., and Pfister, C. (2000). Impacts of interannual climate variability on past and future forest composition. *Regional Environmental Change* **1**, 112-125.

Burga, C. A., and Perret, R. (1998). "Vegetation und Klima der Schweiz seit dem jüngeren Eiszeitalter." Ott Verlag, Thun.

Caldararo, N. (2002). Human ecological intervention and the role of forest fires in human ecology. *Science of the Total Environment* **292**, 141-165.

Carraro, G., Klötzli, F., Walther, G.-R., Gianoni, P., and Mossi, R. (1999). "Observed changes in vegetation in relation to climate warming." vdf, Hochschulverlag AG ETH Zürich, Zürich.

Clark, J. S., Merkt, J., and Müller, H. (1989). Post-glacial fire, vegetation, and human history on the northern alpine forelands, south-western Germany. *Journal of Ecology* **77**, 897-925.

Conedera, M., Marcozzi, M., Jud, B., Mandallaz, D., Chatelain, F., Frank, C., Kienast, F., Ambrosetti, P., and Corti, G. (1996). "Incendi boschivi al Sud delle Alpi: Passato, presente e possibili sviluppi futuri." vdf, Hochschulverlag ETH Zürich, Zürich.

Ellenberg, H. (1996). "Vegetation Mitteleuropas mit den Alpen in ökologischer Sicht." Ulmer, Stuttgart.

Gobet, E., Tinner, W., Hubschmid, P., Jansen, I., Wehrli, M., Ammann, B., and Wick, L. (2000). Influence of human impact and bedrock differences on the vegetational history of the Insubrian Southern Alps. *Vegetation History and Archaeobotany* **9**, 175-178.

Haas, J. N. (1996). Pollen and plant macrofossil evidence of vegetation change at Wallisellen-Langachemoos (Switzerland) during the Mesolithic-Neolithic transition 8500 to 6500 years ago.

Dissertationes Botanicae **267**, 1-67.

Heiri, O. (2001). "Holocene palaeolimnology of Swiss mountain lakes reconstructed using subfossil chironomid remains: Past climate and prehistoric human impact on lake ecosystems." Unpublished PhD thesis, University of Bern, Bern.

Kaltenrieder, P. (1999). "Lokale Vegetationsgeschichte und Holozäne Schwankungen der oberen Waldgrenze: Eine makrorestanalytische Untersuchung am Gouillé Rion (2343 m, VS)." Unpublished Master thesis, University of Bern, Bern.

Keller, F., Lischke, H., Mathis, T., Mohl, A., Wick, L., Ammann, B., and Kienast, F. (2002). Effects of climate, fire, and humans on forest dynamics: Forest simulations compared to the palaeological record. *Ecological Modelling* **152**, 109-127.

Klötzli, F., Walther, G.-R., Carraro, G., and Grundmann, A. (1996). Anlaufender Biomwandel in Insubrien. *Verhandlungen der Gesellschaft für Ökologie* **26**, 537-550.

Küster, H. (1994). The economic use of Abies wood as timber in central Europe during Roman times. *Vegetation History and Archaeobotany* **3**, 25-32.

Küttel, M. (1990). Der subalpine Schutzwald im Urserental: Ein inelastisches Ökosystem. *Botanica Helvetica* 100, 183-197.

Lang, G., and Tobolski, K. (1985). Hobschensee: Late-Glacial and Holocene environment of a lake near the timberline. *Dissertationes Botanicae* **87**, 209-228.

Lotter, A. F., Birks, H. J. B., Hofmann, W., and Marchetto, A. (1997). Modern diatom, cladocera, chironomid, and chrysophyte cyst assemblages as quantitative indicators for the reconstruction of past environmental conditions in the Alps. I. Climate. *Journal of Paleolimnology* **18**, 395-420.

Markgraf, V. (1969). Moorkundliche und vegetationsgeschichtliche Untersuchungen an einem Moorsee an der Waldgrenze im Wallis. *Botanische Jahrbücher* **89**, 1-63.

Ravazzi, C. (2002). Late Quaternary history of spruce in southern Europe. *Review of Palaeobotany and Palynology* **120**, 131-177.

Richoz, I. (1998). Etude paléoécologique du lac de Seedorf (Fribourg, Suisse). Histoire de la végétation et du milieu durant l'Holocène: le rôle de l'homme et du climat. *Dissertationes Botanicae* **293**, 1-177.

Tinner, W., Ammann, B., and Germann, P. (1996). Treeline fluctuations recorded for 12,500 years by soil profiles, pollen, and plant macrofossils in the central Swiss Alps. *Arctic and Alpine Research* **28**, 131-147.

Tinner, W., Hubschmid, P., Wehrli, M., Ammann, B., and Conedera, M. (1999). Long-term forest fire ecology and dynamics in southern Switzerland. *Journal of Ecology* **87**, 273-289.

Tinner, W., Conedera, M., Gobet, E., Hubschmid, P., Wehrli, M., and Ammann, B. (2000). A palaeoecological attempt to classify fire sensitivity of trees in the southern Alps. *The Holocene* **10**, 565-574.

Tinner, W., and Ammann, B. (2001). Timberline paleoecology in the Alps. *PAGES News* **9**, 9-11.

Tinner, W., and Lotter, A. F. (2001). Central European vegetation response to abrupt climate change at 8.2 ka. *Geology* **29**, 551-554.

Tinner, W., and Theurillat, J.-P. (in press). Uppermost limit, extent, and fluctuations of the timberline and treeline ecocline in the Swiss Central Alps during the past 11,500 years. *Arctic, Antarctic, and Alpine Research.*

Wegmüller, S., and Lotter, A. F. (1990). Palynostratigraphische Untersuchungen zur spät- und postglazialen Vegetationsgeschichte der nordwestlichen Kalkvoralpen. *Botanica Helvetica* **100**, 37-73.

Welten, M. (1982). Vegetationsgeschichtliche Untersuchungen in den westlichen Schweizer Alpen: Bern-Wallis. *Denkschriften Schweizerische Naturforschende Gesellschaft* **95**, 1-104.

Welten, M. (1944). Pollenanalytische, stratigraphische und geochronologische Untersuchungen aus dem Faulenseemoos bei Spiez. *Veröffentlichungen Geobotanisches Institut Rübel Zürich* **21**, 1-201.

Wick Olatunbosi, L. (1996). "Spät- und postglaziale Vegetationsgeschichte in den Südalpen zwischen Comersee und Splügenpass (Norditalien)." Inaugural dissertation University of Bern, Bern.

Wick, L., van Leeuwen, J. N. F., van der Knaap, W. O., and Lotter, A. F. (in press). Holocene vegetation development in the catchment of Sägistalsee (1935 m asl), a small lake in the Swiss Alps. *Journal of Palaeolimnology*.

Climate Fluctuations Derived from Tree-rings and Other Proxy-records in the Chilean Andes: State of the Art and Future Prospects

Antonio Lara[1*], Alexia Wolodarsky-Franke[1], Juan Carlos Aravena[2], Ricardo Villalba[3], Maria Eugenia Solari[4], Liliana Pezoa[1], Andrés Rivera[5], and Carlos Le Quesne[1]

[1]*Instituto de Silvicultura, Facultad de Ciencias Forestales, Universidad Austral de Chile, Casilla 567, Valdivia, Chile*

[2]*Centro de Estudios Cuaternarios, Universidad de Magallanes - Department of Geography, University of Western Ontario*

[3]*Departemento de Dendrocronología e Historia Ambiental, IANIGLA, Mendoza, Argentina*

[4]*Instituto de Ciencias Sociales, Universidad Austral de Chile*

[5]*Departamento de Geografía, Universidad de Chile - School of Geographical Sciences, University of Bristol*

**phone ++56 63 221228, 226302, fax ++56 63 221230, email alara@uach.cl*

Keywords: Andes, Climate fluctuations, Glacier fluctuations, Patagonia, Proxy-records, Tree-ring

1. Introduction

Treeline and high elevation sites in the central and southern Chilean Andes (32°39' to 55°S) have shown to be an excellent source of paleoenvironmental records because their physical and biological systems are highly sensitive to climatic and environmental variations. In addition, most of these sites have been less disturbed by logging and other human induced disturbances, which enhances the climatic signals present in the proxy records (Luckman 1990; Villalba et al. 1997).

Current studies on tree-ring, glacier and documentary records in the Chilean Andes have led to important progress in the understanding of climate fluctuations in the last 1000 years against which current changes can be assessed. The integration of

U. M. Huber et al. (eds.), Global Change and Mountain Regions, 145-156.

these proxies provides an opportunity to study climatic signals across a wide range of spatial and temporal scales. In this contribution, we discuss the state of the art in the reconstruction of climate variability from tree-rings and other proxy records in three main regions along the Chilean Andes: the Central Andes (32°40'-39°S), Northern Patagonia (39°-48°S) and Southern Patagonia (48°-55°S). We also discuss future needs and challenges for global change research in the Chilean mountain ranges.

2. Research in Chilean mountain environments: Progress and gaps

Tree-ring studies at high elevation sites have produced a network of tree-ring chronologies, covering a 2530 km latitudinal gradient in the Chilean Andes (Fig. 1). This research has rendered important progress in the understanding of climate and environmental fluctuations on inter-annual to century timescales and of the potential mechanisms behind this variability.

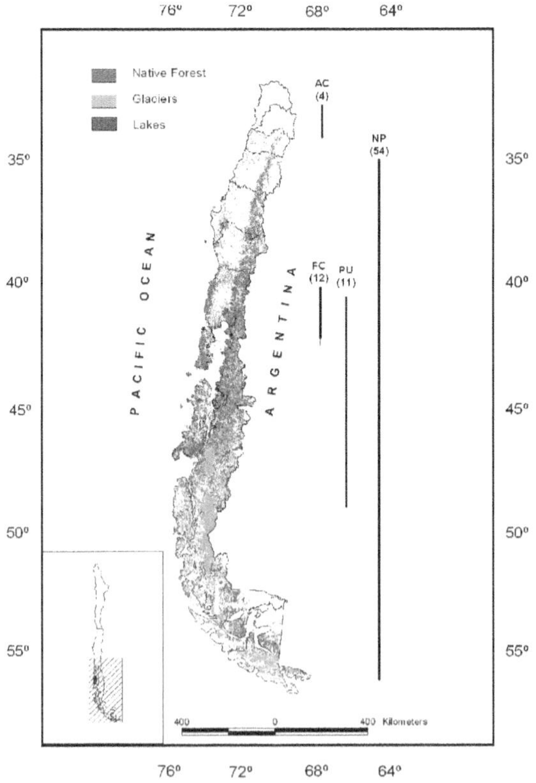

Figure 1: Map showing the latitudinal distribution of tree-ring chronologies for different native species. AC: *Austrocedrus chilensis*; NP: *Nothofagus pumilio*; FC: *Fitzroya cupressoides*; PU: *Pilgerodendron uvifera*. The numbers in brackets indicate the number of existing chronologies for each species in Chile.

2.1 Central Andes (32°40'- 39°S)

This region has a climate of Mediterranean type. Rain and snow are concentrated in fall and winter, when over 75% of the annual precipitation occurs, with a north to south increase in rainfall. There is also an important west-east moisture gradient, with a distinct rainshadow effect east of the highest Andean peaks. Interannual climate variability is mainly driven by El Niño-Southern Oscillation (ENSO) events and the changes in the intensity and latitudinal position of the Southeast Pacific anticyclone (Miller 1976).

In the northernmost portion of the Central Chilean Andes (32°40'-35°40'S), the long-lived conifer *Austrocedrus chilensis* has been used for dendroclimatic studies. These moisture-limited sites yielded a precipitation reconstruction for the last 1000 years. *Austrocedrus* growth is positively related to winter precipitation, and one of the main features of the reconstruction are two long periods of droughts, one between AD 1270-1450, and the other between AD 1600-1650 (LaMarche et al. 1979; Boninsegna 1988). Instrumental precipitation records between 33° and 36° S show no increasing or decreasing trend in precipitation (Pezoa 2003) for the 1931-2001 period. Interannual precipitation variability is high with a variation coefficient ranging between 41% and 43% for most stations (Pezoa 2003). Although temperature reconstructions from tree-rings have not been developed for this region, instrumental records between 33° and 36° S show a steady increase in mean annual temperatures for the period 1965-2001 (Pezoa 2003).

Dendroclimatic research has also been conducted in the northernmost portion of the deciduous *Nothofagus pumilio* forests at upper treeline and high elevation sites (1500-1720 m asl), ranging from 35°37' to 37°30' S. Here, eight tree-ring chronologies have been developed. Results indicate a positive correlation of tree-ring width with late-spring and early-summer precipitation and a negative correlation with temperature (Lara et al. 2001). In this region, high temperatures in spring and summer, which enhance evapotranspiration and decrease water availability, appear to reduce radial growth. In contrast, at wetter high elevation sites located further south (38°37' S), radial growth is negatively correlated with late-spring and early-summer precipitation. From this set of chronologies, a reconstruction of November-December precipitation, accounting for 37% of the instrumentally recorded precipitation variance, was produced (Fig. 2a). This reconstruction shows that the twentieth century has the most extreme intervals of both summer drought and wetness since 1837, and that the driest and wettest 25-year periods are 1890-1914 and 1917-1941, respectively (Fig. 2a; Lara et al. 2001).

Progress has also been made in the inventory of glaciers and the analysis of their fluctuations in relation to regional climate change (Rivera et al. 2000). Studies based on the interpretation of aerial photographs show a general glacier retreat in the Central Andes (Rivera et al. 2000). A local study at Los Cipreses Glacier (34° 33') based on historical documents, sketches and maps shows an average retreat of 10 m/year between 1858-1888, which increases to 30 m/year for 1888-1968 (Le Quesne and Acuña 2003).

Despite the progress in tree-ring studies, there is an important geographic gap in the area north of the Central Andes, ranging from 17°40' to 32°40' S. This gap includes the Chilean *Altiplano,* from 19° to 22° S (the high Andean plateau shared between Perú, Bolivia, Argentina, and Chile), and the region located between 22° and 32° 40' S. Filling this gap is a priority, and currently we are working on the development of tree-ring chronologies of *Polylepis tarapacana*, collected from 20 sites located at 4151-4781 m asl, along its entire latitudinal range in Chile (17°40'-21°20'). Research in the adjacent Bolivian and Argentinean Altiplano has rendered four chronologies, the longest starting in AD 1297 (Argollo et al. 2003). The chronologies located between 18° and 22° S show a strong positive correlation with summer precipitation (Soliz et al. 2003; Argollo et al. in preparation). Further south, between 21° and 33° S, the potential for dendroclimatological studies of *Prosopis spp.* and other species growing in scattered populations through a vast desert region should be investigated to fill the remaining gap spanning 1400 km between the *Polylepis* and the *Austrocedrus* populations in Chile. Sample collection of *Prosopis chilensis* at various sites in northern Chile started in 2003. A tree-ring chronology of *Prosopis ferox* has been developed in northwestern Argentina, indicating high dendroclimatic potential for this species distributed between 20° and 25° S in Bolivia and Argentina, reaching 240-280 km west of the border with Chile (Morales et al. 2001). This chronology, developed for a site at 3500 m asl, has a positive correlation with precipitation and a negative correlation with temperature, interpreted as a positive response to soil water availability.

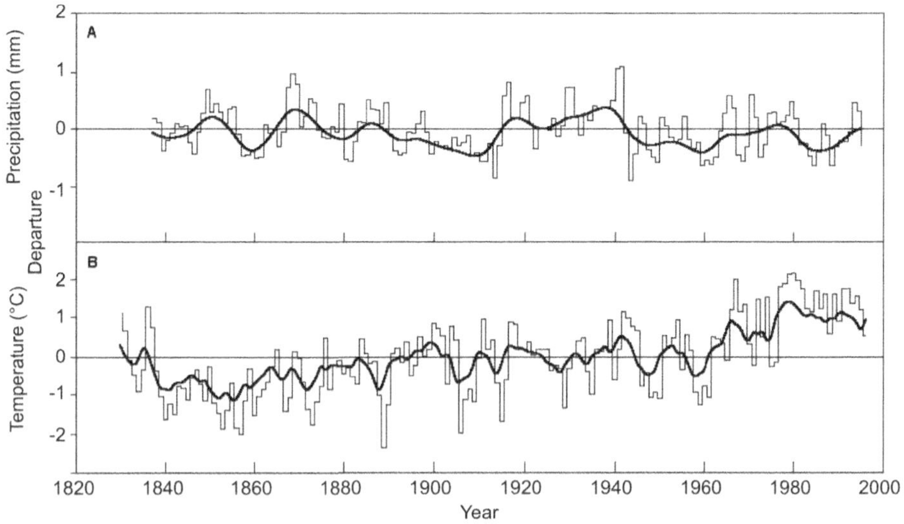

Figure 2: A) Tree-ring based reconstruction of November-December precipitation for the Central Andes of Chile (35°-39°S) from 1837 to 1995. A trend line drawn to emphasize the low-frequency variation was obtained using a cubic spline (Cook and Peters 1981). B) Tree-ring based reconstruction of minimum annual temperature for Chilean Southern Patagonia from 1829 to 1996. A trend line drawn to emphasize the low-frequency variation was obtained using an exponential filter (Essenwanger 1986).

2.2 Northern Patagonia (39°-48° S)

South of 39°S, strong westerlies are remarkably persistent throughout the year, occurring at least 75% of the time along the entire coast. Temperature patterns are strongly influenced by latitude and elevation. The increasing influence of the westerlies is reflected in more abundant rainfall and a reduced summer season towards the south. Precipitation over the Andes increases with elevation, reaching a maximum close to the crest of the range. Precipitation is closely related to the north-south seasonal shifts of the Southeastern Pacific anticyclone. In northern Patagonia, warm El Niño events are generally associated with cooler/wetter winter-springs and warmer/drier summers, whereas most La Niña events correspond to warmer/drier winter-springs and cooler/wetter summers (Villalba et al. 2003).

Dendroclimatological research has been quite extensive in this region, using *Fitzroya cupressoides* and *Nothofagus pumilio,* and to a lesser extent *Pilgerodendron uvifera*. These studies have produced a network of 23 tree-ring chronologies for *Fitzroya* (40°10' to 43°30' S), both in Chile and Argentina, 19 of which are >1000-year long (Lara et al. 2000), and recently a 5666-year long chronology has been generated (Wolodarsky-Franke 2002). Based on the significant negative correlation between *Fitzroya* tree-ring widths and previous summer (December to March) mean temperatures, a 3622-year temperature reconstruction was developed (Lara and Villalba 1993). The longest period of slow growth, and therefore above-mean reconstructed temperatures, in this chronology was from 80 BC to AD 160. Current research is analyzing this period in more detail, using tree-ring records from other sites, as well as lake sediments and pollen records for a multi-proxy approach.

Growth patterns and climatic response of *Pilgerodendron uvifera*, reconstructed from a network of nine chronologies situated between 39°36' and 49°15' S, have been studied (Szeics et al. 2000). Sites located in the southern portion of this range show a strong negative response to summer temperatures. Conversely, a positive response to summer and annual temperatures is reported for two sites at or near treeline in the Coastal Archipelagoes (46°10', 700 m asl). A marked increase in tree-growth since the mid-20th century, attributed to an increase in mean annual temperature in this period, is reported (Szeics et al. 2000). However, none of the other *Pilgerodendron* sites show evidence of a warming trend during the 20th century, following a similar pattern as the one described for the 3,622-year temperature reconstruction from *Fitzroya* tree-rings (Lara and Villalba 1993).

Recent studies of instrumental records in the Southern Andes of Argentina and Chile report a negative trend in mean annual temperature with a marked cooling period from 1950 to 1975 in the region between 37° and 43° S (Rosenblüth et al. 1997; Pezoa 2003; Villalba et al. 2003). This is described as one of the main regional patterns in the Southern Andes from 37°-56° S (Villalba et al. 2003). The second dominant pattern, which followed this cooling trend, is a pronounced and widespread increase of temperatures starting in 1976 in both the Chilean and Argentinean North Patagonian Andes (Villalba et al. 2003).

Temperature reconstructions from composite tree-ring chronologies across Northern and Southern Patagonia in both Chile and Argentina, using standardization methods to preserve low-frequency (decadal to centennial) variations have been recently accomplished (Villalba et al. 2003). This study indicates that temperatures during the 20th century have been anomalously warm across northern and southern Patagonia in both Argentina and Chile (37°-55° S), compared to the reconstructed temperatures for the past 360 years (Villalba et al. 2003). These reconstructions show a well-defined cold period from 1640-1850, which is roughly synchronous with the "Little Ice Age" in both Hemispheres (Villalba et al. 2003). The warming trend during the 20th century is distinguishable, despite the cooling trend from 1950-1975, observed between 37°-43° S. The increase in temperature since 1976 is most evident in summer and may be associated with a shift in the Pacific Decadal Oscillation (PDO) at that time (Graham 1994). Sea Surface Temperature (SST) anomalies associated with this decadal mode reached into mid-latitudes in both hemispheres. Long-term dominant modes of atmospheric circulation in the Southern Hemisphere are connected to global changes in SST (Mo 2000). The 20th century increase in temperatures for the Southern Andes is the local atmospheric response to large-scale ocean-atmosphere changes (Villalba et al. 2003).

Most glaciers are retreating in Northern Patagonia and the reduction of ice-masses that have accumulated over preceding centuries has continued (or accelerated) to the present. Studies based on the interpretation of air photos and satellite images indicate that land-based glaciers and most calving glaciers in the Northern and Southern Patagonian Icefields have retreated since 1945, in some cases at an increasing rate (Naruse and Aniya 1992; Cassasa et al. 2002; Rivera et al. 2000; 2002).

Climatic reconstructions using historical records are in progress for the area located between 39° to 41° S. These records include documents, maps and sketches prepared by travelers and surveyors. Data are more abundant after 1850 and the first photographs were taken in 1901. These records also show a steady and distinct glacier retreat through present. A general glacial retreat in Northern Chilean Patagonia is probably the result not only of increased temperatures since 1850, but also of a regional decrease in precipitation registered at least since 1931 (Rosenblüth et al. 1995). The steady precipitation decrease in the 1931-2001 period, recorded at most of the weather stations between 37° and 45° S, is the dominant regional precipitation pattern (Pezoa 2003).

2.3 Southern Patagonia (48° to 55° S)

The climate of southern Chilean Patagonia is characterized by a dramatic decrease in precipitation and humidity from west to east. In southernmost Patagonia, this gradient shifts from a west to east orientation to a south-west to north-east pattern, following the dominant position of the Andean range (Aravena et al. 2002). Annual rainfall ranges from 8000 mm in the western archipelagoes to 250 mm and less on the plateau east of the Andes. Precipitation is mainly affected by the influence of the strong and persistent westerly winds, which dominate the whole region throughout the year, and there is almost no seasonality in precipitation distribution (Carrasco et al. 1998).

Nothofagus pumilio tree-ring growth studied at 21 high elevation and upper treeline sites (500-980 m asl) in southern Chilean Patagonia shows a significant positive correlation with summer temperature (December-January) of the current growing season at most sites (Aravena et al. 2002). Some sites show a significant positive correlation between tree growth and fall-early winter temperature of the previous season. In addition to this temperature response, most chronologies located towards the southern limit of *N. pumilio* on Navarino Island (55° S) have a negative correlation with precipitation (Aravena et al. 2002). Tree-growth may be negatively affected by more extended snowfall seasons and snow accumulation through the spring, associated with higher precipitation years at these relatively cooler and wetter southern sites. This effect has also been described for the Argentinean portion of Northern Patagonia at 41° S (Villalba et al. 1997).

The significant correlation between *N. pumilio* tree-ring widths and temperature allowed the reconstruction of minimum annual temperature fluctuations since 1829 (Fig. 2b; Aravena et al. 2002). During most of the 19th century, minimum annual temperatures remained below average and increased to values fluctuating around the mean during the period 1900-1960, followed by a clear trend towards above-average values after 1963 (Fig. 2b). This warming trend since 1963 coincides with the patterns described from instrumental records for the extreme south of South America (45°-55° S), but contrasts with the cooling trend between 1950 and 1975 for Northern Patagonia (37°-43° S), described above (Rosenblüth et al. 1997; Villalba et al. 2003). Precipitation patterns from the only two available weather stations in Southern Chilean Patagonia show a steep increase since 1983 for Faro Evangelistas, located at 52°24' S at the western fringe of the Archipelagoes, and a slighter increase for Punta Arenas (53° S) between 1983 and 2001, following two decades of below-mean precipitation (Carrasco et al. 2002; Pezoa 2003). This recent precipitation increase is in contrast to the dominant pattern described for Northern Patagonia.

Glaciers in the Southern Patagonian Icefield show a substantial and rapid retreat, following the pattern described for the Central and North Patagonian Andes (Naruse and Aniya 1992; Cassasa et al 2002; Rivera et al. 2002). Pío XI Glacier is an exception to this general trend, showing a distinct advance due to local ice dynamic factors (Rivera et al. 1997). Current estimates indicate that the general retreat observed in the Southern Patagonian Icefield accounts for 6% of the global sea level rise (1-2 mm/year during the last 100 years; Rivera et al. 2002).

Historical documents for the Magellan region (50°-56° S) date to the time of its discovery by Europeans in 1520 and historical sources exist since then (Prieto and Herrera 1998). These data indicate a cold period between 1520 and 1670, synchronous with a cold interval identified by tree ring records (Villalba 1994). Photographs taken by Agostini (mainly in Southern Patagonia in 1923, 1945, 1949 and 1955) show a general retreat of glaciers during the 20th century (Solari et al. 2003).

3. Future needs and challenges for global change research in the Chilean mountain ranges

Research in the Chilean Andes (latitudes 32°40'-55° S), using tree-rings and other climate proxy records, has demonstrated a high potential for understanding the environmental and climate variability of the last millennia in this region. Instrumental records, as well as proxy-records from tree-rings, glaciers and historical documents, show a consistent increase in temperature during the 20th century, compared to the previous 360 years. This warming trend documented for the Chilean and Argentinean Andes (37°-55° S) is a response to large-scale ocean-atmosphere changes expressed in increasing Sea Surface Temperature (SST) of the Southern Pacific and Atlantic Oceans (Villalba et al. 2003; Villalba et al., this volume). Further research on the relationships between SST and atmospheric temperatures over southern South America will improve our knowledge of global change mechanisms and responses in this region.

A warming trend in the Central Andes since 1858 is evident from glacier records and is a dominant pattern at least since 1965 based on instrumental records located between 33° and 36° S. This 20th century warming trend is also clear in Southern Patagonia, as indicated by tree-ring, glacier and instrumental records. Conversely, in Northern Patagonia a cool period between 1950 and 1975 is a dominant feature, and some tree-ring records do not show a temperature increase in recent decades. Nevertheless, glaciers show a strong retreat in Northern Patagonia during the 20th century. Glacier records give a good estimate of overall temperature and/or precipitation changes from one century to the next (Luckman and Villalba 2001). In contrast, tree-rings generally show a relatively strong inter-annual climatic signal, as well as a decadal or centennial signal of variable strength, depending on species, site, length of the individual tree-ring series and standardization methods (Briffa et al. 1996). These differences between climatic responses probably explain the discrepancies between the glacier records and some of the tree-ring records in Northern Patagonia. In Southern Patagonia, the scarcity of instrumental records between 46° to 55° S, in a vast and climatically highly variable area, is a limitation for understanding temperature and precipitation patterns.

Future research should address these and other limitations to improve our knowledge of the long-term spatial and temporal patterns of climatic variability on both the western and the eastern slopes of the Andes. A main objective of future research in mountain areas in Chile should be to continue the development of millennial tree-ring chronologies of long-lived native species, such as *Fitzroya cupressoides, Austrocedrus chilensis, Pilgerodendron uvifera,* and *Araucaria araucana,* to detect decadal to centennial signals in climatic variations and to distinguish between natural and human-induced climatic changes. Since *Pilgerodendron uvifera* is the only conifer growing between 43°30' and 54° S, which shows both a temperature and precipitation signal, future research on this species seems promising. Streamflow reconstructions in Northern Patagonia using *Pilgerodendron* tree-rings are in progress (Urrutia et al. 2003). These types of studies are needed for a better understanding of the hydrological response to the observed precipitation decrease in Northern Patagonia and the widespread temperature increase throughout the Andes. Reconstructing long-term streamflow as a basis for predicting future trends and their possible economic impacts should receive high priority, since water availability is a key factor for future development throughout the Andes. Hydroelectricity, fish farms, sports fishing and tourism are major economic activities, which strongly depend on water availability (Urrutia et al. 2003).

The development of tree-ring chronologies for the Northern Chilean Andes (17°40'-32°40'), which has already been started for *Polylepis tarapacana* and *Prosopis chilensis*, should include new species and sites and must also be regarded as a high research priority. These studies will document precipitation patterns in this vast subtropical area, providing critical information for a better understanding of ENSO, the influence of the Easterlies, and changes in SST of the Atlantic Ocean.

Since fire is one of the main disturbances throughout the Chilean Andes, research on the links between fire regimes and inter-annual and decadal scale climate variability should be stressed, enhancing and broadening the studies already completed or in progress in *Austrocedrus*, *Fitzroya* and *Araucaria* forests (Lara et al. 1999; 2003; González 2002; Aravena et al. 2003). A network of fire histories would provide a better understanding of the effects of climate variability and seasonality on fire regimes in different ecosystems along the major environmental gradients. The study of the climatic and ecological effects of volcanism reconstructed from tree-rings is also a relevant topic. Research on the increase in interannual climate variability and its influence on forest dynamics (e.g. tree mortality, seedling establishment, fires ignited by increased lightning occurrence) should be developed in Chile, complementing research in Northern Argentinean Patagonia in *Austrocedrus* and *Nothofagus dombeyi* habitats (Villalba et al., this volume).

Studies on the integration of tree-ring and stratigraphic records (pollen, charcoal and tephras from lake sediments and peat bogs) to decipher the patterns, rates and directions of changes in vegetation, climate and fire regimes since the Last Glacial Maximum (c. 17,000 years B.P.) have started. A recent study integrates high-resolution pollen and charcoal records from small lakes with *Pilgerodendron* tree-rings in the Chonos Archipelago (44°20' S, Szeicz et al. 2003). This reconstruction of the impacts of climate change and fire on vegetation shows promising results. Research using *Fitzroya* tree-rings from sub-fossil wood yielded a floating chronology that was radiocarbon dated to 50,000 ^{14}C yr B.P. (Roig et al. 2000), indicating a potential for the integration of records into the Pleistocene. Multi-proxy approaches require a considerable amount of effort, collaboration and funding, but provide a unique opportunity to improve the understanding of the long-term spatial and temporal patterns of climate, fire and volcanism, and therefore should be given a high priority.

Collaboration between researchers along the South American Andes, including Perú, Bolivia, Argentina, and Chile, as well as training of young scholars, is crucial to make effective progress in the study of climate change in the region. An ongoing collaborative research initiative on climate variability in the Americas from treeline environments (CRN03 project of the Inter American Institute for Global Change Research, IAI) is making a significant contribution to our understanding of large-scale climate change mechanisms along a transect spanning from Alaska to Tierra del Fuego. Such initiatives should be broadened in geographic range and their long-term continuation should be assured. High quality datasets from paleo-climate records covering a wide geographic area in the Americas, developed through a focused and effective collaboration, can be used to validate Global Climate Models (GCMs). Such studies have the potential to achieve major breakthroughs in the improvement of the resolution and quality of GCMs, as well as in the understanding of global change patterns and mechanisms. The improved predictive capacity of climate models is relevant to the planning of natural resource management as well as for policy making.

4. Acknowledgements

This review is largely based on research supported by FONDECYT Projects 1000445, 1010200, and 1030766, the CRN03 project of the Inter American Institute for Global Change Research IAI, and Mideplan, through its Iniciativa Cientifica Milenio (ICM). We thank Museo Histórico-Antropológico U. Austral (Valdivia), Museo Salesiano (Punta Arenas), Club Andino Alemán (Santiago) and the Ministerio de Obras Públicas (Santiago) for providing historical and photographic records. Dirección Meteorológica de Chile contributed with the temperature and precipitation records. E. Neira and P. Romero drew the figures. A. Lara acknowledges support from a Bullard Fellowship from the Harvard Forest (Harvard University). We are grateful to the editors for the invitation to contribute with this chapter and for their careful review of earlier versions of our paper.

5. References

Aravena, J. C., Lara, A., Wolodarsky-Franke, A., Villalba, R., and Cuq, E. (2002). Tree-ring growth patterns and temperature reconstruction from *Nothofagus pumilio* (Fagaceae) forests at the upper tree line of southern Chilean Patagonia. *Revista Chilena de Historia Natural* **75**, 361-376.

Aravena, J.C., LeQuesne, C., Jiménez, H., Lara, A., and Armesto, J. J. (2003). Fire history in central Chile: tree ring and modern records. *In* "Fire and climatic change in temperate ecosystems of the Western Americas." (T. T. Veblen, T. Swetnam, and G. Montenegro, Eds.), pp. 343-356. Springer, New York.

Argollo, J., Soliz, C., Villalba, R., and Montevilla, J. (2003). Dendrochronology and dendroclimatology in Bolivia. *In* "Proceedings of the Fourth Annual Science Meeting, IAI CRN 03," Mendoza, Argentina October 10-16, p. 4.

Argollo, J., Soliz, C., and Villalba, R. (in preparation). Potencialidad dendrocronológica de *Polylepis tarapacana* en los Andes Centrales de Bolivia.

Boninsegna, J. A. (1988). Santiago de Chile winter rainfall since 1220 as being reconstructed by tree rings. *Quaternary of South America and Antarctic Peninsula* **6**, 67-87.

Briffa, K. R., Jones, P. D., Bartolin, T. S., Schweingruber, F. H., Karlen, W., and Shiyatov, S. G. (1996). Tree-ring variables as proxy-climate indicators: Problems with low-frequency signals. *In* "Climate variations and forcing mechanisms of the last 2000 years." (P. D. Jones, R. S. Bradley, and J. Jouzel, Eds.), pp. 9-41. NATO ASI Series, Vol. 141, Springer, Heidelberg.

Carrasco, J. F., Casassa, G., and Rivera, A. (1998). Climatología actual del Campo de Hielo Sur y posibles cambios por incremento del efecto invernadero. *Anales Instituto de la Patagonia, Serie Ciencias Naturales* **26**, 1109-1128.

Carrasco, J. F., Casassa, G., and Rivera, A. (2002). Meteorological and climatological aspects of the Southern Patagonia Icefield. *In* "The Patagonian Icefields: A unique natural laboratory for environmental and climate change studies." (G. Casassa, F. Sepúlveda, and R. Sinclair, Eds.), pp. 29-41. Series of the Centro de Estudios Científicos. Kluwer Academic/Plenum Publishers, Dortrecht.

Casassa, G., Rivera, A., Aniya, M., and Naruse, R. (2002). Current knowledge of the Southern Patagonia Icefield. *In* "The Patagonian Icefields: A unique natural laboratory for environmental and climate change studies." (G. Casassa, F. Sepúlveda, and R. Sinclair, Eds.), pp. 67-83. Series of the Centro de Estudios Científicos. Kluwer Academic/Plenum Publishers, Dortrecht.

Cook, E. R., and Peters, K. (1981). The smoothing spline: A new approach to standardizing forest interior ring-width series for dendroclimatic studies. *Tree-Ring Bulletin* **41**, 45-53.
Essenwanger, O. (1986). Elements of statistical analysis. World survey of climatology. Vol. 1B, Elsevier, Amsterdam.
González, M. E. (2002). "Fire history of *Araucaria-Nothofagus* forests in the Andean Cordillera of South-Central Chile." Ph.D. thesis, University of Colorado, Boulder.
Graham, N. E. (1994). Decadal-scale climate variability in the 1970s and 1980s: Observations and model results. *Climate Dynamics* **10**, 135-162.
La Marche, V., Holmes, R. L., Dunwiddie, P., and Drew, L. (1979). Tree-ring chronologies of the Southern Hemisphere. Vol. 2: Chile. Chronology Series V, Arizona University of Arizona.
Lara, A., and Villalba, R. (1993). A 3620-year temperture record from *Fitzroya cupressoides* tree rings in southern South America. *Science* **260**, 1104-1106.
Lara, A., Aravena, J. C., Fraver, S., and Wolodarsky-Franke, A. (1999). Fire and the dynamics of alerce *(Fitzroya cupressoides)* forests of Chile's Cordillera Pelada. *Ecosience* **6**, 100-109.
Lara, A., Villalba, R., Aravena, J. C., Wolodarsky-Franke, A., and Neira, E. (2000). Desarrollo de una red de cronologías de *Fitzroya cupressoides* (alerce) para Chile y Argentina. *In* "Dendrocronología en América Latina." (F. Roig, Ed.), pp. 217-244. Editorial Nacional de Cuyo, Mendoza.
Lara, A., Aravena, J. C., Wolodarsky-Franke, A., Villalba, R., Luckman, B., and Wilson, R. (2001). Dendroclimatology of high-elevation *Nothofagus pumilio* forests in the Central Andes of Chile. *Canadian Journal Forestry Research* **31**, 925-936.
Lara, A., Wolodarsky-Franke, A., Aravena, J. C., Cortés, M., Fraver, S., and Silla, F. (2003). Fire regimes and forest dynamics in the Lake Region of South-Central Chile. *In* "Fire an climatic change in temperate ecosystems of the Western Americas." (T. T. Veblen, T. Swetnam, and G. Montenegro, Eds.), pp. 316-336. Springer, New York.
LeQuesne, C., Aravena, J. C., Alvarez García, M. A., and Fernández Prieto, J. A. (2000). Dendrocronología de *Austrocedrus chilensis* en Chile Central. *In* "Dendrocronología en América Latina." (F. Roig, Ed.), pp. 159-175. Editorial Nacional de Cuyo, Mendoza.
Le Quesne, C., and Acuña, C. (2003). Fluctuaciones históricas del ventisquero Cipreses y su relación con el registro de anillos de *Austrocedrus chilensis* en Chile Central (Glacier (34°33'S 70°22'W). *In* "Symposium on Global Change: Towards a systemic view held at the meeting of the IGBP Scientific Steering Committee," Punta Arenas, Chile.
Luckman, B. H. (1990). Mountain areas and global change: a view from the Canadian Rockies. *Mountain Research and Development* **10**,183-185.
Luckman, B. H., and Villalba, R. (2001). Assessing the synchronicity of glacier fluctuations in the Western Cordillera of the Americas during the last millennium. *In* "Interhemispheric climate linkages." (V. Markgraf, Ed.), pp.119-140. Academic Press, San Diego, CA.
Mo, K. (2000). Relationships between low-frequency variability in the Southern Hemisphere and sea surface temperature anomalies. *Journal of Climate* **13**, 3599-3610.
Morales, M. S., Villalba, R., Grau, H. R., Villagra, P., Boninsegna, J. A., Ripalta, A., and Paolini, L. (2001). Potencialidad de *Prosopis ferox* Griseb (Leguminosae, subfamilia: Mimosoideae) para estudios dendrocronológicos en desiertos subtropicales de alta montaña. *Revista Chilena de Historia Natural* **74**, 865-872.
Miller, A. (1976). The climate of Chile. *In* "World survey of climatology. Climates of Central and South America." (W. Schwerdtfeger, Ed.), pp. 113-131. Elsevier, Amsterdam.
Naruse, R., and Aniya, M. (1992). Outline of Glacier Research Project in Patagonia, 1990. *Bulletin of Glacier Research* **10**, 31-38.
Pezoa, L. S. (2003). "Recopilación y analisis de la variación de las temperatures (período 1965-2001) y las precipitaciones (período 1931-2001) a partir de la información de estaciones meteorológicas de Chile entre los 33° y 53° de latitud sur." Tesis de grado Escuela de Ingeniería Forestal. Universidad Austral de Chile.
Prieto, M. R., and Herrera, R. (1998). Naos, clima y glaciares en el Estrecho de Magallanes durante el siglo XVI. *Anuario de Estudios Americanos*, Tomo LV-2: 413-439.
Rivera, A., Aravena, J. C., and Casassa, G. (1997). Recent fluctuations of glaciar Pío XI, Patagonia: Discussion of a glacial surge hypothesis. *Mountain Research and Development* **17**, 309-322.
Rivera, A., Casassa, G., Acuña,C., and Lange, H. (2000). Variaciones recientes de glaciares en Chile. *Investigaciones Geográficas* **34**, 25-52.

Rivera, A., Acuña, C., Casassa, G., and Bown, F. (2002). Use of remotely sensed and field data to estimate the contribution of Chilean glaciers to eustatic sea-level rise. *Annals of Glaciology* **34**, 367-372.

Roig F, Le Quesne, C., Bonisegna, J. J., Briffa, K., Lara, A., Grudd, N., Jones, P., and Villagrán, C. (2000). Climate variability 50,000 years ago in mid-latitude Chile as reconstructed from tree rings. *Nature* **410**, 567-570.

Rosenblüth, B., Casassa, G., and Fuenzalida, H. A. (1995). Recent climate changes in western Patagonia. *Bulletin of Glacier Research* **13**, 127-132.

Rosenblüth, B., Fuenzalida, H. A., and Aceituno, P. (1997). Recent temperature variations in southern South America. *International Journal of Climatology* **17**, 67-85.

Solari, M. E., Prieto, M. R., Gutiérrez, A. G., and Araya, C. (2003). Caracterización de la variabilidad a través de registros históricos del sur de Chile (41°-51° S) entre los años 1850 y 1950. *In* "Symposium on Global Change: Towards a systemic view held at the meeting of the IGBP Scientific Steering Committee," Punta Arenas, Chile.

Soliz, C., Argollo, J., Villalba, R., and Stahle, D. (2003). Climatic correlation of *Polylepis tarapacana* tree-ring chronology at Caquella Volcano in the Bolivian altiplano. *In* "Fourth Annual Scientific Meeting IAI CRN 03," Mendoza, Argentina.

Szeicz, J., Lara, A., Díaz, S., and Aravena, J. C. (2000). Dendrochronological studies of *Pilgerodendron uviferum* in southern South America. *In* "Dendrocronología en América Latina." (F. Roig, Ed.), pp. 245-270. Editorial Nacional de Cuyo, Mendoza.

Szeicz, J., Haberle, M., Simon, G., and Bennett, K. (2003). Dynamics of North Patagonian rainforests from fine-resolution pollen, charcoal and tree-ring analysis, Chonos Archipelago, Southern Chile. *Austral Ecology* **28**, 413-422.

Urrutia, R., Lara, A., Villalba, R., Pezoa, L., LeQuesne, C., Cuq, E., and Wolodarsky-Franke, A. (2003). Streamflow reconstruction from tree-ring chronologies of *Austrocedrus chilensis* and *Pilgerodendron uviferum* in the Xth Region. *In* "Fourth Annual Scientific Meeting IAI CRN 03," Mendoza.

Villalba, R. (1994). Tree-rings and glacial evidence for the Medieval Warm Epoch and the Little Ice Age in Southern South America. *Climatic Change* **30**, 1-15.

Villalba, R, Boninsegna, J. A., Veblen, T. T., Schmelter, A., and Rubulis, S. (1997). Recent trends in tree-ring records from high elevation sites in the Andes of northern Patagonia. *Climatic Change* **36**, 425-454.

Villalba, R., Lara, A., Boninsegna, J. A., Masiokas, M., Delgado, S., Aravena, J. C., Roig, F. A., Schmelter, A., Wolodarsky, A., and Ripalta, A. (2003). Large-scale temperature changes across the Southern Andes: 20th century variations in the context of the past 400 years. *Climatic change* **59**, 177-232.

Wolodarsky-Franke, A. (2002). Fluctuaciones ambientales de los últimos 1000 años a partir de anillos de crecimiento de *Fitzroya cupressoides* en el área del Volcán Apagado, X Región, Chile. Master thesis, Facultad de Ciencias, Universidad Austral de Chile.

Biogeographical Consequences of Recent Climate Changes in the Southern Andes of Argentina

Ricardo Villalba[1*], Mariano H. Masiokas[1], Thomas Kitzberger[2], and José A. Boninsegna[1]
[1]*Departamento de Dendrocronología e Historia Ambiental, IANIGLA - CRICYT, Mendoza, Argentina*
[2]*Departamento de Ecología, Universidad Nacional del Comahue, Bariloche, Argentina*
**phone +54-261-4287029 ext. 48, fax +54-261-4285940, e-mail Ricardo@lab.cricyt.edu.ar*

Keywords: Climate changes, Dendrochronology, Fire regimes, Forest dynamics, Glacier fluctuations, Southern Andes.

1. State of mountain research in the Southern Andes of Argentina

Long-term trends of temperature variations across the Southern Andes (37-55°S) have been recently examined using a combination of instrumental and proxy records. Tree-ring based reconstructions indicate that the annual temperatures during the 20th century have been anomalously warm across the Southern Andes in the context of the past four centuries. The mean annual temperatures for northern and southern Patagonia during the interval 1900-1990 are 0.53°C and 0.86°C above the AD 1640-1899 means, respectively. Increased temperatures are seriously impacting the physical and biological systems across the Southern Andes.

Non-forested glacial deposits around most Patagonian glaciers provide evidence of glacier retreat since the late Neoglacial advance dated between AD 1520 and 1850, approximately. The reduction of ice masses accumulated over preceding centuries has continued and accelerated to the present. Indeed, all land based glaciers and most calving glaciers in northern and southern Patagonia are presently retreating and thinning at a fast rate. The reconstructed temperature records suggest that the

U. M. Huber et al. (eds.), Global Change and Mountain Regions, 157-166.

generalized glacier retreat is largely due to the temperature increase during the 20th century across the southern Andes.

Events of episodic tree mortality have been recognized during the 20th century for *Austrocedrus chilensis* growing along the xeric forest-steppe border in northern Patagonia. These regional events of mortality are associated with extreme dry-warm climatic conditions during single summers, or more commonly during two consecutive summers in the 20th century. More recently, the enhancement of warm-dry conditions, resulting from the 1997-98 El Niño-Southern Oscillation event, caused the extensive mortality of *Nothofagus dombeyi*, a tree species growing in mesic forests of northern Patagonia. On the other hand, climatic fluctuations on decadal time scale appear to influence tree establishment and long-term changes in fire regime. In particular, warmer temperatures in the northern Patagonian Andes since the mid 1970s, due to changes in the Pacific Decadal Oscillation (PDO) modes, have increased the occurrence of lightning-induced fires. It is crucial to understand the effects of recent climate variations on both physical and biological systems to properly predict ecosystem responses to future climatic changes.

2. Temperature changes of the past four centuries across the Southern Andes

The Southern Andes spread over a wide latitudinal range (37-56°S) on the western side of southern South America, encompassing a wide variety of climate regimes from the Mediterranean type with a dry summer in the north to an all year rainy type in the south (Miller 1976). This chain of mountains is an excellent source of paleoenvironmental records because its physical and biological systems are highly sensitive to climatic variation, and they provide complementary records across a range of spatial and temporal resolutions. A network of more than 100 chronologies from *Nothofagus pumilio* at alpine treeline has recently been developed across the Southern Andes of Argentina and Chile (Lara et al. 2001; Aravena et al. 2002; Villalba et al. 2003). With this increasing number of tree-ring collections from temperature-sensitive sites, the use of a spatial approach to reconstruct the large-scale temperature patterns across 2000-km in the southern Andes now appears to be feasible.

Based on a 60-year interval (1930-1990), the dominant patterns of temperature variations were derived from the instrumental records available in the Southern Andes. Two dominant patterns of temperature variations emerge from a principal component analysis of the meteorological records (Villalba et al. 2003). The first pattern, which is associated with the stations in the northern sector of the southern Andes (Temuco, Valdivia and Puerto Montt), is characterized by a cooling trend from the 1930-1940s to the mid 1970s followed by temperature increases during the 1980s and 1990s. The second pattern represents the temperature changes observed in the southern stations, such as Punta Arenas, Río Gallegos and Ushuaia. At the southern tip of South America, a slight cooling, recorded between 1930 and 1950, was followed by steadily increasing temperatures during the past four decades (Rosenblüth et al. 1995; Villalba et al. 2003).

Tree-ring based reconstructions of the two dominant patterns of annual temperature have been developed using standard dendroclimatological methods (Fritts 1976). Temperature reconstructions, which cover the interval AD 1640-1990, explain 55% and 44% of the total variance in the northern and southern temperature patterns, respectively. Co-spectral analyses between actual and reconstructed data indicate that the reconstructions reproduce the low frequency components present in the instrumental records with a high degree of fidelity. Both reconstructions show that temperatures during the 20th century have been anomalously warm in the context of the past four centuries (Fig. 1). The mean annual temperatures for the northern and southern sectors during the interval 1900-1990 are 0.53°C and 0.86°C above the AD 1640-1899 means, respectively.

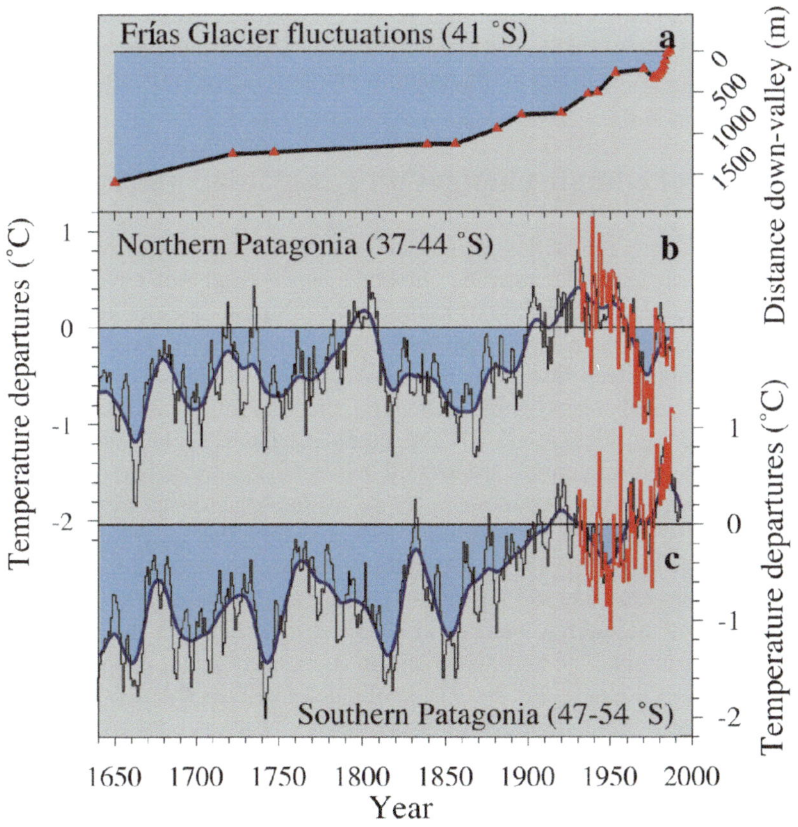

Figure 1: Glacier and tree-ring evidence for temperature changes across the Southern Andes of Argentina during the past four centuries: a) Ice front positions (relative to the glacier terminus in summer of 1986) of Frías Glacier reconstructed from a combination of historical records, aerial photographs and tree-ring records. Fluctuations in ice front positions were measured during the period 1975-1986; b) and c) Annual temperature reconstructions for the northern and southern Patagonian Andes, respectively. To emphasize low frequency variations, the data series (annual values) were smoothed with a cubic spline of 25 years (blue line). The actual values (red) have been superimposed in the most recent portion of the reconstructions.

Spatial correlations between the reconstructions and annual variations in sea-surface temperature (SST) in the South Atlantic and Pacific indicate that the dominant patterns of temperature variations across Patagonia are related to large-scale SST patterns (Villalba et al. 2003). Temperatures in northern Patagonia are significantly correlated with SST over the Pacific from the date line to the subtropical coast (40°S) of South America. This spatial pattern of correlation resembles the South Pacific counterpart of the "Pacific Interdecadal" mode of SST recently identified by Enfield and Mestas-Nuñez (2000). In contrast, temperature variations in the southern sector of the Andes are related to SST over both the South Atlantic and South Pacific. The associated SST spatial pattern resembles the "Global Warming" mode of SST variability, characterized by a steady warming across the South Atlantic and at higher latitudes in the Southern Hemisphere (Enfield and Mestas-Nuñez 2000). A rotated empirical orthogonal function analysis on SST across the South Pacific and South Atlantic Oceans shows that four discrete modes of SST variability explain a third of the total variance in temperature fluctuations across the Southern Andes (Villalba et al. 2003).

3. Impacts of climate changes on Patagonian glaciers

In general, glacier retreat has been observed in the 20th century across the Southern Andes in response to the documented climatic changes. Eight ice-covered areas have been selected to reconstruct past positions and to monitor the present dynamics of glaciers in the Argentinean Patagonian Andes (a similar strategy for the study of glacier variations on the Chilean side of the Southern Andes is conducted by Lara et al., this book). Four glaciated areas (Lanín, Tronador, Esperanza and Tunel) are located in the northern Patagonia Andes (39-43°S). The remaining glaciers (San Lorenzo, Narváez, Piedras Blancas and Ameghino) are located between 47° and 50°S.

Frías Glacier, the northernmost ice body in the Argentinean portion of Mount Tronador (41°S), reached its maximum extent during the Last Neoglacial Event around AD 1640-1660 (Fig. 1). Since that time, the glacier retreated at an average rate of 2.5 m yr^{-1} from AD 1650 to 1850. Mean retreat rates increased after 1850, reaching more than 7 m yr^{-1} between AD 1850 and 1900, 10 m yr^{-1} between 1910 and 1940, and 36 m yr^{-1} for the interval 1976-1986, when annual measurements of front ice positions were conducted (Villalba et al. 1990). Loss of glacier volume was slow during the 17th and 18th century, more continuous from the mid-19th century to the mid-1970s, and followed by a rapid loss and further acceleration during the past three decades. This is consistent with the reconstructions of temperature for northern Patagonia showing cold conditions from AD 1640 to 1850, followed by a warming trend from the 1850s to 1920s, a period of moderate cooling between the 1940s and 1970, and a return to warmer conditions after the mid-1970s (Fig. 1).

Information on glacial history during the past 1000 years for the southern sector of the Southern Andes is still inadequate to properly reconstruct the full temporal sequences of events related to the latest Neoglacial event. Ring counts from trees growing on the moraines indicate that the most external moraine associated with the last Neoglacial event was formed before AD 1700. Our temperature reconstruction for the southern sector of the Southern Andes shows several cold intervals between AD 1640 and1850 (Fig. 1). The persistence of cold conditions during this long-term interval in combination with the severe cold peaks centered on AD 1820 and 1850 may be responsible for the formation of a second massive moraine, a common feature to most glaciers between 47 and 50°S. Consequently, the glacier expansions in the southern sector during the latest Neoglacial might have consisted of two major events, one before AD 1700 and a second during the 19th century. In contrast to the glaciers at lower latitudes in Patagonia, the 19th century advance in the southern sector was similar in magnitude to the first glacial event of the latest Neoglacial (Masiokas et al. 2001). Ice retreat rates substantially increased in both regions due to increasing temperatures during the 20th century (Fig. 1). However, the glacier retreat must also have been affected by the documented regional decrease in precipitation along the Pacific coast of southern South America (Rosenblüth et al. 1995).

The rise in air temperature, suggested by the retreat of temperature-sensitive glaciers along Patagonia, is somewhat greater than the global average temperature rise derived primarily from instrumental records at lower elevations. The present glacier shrinkage is a general phenomenon across the southern Andes (except for a few calving glaciers) and the recession of glaciers has taken place in several stages during the past four centuries with increasing intensity since the mid-1970s. This higher rate of retreat since the mid-1970s corresponds with a shift in the PDO, which has been reported in several recent publications (Ebbesmeyer et al. 1991; Graham 1994; Villalba et al. 2001). Changes in glacier sizes and regional temperatures during the 20th century across the Southern Andes appear to be unprecedented in the context of the past 400 years. The increasing rate of glacier retreat in the past decades bears examination because it may provide insights into the relationships between climate and glaciers in the near future.

4. Impacts of climate changes on Patagonian forests

Climate variability is a major influence on forest dynamics both indirectly through effects on climatically related disturbances, such as insect outbreaks and fires, and directly through influences on tree establishment and mortality (Archer 1994). Evidence from dendroecological studies in northern Patagonia suggests that climate change during most of the 20th century has already been affecting the structure and dynamics of native forests. Moreover, climate conditions at different temporal scales strongly affect fire occurrence in northern Patagonia through the alteration of ignition opportunities and fuel production.

Austrocedrus chilensis establishment at the forest-steppe ecotone in northern Patagonia appears to be episodic in relation to decade-scale climatic variations. Following the warm-dry interval from 1956 to 1962, the climatic pattern changed drastically. Relatively cold and wet conditions started in 1963 and prevailed until the mid-1970s (Fig. 1). A regional age structure analysis along 400 km in the eastern foothills of the Andes shows widespread *Austrocedrus* establishment during this interval. Peaks of tree recruitment coincide with significantly cooler and wetter summers in 1964-66 and 1973-75. With the exception of the summer of 1984, warm-dry summers prevailed throughout the 1980s. The abrupt decrease in tree establishment since 1978 and the near absence of trees dating from the 1980s coincides with this decade-scale change towards a warmer and drier climate in northern Patagonia (Villalba and Veblen 1997).

Interannual variability of climate and the occurrence of extreme events can also have strong effects on the structure and function of forests. Dendrochronological methods were used to date the outermost ring on dead standing and fallen trees to estimate the dates of tree death at nine stands of *Austrocedrus chilensis* along the forest-steppe border (37-43°S). Episodes of exceptionally high tree mortality coincide with exceptionally dry springs and summers during the 1910s, 1942-43, and the 1950s. Droughts and associated mortality events of *Austrocedrus chilensis* during the 20th century were found to be strongly related to above-average sea level pressure off the coast of Chile at the same latitudes. Temperature and precipitation in northern Patagonia are strongly influenced by the intensity and latitudinal position of the southeastern Pacific anticyclone, which, in turn, is greatly affected by the Southern Oscillation events (Villalba and Veblen 1998).

In response to the climatic anomalies imposed by the severe 1997-98 El Niño-Southern Oscillation (ENSO) event, intense droughts occurred in the austral summers 1997/98 and particularly 1998/99. These events caused extensive mortality and dieback in the mesic *Nothofagus dombeyi* forests in the eastern part of its distribution in northern Patagonia (Fig. 2). An evaluation of the regional impact of this drought indicates that 41.8% of the pure *Nothofagus* or mixed *Nothofagus-Austrocedrus* forests located in the Nahuel Huapí National Park were damaged by this extreme climatic event (Bran et al. 2001). These mortality events in northern Patagonia show that forest composition can be strongly affected not only by long-term (decade-scale or larger) fluctuations in climate, but also by interannual climate variability. Tree-demography processes appear to respond differentially to high- versus low-frequency variations in climate. In the *Austrocedrus chilensis* and *Nothofagus dombeyi* forests on the eastern slopes of the Andes, seasonal- to annual-scale droughts increase mortality rates, but decade-scale cool-wet spells are required for widespread increases in tree establishment.

Climate variations at different temporal scales also affect disturbance regimes differentially. Conditions affecting fire, such as ignition opportunities and fuel desiccation, are strongly associated with year-to-year climate variations in northern Patagonia (Kitzberger et al. 1997; Veblen et al. 1999). Years of extreme fire occurrence are associated both with dry winter-springs and with warm summers related to the cold and warm phase of the Southern Oscillation, respectively (Kitzberger and Veblen 1997). Years in which the southeast Pacific anticyclone is intense and located farther south than normal are years of enhanced fires.

Both the number of lightning ignitions and the area burned increase exponentially with summer temperature. Years in which lightning-ignited fires occur in northern Patagonia are associated with above-average sea level pressure at mid to high latitudes (45-55°S) east of southern South America during summer. During summers of anomalously high pressure over the Atlantic Ocean, the Atlantic portion of the subtropical high-pressure belt is located farther south, which allows Atlantic

Figure 2: Large-scale mortality event in the old-growth forests of *Nothofagus dombeyi* in northern Patagonia caused by the severe droughts in the summers of 97/98 and 98/99. Summer (January to March) temperature and precipitation variations for the interval 1984-2001 are shown in the upper right-hand corner. Climate data were provided by INTA, Bariloche, Argentina.

subtropical air to flow southwestwards into Northern Patagonia (van Loon et al. 1993). These moist, warm air masses spawn thunderstorms and lightning through convective uplift over the heated Patagonian plains or by advective or orographic uplift when they reach cooler air in the Andean foothills (Komarek 1966).

At a decade-scale, the frequency of lightning ignitions closely track changes in summer temperature (Fig. 3). Mean summer temperatures in northern Patagonia increased after the mid-1970s by ca. 1°C when compared to the previous 1938-1976 period ($P < 0.01$). This warming trend has been accompanied by a three-fold increase in the rate of lightning ignitions (from 0.6 ignitions/year to 1.95 ignitions/year; $P < 0.02$). Recent warming trends observed in the South Atlantic and South Pacific Oceans may be influencing long-term changes in fire ignitions in northern Patagonia.

5. Final remarks

The temperature reconstructions for northern and southern Patagonia show substantial inter-centennial changes and point out how unusual climate conditions during the 20th century have been in the context of the past 400 years. Warmer conditions during the 1930-1940s and after the mid-1970s across the northern sector of the Southern Andes were unprecedented in the 360-year reconstruction (Fig. 1).

Similarly, temperatures in the southern sector since the mid-1970s have reached unparalleled levels in the past 360 years. Recent surveys of past environmental changes indicate that the anomalous climate of the past century has affected both the biotic and abiotic components of the landscape in the Southern Andes. Due to the steeper precipitation and temperature gradients imposed by the local mountains, the physical and biological systems in the Southern Andes are particularly sensitive to relatively minor changes in climate.

Loss of glacier volume in the Southern Andes started in the middle of the 17th century and continued in several stages of increasing rates during the 20th century, particularly in the past decades. Melting of ground ice has markedly accelerated since the mid-1970s, concurrent with large-scale atmospheric changes related to the Pacific Decadal Oscillation. It is clear, however, that carefully measured glacier data are extremely limited in the Southern Andes. Existing records need to be expanded both temporally and spatially to better understand the relationships between glacier fluctuations and regional climatic forcings.

Our dendroecological analyses suggest that some tree species are already responding to the anomalous climate during the 20th century. Increased numbers of mortality events and lightning-ignited fires are the responses to warmer summers, consistent with the persistence of atmospheric circulation modes that might have been unusual in the previous three centuries. It is in the context of global warming that the science community is challenged to bridge the traditional local spatial scale of ecology and geomorphology to the regional scale of climatology by linking ecosystem responses to variations in the large-scale synoptic forcings of regional climate. For reaching such a goal, several tasks need to be accomplished in the Southern Andes.

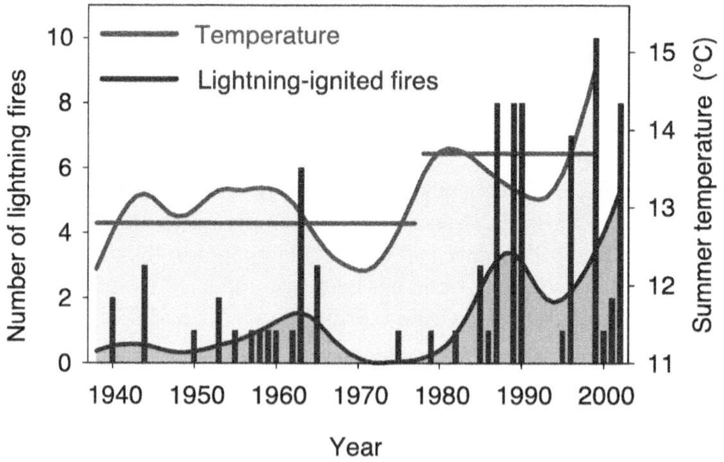

Figure 3: Lightning-ignited fires (vertical bars) reported in Lanín, Nahuel Huapí, Lago Puelo and Los Alerces National Parks between 1945-2002 (Bruno and Martin 1982; Administración de Parques Nacionales, unpubl. data) and mean summer (December to March) temperatures (red line), based on Bariloche Airport weather station. To emphasize decade-scale variations in temperature and lightning ignition, both series were smoothed with a 15-year cubic spline. Horizontal lines are long-term mean summer temperatures for the 1938-1976 and 1977-1999 periods.

Time series of environmental processes in the Southern Andes are short, fragmentary, and suffer from serious drawbacks, due to undocumented changes in instrumentation and station locations. A substantial commitment should be made to rescue, maintain and expand current meteorological and hydrological data collections. It is strongly advised to establish high-resolution maps of climatic, hydrologic, topographic, vegetation and soil property attributes for the Southern Andes. Free and open access to the Andean data sets is essential to future progress. Coordination between local and international research programs is also critical.

There are numerous gaps in our current understanding of basic environmental principles and processes in the Southern Andes. Interdisciplinary studies linking current and paleoenvironmental changes should be fostered to assemble a more complete understanding of the Andes system and its future response to Global Change. Investments in long-term, process-based environmental studies should be encouraged. In addition, we need to address the role of humans in driving or triggering past and future environmental changes in the Southern Andes, and reach a comprehensive understanding of the impact of climate variations on local and regional human activities. The studies reviewed in this paper represent the first attempts to link local-scale physical and biological processes in the Southern Andes with current variations in large-scale atmospheric conditions. New research devoted to establishing linkages between climate variations, physical-biological systems, and human activities in the Southern Andes is strongly advised.

6. Acknowledgements

This review is largely based on studies supported by the CRN03 project from the Inter American Institute for Global Change Research (IAI) and the Argentinean Agency for Promotion of Science and Technology (PICT 07-03093). We are grateful to the editors for inviting us to participate in this book and for their careful reviews of an early draft of our paper.

7. References

Aravena, J. C., Lara, A., Wolodarsky-Franke, A., Villalba, R., and Cuq, E. (2002). Tree-ring growth patterns and temperature reconstruction from *Nothofagus pumilio* (Fagaceae) forests in the upper tree line of Southern Chilean Patagonia. *Revista Chilena de Historia Natural* **75**, 361-376.

Archer, S. (1994). Regulation of ecosystem structure and function: Climatic versus non-climatic factors. *In* "Handbook of agricultural meteorology." (J. E. Griffiths, Ed.). pp. 245-255. Oxford University Press, Oxford.

Bran, D., Pérez, A., Ghermandi, L., and Barrios Lamuniére, S. D. (2001). Evaluación de poblaciones de cohiue (*Nothofagus dombeyi*) del Parque Nacional Nahuel Huapí, afectadas por las sequía 98/99, a escala paisaje (1:250.000). *In* "I Reunión Binacional de Ecología", Abstract Volume, p. 63.

Bruno, J., and Martín, C. (1982). Los incendios forestales en los Parques Nacionales. Administración de Parques Nacionales, Buenos Aires.

Ebbesmeyer, C. C., Cayan, D. R., McLain, D. R., Nichols, F. H., Peterson, D. H. and Redmond, K. T. (1991). 1976 step in the Pacific climate: Forty environmental changes between 1968-75 and 1977-84. *In* "Proceedings of the 7th Annual Pacific Climate Workshop". (J. L. Betancourt, and V. L. Tharp, Eds.), pp. 115-126. California Department of Water Resources, Interagency Ecological Studies Program, Report 26.

Enfield, D. B., and Mestas-Nuñez, A. M. (2000). Global modes of ENSO and non-ENSO sea surface temperature variability and their associations with climate. *In* "El Niño and the Southern Oscillation, multiscale variability and global and regional impacts." (H. Diaz, and V. Markgraf, Eds.), pp. 89-112. Cambridge University Press, Cambridge.

Fritts, H. C. (1976). "Tree rings and climate." Academic Press, London.

Graham, N. E. (1994). Decadal-scale climate variability in the 1970s and 1980s: Observations and model results. *Climate Dynamics* **10**, 135-162.

Kitzberger, T., and Veblen, T. T. (1997). Influences of human and ENSO on fire history of *Austrocedurs chilensis* woodlands in northern Patagonia, Argentina. *Ecoscience* **4**, 508-520.

Kitzberger, T., Veblen, T. T., and Villalba, R. (1997). Climatic influences on fire regimes along a rainforest-to-xeric woodland gradient in northern Patagonia, Argentina. *Journal of Biogeography* **23**, 35-47.

Komarek, E. V. (1966). The meteorological basis for fire ecology. *In* "Proceeding of the fifth Tall Timbers Fire Ecology Conference." Tallahasse, Florida.

Lara, A., Aravena, J. C., Villalba, R., Wolodarsky-Franke, A., Luckman, B., and Wilson, R. (2001). Dendroclimatology of high-elevation *Nothofagus pumilio* forests at their northern distribution limit in the central Andes of Chile. *Canadian Journal of Forest Research* **31**, 925-936.

Masiokas, M. H., Villalba, R., Trombotto, D., Delgado, S., Luckman, B., Ripalta, A., and Hernandez, J. (2001). Dendrogeomorphological reconstruction of glacier variations in Patagonia during the past 1000 years. *In* "International Conference on Tree Rings and People," (M. Kaennel Dobbertin, and O. U. Bräker, Eds.), Abstracts. Swiss Federal Research Institute WSL, Birmensdorf, Switzerland.

Miller, A. (1976). The climate of Chile. *In* "World survey of climatology 12: Climates of Central and South America." (W. Schwerdtfeger, Ed.), pp. 113-131. Elsevier, Amsterdam.

Rosenblüth, B., Casassa, G., and Fuenzalida, H. (1995). Recent climatic changes in western Patagonia. *Bulletin Glacier Research* **13**, 127-132.

van Loon, H., Kidson, J. W., and Mullan, A. B. (1993). Decadal variation of the annual cycle in the Australian data set. *Journal of Climate* **6**, 1227-1231.

Veblen, T. T., Kitzberger, T., Villalba, R., and Donnegan, J. (1999). Fire history in northern Patagonia: The roles of humans and climatic variations. *Ecological Monographs* **69**, 47-67.

Villalba, R., and Veblen, T. T. (1997). Regional patterns of tree population age structure in northern Patagonia: Climatic and disturbance influences. *Journal of Ecology* **85**, 113-124.

Villalba, R., and Veblen, T. T. (1998). Climatic influences on episodic tree mortality at the forest-steppe ecotone in northern Patagonia. *Ecology* **79**, 2624-2640.

Villalba, R., Leiva, J. C., Rubulis, S., Suarez, J. A., and Lenzano, L. (1990). Climate, tree rings and glacier fluctuations in the Frías valley, Río Negro, Argentina. *Arctic and Alpine Research* **22**, 150-174.

Villalba, R., D'Arrigo, R. D., Cook, E. R., Wiles, G, and Jacoby, G. C. (2001). Decadal-scale climatic variability along the extratropical western coast of the Americas: Evidences from tree-ring records. *In* "Inter-hemispheric climate linkages." (V. Markgraf, Ed.), pp. 155-172. Academic Press, San Diego, Califormia.

Villalba, R., Lara, A., Boninsegna, J. A., Masiokas, M., Delgado, S., Aravena, J. C., Roig, F.A., Schmelter, A., Wolodarsky, A., and Ripalta, A. (2003). Large-scale temperature changes across the southern Andes: 20th-century variations in the context of the past 400 years. *Climatic Change*.

Part II: Cryospheric changes

Mountain glaciers, snow cover and permafrost may be the best icons for global climate change. They are exquisitely emblematic of the balance between heat and cold on the planet. Glaciers and snow are crucial water reservoirs throughout the world, damping the variation in water supply over the course of a year. Snow and ice are not however always boons. The annual cycles of freezing and thawing, of deposition and ablation create local hazards while the on-going loss of glacial ice contributes significantly to rising sea levels. Glaciers, snow cover and permafrost mediate many of our concerns arising from climate change.

In the first contribution, *Haeberli* presents a broad overview of glacial monitoring throughout the world He explains the necessity of a tiered observing system based on focused energy and mass balance measurements within a broader context of glacial size and even presence/disappearance data. He notes data gaps in both the Andes and the Himalayas but holds out hope in the form of more complete data from remote sensing.

The increasing loss of glacial mass worldwide leads *Dyurgenov* to highlight the potential extinction of mountain glaciers. These losses account for an important fraction of the rise in mean sea level. Advances in technology offer some hope that important data on large and poorly sampled glaciers might still be gathered in these times of shrinking budgets.

The next three papers focus on tropical glaciers. *Kaser et al.* remind us that a glacier is a completely physical system, the mass of which is an immediate result of energy exchange and mass inputs. While in the mid-latitudes, temperature tends to dominate the mass balance equation, in tropical regions other meteorological variables such as humidity, cloudiness and precipitation are equally important. Their observations regarding Kilimanjaro, an important media story, are very thought provoking. *Francou et al.* focus on glaciers in the tropical Andes. Here they find that the ongoing loss of glacial mass is explained largely through changes occurring in the wet season, particularly as those seasons are themselves changed by ENSO. *Mark* and *Seltzer* examine the historical records and the likely impacts of glacial recession in the Peruvian Andes. They emphasize changes associated with lower precipitation and higher influx of radiation. And while high rates of recession have occurred in the past, the present rates will have significant hydrologic impacts on drier western slope economies.

The next two papers examine how climate change may affect the hazards in mountain areas mediated by permafrost, snow and glaciers. *Harris* focuses on permafrost and notes that even slight warming can lead to very large changes in the extent of frozen soil and rock with considerable increases in the risk of slope failure. *Kääb et al.* look at a wider range of potential hazards, including glacial lake outbursts, floods, glacial surges and slope failures, as well as yet more destructive combinations of events. Both papers recommend broad geographic assessment of hazards with spatial modeling in GIS coupled with more detailed site specific analysis or laboratory experiments.

Martin and *Etchevers* estimate the impacts of climate change on snow cover in the French Alps. Climate change could lead to a considerable shortening of the period of snow cover, with great impacts on the skiing industry as well as on timing and amount of water in major Alpine rivers. Impacts on avalanche hazard are much harder to predict as these are associated with extreme weather events, which are difficult to predict in climate change scenarios.

Finally *Hock et al.* examine how climate change will drive discharge from mountain glaciers. They emphasize that discharge responds at a variety of time scales related to both energy balances and extents of glaciation in the watershed. While the physics of melting have been studied extensively, much less is known about the routing of melt water through the glacier, key to understanding discharge at shorter diurnal and seasonal time frames.

Mountain Glaciers in Global Climate-related Observing Systems

Wilfried Haeberli
World Glacier Monitoring Service, Glaciology and Geomorphodynamics Group, Geography Department, University of Zurich, Zollikerstr. 107, CH-8008 Zurich, Switzerland
phone +41-1-635 51 20, fax +41-1-635 68 48, email haeberli@geo.unizh.ch

Keywords: Atmospheric warming, Climate change, Environment, Glaciers, Monitoring, Mountains.

1. Introduction

Fluctuations of glaciers and ice caps in cold mountain areas have been systematically observed for more than a century in various parts of the world and are considered to be highly reliable indications of worldwide warming trends (cf. Fig. 2.39a in IPCC 2001). Mountain glaciers and ice caps are, therefore, key variables for early-detection strategies in global climate-related observations. Advanced monitoring strategies integrate detailed observations of mass and energy balance at selected reference glaciers with more widely distributed determinations of changes in area, volume and length; repeated compilation of glacier inventories enables global representativity to be reached (IAHS(ICSI)/UNEP/UNESCO 1989; 1998; 2001; cf. Haeberli et al. 2000; 2002).

Long-term mass balance measurements (Fig. 1) provide direct (undelayed) signals of climate change and constitute the basis for developing coupled energy-balance/flow models for sensitivity studies. These investigations explore complex feed-back effects (albedo, surface altitude, dynamic response) and can be used in conjunction with coupled ocean-atmosphere general circulation models (e.g. model validation, hydrological impacts at regional and global scales; cf. Beniston et al. 1997). They combine direct glaciological and geodetic/photogrammetric methods in order to determine changes in volume/mass of entire glaciers (repeated mapping)

U. M. Huber et al. (eds.), Global Change and Mountain Regions, 169-175.

with high spatio-temporal resolution (annual measurements at stakes and pits). Laser altimetry combined with a kinematic Global Positioning System (GPS) is applied for monitoring thickness and volume changes of very large glaciers which are the main meltwater contributors to ongoing sea-level rise (Arendt et al. 2002).

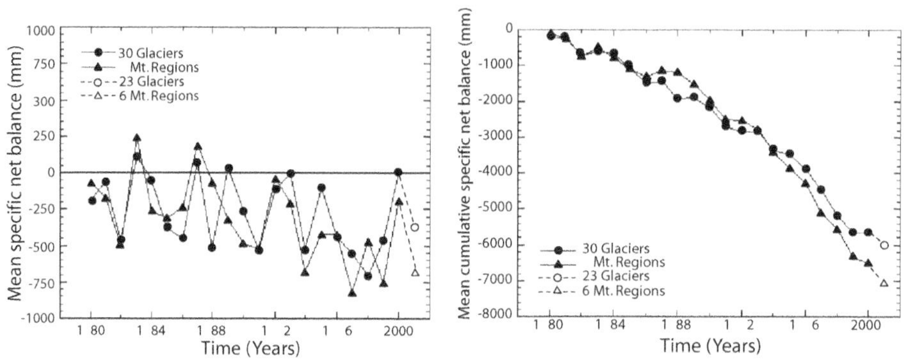

Figure 1: Annual (left) and cumulative (right) glacier mass balances as collected by the World Glacier Monitoring Service.

Change in glacier length (Fig. 2) is a strongly enhanced and easily measured but indirect, filtered and delayed signal of climate change (Oerlemans 2001). It represents an intuitively understood and most easily observed phenomenon to illustrate the reality and impacts of climate change. Work on glacier recession has considerable potential to support or qualify the instrumental record of temperature change and to cast further light on regional or worldwide temperature changes before the instrumental era. Studies of past fluctuations in glacier length are particularly useful for the reconstruction of late Quaternary climate variability (Haeberli et al. 1999). Some records are from more remote areas where there are few, if any, meteorological observations and, on average, glaciers exist at significantly higher altitudes than meteorological stations, which may be very useful in increasing our understanding of the differences in temperature change at different levels of the atmosphere.

Modern glacier inventories are compiled by using a combination of remote sensing and GIS technologies. Surveys take place at time intervals of a few decades – the characteristic dynamic response time of medium-sized mountain glaciers. Length and area changes can be measured for a great number of ice bodies. Area changes mainly enter calculations of glacier contributions to sea-level rise and of regional hydrological impacts (cf. Meier and Bahr 1996). In contrast, cumulative length changes not only influence landscape evolution and natural hazards (especially from ice- and moraine-dammed lakes) but can also be converted to average mass balance over decadal time intervals and, thus, help to establish the representativity of the few direct mass balance observations (Haeberli and Hölzle 1995; Hölzle et al. 2003).

In addition to being excellent indicators of climate change, glaciers and ice caps are observed in connection with climate and earth-system modelling, water resource management, sea-level observations (large glaciers are expected to contribute substantially to sea level rise over the next century; Dyurgerov and Meier 1997; Dyurgerov, this volume), natural hazard assessments and community planning with respect to tourism and recreation.

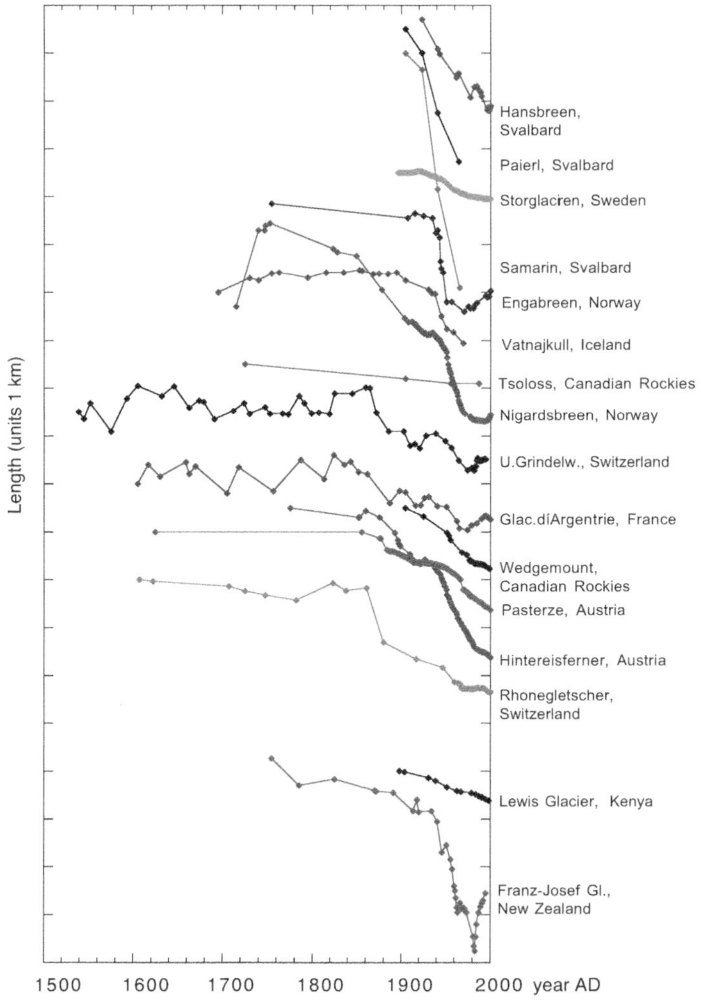

Figure 2: Cumulative length changes of selected glaciers worldwide. Compiled by H. Oerlemans from WGMS data with some additions.

2. Global observation network of glaciers

Records of glacier mass balance and of changes in glacier length, as well as a worldwide but rather preliminary glacier inventory, have been compiled by the World Glacier Monitoring Service (WGMS) in Zurich, Switzerland (http://www.geounizh.ch/wgms/index1.htm). This organization coordinates worldwide glacier monitoring, annually reports mass balance data for about 50 glaciers and publishes comprehensive information about length, area and volume changes of roughly 500 glaciers every 5 years. The WGMS mandate is to continuously upgrade, collect and periodically publish glacier inventory and fluctuation data, as well as to include satellite observations of remote glaciers and to assess ongoing changes. The WGMS maintains data exchange with the International Council of Scientific Unions (ICSU) World Data Center A (WDC-A) for Glaciology (http://nsidc.org/NOAA/index.html).

The Terrestrial Observation Panel for Climate (TOPC) now created a glacier observation network (Haeberli et al. 2000) to meet the needs of the Global Terrestrial Observing System (GTOS) and the Global Climate Observing System (GCOS). This network was developed by matching the observing system of WGMS sites with the concept of a Global Hierarchical Observing Strategy (GHOST). This integrated approach adopts a tier system which links investigations across a range of scales. Observations include detailed process studies at one extreme and global satellite imagery at pixel resolution at the other. The following elements are part of this tiered approach:

- extensive and process-oriented glacier mass balance and flow studies within the major climatic zones for calibrating numerical models (there are about 10 glaciers with intensive research and observation activities that represent such sites - Storglaciären in northern Sweden is an example, see Hock et al., this volume);
- regional glacier mass changes within major mountain systems, observed with a limited number of strategically selected index stakes combined with precision mapping at about decadal intervals (on about 50 glaciers annual mass balance studies reflect regional patterns of glacier mass changes. However, spatial coverage still shows gaps in some of the major mountain systems such as, for instance, the Himalayas, the New Zealand Alps or Patagionia);
- long-term observations of glacier length changes (a minimum of about 10 sites within each major mountain range should be selected to represent different glacier sizes and dynamic responses). The representativity of glaciers is assessed by intercomparison of geometrically comparable glaciers. Investigations include the dynamic fitting of glacier flow models to long time series of measured cumulative length change (Oerlemans et al. 1998), and mass-change reconstructions using concepts of mass conservation (Hölzle et al. 2003);
- glacier inventories repeated at time intervals of a few decades by using satellite remote sensing (continuous upgrading and analyses of existing and newly available data, modelling of data following the scheme developed by Haeberli and Hölzle 1995).

A number of additional glaciers are planned to be selected for mass balance measurements. A resolution of 0.01 to 0.1 m is required for assessing mass change, of 1 to 10 m for length change and of 10-100 m for model validation with inventory parameters. The time resolution for measurements is 1 year for mass balance, 1 to 10 years for length changes and a few decades for inventories. The recently launched US Geological Survey-led ASTER/GLIMS project (Advanced Spaceborne Thermal Emission and Reflection Radiometer/Global Land Ice Measurements from Space) attempts to compile a worldwide glacier inventory for the time slice around the year

2000. Corresponding pilot studies are well underway (Kieffer et al. 2000; Kääb et al. 2002; Paul et al. 2002).

3. The response of glaciers to climate change

The global retreat of mountain glaciers during the 20th century is striking. Trends in long time series of cumulative glacier-length and volume changes represent convincing evidence of fast climatic change at a global scale.

Since 1990, the Intergovernmental Panel on Climate Change (IPCC) has documented such changes as evidence of the existence of global warming, independent of the various surface temperature datasets. This is considered valid because a worldwide retreat is unlikely to be related to a reduction in global mountain precipitation.

Characteristic average rates of glacier thinning are a few decimeters per year for temperate glaciers in humid climates and centimeters to a decimeter per year for glaciers in continental climates with firn areas below melting temperature.

The total retreat of glacier termini is commonly measured in kilometers for larger glaciers and in hundreds of meters for small ones. The apparent homogeneity of the signal at the secular time scale, however, contrasts with great variability at local/ regional scales and over shorter time periods of years to decades. Intermittent periods of mass gain and glacier advance during the second half of the 20th century have been reported from various mountain chains, especially in areas of abundant precipitation, such as southern Alaska, Norway and New Zealand.

Analyses of repeated glacier inventory data show that the European Alps, for instance, have lost 30 to 40% in glacierized surface area and around 50% in ice volume between about 1850 and 1970 (Haeberli and Hölzle 1995). A further 25% of the remaining volume is estimated to have been lost since then. The recent emergence of a stone-age man from cold ice on a high-altitude ridge of the Oetztal Alps is a striking illustration of the fact that the extent of Alpine ice is probably less today than during the past 5000 years (Haeberli et al. 1999).

Air temperature and, to a lesser degree, precipitation are considered to be the most important factors reflected in glacier changes. Detailed data interpretation, however, is not straightforward and must be assisted by numerical modelling of physical aspects to take into account site-specific processes

Cumulative mass balances reflect not only regional climatic variability but also marked differences in the sensitivity of the observed glaciers. Sensitivities of temperate glaciers in maritime climates are generally up to an order of magnitude higher than the sensitivity of polythermal to cold glaciers in arid mountains. Spatial correlations of mass-balance values typically have a critical range of about 500 km and tend to markedly increase with growing length of the considered time interval (Letréguilly and Reynaud 1990; Cogley and Adams 1998). Decadal to secular trends are comparable beyond the scale of individual mountains. Besides individual hypsometric effects, continentality of the climate appears to be the main factor influencing long-term trends worldwide.

The frequency and scale of glacier response to climate fluctuations depends on glacier size (Oerlemans 2001). Small cirque glaciers and glacierets provide annual signals wheras the tongue reaction of medium-sized and long valley glaciers undergoes decadal to secular smoothing. Varying and predominantly slope-dependent dynamic response times of individual glaciers must be considered in analyses of glacier retreat as compared to instrumental proxy temperature records. Glacier advance and retreat may, in fact, lag climatic changes to varying degrees. Surging, heavily debris-covered and calving glaciers have strong non-climatic driving mechanisms.

4. Future perspectives

Most major mountain ranges of the world are represented in studies of glaciers and ice caps. A key priority is to continue long-term mass balance observations and expand these into additional regions, such as Patagonia, the Andes (Kaser et al., Mark and Seltzer, Francou et al., all this volume) and the mountains of New Zealand. More numerous observations of glacier area, thickness and length changes by application of remote sensing technologies (laser altimetry, aerial photography, high-resolution satellite, visible and infrared imagery from systems such as ASTER and Landsat) must be coordinated with the *in situ* measurements traditionally collected by the WGMS.

Numerical modelling studies (Oerlemans et al. 1998) confirm that many, if not most, glaciers of the presently existing worldwide mass balance network could disappear within decades if warming trends continue or even accelerate. An appropriate strategy for dealing with this problem will have to be developed.

Considering the potential impact on sea-level rise, the effects of (a) firn warming in presently cold subarctic and high-mountain accumulation areas, (b) possible runaway trends of mass balance/altitude feed-backs on large/flat glaciers with long dynamic response times, and (c) large ice volumes below sea level in the case of many important meltwater producers in maritime environments must be considered.

Most importantly, worldwide glacier monitoring must receive adequate funding and a new enlarged and internationally organized leading structure in view of the increasing public interest and new data formats. The opportunity provided by the current ASTER/GLIMS project should be used to further develop links with the remote sensing community.

5. References

Arendt, A., Echelmeyer, K., Harrison, W. D., Lingle, G., and Valentine, V. (2002). Rapid wastage of Alaska Glaciers and their contribution to rising sea level. *Science* **297**, 382-386.

Beniston, M., Haeberli, W., Hölzle, M., and Taylor, A. (1997). On the potential use of glacier and permafrost observations for verification of climate models. *Annals of Glaciology* **25**, 400-406.

Cogley, J. G., and Adams, W. P. (1998). Mass balance of glaciers other than the ice sheets. *Journal of Glaciology* **44**, 315-325.

Dyurgerov, M. B., and Meier, M. F. (1997). Year-to-year fluctuations of global mass balance of small glaciers and their contribution to sea level. *Arctic and Alpine Research* **29**, 392-402.

Forel, F.-A. (1895). Les variations périodiques des glaciers. Discours préliminaire. Archives des Sciences physiques et naturelles, Genève, XXXIV, 209 - 229.

Haeberli, W., and Hölzle, M. (1995). Application of inventory data for estimating characteristics of and regional climate-change effects on mountain glaciers: A pilot study with the European Alps. *Annals of Glaciology* **21**, 206-212. Russian translation in „Data of Glaciological Studies 82." pp. 116-124. Moscow.

Haeberli, W., Hölzle, M., and Suter, S., Eds. (1998). Into the second century of worldwide glacier monitoring: Prospects and strategies. A contribution to the International Hydrological Programme (IHP) and the Global Environment Monitoring System (GEMS). *UNESCO - Studies and Reports in Hydrology* **56**.

Haeberli, W., Frauenfelder, R., Hölzle, M., and Maisch, M. (1999). On rates and acceleration trends of global glacier mass changes. *Geografiska Annaler* **81A**, 585-591.

Haeberli, W., Barry, R., and Cihlar, J. (2000). Glacier monitoring within the Global Climate Observing System. *Annals of Glaciology* **31**, 241-246.

Haeberli, W., Maisch, M., and Paul, F. (2002). Mountain glaciers in global glimate-related observation networks. *WMO Bulletin* **51**, 18-25.

Hölzle, M., Haeberli, W., Dischl, M., and Peschke, W. (2003). Secular glacier mass balances derived from cumulative glacier length changes. *Global and Planetary Change* **36**, 77-89.

IAHS (ICSI)/UNEP/UNESCO (1989). „World Glacier Inventory - Status 1988." (W. Haeberli, H. Bösch, K. Scherler, G. Østrem, and C. C. Wallén, Eds.), Nairobi.

IAHS (ICSI)/UNEP/UNESCO (1998). "Fluctuations of glaciers 1990–95." (W. Haeberli, M. Hölzle, S. Suter, and R. Frauenfelder, Eds.). World Glacier Monitoring Service, University and ETH Zurich.

IAHS(ICSI)/UNEP/UNESCO (2001). „Glacier mass balance bulletin no. 6." (Haeberli, W., Frauenfelder, R., and Hölzle, M., Eds.). World Glacier Monitoring Service, University and ETH Zurich.

IPCC (2001). Climate change 2001: The scientific basis. Contribution of Working Group I to the Third Assessment Report of the Intergovernmental Panel on Climate Change, Cambridge University.

Kääb, A., Paul, F., Maisch, M., Hölzle, M., and Haeberli, W. (2002). The new remote-sensing-derived Swiss glacier inventory: II. First results. *Annals of Glaciology* **34**, 362-366.

Kieffer, H., Kargel, J. S., Barry, R., Bindschadler, R., Bishop, M., MacKinnon, D., Ohmura, A., Raup, B., Antoninetti, M., Bamber, J., Braun, M., Brown, I., Cohen, D., Copland, L., DueHagen, J., Engeset, R. V., Fitzharris, B., Fujita, K., Haeberli, W., Hagen, J. O., Hall, D., Hölzle, M., Johansson, M., Kääb, A., Koenig, M., Konovalov, V., Maisch, M., Paul, F., Rau, F., Reeh, N., Rignot, E., Rivera, A., de Ruyter de Wildt, M., Scambos, T., Schaper, J., Scharfen, G., Shroder, J., Solomina, O., Thompson, D. van der Veen, K., Wohlleben, T., and Young, N. (2000). New eyes in the sky measure glaciers and ice sheets. *In* „EOS, Transactions, American Geophysical Union, 81/24," June 13, 265, 270-271.

Letréguilly, A., and Reynaud, L. (1990). Space and time distribution of glacier mass balance in the northern hemisphere. *Arctiv and Alpine Research* **22**, 43-50.

Meier, M. F., and Bahr, D. B. (1996). Counting glaciers: Use of scaling methods to estimate the number and size distribution of the glaciers on the world. *In* „Glaciers, ice sheets and volcanoes: A tribute to Mark F. Meier." (S. C. Colbeck, Ed.), pp. 1-120, CRREL Special Report 27.

Oerlemans, J. (2001). „Glaciers and climate change." Balkema Publishers, Rotterdam.

Oerlemans, J., Anderson, B., Hubbard, A., Huybrechts, P., Johannesson, T., Knap, W. H., Schmeits, M., Stroeven, A. P., van de Wal, R. S. W., Wallinga, J., and Zuo, Z. (1998). Modelling the response of glaciers to climate warming. *Climate Dynamics* **14**, 267-274.

Paul, F., Kääb, A., Maisch, M., Kellenberger, T., and Haeberli, W. (2002). The new remote sensing-derived Swiss Glacier Inventory: I. Methods. *Annals of Glaciology* **34**, 355-361.

Mountain Glaciers are at Risk of Extinction

Mark B. Dyurgerov
Institute of Arctic and Alpine Research, University of Colorado at Boulder, 1560 30th Street, Campus Box 450, Boulder CO 80309-0450, USA
phone 1-303-492-5800, fax 1-303-492-6388, e-mail dyurg@tintin.colorado.edu

Keywords: Contribution to sea-level, Mass balance, Mountain glaciers, Prediction of sea-level rise, Volume change.

1. Why do we need to study glaciers?

Mountain glaciers are a product of climate and are important environmental components of local, regional and global water cycles. Glaciers are sources of beauty in the mountain landscapes and, in many cases, have been among the primary agents responsible for forming these landscapes. Glacier mass balance data have received increasing attention in recent years because of their usefulness in detecting climate change and explaining rising sea level (Meier 1984; Church et al. 2001). Understanding changes in glacier volume is important for regional water supply and power generation. In addition, observations made by the scientific community, tourists and climbers have shown that alpine glaciers are disappearing from mountain ranges around the globe. These changes have profound implications for sources of fresh-water on land, cause sea-level rise and make mountains less attractive, and more difficult and less appealing to climb (Bowen 2001; Meier and Wahr 2002; Meier et al. 2003).

Ironically, the resurgence of interest in mountain glaciers coincides with declining funds for programs to monitor glacier mass balance. Many glaciological stations, with observational records of 20 years and longer, were closed in the former Soviet Union during the 1980-1990s, e.g. Fedchenko and Abramov Glaciers in the Pamir, Shumskiy Glacier in Dzhungaria, Golubina Glacier in the Tien Shan, Obrucheva and IGAN glaciers in the Polar Ural, and Vavilova Ice Cap in Severnaya Zemlya. Several

U. M. Huber et al. (eds.), Global Change and Mountain Regions, 177-184.

glaciological stations were also closed in the U.S.A. (e.g. Blue Glacier in the Olympic Mountains), in Canada (e.g. Sentinel Glacier in the Rocky Mountains), and in East Africa (e.g. Lewis Glacier, Kenya). Despite the importance of glacier studies, funds for glacier monitoring by labor-intensive routine methods are declining. This necessitates the incorporation of new technologies, such as air- and space-born laser altimetry and remote sensing, to monitor changes of large glaciers and glacier systems.

2. The long-term changes

The major and worldwide retreat of mountain glaciers started at the end of the Little Ice Age (LIA), in approximately the mid-19th century. Table 1 provides a summary of data relating to changes in glacier area and volume of selected mountain regions, from the tropics to the high latitudes. Some data sets cover the period from the end of the LIA (ca. AD 1850) until the middle or end of the 20th century. Other data sets show the changes from about the middle to the end of the 20th century. Both periods reveal that glacier wastage has been pervasive and global in scale. Glaciers in regions with more maritime climate conditions have generally experienced larger wastage (e.g. Alps) than the drier and colder regions (e.g. Altai and high Arctic). Also, glacier wastage increased distinctly since the mid-1950s, specifically in the tropics, in the Alps, and in all mountain regions of North America. Since the late-1970s, this acceleration appears to be in tune with global temperature trends. Exceptionally strong wastage has been observed in Alaska (Arendt et al. 2002) and in some regions of Central Asia (Pamir, Tien Shan, and Dzhungaria). In Alaska, the wastage of glaciers can be attributed to the increase in both annual and summer air temperatures in this region of the Northern Hemisphere. In the north-western parts of Central Asia, where glacier mass balance observations were carried out over the last decades, glacier wastage may be explained by a simultaneous increase in summer air temperature and decrease in summer

Table 1. Long-term changes in area and volume of glaciers in selected mountain regions.

Region	Time period	Loss of area %	Loss of vol. % (km^3)	References
Tropics				
Mount Kenya	1850-1993	-74		Kaser et al. 2002
Rwenzory	1906-1990	-74		Kaser et al. 2002
Irian Jaya	ca. 1850-1990	-93		Kaser et al. 2002
Cordillera Real (Bolivia)	1920-1970	-12		Kaser et al. 2002
Pico Bolivar (Venezuela)	1910-1972	-80		Kaser et al. 2002
European Alps	ca. 1850-1994	-35	-50	Meier et al. 2003
Caucasus	1894-1970	-29	-50	Meier et al. 2003
Elbrus (Central Caucasus)	1887-1997	-14		Zolotarev et al. 2002
Spitzbergen	ca. 1850-1973	-6	-13	Meier et al. 2003
New Zealand	ca. 1850-1990	-26		Chin 1996
Tien Shan	1955-1995	-15	-22	Meier et al. 2003
Zailisky Alatau (W. Tien Shan)	1955-1990	-29	-32	Vilesov et al. 2001
Akshirak (InternalTien Shan)	1943-2001	-26		Khromova et al. 2002
Gissar-Alai (Pamirs)	1957-1980	-16		Shetinnikov 1998
Malaspina Glacier, Alaska	ca. 1974-1995	-1.5	-1.9	Meier et al. 2003
Southern Patagonia Icefield	1945-1986		(-200 km^3)	Anya et al. 1997

precipitation (Khromova et al. 2002). The decrease in summer snowfall is crucial for glacier regimes in this region, as summer precipitation (June-September) constitutes more than 50% of their annual accumulation (Ageta and Higuchi 1984; Dyurgerov et al. 1995).

3. Most recent variability and change

Since the 1960s, data on glacier volume changes have become available with annual resolution (Cogley 2002; Dyurgerov 2002). These data have been used to analyze year-to-year fluctuations and trends (Dyurgerov 2002; Meier and Wahr 2002). The results of this analysis confirm, in general, that the global glacier wastage that started after the LIA continued until the end of the 20th century. In addition to these changes, a close relationship between the magnitude of both long-term trends in glacier volume and interannual fluctuations in mass balance has been identified. In some regions, large long-term changes in glacier volume correspond to large interannual fluctuations in mass balance (Fig. 1). This pattern is, for example, characteristic for the Rocky Mountains, southwestern Alaska, and the northwestern parts of Central Asia. In other regions (e.g. Altai and Canadian Archipelago, Axel Heiberg Island), small interannual fluctuations in mass balance correspond with small long-term changes in glacier volume. These differences in spatial and temporal mass balance patterns have been related to fluctuations and changes in regional climate (Hodge et al.1998; McCabe et al. 1999). However, mass balance patterns at the scale of continents and hemispheres have not yet been properly modeled and are still poorly understood. For example, studies on the teleconnections between mass balances of distant glaciers, such as in Central Asia and Canada, have only recently been initiated (McCabe et al. 1999; Meier et al. 2003). In general, the colder and drier regions show smaller interannual fluctuations in mass balance and less pronounced long-term trends. This pattern is most distinct in the Canadian Arctic and in the Dry Valley, West Antarctica.

We have also learned from high-resolution data analysis that the changes in glacier volume have occurred in several distinct shifts, or abrupt changes. The first shift in volume change was observed in the mid-1970s (McCabe and Fountain 1995; Cao 1998). At the end of the 1980s, rates of ice wastage had further accelerated and enveloped additional regions, such as Alaska, the Rocky Mountains, and the Caucasus (Dyurgerov and Meier 2000). Glaciers in the tropics appear to be in danger of complete extinction (Kaser and Ostmaston 2002); several decades would be enough for their complete disappearance from some mountain ranges (Thompson et al. 2003). Large glaciers in Alaska, despite their dramatic wastage observed over the last decades (Arendt et al. 2002), are likely to survive over hundreds of years.

4. Global consequences

Ice melt is an important component of relative sea-level rise (RSL). If current rates of RSL were to last for decades or hundreds of years, the socioeconomic effects and environmental consequences would be dramatic (Warrick et al.1995; Trenberth 1999).

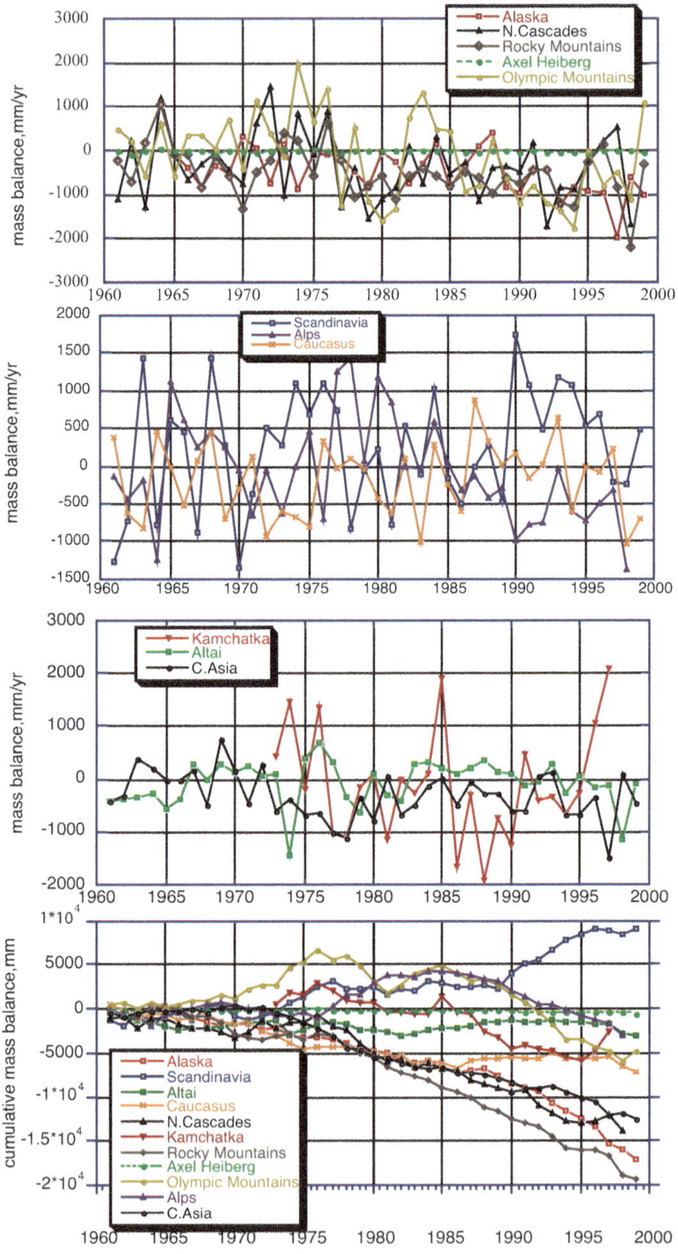

Figure 1: a) - c) The amplitude of year-to-year mass balance fluctuations of glaciers is largest in the most humid regions, such as the NW USA, Scandinavia, the Alps, and Kamchatka, and decreases in the regions with lower annual precipitation, such as the Caucasus, the Altai, the Central Asia ranges and Axel Heiberg Island in the Canadian Arctic; d) Long-term trends in glacier volume changes (cumulative mass balance) correspond, in general, with the magnitude of interannual variability in mass balance; lower interannual variability corresponds to smaller long-term changes.

The estimate of RSL, provided by the 2001 Intergovernmental Panel of Climate Change (IPCC), ranges from 1.0 to 2.0 mm/yr with a central value of 1.5 mm/yr (Church et al. 2001). This increase is partly caused by negative glacier mass balances of mountain glaciers (Meier 1984; Warrick et al. 1995). The globally averaged mass balance can be converted into units of sea-level change (361 km^3 of water equivalent increase sea-level by 1 mm). There are many problems in converting mass balances of individual glaciers to RSL (Warrick et al. 1995). An updated global compilation of continuous mass balance time series was recently assembled by Dyurgerov (2002) and this is one of the necessary steps for making such calculations. Based on the analysis of this comprehensive data set, area-weighted average annual mass balance changed from -82 mm/yr (-56 km^3/yr) during the period 1961-1976, to -125 mm/yr (-85 km^3/yr) during the 1977-87 period and to -217 mm/yr (-147 km^3/yr) during the 1988-98 period. These mass balance changes result in RSL of 0.15±0.05 mm/yr (10%), 0.24±0.05 mm/yr (16%) and 0.41±0.12 mm/yr (27%) for these three periods. Thus, the contribution of mountain and subpolar glaciers to RSL has been accelerating. For the entire period 1961-1998, glaciers contributed about 10 mm, 20% of the observed RSL. These estimates were used in IPCC-95 and IPCC-2001, but our and IPCC's predictions for future glacier contribution to sea-level change are different.

5. Can mass balance observations be used to predict RSL?

In this paper, an attempt has been made to use observed mass balance time series to predict glacier volume losses and estimate the effects of glacier wastage on RSL at the end of the 21st century. Extrapolations beyond the range of observational data are based on best-fit polynomial curves (Fig. 2). In our opinion, such extrapolations should not be derived from the entire time series (1961-1998), because of an abrupt shift in mass balance at the end of the 1980s. Due to this shift, the time series can no longer be considered homogeneous. We therefore split the time series into two periods (1961-87 and 1988-98) to derive extrapolations (Fig. 2). Global mass balance for the time period after 1998 could not be estimated, because the World Glacier Monitoring Service (WGMS) has not yet completed data processing for this time interval.

It needs to be emphasized that the calculation of the expected glacier contribution to RSL for the year 2100 is not a forecast but a possible scenario. This scenario has to be constantly adjusted, using the results of new measurements. Thus, the continuous and standardized observations on mass balance, established in the mid-20th century, are important for monitoring and forecasting of glacier contributions to RSL.

Arrows in Figure 2 show how our estimates of glacier contribution to RSL until the year 2100 compare to those predicted by IPCC-95 (Warrick et al. 1996) and IPCC-2001 (Church et al. 2001). The extrapolated trend from 1961 to 1988 shows substantially lower values compared to those predicted by IS92 Scenario (IPCC-95), but it falls within the middle of the range of the IPCC-2001 prediction. However, when extrapolations are based on the more recent trend between 1988-98, expected glacier contribution to the RSL may be substantially larger than the upper value predicted by IPCC-2001 (Church et al. 2001). The difference between the IPCC-2001 prediction and the estimates presented in this paper can be explained by the fact that the IPCC predictions are based on climate modeling (see IPCC-2001), whereas our estimates are based on observational data.

Every additional year of observational data will refine the forecast. Thus, it is important to process and disseminate observational data in a timely manner. The results of observations from 1998 until 2002, which up to now are only partly available, may change the calculations presented here substantially.

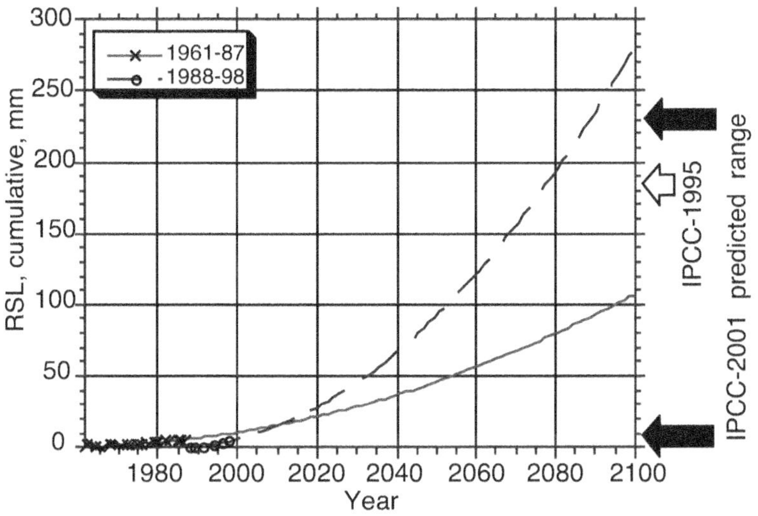

Figure 2: Expected trends in glacier contribution to sea level rise, calculated by statistical extrapolation of the time series 1961-87 (line with crosses) and 1988-98 (open circles) to the year 2100. Open arrow shows the projected glacier contribution based on the IS92 Scenario (IPCC-1995); solid arrow shows the range of glacier contribution to relative sea-level rise (RSL) projected by IPCC-2001.

6. A priority for the future

Widely applied new methods of remotely assessing glacier volume changes, such as laser altimeter photogrammetry (Echelmeyer et al. 1996) and global positioning systems provide new opportunities for evaluating glaciers that are too difficult to measure directly, either because of their large size and/or the difficult terrain. It is important to test these methods against traditional approaches on benchmark glaciers where long-term mass balance time series exist.

In particular, it is crucial to continue long-term glacier mass balance programs. The value of this data increases proportionally to the length of the record. Long-term records are needed to understand the response of glaciers to climatic variations and the subsequent effects on natural hazards, runoff, and sea level change. Without the commitment of the many field scientists who maintained these programs through financially difficult times, many of our current insights would not be possible. We can only hope that the mass balance programs are continued and expanded into the future, thereby allowing us to gain new insights that we cannot yet foresee (Fountain et al. 1999).

Our present knowledge of glacier regimes is biased towards small- and mid-sized glaciers. It has been shown that these glaciers are disappearing at an accelerating rate. Estimates of the global contribution of glaciers to sea level rise are traditionally based on labor-intensive mass-balance measurements on the surface of relatively small glaciers. In contrast, large glaciers in Central Asia, Alaska, the Patagonian Ice Fields and the Arctic Archipelagos that would be of particular importance in understanding RSL are poorly sampled. The launch of two new environmental satellites (Landsat-7 and Terra) by NASA in 1999 marks a leap in the capabilities of mapping and measuring large glaciers in these regions. These data need to be managed with Geographic Information Systems (GIS) and supported by strong field-based programs for the validation of satellite measurements. Remotely sensed, field-validated studies will be the state of the art in glaciology in the future. Studies of this nature will require even closer cooperation within the international community than has been the case in the past. It seems that the NASA program "Global Land Ice Measurements from Space" (GLIMS) is the ideal platform for the establishment of such an international glaciological consortium.

7. Acknowledgements

I thank the reviewers for help with editing of this paper and many important suggestions. This work was financially supported by NSF grants: OPP-0081379, OPP/HARC-0100120 and BCS-0099236.

8. References

Ageta, Y., and Higuchi, K. (1984). Estimation of mass balance components of a summer-accumulation type glacier in Nepal Himalaya. *Geografiska Annaler* **66A**, 249-255.

Anya, M., Sato, H., Naruse, R., Skvarca, P., and Casassa, G. (1997). Recent glacier variations in the Southern Patagonia Icefield, South America. *Arctic and Alpine Research* **29**, 1-12.

Arendt, A., Echelmeyer, K., Harrison, W. D., Lingle, G., and Valentine, V. (2002). Rapid wastage of Alaska Glaciers and their contribution to rising sea level. *Science* **297**, 382-386.

Bowen, M. (2001). Canaries in a coal mine. *Climbing* **208**, 91-97, 138-139.

Cao, M. S. (1998). Detection of abrupt changes in glacier mass balance in the Tien Shan Mountains. *Journal of Glaciology* **44**, 352-358.

Chinn, T. (1996). How much ice has been lost? *New Zealand Alpine Journal* **39**, 88-95.

Church, J. A., Gregory, J. M., Huybrechts, P., Kuhn, M., Lambeck, K., Nhuan, M. T., Qin, D., Woodworth, P. L. (2001). Changes in sea level. *In* "Climate change 2001, the scientific basis. Contribution of Working Group 1 to the Third Assessment Report of the Intergovernmental Panel on Climate Change." (J. T. Houghton et al., Eds.), pp. 641-693. Cambridge University Press, Cambridge.

Cogley, G. (2002). http://www.trentu.ca/geography/glaciology.htm

Dyurgerov, M., Liu Choahai, and Xie Zichu, Eds. (1995). "Oledenenie Tyan Shanya (Trien-Shan Glaciers)." Publishing House Nauka, Moscow (Russian).

Dyurgerov, M. B., and Meier, M. F. (2000). Twentieth century climate change: Evidence from small glaciers. *In* "Proceedings of the National Academy of Sciences 97," pp. 1406-1411.

Dyurgerov, M. B. (2002). Glacier mass balance and regime: Data of measurements and analysis. *INSTAAR Occasional Paper* **55** (INSTAAR: ttp://instaar.colorado.edu/other/occ_papers/html).

Echelmeyer, K., Harrison, W., Larsen, C., Sapiano, J., Mitchell, J., DeMallie, J., and Rabus, B. (1996). Airborne elevation profiling of glaciers: A case study in Alaska. *Journal of Glaciology* **42**, 538-547.

Fountain, A. G., Jansson, P., Kaser, G., and Dyurgerov, M. B. (1999). Summary of the workshop on methods of mass balance measurement and modeling, Tarfala, Sweden, August 10-12, 1998. *Geografiska Annaler* **81A**, 461-465.

Hodge, S. M., Trabant, D. C., Krimmel, R. M., Heinrichs, T. A., March, R. S., and Josberger, E. G. (1998). Climate variations and changes in mass of three glaciers in western North America. *Journal of Climate* **11**, 2161 – 2179.

Kaser, G., and Osmaston, H. (2002). "Tropical Glaciers." Cambridge University Press, Cambridge.

Khromova, T. E., Dyurgerov, M. B., and Barry, R. G. (2002). Application of ASTER images and historical data to estimate the disintegration of a glacier system. *Geophysical Research Letters* (submitted).

McCabe, G. J., and Fountain, A. G. (1995). Relations between atmospheric circulation and mass balance of South Cascade Glacier, Washington, U.S.A. *Arctic and Alpine Research* **27**, 226–233.

McCabe, G. J., Fountain, A. G., and Dyurgerov, M. B. (1999). Effects of the 1976-77 climate transition on the mass balance of Northern Hemisphere glaciers. *Arctic, Antarctic and Alpine Research* **3**, 64-72.

Meier, M. F. (1984). Contribution of small glaciers to global sea level. *Science* **226**, 1418-1421.

Meier, M. F., and Wahr, J. M. (2002). Sea level is rising: Do we know why? *Proceedings of National Academy of Sciences* **99**, 6524-6526.

Meier, M. F., Dyurgerov, M. B., and McCabe, G. J. (2003). "The health of glaciers - Recent changes in glacier regime." Climate Change, Kluwer. Special Issue.

Shetinnikov, A. S. (1998). "Morfologiya i rezhim lednikov Pamiro-Alaya (The morphology and regime of Pamir-Alai glaciers), Central Asia." Hydro-Meteorological Institute, Tashkent (Russian).

Thompson, L. G., Mosley-Thompson, E., Davis, M. E., Lin, P.-N., Henderson, K., and Mashiotta, T. A. (2003). "Tropical glacier and ice core evidence of climate change on annual to millennial time scales." Climate Change, Kluwer. Special Issue.

Trenberth, K. T. (1999). The extreme weather events of 1997 and 1998. Consequences. *The Nature and Implications of Environmental Change* **5**, 3-15.

Vilesov, E. N., and Uvarov, V. N. (2001). "Evolyutsiya sovremenngo oledeneniya Zailiyskogo Alatau v XX veke (The evolution of modern glaciation of the Zailiyskiy Alatau in the XXth century)." Kazakh State University, Almaty (Russian).

Warrick, R. A., LeProvost, C., Meier, M. F., Oerlemans, J., and Woodworth, P. L. (1996). Changes in sea level. *In* "Climate change 1995. The science of climate change. Contribution of Working Group 1 to the Second Assessment. Report of the Intergovernmental Panel on Climate Change (IPCC)."(J. T. Houghton, L. G. Meira Filho, B. A. Callander, N. Harris, A. Kattenberg, and K. Maskell, Eds.). Cambridge University Press, Cambridge.

Zolotarev, E. A., and Khar'kovets, E. G. (2000). Oledenenie El'brusa v kontse XX v. (Glaciation of Elbrus at the end of the XX century; digital orthophoto map of Elbrus for 1997). *Materialy Glyatsiologicheskikh Issledovaniiy* (Data of Glaciological Studies) **89**, 69-76.

Low Latitude Glaciers: Unique Global Climate Indicators and Essential Contributors to Regional Fresh Water Supply. A Conceptual Approach.

Georg Kaser*, Christian Georges, Irmgard Juen, and Thomas Mölg
Tropical Glaciology Group, Department of Geography, Innsbruck University, 6020 Innsbruck, Austria
**phone +43 512 507 5407, fax +43 512 507 2895, e-mail georg.kaser@uibk.ac.at*

Keywords: Glacier-climate relationships, Low latitudes, Modeling, Runoff

1. Introduction

Greenhouse gases in the atmosphere trap energy and, if their concentrations increase, e.g. from anthropogenic sources, the aggregate energy of the earth system increases as well. As a consequence, intensities of fluid dynamic processes (atmosphere and oceans), phase changing processes, biochemical processes, and the thermal status of the system will change in a complex and highly interactive manner. Manifold changes in local, regional and global climate are therefore to be expected, but are anything but easy to detect because: Firstly, climate itself is characterised by multi-scale dynamic variability of interacting processes and states. Thus, trends, fluctuations or changes can only be analysed for selected parameters and must be extracted from noise. Secondly, instrumental records, which concentrate on isolated parameters, are limited in time, and proxy-indicators, although covering longer time scales, show complex dependencies on climate, which can be difficult to interpret unequivocally. This paper emphasizes the role of low-latitude glaciers as i) climate proxies and ii) climate-dependent freshwater sources.

Among climate proxies, glaciers play an exceptional role. They exclusively follow physical laws in their response to climate and its fluctuations, and these laws are the

U. M. Huber et al. (eds.), Global Change and Mountain Regions, 185-195.

same, independent of where the glaciers are located. The gain and loss of ice masses, the so-called mass balance, is the immediate result of weather conditions. In practice, the physical processes governing the mass balance of glaciers must be linked to climate variables by parameterisation. One of these variables is air temperature, which has the highest correlation with glacier fluctuations in the mid and high latitudes. This correlation is so strong, that simple, black box statistical models, which include ablation period temperature and winter precipitation, produce highly satisfactory simulations of glacier mass balances (e.g. Braithwaite and Zhang 2000; Oerlemans 2001). Without doubt, changes in air temperature can also be expected to have a distinct impact on fluctuations of low latitude glaciers. However, since seasonality in low-latitude climate is solely a function of the annual cycle of air humidity and thermal seasons are absent, low-latitude glaciers are particularly sensitive to a broad spectrum of climate variables, not merely air temperature. In order to examine the relationship between mass balance and different climate parameters on low-latitude glaciers, we first outline the general features of glacier mass balance (section 2) and low latitude climate (section 3), then present three examples where observed differential glacier retreats indicate an impact of changing humidity (section 4), and, based on these observations, propose a concept for an analytical model (section 5). Lastly, we discuss the characteristics of low-latitude glacier runoff and its importance for water supply (section 6) and address future research perspectives (section 7).

2. Glacier mass balance

The mass balance of a glacier is the immediate result of weather conditions and, in the long term, of climate. Changes in mass balance, in turn, directly impact glacier runoff. On the income side, the mass balance of a glacier is due to the accumulation of solid precipitation, whereas the expenditure side depends on the availability of energy for ablation. Ablation can be due to fusion and/or sublimation. The energy balance equation at a melting glacier surface can be defined as

$$SW_{in}\,(1\text{-}\alpha) + LW_{in} + LW_{out} + H + C + L_s\,S + L_m M = 0 \qquad (1)$$

where SW_{in} is the incoming short-wave radiation, α the short-wave albedo of the surface, LW_{in} the atmospheric long-wave emission toward the ground, LW_{out} the emission from the glacier surface, H the sensible heat flux, and C the conductive heat flow in the subsurface. $L_s\,S$ and $L_m M$ are the mass consuming terms of equation 1, where L_s and L_m are the latent heat of sublimation and melting, and S and M are the rates of sublimation and melt, respectively. Conventionally, energy fluxes towards the surface are assumed to be positive. Most of the energy budget terms are closely linked to air temperature and/or air humidity. Air temperature controls most of the energy fluxes and is, in turn, primarily determined by the energy budget of the surrounding surfaces (Ohmura 2001). However, air humidity and atmospheric moisture content also exert a great influence on the energy budget terms:

- The *incoming short-wave radiation, SW_{in}*, is markedly controlled by cloud cover.
- The reflectivity (*albedo, α*) of the surface strongly depends on snow cover.
- The *albedo* is higher on a sublimating surface than on a melting surface.
- Under constant temperature conditions, the *incoming long-wave radiation, LW_{in}*, emitted from the atmosphere, increases with atmospheric moisture content.
- The *latent heat flux of sublimation, L_s S,* depends on the vapour pressure gradient between the glacier surface and the air. Sublimation either consumes a surplus of energy or takes it from the uppermost snow or ice layers (*C*) by cooling the glacier surface. If a surplus of energy is consumed by sublimation, this energy is no longer available for melting. If, however, humid conditions reduce sublimation, an energy surplus becomes available for melting, which is eight times more effective at removing ice than sublimation.
- If sublimation cools the surface, *long-wave emission, LW_{out}*, from the glacier surface is reduced.

These energy fluxes are of particular concern when considering the response of glaciers to climate factors other than air temperature. Note that, contrary to changes in air temperature, a given change in air humidity can cause glacier responses of different signs. The ultimate response of a glacier to a change in air humidity is closely linked to the initial glacier-climate conditions.

3. Low-latitude climate characteristics

Compared to the mid and high latitudes, the low-latitude atmosphere is characterized by a pronounced thermal homogeneity where frontal activities are practically absent and daily temperature variations are by far higher than annual variations. Two principle features of global zonal circulation, the Inter-Tropical Convergence, ITC, and the trade winds, characterise three different climate regimes. Whereas the ITC is part of a wet climate, trade winds are connected with particularly dry conditions. The respective climate zones are known as the *tropics* and the *subtropics*. The annual cycle of solar radiation leads to an oscillation in the dynamic atmospheric circulation patterns, leaving only the *inner tropics* with continuously wet conditions. The *outer tropics* can be treated as an intermediate zone between the tropics and the subtropics; during the wet season, they have tropical conditions (wet), during the dry season, subtropical conditions (dry) (Kaser 2001). Most of the low-latitude glaciers are found in the outer tropical regime, mainly because of the meridional lengthening of the austral summer ITC over South America (Kaser and Osmaston 2002).

In essence, low latitude seasonality is clearly related to hygric cycles whereas thermal characteristics are highly homogeneous and annual temperature variations small (Kaser and Osmaston 2002). Thus, ablation on low-latitude glaciers occurs all year round whereas accumulation prevails during the wet season. Beyond this, the above mentioned peculiarities of humidity-driven mass balance processes do not only appear from time to time and scattered over the ablation season, as for example in the Alps, but characterise entire seasons. During wet seasons, SW_{in} is usually reduced by more or less persistent cloud cover and α can be high due to snow cover. On the other hand, LW_{in} almost balances LW_{out} (Francou et al. 2003), sublimation is suppressed, and, as a consequence, melting is highly effective. This and the occurrence of precipitation induce a pronounced mass turnover during the wet season. During the dry season, the long-wave balance is negative and, additionally, sublimation consumes most of the available energy, which reduces ablation markedly (Wagnon et al. 1999a,b; 2001). The dry season is thus characterised by a strongly reduced mass turnover. Due to these processes, air humidity, atmospheric moisture content, and the length of the respective seasons dominate the mass balance characteristics of a low-latitude glacier (Kaser 2001; Francou et al. 2003).

A highly mass consuming scenario can be imagined as follows: if, after a dry season, the onset of the wet season occurs rather gradually, air humidity may rise, quickly turning sublimation into melt. Also, the atmospheric emissivity may increase and ablation will become considerably higher. At the same time, the reduction of SW_{in}, due to increasing cloud cover, is comparatively small. The extreme ablation rates will therefore only be stopped by precipitation and consequent albedo increase. Hence, the immediate occurrence of precipitation at the beginning of the wet period can be crucial for the positive mass balance of a low latitude glacier (Francou et al. 2003). In another scenario, minor hygric changes can have a highly mass conserving effect: Precipitation from isolated storms during the dry season can increase albedo for weeks or even months, thereby decreasing ablation rates on the glacier. The frequency of such events can have a marked impact on glaciers.

4. Interpretation of differential glacier retreat: Three examples from the tropics

4.1 Peruvian Cordillera Blanca (Kaser and Georges 1997)

The Peruvian Cordillera Blanca stretches 180 km from 8°30′ to 10°S, reaches 6000 m asl at several summits, and is a massive barrier against persistent lower-troposphere easterlies, causing wetter conditions on the Amazon side and drier conditions on the Pacific side. The climate is typical for the outer tropics; the wet seasons (October - April) bring 70-80 % of the annual precipitation. In the far north, the Santa Cruz - Pucahirca group shows the largest east-west extension within the Cordillera Blanca. There, a significant retreat of glaciers between two quasi-stationary positions around 1930 and 1950 was reconstructed from air photographs.

The derived equilibrium line altitudes (ELAs), as well as their shift $\Delta ELA_{1930\text{-}1950}$, show marked spatial differences among the six glacier areas analysed, with a regional gradient from low values in the east to high values in the west. In the drier westernmost mountains, smaller scale patterns are superimposed on this east-west gradient, with high values of ELAs and $\Delta ELAs_{1930\text{-}1950}$ on the slopes exposed to the morning sun. The observed spatial patterns of ELAs can best be explained in terms of humidity: i) Conditions are wetter on the windward Amazon side than on the Pacific side of the Andean chain, causing the east-west trend, and ii) in the drier western mountains, the diurnal cycle of convective clouds leads to shading of the western slopes during the afternoon, causing a local decrease in effective solar radiation. The differential $\Delta ELAs_{1930-1950}$ cannot be explained by differences in air temperature changes which are unlikely to be very variable across a few tens of kilometers. However, temperature changes may have played a secondary role. The maximum temperature contribution to ELA changes can be determined by assessing $\Delta ELA_{1930-1950}$ of the wet eastern glaciers, which are minimally affected by changes in humidity. The combined effects of a differential decrease in precipitation and a differential increase in effective solar radiation are likely the primary drivers behind the observed spatial differences in $\Delta ELAs_{1930-1950}$. ELA changes are probably associated with a decrease in moisture advection from the Amazon basin, which would have affected humidity-related processes along the east-west gradient and would have weakened the diurnal convective cloud cycle.

We used a glacier-climate model (Kuhn 1980) to estimate which combination of climate variables could have caused the observed $\Delta ELAs_{1930-1950}$. Based on the assumption that $\Delta ELAs_{1930-1950}$ of the wet eastern glaciers provide a maximum estimate of the temperature contribution to ELA changes in the Cordillera Blanca (see above), we calculated a temperature increase of 0.12°C. This temperature change can only explain 33 % of the largest observed ELA rise in the western mountains. The remaining change in ELA was estimated to be caused by a combination of a 2.7% (0.4 MJ m^{-2} d^{-1}) increase in SW_{in} $(1\text{-}\alpha)$, and a 13 % (155 mm a^{-1}) decrease in accumulation. Potential changes in sublimation were not calculated but probably played a significant role as well.

4.2 Rwenzori Mountains, Uganda/Congo (Mölg et al. 2003a)

In the Rwenzori Mountains, Uganda/Congo, glacier retreat throughout the 20th century also shows a striking spatial variability. Spatially differential glacier retreat between 1906 and 1955, as well as between 1955 and 1990, can be observed across all altitude ranges up to the crests. This spatial pattern rules out temperature increases as the primary driver for the same reasons as discussed for the 1930-1950 glacier retreat in the Cordillera Blanca. Small-scale variability in glacier retreat can only be due to changes in a limited number of mass balance terms, all of which are related to atmospheric moisture content. A closer examination of glacial and climatic evidence leads to the following hypothesis: owing to a drier atmosphere since the end of the 19th century (Kruss 1983, 1984; Hastenrath 1984, 2001; Nicholson et al. 2000; Nicholson and Yin 2001), both accumulation and convective cloud activity have decreased in East Africa. Consequently, increased incoming short-wave radiation, especially during the morning hours, induced differentially increased ablation on slopes exposed to the morning sun that could not be compensated for by mass advection. The results obtained from a combined radiation-terrain model, run for both a wetter and drier climate scenario, confirm this hypothesis. Model simulations closely reproduce observed spatial patterns of $\Delta ELAs_{1930-1950}$, with strong local losses in glacier surface area where incoming short-wave radiation is high.

4.3 Kilimanjaro, Tanzania/Kenya (Kaser et al. in press, Mölg et al. 2003b)

The summit glaciers on Kilimanjaro, Tanzania/Kenya, are located between 5,800 and 5,900 m asl and are characterised by vertical ice cliffs along the glacier margins, sharp-edged morphology, and penitentes. From a physical viewpoint, all these ice features indicate that net short-wave radiation SW_{in} (1- α) and latent heat flux L_sS provide most of the energy for ablation (e.g. Kraus 1972). Ablation from positive air temperature (affected by sensible heat flux, *H,* and incoming long-wave radiation, LW_{in}) is likely absent, which is supported by measured air temperature on the summit's Northern Icefield (climate station operated by University of Massachusetts, Department of Geosciences), which never exceeded – 1.6°C since the year 2000. An additional indication is the presence of permafrost at 4,700 m, lower than the tongue of the lowest-reaching slope glacier. The pronounced east-west orientation of the summit ice bodies, and the primary north or south orientation of the fast retreating ice walls both point to solar radiation as the main factor in maintaining glacier recession on the summit. In contrast to the stagnant summit glaciers, the convex-shaped slope glaciers still show dynamics but have retreated far above the altitude of their thermal readiness. The combined evidence from summit and slope glaciers strongly supports that modern glacier recession on Kilimanjaro, which began around 1880 (Hastenrath 1984), is governed by atmospheric conditions that have been much drier than previously when glaciers experienced growth. This is consistent with glacier retreat on the two other glaciated massifs in East Africa, Rwenzori and Mount Kenya, which are both strongly affected by drier climate since the end of the 19th century (Kruss and Hastenrath 1987; Mölg et al. 2003b). Decreased atmospheric moisture since this time interval is indicated by climatic proxy data, such as historical accounts of lake levels, water balance models of lakes, paleolimnological data and wind and current observations in the Indian Ocean, which strongly impact East African rainfall (Hastenrath 1984, 2001; Verschuren et al. 2000; Nicholson and Yin 2001).

To evaluate the role of solar radiation in maintaining the rapid retreat of vertical ice walls on the summit glaciers of Kilimanjaro, we applied an ice-radiation geometry model to an idealized representation of the 1880 ice cap. Simulation results confirm that solar radiation is the primary climatic driver behind the retreat of the summit ice walls under drier climate conditions since ca. 1880 (Mölg et al. 2003b).

5. An analytical model approach for investigating low-latitude climate-glacier relations

In the mid latitudes, temperature, moisture advection, and air humidity are strongly connected to high-frequency frontal activities. In contrast, in the low latitudes, temperature is quite constant throughout the year and moisture advection, as well as air humidity, is linked to persistent features of atmospheric circulation. Hence, low-latitude glacier history offers a promising view into the history of long-distance advection and convective transport of moisture, and provides information concerning past atmosphere and ocean dynamics and evaporation conditions on a regional to global scale. This is impressively shown by an analysis of glaciological data from the Central Andes (29°S-30°S), at the boundary between the extratropical westerlies and tropical circulation patterns (Kull and Grosjean 2000). Changes in past glacier extent in this region appear to have been strongly linked to changes in precipitation patterns on millennial timescales. More recently, Francou et al. (2003) have successfully correlated a 10-year measured glacier mass balance series from Glaciar Chacaltaya, Cordillera Real, Bolivia, with atmospheric circulation characteristics derived from reanalysed atmospheric data.

Recently, a process-based glacier model was developed, which incorporates the unique aspects of low-latitude climate-glacier interactions as a function of the different mass and energy balance processes (Kaser 2001). The model calculates one of the most characterising features of glacier-climate relations, the vertical profile of the glacier mass balance, *VBP,* for both wet and dry conditions in the low latitudes. The key terms in the model are i) the duration of the ablation period, which usually lasts all year in the low latitudes, and ii) a parameter describing the contribution of melting and sublimation to ablation under a given availability of energy. Under dry climate conditions, energy is assigned preferably to sublimation, under wet conditions to melting. The *VBPs* for the mid latitudes, the wet tropics, and the dry subtropics, as calculated by the model, are shown in Figure 1. Modelled VBPs of the dry and wet season in the outer tropics correspond closely with those measured on Glaciar Uruashraju in the Cordillera Blanca (Kaser 2001). The differences between the dry and the wet VBP are striking and clearly indicate how changes in the intensity and duration of seasons affect the state of a glacier. Mass turnover is extremely reduced under dry conditions and the vertical balance gradients are weak. Thus, small changes in mass balance have marked effects on the *ELA* and thus on the extent of the glacier. The wet VBP is characterized by a strong balance gradient on the tongue but a weak one in the accumulation area. Consequently, tongues are shorter than on mid latitude or dry glaciers.

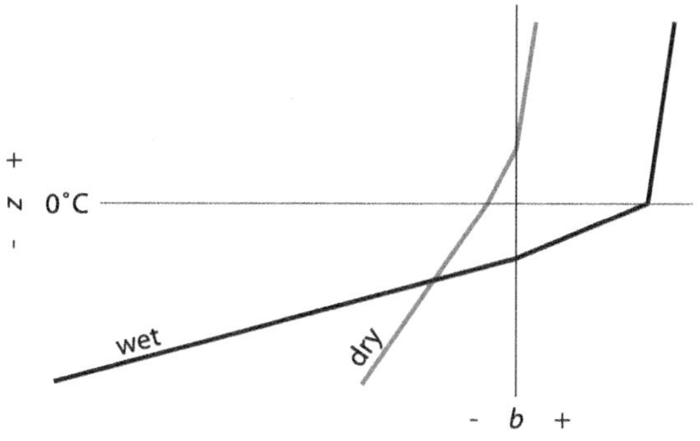

Figure 1: Vertical profiles of specific mass balance on glaciers, as modeled for different climate regimes (after Kaser 2001). *z* indicates the elevation and *b* the specific mass balance in kg m^{-2}; positive mass balance (+) and negative mass balance (-). The inclination of the profiles indicates the mass turnover of a glacier where steep profiles (Subtropics/tropical dry seasons) represent weak mass turn over. Note that a mean annual profile from the outer tropics, averaged over a wet and a dry season, may look similar to the mid latitude one. However, despite this similarity, sensitivities to different climate variables are different.

In addition to sensitivity studies for testing the potential impact of different climate variables on glacier mass balance, the model can also be used to investigate the impact of seasonality on mass balance from a process-based point of view. Seasonality characteristics do not only vary spatially related to the location of atmospheric circulation features, but may also change with time, because circulation features change. Hence, process-based glacier-climate studies in the low latitudes promise interesting reconstructions of global climate history, including changes of the thermal status of the earth system as well as dynamic and phase-changing (e.g. condensation, evaporation) processes. First preliminary model applications for tropical glaciers show promising results and open up new perspectives for both sensitivity analyses and the reconstruction of climate history from past glacier extents. For instance, the retreat of glaciers on Cordillera Blanca during the Younger Dryas (12,900 to 11,600 cal yr BP), due to cold but probably dry conditions (Rodbell and Seltzer 2000), is well simulated by the model (Juen et al. 2002).

6. The impact of glaciers on runoff

The lack of thermal seasonality in low latitudes does not allow the accumulation of a long-lasting snow cover outside the glaciers, leaving glaciers as the major seasonally changing water reservoirs in the low latitude mountains (Ribstein et al. 1995; Kaser et al. 2003b). If the dry seasons are pronounced, as is the case in most of the glacierized low latitude mountains, glacier meltwater, although reduced by sublimation, is the only significant contribution to runoff. Contrary to the mid and high latitudes, glacier melt smoothes the seasonal variation of runoff (Kaser and Osmaston 2002; Mark and Seltzer, this volume). Analyses from the Peruvian Cordillera Blanca show a high correlation between the reservoir capacity of catchment basins - i.e. the monthly or seasonal capacity of catchment basins to store precipitation - and their rate of glacierization (Kaser et al. 2003b). Future resource management in quickly developing societies with increasing demands on freshwater has to take into consideration that a considerable amount of the water supply in the seasonally dry tropics is dependent on the consumption of water previously stored in ice. If, as in some less glaciated catchment basins (e.g. in the southern Cordillera Blanca), the "savings" will soon be consumed, there is no credit to be called upon and runoff will strongly follow the rainy seasons (Fig. 2). But even if glaciers were to advance under colder conditions, savings would be stored in glacier ice and less dry season runoff would be available, unless the amount and seasonal distribution of precipitation were to change markedly. Thus, less runoff has to be expected during future dry seasons and water management strategies have to increasingly concentrate on water quality preservation.

7. Future perspectives

Major climate changes include shifts of the large-scale atmospheric circulation patterns, such as the equatorial easterlies, the trade winds and the westerlies, all inducing strong moisture related gradients when crossing barriers such as the South

American Andes (e.g. Kaser 2002). If the energy content of the earth system changes, these circulation patterns will change their intensity and position with potential major effects on spatial humidity patterns and glacierization (Kull and Grosjean 2000). In order to monitor such changes, a large-scale network of glacier mass balance measurements along climate gradients will be necessary. Ideally, glaciers should be monitored along the north-south chain of the Andes to assess shifts in large-scale atmospheric circulation. Along these transects, pairs of glaciers, from both the west and the east side of the mountain chain, should be investigated to assess the orographic effects of the Andes on the glacier mass balance response to climate change. Similar monitoring networks along climate gradients in the Himalayas could lead to a better understanding of fluctuations in monsoon circulation (Kaser et al. 2003a).

The glaciers on Kilimanjaro's summit plateau offer a particularly interesting tool for investigating climate change in the low latitudes. Under present atmospheric conditions, retreat on the vertical cliffs of these glaciers is unstoppable and they will vanish sooner or later. This raises questions concerning the former vertical structure, and thermal and dynamic conditions of the tropical atmosphere, which allowed glaciers to accumulate prior to the 1880s. The study of present climate-glacier interaction and the reconstruction of past glacier mass balances can help to understand convective dynamic processes at a local scale. In addition, these mass balances can be used to reconstruct patterns of atmospheric circulation and moisture conditions over East Africa and the global tropics.

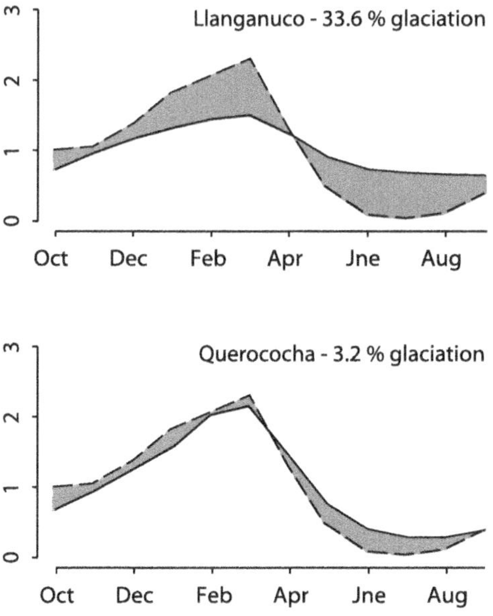

Figure 2: Mean seasonal variations of precipitation (stippled lines) and runoff (solid lines) coefficients in catchments with different glacier coverage in the Cordillera Blanca, Peru. The coefficients indicate mean monthly deviations from long-term means where 1 is the mean. The shaded area is equivalent to the value of the reservoir capacity of the respective catchment area (after Kaser et al. 2003a).

8. Acknowledgments

The authors are grateful to Wilfried Haeberli and two additional reviewers as well as to the editors of the book who all provided constructive and useful comments on an earlier version of the present contribution. Field studies in the Cordillera Blanca, Peru, were carried out within the FWF (Austrian Science Foundation) project P-13567 GEO and were supported by INRENA (Instituto Nacional de Recursos Naturales), Peru and SALEWA mountaineering equipment.

9. References

Braithwaite, R., and Zhang, Y. (2000). Sensitivity of mass balance of five Swiss glaciers to temperature changes assessed by tuning a degree-day model. *Journal of Glaciology* **46**, 7-14.

Francou, B., Vuille, M., Wagnon, P., Mendoza, J., and Sicart, J.-E. (2003). Tropical climate change recorded by a glacier in the central Andes during the last decades of the 20th century: Chacaltaya, Bolivia, 16° S. *Journal of Geophysical Research - Atmospheres* **108**, 4154-4165.

Hastenrath, S. (1984). "The glaciers of Equatorial East Africa." Reidel, Dordrecht.

Hastenrath, S. (2001). Variations of East African climate during the past two centuries. *Climatic Change* **50**, 209-217.

Juen, I., Georges, C., and Kaser, G. (2002). Modelling Younger Dryas glacier extents in the tropical Cordillera Blanca. European Geophysical Society XXVII General Assembly. Nice, France, 21-26 April 2002 (www.copernicus.org/EGS/egsga/nice02/programme/overview.htm).

Kaser, G. (2001). Glacier-climate interaction at low-latitudes. *Journal of Glaciology* **47**, 195-204.

Kaser, G. (2002). Glacier mass balance and climate in the South American Andes: An example from the tropics and a long term and large scale concept for the Southern Patagonian Icefield. *In* "The Patagonian icefields: A unique natural laboratory for environmental and climate change studies." (G. Casassa, F. Sep·lveda, and R. Sinclair, Eds.), pp. 89-99. Series of the Centro de Estudios Científicos. Kluwer, New York.

Kaser, G., and Georges, Ch. (1997). Changes in the equilibrium line altitude in the tropical Cordillera Blanca (Perú) between 1930 and 150 and their spatial variations. *Annals of Glaciology* **24**, 344-349.

Kaser, G., and Osmaston, H. (2002). "Tropical Glaciers." International Hydrological Series. UNESCO-IHP/Cambridge University Press.

Kaser, G., Fountain, A., and Jansson, P. (2003a). "A manual for monitoring the mass balance of mountain glaciers with particular attention to low latitude characteristics." A contribution from the International Commission on Snow and Ice (ICSI) to the UNESCO HKH-FRIEND program. UNESCO technical paper.

Kaser, G., Juen, I., Georges, Ch., Gómez, J., and Tamayo, W. (2003b). Glaciers and Hydrology in the Tropical Cordillera Blanca, Perú. *Journal of Hydrology* **282**, 130-144.

Kaser, G., Hardy, D. R., Mölg, T., Hyera, T., and Bradley, R. S. (in press). Modern glacier retreat on Kilimanjaro as evidence of climate change: Observations and facts. *International Journal of Climatology*.

Kraus, H. (1972). Energy exchange at air-ice interface. *IAHS Publication* **107**, 128-164.

Kruss, P. (1983). Climate change in East Africa: A numerical simulation from the 100 years of terminus record at Lewis glacier, Mount Kenya. *Zeitschrift für Gletscherkunde und Glazialgeologie* **19**, 43-60.

Kruss, P. D. (1984). Terminus response of Lewis Glacier, Mount Kenya, to sinusoidal net balance forcing. *Journal of Glaciology* **30**, 212-217.

Kruss, P. D., and Hastenrath, S. (1987). The role of radiation geometry in the climate response of Mount Kenya's glaciers, part 1: Horizontal reference surfaces. *International Journal of Climatology* **7**, 493-505.

Kuhn, M. (1980). Climate and glaciers. Sea level, ice and climate change. *In* "Proceedings of the Camberra Symposium, December 1979." *IAHS Publications* **131**, 3-20.

Kull, Ch., and Grosjean, M. (2000). Late Pleistocene climate conditions in the north Chilean Andes drawn from a climate-glacier model. *Journal of Glaciology* **46**, 622-632.

Mölg, T., Georges, C., and Kaser, G. (2003a). The contribution of increased incoming shortwave radiation to the retreat of the Rwenzori Glaciers, East Africa, during the 20th century. *International Journal of Climatology* **23**, 291-303.

Mölg, T., Hardy, D. R., and Kaser, G. (2003b). Solar radiation-maintained glacier recession on Kilimanjaro drawn from combined ice-radiation geometry. *Journal of Geophysical Research* **108**, 4731 (doi: 10.1029/2003JD003546) (in press).

Nicholson, S. E., Yin, X., and Ba, M. B. (2000). On the feasibility of using a lake water balance model to infer rainfall: An example from Lake Victoria. *Hydrological Science-Journal-des Sciences Hydrologiques* **45**, 75-95.

Nicholson, S. E., and Yin, X. (2001). Rainfall conditions in Equatorial East Africa during the nineteenth century as inferred from the record of Lake Victoria. *Climatic Change* **48**, 387-398.

Oerlemans, J. (2001). "Glaciers and climate change." Balkema, Lisse.

Ohmura, A. (2001). Physical basis for the temperature/melt-index method. *Journal of Applied Meteorology* **40**, 753-761.

Ribstein, R., Tiriau, R., Francou, B., and Saravia, R. (1995). Tropical climate and glacier hydrology: A case study in Bolivia. *Journal of Hydrology* **165**, 221-234.

Rodbell, D. T., and Seltzer, G. (2000). Rapid ice margin fluctuations during the Younger Dryas in the tropical Andes. *Quaternary Research* **54**, 328-338.

Verschuren, D., Laird, K. R., and Cumming, B. F. (2000). Rainfall and drought in equatorial east Africa during the past 1,100 years. *Nature* **403**, 410-414.

Wagnon, P., Ribstein, P., Kaser, G., and Berton, P. (1999a). Climate variability, energy balance and runoff on a tropical Glacier. *Global and Planetary Change* **22**, 49-58.

Wagnon, P., Ribstein, P., Francou, B. and Pouyaud, B. (1999b). Annual cycle of energy balance of Zongo Glacier, Cordillera real, Bolivia. *Journal of Geophysical Research* **104**, 3907-3924.

Wagnon, P., Ribstein, P., Francou, B., and Sicart, J. (2001). Anomalous heat and mass budget of Glaciar Zongo, Bolivia, during the 1997/98 El Niño year. *Journal of Glaciology* **47**, 21-28.

Glaciers of the Tropical Andes: Indicators of Global Climate Variability

Bernard Francou[1*], Pierre Ribstein[2], Patrick Wagnon[1], Edson Ramirez[3], and Bernard Pouyaud[2]
[1]*Institut de Recherche pour le Développement (IRD), Laboratoire de Glaciologie et de Géophysique de l'Environnement (LGGE), BP 96, F-38402 Saint-Martin d'Hères, France*
[2]*IRD, Maison des Sciences de l'Eau, BP 64501, F-34394 Montpellier, France*
[3]*Universidad Mayor de San Andrés, IRD, CP 9214, La Paz, Bolivia*
[*]*phone +33-4-76824251, fax 591-2-278 4925, e-mail francou@glaciog.ujf-grenoble.fr*

Keywords: Andes, Energy balance, ENSO, Global warming, Mass balance, Tropics.

1. Introduction: A network of glacier monitoring along the tropical Andes

Over the last decade, mass balance has been monitored on several glaciers of the tropical Andes by the Institute of Research for Development (IRD, France) in collaboration with South American partners. This network includes glaciers in the Cordillera Real of Bolivia, Zongo and Chacaltaya (16°S), glaciers in the Cordillera Blanca of Peru, Yanamarey and Artezonraju (9°S), and glaciers in the eastern and western cordilleras of Ecuador, Antizana (0°28'S) and Carihuayrazo (1°S) (Fig. 1). Some of these have been listed as benchmark glaciers by the Word Glacier Monitoring Service (WGMS 2001), and the data are accessible to the scientific community. This network is designed to capture the effects of climate change, and especially ENSO variability, both in the outer (Bolivia, Peru) and the inner (Ecuador) tropical Andes. Glaciers have been selected to be representative of the regional glacierization. Each monitoring programme includes two glaciers, a large one (1 km^2 or more) with a substantial accumulation zone, and a small one that is more directly sensitive to ablation processes. Information about the long-term evolution of some of these

U. M. Huber et al. (eds.), Global Change and Mountain Regions, 197-204.

glaciers has been extracted from aerial photographs, available for the last five decades (Francou et al. 2000; Ramirez et al. 2001). The particular nature of climate in the Tropics allows ablation to occur at anytime throughout the year in the lowest part of glaciers. Thus, the ablation zone has been surveyed in monthly intervals at several sites, providing interesting details about the seasonal response of tropical glaciers (Francou et al. 2003). In some cases, the hydrological balance can be obtained from rain gauges and runoff stations that are installed close to the glacier limits (Ribstein et al. 1995). This information offers the possibility to calculate mass balance from two independent sources (Sicart 2002). In two cases, Zongo and Antizana, radiative and turbulent fluxes have been measured year round by automatic weather stations on the glacier surface, making it possible to estimate the energy balance on one or several points (Wagnon et al. 1999; Sicart 2002). The knowledge of the processes that control melting and sublimation is a necessary pre-condition for understanding the response of glaciers to atmospheric changes over the Andes.

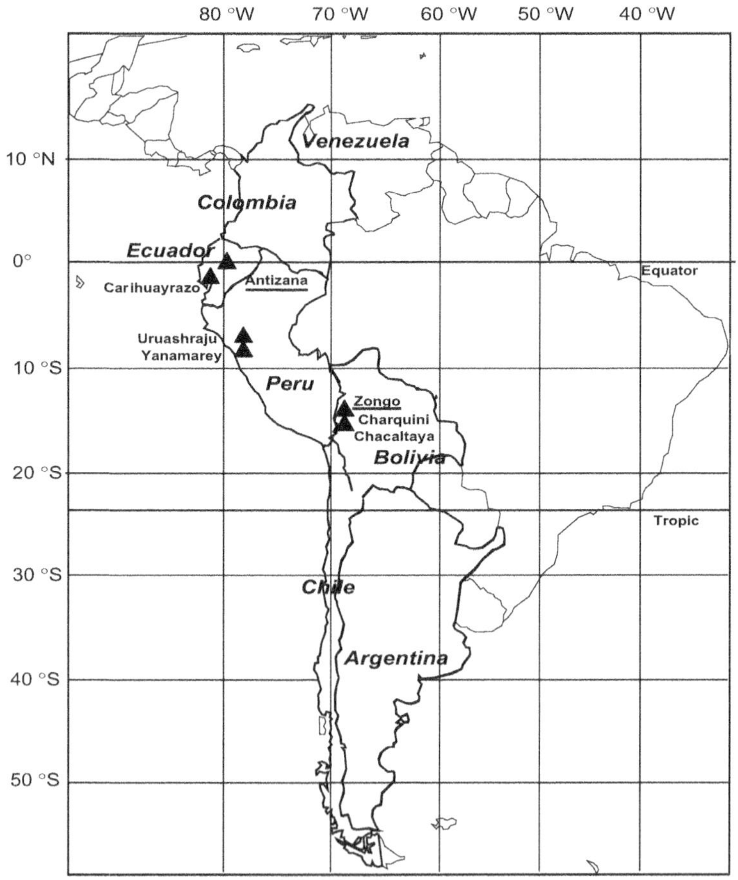

Figure 1: Glaciers monitored by IRD and South-American Partners in the Tropical Andes. The glaciers where energy balance is estimated at year time-scale from automatic weather stations are underlined.

2. Principal results

The data presented below focus only on the main results from our research on mass balance of tropical Andean glaciers.

1. During the last decade, mass balance in this region has experienced a strongly negative trend. Large glaciers, such as Zongo and Antizana, lost 3-5 m of water-equivalent, whereas small ones, such as Chacaltaya in Bolivia, have retreated even more dramatically, with a deficit as high as 13 m of water-equivalent (Fig. 2). The glaciers in the tropical Andes have been retreating in a coherent way, which suggests a common response to a global climate forcing along the mountain chain.

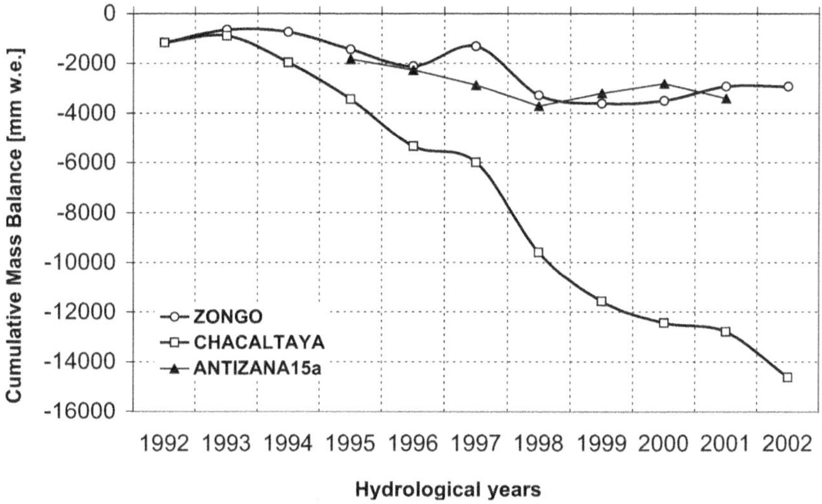

Figure 2: Cumulative mass balance of three glaciers in the tropical Andes. For Zongo and Chacaltaya (Bolivia), the hydrological year is September-August (1992=1991-1992), for Antizana15 (Ecuador) January-December. Note the coherence in the evolution of mass balances. Nevertheless, small glaciers, such as Chacaltaya, have experienced much higher deficits over the observed decade (-1.3 $m.a^{-1}$ w.e. for Chacaltaya vs –0.3 $m.a^{-1}$ w.e. for Zongo).

2. The glacier shrinkage observed during the 1991-2001 period indicates a clear acceleration of the recession trend: Ablation rates have been three to five times higher than during former decades. The small glaciers, such as Chacaltaya, have shown increased recession since at least 1983 (Fig. 3). The fluctuations in mass balance, measured prior to 1983 on several glaciers of the Cordillera Blanca, indicate that the acceleration in glacier retreat had already begun after 1976/1977 (Ames and Francou 1995; Kaser 1999).
3. If the observed rate of recession continues, many small glaciers are expected to disappear in the near future. A model simulation was undertaken for Chacaltaya Glacier (0.06 km^2) to estimate the time required for the complete disappearance

of the glacier. This prediction was based on an average ice deficit of 1.07 m yr^{-1}, which was the rate measured in the 1983-1998 period, and on direct measurements of ice thickness with ground-penetrating radar. The modelling results predict a complete disappearance of the glacier within the next 10-15 years (Ramirez et al. 2001). Water resource managers are concerned by this scenario because most of the glaciers in the Andes are less than 1 km^2 in size and many high-elevation basins with small-sized glaciers are likely to experience a significant decrease in water discharge (Ribstein et al. 1999).

Assuming that ice is lost only through melting, the recession of Chacaltaya since 1983 translates into an average increase in heat supply for glacier melting of 10 W m^{-2} (Ramirez et al. 2001). Similar to Yanamarey glacier in Peru (Hastenrath and Ames 1995), a sensitivity analysis suggests that the stabilization of the glacier would require an increase in cloudiness of less than one tenth, an air temperature decrease of about 1.5°C, a decrease in specific humidity of about 0.5 g kg^{-1}, or some combination of these processes. Such a stabilization would result in a drop of the present average equilibrium line altitude (ELA) by -200 m. Hence, with the above climate changes ELA would decrease from 5400 m asl (decadal average: 1991-2001) to 5220 m asl, which is close to the average ELA observed during the 1940-1963 period, prior to accelerated glacier recession.

4. A better knowledge of the seasonal evolution of mass balance is important for understanding the combination of processes that operate at the glacier surface. In Bolivia, where precipitation seasonality is very strong, the effect of climate variability on mass balance is concentrated in the summer months (October-April), which correspond to the wet season. The wet season explains more than 95% of the variance in the yearly mass balance of Chacaltaya glacier (based on a regression of data from 120 months). In the summer, the most important months are December-February (DJF), which account for 78% of the total variance

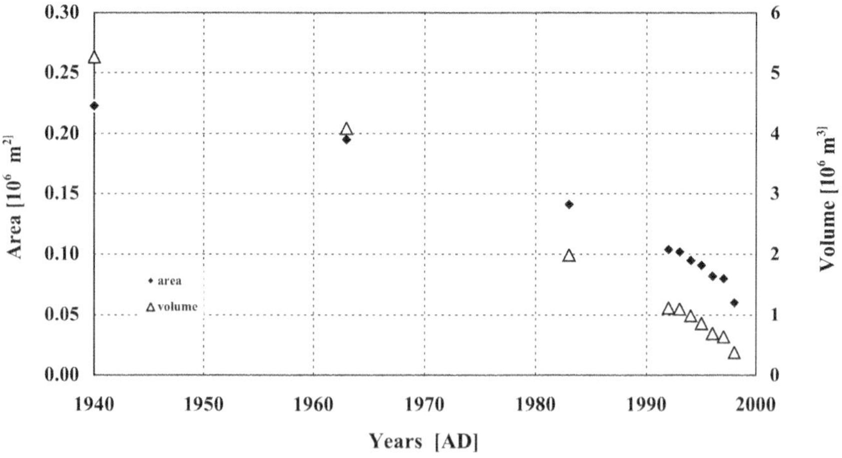

Figure 3: Evolution of Chacaltaya glacier (area and ice volume) over the last decades (from Francou et al. 2000).

(Francou et al. 2003). Between September and December, precipitation (snowfall) is not very frequent and the glacier surface is characterized by bare ice with low albedo. Therefore, the glacier absorbs a great part of the solar radiation, leading to the highest melt rates of the year. At the beginning of January, snowfalls start to become increasingly important and frequent enough to maintain a permanent snow cover of high albedo at the glacier surface. However, due to enhanced cloudiness, incoming long wave radiation is high and thus, there is still a large amount of energy available for melting at the glacier surface. During the dry season (May-September), the incoming long wave radiation is greatly reduced due to decreased cloudiness and therefore, the energy available at the glacier surface decreases. In addition, with enhanced katabatic winds and low air humidity, almost all of this energy is used for sublimation, and melting rates are the lowest of the year. In conclusion, on glaciers of the outer tropics, such as Zongo Glacier, that are characterized by an absence of major thermal seasonality, precipitation, which has a strong feedback on albedo, cloudiness, which controls the incoming long-wave radiation, and wind speed and air humidity, which are responsible for sublimation, are the key variables explaining the seasonal variation of melting and therefore mass balance of the glacier (Wagnon et al. 1999; Sicart et al. in review). In contrast, when seasonality is weak, as in the inner-tropical mountains of Ecuador, melting can occur at any time of the year at low elevation, with the highest rates during the equinoxes, when incoming radiation is at its maximum and albedo is low (due to low snowfall or the occurrence of liquid precipitation) (Francou et al. in review).

5. During the last decade, both in Bolivia and Ecuador, ablation rates have increased significantly during the Pacific warm ENSO phases (El Niño) and decreased during the cold phases (La Niña). Wagnon et al. (2001) applied an energy balance approach to simulate melting processes during the extreme phases of ENSO. This analysis revealed that net all-wave radiation, which is modulated by albedo, is the main factor governing ablation. Albedo, in turn, is dependent upon snow cover, and the snowfall deficit in the early wet season (DJF), generally observed during warm ENSO events, maintains low albedo surfaces of bare ice. Ice surfaces are in direct contact with the atmosphere, which leads to an enhanced absorption of solar radiation and thus to increased melting
6. On interannual time-scales, a close correlation between mass balance and (reanalyzed) air temperature at 500 hPa exists (Francou et al. 2003) (Fig. 4). During El Niño events, near-surface air temperature in the Andes is 0.7-1.3°C higher than during La Niña (Vuille et al. 2003). However, unlike on mid-latitude glaciers, it is not the variation of atmospheric temperature, through the sensible heat flux, which explains the observed variations in mass balance. Other factors play a more important role, in particular humidity, which governs sublimation, and cloudiness, which controls the incoming long-wave radiation (Sicart 2002). Temperature integrates the main energy fluxes that operate at the glacier surface and is thus a relevant variable to explain glacier mass balance evolution on long time-scales (month, year).

7. Based on information from meteorological stations and reanalyzed data from the NCEP- NCAR climate data set, temperature in the tropical Andes has risen by 0.1°C decade^{-1} over the last 50 years and more than 0.2°C decade^{-1} during the last three decades (Vuille et al. 2003). The results of our investigations on energy balance lead us to consider temperature as an important factor behind glacier shrinkage in the tropics, because it is closely interconnected with other relevant climate variables, such as humidity and cloudiness.
8. Indeed, the strong linkage between high melting rates and increased atmospheric humidity is now evident (Hastenrath and Kruss, 1992; Wagnon et al. 1999). Observational and modelling results indicate that relative humidity and water vapour increased by 0.5-1.0% decade^{-1} and 0.1-0.2 hPa decade^{-1} respectively between 1950 and 1995 in both the inner and outer tropics (Vuille et al. 2003). Air humidity variations play an important role in the transfer of energy from sublimation to melting and increased humidity can explain an important part of the observed ice deficit, particularly in the outer tropics.

   Glacier evolution in the central Andes is to a large extent controlled by tropical Pacific sea surface temperatures (SST). The higher SST observed since the mid 1970s in the eastern equatorial Pacific, particularly in the Niño1+2 region, have been associated with a higher frequency and intensity of El Niño events and changes in the spatio-temporal evolution of El Niño. These changes have contributed to the

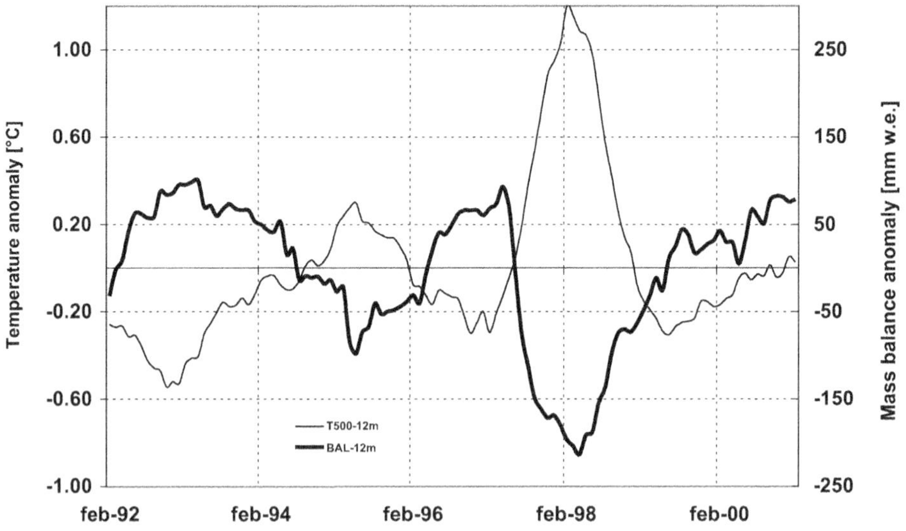

Figure 4: Temperature anomaly at 500 hPa (T500-12m in the figure) and mass balance anomaly (BAL-12m) at Chacaltaya Glacier during 120 months from September 1991 to August 2001. The reanalysed temperature was extracted from the NCEP-NCAR data set. Both time series have been smoothed with a 12-month-averaging filter to remove seasonality. The means of temperature and mass balance are –5.8°C and -107 mm respectively. The two positive peaks of temperature relate to the 1994-1995 and 1997-1998 ENSO warm events, whereas the two negative excursions, 1995-1996 and 1999-2000, closely match ENSO cold events. The 1992-1993 El Niño cold anomaly is associated with the cooling effect of the Pinatubo eruption (from Francou et al. 2003).

dramatic acceleration of glacier retreat in the central Andes (Francou et al. 2003). Whether or not global warming is affecting ENSO evolution is still controversial, but several recent studies have pointed out a strengthening of the tropical general circulation during the last decade, which might be due to natural variability on decadal or multi-decadal time scales (e.g. Wielicki et al. 2002).

These results confirm that tropical glaciers are not just reflecting local-scale climatic conditions but are largely governed by large-scale forcing. It is clear that ENSO variability in the Pacific domain is the main forcing factor behind glacier mass balance variability in the Tropical Andes. An increased frequency of long and intense warm ENSO events considerably accelerated the glacier retreat since the late 1970s, whereas short and less pronounced cold events (La Niña) allowed glaciers to briefly save or gain mass. It is probable that the effects of El Niño on glaciers are superimposed on the effects of global warming, but because of the shortness of the records it has not been possible to tease apart the impacts of these two forcing mechanisms. Global warming might potentially be the primary driver behind the observed glacial retreat, if changes in ENSO structure since the late 1970s are related to global warming. However, the potential impact of global warming on ENSO is still debated among climatologists.

3. Conclusion and future developments

Tropical glaciers are of great interest because of their high sensitivity, which makes them excellent indicators of climate change at low latitudes (see Kaser and Omaston 2002). Tropical regions are known for the spatial consistency of temperature and the low thermal seasonality (Hastenrath 1996). This consistency is reflected in the glaciers since their mass balance tends to show a common response to global climate variability from Bolivia to Ecuador. Considering the effects of ENSO events on glaciers, we can stress that an important part of this variability is driven from the tropical Pacific domain. Nevertheless, to gain the status of "excellent indicators" of global climate change, glaciers have to be very carefully analyzed and the linkages with climate have to be clearly assessed. A better understanding of the evolution of tropical glaciers has to be supported with three kinds of basic information:

1. A permanent observation system along the Andean chain –inner and outer tropics– to provide the scientific community with high-resolution data on mass balance. International programmes (GTN-G, GCOS, WGMS) have developed strategies for mass balance measurements and define investigations in the southern hemisphere as a high priority.
2. Energy balance measurements: Data on energy fluxes are of great interest for understanding the processes operating at the glacier surface, because temperature-based models (i.e. degree-day models) are not suitable for tropical glaciers where ablation is mainly controlled by radiative fluxes.
3. Climate data from high elevation to link the response of glaciers to large-scale atmospheric forcing.

In the last decade, the establishment of large-scale monitoring programmes, which also include a network of surveyed glaciers in various regions of the tropical Andes, has greatly improved our knowledge on tropical glaciers. Nevertheless, much remains to be done to make this network completely operational in the long term.

4. References

Ames, A., and Francou, B. (1995). Cordillera Blanca, Perú. Glaciares en la Historia. *Bulletin de l'Institut Français d'Etudes Andines* **24**, 37-64.

Francou, B., Ribstein, P., Tiriau, E., and Saravia, R. (1995). Monthly balance and water discharge on an inter tropical glacier. The Zongo Glacier, Cordillera Real, Bolivia,16°S. *Journal of Glaciology* **42**, 61-67.

Francou, B., Ramirez, E., Cáceres, B., and Mendoza, J. (2000). Glacier evolution in the tropical Andes during the last decades of the 20th century. Chacaltaya, Bolivia, and Antizana, Ecuador. *Ambio* **29**, 416-422.

Francou, B., Vuille, M., Wagnon, P., Mendoza, J., and Sicart, J. E. (2003). Tropical climate change recorded by a glacier of the central Andes during the last decades of the 20th century: Chacaltaya, Bolivia, 16°S. *Journal of Geophysical Research* **108**, 4154.

Francou, B., Vuille, M., Favier, V., and Caceres, B. (in review). ENSO climate variability impacting glacier mass balance at low latitude: Antizana 15, Andes of Ecuador, 0°28'S.

Hastenrath, S. (1996). "Climate dynamics of the Tropics". Kluwer, Dordrecht.

Hastenrath, S., and Ames, A. (1995). Diagnosing the imbalance of Yanamarey Glacier in the Cordillera Blanca of Peru. *Journal of. Geophysical Research* **100**, 5105-5112.

Hastenrath, S., and Kruss, P. D. (1992). The dramatic retreat of Mount Kenya's glaciers between 1963 and 1983: Greenhouse forcing. *Annals of Glaciology* **16**, 127-133.

Kaser, G. (1999). A review of the modern fluctuations of tropical glaciers. *Global Planetary Change* **22**, 93-103.

Kaser, G., and Omaston, H. (2002). "Tropical Glaciers." International Hydrological Series. UNESCO-IHP/ Cambridge University Press.

Ramirez, E., Francou, B., Ribstein, P., Descloitres, M., Guerin, R., Mendoza, J., Gallaire, R., Pouyaud, B., and Jordan, E. (2001). Small glaciers disappearing in the tropical Andes: A case study in Bolivia: Glaciar Chacaltaya (16°S). *Journal of Glaciology* **47**, 187-194.

Ribstein, P., Tiriau, E., Francou, B., and Saravia, R. (1995). Tropical climate and glacier hydrology: A case study in Bolivia. *Journal of Hydrology* **165**, 221-234.

Ribstein, P., Pouyaud, B., Sicart, J. E., Wagnon, P., Ramirez, E., and Francou, B. (1999). Variabilité climatique et fonctionnement hydrologique d'un glacier tropical. In "Comité National Français de Géodésie et Géophysique. Rapport Quadriennal, Section 6." XXIIe Assemblée Générale de l'Union Géodésique et Géophysique Internationale, Birmingham, July 18-30 1999, pp. 279-287.

Sicart, J. E. (2002). "Contribution à l'étude des flux d'énergie, du bilan de masse et du débit de fonte d'un glacier tropical: Le Zongo, Bolivie." Unpublished Thèse d'Université, Université de Paris 6, Paris.

Sicart, J. E., Wagnon P., and Ribstein, P. (in review). On the relation between meteorological variables and the melting of outer tropics' glaciers. *Journal of Geophysical Research*.

Vuille, M., Bradley, R. S., Werner, M., and Keimig, F. (2003). 20th century climate change in the tropical Andes. *Climatic Change* (in press).

Wagnon, P., Ribstein, P., Francou, B., and Pouyaud, B. (1999). Annual cycle of energy balance of Zongo glacier, Cordillera Real, Bolivia. *Journal of Geophysical Research* **104**, 3907-3923.

Wagnon, P., Ribstein, P., Francou, B., and Sicart, J. E. (2001). Anomalous heat and mass budget of Glaciar Zongo, Bolivia, during the 1997-98 El Niño year. *Journal of Glaciology* **47**, 21-28.

Wielicki, B. A., Wong, T., Allan, R. P., Slingo, A., Kiehl, J. T., Soden, B. J., Gordon, C. T., Miller, A. J., Shi-Keng Yang, Randall, D. A., Robertson, F., Susskind, J., and Jacobowitz, H. (2002). Evidence for large decadal variability in the tropical mean radiative energy budget. *Science* **295**, 841-844.

WGMS (2001). "Glaciers Mass Balance Bulletin 6 (1998-1999)." A contribution to the GCOS, the GTOS, the GEMS and the IHP. Compiled by the World Glacier Monitoring Service (IAHS-ICSI, UNEP, UNESCO, WMO), Zurich.

Glacier Recession in the Peruvian Andes: Climatic Forcing, Hydrologic Impact and Comparative Rates Over Time

Bryan G. Mark[1*] and Geoffrey O. Seltzer[2]
[1]*Max Planck Institute for Biogeochemistry, Jena 07745, Germany*
c/o Department of Geography, University of Glasgow, Glasgow G12 8QQ, United Kingdom
[2]*Department of Earth Sciences, Syracuse University, Syracuse, NY 13244, USA*
**phone +44-141-330-2287, fax +44-141-330-4894, email bmark@geog.gla.ac.uk*

Keywords: Climate forcing, Glacier recession, Hydrology, Peruvian Andes, Rates of deglaciation, Terrain modelling.

1. Introduction

Tropical glaciers are intriguing and apparently rapidly disappearing components of the cryosphere that literally crown a vast ecosystem of global significance. Half of the Earth's surface area lies between the tropics of Capricorn and Cancer, wherein a staggering 75% of the global population resides (Thompson 2000). Tropical glaciers are highly sensitive to climate changes over different temporal and spatial scales, notably ENSO, and are important hydrological resources in tropical highlands (Francou et al. 1995; 2000; this volume; Wagnon et al. 2001; Kaser and Osmaston 2002). Moreover, resolving the complex dynamics and variability of the tropical climate over longer time periods presents important goals to the global modelling community. Compiling an accurate understanding of the timing and climate response of tropical glaciers in the past is a crucial source of palaeoclimatic information for the validation and comparison of climate models (e.g. Farrera et al. 1999; Hostetler and Clark 2000; Porter 2001; Harrison et al. 2002; Seltzer et al. 2002). Deciphering the relative strength of different climatic forcing mechanisms on tropical glacier behaviour and quantifying hydrological changes associated with glacier recession are

U. M. Huber et al. (eds.), Global Change and Mountain Regions, 205-214.

therefore relevant to interpreting the past climate and predicting the impact of future climate changes. Much scientific, social and political attention now concerns future changes in climate, with temperature change predominant. As we look to the future of global change research, specific issues of data access, sharing, and logistics that challenge workers in tropical South America are also common to other such regions globally, particularly those situated in developing countries.

We have studied both present-day glacier recession and field evidence of past episodes of deglaciation in Perú to test hypotheses related to this important climatically forced process in the tropical Andes. Modern glacier recession raises the issues of the nature of climatic forcing and the impact on surface water runoff. While rates of contemporary glacier recession appear to be accelerating (e.g. Thompson et al. 2000), careful analysis of the timing and volumetric extent of deglaciation from Late Glacial and Holocene moraine positions provides a historical comparison with important implications for understanding glacial-to-interglacial transitions. Our research incorporates three specific parts: (1) an analysis of the spatial variability of late 20th century glacier recession in the Queshque massif of the southern Cordillera Blanca, Perú; (2) an evaluation of the hydrological significance of glacial meltwater with respect to streamflow in the Cordillera Blanca region; and (3) an evaluation of the rate and extent of deglaciation during the late-Pleistocene and Holocene compared to modern glacier recession in the Cordillera Vilcanota/Quelccaya.

2. Review of recent research

2.1 Late 20th century glacier recession of Queshque massif, Cordillera Blanca

To investigate the forcing mechanisms behind recent tropical glacier recession, we have coupled our research with the growing number of empirical and theoretical studies in the most-extensively glaciated region of the tropics, the Cordillera Blanca, Perú. Progressive glacier recession in the Peruvian Andes throughout the 20th century (e.g. Petersen et al. 1969; Hastenrath and Ames 1995a; Kaser and Osmaston 2002) is consistent with the well-documented global retreat of alpine glaciers (e.g. Dyurgerov and Meier 2000; Haeberli, Dyurgerov, both this volume). Previous inferences into the climatic forcing of this recession have compensated for the lack of historical climate data by using simple quantitative expressions of mass balance, but they have been limited in temporal and spatial coverage, and have also been challenged in distinguishing between the effects of changes in temperature versus precipitation. An equilibrium mass balance model with assumed constants was applied to suggest alternative simplified climatic changes compatible with the changes in glacier extent observed in the Cordillera Blanca between 1920 and 1970 using aerial photography. The climatic forcing of this early to mid 20th century glacier recession was thus best accounted for by a decrease in atmospheric humidity, resulting in increased global radiation due to lack of clouds, and a decrease in precipitation (Kaser and Georges 1997). Temperature changes were hypothesized to have had a lesser effect, given a distinctive spatial pattern of observed recession. Another project based on carefully mapped observations of the Yanamarey Glacier from 1973 through 1988 estimated that a cloudiness increase of less than 10%, a temperature decrease of 2°C, an increase in specific humidity of 0.1 g kg^{-1} or a combination of the three climate variables would be needed to stabilize the observed negative mass balance (Hastenrath and Ames 1995b). However, this work focused only on a single glacier, and lacked an assessment of the spatial distribution of ice loss. Our investigation was designed to evaluate the most recent four decades of glacier mass loss, spatially distributed over three different glaciers, so as to test specifically the relative influence of temperature and solar radiation.

Solar radiation has been shown to be the largest component of the glacier surface energy balance affecting ablation for mountain glaciers in general (Oerlemans and Knap 1998), and more specifically for tropical Andean glaciers (Wagnon et al. 1999). Solar energy receipt on glaciers is influenced by glacier aspect, surrounding topography, and the glacier surface albedo. Solar radiation and any related climatic phenomenon such as cloudiness should have a differential effect on ablation based on solar geometry, while a uniform temperature rise could potentially override this spatially heterogeneous response on a regional scale (e.g. Hastenrath and Kruss 1992). The central questions we consider are: What has the spatial pattern of the most recent glacier recession been, and how does this reflect the relative influence of solar radiation versus other climatic variables? In other words, was recession a) spatially homogenous across glaciers with different aspects and therefore due to a temperature increase, or b) spatially heterogeneous and therefore related to solar radiation. It is important to consider how the actual volume (and thus mass) of these glaciers with different aspects changed relative to solar insolation, so that appropriate mass balance modelling can be used to quantify the influence of different climatic variables.

We use digital renditions of glacier surfaces based on available aerial photogrammetry and differential Global Positioning System (GPS) measurements to quantify the volume of ice loss between 1962 and 1999 from three glaciers with different aspects from the Queshque massif in the Southern Cordillera Blanca (9°52'30" S, 77°15'00" W). To test the influence of the solar geometry on the glacier recession, we used a simple insolation model to simulate solar energy receipt over the digital elevation model of the Queshque massif and the surrounding topography. The three Queshque glaciers flow to the east, southwest and south, demonstrating a first order structural control by the NW-SE strike of this part of the Peruvian Andes. The 1962 terminus elevations were lower on the southwest-facing slopes, which indicates higher sensitivity to solar radiation than to regional precipitation gradients. However, our modelling results show that glaciers on the southwest aspect experienced the largest relative mass loss since 1962. This negative relationship between simulated solar radiation and relative mass loss by aspect indicates that forcing by solar radiation or related shading was not predominant. Using a sensitivity analysis of the observed glacier mass loss and time series analyses of historical climate data, we present compelling evidence that increased temperatures, in the form of sensible heat transfer, with an accompanying increase in humidity have been the predominant forcing behind recent glacier recession in the southern Cordillera Blanca (Mark and Seltzer in prep).

2.2 Hydrologic impact of glacier meltwater, Cordillera Blanca

Another widely cited concern related to future glacier recession is water supply, particularly for developing Andean nations such as Perú (Chen and Ohmura 1990; Barry and Seimon 2000). In these low-latitude regions where rainfall is strongly seasonal and the persistently high temperature conditions prevent significant snow accumulation, glaciers are potentially significant water reservoirs that serve to buffer stream discharge by providing meltwater year-round. It is not clear how much runoff presently receding glaciers are contributing to stream flow, although recent estimates from Bolivia indicate up to 30% of the annual discharge from streams fed by the Chacaltaya glacier is comprised of glacier melt (Francou et al. 2000). Understanding how much of surface water originates from glacial melt is important for planning purposes, especially in the context of rapidly retreating glaciers. The glacial-fed Río Santa is a valuable hydrological resource for the country of Perú. Of the rivers draining to the Pacific coast of Perú, the Río Santa has the second largest discharge volume and maintains the most regular flow, featuring the least variability in monthly runoff over the year. Four hydroelectric plants supply power to mines and villages along its banks and to coastal cities with large steel mills, anchovy fishery industries, and sugar plantations. The upper Río Santa watershed, draining the western side of the glaciated Cordillera Blanca, is referred to as the Callejón de Huaylas, where we have focused a second component of our research.

We focus this part of our research on a very practical question: how much of the annual Río Santa discharge in the Callejón de Huaylas is not dependent on the highly seasonal precipitation, but derives from melting of glacier ice? To answer this question, we first estimate the percent contribution of glacier melt to two individual glacier catchments in the Cordillera Blanca over the 1998-99 hydrological year by using monthly observations of precipitation, discharge, and the ^{18}O of stream water. We then develop a mixing model based on the hydrochemistry of glaciated and non-glaciated streams to trace the impact of glacier melt water downstream in the larger streams. Individual pro-glacial lakes are manageable settings for measuring the dynamics of the annual glacier-hydrological regime, whereas analyses of historical runoff data for larger glaciated tributary streams of the Río Santa address the impact of glacier melt to water supply at the regional scale.

The results not only show that glacier melt provides significant (10-20%) discharge to the Río Santa annually, but also indicate that maximum melt occurs in the austral spring, and that glaciers effectively buffer seasonal differences in discharge (Fig. 1). Tributary watersheds to the Río Santa with larger fractions of glacier cover have less variable runoff and enhanced discharge, demonstrating that the glaciers effectively buffer stream discharge seasonally. With continued glacier melting, streamflow will likely become more variable, and there will be less dry season runoff. This could reduce agricultural yields that rely on irrigation, decrease power production, and cause potential seasonal shortages in water supply. Mitigation efforts such as constructing dams and reservoirs have been considered, but are problematic in this tectonically active region where slope stability is further compromised by ongoing glacier recession (e.g. Ames 1998).

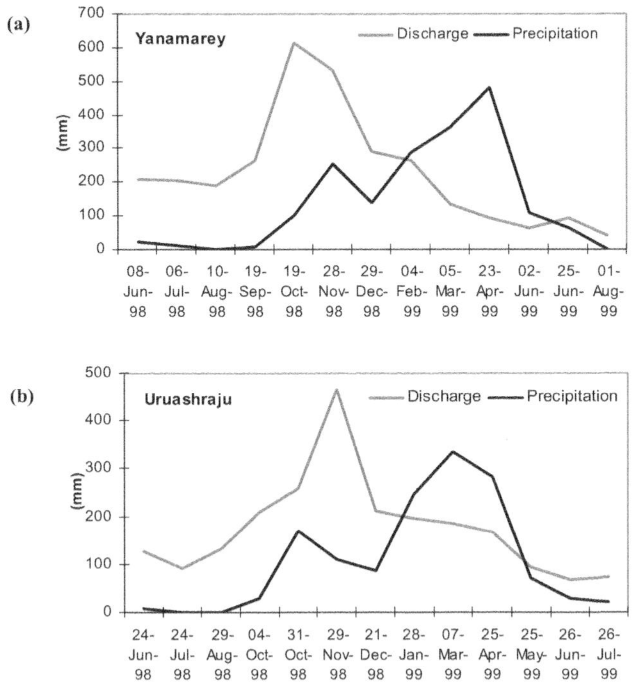

Figure 1: Monthly measurements of discharge ($m^3 s^{-1}$) plotted with the monthly precipitation totals (mm) as observed over the 1998-99 hydrological year for two glacier catchments in the southern Cordillera Blanca, Perú: (a) the Yanamarey glacier; (b) the Uruashraju glacier (from Mark and Seltzer accepted). The data show the importance of glaciers in buffering strongly seasonal discharge, and the significant seasonal character to meltwater production, with a maximum in austral spring.

2.3 Comparative rates of deglaciation, Cordillera Vilcanota/Quelccaya Ice Cap region

Finally, we have conducted an innovative terrain modelling exercise to compare the volumetric rates of late-Glacial and Holocene deglaciation in the Cordillera Vilcanota/Quelccaya Ice Cap region of SE Perú (Mark et al. 2002). The glacial moraines of this region were dated and mapped by previous workers (Goodman et al. 2001), and the site was notable for supplying one of only two maximum limiting ages for the Last Glacial Maximum in the central Andes of Perú and Bolivia (Seltzer 1990). We combine the traditional geomorphological maps and chronologies with a digital elevation model to recreate paleoglacier volumes for distinct periods of time. Within the limitations of the model and chronology, we then compute different rates of deglaciation. The results show that, despite the much smaller volumes of ice involved, the most recent late Holocene rates of deglaciation are larger than late-Glacial rates. However, the rates fall within the range of late 20th century rates measured from the nearby Qori Kalis glacier (Brecher and Thompson 1993; Thompson 2000), indicating that deglacial rates vary significantly over time, and high rates of glacier recession may not be exclusive to the late 20th century (Fig. 2). This has important implications for understanding how topography controls glacier mass distribution on the Altiplano, and thus influences the rate of deglaciation. The large, flat tongues of ice that formed during late-glacial and Holocene advances would have melted rapidly with relatively small changes in temperature and resultant rise in equilibrium line altitude (ELA). Similarly, the flat-lying Quelccaya Icecap is rapidly wasting as temperatures rise and expose the bulk of ice volume below the regional ELA. Nevertheless, up-to-date measurements of the Qori Kalis glacier do confirm the reality of recent global warming, as the rate of terminus retreat continues to accelerate, exceeding late-glacial and Holocene rates (Thompson et al. 2000).

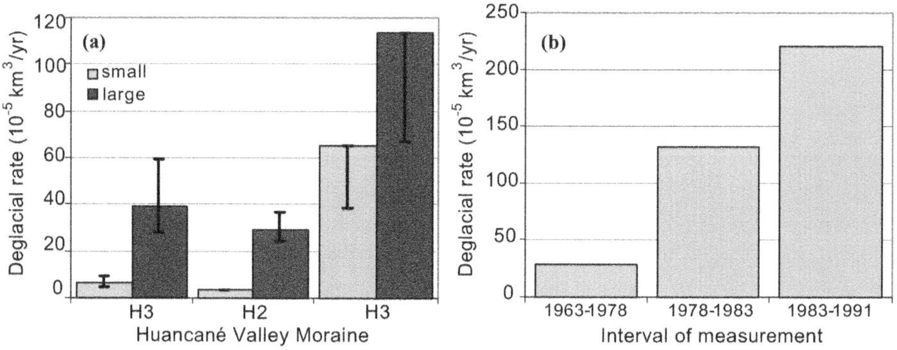

Figure 2: (a) Comparative rates of deglaciation for different moraine positions in the Huancané valley, Quelccaya Ice Cap region of south-eastern Perú, with moraines labelled by age from youngest to oldest as H1 (~270 yrs BP), H2 (~13,000 yr BP), and H3 (>14,300 yr BP). For each moraine position, a "large" and "small" rate are presented to indicate the range of uncertainty in terrain-modelled glacier volumes and moraine chronology. Hence, the "large" rate is the maximum conceptual glacier volume divided by the shortest possible time interval of deglaciation, or smallest difference in sequential moraine ages. Likewise, the "small" rate is the minimum conceptual glacier volume divided by the longest possible time interval. Error bars on each estimate represent the error range associated with a one-sigma range in calibrated radiocarbon ages used in the moraine chronology. All rates are presented as 10^{-5} km^3yr^{-1}; (b) Rates of glacier recession for the Qori Kalis glacier, the largest outlet glacier from the Quelccaya Ice Cap (from Brecher and Thompson 1993). The calculations were made from terrestrial and aerial photogrammetry for three different time intervals: 1963-1978; 1978-1983; and 1983-1993. A clear acceleration in rate towards the present is apparent. The rates are presented as 10^{-5} km^3yr^{-1}, to enable the comparison with rates calculated for other moraine positions in the area. The rate for the H1 moraine position falls within the range of values observed for the Qori Kalis over the late 20th century.

3. Expanding and integrating future research

Although dramatic 20th century glacier recession in tropical mountains has caught media attention as an indication of global climate warming (e.g. Whitfield 2001), a lack of observational data continues to hamper dynamical understanding of these climatically sensitive systems. Glaciated mountain regions in the tropics remain much less intensely researched than similar regions in the middle latitudes, even though on-going glacier recession poses significant hazards to public safety and water supply

(Kaser and Osmaston 2002). An extensive network of mass balance profiles in the Andes coupled with appropriate spatial and climatic modelling would contribute much needed information on the status of these glaciers, both how and why they might be changing. The development of mass balance profiles is the most effective approach for characterizing the climate-glacier interaction. Furthermore, greater integration and data sharing between different agencies and institutions would enhance the value and application of such research. Mountain-based research is motivated not only by the academic rationale to understand better the fundamental mechanisms of tropical mass balance, but also by the very practical issues of water resources and hazards due to glacier recession. However, it is likely that the most important application of research on tropical glaciers in the coming decades will be associated with the use of tropical glaciers as indicators of climate change.

Additional basic observations are needed to formulate theoretical and empirical understanding of glacier-climate interactions from the local watershed scale to the global tropics. Observing conventional measurement protocol and establishing mass-balance monitoring networks across regional climate gradients are important considerations (see Kaser et al., this volume). It is clear that more variables need to be measured, notably humidity, radiation, cloudiness, and surface albedo (see Francou et al., this volume). In many cases, instrumentation installed in the middle part of the 20th century now needs updating, especially stream gauges and weather stations. As always, data would be improved by a denser network of observations, especially for highly spatially-variant parameters such as precipitation. Also, installing more stream gauges at different spatial scales from the small glacier catchment to the larger watersheds, with measurements taken at finer time intervals, would facilitate a more complete understanding of the role of glacier recession in regulating discharge. Finally, realistic long-range data gathering in the often remote areas of the tropical highlands requires a sensible balance between technological sophistication and sustainability. The lower costs of ever more sophisticated technological instrumentation will facilitate elegant and lightweight monitoring stations. However, while lower costs make high-tech operations attractive, they often also involve more complex demands on maintenance that can be problematic in remote regions, and should be considered carefully to insure maximum data recovery. International efforts to centralize glacier observations and distribute data, such as the Global Terrestrial Network for Glaciers, managed by the World Glacier Monitoring Service as part of the Global Climate Observing System, are important initiatives to be complemented by future work in the Andes. Such efforts have already recognized the importance of prioritizing mass balance measurements in the southern hemisphere.

While instrumented studies of energy and mass flux at small scales over specific glaciers are critical, and should be continued, they also need to be accompanied by a greater spatial and temporal coverage. As shown by our results, studying the spatial distribution of glacier mass is critical to understanding the relative role of different forcing mechanisms. Disentangling the climatic signal from other influences on mass balance requires sampling different types of glaciers in different contexts. While the actual ice mass balance is what we are after, mass balance measurements are costly and difficult to maintain. Further, because topography fundamentally controls ice distribution (e.g. Kerr 1993), a better representation of topography with more precise digital elevation models will enhance our ability to link mass balance to glacier extent, providing a link between spatial scales. Advances in geospatial modelling, GPS and remote sensing with Landsat, SPOT and ASTER satellites all hold promise. The Global Land Ice Measurements from Space (GLIMS) program has been initiated with the expressed purpose of utilizing the ASTER data to monitor glaciers, and the Peruvian Andes are identified among the limited number of high priority sites. Moreover, satellites have potential to provide wide-scale measurements of other variables that are not reliably provided by ground measurements, such as humidity flux, radiation, and precipitation. Advances in GIS technology make integration of these various spatial components over different scales much more operationally practical. Likewise, progress in understanding the processes of glacier-climate dynamics over longer time scales of the Quaternary is being enhanced by better chronologies and the incorporation of geospatial modelling (Mark et al. 2002).

All these concerns highlight both the interdisciplinary nature of tropical glacier research and the broad potential for creative, international educational initiatives. We envision the Cordillera Blanca as an ideal site to establish an international field camp. Combined with a relatively rich heritage of previous work in this most glaciated region of the tropics, the context of deglacial impact to water supply and public safety, as well as to other social, economic and natural components of the ecosystem, make the Cordillera Blanca an ideal site. For example, there is an ongoing need to study and map the geological hazards presented by debris flows and moraine-dammed lake outbursts that naturally complement glacier monitoring and stream gauging. An international educational facility like the Juneau Icefields Research Program has long been discussed informally, but has not received deliberate discussion. Such an enterprise would set a precedent for applied, multi-disciplinary, and international collaboration. Better international cooperation would assist in compensating for inconsistent funding, and help the currently small and dispersed community of researchers overcome logistical and bureaucratic obstacles of data access and application. Some small-scale advances have been made in this context: annual summer field camp expeditions are run by the Geology Department at Union College, USA; ongoing work and contractual agreements with local Peruvian agencies have been established by both IRD, France, and the University of Innsbruck; and the Lima-based Andean Institute of Glaciology and Environmental Geology (INAGGA) has organized international glaciological field camps in 2000 and 2002. Future research will benefit by continuing to build upon such endeavours in this mountain region with a long history of human-environmental interactions.

4. Acknowledgements

Our research has been sponsored by grants from the National Science Foundation, the Geological Society of America, the US Fulbright Commission, and the Graduate School of Syracuse University. We acknowledge the helpful collaboration with Peruvian colleagues in the Instituto Geofísico del Perú (IGP), INAGGA, Servicio Nacional de Meteorología y Hidrología (SENAMHI), Egenor S.A., and Instituto

Nacional de Recusos Natuales (INRENA). In particular, we thank Jesús Gómez, Jose 'Pepe' Ames, Chris Hopkinson, Bob Seavey and the INRENA office in Huaraz for help with field work, Abel Rodríguez of Egenor S.A. for contributing historical stream discharge data, Henry Brecher of the Byrd Polar Research Center for assisting with photogrammetry, and Alcides Ames for his logistical assistance and wise counsel. Glacial geological research has been conducted in collaboration with Donald Rodbell and students at Union College, and enhanced with satellite imagery supplied by Cornell University.

5. References

Ames, A. (1998). A documentation of glacier tongue variations and lake developments in the Cordillera Blanca, Peru. *Zeitschrift für Gletscherkunde und Glazialgeologie* **34**, 1-36.

Barry, R. G., and Seimon, A. (2000). Research for mountain area development: Climatic fluctuations in the mountains of the Americas and their significance. *Ambio* **29**, 364-370.

Brecher, H., and Thompson, L. G. (1993). Measurement of the retreat of Qori Kalis glacier in the tropical Andes of Peru by terrestrial photogrammetry. *Photogrammetric Engineering and Remote Sensing* **59**, 1017-1022.

Chen, J., and Ohmura, A. (1990). Estimation of Alpine glacier water resources and their change since the 1870s. "Hydrology in mountainous regions," *International Association of Hydrological Sciences Publication* **193**, 127-135.

Dyurgerov, M. B., and Meier, M. F. (2000). Twentieth century climate change: Evidence from small glaciers. *Proceedings of the National Academy of Science* **97**, 1406-1411.

Farrera, I., Harrison, S. P., Prentice, I. C., Ramstein, G., Guiot, J., Bartlein, P. J., Bonnefille, R., Bush, M., Cramer, W., von Grafenstein, U., Holmgren, K., Hooghiemstra, H., Hope, G., Jolly, D., Lauritzen, S.-E., Ono, Y., Pinot, S., Stute, M., and Yu, G. (1999). Tropical climates at the Last Glacial Maximum: A new synthesis of terrestrial palaeoclimate data: I. Vegetation, lake-levels and geochemistry. *Climate Dynamics* **15**, 823-856.

Francou, B., Ribstein, P., Savaria, R., and Tiriau, E. (1995). Monthly balance and water discharge of an inter-tropical glacier: Zongo Glacier, Cordillera Real, Bolivia, 16°S. *Journal of Glaciology* **41**, 61-67.

Francou, B., Ramirez, E., Cáceres, B., and Mendoza, J. (2000). Glacier evolution in the tropical Andes during the last decades of the 20th century: Chacaltaya, Bolivia, and Antizana, Ecuador. *Ambio* **29**, 416-422.

Goodman, A. Y., Rodbell, D. T., Seltzer, G. O., and Mark, B. G. (2001). Subdivision of glacial deposits in southeastern Peru based on pedogenic development and radiometric ages. *Quaternary Research* **56**, 31-50.

Harrison, S. P., Braconnot, P., Joussaume, S., Hewitt, C., and Stouffer, R. J. (2002). Comparison of palaeoclimate simulations enhances confidence in models. *Eos* **83**, 447.

Hastenrath, S., and Ames, A. (1995a). Recession of Yanamarey glacier in the Cordillera Blanca, Peru, during the 20th century. *Journal of Glaciology* **41**, 191-196.

Hastenrath, S., and Ames, A. (1995b). Diagnosing the imbalance of Yanamarey glacier in the Cordillera Blanca of Peru. *Journal of Geophysical Research* **100**, 5105-5112.

Hastenrath, S., and Kruss, P. D. (1992). The dramatic retreat of Mount Kenya's glaciers between 1963 and 1987: Greenhouse forcing. *Annals of Glaciology* **16**, 127-133.

Hostetler, S., and Clark, P. U. (2000). Tropical climate at the Last Glacial Maximum inferred from glacier mass-balance modelling. *Science* **290**, 1747-1750.

Kaser, G., and Georges, Ch. (1997). Changes in the equilibrium-line altitude in the tropical Cordillera Blanca, Peru, 1930-50, and their spatial variations. *Annals of Glaciology* **24**, 344-349.

Kaser, G., and Osmaston, H. (2002). "Tropical Glaciers." International Hydrology Series, Cambridge University Press.

Kerr, A. (1993). Topography, climate and ice masses: A review. *Terra Nova* **5**, 332-342.

Mark, B. G., Seltzer, G. O., Rodbell, D. T., and Goodman, A. Y. (2002). Rates of deglaciation during the Last Glaciation and Holocene in the Cordillera Vilcanota-Quelccaya Ice Cap Region, Southeastern

Peru. *Quaternary Research* **57**, 287-298.
Mark, B. G., and Seltzer, G. O. (accepted). Tropical glacial meltwater contribution to stream discharge: A case study in the Cordillera Blanca, Peru. *Journal of Glaciology*.
Mark, B. G. and Seltzer, G. O. (in prep.). Forcing mechanisms behind recent deglaciation in the Cordillera Blanca, Peru, from 1962-1999.
Oerlemans, J. and Knap, W. H. (1998). A 1 year record of global radiation and albedo in the ablation zone of Morteratschgletscher, Southwest Switzerland. *Journal of Glaciology* **44**, 231-238.
Petersen, U., Sassarini, L., and Plenge, R. (1969). Glaciar Yanasinga (Central Peru): 24 years of measurements. *Journal of Glaciology* **8**, 487-489.
Porter, S. (2001). Snowline depression in the tropics during the Last Glaciation. *Quaternary Science Reviews* **20**, 1067-1091.
Seltzer, G. O., Rodbell, D. T., Baker, P. A., Fritz, S. C., Tapia, P. M., Rowe, H. D., and Dunbar, R. B. (2002). Early warming of tropical South America at the last glacial-interglacial transition. *Science* **296**, 1685-1686.
Seltzer, G. O. (1990). Recent glacial history and paleoclimate of the Peruvian-Bolivian Andes. *Quaternary Science Reviews* **9**, 137-152.
Thompson, L. G. (2000). Ice core evidence for climate change in the Tropics: Implications for our future. *Quaternary Science Reviews* **19**, 19-35.
Thompson, L. G., Mosley-Thompson, E., and Henderson, K. A. (2000). Ice-core palaeoclimate records in tropical South America since the Last Glacial Maximum. *Journal of Quaternary Science* **15**, 377-394.
Wagnon, P., Ribstein, P., Francou, B., and Pouyard, B. (1999). Annual cycle of energy balance of Zongo Glacier, Cordillera Real, Bolivia. *Journal of Geophysical Research* **104**, 3907-3923.
Wagnon, P., Ribstein, P., Francou, B., and Sicart, J. E. (2001). Anomalous heat and mass budget of Glaciar Zongo, Bolivia, during the 1997/98 El Niño year. *Journal of Glaciology* **47**, 21-28.
Whitfield, J. (2001). "Tropical glaciers in retreat." Nature Science Update, February 19, 2001 (http://www.nature.com/nsu/010222/010222-14.html).

Climate Change, Mountain Permafrost Degradation and Geotechnical Hazard

Charles Harris
Department of Earth Sciences, Cardiff University, P.O. Box 914, Cardiff CF10 3YE, United Kingdom
phone +44-2920 874 336, fax +44-2920 874 326, email harrisc@cardiff.ac.uk

Keywords: Climate change, Hazard assessment, Modelling, Monitoring, Permafrost.

1. Introduction

The IPA Circum-Polar Permafrost Map (Brown et al. 1997) shows discontinuous and sporadic permafrost in the mountains of Europe, including Scandinavia, the Alps, the Pyrenees, and further east in the Urals. In general, the lower altitudinal limit of mountain permafrost increases with decreasing latitude, from sea level in Svalbard, to around 1500 m in Southern Norway, to above 2500 m in the southern Swiss Alps. Many of these low-latitude mountain regions have permafrost temperatures that are only a few degrees below zero, so that a slight shift in energy flux at the ground surface is likely to cause a significant increase in the depth of summer thawing and, in consequence, widespread permafrost degradation. Where permafrost is ice-rich, degradation caused by global warming is likely to be associated with increased magnitude and frequency of mountain slope instability (Harris et al. 2001a). Traditional landslide hazard assessment approaches, based on forward projection of historical data on distribution and magnitude-frequency relationships (Varnes 1984), may therefore become increasingly inappropriate if climate change leads to a significant change in the thresholds of processes within the permafrost geomorphic system. In this paper, approaches to the assessment of geotechnical hazards associated with mountain permafrost in a warming climate are outlined in the context of recent European collaborative research. A critical first stage is the early detection of

U. M. Huber et al. (eds.), Global Change and Mountain Regions, 215-224.

permafrost responses to climate change through integrated monitoring systems.

2. Monitoring mountain permafrost

A Terrestrial Network for Permafrost (GTN-P) has been established under the international Global Terrestrial Observing System (GTOS) initiative to organise and manage a global network of permafrost observations, most importantly of changes in frozen ground temperature (Burgess et al. 2000). The establishment of long-term mountain permafrost monitoring networks is a high priority, since potential hazards associated with permafrost degradation in high mountains may be severe (see Kääb, this volume). The European PACE (Permafrost and Climate in Europe) Project transect of instrumented permafrost boreholes is an example of such a network. It extends from Janssonhaugen, in Svalbard, to the Stelvio Pass, in the Italian Alps, including sites in northern Sweden, Southern Norway and the Swiss Alps (Harris and Vonder Mühll 2001). Permafrost monitoring networks should form the focus for complementary investigations, and in the case of the PACE programme these included geophysical surveys, microclimatic investigations, numerical modelling of permafrost distribution, and physical modelling of permafrost-related slope instability. Results were integrated with the aim of improving assessment of permafrost-related hazards in the context of changing global climates.

3. Mountain permafrost distribution

Mountain permafrost has a complex spatial distribution that depends largely on atmospheric temperatures, radiation and snow distribution in space and time. Clearly, permafrost distribution is critically important to geotechnical hazard assessment. A distinction is necessary between regional-scale assessments, where maps of permafrost distribution form a critical element, and site-specific hazard assessments, where detailed field investigations may be employed to detect those areas underlain by permafrost.

The development of numerical models of permafrost distribution in high mountains has advanced considerably in recent years (e.g. Etzelmüller et al. 2001; Hoelzle et al. 2001), but much remains to be done. Two approaches may be identified: *Empirical-statistical* models and *process-oriented* models. The former relate observed permafrost distribution to topoclimatic factors, such as altitude, slope, aspect, mean air temperature, solar radiation etc. Such models are therefore calibrated within a specific geographical area, and they can be applied within this area using GIS techniques to give a spatial representation of permafrost distribution (Fig. 1) (e.g. Hoelzle 1996; Ishikawa and Hirakawa 2000).

Process-oriented models on the other hand, demand a more detailed understanding of the energy fluxes between the atmosphere and permafrost. They parameterise radiation and sensible heat fluxes in relation to ground characteristics, including surface properties. Accurate modelling of snow cover is particularly critical, and often particularly challenging (Hoelzle et al. 2001). Process-based numerical models are, therefore, complex and require large amounts of precisely measured or computed data. Output is generally mean annual ground surface temperatures from which it is necessary to estimate mean annual permafrost temperatures. This requires a thermal-offset model (Burn and Smith 1988), which in mountain regions is difficult to establish (e.g. Stocker-Mittaz et al. 2002). The thermal offset is the difference in mean annual temperature between the top of the permafrost and the ground surface, and depends largely on the active-layer composition, moisture status and the potential for vertical and lateral heat advection either by water or, in coarse deposits, by air drainage. The final aim is the prediction of changes in the spatial distribution pattern of mountain permafrost in response to climate change, especially in the thermally sensitive marginal zones. Difficulties arise both from the thermal complexities of the boundary layer discussed above, and the fact that the permafrost-climate thermal relationship is unlikely to be in equilibrium, so that the present permafrost thermal regime is in part a function of climate over past decades or centuries.

Over small areas, the distribution of permafrost may be investigated by measuring the basal temperature of dry late winter snow (BTS) (Haeberli 1973). This approach relies on the assumption that snow insulates the soil from atmospheric energy transfers. When the snow cover is at least 1 m thick and melting has not yet begun, the BTS will remain nearly constant, depending largely on heat flux from below, which in turn is strongly influenced by the presence or absence of permafrost (Haeberli and

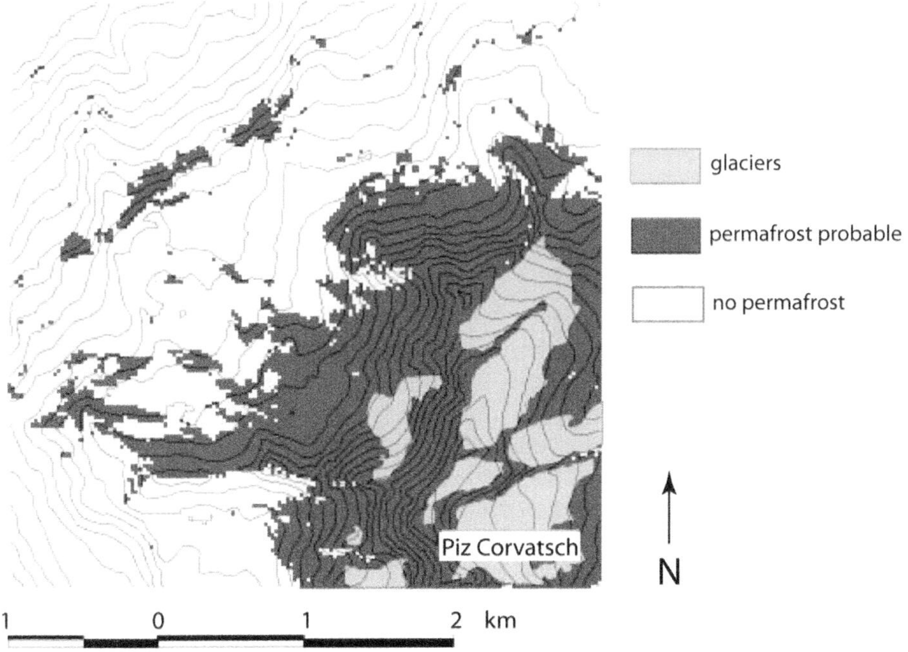

Figure 1: Permafrost distribution modelled with the programme PERMAMAP (Hoelzle and Haeberli 1995) for the local area of Corvatsch-Furtschells, Upper Engadin, Switzerland (from Hoelzle et al. 2001).

Patzelt 1982). Haeberli (1973) proposed that when permafrost is present the BTS is likely to be -3°C or colder, and when the ground is permafrost-free, the BTS is likely to be -2°C or warmer. Calibration of both empirical-statistical and process-oriented numerical models has generally relied on BTS measurements.

Although some care is needed in interpreting BTS data, since timing of winter snowfall can cause inter-annual variations and BTS-permafrost relationships may change with latitude, the technique provides a simple, cheap and effective approach to the identification of discontinuous mountain permafrost within small site-specific areas. A future promising line of research is the use of air-borne radiometry for rapid automatic BTS measurement (Vonder Mühll et al. 2001).

4. Permafrost and slope instability

4.1. Slopes formed in frozen soils

The physical stability of permafrost terrain, especially on steep mountainsides, is highly sensitive to thermal changes, since thawing reduces the strength of both ice-rich sediments and frozen jointed bedrock. Ice-rich soils undergo thaw consolidation during melting, with resulting elevated pore water pressures, so that formerly frozen sediment-mantled slopes may become unstable (e.g. Morgenstern and Nixon 1971). The nature of the resulting slope failure ranges from shallow translational landslides in finer-grained sediments to rapid mudflows and debris flows. Warming-induced permafrost degradation is, therefore, likely to lead to increasing scale and frequency of such slope failures and may also cause thaw settlement damage to foundations (e.g. Haeberli 1992; King and Kalisch 1998).

4.2. Slopes formed in frozen bedrock

Destabilisation of frozen rock slopes by increasing ambient temperature has been attributed to the melting of the ice that bonds joints (e.g. Dramis et al. 1995). Release of meltwater may also cause elevated water pressures in the joint, reducing frictional strength, and migration of groundwater into the formerly frozen rock may further raise pore pressures. In fact, it appears that warming ice-bonded rock slopes may be even more vulnerable to slope failure immediately prior to ice melt, since laboratory tests have shown a decrease in both stiffness and strength of ice with increase in temperature. Davies et al. (2002) conducted a series of direct shear box tests, using simulated ice-bonded rock joints, and demonstrated clearly the loss of strength as temperatures rose. Slope stability analysis showed that a warming frozen 70° model test slope containing planar ice-bonded joints was most vulnerable to rock sliding at a temperature of around – 0.5°C, *before* the joints began to thaw. This hypothesis was elegantly validated in scaled physical modelling experiments using the geotechnical centrifuge. The stability of frozen rock slopes may, therefore, be much more sensitive to changes in temperature than previously envisaged, and failure may occur before significant melting begins.

5. Laboratory modelling of geomorphic process-response

Field studies of landslide triggering mechanisms on thawing slopes in the mountain permafrost zone is extremely challenging, not least because of the difficulty in predicting exactly where and when slope failures will occur. As indicated above, an alternative approach is laboratory physical modelling in which it is possible to control geotechnical and cryogenic properties of the slope materials and their thermal regime, and to accurately monitor process variables. The basis for such physical modelling is that geometric and dynamic similarity between model and prototype (full-scale equivalent) allows the response of the prototype to be interpolated from the observed response of the model. Correct scaling of self-weight stresses within a small-scale model is possible under an elevated gravitational field (where reduced model linear dimensions are compensated for by increased gravitational acceleration), and this principal has been widely applied in geotechnical centrifuge modelling studies (Scholfield 1980). Recently, scaled centrifuge modelling has been used not only to investigate warming frozen rock slopes, but also to investigate slope instability during thawing of ice-rich frozen soils (e.g. Harris et al. 2002).

Critically important in thawing ice-rich soils is the rate of thawing (and hence release of melt water) in relation to the rate of water seepage away from the thaw front. The thaw consolidation ratio (Morgenstern and Nixon 1971) expresses the ratio of these factors as:

$$R = \frac{\alpha}{2\sqrt{C_v}} \tag{1}$$

Where α is the thaw factor, defined as

$$\alpha = \frac{X}{\sqrt{t}} \tag{2}$$

X is depth of thaw in time t and C_v is the coefficient of consolidation.

Similitude in soil properties between prototype and model demands that the same soil is used in the model, that it should have the same stress history and be subject to the same stress conditions as those that apply at prototype scale. It is critically important, therefore, that no conflicts arise in the thaw consolidation ratio when scaling from model to prototype. It may be readily demonstrated that the scaling ratio of both α and C_v is 1:1, so that when thawing a scale model of ice-rich frozen soil under elevated gravity in a geotechnical centrifuge, thaw consolidation processes accurately replicate those in the equivalent prototype. In centrifuge modelling, linear dimensions scale as 1/N, area $1/N^2$, volume $1/N^3$, stress 1/1, temperature 1/1, time for seepage force similarity $1/N^2$ and time for conductive and advective heat transfer $1/N^2$ (Ketcham and Black 1995). Centrifuge modelling of the thaw consolidation process therefore involves time scaling of $1/N^2$ (e.g. at 10 times gravitational acceleration (10g), 1 hr of model time is equivalent to 100 hr (over 4 days) in the full-scale prototype). An illustration of the effectiveness of this approach and its potential for investigating triggering mechanisms is provided by Harris et al. (2002) who described 1/20 scale centrifuge modelling of shallow translational landsliding on a simple 24° planar slope formed in overconsolidated silty clay soil (Fig. 2). Ice segregation during model freezing led to frost heave, with a heaving ratio (ratio of frost heave to frozen soil thickness) of 0.14. The average thaw rate, scaled to prototype time was 1.15 mm per hour, giving a thaw consolidation ratio of 0.71. The model was thawed in

(a)

(b)

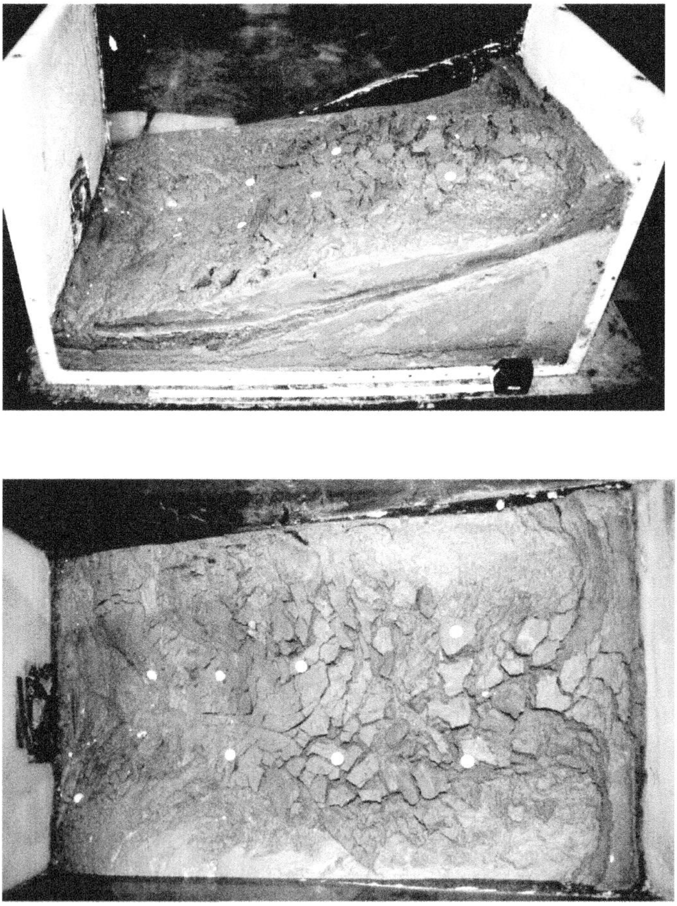

Figure 2: 1/20 scale centrifuge model of a shallow translational landslide developed during thaw of a 24° slope formed in frozen overconsolidated silty clay. Thawing took place under gravitational acceleration of 20g. Test box length is 750 mm, which scales to a prototype length of 15 m. The model thickness is 90 mm, scaling to 1.8 m and the slip surface depths are approximately 30 mm and 50 mm, equivalent to 0.6 m and 1.0 m at prototype scale. (a) Side view. Note two adjacent arcuate slip scars on right of photograph and corresponding lobate toe zones on the left of the photograph; (b) Plan view. The twin arcuate slip scars are clearly visible on the right of the photograph. White discs are plastic displacement markers in the central part of the model where displacements were least.

the centrifuge from the surface downwards at 20g during which time pore pressures, soil temperatures and surface displacements were monitored. The resulting planar landslide was triggered as pore pressures rose. Sliding took place in a sequence of slip-stick events over two distinct shear surfaces, each slip event being triggered by pore pressure rise, with excess pore pressures rapidly dissipating during landslide displacement. Such investigations of geomorphic process variables and triggering mechanisms provide important inputs into hazard assessment techniques in mountain permafrost terrain.

6. Geotechnical hazard assessment

Hazard mapping in mountain terrain has in the past largely focused on the spatial classification of terrain units in terms of the geoenvironmental factors considered conducive to the promotion of slope instability, debris flows, avalanches etc. These have frequently been combined with historical magnitude and frequency data (e.g. Ives and Messerli 1981; Anbalagan 1992). Few studies have specifically addressed the role of permafrost, and none have considered the potential changes in nature, frequency and magnitude of slope instability that might arise from widespread permafrost degradation in a warming global climate. This problem was addressed by the European Union PACE Project and Harris et al. (2001b) reported a structured, two-phased approach designed specifically for hazard assessment in the zone of discontinuous mountain permafrost. The authors emphasised the need to understand triggering mechanisms and thresholds for slope failure and thaw subsidence in a warming mountain permafrost environment. The recommended assessment protocols represent an initial simple procedure designed for the use of practitioners.

Phase 1 of the PACE Protocols consists of a desk study and site reconnaissance to assess the potential for and character of permafrost-related hazard. If, as a result of this desk study, it is considered that permafrost is likely, or cannot be excluded within the site, Phase 2 investigations are necessary, involving detailed ground investigations. Typically, data requirements for Phase 1 will include permafrost distribution, topographic information, slope elements, and geological information (bedrock and surface materials). Combining permafrost distribution, slope elements and geological criteria allows potential permafrost hazard zones to be identified, since permafrost, slope gradient and substrate character largely determine potential thaw consolidation and potential thaw-related slope instability. Hazard characterisation should be undertaken by differentiating between permafrost developed in bedrock and permafrost developed in superficial sediments and determination of potential ice content of the substrate. Integrating these geological factors with topography (defined largely by slope) allows the identification of areas potentially susceptible to permafrost-related slope failures (Table 1).

Phase 2 investigations include detailed large-scale geological mapping and preliminary materials classification within the zone of influence of the proposed project. The aim is to validate and refine the initial hazard assessment matrix (Table 1), relating slope gradient to subsurface materials (bedrock/sediment character), and to identify potentially thaw-sensitive areas where further ground investigations are necessary. Steep bedrock slopes within the potential permafrost zone require standard engineering rock mass descriptions and discontinuity surveys (Hoek and Bray 1981). Analysis of potential rock slope instability should include the destabilisation effects of permafrost warming discussed earlier in this chapter. Investigation of permafrost character, including ice content and ground temperatures, may require drilling of boreholes, using cold compressed air rather than water as the flushing fluid. Spatial integration of permafrost conditions between boreholes may be achieved by using a range of geophysical techniques (see Hauck 2001 for details). If boreholes are to be drilled as part of the ground investigation, it is strongly recommended that monitoring equipment be installed. The thermal and physical condition of permafrost may then be monitored throughout the lifetime of the project.

Table 1: Matrix of potential hazards related to permafrost degradation (Harris et al. 2001b).

	Non-competent lithologies (shales, soft mudstones etc.)	*Competent well-jointed lithologies*	*Competent massive lithologies*	*Fine Grained (silts, clays, some tills)*	*Coarse Grained (screes, gravels, sands)*
≥75	rock fall	rock fall	occasional rock fall	-	-
30-74	debris flows and landslides (including deep-seated failures)	Rockslides, debris flows	-	debris flows	debris flows
15-29	Landslides, thaw subsidence	rockslides	-	landslide/ mudflow	accelerated permafrost creep (rock glaciers)
<15	thaw settlement	-	-	thaw subsidence, solifluction, mudslides on steeper slopes	accelerated permafrost creep
0	thaw settlement	-	-	thaw settlement	-

7. The future

Three aspects of the mountain permafrost hazard assessment methodologies outlined above appear to offer the potential for significant progress in our predictive capabilities. Firstly, the development of more robust permafrost distribution models based on computation of energy balance at the ground surface, and between the ground surface and permafrost via the active layer, will allow not only estimates of the response of permafrost to a given climatic change, but also time-scales of response to be predicted. Such modelling will require increasingly sophisticated inputs of atmospheric and geothermal parameters, and probably rely heavily on synthetic data derived from calibrated remote sensing sources. Secondly, systematic investigation of the geomorphic process-response to thermal forcing is required to better predict thresholds and event magnitude/frequency relationships. Field monitoring programmes are vital in this respect, but are unlikely to yield sufficient data to allow quantification of the significance of a range of topographical, geological and geocryological factors. In this regard, laboratory simulation studies may allow the influence of a wider range of site parameters to be quantified. Finally, integration of spatial data in a GIS environment and the derivation of permafrost-related terrain factors from remotely sensed data are likely to yield new and novel approaches to mapping geomorphic responses to climate change (see for instance, Etzelmüller et al. 2001). It is critically important that calibration methods and sensitivity analyses keep pace with developments in GIS-based data manipulation, to ensure the accuracy of predictions. However, combination of these three research themes is likely to yield a new generation of mountain permafrost hazard assessment tools, with applications in geoenvironmental management, land-use planning, engineering and geomorphological research.

8. References

Anbalagan, R. (1992). Landslide and hazard evaluation and zonation mapping in mountainous terrain. *Engineering Geology* **32**, 269-277.

Brown, R. J., Ferrians, O. J., Heginbottom, J. A., and Melnikov, E. S. (1997). Circum-arctic map of permafrost and ground ice conditions. International Permafrost Association, US Geological Survey.

Burgess, M. M., Smith, S. L., Brown, J., Romanovsky, V., and Hinkel, K. (2000). Global Terrestrail Network for Permafrost (GTN-P): Permafrost monitoring contributing to global climate observations. Current Research 2000-E14, Geological Survey of Canada, 1-8.

Burn, C. R., and Smith, C. A. S. (1988). Observations of the 'thermal offset' in near-surface mean annual ground temperatures at several sites near Mayo, Yukon Territory Canada. *Arctic* **41**, 99-104.

Davies, M. C. R. D., Hamza, O., and Harris, C. (2002). Physical modelling of the effect of climate change on rock slope stability. *In* "Physical modelling in geotechnics: ICPMG." (R. Phillips, P. J. Guo, and R. Popescu, Eds.), pp. 303-308.

Dramis, F., Govi, M., Guglielmin, M., and Mortara, G. (1995). Mountain permafrost and slope instability in the Italian Alps: The Val Pola landslide. *Permafrost and Periglacial Processes* **6**, 73-82.

Etzelmüller, B., Ødegård, R. S., Berthling, I., and Sollid, J. L. (2001). Terrain parameters and remote sensing data in the analysis of permafrost distribution and periglacial processes: Principles and examples from southern Norway. *Permafrost and Periglacial Processes* **12**, 79-92.

Haeberli, W. (1973). Die Basis Temperatur der winterlichen Schneedecke als möglicher Indikator für die Verbreitung von Permafrost. *Zeitschrift für Gletscherkunde und Glazialgeologie* **9**, 221-227.

Haeberli, W. (1992). Construction, environmental problems and natural hazards in periglacial mountain belts. *Permafrost and Periglacial Processes* **3**, 111-124.

Haeberli, W., und Patzelt, G. (1982). Permafrostkartierung im Gebiet der Hochebenkar-Blockgletscher, Obergurgl, Ötztaler Alpen. *Zeitschrift für Gletscherkunde und Glazialgeologie* **18**, 127-150.

Harris, C., and Vonder Mühll, D. (2001). Permafrost and climate in Europe: Climate change, mountain permafrost degradation and geotechnical hazard. *In* "Global change and protected areas." (G. Visconti, M. Beniston, E. D. Iannorelli, and D. Barba, Eds.), pp. 71-82. Kluwer, Dordrecht.

Harris, C., Haeberli, W., Vonder Mühll, D., and King, L. (2001a). Permafrost monitoring in the high mountains of Europe: The PACE Project in its global context. *Permafrost and Periglacial Processes* **12**, 3-12.

Harris, C., Davies, M. C. R., and Etzelmüller, B. (2001b). The assessment of potential geotechnical hazards associated with mountain permafrost in a warming global climate. *Permafrost and Periglacial Processes* **12**,145-156.

Harris, C., Davies, M. C. R., and Rea, B. (2002). Centrifuge modelling of slope processes in thawing ice-rich soil. *In* "Physical modelling in geotechnics: ICPMG." (R. Phillips, P. J. Guo, and R. Popescu, Eds.), pp. 297-302.

Hauk, C. (2001). "Geophysical methods for detecting permafrost in high mountains." Versuchsanstalt für Wasserbau Hydrologie und Glaziologie der Eidenössischen Technischen Hochschule Zürich, 171.

Hoek, E., and Bray, J. (1981). "Introduction to rock mechanics." Wiley, New York.

Hoelzle, M. (1996). Mapping and modelling of mountain permafrost distribution in the Alps. *Norsk Geografisk Tidsskrift* **50**, 11-16.

Hoelzle, M., and Haeberli, W. (1995). Simulating the effects of mean annual air temperature changes on permafrost distribution and glacier size. An example from the Upper Engadin, Swiss Alps. *Annals of Glaciology* **21**, 400-405.

Hoelzle, M., Stocker-Mittaz, C., Etzelmüller, B., and Haeberli, W. (2001). Surface energy fluxes and distribution models relating to permafrost in European Mountain areas: An overview of current developments. *Permafrost and Periglacial Processes* **12**, 53-68.

Ishikawa, M., and Hirakawa, K. (2000). Mountain permafrost distribution based on BTS measurements and DC resistivity soundings in the Daisetsu Mountains, Hokkaido, Japan. *Permafrost and Periglacial Processes* **11**, 109-123.

Ives, J. D., and Messerli, B. (1981). Mountain hazard mapping in Nepal – Introduction to an applied mountain research project. *Mountain Research and Development* **1**, 223-230.

Ketcham, S. A., and Black, P. B. (1995). Initial results from small-scale frost heave experiments in a centrifuge. U.S. Cold Regions Research and Engineering Laboratory, Report 95-9.

King, L., and Kalisch, A. (1998). Permafrost distribution and implications for construction in the Zermatt area, Swiss Alps. *In* "Seventh International Conference on Permafrost Proceedings." (A. G. Lewkowicz, and M. Allard, Eds.), Collection Nordicana. Centre d'Etudes Nordiques, Universite Laval, Quebec, PQ, Canada, 569-574.

Morgenstern, N. R., and Nixon, J. F. (1971). One-dimensional consolidation of thawing soils. *Canadian Geotechnical Journal* **8**, 558-565.

Scholfield, A. N. (1980). Cambridge geotechnical centrifuge operations, 20th Rankine Lecture, *Géotechnique* **30**, 227-269.

Stocker-Mittaz, C., Hoelzle, M., and Haeberli, W. (2002). Modelling Alpine permafrost distribution based on energy-balance data: A first step. *Permafrost and Periglacial Processes* **13**, 271-282.

Varnes, D. J. (1984). "Landslide hazard zonation: A review of principles and practice." UNESCO, Paris.

Vonder Mühll, D., Hauck, C., Gubler, H, McDonald, R., and Russill, N. (2001). New geophysical methods of investigating the nature and distribution of mountain permafrost with special reference to radiometry techniques. *Permafrost and Periglacial Processes* **12**, 27-38.

Glacier and Permafrost Hazards in High Mountains

Andreas Kääb[1*], John M. Reynolds[2], and Wilfried Haeberli[1]
[1] *Department of Geography, University of Zurich-Irchel, Winterthurerstr. 190, CH-8057 Zurich, Switzerland*
[2] *Reynolds Geo-Sciences Ltd, 2 Long Barn, Pistyll Farm, Nercwys, Mold, UK-Flintshire, CH 7 4 EW*
**phone +41-1-635 51 46, fax +41-1-635 68 48, email kaeaeb@geo.unizh.ch*

Keywords: Climate change, Debris flow, Flood, Glacier, Ice avalanche, Permafrost.

1. Glacier and permafrost related hazards

Glacier- and permafrost-related hazards represent a continuous threat to human lives and infrastructure in high mountain regions. Related disasters can kill hundreds or even thousands of people at once and cause damage with a global sum on the order of 10^8 Euro annually. Glacier and permafrost hazards in high mountains include:

- outbursts of glacier lakes, causing floods and debris flows;
- ice break-offs and subsequent ice avalanches from steep glaciers;
- stable and unstable glacier length variations;
- destabilisation of frozen or unfrozen debris slopes;
- destabilisation of rock walls; and
- combinations or chain reactions of these processes.

Generally, glacier floods represent the glacial risk with the highest potential for disaster and damages (up to 3 km^3 flood volume, and, in exceptional cases, up to 40,000 m^3/s runoff and over 1,200 km run-out distances). Glacier floods occur in most glacierised mountains of the world and are triggered by the outburst of water reservoirs in, on, underneath and at the margins of glaciers. Most reservoir types develop slowly and can be identified at the surface, a precondition that favours the

U. M. Huber et al. (eds.), Global Change and Mountain Regions, 225-234.

application of remote sensing techniques for monitoring glacial and periglacial lakes. Floods from ice-dammed lakes and proglacial moraine-dammed lakes, in particular, represent a severe and sometimes recurring danger. Thus, various studies have focused on these glacier lakes, which occur in many mountain ranges of the world (Fig. 1; e.g. Richardson and Reynolds 2000a).

Compared to the distances covered by glacier floods, ice avalanches often affect much smaller areas. Corresponding disasters are generally restricted to densely populated high-mountain regions. However, in combination with other glacier hazards, ice avalanches have the potential for far-reaching disasters. In zones with high seismic activity and geothermal heat flow, the risk of major ice break-offs is greatly increased, as was demonstrated dramatically by one of the most destructive glacier catastrophes, the Huascarán disaster in 1970, with a loss of over 18,000 lives (Plafker et al. 1971). Also, the extraordinary 20 September 2002 rock/ice avalanche

Figure 1: Belvedere Glacier, Monte Rosa massive, Italian Alps, as seen on 13 October 2002. In 1979, an outburst flood from the Lago delle Locce breached the lateral moraines of Belvedere Glacier and caused a severe debris flow, which among other things destroyed a chair lift. To prevent further lake outbursts, the lake level was lowered and controlled by an artificial channel. In summer 2001, the glacier started an untypical surge-type movement with glacier speeds increased by one order of magnitude compared to previous decades. As a consequence, the glacier became heavily crevassed and its tongue advanced over the Little Ice Age moraines, destroying forest and tourism infrastructure. In spring 2002, a so-called supraglacial lake of 3 million m^3 developed on the glacier, which became a severe flood risk for the village of Macugnaga within a couple of weeks. The Italian Civil Defense Department and the scientists involved initiated emergency actions. These included continuous lake level monitoring, evacuation of certain parts of the village of Macugnaga, an automatic alarm system, the installation of pumps and detailed scientific investigations. From summer to autumn 2002 the lake lowered naturally. (Haeberli et al. 2002; Kääb et al. 2003) (Photo: Christine Rothenbühler).

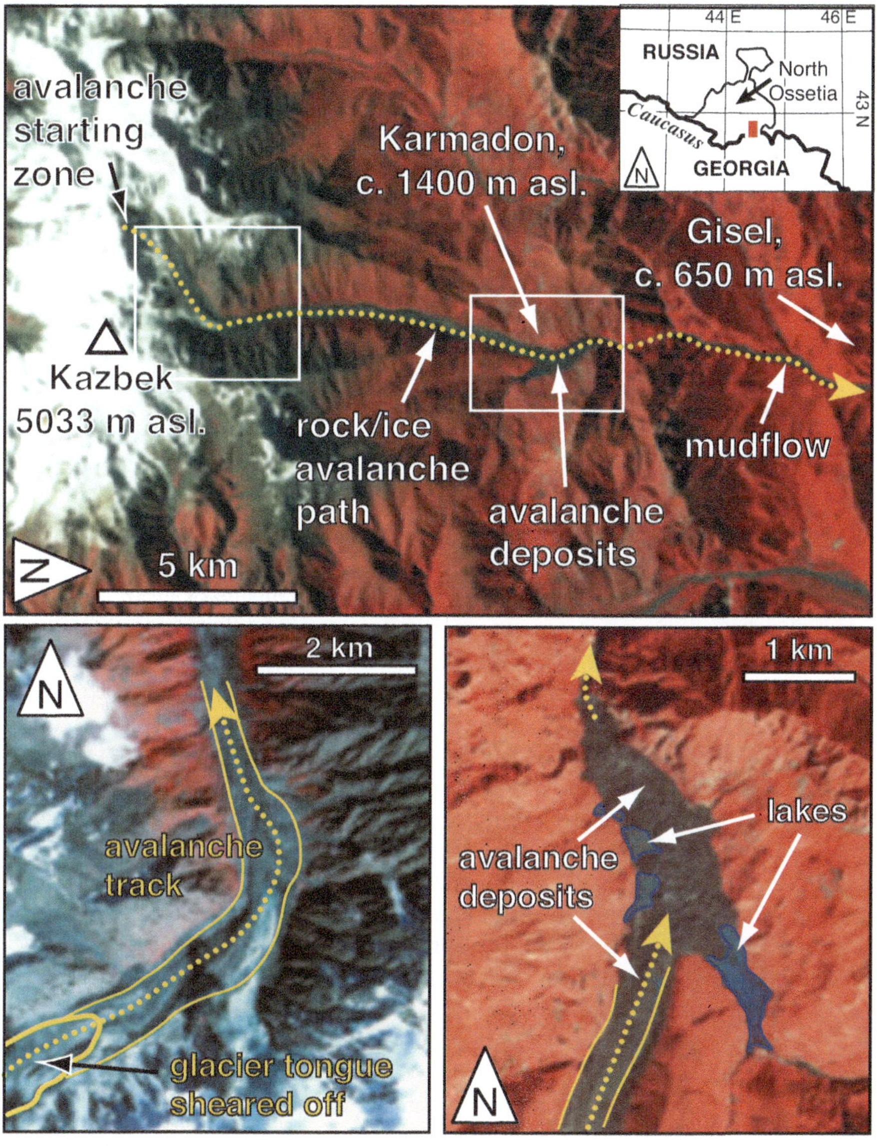

Figure 2: ASTER false color satellite image of 6 October 2002, showing traces and deposits of the 20 September 2002 Kolka-Karmadon rock/ice avalanche that killed over 120 people. The red rectangle in the inset indicates the location of the satellite image. The two white rectangles mark areas that are displayed in more detail in the small pictures. A rock/ice avalanche of several million m^3 started on the north face of the Dzimarai-khokh in the Kazbek massive. Its impact sheared off the entire tongue of Kolka glacier. The subsequent 80 million m^3 avalanche destroyed parts of the village Karmadon before it was stopped by the narrowing valley flanks of the Karmadon gorge. A devastating mudflow was triggered by the rock/ice avalanche and continued 15 km further down-valley but did not reach the village of Gisel. The rivers entering the gorge were progressively dammed by the avalanche deposits and formed huge unstable lakes (Kääb et al. 2003).

at Kolka/Karmadon (Caucasus), a combination of rock and ice destabilisation killing over 100 people, drastically underlines the devastating potential of ice avalanches (Fig. 2; Kääb et al. 2003).

Advancing and retreating glaciers can pose a direct risk to mountain infrastructures. From a global point of view, the prediction of glacier length variations is complicated by the fact that glaciers can vary in a continuous (stable) or unstable way (i.e. glacier surges; e.g. Haeberli et al. 2002). Surging glaciers are able to rapidly destroy installations or induce other hazards, such as ice- and rock-falls. They can temporarily dam lakes, which, when these dams fail, produce some of the largest known outburst floods. Within the last 100 years, for example, a large ice dam in the Upper Indus Basin in Pakistan failed, resulting in the total drainage of 3 km^3 of stored water within 48 hrs.

A widespread risk in high mountains is related to accumulations of loose sediments on steep slopes, which represent potential sources of debris flows (Zimmermann and Haeberli 1992). Such debris accumulations can occur in the form of moraines, moraine dams, or steep valley flanks uncovered by retreating glaciers. In other words, they are strongly connected to glacier development. Another important factor in this context is permafrost, which influences the stability and hydrology of debris slopes (see Harris, this volume). Whilst the trigger mechanisms of these frequently unexpected debris flows often remain unclear (e.g. melting dead ice, permafrost-hydrology interactions) and are therefore difficult to predict in individual cases, the respective hazard potential seems to be connected to the presence of permafrost and its changes. The instability of rock slopes can also be connected to glacier variations. Glacier retreat leads to stress redistribution within adjacent valley flanks, which can cause mass movements such as rock-slides (Kääb 2002). For example, late-glacial ice retreat was associated with a large number of landslides and these events have also been observed during the present glacier retreat since the end of the Little Ice Age. Changes in the thermal regime of cold rock walls and related effects on rock stability (Davies et al. 2001) are still poorly understood processes but are of increasing concern in view of recent catastrophes (Giani et al. 2001).

Glacier floods, ice avalanches and glacier length variations, represent relevant hazard potentials in high mountains on an individual basis. However, combinations and interactions between these or other hazard types are of similar or even greater importance. In fact, many of the largest known glacier catastrophes are characterised by hazard combinations and/or process chains. For instance, if ice and/or rock avalanches enter natural or artificial (i.e. reservoir) lakes, they are able to trigger flood waves and, as a consequence, can lead to overflowing and breaching of natural (or artificial) dams with corresponding flood and debris flow disasters. Ice avalanches are a special risk in the winter season when the run-out distance increases considerably due to strongly reduced friction on snow. In addition, ice avalanches are able to trigger large snow avalanches, thereby greatly enhancing the avalanche volume (Alean 1985). Of much more importance than the direct impact of glacier length variations is the indirect risk of triggering glacier floods and ice avalanches (e.g. Grove 1987). Advancing glaciers are able to dam rivers and create lakes. Many known glacier floods have their origin in such ice-dammed lakes (e.g. Bruce et al. 1987). On the other hand, retreating glaciers often leave behind moraine-dammed lakes. These moraines can be breached resulting in floods and also represent an important sediment source for debris flows (e.g. Haeberli 1983). Glaciers retreating or advancing over a topographic break in slope have a greatly increased risk of ice breaking off their tongue (cf. the 1965 catastrophe of Allalin Glacier, Swiss Alps). Shock waves related to ice- and rock-fall impacts are able to destabilise glaciers and other high mountain terrain.

While the above interactions represent trigger-chains that influence hazard situations on short timescales, many environmental changes in glacial environments can impact the hazard potential on longer timescales of decades or centuries. For example, retreating glaciers may destabilise steep slopes and cause mass movements, or uncover steep debris reservoirs that are sources of periglacial debris flows for decades or centuries after the actual retreat (Zimmermann and Haeberli 1992; Haeberli et al. 1997; Kääb 2000). Such system interactions in high mountains clearly show the urgent need for integrated hazard assessments to account for a variety of relevant processes and their linkages on different timescales.

2. Climate change, human activities and related shifts of hazard zones

Mountain regions are particularly sensitive to climate change. Changes in glaciers, snow and permafrost and corresponding impacts on natural hazards in high-mountain systems could, in fact, be among the most directly visible signals of global warming and may seriously affect human activities (Haeberli and Beniston 1998). For example, in the Himalayas in Nepal and Bhutan, glacier lake outburst floods occurred at a frequency of roughly one flood per decade in the 1950s; this rate has since increased to one flood every three years in the 1990s and it is anticipated that event frequency could further increase to one significant glacier outburst flood each year by 2010 (Richardson and Reynolds 2000a). It is predicted that both the number and size of glacial lakes will increase as climate changes. Coupled with increasing rural development and investments in infrastructure, particularly in hydropower, the vulnerability of mountain communities to outburst floods is growing rapidly. Furthermore, for those rivers fed largely by ice melt, reduction in glacier volumes will have a particularly strong impact on dry-season river flows, and on the provision of downstream water for hydropower, irrigation and potable water supplies. In some regions, most noticeably in Pakistan, climate change will increase environmental, economic and social vulnerability for tens of millions of people. Whilst catastrophic floods (too much water too quickly) are a very palpable hazard, so too are "soft" hazards, such as reduced glacier water during the dry season. Consequently, hazards in high mountains must be considered in relation to water resource management and cannot be seen in isolation. As an example, in Peru, hazards associated with high altitude glacial lakes are being mitigated, using methods that control the lake water volume and ensure safe water reservoirs.

Marked changes in glacier extent due to climate change may be accompanied by both the formation and disappearance of ice- and moraine-dammed lakes, and steep hanging glaciers may become less stable. On the other hand, steep glacier tongues with their present-day potential for large ice avalanches could disappear. Re-vegetation of deglaciated terrain is slow and leaves morainic deposits unprotected against erosion over extensive time periods of several decades and more. On steep slopes, freshly exposed or thawing non-consolidated sediments can become unstable, resulting in debris flows and landslides of varying magnitudes. Once one event has occurred in a particular valley, the remaining slopes may become destabilised even further. The risk of secondary damming of rivers by debris flows also needs to be considered. In places of pronounced glacier retreat, changes in stress distribution and surface conditions of rock walls in deeply cut glacier troughs could induce large mass instabilities. The general tendency is towards a shifting of hazard zones with considerable changes in the processes involved and a widespread decrease in the stability of high-mountain slopes. Special measures are needed to ensure the structural stability and durability of installations for tourism, transportation and telecommunication in permafrost areas. Similarly, detailed hazard assessments must be undertaken routinely and regularly to avoid damage to hydropower installations due to the impact of glacier-derived floods, which can cost many tens of millions of Euros. If, in fact, environmental conditions in high-mountain regions were to evolve beyond the range of Holocene and historical variability, hazard assessments may become increasingly difficult because estimates of hazard potential based on empirical data from the past (historical documents, statistics, geomorphological evidence) will not be directly applicable under new conditions.

3. Modern methods of hazard assessment

Historical data on glacier and permafrost hazards can be used to test spatial models based on new earth observation and geo-informatics techniques. Such modern methodologies provide powerful tools to assist hazard assessments in complex mountain systems, which are experiencing increasing change and divergence from equilibrium conditions.

The assessment of glacier and permafost hazards requires systematic and integrative approaches. Presently, the most successful strategy is based on the combination of remote sensing, modelling with Geographical Information Systems (GIS), geophysical soundings and other local field surveys (Richardson and Reynolds, 2000b). These methods are best structured in a downscaling approach from area-wide first-order assessments for systematically detecting hazard potentials (i.e. the domain of space-borne remote sensing and GIS-techniques) to detailed ground-based or air-borne local investigations in high-risk areas (i.e. the domain of geophysics, surveying, and air-borne and close-range remote sensing).

Air- and space-borne optical and microwave data can be applied to automatically classify glaciers, lakes, debris and other terrain types relevant to glacier and permafrost hazards (Kääb et al. 2002). Furthermore, some of this data can be used to derive digital terrain models (DTM), an invaluable prerequisite for analysing hazard potential in high-mountains and for related GIS-modelling (Huggel et al. 2002; Kääb

2002). Even ice flow and terrain displacements can be measured with high accuracy from repeated remote sensing data (Kääb 2002). With these methods, the terrain cover, geometry and dynamics of an area can be fully investigated without direct access. This can be especially beneficial in mountain areas where the potential sources of glacial hazards lie in geopolitically unstable regions (e.g. Kashmir, Afghanistan) but where the principal impact zones lie significantly downstream. Remote sensing can also be of great use in assessing glacial hazards across international borders where glaciers in an inaccessible part of a country extend their impact into a neighbouring country (e.g. China into Bhutan, India and Nepal).

A further step towards an integrative hazard assessment consists in the application of GIS and other numerical models for simulating processes that are too complex or undetectable by remote monitoring. Glacier lake outburst floods, ice avalanches or debris flows can be modelled with a GIS (e.g. Huggel et al. 2002). Also, permafrost distribution, approximate ground-, firn- and ice-temperatures, or various other terrain parameters that have an impact on natural hazards can be computed. Especially the fusion of remote sensing results with numerical process models provides a promising base for the assessment of hazard potentials (e.g. Huggel et al. 2002).

A more detailed analysis of the hazard sources detected by remote sensing often involves ground-based methods. Geophysical investigations, employing electrical resistivity tomography and ground penetrating radar (Reynolds 1997), in particular, have been used to develop three-dimensional maps of geological structures and have provided information on instability zones such as buried ice bodies within moraine dams, which could lead to breaches in the dam if the ice were to melt (Pant and Reynolds 2000; Richardson and Reynolds 2000b). Furthermore, the use of geophysical methods can provide information about the prevalent physical processes behind glacial and periglacial hazards and can lead to a better understanding of the behaviour of natural dams and their potential to fail. Terrestrial surveying, using laser ranging or Global Positioning Systems (GPS), is needed for accurate mapping and detection of terrain dynamics with high spatial and temporal resolution.

While many glacial and periglacial hazards may develop into major potential hazards, if left unchecked, there are many examples in the Alps, in Nepal, Bhutan and especially Peru, where lakes with high hazard potential have been re-mediated very successfully (Fig.3; e.g. Reynolds et al. 1998; Haeberli et al. 2001). The remote monitoring of changes in glacial lakes is crucial in order to help prioritise which lakes should be re-mediated first and when it would be most expedient.

4. The near future and its challenges

Glacierized mountain areas would be among the most heavily affected parts of the world in the event of accelerated future warming. Due to the complex interactions of the different variables of the energy balance in such areas, potential future changes can only be estimated very roughly. Empirical methods and energy balance considerations indicate that a large fraction (about one-third to one-half) of the presently existing mountain glacier mass on earth could disappear over the next 100 years with anticipated atmospheric changes. With an associated upward shift

Figure 3: The nearly completed spillway at Tsho Rolpa (4,500 m), Rolwaling Himal, Nepal, in 2000. The lake level was lowered successfully by 3.5 m following construction of the new spillway and sluice gates. It took from April 1999 to July 2000 to complete the works at a cost of ca $3 million. Heavy machinery had to be flown to the site in bits and reconstructed at site. Glacier ice is still present within the moraine dam (right-hand side of picture). Although the glacial hazard has been reduced significantly, it has been recommended that the lake level should be drawn down at least a further 11.5 m to ensure an adequate factor of safety for the 150-m high moraine dam (Photograph © Reynolds Geo-Sciences Ltd, 2000).

of the equilibrium line by some 200 to 300 meters, yearly thickness losses of 1 to 2 meters would have to be expected for temperate glaciers, and many low-latitude mountain ranges would lose major parts of their glacier cover within decades. The consequences would include changes in hazard situations, but also in the water cycle and in landscape evolution.

Under such circumstances, the concept of sustainable development in the highest belts of cold mountain areas becomes questionable, because large-scale climatic forcing would by far outweigh any local environmental influences. The main challenge would, in fact, be to adapt to high and accelerating rates of environmental change (Haeberli and Beniston 1998). Empirical knowledge would have to be increasingly replaced by improved process understanding, especially concerning runoff formation and slope stability. Robust numerical models would have to help with the design of hazard mitigation measures at high altitudes. The intensive research on glacier hazards carried out in Switzerland during the past decades (e.g. Haeberli et al. 2001) can illustrate possibilities and limitations of hazard assessments and mitigation.

The above-mentioned recent catastrophes clearly demonstrate the key issues with respect to assessing and mitigating glacier and permafrost hazards:

- The large potential for hazard assessment based on remote sensing and numerical modelling has to be fully exploited, and knowledge has to be transferred to affected regions in the second and third world;
- Scientifically objective criteria need to be developed to assess the hazard potential of glacial lakes and other glacial and periglacial hazards;
- Scientists should work towards a greater transfer of information and improved communication between the scientific and political communities to raise the

awareness and willingness of the responsible authorities to use the available information and knowledge basis on glacial and periglacial hazards;

- The impacts of environmental change on hazard potential need to be continually monitored and a rapid transfer of this information is critical for the successful mitigation of hazards in highly sensitive high-mountain environments.

The sudden and unexpected surge-type flow acceleration and advance of Ghiacciaio del Belvedere (Italian Alps; Haeberli et al. 2002; Kääb et al. 2003) and the Kolka/Karmadon rock/ice avalanche (Caucasus), which has been without internationally known precedent (Kääb et al. 2003), clearly shows that future surprises cannot be excluded. The learning process must continue and an open exchange of knowledge and experience must guarantee high-quality research on glacier and permafrost hazards and their mitigation in high mountain ranges of the world.

5. References

Alean, J. C. (1985). Ice avalanches: Some empirical information about their formation and reach. *Journal of Glaciology* **31**, 324-333.

Bruce, R., Cabrera, G. A., Leiva, J. C., and Lenzano, L. E. (1987). The 1985 surge and ice dam of Glaciar Grande del Nevado del Plomo, Argentina. *Journal of Glaciology* **33**, 131-132.

Davies, M. C. R., Hanza, O., and Harris, C. (2001). The effect of rise in mean annual temperature on the stability of rock slopes containing ice-filled discontinuities. *Permafrost and Periglacial Processes* **12**, 137-144.

Giani, G. P., Silvano, S., and Zanon, G. (2001). Avalanche of 18 January 1997 on Brenva glacier, Mont Blanc Group, Western Italian Alps: An unusual process of formation. *Annals of Glaciology* **32**, 333-338.

Grove, J. M. (1987). Glacier fluctuations and hazards. *Geographical Journal* **153**, 351-369.

Haeberli, W. (1983). Frequency and characteristics of glacier floods in the Swiss Alps. *Annals of Glaciology* **4**, 85-90.

Haeberli, W., and Beniston, M. (1998). Climate change and its impacts on glaciers and permafrost in the Alps. *Ambio* **27**, 258-265.

Haeberli, W., Wegmann, M., and Vonder Mühll, D. (1997). Slope stability problems related to glacier shrinkage and permafrost degradation in the Alps. *Eclogae Geologicae Helvetiae* **90**, 407-414.

Haeberli, W., Kääb, A., Vonder Mühll, D., and Teysseire, P. (2001). Prevention of outburst floods from periglacial lakes at Grubengletscher, Valais, Swiss Alps. *Journal of Glaciology* **47**, 111-122.

Haeberli, W., Kääb, A., Paul, F., Chiarle, M., Mortara, G., Mazza, A., Deline, P., and Richardson, S. (2002). A surge-type movement at Ghiacciaio del Belvedere and a developing slope instability in the east face of Monte Rosa, Macugnaga, Italian Alps. *Norwegian Journal of Geography* **56**, 104-111.

Huggel, C., Kääb, A., Haeberli, W., Teysseire, P., and Paul, F. (2002). Remote sensing based assessment of hazards from glacier lake outbursts: a case study in the Swiss Alps. *Canadian Geotechnical Journal* **39**, 316-330.

Kääb, A. (2000). Photogrammetry for early recognition of high mountain hazards: New techniques and applications. *Physics and Chemistry of the Earth, Part B* **25**, 765-770.

Kääb, A. (2002). Monitoring high-mountain terrain deformation from repeated air- and spaceborne optical data: Examples using digital aerial imagery and ASTER data. *ISPRS Journal of Photogrammetry and Remote Sensing* **57**, 39-52.

Kääb, A., Paul, F., Maisch, M., Hölzle, M., and Haeberli, W. (2002). The new remote-sensing-derived Swiss glacier inventory: II. First results. *Annals of Glaciology* **34**, 362-366.

Kääb, A., Wessels, R., Haeberli, W., Huggel, C., Kargel, J. S., and Khalsa, S. J. S. (2003). Rapid ASTER imaging facilitates timely assessment of glacier hazards and disasters. *EOS Transactions, American Geophysical Union* **84**, 117, 121.

Pant, S. R., and Reynolds, J. M. (2000). Application of electrical imaging techniques for the investigation of natural dams: An example from Thulagi Glacier Lake, Nepal. *Journal of Nepal Geological Society*

22, 211-218.

Plafker, G., Ericksen, G. E., and Fernandez, J. (1971). Geological aspects of the May 31, 1970, Peru earthquake. *Bulletin of the Seismological Society of America* **1**, 543-578.

Reynolds, J. M. (1997). "An introduction to applied and environmental geophysics." John Wiley & Sons, London.

Reynolds, J. M., Dolecki, A., and Portocarrero, C. (1998). The construction of a drainage tunnel as part of glacial lake hazard mitigation at Hualcán, Peru. *In* "Geohazards in engineering geology." (J. G. Maund, and M. Eddleston, Eds.), Geological Society, London, Engineering Geology Special Publications 15, 41-48.

Richardson, S. D., and Reynolds, J. M. (2000a). An overview of glacial hazards in the Himalayas. *Quaternary International* **65/66**, 31-47.

Richardson, S. D., and Reynolds, J. M. (2000b). Degradation of ice-cored moraine dams: Implications for hazard development. *Debris Covered Glaciers* (Proceedings of a workshop held at Seattle, Washington, USA, September 2000). *IAHS Publication* **264**, 187-197.

Zimmermann, M., and Haeberli, W. (1992). Climatic change and debris flow activity in high mountain areas: A case study in the Swiss Alps. *Catena Suppl*ement **22**, 59-72.

Impact of Climatic Changes on Snow Cover and Snow Hydrology in the French Alps

Eric Martin* and Pierre Etchevers
Centre d'études de la neige, 1441 rue de la piscine, 38406 St Martin d'Hères CEDEX, France
phone +33 4 76637900, fax +33 4 76515346, email eric.martin@meteo.fr

Keywords: Climatic change, French Alps, Hydrology, Snow cover

1. Introduction

A better understanding of the potential effects of climate change on snow cover is critical, considering the far-reaching environmental and socio-economic implications on water resources, winter tourism, ecology as well as local changes in climate. The snow coverage of the French Alps depends on weather conditions in a rather complex way. For a given winter, snow cover is the consequence of the various meteorological events encountered (frequency and intensity of snowfall events, atmospheric circulation patterns, cold and warm periods). Simple relationships between snow cover and averaged climate variables, such as mean temperature and precipitation, can therefore not adequately explain interannual variability of snow cover. Models can be used to gain a better understanding of the complex interactions between different climate variables and their effects on snow cover. In addition, models are of great interest for the assessment of potential climatic change impacts.

This paper discusses the observed interannual variability in snow cover in the French Alps. In addition, numerical models are used to test the sensitivity of snow cover changes to global warming and to assess the potential impact of these changes on avalanche activity and mountain hydrology.

U. M. Huber et al. (eds.), Global Change and Mountain Regions, 235-242.

2. Impact of climate warming on snow cover

2.1 Observations of recent changes

There are only few long-term data series of snow cover changes in France. Individual records can only provide limited information on regional trends because of local effects, such as altitude and aspect. However, a single data series can give useful information on interannual variations in snow depth and snow cover duration at a given site. The Centre d'Études de la Neige has monitored snow cover in Col de Porte since 1960. The Col de Porte (David and Martin 1999) is situated at 1320 m asl in the Chartreuse massif (northern Pre-Alps). Mean snow depth, measured annually between February 11-20, clearly shows distinct interannual variability (Fig. 1). Particularly thin snow cover was characteristic for the winter of 1964 (which was used as a reference year until recently) and winters in the late 1980s and early 1990s. Winters characterized by substantial snow-pack became less common in the last decade. Snow depth exceeded 1.5 m only once during the last decade, as opposed to three to four times in each of the previous decades. Snow cover duration also shows high interannual variability. Winters with the longest snow cover duration occurred between the mid 1970s and mid 1980s. The slight decrease in snow depth and snow cover duration since the start of the observation period might be linked to the observed increase in winter temperature (+2°). A negative trend of approximately -0.6 days year^{-1} can also be observed for the number of days with snow cover on the ground (Fig. 1).

2.2 Snow models

The two numerical models SAFRAN and CROCUS have been used to simulate changes in snow cover (Durand et al. 1999). The meteorological analysis system SAFRAN incorporates a range of meteorological data sources (mesoscale model outputs, radio-soundings, standard and automatic observation networks) and simulates meteorological variables at hourly time steps. These models are able to resolve the spatial scale of the massif (between 500 and 1000 km^2, e.g. the Chartreuse massif), which is considered to be homogeneous from a meteorological point of view. SAFRAN provides the input data for the snow model CROCUS.

CROCUS is a multi-layer snow model that simulates the evolution of snowpack characteristics as a function of weather conditions (Brun et al. 1989; 1992). The input data of the model are: air temperature, humidity, wind velocity, short- and long-wave incoming radiation, and amount and phase of precipitation. The model derives the internal conditions of the snowpack by estimating variables such as temperature, liquid water content, density and snow type. To calculate the different variables, the snowpack is divided into different layers that are oriented parallel to the slope. Energy transfers inside the snowpack are projected perpendicular to the slope. In this study, snow cover is calculated assuming level terrain.

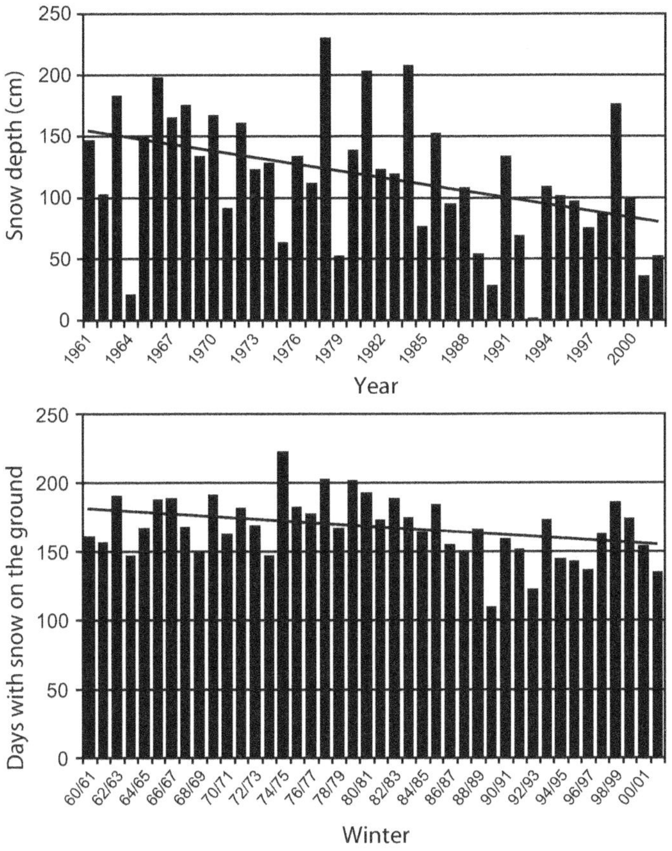

Figure 1: Interannual variability in mean snow depth (11-20 February, top panel) and snow cover duration (days with snow on the ground, bottom panel) at the Col de Porte (1320 m asl, French Alps). A linear trend (black line) has been calculated for each dataset.

2.3 Simulation results

One of the major advantages of the SAFRAN/CROCUS simulations of snow cover is that the modelled datasets can be easily compared between regions, because they are less sensitive to local effects. They account for variables such as altitude, orientation, slope and landscape features. The CROCUS model successfully simulates present snow conditions (snow depth, snow cover duration) in the French Alps. The regional patterns of mean snow cover duration at 1500 m asl (Fig. 2) closely reflect the precipitation gradient across the French Alps. Snow cover duration clearly decreases from 170 days/year in the wet regions of the Northwest to 100-140 days/year in the dry regions of the Southeast. One exception to this pattern is the Mercantour massif in the extreme Southeast corner of the Alps. This region, located in the immediate vicinity of the Mediterranean Sea, is characterized by relatively long snow cover duration, which can be attributed to the impact of low-pressure systems from the South. Gradients in snow cover duration can also be observed in the Pyrenees, with longer snow cover duration in the wet areas of the Northwest (where total precipitation amounts are comparable to the northern Alps). As mean winter temperature in the Pyrenees is higher than in the northern Alps, the mean snow cover duration is shorter (70 to 124 days/year) and comparable to the southern Alps. At high elevation, the snow cover conditions are similar in both regions, because increased precipitation in the South balances the effects of higher temperature.

In order to assess the impact of climatic warming on snow cover duration, a simple doubled CO_2 scenario has been investigated. Based on the results from a General Circulation Model run (the model ARPEGE developed by Météo-France), mean annual temperature was increased by 1.8°C across the French Alps, whereas precipitation was kept constant (Martin et al. 1994). Although the total amount of precipitation did not change in the simulation, the proportion of rainfall relative to snowfall increased due to higher temperatures. The magnitude of the temperature impact on snow cover appears to be spatially differentiated and is primarily related to elevation. Because of the cold climate conditions above 2500 m asl, a small temperature increase has only moderate effects on snow cover. The impact at these altitudes includes a delay in the beginning of continuous snow cover, an increase in the melting rate in the spring, a

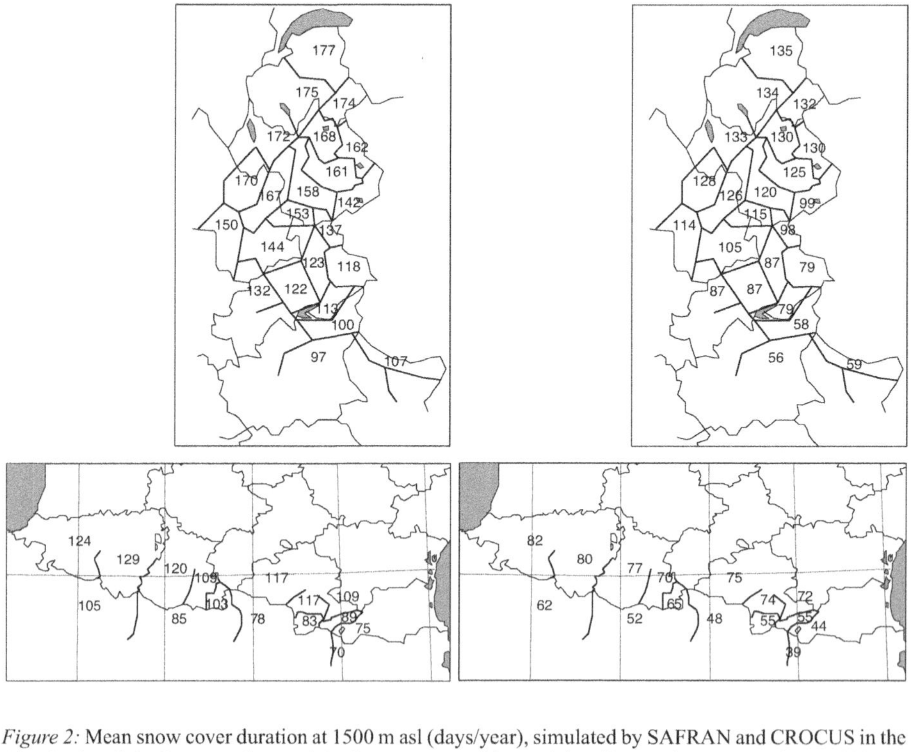

Figure 2: Mean snow cover duration at 1500 m asl (days/year), simulated by SAFRAN and CROCUS in the Alps (top) and the Pyrenees (bottom). Left: present conditions; right: climate change scenario (+1.8°C).

decrease in snow cover duration by 10-15 days and a small decrease in snow depth. Temperature impacts clearly increase with decreasing elevation (and increasing temperature) and appear to be particularly pronounced at approximately 1500 m asl. At this elevation, snow cover duration decreases by approximately one month (Fig. 2) and snow depth shows a decline of 40 to 20 cm. There are also distinct regional differences in the temperature response of snow cover. Snow cover duration decreases from five to four months in the northern French Alps and from three to two months in the southern Alps and the Pyrenees. Changes of this magnitude could have a major impact on the skiing industry in the French Alps, especially at lower elevations. The length of the skiing season could decrease substantially and ski areas may increasingly move into less temperature-sensitive high-elevation areas.

The described results are based on the assumption that temperature changes were not associated with precipitation changes, which might not be valid. Some climate change scenarios anticipate an increase of winter precipitation in a doubled CO_2 climate. This increase may compensate, at least partially, for a decrease in snow depth at high elevation. If warming is in fact associated with increased precipitation, snow depth may increase at high elevations and decrease at low elevations. Future studies require the use of sophisticated downscaling methods in order to link changes in atmospheric patterns to changes in snow cover. A study (Martin et al. 1997) in the French Alps showed that, with a statistical downscaling approach, it is possible to render a more spatially precise scenario of global change effects on snow cover, with regionally differentiated impacts across the French Alps massifs. However, it appears that the uncertainties of the climate change scenarios remain significant.

3. Impact of climate warming on avalanche activity

3.1 Avalanche models

The model used in this study is MEPRA (méthode experte pour la prévision du risque d'avalanche). This model assesses the stability of the snow cover, simulated by CROCUS, on a given slope (Giraud 1993). MEPRA deduces additional mechanical characteristics (e.g. additional grain characteristics, mechanical resistance) and adds this new information to the snow profile. After classifying the different profiles, the model predicts the "natural" mechanical stability of the snowpack (i.e. the risk of spontaneous avalanches without human overloading) by calculating the ratio of shear strength/shear stress for each layer. This estimation is completed by a classification into different avalanche types.

3.2 Simulation results

Natural avalanche activity is primarily related to extreme precipitation events. Global warming may impact avalanche activity by affecting mean snowfall and, more importantly, the frequency and magnitude of extreme events. A modelling study (Martin et al. 2000) in the French Alps shows that an increase in mean winter temperatures is associated with a slight decrease of avalanche hazard. Increased temperatures in this region appear to be associated with a significant decrease of the proportion of snowfall at middle elevation and a more stable structure of the snow cover. However, model results also indicate that, if increases in temperature are accompanied by an increased frequency and magnitude of extreme precipitation events, avalanche hazard may actually increase slightly. When interpreting these model results, one needs to keep in mind that General Circulation Models poorly simulate extreme precipitation events and that the impact on avalanche activity may be more pronounced than indicated by the model. The possible impact of a warmer climate on extreme precipitation events and their effect on avalanche activity requires further study.

A study of past avalanche activity in Switzerland (Laternser et al. 1997), based on observational data since the 15th century, shows no clear trend. This may be at least partially related to difficulties associated with avalanche observations. Also, high short-term variability of avalanche activity might mask potential long-term trends associated with global warming.

4. Impact of climate warming on the flow regimes of mountain rivers

In the framework of the French research project GEWEX Rhône (Etchevers et al. 2001), the potential impacts of climate change on the hydrology of the French part of the Rhône Basin (80,000 km^2) was assessed, using a coupled meteorological-hydrological model (Etchevers et al. 2002). The rivers Durance and Isère are two of the major alpine tributaries of the Rhone River in this area and the hydrological response of these rivers was studied in detail.

4.1 Meteorological-hydrological models

In this study, the SAFRAN and CROCUS models, presented above, were coupled with the ISBA (Interaction Soil-Biosphere-Atmosphere) soil vegetation atmosphere transfer model (Noilhan and Planton 1989) and the hydrological model MODCOU (Ledoux et al. 1989). This integrated approach allowed a comprehensive simulation of the various components of the hydrological budget for the whole basin.

4.2 Simulation results

A total of six doubled CO_2 climate scenarios were investigated, using various GCMs (Arpège-Centre national de recherches météorologiques, ECHAM - Max Planck Institute, UKMO -United Kingdom Meteorological Office, LMD - Laboratoire de Météorologie Dynamique, HC - Hadley Center, UR - University of Reading, Etchevers, unpublished data). GCM results indicate that the temperature changes are relatively homogeneous across the French Alps. Warming is most pronounced in December (around 2°C) and moderate from January to April (1 to 2°C). A comparison of these different scenarios gives some estimate of the uncertainties of GCM predictions. Precipitation changes are highly variable from one model to the other. In some of the models, total precipitation increases (up to +50%). Because precipitation increases are accompanied by increased temperature, snowfall does not increase in the Isère and Durance basins. Snow cover simulations reflect these changes. In the model scenarios, snow cover extent in January and February is drastically diminished from 24,000 km^2 under present conditions to 6000-10,000 km^2.

The modelled changes in snow cover are reflected in the hydrological response of the Rhone River Basin. Seasonal river discharge appears to be greatly affected in all climate change scenarios. Model results for the Isère Basin indicate that the timing of maximum snow cover (in water equivalent) advances from April to March. Winter discharge increases by 10%, whereas spring discharge decreases by ≤15%. In addition, the time of maximum spring discharge advances. The most important reduction in discharge is observed in July, because of increased evapotranspiration and a decreased contribution of snowmelt. Annual river discharge increases in three of the four scenarios and decreases in one. Model simulations show changes of similar magnitude in the High-Durance Basin. However, the decrease in annual river discharge is slightly more pronounced. In the Doubs Basin, Jura mountains, snow cover disappears almost completely, greatly decreasing the snowmelt contribution to river discharge in the spring.

5. Conclusions and future research directions

Our modelling results indicate that snow cover is highly sensitive to increased temperature in a doubled CO_2 climate. This temperature impact can to some extent be offset by an increase in precipitation. Our simulations show that temperature sensitivity is particularly high at mid elevations (around 1000-1500 m asl). Snow cover extent in this altitudinal range can decrease by 50% in winter. The hydrological consequences of these changes are increased winter discharge and decreased summer discharge. These changes would increase the vulnerability of both mountain regions and adjacent lowlands to summer droughts. It must be noted that the uncertainties of climate predictions are still very large, as highlighted by the range of climate change predictions from different models.

Future research in mountain regions should therefore focus on:

- Improving model simulations of climate change impacts on snow cover, avalanche activity, and water resources in mountain regions. In this context, special efforts should be put towards the downscaling of GCM outputs. Currently, most GCMs have insufficient resolution to adequately resolve even major mountain ranges. This is a limitation of most impact studies, as some important characteristics of the climate (variability, extreme events, precipitation amounts) are not well simulated.
- Assessing the integrated impact of environmental change at the scale of watersheds by using coupled meteorological-hydrological models. Such research should include the effects of human impact (e.g. land use changes) on the hydrologic cycle.

- Assessing changes in the frequency and magnitude of extreme events and their impacts on avalanches and floods.
- As snow cover is expected to decrease in temperate regions in response to the temperature increase predicted for this century, there is a need for snow monitoring at larger scales (e.g. the European mountains). Monitoring should be based on existing networks, but also on remote sensing.

6. References

Brun, E., Martin, E., Simon, V., Gendre, C., and Coléou, C. (1989). An energy and mass model of snow cover suitable for operational avalanche forecasting. *Journal of Glaciology* **35**, 333-342.

Brun, E., David, P., Sudul, M., and Brunot, G. (1992). A numerical model to simulate snow-cover stratigraphy for operational avalanche forecasting. *Journal of Glaciology* **38**, 13-22.

David, P., and Martin, E. (1999). "Le laboratoire du Col de Porte pour l'étude de la neige: histoire et climatologie." *La Météorologie* **28**, 23-34.

Durand, Y., Giraud, G., Brun, E., Mérindol, L., and Martin, E. (1999). A computer-based system simulating snowpack structures as a tool for regional avalanche forecasting. *Journal of Glaciology* **45**, 469-484.

Etchevers, P., Golaz, C., and Habets, F. (2001). Simulation of the water budget and the river flows of the Rhône basin from 1981 to 1994. *Journal of Hydrology* **244**, 60-85.

Etchevers, P., Golaz, C., Habets, F., and Noilhan, J. (2002). Impact of a climate change on the Rhone river catchment hydrology. *Journal of Geophysical Research* **107** (10.1029/2001JD000490).

Giraud, G. (1993). "MEPRA: An expert system for avalanche risk forecasting," (R. Armstrong, Ed.). International snow science workshop, 4-8 October, 1992, Breckenridge, Colorado.

Laternser, M., Schneebeli, M., Foehn, P., and Amman, W. (1997). "Climat, neige et avalanches," Argumment de la recherche 13/97, Swiss Federal Institute for Forest, Snow and Landscape Research.

Ledoux, E., Girard, G., de Marsilly, G., and Deschenes, J. (1889). Spatially distributed modelling: Conceptual approach, coupling surface water and ground water. *In* "Unsaturated hydrologic modelling: Theory and practice." (H. J. Morel-Seytoux, Ed.), pp. 435-454. NATO Scientific Series C, 275.

Martin, E., Brun, E., and Durand, Y. (1994). Sensitivity of the French Alps snow cover to the variation of climatic variables. *Annales Geophysicae* **12**, 469-477.

Martin, E., Giraud, G., Lejeune, Y., and Boudart, G. (2000). Impact of climate change on avalanche hazard. *Annals of Glaciology* **32**, 163-167.

Martin, E., Timbal, B., and Brun, E. (1997). Downscaling of general circulation models outputs: Simulation of the snow climatology of the French Alps. Sensitivity to climate changes. *Climate Dynamics* **13**, 45-56.

Noilhan, J., and Planton, S. (1989). A simple parameterisation of land surface processes for meteorological models. *Monthly Weather Review* **117**, 536-549.

Modelling the Response of Mountain Glacier Discharge to Climate Warming

Regine Hock[1*], Peter Jansson[1], and Ludwig N. Braun[2]
[1]*Department of Physical Geography and Quaternary Geology, Stockholm University, SE-106 91 Stockholm, Sweden*
[2]*Commission for Glaciology, Bavarian Academy of Sciences, Marstallplatz 8, D-80539 Munich, Germany*
[*]*phone +46-8-164784, fax +46-8-164818, e-mail regine.hock@natgeo.su.se*

Keywords: Climate change, Glacier discharge, Glacier mass balance, Modelling, Storglaciären, Vernagtferner.

1. Introduction

Glaciers are characteristic features of mountain environments but are often not recognized for their strong influence on catchment runoff quantity and distribution. Such modification occurs with glacierization of only a few percent of the total catchment area, and affects adjacent lowlands far beyond the limits of mountain ranges. The main impact occurs because glaciers temporarily store water as snow and ice on many different time scales (Jansson et al. 2003), the release from storage being controlled by both climate and internal drainage mechanisms.

Mountain glaciers have generally experienced a worldwide retreat and thinning since the beginning of the 20th century in response to a ~0.6°C increase in mean global temperature (IPCC 2001). General Circulation Models (GCMs) predict enhanced global warming in the coming decades due to anthropogenically-induced greenhouse warming (IPCC 2001), which will likely accelerate the current glacier decline. Consequently, additional water is expected to be released from glacier storage thus modifying current streamflow regimes further. Any change in storage and release of water by glaciers is important for all aspects of watershed management, including the

U. M. Huber et al. (eds.), Global Change and Mountain Regions, 243-252.

operation of hydroelectric facilities and flood forecasting, and hence such changes have direct economic implications in many parts of the world. In this paper, we discuss the specific characteristics of glacier discharge and the expected changes from climate warming. We emphasize a need to maintain and establish long-term monitoring programmes combining glacier mass balance and discharge measurements. In addition, further studies are needed to quantitatively assess the impact of climate change on runoff by fully exploiting the growing availability of modern, sophisticated modelling tools.

2. Characteristics of glacier discharge

The distinct modification of streamflow by glaciers is well documented in the literature (e.g. Meier and Tangborn 1961; Stenborg 1970; Fountain and Tangborn 1985; Lang 1986; Braithwaite and Olesen 1988; Chen and Ohmura 1990; Collins and Taylor 1990; Hopkinson and Young 1998). Five specific characteristics can be identified (Table 1):

1. *Specific runoff*: Total streamflow is reduced in years of positive glacier net balance, when water is withdrawn from the annual hydrological cycle and put into glacier storage. The opposite occurs in years of negative glacier mass balance since water is released from long-term glacier storage, thereby increasing streamflow.
2. *Seasonal variation*: In climatic regions subject to distinct accumulation and ablation periods, most annual runoff is concentrated during the summer melt season, while runoff is negligible during winter when most precipitation is stored as snow. Kuhn and Batlogg (1998) found the ratio of maximum monthly runoff to mean monthly runoff to increase linearly with increasing glacierization above a 5% level. In contrast to snow-covered basins subject to spring meltwater peaks, the seasonal glacier-meltwater runoff-peak is delayed due to refreezing and firn saturation at the onset of melt. In addition, melt rates are generally higher in summer due to larger sun altitude angles, higher air temperatures and lower albedo of ice compared to snow.
3. *Diurnal variation*: Glacier discharge is characterized by large diurnal fluctuations caused by the pronounced diurnal cyclicity in meltwater production. Daily peak discharges increase by up to several hundred percent of daily minimum flows during precipitation-free days.
4. *Year-to-year runoff variability*: Glaciers tend to dampen annual streamflow variations, generally referred to as the *glacier compensation effect* (Lang 1986), where ablation variations offset precipitation variations. A moderate fraction of 10 to 40% glacierization reduces year-to-year variability to a minimum. The variability becomes larger at both higher and lower glacierization-levels.
5. *Runoff correlation*: Finally, there is a tendency for runoff to correlate positively with temperature and negatively with precipitation with increasing glacierization, while glacier-free basins show positive correlations between runoff and precipitation. This indicates the dominance of glacier melt water over precipitation as contributor to runoff in highly glacierized basins.

Table 1: Schematic summary of discharge characteristics in highly glacierized basins and expected changes when balanced glacier mass balances turn negative for prolonged periods due to a warming climate. Responses are shown both for an initial phase and a later stage when glacier sizes and volumes have been significantly reduced by long-term glacier mass loss.

Variable	Characteristic	Expected change of variable under a climate warming	
		Initial	*Later stage*
Specific runoff	Decrease for positive mass balances Increase for negative mass balances	Increase →	Decrease
Seasonal variation	Runoff concentration during melt season	Prolongation of melt season, reduced runoff concentration	
Diurnal fluctuation	Pronounced diurnal cyclicity	Increase →	Decrease
Year-to-year variability	Glacier compensation effect: reduced variability at moderate glacierization	Increase or decrease depending on initial glacierization	Increase
Runoff correlation	Positive correlation with temperature Negative correlation with precipitation	Increase →	Decrease

3. Effects of climate warming

Various studies have investigated the effects of climate warming on glacier discharge by analysing historic discharge data from glacierized and non-glacierized basins (Fountain and Tangborn 1985; Chen and Ohmura 1990) or by runoff modelling (Singh and Kumar 1997; Kuhn and Batlogg 1998; Braun et al. 2000). The primary effect of climate warming on runoff is increased streamflow caused by larger glacier melt rates. These are further accelerated by positive feedback mechanisms, such as enlargement of bare ice areas that reduce albedo compared to that of snow-cover. Runoff distribution will also be modified by changes in meltwater routing through the glacier, caused by a reduction in the thickness and extent of firn and snow cover. Since water flow velocities through ice considerably exceed those through snow and firn (Fountain and Walder 1998), a reduced firn body and snow cover will reduce the water retention capacity and will thus speed up average transport of water on and through the glacier. Enhanced diurnal discharge peaks and amplitudes are a direct consequence (Fig. 1).

The response of glacier runoff to climate warming is a matter of timescale. Although the response is immediate, some response-variables will change sign at a later stage when enhanced melt rates have caused glacier volume to decrease significantly (Table 1). Assuming unaltered precipitation conditions, specific runoff will initially increase due to release of water from storage when mass balances are negative. However, prolonged long-term mass loss will reduce glacier volume, which in turn will lead to reduced water yields (Braun et al. 2000; Jansson et al. 2003). For basins in Washington State, Fountain and Tangborn (1985) found that, for glacier mass balances of −1 m, a mean annual runoff of 2 m and glacierization of 20%, streamflow is enhanced by about 10% due to glacier mass loss. Singh and Kumar (1997) and Kuhn

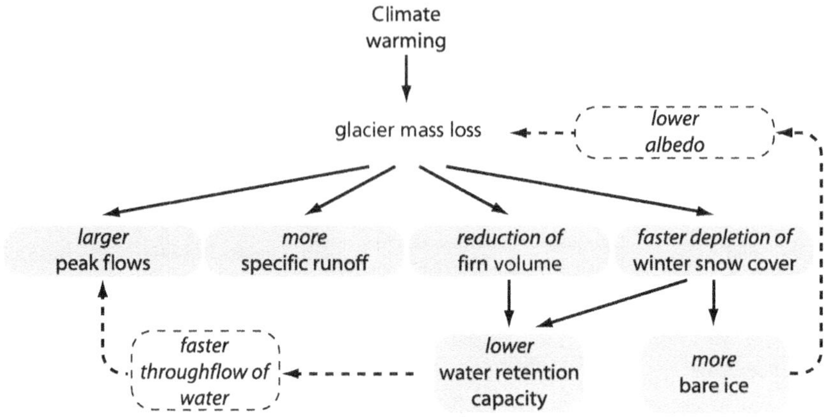

Figure 1: Short-term effects of climate warming on glacier discharge including feedback mechanisms.

and Batlogg (1998) modelled enhanced specific runoff owing to increased glacier melt for Himalayan and Austrian Alpine basins, respectively. Singh and Kumar's study is based on daily time steps and a GCM-output for a future climate forcing, while Kuhn and Battlogg's study uses monthly time steps and assumes uniform temperature shifts. Both assume unchanged glacier size, thus failing to model potential runoff reduction due to glacier retreat. Braun et al. (2000) modelled the effect of a 2xCO_2 scenario, obtained from a regional climate model, on daily discharge of a 40% glacierized basin in the Austrian Alps using the conceptual HBV3-ETH9 runoff model and employing different assumptions on deglaciation. Current water yields during summer are more than doubled using present-day glacier boundaries, while water yields are drastically reduced when a markedly reduced glacierized area is prescribed. In this case, the runoff regime approaches the characteristics of present-day nivo-pluvial regimes of lower-Alpine areas. The long-term effect of runoff reduction, resulting from a decrease in ice-covered area, was also detected by Chen and Ohmura (1990), who analyzed multi-decadal discharge records in the Swiss Alps.

Climate warming is expected to prolong the melt season, thus reducing seasonal runoff concentration, i.e. the ratio of maximum monthly to mean monthly discharge. In addition, diurnal discharge fluctuation will, at least in an initial phase, be amplified by enhanced daily melt water production and more efficient water transport through the glacier (Braun et al. 2000; Willis et al. 2002). Figure 2 shows the temporal development of bare ice surface at the end of the melt season and the amplitude of the diurnal variation of discharge of Vernagtferner, Austria, since the beginning of measurements in 1974. In the 1970s, when mass balance was positive (Fig. 3) and 10–30% of the glacier area consisted of bare ice at the end of the balance year, typical mean diurnal variation of discharge during the melt season was 0.5–1 m^3s^{-1} with maximum amplitudes of ~5 m^3s^{-1}, based on hourly mean values. In the 1990s, bare ice area was as large as 90% as a result of strongly negative mass balances (Fig. 3). Mean diurnal variation of summer discharge was up to 3 m^3s^{-1} with maximum amplitudes exceeding 10 m^3s^{-1}. Preliminary modelling results from Storglaciären (3 km²), northern Sweden, on the effects of a temperature rise on hourly discharge are shown in Figure 4. Melt is modelled with a grid-based energy balance model, based on meteorological data collected on the glacier (Hock 1998). Water is routed through the glacier by a linear reservoir approach. After calibration, the observed temperature record is raised by 2°C. Results show increased runoff, more pronounced daily amplitudes and a prolonged runoff period.

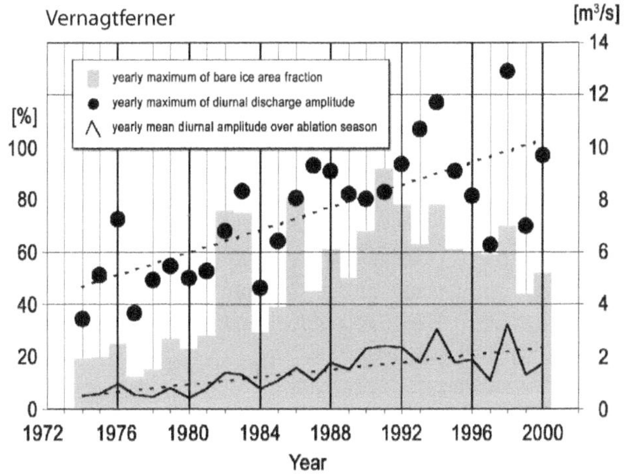

Figure 2: Bare ice area fraction of Vernagtferner (Oetztal Alps, Austria) at the end of the ablation season, and mean (May-October, solid line) and maximum diurnal amplitude of discharge (m^3s^{-1}) at gauging station Vernagtbach (basin area 11.4 km²), for each year between 1974 and 2000, based on hourly means. Dashed lines denote linear trends.

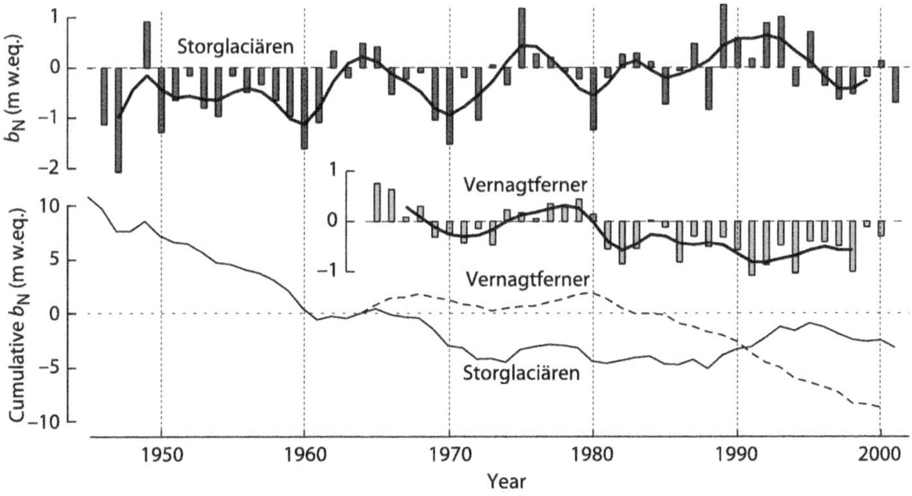

Figure 3: Mass balance (b_N) of Storglaciären, northern Sweden, (1946–2001) and Vernagtferner (1964–2000). Running 5-year moving averages are shown as solid lines. Cumulative net balance for both glaciers with a common point at 0 m w.eq. at 1964 (the start of the Vernagtferner mass balance program).

The impact of climate warming on year-to-year runoff variability will depend on initial glacierization, i.e. whether or not glacierization is above or below the ~10 to 40% yielding minimum variability. When glacierization falls below this threshold, runoff variability tends to increase. In the long term, runoff will correlate positively with precipitation instead of air temperature because the percentage of temperature-dependent melt-derived glacier runoff decreases with decreasing glacierization. However, initially, correlation with air temperature may increase since meltwater yield is increased due to rising temperature.

Another issue of concern is how climate change will affect glacier runoff in different climatic regions. Hagg and Braun (2003) made a comparison between the Alps and the continental mountain ranges of the Tien Shan, Central Asia, and showed that the effects of a doubled CO_2 climate on glacier runoff are similar in both areas. In an initial phase, runoff is greatly enhanced due to higher temperature, but with decreasing glacier area, water yield will gradually diminish. Once glaciers have disappeared, summer runoff is greatly reduced. Since melt water from glaciers is the only source for water during dry seasons in the continental areas of Central Asia, loss of glaciers will affect lowland areas even more drastically in Asia than in, e.g. Central Europe.

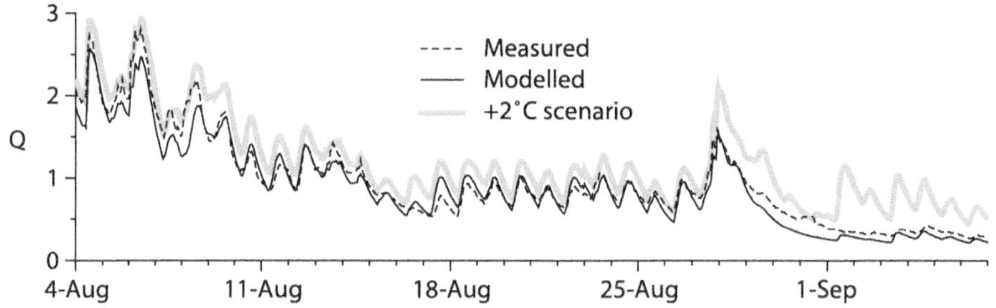

Figure 4: Measured and simulated hourly discharge (Q) from Storglaciären, Sweden, 4 August to 6 September 1994. Measured air temperature is linearly shifted by 2°C in the scenario simulation, leaving all other variables and parameters unchanged. Discharge shows the pronounced melt-induced diurnal cyclicity typical of glacier discharge regimes.

4. Research gaps and strategies for future research

4.1 Measurements

While previous research has focussed on the impact of climate change on glacier mass balance (e.g. Braithwaite and Zhang 1999) and glacier size variations (e.g. Oerlemans 1994), far less attention has been paid to the effects on glacier discharge. Glaciers have generally been recognized as sensitive indicators of climate change and contributors to sea level change (Meier 1984). More than 280 glaciers have been subject to annual mass balance measurements at one time or another since 1946 (Dyurgerov 2002). Long-term measurements of glacier discharge have often been promoted by hydroelectric power schemes, but have rarely been assessed from a broader water resources aspect. Only very few glaciers are subject to simultaneous mass balance and discharge monitoring programmes. This may partly be due to practical difficulties associated with measuring stream discharge in the glacier environment, which is characterized by considerable turbulence, high sediment load and frequently changing channel geometry. Glaciers subject to such programmes are, e.g. Vernagtferner, Austria (Escher-Vetter and Reinwarth 1994), and South Cascade Glacier, USA (Krimmel 2001), where streamflow monitoring has been combined with annual mass balance measurements since 1974 and 1958, respectively. Storglaciären, Sweden, has the longest, detailed mass balance record in the world (Fig. 3, Holmlund and Jansson 1999), reaching back to 1945/46. However, proglacial discharge records are only available for a few melt seasons and further downstream (21% glacierization) since 1968.

Obviously, more combined long-term measurements of mass balance and discharge are needed to detect and analyze the effects of climate change on runoff and as a valuable database for modelling experiments. A monitoring network should emphasize geographical distribution in order to encompass different climate settings. For example, Storglaciären experienced positive mass balances for almost a decade due to enhanced winter accumulation, while Vernagtferner showed strongly negative mass balances during the same period (Fig. 3). Consequently, glaciers in different regions respond, at least initially, differently to climate warming, which in turn will trigger different hydrological responses.

4.2 Modelling

Although the characteristics of glacier discharge and their importance for catchment runoff have long been recognized, surprisingly few studies have attempted to quantitatively assess the effects of climate change on glacier discharge using numerical models (e.g. Braun et al. 2000). This dearth of studies is surprising considering the volume of investigations that focus on mass balance changes, although it is primarily their hydrological consequences that are immediate and entail direct economic impacts. To capture the full range of influences of climate warming on glacier discharge, better tools than those previously adopted need to be developed.

Modelling of glacier discharge involves two principal steps: (1) modelling of glacier mass balance, i.e. snow accumulation and glacier melt, and (2) discharge routing of melt and rain water through the glacier, i.e. transformation of water inputs into a discharge hydrograph. A large number of energy balance and temperature-index melt-models have been developed to compute melt. The current trend of these models is towards distributed modelling (summaries in Hock 1998; 2003). Temperature-index models are widely used, promoted by ease of application and low data input requirements. However it remains unclear how model parameters will change under a different climate, a limitation that needs further research. Hence, more physically based energy balance models provide a more reliable tool for climate impact studies, although application is restricted to areas with a sufficiently detailed database, including data on air temperature, humidity, wind speed and radiation. A critical parameter is to determine whether precipitation falls as rain or snow, e.g. a strong summer snowfall can effectively shut off glacier discharge for an extended period of the melt season. Currently, many conceptual runoff models do not include explicit routing routines for water transport through the glacier, taking into account the different hydraulic properties of snow, firn and ice with respect to throughflow velocities. Such routines are necessary to accurately capture the effects of amplified diurnal discharge cycles, resulting from accelerated runoff generation in response to a reduced extent of firn and snow cover under a warming climate.

It is obvious that the impact of climate warming on glacier discharge is complex and differs depending on the time-scale considered, in particular whether or not the glacier varies in size. Hence, modelling strategies need to be adjusted to the purpose of the study. When assessing short-term effects, such as amplified diurnal discharge fluctuations, changes in glacier size can be neglected. However, a high temporal resolution, e.g. hourly time steps, is necessary to capture peak flows, especially when considering the enormous diurnal discharge amplitudes typical of glacier regimes. This effect is important for watershed management and flood forecasting in areas close to glacier headwaters, particularly when peak melt rates and intense rain showers coincide. Modelling long-term effects, such as a decline in streamflow contribution caused by a decrease in glacier size, inevitably requires consideration of changes in glacier area. Thus, it is necessary to couple melt models to flow models or other tools capable of adjusting glacier sizes to climate change. A coarser time resolution, such as monthly or annual time steps, is sufficient to capture area changes. The long-term decrease in specific runoff to be expected from continuous glacier retreat is of major concern, especially in arid to semi-arid areas that are fed by rivers originating from remote glacierized mountain areas (e.g. Zhenniang and Xiaogang 1992). Accurate modelling of this effect is essential for these regions since the risk for water shortage will increase with a warming climate. In summary, the full range of responses to climate change should be addressed by a nested approach, whereby discharge is computed with high resolution, e.g. hourly, while glacier size changes are adjusted only on annual to decadal time scales.

To assess glacier sensitivity to climate change, models are often forced by simple linear shifts in climate data records (e.g. Kuhn and Batlogg 1998; Braithwaite and Zhang 1999). Such studies are useful to investigate sensitivities, but preclude the prediction of the "real" responses to climate change for a specific catchment, since future climate changes will not be homogeneous. Hence, the direct use of results from climate models is preferable in impact studies. General Circulation Models are currently too coarse in resolution, but major advances in regional climate modelling have been made to capture the effects of complex mountain topography in a more realistic way. For glacier hydrology, it is important to consider seasonal variations in predicted climate changes to account for seasonality in snow accumulation and melt.

We conclude that more sophisticated and detailed modelling studies are required to predict the full range of climate change impacts on glacier discharge on all time scales. We emphasize the need to model short-term as well as long-term effects including consideration of glacier size changes and direct use of results from climate models for scenario runs. Such studies should adopt more holistic views and an interdisciplinary approach, encompassing the disciplines glaciology, hydrology and meteorology.

5. References

Braithwaite, R. J., and Olesen, O. B. (1988). Effect of glaciers on annual run-off, Johan Dahl Land, south Greenland. *Journal of Glaciology* **34**, 200-207.
Braithwaite, R. J., and Zhang, Y. (1999). Modelling changes in glacier mass balance that may occur as a result of climate changes. *Geografiska Annaler* **81A**, 489-496.
Braun, L. N., Weber, M., and Schulz, M. (2000). Consequences of climate change for runoff from Alpine regions. *Annals of Glaciology* **31**, 19-25.
Chen J., and Ohmura, A. (1990). On the influence of Alpine glaciers on runoff. *In* "Hydrology of mountainous regions I." (H. Lang, and A. Musy, Eds.), Proceedings of two Lausanne Symposia, 1990: IAHS Publ. 193, 117-126.
Collins, D. N., and Taylor, D. P. (1990). Variability of runoff from partially-glacierised Alpine basins. *In* "Hydrology of mountainous regions I." (H. Lang, and A. Musy, Eds.), Proceedings of two Lausanne Symposia, 1990: IAHS Publ. 193, 365-372.
Dyurgerov, M. (2002). "Glacier mass balance and regime: Data of measurements and analysis." Institute of Arctic and Alpine Research, University of Colorado. Occasional Paper No. 55.
Escher-Vetter, H., and Reinwarth, O. (1994). Two decades of runoff measurements (1974 to 1993) at the Pegelstation Vernagtbach/Oetztal Alps. *Zeitschrift für Gletscherkunde und Glazialgeologie* **30**, 53-98.
Fountain, A., and Tangborn, W. (1985). The effect of glaciers on streamflow variations. *Water Resources Research* **21**, 579-586.
Fountain, A. G., and Walder, J. S. (1998). Water flow through temperate glaciers. *Reviews of Geophysics* **36**, 299-328.
Hagg, W., and Braun, L. N. (2003). The influence of glacier retreat on water yield from high mountain areas, comparison Alps – Central Asia. *In* "Climate and hydrology in mountain areas." (C. De Jong, D. Collins, and R. Ranzi, Eds.), John Wiley and Sons, London (submitted).
Hock, R. (1998). "Modelling of glacier melt and discharge." Zürcher Geographische Schriften 70, 140 pp.
Hock, R. (2003). Temperature index melt modelling in mountain regions. *Journal of Hydrology* (in press).
Holmlund, P., and Jansson, P. (1999). The Tarfala mass balance programme. *Geografiska Annaler* **81A**, 621-631.
Hopkinson C., and Young, G. (1998). The effect of glacier wastage on the flow of the Bow river at Banff, Alberta, 1951-1993. *Hydrological Processes* **12**, 1745-1762.
IPCC (2001). *In* "Climate Change 2001: The scientific basis". (J. T. Houghton, Y. Ding, D. J. Griggs, M. Noguer, P. J. Van der Linden, X. Dai, K. Maskell, and C. A. Johnson, Eds.). Cambridge University Press, Cambridge.
Jansson, P., Hock, R., and Schneider, T. (2003). The concept of glacier water storage: A review. *Journal of Hydrology* (in press).
Kuhn, M., and Batlogg, N. (1998). Glacier runoff in Alpine headwaters in a changing climate. *In* "Hydrology, water resources and ecology in headwaters." (K. U. Kovar, N. Tappeiner, E. Peters, and R. G. Craig, Eds.), pp. 78-88. IAHS Publication 248.
Krimmel, R. (2001). "Water, ice, and meteorological measurements at South Cascade Glacier, Washington, 2000-01 balance years." U.S. Geological Survey, Water-Resources Investigations Report 02-4165.
Lang, H., (1986). Forecasting meltwater runoff from snow-covered areas and from glacier basins. *In* "River flow modelling and forecasting." (D. A. Kraijenhoff, and J. R. Moll, Eds.), pp. 99-127. D. Reidel Publishing, Dordrecht.
Meier, M. F., and Tangborn, W. V. (1961). Distinctive characteristics of glacier runoff. U.S. Geological Survey Professional Paper 424-B, 14-16.
Meier, M. F. (1984). Contributions of small glaciers to global sea level. *Science* **226**, 1418-1421.
Oerlemans, J. (1994). Quantifying global warming from the retreat of glaciers. *Science* **264**, 243-245.
Singh, P., and Kumar, N. (1997). Impact assessment of climate change on the hydrological response of a snow and glacier melt runoff dominated Himalayan river. *Journal of Hydrology* **193**, 316-350.
Stenborg, T. (1970). Delay of runoff from a glacier basin. *Geografiska Annaler* **52A**, 1-30.
Willis, I. C., Anold, N. S., and brock, B. W. (2002). Effect of snowpack removal on energy balance, melt

and runoff in a small supraglacial catchment. Hydrological Processes 16, 2721-2749.

Zhenniang, Y., and Xiaogang, H. (1992). Study of glacier meltwater resources in China. *Annals of Glaciology* **16**, 141-145.

Part III: Hydrological changes

"Climate change" is often a synonym for "climate warming." However, according to *Schär & Frei*, the intrinsic link between the energy and hydrological cycle of the climate system justifies the equation with "climate moistening." In terms of energy, the moistening is more relevant than the warming. The authors assume that orographic precipitation could increase in mid and high latitudes at a similar rate as the atmospheric moisture content. The frequency and intensity of extreme events could even be more important as they may even increase independent of a decrease in mean precipitation amounts, which is especially well known in arid and semi-arid regions.

Diaz discusses the network of mountain observatories in the western US that are documenting the temperature increase since the last century. In stark contrast, information on the temperature change above 2000 m is scarce, as the present observatories cover only less than 10% of the territory. However, changes in the alpine cryosphere do not only signal large scale climate change, but - being directly coupled with the run-off regime - also represent important decision-making tools for managers and policy makers. The author proposes an adequate monitoring of critical and environmental variables in the mountains of the western US as a part of a global change mountain network.

Investigation of the spatial variability of snow cover and evaporation in relation to temperature, precipitation and rate of drainage in altitudes between 1500 and 2000 m in the Swiss Alps is the topic of *Menzel & Lang* showing astonishing differences between Eastern and Western, Northern and Southern Alps. Based on IPCC scenarios, a drastic reduction of the snow cover could be possible in the next 50 to 100 years with all its consequences for the hydrological regime.

Runoff sub-components under different environmental conditions, discussed by *Becker*, show the difficulties and uncertainties in the understanding of hydrological processes in heterogeneous mountain areas. The different response time between the runoff generating event and the corresponding increase in surface flow, interflow and baseflow may show that our knowledge about these complex processes in mountain catchments is still very limited. For two investigated hillslope/spring systems in the

Southern Black Forest Mountains, Germany, *Uhlenbrook et al.* demonstrate that the processes and hydrological responses in mountainous landscapes can be very diverse, even on relatively similar systems. The spatial heterogeneity appears to be related to highly variable soil structure overlain by land use and vegetation patterns, which will have a significant influence on the future recharge of springs and the discharge and composition of runoff components and their hydrochemical composition. Consequently, multi-disciplinary approaches are required to decode the complex hillslope system with its non-linear processes and numerous feedback mechanisms. In the same sense, *Kirnbauer et al.* identify space-time patterns of runoff generation in the Austrian Alps in a small, well equipped catchment. If typical and well defined runoff events, identified in a small to very small experimental catchment, are also applicable to medium-size catchments, remain an open question: Different threshold conditions in a complex system may create different runoff conditions.

Analyzing their results from the Spanish Pyrenees, including paleodata about climate variability and land use and land cover changes with a special focus on deforestation, *Garcia-Ruiz et al.* consider three scales: Small experimental plots to understand the impact of different land use types on runoff and erosion; small experimental catchments to get the information about discharge and sediment transport; the basin or large scale study including the growth of the Ebro delta since the Roman time. This integrative and comparative analysis of different scales can lead to a more precise interpretation of changes and to a better differentiation of natural and human driving forces.

McGlynn's article investigates the role of the riparian zones in steep mountain watersheds in very different catchments from New Zealand to the US. Riparian zones are narrow in the headwaters and can increase in lower areas to a valley bottom and even to a flood plain. The author discusses different scales and the problem of modeling complex mountain watersheds, as it is done also in the contribution of *Gurtz et al.* about the use of hydrological models for the simulation of climate change impacts on mountain hydrology in different catchments of the Swiss Alps. Interesting is the statement, that the interflow is more or less the most important runoff component in mountain regions, when the necessary storage capacity with adequate soil and slope conditions are existing.

Effects of climate variability and change on mountain water resources in the western US, discussed by *Leung,* threaten the urbanised and industrialised region with its export oriented and irrigated agriculture. The variations of ENSO play an important role for the precipitation regime and for the precipitation anomalies in the different mountain areas. Warmer temperatures effects are expressed mainly in the reduction of snowpack along the coastal mountains. In contrast, in the Rocky Mountains, snowpack change is governed by temperature and precipitation changes, so that the reduction in this continental climate is limited. High-resolution modelling will be important in future research programs, but higher spatial resolution improves the simulation of temperature and snow, but does not necessarily improve the simulation of precipitation.

Orographic Precipitation and Climate Change

Christoph Schär* and Christoph Frei
Institute for Atmospheric and Climate Science, Swiss Federal Institute of Technology, ETH Zurich, CH-8057 Zurich, Switzerland
**phone +41-1-635-5199, fax +41-1-362-5197, e-mail christoph.schaer@iac.umnw.ethz.ch*

Keywords: Climate change, Climate models, Extreme events, Precipitation, Runoff, Scenarios.

1. Introduction

More than half of the accessible freshwater is used directly or indirectly by humankind, and much of this precious resource has its origin in mountainous regions, ultimately in the form of orographic precipitation. In many areas, mountains function as "water towers" for the surrounding regions. Melt from snow cover and glaciers represents an important contribution to runoff in the surrounding areas, especially during seasons when precipitation is sparse or completely absent. Mountain freshwater resources are heavily utilized for agricultural purposes (e.g. irrigation) and for the generation of hydropower, thus being of great socio-economic importance. Yet, heavy orographic precipitation events also represent a potential hazard, as they may lead to floods, avalanches and mudslides that often cause countless loss of life and tremendous damage. The potential consequences of such events may be extreme. For instance, a single catastrophic mudslide event that took place in Venezuela on December 15, 1999, is estimated to have caused more than 20,000 casualties according to re-insurance estimates.

Extreme events are a characteristic property of mountain climates. On geological timescales, heavy precipitation events, floods, water erosion, avalanches and mudslides have contributed towards shaping the landscape and environment. Civilizations in mountainous regions have for a long time adjusted their infrastructure to this challenge (e.g. location of settlements) and have in many regions successfully

U. M. Huber et al. (eds.), Global Change and Mountain Regions, 255-266.

undertaken measures to protect themselves against natural hazards (e.g. damming of major rivers). In many mountain ranges, a wide range of planning measures is underway to mitigate and adapt to the threat of extreme weather and hydrological conditions (e.g. building codes, dimension of bridges, operation of dams). In general, the planning of such systems is based on the assumption of a stationary climate. However, this principle is likely to become obsolete due to climate change. Indeed, in terms of their socio-economic implications, future changes in the frequency and character of extreme events are likely to be more relevant than the comparatively slow shifts in mean temperature and precipitation.

According to the Third Assessment Report of the Intergovernmental Panel on Climate Change (IPCC), there is increasing evidence that the climate system is experiencing pronounced changes. A substantial fraction of the warming observed in the last 50 years is attributed to anthropogenic greenhouse gases (IPCC 2001). The global mean surface temperature has increased by ~0.6°C since the late 19th century, and the 1990s are believed to be the warmest decade of the last millennium in the Northern Hemisphere. In addition to the warming, substantial changes of the hydrological cycle have been detected. In the northern hemisphere, the overall signature is an increase in total precipitation amounts, except for the sub-tropics. In the northern middle and high latitudes, annual land precipitation has increased by 0.5 to 1%/decade during the last 100 years, while over the northern tropics and sub-tropics (0°N to 40°N) there is a tendency for a slight precipitation decrease (New et al. 2001). Where time series are available, changes in annual streamflow are often observed to relate well to changes in total precipitation. Consistent with the observed warming, there has been an increase of the total atmospheric water vapour content.

Climate change scenarios that account for future greenhouse gas and aerosol emissions suggest that the observed warming will accelerate and the hydrological cycle will intensify. Such changes would imply important repercussions for orographic precipitation and mountain climates. However, at present there are major uncertainties in particular regarding the regional patterns of climate change (IPCC 2001). Here, a short overview is presented on the role of climate change on orographic precipitation, including considerations of climate-induced changes in the frequency of extreme weather and hydrological conditions. The paper will cover observations, processes and scenarios and will conclude with a broad outlook.

2. Observations

Monitoring orographic precipitation on seasonal to decadal time scales is a challenging issue for several reasons. Records from conventional rain gauges suffer from systematic measurement biases (Yang et al. 1999) and inhomogeneities due to changes in observation practice and station displacements (Peterson et al. 1998). These errors are particularly large in mountain regions, due to the higher proportion of snowfall and the large spatial variability of precipitation. Monitoring orographic precipitation requires particularly dense networks, which is difficult to achieve considering the technical problems encountered in these remote areas. As a result, analyses of long-term precipitation variations that resolve the prominent topographical imprints are available for a few mountain areas only.

A notable example is the region of the European Alps, where data from present-day high-resolution networks (Frei and Schär 1998) and sparser homogenized long-term records (Schmutz et al. 2003) can be used for the reconstruction of precipitation variability and trends back to the beginning of the 20th century (Fig. 1; Schmidli et al. 2000; 2001): During the winter season, mean precipitation has increased, particularly in the north-western parts of the Alps, whilst a decreasing trend was detected for the southern and eastern parts of the ridge during autumn. These observed trends are statistically significant and amount to 30% per 100 years. Their spatial distribution shows a relation to some of the topographic characteristics of the region. Trends during spring and summer are smaller and barely significant. Some of these observed changes are in qualitative agreement with climate change scenarios (see below), however, the extent to which these changes are related to anthropogenic climate change is unknown.

The identification of trends in heavy precipitation is confronted with even more obstacles. Firstly, it is difficult to recover high-quality station records with daily resolution over sufficiently extended periods. Secondly, there are fundamental limitations in our ability to distinguish between systematic trends and merely random occurrence in records of rare extremes. The ability to identify trends in extreme events can be described by the "detection probability" (Frei and Schär 2001). The detection probability decreases with increasing rareness of events and becomes negligible for events with a return period of several years or more, even if the change corresponds to a doubling of the event frequency. It follows that care must be exercised when inferring the absence of a trend from the absence of statistical significance, and trend analysis should preferably focus on more frequent moderate intensity (rather than rare

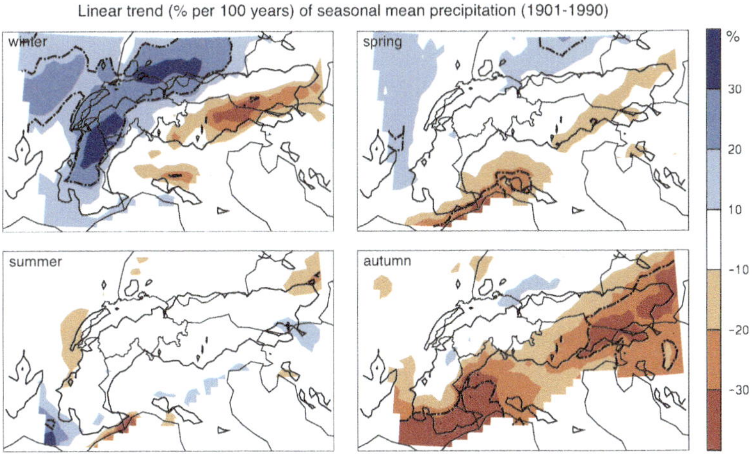

Figure 1: Alpine precipitation trends stratified according to season (Schmidli et al. 2002). Colors: Trend magnitude in % change per 100 years; Bold contours: 850 m topographic contour; Thick dash-dotted contours: Statistical significance at the 10% level.

extreme) events.

In the Swiss portion of the Alpine region, the frequency of intense daily precipitation events (days exceeding the 30-day return period threshold) has increased in winter and autumn by 20–80% during the 20th century (Frei and Schär 2001). In autumn, the increase in precipitation intensity is accompanied by a decrease in rainday frequency and these trends mutually compensate to yield no trend in mean precipitation. A similar compensation was found for the Italian part of the Alps (Brunetti et al. 2001). Recent climate change scenarios for the Alps predict changes of intense precipitation with similar seasonal characteristics (e.g. Durman et al. 2001).

Long-term trends in the components of the water cycle have been reported for several other mountain regions of the world. For Scandinavia, analyses of homogenised precipitation series indicate that mean precipitation has increased in coastal Norway, Sweden and Denmark (Hanssen-Bauer and Forland 2000; Schmith 2000). In winter, the 20th century increase amounts to 15-20%, and is related to particularly wet conditions since ~1980. During the same period, the atmospheric circulation regime of the North Atlantic (the so-called North Atlantic Oscillation, NAO, see e.g. Wanner et al. 2001) was in a prolonged phase of strong westerlies (i.e. high NAO-index states), characteristic of enhanced transport of warm and moist airmasses towards the European continent. Some fraction of the observed precipitation trend in Scandinavia is therefore related to the recent NAO anomalies (Hurrell 1995). However, attempts to quantify this contribution were not successful in explaining the full magnitude of the observed precipitation trend (Hanssen-Bauer and Forland 2000; Schmith 2000).

In the North American Rocky Mountains, an increase of mean precipitation was found from September through December and this was accompanied by an increase in annual stream flow (Lettenmaier et al. 1994). More than half of the 20th century increases in mean annual precipitation is due to more frequent heavy precipitation in the upper 10 percentiles of the distribution (Karl and Knight 1998). At the same time, the frequency of convective cloud types was found to have increased over the last 40-50 years (Sun et al. 2001). However, the observed increase in heavy precipitation events is not accompanied by an upward trend in peak stream flow in the Rockies (unlike in the rest of the country). This might be explained by the observed decrease in the extent of spring-time snow cover, which compensates for the precipitation increase (Groisman et al. 2001). Long-term trends in mean precipitation, showing strong regional variations, were also reported from Alaska (Stafford et al. 2000).

For the tropics, observations demonstrate that the prime mode of variability is associated with El Niño (e.g. Dai et al. 1997). This phenomenon yields variability on time-scales of a few years, which is likely to dominate the system for some time into the future. Nevertheless, long-term trends are also reported for some tropical land areas, especially the latitude band 0-20°N, where mean annual land precipitation decreased during the second half of the 20th century (see New et al. 2001).

3. Processes and feedbacks

On the global scale, the term "climate change" is often equated with the term "climate warming". However, the energy cycle of the climate system is intrinsically linked with the hydrological cycle. To a first approximation, it would indeed be more appropriate to equate "climate change" with "climate moistening", as in terms of energy the moistening is more relevant than the warming. One of the primary drivers behind the associated intensification of the hydrological cycle is the Clausius-Clapeyron relationship, which represents the dependency of the saturation vapor pressure upon temperature. In quantitative terms, the ability of air to hold water vapor increases by ~6% per degree Kelvin. There are strong indications – from physical arguments, climate observations and climate models (see e.g. DelGenio et al. 1991; Allen and Ingram 2002) – that the relative humidity of the atmosphere will in the long-term and large-scale mean not change much in a future climate (with the possible exception of tropical regions). The Clausius-Clapeyron relation thus implies that the total moisture content of the atmosphere must increase by ~6% per degree warming.

The simplest picture of the intensified hydrological cycle in a warmer world is thus one of a percent-wise intensification, where the atmospheric moisture content, evapotranspiration and precipitation increase simultaneously at a rate comparable to the aforementioned ~6%/K. However, such a scenario is not likely to come true. Rather, GCM scenarios of greenhouse gas-induced warming suggest that, although the atmospheric moisture content will increase by ~6%/K, precipitation and evapotranspiration will increase at a slower rate of 1-3%/K (IPCC 2001). This peculiar behavior is due to the large-scale balances between atmospheric and terrestrial radiation, evaporative cooling and condensational heating (Boer et al. 1993; Hartmann 1994; Trenberth 1999), but the underlying processes are still not fully understood. As a result, we are facing a climate in which the intensification of the global mean hydrological cycle is substantially weaker than the increase of the atmospheric water content. The unequal increase of these critical factors implies an increase in the mean atmospheric residence time of water molecules (Trenberth 1999). This will likely be accompanied by changes in the intensity of rainfall events and shifts in the geographical distribution of climate zones. In addition, the increase of the atmospheric moisture content implies important changes in the radiation balance, as water vapor is the most important greenhouse gas of our atmosphere.

The expected increase of global mean precipitation by 1-3%/K alone would probably be of comparatively minor concern. This increase may however be amplified in certain regions and over complex topography:

- First, the increase in precipitation will not occur uniformly but changes will be associated with specific geographical patterns and will vary with seasons. More specifically, the mid and high latitudes are expected to experience a higher relative increase in total precipitation in particular during winter, while there is evidence that some sub-tropical and semi-arid regions might experience an increased risk of summer droughts (e.g. Weatherald and Manabe 1995). These may arise where the local increase in evapotranspiration exceeds the increase in precipitation.
- Second, the frequency of heavy precipitation events is not directly linked to mean precipitation amounts. Frequency changes at the extreme tails of the distribution can take on large magnitudes even if mean changes are small (Frei et al. 1998). Also, several climate model scenarios suggest that the frequency

of heavy precipitation events may increase irrespective of a decrease in mean precipitation amounts (e.g. Durman et al. 2001; Semenov and Bengtsson 2002). Recent observations have revealed such trends, for instance in the Alps, where total autumn precipitation has changed little or has even decreased, but intense precipitation events have increased (Frei and Schär 2001; Schmidli et al. 2002).

- Third, orographic precipitation is likely to increase at a similar rate as the atmospheric moisture content (~6%/K), rather than at the rate of the global mean precipitation increase (1-3%/K). This particular characteristic of orographic precipitation is due to the fact that mountains effectively extract a fraction of the atmospheric moisture flux, which increases at a similar rate as the atmospheric water content. Again, there are indications for this kind of behavior. For instance, Figure 1 shows the most pronounced trends in Alpine precipitation (of either sign) in the immediate vicinity of the topography.

Detailed analysis suggests that the anticipated intensification of the hydrological cycle may be particularly effective in the mid and high latitudes, in relation to heavy precipitation events, and in mountainous regions. With regard to runoff and flooding in mountainous regions, the increase in the frequency and intensity of extreme precipitation events is of concern, as this factor works in the same direction as the anticipated increase in the fraction of liquid precipitation at the expense of snowfall. These hydrological factors are discussed in other chapters of this book.

The above argumentation is in its core a thermodynamic one and therefore applies at best to the large-scale mean and not necessarily to individual mountain ranges. Indeed, there is broad evidence that interannual variations of orographic precipitation are largely controlled by planetary and synoptic-scale atmospheric circulation anomalies (e.g. Massacand et al. 1998). Hence, a thermodynamic argument alone is insufficient to derive realistic and spatially specific climate change scenarios.

4. Scenarios

The overall changes in the hydrological cycle will be highly complex and will involve a number of factors, such as changes in storm track dynamics, soil moisture conditions, and cloud formation processes. The complexity of these interactions calls for a detailed numerical assessment using general circulation models (GCMs). Most GCM studies of increased greenhouse gas scenarios show a pronounced increase in the frequency and intensity of heavy precipitation events. For instance, for equilibrium doubling of carbon dioxide, Hennessy et al. (1997) find that, for a specified return period of 1 year (corresponding to an event size that is exceeded once every year) there is an increase in precipitation intensity of 10 to 25% in Europe, North America, Australia and India. McGuffie et al. (1999) confirm this conclusion when comparing the results from five different GCMs. From an ensemble of transient GCM experiments, Kharin and Zwiers (2000) compute the change in the 20-year return period of daily rainfall for the next 100 years. They find an increase in precipitation intensity almost everywhere on the globe, with the relative change in precipitation extremes exceeding that in mean precipitation. Semenov and Bengtsson

(2002) find increases in mean precipitation intensity, even in regions where mean annual precipitation decreases. Using a probabilistic analysis of 19 GCM simulations, Palmer and Räisänen (2002) estimate that the probability of total winter precipitation exceeding two standard deviations above normal will increase by a factor of five over parts of the UK during the next 100 years. They find similar increases in probability for the Asian monsoon region, with potentially serious implications for flooding in Bangladesh.

The aforementioned studies serve to demonstrate the high sensitivity of precipitation to global warming scenarios. Currently, however, most coupled atmosphere-ocean models have a horizontal resolution of ~300 km, and this is insufficient to properly resolve even major mountain ranges. As a result, coupled atmosphere-ocean GCMs drastically underestimate the intensity of extreme precipitation events in mountainous region. In response to this model limitation, a wide range of downscaling methodologies has been developed. These include both statistical approaches (which are calibrated with observational data from the past) and numerical approaches (which are more closely based on physical laws). Succinct reviews of these methodologies can be found in Giorgi and Mearns (1991) and Wilby and Wigley (1997).

Numerical downscaling procedures typically employ a sequence of nested models, ranging from a low-resolution GCM to a high-resolution impact model. In the example illustrated in Figure 2, two regional climate models (RCMs) are utilized to bridge the scale-gap between the low-resolution GCM and a high-resolution runoff model covering the Rhine basin upstream of Cologne (Kleinn et al. 2002). At each step in the chain, models are forced at their lateral boundaries (and in terms of their sea-surface temperatures) by the results of larger-scale models that are one step up in the spatial hierarchy. With increasing model resolution, the weather evolution and the hydrological cycle are simulated with increasing spatial detail and accuracy.

The quality of climate change scenarios critically depends upon the quality of both the global model component (which must accurately describe the large-scale climatology) and that of the downscaling procedure (which must adequately represent smaller-scale processes). In particular, downscaling cannot correct for systematic errors in the large-scale forcing. The quality of a downscaling procedure can be assessed by testing – under current climatic conditions – its ability to represent the mean climate, the frequency of rare events, natural interannual variations, as well as specific meteorological processes (e.g. Christensen et al. 1996; Frei et al. 2003; Vidale et al. 2003; Wild et al. 2001; respectively).

The credibility of a specific scenario depends upon a wide range of factors, among them the geographical region and season under consideration, and the GCM and RCM under use. Recent studies confirm that increasing the computational resolution does indeed improve the representation of the hydrological cycle, in particular in mountainous regions (e.g. Jones et al. 1995; Leung and Ghan 1999; Durman et al. 2001; Giorgi et al. 2001). Despite this progress, current models entail substantial uncertainties. For instance, difficulties exist concerning the simulation of large-scale quasi-stationary atmospheric circulation patterns, the terrestrial hydrological cycle in semi-arid regions, and convective precipitation processes. In addition, some aspects of the climate system are not predictable, as the atmosphere is an inherently chaotic dynamical system (Palmer 2000). Climate change scenarios (from numerical as well as statistical approaches) should thus be used in a process/sensitivity study mode, rather than be taken literally as predictions. In particular, the term "scenario" is defined as a consistent evolution of a system into the future, but *without* specified probability. Thus, future research in this area should aim towards the construction of probabilistic climate change scenarios, a task that will require the quantification of all relevant uncertainties.

5. Outlook

While there is ample evidence to reject the assumption of a stationary precipitation climate, current climate models can at most simulate the general direction of mean precipitation changes, and it appears unrealistic at present to make quantitative statements about the geographical distribution of climate changes in specific mountainous regions. For actual planning purposes relating to flood protection, agriculture and water resource infrastructure, the prime implication of climate change is thus to increase the uncertainty. This uncertainty – in concert with the growing population pressure, changes in land-use and settlement structure, and per-capita increases in freshwater use – implies that the role of mountains as a freshwater source will become more critical and more difficult to assess.

A better understanding of climate and climate variability, the underlying atmospheric and hydrological processes and their interaction with topography is thus urgently needed. A specific list of key topics is discussed in the recent IPCC report (IPCC 2001). Regarding climate and climate change in mountainous regions, there is a particular need to better understand diurnal mountain circulations, land-surface processes, orographic precipitation processes, and their interactions. The investigation

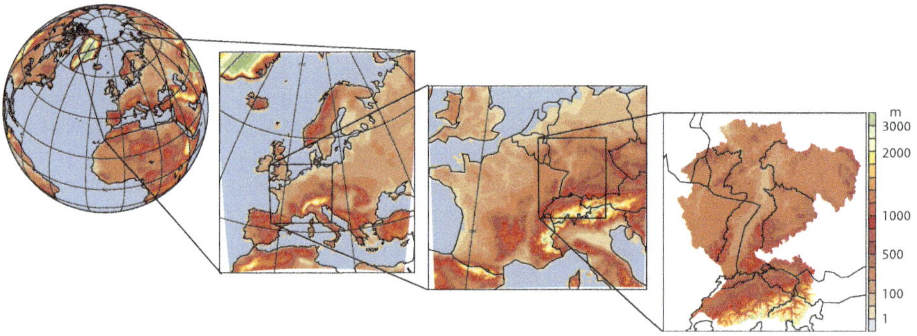

Figure 2: An example of a downscaling model chain used for the assessment of climate-change impacts upon the runoff of the river Rhine (Kleinn et al. 2002). The downscaling chain includes a coupled atmosphere/ocean GCM with a horizontal resolution of ~300 km, an atmosphere-only GCM with a resolution of ~120 km, two regional climate models with resolutions of 56 and 14 km, respectively, and a hydrological runoff model with a resolution of 1 km. The panels show the topography [m] for the different model components.

of these issues is coordinated under the umbrella of large international cooperations such as the World Climate Research Program (WCRP), the Climate Variability and Predictability (CLIVAR) and the Global Energy and Water Cycle (GEWEX) programs of the World Meteorological Organisation (WMO) and the International Council of Scientific Unions (ICSU).

It is, however, important to realize that weather, climate and hydrology of many mountain ranges are poorly understood, even under current climatic conditions. It is the authors' opinion that in such circumstances climate change assessments are of limited use. Establishing the current climatology, assessing its natural seasonal and interannual variations, understanding the underlying key processes on short (meteorological and hydrological) time scales, and pinpointing the vulnerability of settlements and infrastructure under current climatic conditions are preconditions for climate change assessments. Improving the understanding of these issues is also of practical benefit for weather and hydrological forecasting on daily to seasonal time scales. Key aspects of this avenue are:

Observations: Although considerable progress has been made in the collation of long-term data and the establishment of global surface-climate analyses (e.g. Dai et al. 1997; Huffman et al. 1997; New et al. 2001), the picture of the Earth's mountain climates offered by these analyses is still very limited. The available station data is mostly confined to dedicated WMO stations, and current analyses can rarely provide the spatial resolution that is necessary for many applications. Therefore, there is an urgent need to improve the accessibility of existing observational data for the scientific community by relaxing restrictions on data exchange (see also Hulme 1994). Considering emerging tendencies of network optimization, there is also a need to preserve the quality of existing networks with regard to spatial resolution and temporal homogeneity (e.g. Karl et al. 1995).

Assimilated Data: For remote mountain areas, better use of existing data will only marginally improve the situation, due to the sparse coverage with climate stations. One way to overcome the irrecoverable lack of data in these cases is the use of model-aided data from atmospheric data assimilation and reanalysis systems (Kalnay et al. 1996; Gibson et al. 1997; Rabier et al. 2000). Such systems are able to assimilate meteorological data from a wide range of sources – among them conventional radiosonde and surface station data, satellite information and data from commercial aircrafts – and exploit numerical models to estimate poorly observed variables (such as precipitation) in remote areas. In many large-scale mountainous river catchments, such data – in conjunction with existing runoff data and satellite information on snow cover extent – is the only hope to better understand the natural interannual and inter-decadal variations of the atmospheric water cycle and the hydrological response.

Numerical Weather Prediction: Twenty or thirty years ago, numerical weather prediction in mountainous regions required a tremendous effort that appeared only feasible in industrialized countries. With the advent of high-resolution atmospheric prediction models (e.g. Simmons and Hollingsworth 2002), it has become feasible to utilize global forecasting products even at remote locations and in regions where data is sparse. Furthermore, with the decreasing price of high-speed computers and the increasing speed of Internet connections, it has become feasible to run purpose-designed high-resolution limited-area forecasting systems at comparatively small costs. The use of such procedures appears particularly promising as regards quantitative precipitation forecasting in mountainous regions. Such applications allow a direct coupling with hydrological runoff and water resource assessment models. In addition, there are promising prospects regarding the prediction of extreme events (e.g. Bougeault et al. 2001), which enables the implementation of short-term warning systems.

Seasonal Forecasting: For many tropical mountain ranges, the future climate will continue to be dominated by El Niño type natural variability (although global warming may impact ENSO frequency and intensity). Such variability is to some extent predictable with lead times of 3 to 6 months (e.g. Stockdale et al. 1998), thus providing important information for the management of water resources over seasonal time scales. In the extratropics, however, the predictability of seasonal variability is much poorer, and it remains to be seen whether seasonal forecasting is of much practical applicability in these areas.

Climate Change Scenarios: A better understanding of the above issues will also help to improve methodologies for constructing climate change scenarios. As almost all mountain ranges are characterized by highly complex topography, it is not feasible to directly rely upon output from coupled atmosphere/ocean GCMs, but rather some downscaling procedure must be applied. Today, RCMs are considered the only foreseeable downscaling tool that is able to adequately represent the inherent non-linearities of regional climates. When applied in mountainous regions, the high spatial resolution of RCMs is of particular advantage. Statistical downscaling methods may also serve their purpose and these are much simpler and cheaper to apply. In the medium term, methodologies will be needed that provide probabilistic climate predictions (rather than scenarios of unknown probability).

6. Acknowledgments

This study was supported by the Swiss National Science Foundation (NCCR Climate, Project 2.2). Special thanks to Jürg Schmidli and Jan Kleinn for their help with the figures. The authors are indebted to a reviewer and several colleagues for useful comments on an earlier version of the manuscript.

7. References

Allen, M., and Ingram, W. (2002). Constraints on future changes in climate and the hydrologic cycle. *Nature* **419**, 224-232.

Boer, G. J. (1993). Climate change and the regulation of the surface moisture and energy budgets. *Climate Dynamics* **8**, 225-239.

Bougeault, P., Binder, P., Buzzi, A., Dirks, R., Houze, R., Kuettner, J., Smith, R. B., Steinacker, R., and Volkert, H. (2001). The MAP special observing period. *Bulletin of the American Meteorological Society* **82**, 433-462.

Brunetti, M., Maugeri, M., and Nanni, T. (2001). Changes in total precipitation, rainy days and extreme events in northeastern Italy. *International Journal of Climatology* **21,** 861-871.

Christensen, J. H., Machenhauer, B., Jones, R. G., Schär, C., Ruti, P. M., Castro, M, and Visconti, G. (1996).

Validation of present-day regional climate simulations over Europe: LAM simulations with observed boundary conditions. *Climate Dynamics* **13**, 489-506.

Dai, A., Fung, I. Y., and DelGenio, A. D. (1997). Surface observed global land precipitation variations during 1900-88. *Journal of Climate* **10**, 2943-2962.

DelGenio, A. D., Lacis, A. A., and Ruedy, R. A. (1991). Simulations of the effect of a warmer climate on atmospheric humidity. *Nature* **251**, 382-385.

Durman, C. F., Gregory, J. M., Hassell, D. C., Jones, R. G., and Murphy, J. M. (2001). A comparison of extreme European daily precipitation simulated by a global and a regional climate model for present and future climates. *Quaterly Journal of the Royal Meteorological Society* **127**, 1005-1015.

Frei, C., and Schär, C. (1998). A precipitation climatology of the Alps from high-resolution rain-gauge observations. *International Journal of Climatology* **18**, 873-900.

Frei, C., Schär, C., Lüthi, D., and Davies, H. C. (1998). Heavy precipitation processes in a warmer climate. *Geophysical Research Letters* **25**, 1431-1434.

Frei, C., and Schär, C. (2001). Detection probability of trends in rare events: Theory and application to heavy precipitation in the Alpine region. *Journal of Climate* **14**, 1564-1584.

Frei, C., Christensen, J. H., Déqué, M., Jacob, D., Jones, R. G., and Vidale, P. L. (2003). Daily precipitation statistics in regional climate models: Evaluation and intercomparison for the European Alps. *Journal of Geophysical Research-Atmospheres* **108** (D3), art. no. 4124.

Gibson, J. K., Kallberg, P., Uppala, S., Nomura, A., Hernandez, A., and Serrano, A. (1997). "ERA description." ECMWF Re-Analysis Project Report Series, European Center for Medium-Range Weather Forecast, Reading.

Giorgi, F., and Mearns, L. O. (1991). Approaches to the simulation of regional climate change: A review. *Reviews in Geophysics* **29**, 191-216.

Giorgi, F., Whetton, P. H., Jones, R. G., Christensen, J. H., Mearns, L. O., Hewitson, B., vonStorch, H., Francisco, R., and Jack, C. (2001). Emerging patterns of simulated regional climatic changes for the 21st century due to anthropogenic forcings. *Geophysical Research Letters* **28**, 3317-3320.

Groisman, P. Y., Knight, R. W., and Karl, T. R. (2001). Heavy precipitation and high streamflow in the contiguous United States: Trends in the twentieth century. *Bulletin of the American Meteorological Society* **82,** 219-246.

Hanssen-Bauer, I. and Forland, E. J. (2000). Temperature and precipitation variations in Norway 1900-1994 and their links to atmospheric circulation. *International Journal of Climatology* **20**, 1693-1708.

Hartmann, D. L. (1994). "Global physical climatology." Academic Press, San Diego.

Hennessy, K. J., Gregory, J. M., and Mitchell, J. F. B. (1997). Changes in daily precipitation under enhanced greenhouse conditions. *Climate Dynamics* **13**, 667-680.

Huffman, G. J., Adler, R. F., Arkin, P., Chang, A., Ferraro, R., Gruber, A., Janowiak, J., McNab, A., Rudolf, B., and Schneider, U. (1997). The Global Precipitation Climatology Project (GPCP) combined precipitation dataset. *Bulletin of the American Meteorological Society* **78**, 5-20.

Hulme, M. (1994). The cost of climate data: A European experience. *Weather* **49**, 168-174.

Hurrell, J. W. (1995). Decadal trends in the North Atlantic Oscillation: Regional temperature and precipitation. *Science* **269**, 676-679.

IPCC (2001). *In* "Climate change 2001: The scientific basis." (J. T. Houghton, Y. Ding, D. J. Griggs, M. Noguer, P. J. Van der Linden, X. Dai, K. Maskell, and C. A. Johnson, Eds.). Cambridge University Press, Cambridge (available from http://www.ipcc.ch/).

Jones, R. G., Murphy, J. M., and Noguer, M. (1995). Simulation of climate change over Europe using a nested regional climate model. Part I: Assessment of control climate including sensitivity to location of lateral boundaries. *Quaterly Journal of the Royal Meteorological Society* **121**, 1413-1449.

Karl, T. R., and Knight, R. W. (1998). Secular trends of precipitation amount, frequency and intensity in the United States. *Bulletin of the American Meteorological Society* **79**, 231-241.

Karl, T. R., Derr, V. E., Easterling, D. R., Folland, C. K., Hofmann, D. J., Levitus, S., Nicholls, N., Parker, D. E., and Withee, G. W. (1995). Critical issues for long-term climate monitoring. *Climatic Change* **31**, 185-221.

Kharin, V. V., and Zwiers, F. W. (2000). Changes in the extremes in an ensemble of transient climate simulations with a coupled atmophere-ocean GCM. *Journal of Climate* **13**, 3760-3788.

Kleinn, J., Frei, C., Gurtz, J., Vidale, P. L., and Schär, C. (2002). Coupled climate-runoff simulations: A process study of current and warmer climate conditions in the Rhine basin. *In* "16th Conference on Hydrology," January 2002, American Meteorological Society (extended abstract).

Lettenmaier, D. P., Wood, E. F., and Wallis, J. R. (1994). Hydro-climatological trends in the continental United States, 1948-88. *Journal of Climate* **7**, 586-607.

Leung, L. R., and Ghan, S. J. (1999). Pacific Northwest climate sensitivity simulated by a regional climate model driven by a GCM. Part I and II. *Journal of Climate* **12**, 2010-2053

Massacand, A. C., Wernli, H., and Davies, H. C. (1998). Heavy precipitation on the Alpine southside: An upper-level precursor. *Geophysical Research Letters* **25**, 1435-1438.

McGuffie, K., Henderson-Sellers, A., Holbrook, N., Kothavala, Z., Balachova, O., and Hoekstra, J. (1999). Assessing simulations of daily temperature and precipitation variability with global climate models for present and enhanced greenhouse climates. *International Journal of Climatology* **19**, 1-26.

New, M., Lister, D., Hulme, M., and Makin, I. (2002). A high-resolution data set of surface climate over global land areas. *Climate Research* **21**, 1-25.

Palmer, T. N. (2000). Predicting uncertainty in forecasts of weather and climate. *Reports on Progress in Physics* **63**, 71-116.

Palmer, T. N., and Räisänen, J. (2002). Quantifying the risk of extreme seasonal precipitation events in a changing climate. *Nature* **415**, 512-514.

Peterson, T. C. et al. (1998). Homogeniety adjustments of *in situ* atmospheric climate data: A review. *International Journal of Climatology* **18**, 1493-1517.

Rabier, E., Jarvinen, H., Klinker, E., Mahfouf, J. F., and Simmons, A. J. (2000). The ECMWF operational implementation of four-dimensional variational assimilation. I: Experimental results with simplified physics. *Quaterly Journal of the Royal Meteorological Society* **126**, 1143-1170.

Schmidli, J., Frei, C., and Schär, C. (2001). Reconstruction of mesoscale precipitationfields from sparse observations in complex terrain. *Journal of Climate* **14**, 3289–3306.

Schmidli, J., Schmutz, C., Frei, C., Wanner, H., and Schär, C. (2002). Mesoscale precipitation in the Alps during the 20th century. *International Journal of Climatology* **22**, 1049-1074.

Schmith, T. (2000). Global warming signature in observed winter precipitation in Northwestern Europe. *Climate Research* **17**, 263-274.

Schmutz, C. (2003). A quality-tested data base of monthly Alpine long-term (1901-1995) precipitation time series. *Theoretical and Applied Climatology* (in press).

Semenov, V. A., and Bengtsson, L. (2002). Secular trend in daily precipitation characteristics: Greenhouse gas simulations with a coupled AOGCM. *Climate Dynamics* **19**, 123-140.

Simmons, A. J., and Hollingsworth, A. (2002). Some aspects of the improvement in skill of numerical weather prediction. *Quarterly Journal of the Royal Meteorological Society* **128**, 647-677.

Stafford, J. M., Wendler, G., and Curtis, J. (2000). Temperature and precipitation of Alaska: 50 year trend analysis. *Theoretical and Applied Climatology* **67**, 33-44.

Stockdale, T. N., Anderson, D. L. T., Alves, J. O. S., and Balmaseda, M. A. (1998). Global seasonal rainfall forecasts using a coupled ocean-atmosphere model. *Nature* **392**, 370-373.

Sun, B., Groisman, P. Y., and Makhov, I. I. (2001). Recent changes in cloud-type frequency and inferred increases in convection over the United States and the Former Soviet Union. *Journal of Climate* **14**, 1864-1880.

Trenberth, K. E. (1999). Conceptual framework for changes of extremes of the hydrological cycle with climate change. *Climatic Change* **42**, 327-339.

Vidale, P. L., Lüthi, D., Frei, C., Seneviratne, S., and Schär, C. (2003). Predictability and uncertainty in a regional climate model. *Journal of Geophysical Research - Atmospheres* (in press).

Wanner, H., Brönnimann, S., Casty, C., Gyalistras, D., Luterbacher, J., Schmutz, C., Stephenson, D. B., and Xoplaki, E. (2001). North Atlantic Oscillation: Concepts and studies. *Surveys in Geophysics* **22**, 321-382.

Weatherald, R. T., and Manabe, S. (1995). The mechanisms of summer dryness induced by greenhouse warming. *Journal of Climate* **8**, 3096-3108.

Wilby, R. L., and Wigley, T. M. L. (1997). Downscaling general circulation model output: A review of methods and limitations. *Progresses in Physical Geography* **21**, 530-548.

Wild, M., Ohmura, A., Gilgen, H., Morcrette, J. J., and Slingo, A. (2001). Evaluation of downward longwave radiation in general circulation models. *Journal of Climate* **14**, 3227-3239.

Yang, D. Q., Elomaa, E., Tuominen, A., Aaltonen, A., Goodison, B., Gunther, T., Golubev, V., Sevruk, B., Madsen, H., and Milkovic, J. (1999). Wind-induced precipitation undercatch of the Hellmann gauges. *Nordic Hydrology* **30**, 57-80.

Monitoring Climate Variability and Change in the Western United States

Henry F. Diaz
Climate Diagnostics Center, David Skaggs Research Center (DSRC) on the Department of Commerce campus, 325 Broadway, Boulder, Colorado, USA
phone 1-303-497-6649, fax 1-303-497-7013, email henry.f.diaz@noaa.gov

Keywords: Elevational gradient, Station data coverage, Temperature change, Western US Mountains.

1. Introduction

Mountain ecosystems of the western United States are complex, and include cold desert biomes, such as those found in Nevada, subpolar biomes found in the upper treeline zone, and tundra ecosystems, occurring above timberline. Many studies (e.g. Thompson 2000) suggest that high elevation environments, comprising glaciers, snow, permafrost, water, and the uppermost limits of vegetation and other complex life forms are among the most sensitive to climatic changes occurring on a global scale. The stratified, elevationally-controlled vegetation belts found on mountain slopes represent an analogue for the different latitudinally-controlled climatic zones, but these condensed vertical gradients are capable of producing unique hotspots of biodiversity, such as those that serve as habitat for a variety of species ranging from butterflies, frogs and toads, to species of birds, trout and salmon. High relief and high gradients make mountain ecosystems very vulnerable to slight changes of temperatures and to extreme precipitation events (Parmesan 1999; Pounds et al. 1999).

Likewise, the source nature of the mountains in providing life-sustaining water for western U.S. society means that climatic and other environmental changes in the mountains of the western United States will have a large impact, not only on the region, but for the rest of the country as well. In essence, mountain regions provide a discreet quantifiable domain where relatively small perturbations in global processes,

U. M. Huber et al. (eds.), Global Change and Mountain Regions, 267-274.

can cascade down to produce large changes in most or all of the myriad interdependent mountain systems, from their hydrological cycle to their complex fauna and flora, and the people that depend on those resources.

This paper examines aspects of the climate-observing network in the western United States, pointing out the need to establish a comprehensive national effort to adequately monitor the state of the climate system in the mountainous US West.

2. Monitoring climate processes in the Western United States

As in any complex geophysical system, to be able to adequately address questions about the past, present, and future status of mountain environments in the US West, one must focus efforts to monitor and anticipate any ongoing changes, and be able to provide a historical context for the measurements. Information on fundamental processes along with patterns of local and regional change can be used to assess impacts of climate variability and mountain ecosystem vulnerability. This information is vital in order to better manage mountain ecosystems, maintain their biodiversity, sustain the use of mountain resources and ecosystems, and preserve the social and economic well being of mountain communities in the western US.

To meet the challenges of observing, understanding, predicting, and verifying changes in our mountain environments, requires a sustained integrated effort on a national scale. The complex nature of the physical, biophysical, biochemical, and human-mediated processes that operate in large-scale mountain systems requires long-term multidisciplinary and multi-institutional activities. Climate monitoring in mountain regions can be a difficult undertaking. To develop the type of long baseline of observations needed to properly assess environmental changes on multiple time scales requires a long-term commitment to quality and stability. Climate-related signals can be subtle, and are sometimes obscured by short-time scale variability. Hence, changes in variability arising from changes in the observing system can obstruct efforts at detection of climate change.

Figures 1 and 2 illustrate the present coverage of weather observing stations in the western US (area west of the 100th meridian). The first figure gives the percent coverage of stations in the western United States according to elevation. This is simply the relative proportion of stations binned by 1000 ft (~300 m) intervals. The graph also compares station coverage to the proportion of area that is occupied by the equivalent topography (the solid line in Fig. 1) in this region, based on a high-resolution digital elevation model (DEM) data set of surface topography. Of course, while the relative proportion of stations is similar to that of the topography, the actual physical coverage of those stations is rather meager (half of all the station elevations are lower than 3500 ft (1067 m)). In the second figure, the percentage of actual area coverage with meteorological observations in the western United States is calculated, assuming that each station is representative of a 100-km^2 area. This is perhaps overly restrictive with respect to surface temperature, which tends to be correlated to a relatively large radius. However, in regions of complex topography typical decorrelation scales are often less than 100 km. Figure 2 shows clearly, that the alpine regions of the US West are grossly undersampled with respect to monitoring precipitation, and the need is obvious for the development of a comprehensive climate monitoring program that would complement other long-term observations programs, such as, for example, the Long-Term Ecological Research program.

In the US, a program to develop a long-term climate reference network for the purposes of climate change monitoring and detection may provide an opportunity to incorporate into the reference network some of the mountain sites where active research programs currently utilize multidisciplinary data sets of high quality, which are, incidentally, taken from these very pristine environments. The integration of measurement programs at these mountain research sites will ensure that the data are state-of-the-art and continue to meet research requirements for studies of climate variability and change.

We have calculated surface temperature trends in the western US for the period 1950 to 2000 based on the available observation network, which contains 4,469 stations. The results are illustrated in Figure 3. Temperature changes in the western United States are positive, in the range of about 0.5 to 0.6°C over the latter half of the 20th century, with a suggestion of smaller trends above about 2 km. The statistical significance of the trends (not shown) vary from greater than 10% below 7000 ft (~2135 m) to not significant above that. However, the calculation of representative

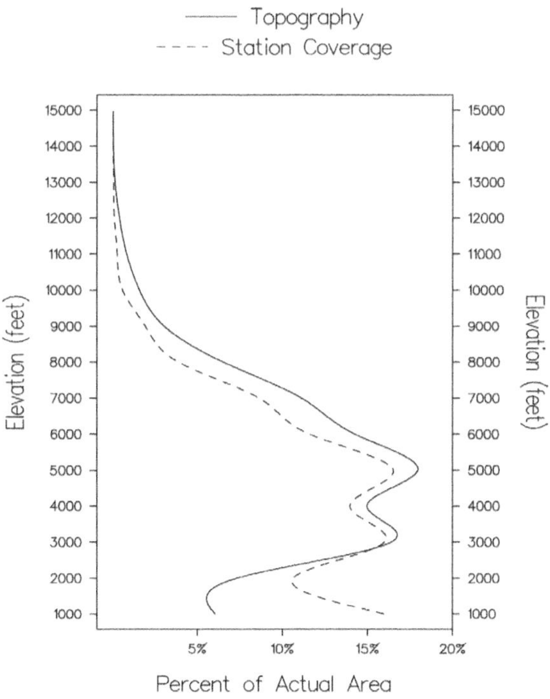

Figure 1: Percent coverage of stations in the western US according to elevation. The dashed line is simply the fraction of all stations within 1000 ft elevation intervals. The solid line represents the proportion of area that is occupied by the equivalent topography in the region.

mean temperature changes above ~ 2 km may not be possible with the current observing network, since it samples less than 10% of the available territory.

Changes in the alpine cryosphere may represent some of the earliest signs of large-scale climate change. The cryospheric variables not only serve as indicators of change but also provide powerful feedbacks through changes in albedo. Timely and detailed knowledge of ongoing changes, coupled with modeling of the effects, will allow managers and policy makers to plan for the impacts arising from such changes. At present there is great uncertainty regarding the amplitude of recent climatic changes and their future course at high elevations of the American Cordillera. Satellite images of the margins of glaciers around the world are being compiled into atlases that will provide the basis for measuring changes in the extents of glaciers through time. Observation and modeling studies of the alpine cryosphere will help document and understand the impacts of global climate change in mountain regions.

Lack of water-flow and water-quality data in critical climate-sensitive areas, such as the mountain regions of the US West impairs our ability to understand and model hydrologic processes governing climatic land-atmosphere-ocean interactions. The information is needed to make reliable projections, and to assess the impacts of variability and change in climate and water resources. Streamflow observations are inadequate and stations are being discontinued. Areas where data are particularly lacking include discharge of freshwater to the oceans and precipitation, snowmelt,

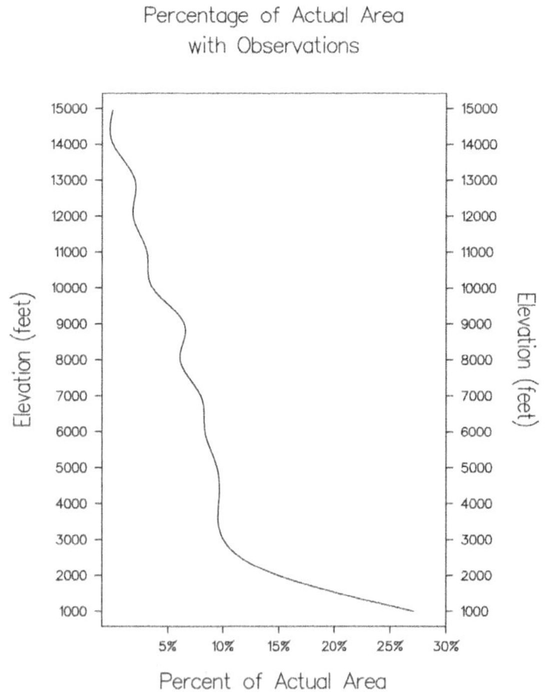

Figure 2: The percentage of actual area coverage with observations in the western US. Values calculated assuming that each station is representative of a 100-km² area.

and runoff in high mountain basins, which contribute disproportionately to the flow of many rivers.

A major problem in carrying out routine observations in high mountains, which are often located in remote areas that require major efforts to visit and keep the measurements going, is often the lack of adequate resources to do the job. Well-established technical means are available, and various innovative technologies are either already available, or in need of only minor development and field-testing for applications. In order to monitor adequately for changes in the natural environment that may be occurring as a result of global climate change, it will be necessary to establish *in situ* streamflow gauging stations, ground-water observation wells, and water-quality measurement sites in selected climatically sensitive basins in the West. The payoff would be improved definition of surface and subsurface flows and transports of water-quality constituents, which will lead to improved understanding of hydrologic processes and how they might respond to global climate changes in the future.

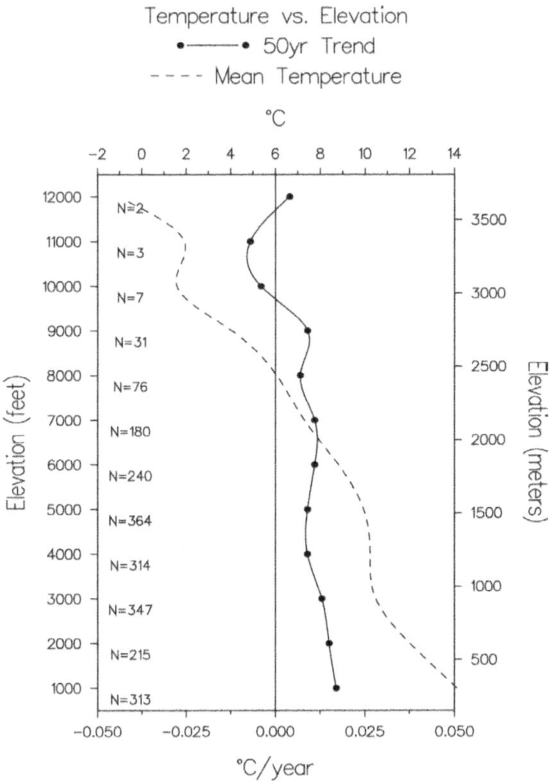

Figure 3: Median temperature trends with elevation for the period 1950–2000 for the western US (west of 100°W). Values are plotted in 1000-ft (~300 m) increments, and the curves are smoothed using a rigid spline. The trend values are plotted against the annual mean temperature of the stations (dashed curve), averaged for each vertical segment, and based on the number of stations shown on the left side of the graph.

Since 1950, springs in the western US have advanced by 1-2 weeks (Cayan et al. 2001) and about 20% of the late spring-summer snowmelt runoff from mid-elevation watersheds has been occurring in earlier months (Dettinger and Diaz 2000). Global change projections indicate that western snowpacks will diminish markedly over the next century and this crucial spring-summer portion of the runoff will be sharply reduced. Besides having direct economic impact, this change in mountain hydroclimate would presumably affect ecosystems, both up and downstream. A better multi-faceted observational system is needed to monitor and understand these changes as they occur.

Continued support of paleoclimate studies is needed to help establish a scale for what can be considered normal variation by looking back at climatic variations in the recent past. Results from paleoclimate studies have shown that during the last 2000 years climate variation has resulted in both warming and cooling events, the Medieval Warm Period at around AD 1000, and the Little Ice Age from around AD 1500 to 1800, which have been accompanied by significant elevation shifts of lake levels and alpine treelines, as well as temperature and rainfall. One advantage of supporting studies with high-resolution paleorecords of these relatively recent events is that by determining the surface spatial distribution of climatic effects it may be possible to infer the past atmospheric driving forces.

Finally, it is noted that temperature changes appear to have been greatest at high latitudes, while changes in precipitation and drought periods appear to be much more important in the tropics over the past 100 years. It is not clear yet whether climate changes in high elevation regions will exhibit an amplification of the global warming signal. It behooves the scientific community to definitively address this important question. To do that, however, a variety of high quality records will be needed. Based on the current sampling network of climate monitoring in the mountainous regions of the western US, it appears that the spatial coverage is inadequate to answer that question.

3. Concluding remarks

Because of the fundamental complexity of mountain regions, progress in understanding the response of both natural and human ecosystems to climatic variation and change will require the integration of various disciplines into a more cohesive intellectual framework. In the mountainous western US, as in other mountainous regions of the world, there is a need to develop a more holistic view of the processes affecting the physical and biological systems comprising the region. The problems must be tackled in the same interconnected manner that the real system operates in.

There are a number of challenges that we face in order to achieve a sustainable long-term effort to comprehensively monitor mountain environments. With the possible exception of some mountain regions in Europe, where long-term observations of climate and other aspects of the physical environment exist, the required efforts apply elsewhere in the world's mountain regions, as they do for the western US. First, the development of improved networks to adequately sample, both spatially and temporally, all the critical elements needed to define the state of the region's climate in order to understand its past, present, and future behavior. A goal of an observing system should be to provide free and open access to real time (or near-real time) data, and access to quality assessment of the data.

The Mountain Research Initiative, which is the theme of this special volume, lists as one of its critical goals that are relevant to the theme of this chapter, the long-term monitoring and analysis of indicators of environmental change in mountain regions. The climate observing system should be linked to ongoing research and be able to support the needs of other users and to accommodate a broad range of uses of the data. The observing system should have the ability to adapt to the use of new technologies as they become available at lower costs, add new variables as needed, etc. Finally, the climate observing system must adhere to the principles for climate monitoring, as outlined in a U.S. National Research Council report (NRC 1999), and to the management guidelines required for implementing them. Adequate monitoring of critical environmental variables and processes are the foundation of understanding global change, variability, and extreme events. Efficient access to comprehensive observational, paleoenvironmental, and model data is also required.

In the United States, a comprehensive national program to monitor the state of its mountain regions is not in place. Some individual efforts by various state and federal agencies has resulted in the development of different observing networks maintained under different operational goals, and with the data sets held in different locations throughout the country. There are a number of efforts currently under way in the US and in other parts of the world to develop a climate-observing network for mountain regions. There have been many previous efforts to bring to the attention of the scientific community the need for special attention to mountain environments and opportunities for establishing a global network of mountain observatories (e.g. Barry 1994; Beniston 1994; Beniston et al. 1997; Diaz et al. 1997; Diaz and Bradley 1997). It is to be hoped that publication of this volume will provide the necessary impetus to jump-start these efforts and to ultimately incorporate the routine monitoring and data archive functions into the existing Global Climate Research Program activities of countries with mountain regions.

4. References

Barry, R. G. (1994). Past and potential future changes in mountain environments. A review. *In* "Mountain environments in changing climates." (M. Beniston, Ed.), pp. 3–33. Routledge, London.

Beniston, M., Ed. (1994). "Mountain environments in changing climates." Routledge, London.

Beniston, M., Diaz, H. F., and Bradley, R. S. (1997). Climatic change at high elevation sites: An overview. *Climatic Change* **36**, 233–251.

Cayan, D. R., Kammerdiener, S. A., Dettinger, M. D., Caprio, J. M., and Peterson, D. H. (2001). Changes in the onset of spring in the western United States. *Bulletin of the American Meteorological Society* **82**, 399–415.

Dettinger, M. D., and Diaz, H. F. (2000). Global characteristics of streamflow seasonality and variability. *Journal of Hydrometeorology* **1**, 289–310.

Diaz, H. F., Beniston, M., and Bradley, R. S. (1997). "Climatic change at high elevation sites." Kluwer, Dordrecht.

Diaz, H. F., and Bradley, R. S. (1997). Temperature variations during the last century at high elevation sites.

Climatic Change **36**, 253–279.
National Research Council (NRC) (1999). Adequacy of climate observing systems. National Academy Press, Washington DC.
Parmesan, C. (1996). Climate and species' range. *Nature* **382**, 765–766.
Parmesan, C. (1999). Poleward shifts in geographical ranges of butterfly species associated with regional warming. *Nature* **78**, 2837-2849.
Pounds, J. A., Fogden, M. P. L., and Campbell, J. H. (1999). Biological response to climate change on a tropical mountain. *Nature* **398**, 611–615.
Thompson, L. G. (2000). Ice core evidence for climate change in the Tropics: Implications for our future. *Quaternary Science Reviews* **19**, 19–35.

Spatial Heterogeneity of Snow Conditions and Evapotranspiration in the Swiss Alps

Lucas Menzel[1*] and Herbert Lang[2]
[1]*Potsdam-Institute for Climate Impact Research (PIK), P.O. Box 601203, D-14412 Potsdam, Germany*
[2]*Institute for Atmospheric and Climate Science, Swiss Federal Institute of Technology (ETH), Winterthurerstrasse 190, CH-8057 Zürich, Switzerland*
**phone +49-331-288 2673, fax +49-331-288 2695, e-mail menzel@pik-potsdam.de*

Keywords: Climate change, Evapotranspiration, Snowcover, Spatially distributed modelling, Swiss Alps.

1. Introduction

In most alpine regions, the presence of snow controls the hydro-climatic situation over a great part of the year. The delayed and long-lasting process of snowmelt guarantees a relatively well-balanced discharge regime of rivers in the spring and summer melting season, even if only a small part of their catchment includes high mountain areas. For the typical alpine weather conditions, this results in high melt water runoff during dry conditions when net radiation and air temperature are high, while, during cooler periods, rainfall compensates for reduced or discontinued melt rates and sustains streamflow at a balanced level. Furthermore, because of the relatively high albedo of snow, changes in alpine snowcover are associated with a feedback to climate, a process that has not yet been very well investigated. For example, a climate-induced decrease in snowcover will reduce surface albedo, which leads to an amplification of the initial warming.

During snow free periods, evapotranspiration is another process, which impacts both water balance and climate over a wide elevational range. The partitioning of radiant energy into sensible and latent heat and hence the amount of energy available

U. M. Huber et al. (eds.), Global Change and Mountain Regions, 275-282.

for both surface warming and evapotranspiration is determined by a variety of local characteristics, such as climatic and topographical conditions, the vegetation structure, different soil types and water availability in the near surface soil zone. These factors may vary considerably with space and time, especially in mountainous areas. The temporal variability of snowcover and evapotranspiration and their spatial heterogeneity across the Alps result in complex dynamics of hydrological processes and marked differences between the hydro-climatic characteristics of the various alpine regions.

The present study aims to investigate the spatial variability in snow cover and evapotranspiration between the different parts of the Swiss Alps, with a special focus on the elevation zone between 1500 and 2000 m asl. So far, relatively few reliable estimates of evapotranspiration in mountainous regions are available (e.g. Hennemuth and Köhler 1984; Bernath 1991). Therefore, we carried out field investigations across a high-alpine valley in Grisons, Switzerland, over three subsequent years (Konzelmann et al. 1997; Menzel and Lang 1998). All energy balance components, as well as additional climatic parameters and soil moisture, were measured at representative sites above and below the timberline. The measurements identified small-scale variations and substantial gradients in the different components of the energy balance in mountainous terrain, which is reflected in the spatial heterogeneity of evapotranspiration. The investigations also served to further develop our model simulations for the spatially distributed determination of evapotranspiration and other components of the water balance in the Swiss Alps (Gurtz et al. 1997; Menzel et al. 1999).

Beside other impacts, climate change in alpine regions modifies the snow (depth and cover) and evapotranspiration conditions and hence the water and energy balance. While, on a global scale, temperature increase was in the range of + 0.6°C over the last century (IPCC 2001), warming was significantly higher in the European Alps (Haeberli and Beniston 1998). The evaluation of meteorological measurements in Switzerland shows that temperature increase was even different in the various regions of Switzerland, ranging from + 1.0°C in the southern alpine region to + 1.6°C in the western parts within the last 100 years (OCCC 2002). Climate model projections suggest that 21st century warming in the European Alps will exceed the global average. Therefore, we also present a simulation study on the possible impact of climate change on snow and evapotranspiration conditions in the investigated elevation zone of the Swiss Alps.

2. Model description

Our study aims to simulate the spatial distribution of evapotranspiration and associated sub-processes over the whole area of Switzerland (Menzel et al. 1999). For this purpose, we applied TRAIN, a model that predicts the major processes at the soil-vegetation-atmosphere interface (Menzel 1997), such as the available radiant energy fluxes, snow cover, soil moisture content, transpiration, interception and interception evaporation for different types of land-cover. The model is generally applicable for large areas, which are subdivided into regular grids of optional cell size.

The simulation of snow accumulation in TRAIN is initiated when air temperature drops below a critical value of + 1.6°C. Then, within a certain temperature interval, the proportion of snowfall relative to total precipitation is assumed to vary linearly with rising or falling temperatures. Below a given threshold temperature (– 0.4°C) only snowfall occurs, which accumulates on the entire land surface. Variations of snow albedo are calculated according to Plüss (1997), with decreasing values during periods without snowfall, or when temperature conditions initiate snowmelt. The simulation of snowmelt is mainly based on a simple temperature-degree approach. The determination of transpiration follows the Penman-Monteith approach (Monteith 1965), with canopy resistances mainly depending on weather conditions, the state of vegetation growth and soil moisture. Interception, i.e., the short-term storage of precipitation in the canopy, and evaporation of intercepted water are simulated using a novel, physically-based approach (Menzel 1997), taking into account the vertical structure of the different vegetation types and their ability to store precipitation in relation to the phenological development.

In order to simulate the spatial patterns of evapotranspiration, Switzerland was divided into 1 km^2 squares. The meteorological data needed for the calculations (precipitation, air temperature, air humidity, wind speed, sunshine duration) were provided by the measuring network operated by Meteo Swiss. A combination of an altitude-dependent regression and the inverse distance method (Schulla 1997) was used for the interpolation of the climatic data on the grid. Other data sets that were required to implement TRAIN include a digital elevation model, with information on slope angles and aspect, and digital maps on land-use, soil depths and soil water storage capacities.

The model was developed and validated at different observation sites, including both pre-alpine and high-alpine climate conditions. TRAIN is usually applied in hourly time steps. Since complete meteorological time series for the whole area of Switzerland were only available on a daily basis, we carried out our simulation studies with daily resolution. For the investigated period 1973-1992, mean annual evapotranspiration over the entire area of Switzerland was found to be 484 mm (Menzel et al. 1999). An independent study carried out by Zierl (2001), only focusing on forested ecosystems, estimates mean annual evapotranspiration over the forested area of Switzerland to be 617 mm. Our investigations on forest evapotranspiration in Switzerland yielded concordant results (616 mm).

3. Significant spatial differences

The present study is focussed on the alpine part of Switzerland. The investigated area is structured into a series of mountain ranges of differing geographic orientation, often subdivided by major valleys and ridges, and with marked differences in climatic properties. Based on these differences, we defined six regions from west to east, the Valais, the Bernese Oberland (including Vaud), the Ticino, Central Switzerland, Grisons and the Engadine. Since our investigations concentrate on both snow-related processes and evapotranspiration, as well as their sensitivity to climate change, we selected the altitude interval between 1500 and 2000 m asl. Furthermore, since most areas within this elevation zone are forested, the focus of our regional comparative study is on this type of land-cover. According to our database, the number of 1 km^2 squares covered with forest within the 1500-2000 m elevation zone ranges from 168 in Central Switzerland and 480 in Grisons. Table 1 presents average data over all investigated squares in the respective regions.

Table 1: Investigated forest regions of the Swiss Alps and their main hydro-climatic characteristics within the 1500-2000 m elevation zone. The regions are ordered along an approximate west-east direction. Data are given as annual averages over the period 1973-1992.

Region name	*Precipitation [mm]*	*Temperature [°C]*	*Evapotranspiration [mm]*	*Drainage rate [mm]*	*Number of days with snow cover*
Valais	1210	4.0	620	590	190
Bernese Oberland and Vaud	1516	4.0	603	913	194
Ticino	1652	3.8	604	1048	195
Central Switzerland	1615	3.7	593	1022	211
Grisons	1273	3.4	575	698	198
Engadine	961	2.2	547	414	198

The spatial variations in mean annual precipitation and mean annual temperature clearly show the different climatic characteristics of the investigated regions. The decrease in mean annual temperature between the western and eastern parts of the Swiss Alps is also reflected in a related gradient in evapotranspiration. A combination of favourable conditions, primarily warmer temperatures and high rainfall, leads to a maximum in forest evapotranspiration in the Valais, while the lowest values occur in the relatively dry and cold Engadine. Drainage rates (the difference between precipitation and evapotranspiration) are highest in the comparatively wet central parts, while they are at a minimum in the Engadine (low precipitation) and the Valais (high evapotranspiration). The mean number of days per year with snow cover only varies over a maximum of 21 days between the six regions.

However, in contrast to the number of days with snow cover, the seasonal progression of snow cover and the absolute values of snow depth show significant regional differences, as the example in Figure 1 demonstrates. This figure presents the mean seasonal development of snow, expressed as snow water equivalent, in the investigated elevation interval in Central Switzerland and in the Engadine. Despite these regional differences in snow conditions, mean annual albedo is 0.5 in all investigated regions. This can be explained both by the fact that snow albedo is largely independent of snow depth and by the assumption that forested areas show a uniform albedo of 0.1 during snow free periods.

Figure 1 also demonstrates the comparatively low temperatures in the Engadine, particularly between December and February. In addition, we can observe an approximately 10 % difference in mean annual forest evapotranspiration between

Central Switzerland and the Engadine, as caused by a combination of both lower evapotranspiration rates during winter and lower peak values during summer in the Engadine. In all investigated regions, the occurrence of evapotranspiration is not limited to the growing season. A combination of interception evaporation and snow sublimation - the latter is especially intensive during snowmelt conditions - leads to considerable evapotranspiration rates during times when transpiration is reduced or at zero values.

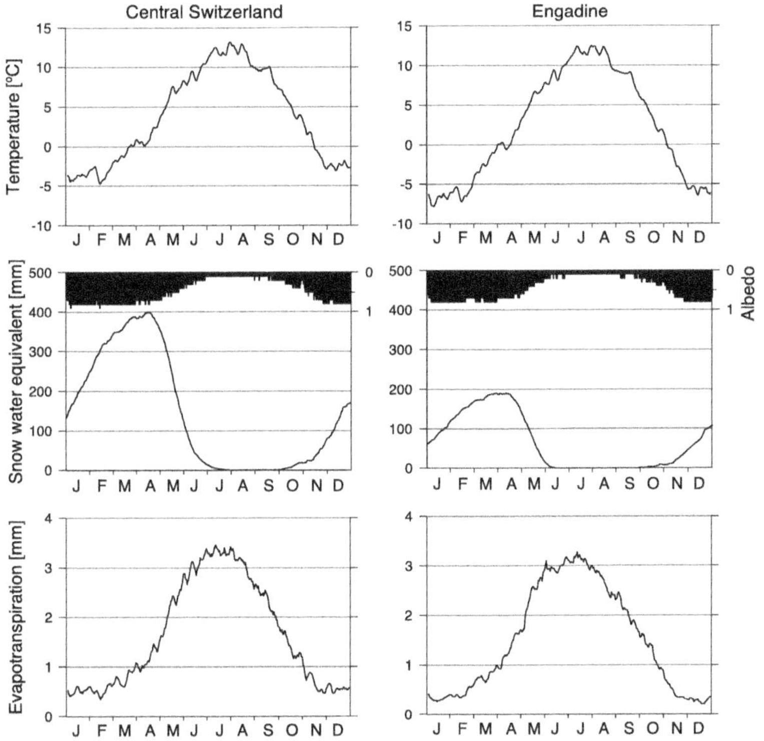

Figure 1: Mean daily data of air temperature, albedo (black bars), snow water equivalent and actual evapotranspiration for forested areas in Central Switzerland (left) and the Engadine (right) within the investigated 1500–2000 m elevation zone. Data refer to the period 1973–1992.

4. Possible impacts of a changing climate

At present, there is still a considerable lack of reliable, small-scale scenario data for mountain regions, because the resolution of General Circulation Models (GCMs) is still inadequate for modelling mountain climates. According to the current emission scenarios of the Intergovernmental Panel on Climate Change (IPCC 2001), GCMs simulate a temperature increase of 3–5°C within the next 50–100 years for the European Alps. It is very likely that the precipitation regime will also change, with increasing winter precipitation and drier summers (OCCC 2002). Since detailed spatial information on future regional changes is not available yet, we applied a climate change scenario in which temperature and precipitation changes were kept constant between the different investigated regions. The applied scenario is based on data given in Parry (2000) and related assumptions. The mean winter temperature increase until 2050, as computed by five different climate models, is given as 3.5°C for the Swiss Alps, a value which was also adopted for the summer period. For our climate change scenario, we assume an increase in winter precipitation by 15 % until 2050 and a reduction of summer precipitation by 10 %. These temperature and precipitation changes were superimposed on the time series of measured data of the reference period 1973-1992. The effects of the climate change scenario on the investigated hydro-climatic conditions in the Swiss Alps are compiled in Table 2.

The data in Table 2 demonstrate that the projected seasonal changes in precipitation will result in an overall increase of mean annual precipitation, which reflects that winter precipitation is dominant in all regions. Evapotranspiration is simulated to increase by 13-20 % under the scenario conditions. A detailed analysis shows that both precipitation and temperature increase lead to an intensification of canopy precipitation storage (interception) and subsequent evaporation. Therefore, the increase in evapotranspiration is mainly due to an increase in interception evaporation. Since evapotranspiration totals are simulated to increase disproportionately relative to the precipitation increase, annual drainage rates are projected to decrease. The reduction of runoff becomes more pronounced from the western towards the eastern regions of the Swiss Alps. This can be ascribed to smaller increases in precipitation totals in the eastern regions, especially in Grisons and the Engadine. According to the climate change scenario, drastic changes in snow conditions can be expected in all regions. Table 2 shows that the number of days per year with snow cover is projected to decrease considerably. This will also have an impact on mean annual albedo of the investigated regions, which is estimated to decrease from a value of 0.5 under present conditions to 0.3 in the future. This change in albedo might be associated

Table 2: Impacts of a climate change scenario on the investigated regions of the Swiss Alps. The results relate to the forested areas within the 1500-2000 m elevation interval. Values in brackets show changes relative to the data of the 1973-1992 reference period (see Table 1).

Region name	*Precipitation [mm]*	*Temperature [°C]*	*Evapotranspiration [mm]*	*Drainage rate [mm]*	*Number of days with snow cover*
Valais	1276 (+ 5.5 %)	7.5 (+ 3.5 °C)	704 (+ 13.5 %)	572 (- 3.1 %)	105 (- 85)
Bernese Oberland and Vaud	1578 (+ 4.1 %)	7.5 (+ 3.5 °C)	719 (+ 19.2 %)	859 (- 5.9 %)	111 (- 83)
Ticino	1693 (+ 2.5 %)	7.3 (+ 3.5 °C)	698 (+ 15.6 %)	995 (- 5.1 %)	93 (- 102)
Central Switzerland	1671 (+ 3.5 %)	7.2 (+ 3.5 °C)	712 (+ 20.1 %)	959 (- 6.2 %)	124 (- 87)
Grisons	1301 (+ 2.2 %)	6.9 (+ 3.5 °C)	680 (+ 18.3 %)	621 (- 11.0 %)	117 (- 81)
Engadine	977 (+ 1.7 %)	5.7 (+ 3.5 °C)	631 (+ 15.4 %)	346 (- 16.4 %)	126 (- 72)

with a positive climate feedback, leading to a regional amplification of the initial temperature increase.

5. Discussion and future directions

The aims of this contribution are multiple. We want to emphasise the considerable spatial differences in principal hydro-climatic characteristics, which exist even in adjacent alpine regions. The focus of this paper is on differences in snow and evapotranspiration conditions in the elevation zone between 1500 and 2000 m asl across the Swiss Alps. Although our results are based on validated, detailed model simulations, attention needs to be drawn towards shortcomings of this type of alpine research. The consideration of hydrological processes in forested areas over large regions and vast elevation zones would require a detailed analysis of differences in soil and vegetation properties. Although our database is comparatively comprehensive, data coverage is still too limited for a more detailed spatial analysis. Furthermore, model restrictions require the assumption of a homogeneous vegetation cover, which only consists of coniferous forests with identical properties over the whole investigated area. However, although a more realistic representation of small-scale patterns in physiography and vegetation would likely increase the spatial heterogeneity of hydrological simulations in each of the study areas, mean spatial differences between the investigated regions would likely be similar to our modelling results.

Further processes, which have not yet been sufficiently investigated, relate to the feedbacks between snow conditions, albedo changes and evapotranspiration at the land surface on one side and weather and climate on the other side. Therefore, further research that considers the interaction between climatic and hydrologic processes in coupled models is urgently needed.

Our investigations concentrate on the 1500-2000 m elevation zone, which is thought to be particularly sensitive to climate change. We applied a relatively simple climate change scenario in order to demonstrate the potential magnitude of changes in snowcover and evapotranspiration. In this scenario, assumed temperature and precipitation changes are superimposed on present-day climate conditions, which were measured over a certain reference period. This is a technique that is widely applied, especially for impact studies in alpine regions. However, changes in other climatic parameters, such as the different components of the radiation balance, air humidity or wind speed, are not taken into account. Also, future projections of small-scale spatial changes in climatic properties are not available so far. This would however be a prerequisite for more realistic impact studies in extremely inhomogeneous alpine terrain. Furthermore, our study – such as a multitude of similar investigations – does not account for changes in vegetation composition and vegetation properties, which certainly occur under climate change. Investigations that consider vegetation change in alpine conditions should be initiated soon and they should be integrated in hydrological models.

The results of our climate change impact study clearly demonstrate the sensitivity of the investigated elevation zone to temperature increase. The projected, drastic decrease of snowcover across all investigated regions will have pronounced effects on both natural processes and economic conditions. For example, winter tourism will concentrate on higher ski areas with safer snow conditions or potential tourists may be discouraged to spend their holidays in the Alps. The combination of reduced water storage during winter, lower summer precipitation and higher evapotranspiration will decrease specific discharge rates in summer, which again leads to reduced runoff in rivers and might raise economical difficulties for the local hydro-electric power industry. Furthermore, water supply problems will occur on a supra-regional scale. In contrast to summer conditions, specific discharge rates will increase in winter, when a higher proportion of precipitation falls as rain. This could not only increase mean flow conditions in rivers but also intensify the occurrence of floods and cause further economic damage.

6. References

Bernath, A. (1991). "Zum Wasserhaushalt im Einzugsgebiet der Rhône bis Gletsch." Zürcher Geographische Schriften 43. Department of Geography ETH, Zurich.

Gurtz, J., Baltensweiler, A., Lang, H., Menzel, L., and Schulla, J. (1997). "Auswirkungen von klimatischen Variationen auf Wasserhaushalt und Abfluss im Flussgebiet des Rheins." Schlussbericht NFP 31. Vdf Hochschulverlag, Zurich.

Haeberli, W., and Beniston, M. (1998). Climate change and its impacts on glaciers and permafrost in the Alps. *Ambio* **27**, 258-265.

Hennemuth, B., and Köhler, U. (1984). Estimation of the energy balance of the Dischma Valley. *Archiv für Meteorologie, Geophysik und Bioklimatologie*, Serie B, **34**, 97-119.

Intergovernmental Panel on Climate Change (IPCC) (2001). "Climate Change 2001: The scientific basis" (J. T. Houghton et al., Eds.). Third assessment report of the IPCC, Cambridge.

Konzelmann, T., Calanca, P., Müller, G., Menzel, L., and Lang, H. (1997). Energy balance and evapotranspiration in a high mountain area during summer. *Journal of Applied Meteorology* **36**, 966-973.

Menzel, L. (1997). "Modellierung der Evapotranspiration im System Boden-Pflanze-Atmosphäre." Zürcher Geographische Schriften 67. Department of Geography ETH, Zurich.

Menzel, L., and Lang, H. (1998). Spatial variation in evapotranspiration in Swiss Alpine regions. *In* "Hydrology, water resources and ecology in headwaters." (K. Kovar, U. Tappeiner, N. E. Peters, and R. G. Craig, Eds.), pp. 115 – 121. IAHS Publication 248, Wallingford.

Menzel, L., Lang, H., and Rohmann, M. (1999). Mean annual actual evaporation. *In* "Hydrological atlas of Switzerland." (Landeshydrologie und –geologie, Ed.), Chapter 4. Geographical Institute of the University Bern, Bern.

Monteith, J. L. (1965). Evaporation and environment. *In* "Proceedings of the 19th Symposium of the Society for Experimental Biology," Cambridge.

Organe consultatif sur les changements climatiques (OCCC) (2002). "Das Klima ändert - auch in der Schweiz. Die wichtigsten Ergebnisse des dritten Wissensstandberichts des IPCC aus Sicht der Schweiz." OCCC, Bern.

Parry, M. L., Ed. (2000). "Assessment of potential effects and adaptations for climate change in Europe: The Europe ACACIA project." University of East Anglia, Norwich.

Plüss, Ch. (1997). "The energy balance over an alpine snowcover." Zürcher Geographische Schriften 65. Department of Geography ETH, Zurich.

Schulla, J. (1997). "Hydrologische Modellierung von Flussgebieten zur Abschätzung der Folgen von Klimaänderungen." Zürcher Geographische Schriften 69. Department of Geography ETH, Zurich.

Zierl, B. (2001). A water balance model to simulate drought in forested ecosystems and its application to the entire forested area in Switzerland. *Journal of Hydrology* **242**, 115–136.

Runoff Processes in Mountain Headwater Catchments: Recent Understanding and Research Challenges

Alfred Becker
Potsdam Institute for Climate Impact Research, PO Box 601203, D-14412 Potsdam, Germany phone +49331 288 2541, fax +49331 288 2642, e-mail becker@pik-potsdam.de

Keywords: Distributed hydrological modelling, Landscape patchiness, Lateral flow components, Response times, Runoff generation, Subsurface stormflow, Travel times

1. Introduction

Runoff generation in mountain catchments is one of the most complex hydrological processes. It is highly variable in space and time, depending on the combination of three main controlling factors: (1) climate, (2) soil and geology, and (3) vegetation. The different combinations of these three factors determine the water balance of landscape units, including soil moisture dynamics, evapotranspiration and runoff generation. When assessing runoff generation, not only the runoff amounts need to be considered, but also the relative streamflow contributions of surface and subsurface runoff, which may differ considerably between areas (Buttle 1998). An overview of runoff mechanisms and components in different environments is given in Uhlenbrook and Leibundgut (1997) and Bonell (1998). The main focus of this paper is on subsurface stormflow, the least understood flow component.

Landscape units with similar combinations of climate, soil and vegetation, and hence similar hydrological and water balance behaviour, are called hydrotopes (see Fig. 1). They are assumed to behave internally similar (quasi-homogeneous) in terms of the different water balance components, including runoff generation. Landscapes are generally composed of a series of hydrotopes, which may differ remarkably. A

U. M. Huber et al. (eds.), Global Change and Mountain Regions, 283-295.

consequence of this "landscape patchiness" is the often pronounced spatial variability of hydrological and related processes, such as runoff generation. For example, in impervious or less permeable areas, such as exposed bedrock, clay soils, or sealed surfaces in urban areas (AIMP in Fig. 1), as well as in water saturated areas (e.g. in shallow soils above the groundwater table (AN) or impervious subsoil), overland flow (RO) is generated during each rainfall and snowmelt event and incorporates nearly 100% of rainfall and/or snowmelt. In contrast, adjacent permeable vegetated areas, especially those with deep groundwater (AG), may never generate any direct surface runoff. Process-based distributed hydrological models are required to assess this spatial variability and allow us to model the hydrological response of the different hydrotopes in the landscape or, alternatively, in evenly-spaced grid cells at a resolution that adequately reflects the patchiness of the landscape (Becker and Braun 1999; Becker et al. 2002).

In these models, the calculated runoff responses of the various hydrotopes or grid cells result in lateral flows, which are superimposed and spatially aggregated in a catchment and then routed downslope along the different surface and subsurface pathways (arrows in Fig. 1) to the nearest channels of the river network, according to their response or travel times. The total outflow of the catchment (basin discharge) is calculated as the aggregate of the different runoff components generated in the catchment. These components will be briefly discussed in the following.

2. Runoff components and their time behaviour

2.1 Three main runoff components

Whenever rainfall and/or snowmelt minus actual evapotranspiration exceeds the infiltration capacity of the soil, or the soil water recharge capacity in a spatial unit such as a hydrotope, "excess water" runs off in one or more of the following flow components (see Fig. 1):

- surface runoff (overland flow RO = rainfall or snowmelt intensity minus infiltration capacity)
- infiltration and percolation through the soil, which either result in contributions to

  (i) interflow (lateral subsurface stormflow RI on less permeable soil layers = infiltration minus soil water recharge and/or percolation to deeper soil layers), or

  (ii) groundwater recharge through percolation, generating increases in base flow RG (groundwater outflow from aquifers into the channel system).

Our understanding of runoff components has evolved remarkably during the second half of the last century (Maidment 1993). Originally, only two runoff components were distinguished: surface and subsurface runoff, which differ remarkably in their response times (see Table 1 in section 2.2). Later, with the increasing application of mathematical runoff models, it became clear that a third component (RI) needs to be considered, in particular in mountainous catchments (Becker 1989; Beven 1989). The response time of this so-called interflow component lies between two extremes, namely:

- the rapid (short-term) response of the surface runoff component (overland flow RO) and
- the slow (long-term) response of the base flow component (RG), which is mainly fed from groundwater in aquifers and other large-scale groundwater bodies and therefore behaves in a rather stable manner, even during longer dry periods.

Accordingly, RI was recognized as a subsurface flow component that responds more rapidly than base flow RG; i.e. the definition of this flow component RI was originally based more on its time behaviour, as understood from hydrograph analysis and flow modelling, rather than on process understanding. RI was later re-named with the more process-oriented term "subsurface stormflow," which in terms of processes and pathways was, and in several aspects still is, not completely understood.

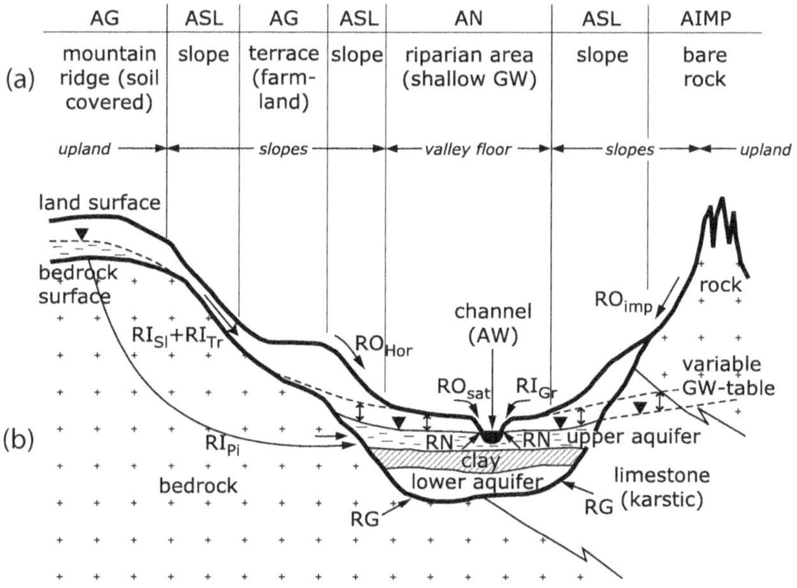

Figure 1: Schematic representation of a valley cross-section showing (a) typical landscape sub-units with similar runoff generation and evaporation characteristics (hydrotopes), (b) different runoff components (R) and subcomponents (abbreviations see Table 2 in Section 3).

Both RI and RO are highly variable in space and time. Because RI is not visible, it is more difficult to measure and model. Gaps in our knowledge on RI need to be overcome in order to simulate this runoff component correctly and to investigate human impacts, including the influence of global change on runoff and associated transports of nutrients and pollutants. For this purpose, comprehensive and integrative experimental and modelling studies need to be carefully designed on the basis of our present understanding of run-off generation (Becker et al. 1999; Uhlenbrook and Leibundgut 2002). A summary of this understanding is given below, with special focus on subsurface stormflow.

2.2 Distinguishing response and travel times of runoff components

Concerning the time behaviour of runoff components, in particular subsurface flow components (RI and RG), it is essential to distinguish (Kendall and McDonnell 1998):

(i) *Response time* T_r, i.e. the time between a runoff-generating event, such as a rainstorm, and the corresponding increase in streamflow (Becker 1989; Maidment 1993). An understanding of T_r is essential for modelling the temporal variability of streamflow, especially in flood analysis and predictions.

(ii) *Transit* or *travel time* T_t (and the related residence time), i.e. the time that water particles and associated substances and solutes (molecules) require in order to "pass" through the relevant system (McDonnell et al. 1999). T_t needs to be known for the analysis and simulation of the transport of nutrients and other solutes and particles that determine water quality and aquatic ecosystem states and thus for environmental impact assessments and hydroecological studies.

Table 1 gives an overview of the average response times T_r and transit times T_t for the three flow components. It should be pointed out that transit times T_t for the subsurface flow components are larger than their response times T_r by at least one order of magnitude (Sklash and Farvolden 1979; McDonnell et al. 1999). This is due to the fact that response times are the result of both the movement of water particles and molecules and pressure wave propagation. The focus in the following will be on response times, although the analysis of travel time behaviour of the different subsurface flow components (e.g. by tracer techniques) plays a key role for improving our understanding of subsurface flow processes, in particular subsurface stormflow RI (Uhlenbrook and Leibundgut 1997).

Table 1: Ranges of response times T_r and transit times T_t of the three main runoff components as derived from results of various studies in different river basisns (see text).

Runoff component	Response time T_r	Transit time T_t
Surface flow	< 1 hour to < 1 day	nearly equal to T_r
Interflow	ca. 1 day to a few weeks	several weeks to > 1 year
Baseflow	ca. 100 days to > 1 year	several years and more

Response times and transit times generally increase with catchment size due to the increase in the length of pathways. However, these increases are relatively small compared to the significant differences between the three flow components listed in Table 1.

T_r of *overland flow* (or surface runoff RO) can vary between less than an hour and less than a day, at least in small (headwater) catchments. For example, the streamflow hydrograph at the outlet of the 1.6 km^2 Schaefertal catchment in the Harz mountains, Central Germany (Fig. 2 in section 4), shows several flashy flood peaks, which can be interpreted as pulse response functions of the surface flow system ($RO_{1,2,3}$) in the catchment (Becker 1989). These hydrographs indicate that in this catchment T_r of the overland flow component RO amounts to approximately three hours. Similar results are obtained at many other places, for example, in a mountainous headwater basin in the crystalline Alps (Tilch et al. 2003), based on both classical and tracer hydrological analyses (see also Kirnbauer et al., this volume). The T_r of overland flow is primarily defined by topography, such as slope angles and the structure and slope of channels and rivers. The response times of this component are relatively well understood and the range of values listed in Table 1 is consistent with the governing physical laws.

T_r of *subsurface stormflow* (RI) is on the order of a day to several weeks (Table 1). Figure 2 (in section 4.) shows an example from the Schaefertal, where T_r can be estimated for the third of the three represented storm events (P3) as about 50 hrs, assuming that the lower and broader secondary peak (RI_3) in the bimodal hydrograph represents subsurface stormflow. This result corresponds closely to response times derived from classical streamflow recession analyses and various modelling studies (e.g. Bazemore et al. 1994; Becker and Mc Donnell 1998; Uhlenbrook et al. 2002). Nevertheless, the response times of subsurface stormflow, especially in mountainous terrain, are still poorly understood because this runoff component often consists of different sub-components the response times of which may vary considerably (see section 3.2).

T_r of *baseflow* (RG) varies between approximately 100 days and more than a year (Table 1). Base flow is basically fed from groundwater in aquifers and in bedrock. T_r is therefore primarily dependent on the hydraulic gradient and transmissivities in the saturated zone. Although an accurate determination of these variables is always difficult, especially in mountain catchments, the range of response times listed in Table 1 is generally accepted, at least as a first approximation (e.g. Kunkel and Wendland 1999).

3. Recent understanding of runoff sub-components in different environments

Table 2 gives a brief summary of the different sub-components of overland flow RO and subsurface stormflow RI and the characteristic conditions under which they occur.

3.1 Overland flow

Infiltration excess overland flow (also called "Horton overland flow": RO_{Hor} in Fig. 1) is mainly generated during high intensity rainstorms when rainfall intensities exceed the soil infiltration capacity. RO_{Hor} represents the dominant runoff component in arid and semi-arid environments where such events typically occur after long dry periods. A special type of this direct surface runoff generation is from impervious areas, such as bare rock and sealed surfaces (RO_{imp} in Fig. 1).

In more humid climates, RO_{Hor} is less prevalent and saturation excess overland flow dominates (RO_{sat} in Fig. 1). RO_{sat} is generated in surface saturation areas, which predominantly occur during and after rainfall and snowmelt events in near-stream zones, wetlands and other areas with shallow groundwater (AN in Fig. 1). As already mentioned, surface saturation is due to rising groundwater tables or transmissivity feedback from less permeable soil layers. Figure 1 illustrates, for example, how a rising groundwater table in the valley floor aquifer (dashed line) can temporarily intersect the soil surface and thus produce dynamically growing saturated areas in the riparian zone (AN) during heavy or long-lasting rainfall and snowmelt events. This can be described by the variable source area concept as first introduced by Hewlett and Hibbert (1967).

Table 2: Storm runoff components and main characteristics of hydrotopes (see Fig. 1).

Overland flow RO	
RO_{Hor}	Infiltration excess overland flow ("Horton" flow) when rainfall or snowmelt intensity exceeds the infiltration capacity of soils (high spatial variability). Characteristic for: Bare soils and croplands, especially in arid and semi-arid regions during high intensity rainstorm events.
RO_{imp}	RO from impervious areas such as bare rocks and sealed areas (paved, built-up, etc.) in all climate zones (nearly constant in areal extent). After an initial surface wetting and evaporation loss of a few millimetres, RO_{imp} amounts to 100% of rainfall or snowmelt in each event.
RO_{sat}	Saturation excess overland flow ("Dunne" flow) from dynamically varying saturated areas, due to rising groundwater tables which intersect the land surface. RO_{sat} also amounts close to 100% of rainfall or snowmelt in each event. It even occurs in case of low-intensity long-lasting rainfall or snowmelt events. Characteristic for: Near-stream riparian areas, flat valley floors with gentle concave slopes, and shallow groundwater areas, mainly in humid and semi-humid regions.
Subsurface stormflow (Interflow) RI occurs as short-term exfiltration of subsurface water to the land surface in depressions or at lower slopes, or directly into channels:	
RI_{Sl}	Subsurface stormflow along preferential pathways (e.g. macropores, pipes), and in highly permeable subsurface layers due to transmissivity feedback from underlying less permeable layers (e.g. bedrock surface).
RI_{Tr}	Subsurface pressure wave translatory flow through temporarily water saturated layers.
RI_{Pi}	Piston flow = subsurface pressure wave translatory flow through the bedrock, from groundwater stored in uplands to valley floor aquifers, after a hydraulic connection is established.
RN	Direct subsurface flow from the valley floor aquifer into the channel system.
RI_{Gr}	Increases in groundwater outflow from the valley floor aquifer due to "groundwater ridging" caused by spatial differences in groundwater recharge, in particular due to spatially heterogeneous inflows from the valley slopes.
Typical landscape sub-units (hydrotopes)	
AG	Areas with deep groundwater table that cannot be reached by plant roots.
AN	Areas with shallow groundwater table, e.g. wetlands, near-stream riparian areas.
AW	Open water surfaces.
ASL	Slope areas with increased potential for the generation of overland flow and interflow.
AIMP	Impervious or less permeable areas, e.g. exposed bedrock, clay soils, sealed areas.

These areas generate not only saturation excess overland flow but also increased subsurface stormflow (RN) into the channel. McDonnell et al. (1999) argue that the saturated areas seem to scale directly with catchment area since the topographic gradient decreases as basin scale increases. In other words, although saturation excess overland flow (RO_{sat}) is a key runoff mechanism in wet landscapes across scales, it plays an increasingly important role in larger basins that are characterized by a higher proportion of surface saturation areas.

3.2 Subsurface stormflow

In mountainous terrain, overland flow contributions are matched, or often exceeded, by rapid subsurface stormflow (RI). Especially in humid regions, this flow component is clearly the dominant source of storm runoff into river channels. However, its generation mechanisms are not yet completely understood. In recent decades, tracer methods, in combination with classical geophysical measurements, have provided insights into flow pathways, source areas and residence times of subsurface stormflow in different hydrological systems (Uhlenbrook et al. 2002). These methods have helped to formulate a set of hypotheses, which may serve as a basis for an improved comprehensive model of subsurface stormflow.

Subsurface stormflow generation, as well as percolation, are initiated after the soil moisture content in the usually unsaturated topsoil has reached and exceeds field capacity (defined as the moisture content of a soil after gravity has removed mobile excess water). From then on, any excess water in the soil zone will run off along one or several of the different subsurface pathways (Table 2), involving the following mechanisms:

a) *Matrix flow:* Percolation through the soil matrix to deeper layers and ultimately to groundwater (groundwater recharge). Matrix flow is relatively slow and therefore does not contribute significantly to rapid subsurface stormflow RI. However, it contributes to the displacement of "old" pre-event soil water by any infiltrating "new" event water.

b) *Bypass flow:* Rapid percolation and lateral (generally downslope) subsurface flow (RI_{SI} in Fig. 1) through preferential flow pathways, such as macropores, soil pipes and highly permeable soil and sediment layers (e.g. debris cover, soil-bedrock interface). This water is a mixture of event water (primary component) and pre-event soil water due to soil water displacement.

c) *Subsurface sheet flow due to transmissivity feedback:* On slopes water saturation may occur above less permeable layers, including the bedrock surface, due to transmissivity feedback. Such water saturation layers usually generate rapid downslope subsurface sheet flow (RI_{SI}). As in the case of bypass flow, the flowing water then accumulates in "subsurface depressions" of the underlying less permeable layers, in particular the soil-bedrock interface, forming a dendritic subsurface channel system. This has been illustrated in an advanced experimental and modelling study in the Panola catchment by McDonnell and his co-workers (1996) where the "channel-like" flow system follows the topography of the bedrock surface and carries most of the subsurface stormflow downslope.

d) *Pressure wave translatory flow* RI_{Tr}. This flow type is assumed to gain increasing importance during heavy and/or long-lasting rainfall events with continuous percolation and runoff generation, which can lead to the built-up of temporarily saturated sub-surface water layers. Such layers can grow in volume and areal extent across entire slopes and finally cover large parts of the catchment. RI_{Tr} follows the sub-surface topography and may be described by a variable source area concept, analogous to the one introduced for overland flow (see section 3.1 and 3.2 c). Pressure wave translatory flow is initiated as soon as continuous hydraulic connections are established across slopes and different elevation zones, linking previously isolated groundwater bodies ("pillows") together (e.g. Uhlenbrook et al., this volume). This type of flow generates the slightly delayed increases in streamflow, as can be seen in Figure 2.

e) *Piston flow:* Alternatively, the hydraulic connection between water bodies in different elevation zones (e.g. between upslope groundwater pillows near mountain ridges and the valley floor aquifer) may be established through the bedrock. This type of pressure wave translatory flow is called piston flow RI_{Pi}.

f) *Increases in groundwater outflow due to "groundwater ridging":* A third mechanism, which can generate pressure wave translatory flow, is related to spatial differences in groundwater recharge in valley floor aquifers. These differences result in temporary "ridges" of the groundwater table, in particular in valley floor aquifers near the lower end of slopes. The increasing hydraulic gradients produce, as a consequence, increased subsurface outflows of pre-event groundwater into the stream.

Among these different mechanisms of subsurface stormflow generation, matrix flow, bypass flow and transmissivity feedback are widely accepted concepts that are supported by a large number of field studies. In contrast, pressure wave translatory flow (d, e, f), although physically sound and plausible, is still to a considerable degree hypothetical. Some of the current experimental evidence for this run-off type will therefore be further discussed in section 4.

3.3 Baseflow

The third runoff component is baseflow, which is varying slowly, and therefore more or less stable even during longer dry periods (RG in Fig. 1, complemented by RN). RG is fed from aquifers within, and sometimes from outside, the catchment (depending on the location of the subsurface water divides). As illustrated in Figure 1, RG generally flows first into the valley floor aquifers and then, by passing through them, into the river system. RG is superimposed on groundwater outflow RN, generated in the valley floor aquifer itself. Together, RG and RN form the total base flow in the river system.

4. Experimental evidence of pressure wave translatory flow: Is it sufficient?

The development of the concept of pressure wave translatory flow has been promoted by the results of numerous tracer studies, which indicate that the contribution of pre-event water to flood streamflow is much greater than expected. This is the case even in mountain catchments, where the potential for event water contributions is highest. Pre-event water contributions to flood streamflow are always greater than 60% and generally even greater than 85% (Bazemore et al. 1994; Becker et al. 1999). Pressure wave translatory flow (d, e, f in section 3.2) is generally comprised of 100% pre-event water. Hence, this process can be used to explain the unexpectedly large contributions of pre-event water from soils and groundwater.

Although the concept of pressure wave translatory flow is physically sound, the experimental evidence on this subsurface runoff type, and on the conditions that control it, is still rather weak. So far, there are only a few examples of streamflow records from small mountain headwater catchments that show evidence for pressure wave translatory flow. To illustrate our current knowledge on this flow type, two experimental studies from Schäfertal (1.4 km^2) and Waldbach (0.3 km^2) in the Harz Mountains, central Germany, will be discussed in more detail. Figure 2 shows the streamflow response of these small headwater catchments to a sequence of three rainfall events in May 1975. These events occurred a few weeks after snowmelt when the initial moisture conditions in the basins were still rather wet (soil moisture near or above field capacity). The streamflow records show immediate short-term rises and peaks during and a few hours after the rainfall events. This response is likely related to the immediately generated overland flow from roads and pathways and from a few impermeable or saturated areas in the basins. In addition, some of the rainfall events are associated with secondary increases in streamflow, with a delay of about 2 days (Waldbach: for all events; Schäfertal: significant only for the third event). These secondary increases are attributed to the delayed response of the interflow component (Becker 1989) and produce the so-called "bimodal type hydrograph".

Groundwater levels in the Schäfertal (Fig. 2b) rose more or less simultaneously with the secondary rise in streamflow. This synchronous increase indicates that the initiation of subsurface stormflow was associated with the establishment, or extension of, a subsurface saturation layer that hydraulically connected the valley slopes with the valley bottom. This saturation layer expanded from the valley floor upslope on both the south- and northside of the valley connecting with previously isolated groundwater pillows along the ridges. The time required for the first establishment of a saturation layer is about 1 day (associated with a marked rise in groundwater levels) and the expansion of this layer takes about another 1 to 1.5 days. As soon as the previously disconnected saturation layers are linked across the whole slope, the hydraulic gradients abruptly increase from a few meters to more than 100 m. As a consequence, pressure wave translatory flow RI_{SI} is either initiated or increases abruptly, causing a secondary rise in streamflow. Differences in the onset of the rise in groundwater levels are probably caused by differences in the subsurface conditions

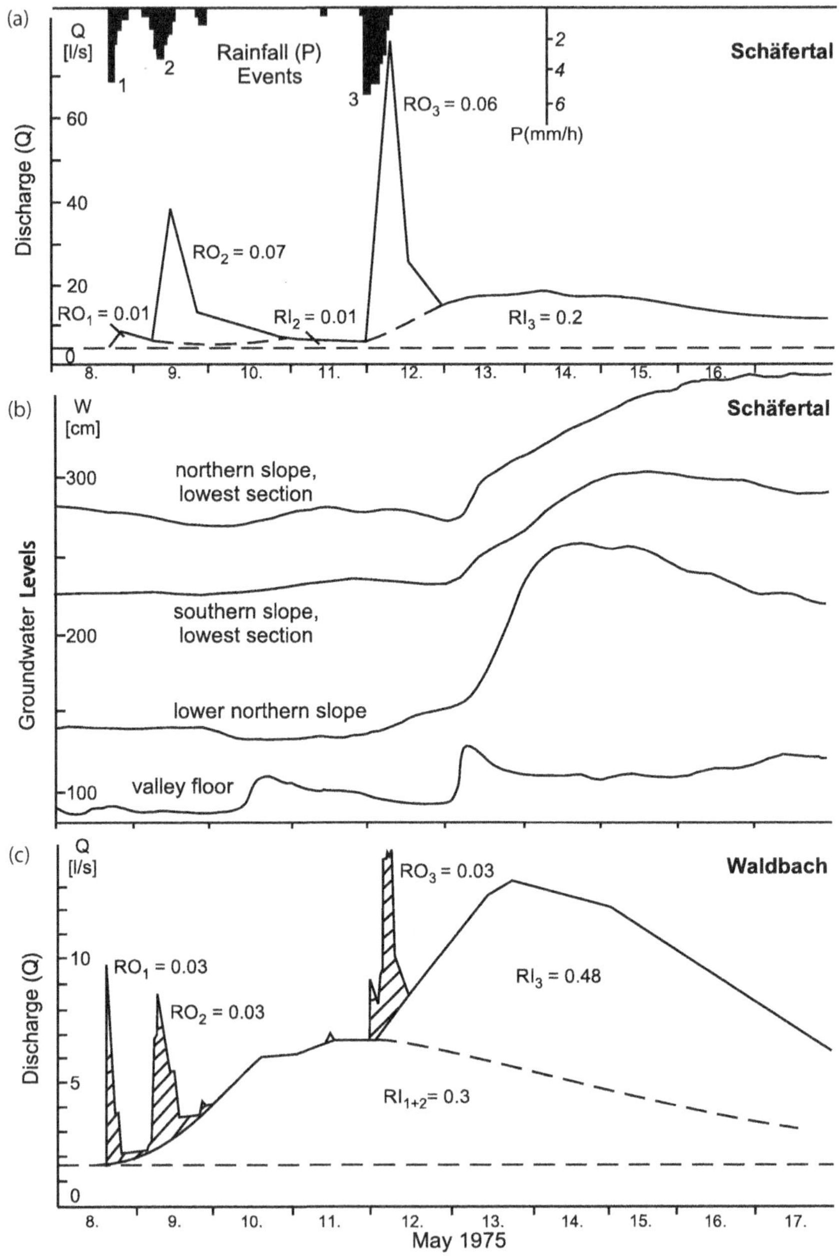

Figure 2: (a) Streamflow hydrograph and (b) groundwater levels from the Schaefertal and (c) streamflow hydrograph from the Waldbach, both in the Harz Mountains, Central Germany (Becker and McDonnell 1998).

across the catchment.

Burt and Butcher (1985) provided additional evidence for this interpretation from observations in a small headwater basin in the UK, where a 1.4 ha hillslope was instrumented with a very dense groundwater observation network. During similar rainfall events, it was observed that as soon as previously disconnected groundwater bodies (pillows) at the bedrock surface merged and formed a continuous saturation layer across the slope, a secondary rise in streamflow occurred. The magnitude of the increase in streamflow and the observed time lag were similar to the observations in the Schäfertal (Becker et al. 1999). Similar observations were also made in other catchments, for example, by Bazemore et al. (1994) and Kirnbauer and Haas (1998). For additional information on ongoing investigations see Kirnbauer et al. (this volume).

Another surprising phenomenon shown in Figure 2 is the remarkable difference in the observed volumes of subsurface stormflow RI_{SI} between the Schäfertal and Waldbach catchments for the same sequence of rainfall events. In the Waldbach, subsurface stormflow responses occurred during all three events, amounting to approximately 30% of precipitation in events 1 and 2, and to nearly 50% in event 3. In the Schäfertal, RI_{SI} was only observed for the second and third event, amounting to about 1.6 and 20% of precipitation, respectively. The difference between the catchments could be explained by different land uses: forest in the Waldbach and agriculture in the Schäfertal. Because of the better-developed and deeper root system in forests, larger amounts of infiltrating water may percolate as bypass flow through macropores to the permeable layer at the soil-bedrock interface. Due to the steeper slopes, this process could generate larger amounts of pressure wave translatory flow to the stream. This illustrates the important impact of topography and land use on subsurface stormflow generation.

Although all these observations, considerations and assumptions are still hypothetical, they point towards a potentially important role of pressure wave translatory flow in subsurface stormflow generation. However, secondary flow peaks, attributed to the occurrence of pressure wave translatory flow, do not occur in the majority of basins and even in the above study catchments they are missing during most events. The reasons for this are still poorly understood. Therefore, additional comprehensive investigations are required to answer critical questions, such as:

What is the structure of the saturation layer (spatial extent in relation to catchment area, elevation range, etc.)?

How can the different components of the perceptual model of pressure wave translatory flow be quantified?

How do topographical, geological and land use characteristics influence the response times of subsurface stormflow components?

Which role does catchment scale play in this context?

It is evident that new measurement techniques need to be developed and widely applied in order to identify the different processes of subsurface storm flow generation.

5. Summary and outlook

This paper gives an overview of the current knowledge on direct runoff generation mechanisms in mountain environments. Based on our recent understanding and available field observations, eight individual runoff components can be distinguished, namely three subcomponents of overland flow RO, and five subcomponents of subsurface stormflow (interflow RI).

The different subcomponents of RO are relatively well understood. However, the heterogeneity of land surfaces in mountain regions causes high spatial and temporal variability in overland flow, which is difficult to assess. Subsurface stormflow is in general the least understood component of run-off, since it is not visible and difficult to measure. Consequently, it is generally modelled in a rather crude and simplified way with the risk that essential components and process specifics are neglected or insufficiently described. Accordingly, a number of uncertainties are involved in runoff simulation results. It is critical to improve our understanding of subsurface stormflow if models are to be applied in Global Change impact studies and in future projections of hydrological and ecological conditions, especially in highly heterogeneous mountain environments.

The still existing deficiencies in our understanding of runoff processes in mountain regions need to be overcome as soon as possible. New measurement techniques and monitoring system designs are needed in order to better assess and describe the complex runoff mechanisms in mountain environments, including isotope and multi-tracer techniques, remote sensing and geophysical exploration techniques (e.g. Becker and McDonnell 1998). In particular, spatial patterns and temporal variations in subsurface flow require a concerted research effort (monitoring and modelling). Subsequently, the results of field studies must be transferred into the modelling practice, especially with view to the requirements of Global Change impact studies in mountain regions. It is an important task for the future to integrate process knowledge into catchment models, as attempted, e.g. by McDonnell et al. (1999), Becker et al. (2002), and Uhlenbrook and Leibundgut (1999; 2002). This will allow us to make "process-realistic" predictions, and thus more accurate impact assessments.

6. References

Bazemore, D. E., Eshleman, K. N., and Hollenbeck, K. J. (1994). The role of soil water in storm-flow generation in a forested headwater catchment: Synthesis of natural tracer and hydrometric evidence. *Journal of Hydrology* **162**, 47-75.

Becker, A. (1989). Specific aspects of runoff formation. *In* "Proceedings of the International Symposium on Headwater Control." Prague, November 1989. Vol. I. Prague University, Prague.

Becker, A., and McDonnell, J. J. (1998). Topographical end ecological controls of runoff generation and lateral flows in mountain catchments. *In* "Hydrology, water resources and ecology in headwaters," (K. Kovar, U. Tappeiner, N. E. Peters, and R. G. Craig, Eds.). Proceedings of the HeadWater '98 Conference, Merano, April 1998. *IAHS Publication* **248**, 199-206.

Becker, A., and Braun, P. (1999). Disaggregation, aggregation and spatial scaling in hydrological modelling. *Journal of Hydrology* **217**, 239-252.

Becker, A., Güntner, A., and Katzenmaier, D. (1999). Required integrated approach to understand runoff generation and flow-path dynamics in catchments. *In* "Integrated methods in catchment hydrology,"

(Ch. Leibundgut, J. J. McDonnell, and G. Schultz, Eds.). Proceedings of the International Symposium, Birmingham/UK, July 1999. *IAHS Publication* **258**, 3-9.

Becker, A., Klöcking, B., Lahmer, W., and Pfützner, B. (2002). The hydrological modelling system ARC/EGMO. *In* "Mathematical models of large watershed hydrology." (V. P. Singh, and D. K. Frevert, Eds.). pp. 321-384. Water Resources Publications, Colorado/USA.

Beven, K. J. (1989). Interflow. *In* "Unsaturated flow in hydrological modelling." (H. J. Morel-Seytoux, Ed.), pp. 191-216. Kluwer, Dordrecht.

Bonell, M. (1998). Selected challenges in runoff generation research in forests from the hillslope to headwater drainage basin scale. *Journal of the American Water Resource Association* **34**, 765-785.

Buttle, J. (1998). Fundamentals of watershed hydrology. *In* "Isotope tracers in catchment hydrology." C. Kendall, and J. J. McDonnell, Eds.), pp. 1-50. Elsevier, Amsterdam.

Burt, T. P., and Butcher, D. P. (1985). Topographic control of soil moisture distribution. *Journal of Soil Science* **36**, 469-486.

Hewlett, J. D., and Hibbert, A. R. (1967). Factors affecting the response of small watersheds to precipitation in humid areas. *In* "International symposium on forest hydrology," (W. E. Sopper, and H. W. Lull, Eds.), pp. 271-275.

Kendall, C., and McDonnell, J. J., Eds. (1998). Isotope tracers in catchment hydrology. Elsevier, Amsterdam.

Kirnbauer, R., and Haas, P. (1998). Observations on runoff generation mechanisms in small Alpine catchments. *In* "Hydrology, water resources and ecology in headwaters." (K. Kovar, U. Tappeiner, N. E. Peters, and R. G. Craig, Eds.), Proceedings of the HeadWater '98 Conference, Merano, April 1998. *IAHS Publication* **248**, 239-247.

Kunkel, R., and Wendland, F. (1999). Das Weg-Zeit-Verhalten des grundwasserbürtigen Abflussanteils im Flusseinzugsgebiet der Elbe. Schriften des FZ Jülich, „Umwelt" Series, Vol. 19.

Maidment, D. R., Ed. (1993). Handbook of hydrology. McGraw-Hill, New York.

McDonnell, J. J., Freer, J., Hooper, R., Kendall, C., Burns, D., Beven, K., and Peters, N. (1996). New method developed for studying flow on hillslopes. *EOS* **77**, 465-472.

McDonnell, J. J., Rowe, L., and Stewart, M. (1999). A combined tracer-hydromatic approach to assessing the effects of catchment scale on water flowpaths, source and age. *IAHS Publication* **258**, 265-274.

Sklash, M. G., and Farvolden, R. N. (1979). The role of groundwater in storm runoff. *Journal of Hydrology* **43**, 45-65.

Tilch, N., Uhlenbrook, S., Didszun, J., Leibundgut, Ch., Zillgens, B., Kirnbauer, R., and Merz, B. (2003). Entschlüsselung von Abflussbildungsprozessen mit Hilfe tracerhydrologischer Ansätze in einem alpinen Einzugsgebiet. *Österreichische Wasser- und Abfallwirtschaft* **55**, 1-9.

Uhlenbrook, S., and Leibundgut, Ch. (1997). Abflussbildung bei Hochwasser in verschiedenen Raumskalen. *Wasser&Boden* **29**, 13-22.

Uhlenbrook, S., and Leibundgut, Ch. (1999). Integration of tracer information into the development of a rainfall-runoff model. *In* "Integrated methods in catchment hydrology," (Ch. Leibundgut, J. J. McDonnell, and G. Schultz, Eds.). Proceedings of the International Symposium, Birmingham/UK, July 1999. *IAHS Publication* **258**, 93-100.

Uhlenbrook, S., Frey, M., Leibundgut, Ch., and Maloszewski, P. (2002). Residence time based hydrograph separations in a meso-scale mountainous basin at event and seasonal time scales. *Water Resources Research* **38**, 1-14.

Uhlenbrook, S., and Leibundgut, Ch. (2002). Process-oriented catchment modelling and multiple-response validation. *Hydrological Processes* **16**, 423-440.

Runoff Generation Processes on Hillslopes and Their Susceptibility to Global Change

Stefan Uhlenbrook, Jens Didszun, and Chris Leibundgut
University of Freiburg, Institute of Hydrology, Fahnenbergplatz, D-79098 Freiburg, Germany phone +49 761 203 35 30, fax +49 761 203 35 94, email stefan.uhlenbrook@hydrology.uni-freiburg.de

Keywords: Hill slope hydrology, Hydrograph separation, Runoff generation, Tracer methods.

1. Introduction

Global change will influence hillslope hydrological processes for a variety of reasons. On the one hand, climate change might alter the hydrological input, i.e. precipitation and snow melt, which might cause an increase or decrease in the intensity of specific hillslope processes. For instance, overland flow might be amplified by increased rain intensities (Horton 1933) or by reduced infiltration due to surface crusts (Yair 1990) or increased hydrophobicity (Doerr et al. 2002), triggered by longer and more pronounced drought periods. However, overland flow could also be significantly influenced by antecedent moisture conditions of the substrate that were either altered due to wetter climate and reduced evapotranspiration at a site or due to different snow and snow melt regimes, changing the hydrological input for a specific precipitation event. On the other hand, global change in the form of land use changes will play a key role in defining the dominant runoff generation processes on hillslopes (cf. summary given in DVWK 1999).

In general, the hydrological response of montane and high-elevation ecosystems is primarily defined by hillslope processes. Thus, changes in these processes play a crucial role in areas that are very sensitive to global change. It has to be noted that many non-linear processes with numerous feedback mechanisms define the hillslope response. Therefore, the implications of global change can be much more dramatic

U. M. Huber et al. (eds.), Global Change and Mountain Regions, 297-307.

than, e.g. a slightly modified hydrological input might indicate. For instance, a slight increase in event-precipitation in specific regions is likely to cause a much higher increase in flood discharge if feedback mechanisms such as an increase of overland flow operate.

Hillslope hydrological processes define how precipitation reaches the stream, how long water is stored in soil and ground water systems before it reaches the stream, and which hydrochemical composition the water has when reaching the stream (see Becker, this volume). To investigate these processes in mountainous basins, different types of field studies have been conducted: (i) Comparisons of the hydrological responses of headwater basins were carried out (e.g. Jones 2000). (ii) Soil physical and hydrometrical studies, using tensiometers and piezometers, were executed at the plot scale (e.g. McDonnell 1990). (iii) Sprinkling experiments at hillslopes, often in combination with tracer tests, were conducted (e.g. Mosley 1982; Flury et al. 1994; Uhlenbrook and Leibundgut 1997a; Lange et al. 2003). (iv) Geophysical measurements proved useful to explore subsurface soil properties, which are crucial for infiltration, percolation and lateral flow processes (e.g. Sherlock et al. 2000). (v) Finally, the use of isotopic tracers (e.g. Sklash and Farvolden 1979; Buttle 1994) in combination with hydrochemical tracers (e.g. Christophersen et al. 1990; Anderson et al. 1997) helped to gain further insights into hillslope processes, in particular into the flowpaths and residence times of water. Each method has its own strengths and shortcomings concerning costs and the temporal and spatial scale at which they can be used. The best and most reliable results can be obtained when different methods are applied at the same test site.

The importance of subsurface flow processes for generating not only base flow but also floods has been shown for many mountainous basins (cf. reviews of Buttle 1994; Uhlenbrook and Leibundgut 1997b; Bonell 1998). However, the way in which dominant subsurface storm flow processes function in different environments and for different circumstances (i.e. input and moisture conditions) is still under discussion. Therefore, the focus of this paper is to summarise our contribution to this research area, and highlight with one case study how the flow processes on two hillslopes, drained by springs, can be explored, using tracer methods.

2. Study area and previous investigations

In recent years, our research team has mainly investigated the meso-scale Brugga catchment (40 km^2), located in the Southern Black Forest Mountains, Germany. This is a pre-alpine mountainous catchment with an elevational range from 434 to 1493 m a.s.l. The bedrock consists of gneiss and is covered by a drift of glacial and periglacial origin of varying depths (0-10 m). Brown soils have mainly developed on this drift cover. The morphology is characterised by moderate to steep slopes (75% of the area), hilltops and hilly uplands (about 20%), and narrow valley floors (less than 5%). The overall average slope is 19°, calculated with a 50x50 m^2 digital elevation model. Forests cover approximately 75% of the basin, and 23% is used as pasture land. The remaining area is used for small settlements. The mean annual discharge amounts to

1220 mm, and is generated by a mean annual precipitation of approximately 1750 mm. Further details about the basin can be found in Uhlenbrook (1999).

Detailed tracer investigations were carried out: (i) Hydrograph separations, using the natural tracers oxygen-18, chloride and dissolved silica (Hoeg et al. 2000; Uhlenbrook et al. 2002), (ii) tracer experiments with artificial tracers on different hillslopes and in the river channel system (Mehlhorn et al. 1998; IHF 2002), and (iii) residence time determinations, using oxygen-18, tritium and CFC concentrations (Uhlenbrook et al. 2002). In agreement with additional field observations during flood formation, the experimental findings led to the development of the following conceptual model of runoff generation for the study site. Three main flow systems were identified:

a. Fast runoff components (surface and near-surface runoff) are generated in sealed or saturated areas, and on steep highly permeable slopes covered by boulder fields where macropore flow is dominant (mean residence time of the water: hours to a few days).
b. Slow base flow components (deep groundwater) originate from the fractured hard rock aquifer and the deeper parts of the weathering zone (mean residence time of the water: six to nine years; determined by tritium and CFC measurements). There is no evidence that these components are important for flood formation.
c. An intermediate flow system originates mainly from the periglacial deposits of the slopes (shallow groundwater). Here a large heterogeneity and variability of processes was observed: soil water displacement takes place and perched groundwater tables can spread above less conductive layers. In addition, lateral macropore flow occurs locally (for process description see, e.g. Bonell 1998). This results mainly in delayed runoff components compared to the surface and near-surface runoffs. However, the shallow groundwater can also contribute to flood formation (mean residence time of the water: two to three years; determined by oxygen-18 measurements). To improve the understanding of shallow groundwater contributions during flood formation, the following experimental research was performed on two hillslopes.

3. Recent results of hillslope investigations in the Brugga Basin

3.1 Problems and Objectives

A long-term hydrograph separation in the Brugga Basin has shown that the shallow groundwater contributes about two-thirds to the total runoff (Uhlenbrook et al. 2002). This finding highlights the importance of this runoff component. Two hillslopes that are drained by a spring at the toe of each hillslope were investigated in further detail to extend the knowledge of runoff generation processes on hillslopes that contribute shallow groundwater. In mountainous areas, springs in these locations are often used for local water supply because they show a persistent discharge. The central question to be examined in this respect is how the land use and the structure of the periglacial debris cover influences the runoff behaviour of the hillslopes, which is observed at the two springs. The chosen springs are more or less similar regarding their altitude, mean discharge, and length and inclination of the hillslope. However, there are differences in the land use and the structure of the periglacial debris cover. Although the structure of the periglacial drift cover in the Brugga basin is complex, due to the heterogeneity of the topography and the formation processes (Tilch et al. 2002), four main layers that reflect the different stages and processes of their formation can be distinguished (Rehfuess 1990). The difference between the two hillslopes with respect to the structure of the drift cover is that the hillslope above spring A has no significant coarse top layer with boulders, but on the hillslope above spring B the coarse material of the top layer is visible at the surface. Furthermore, the land use differs, with pastureland and conifer forest dominating at spring A and B, respectively.

The main objective of this study is to gain further insights into the impacts of the different physiographic and hydro-climatic characteristics of these two hillslopes on runoff and hydrochemical response. This is of particular interest, as these characteristics are subject to global change, i.e. climate and landuse changes, including human impact on soil properties. A thorough understanding of the spatial and temporal variability of hillslope processes in mountain landscapes is essential for simulating the hydrological impact of future changes in land use or climate.

3.2 Methods

During autumn 1999, the discharge during three rainfall events and several low flow periods was monitored continuously. Precipitation (10 minute intervals) was measured at two rain gauges close to springs A and B. Furthermore, water samples were taken at each spring with a 4-hour time interval using automatic samplers. The water samples were analysed for stable isotopes (^{18}O and ^{2}H), dissolved silica and major anions (Cl^-, NO_3^-, SO_4^{2-}) and cations (Na^+, K^+, Ca^{2+}, Mg^{2+}). In addition, classical meteorological parameters were observed at a nearby climate station, and rainwater was sampled every 2 mm and analysed for stable isotopes (^{18}O and ^{2}H), reported in ‰ relative to V-SMOW.

3.3 Results

The results show distinct differences between the two springs that were more obvious than expected. There were clear differences between the springs' hydrographs and between their chemical and isotopic response to rainfall events. Spring A is characterized by a slow and delayed runoff behaviour (Fig. 1). Although the time lag between rainfall and increased runoff is only a few hours, the peak discharge is not reached until two to four days after the beginning of the rainfall, depending on its intensity. In contrast, the time lag at spring B is shorter, the peak discharge is higher and reached about two days earlier, and the recession is considerably steeper than at spring A. However, despite the rapid runoff response, spring B also shows a fairly constant discharge of 0.3 l/s during summer droughts, which suggests that the spring is fed by at least two runoff components, a long-lasting base flow component and a

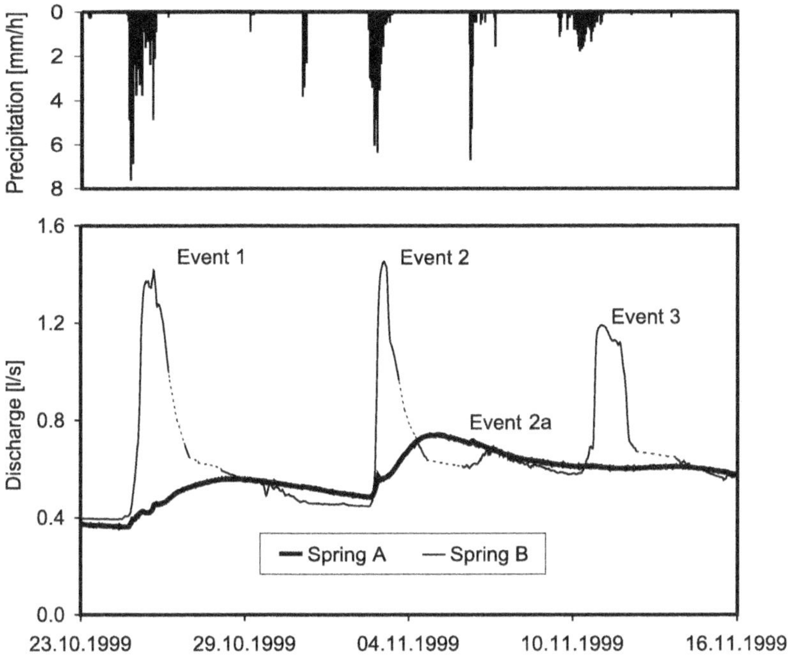

Figure 1: Precipitation and discharge of both springs during autumn 1999. Dotted lines represent interpolations where data was missing due to technical problems.

dynamic storm flow component.

The precipitation and runoff volumes for each event are given in Table 1. Significant differences in precipitation between both springs occur during events 2a and 3. For spring A, the discharge volume during storm events was calculated by summing up the discharge shown in the hydrograph from the beginning of the rising limb to the beginning of the next event. A discharge volume could not be calculated for event 2a, as there is no significant rise in discharge. For event 3, the calculation was also difficult because of the small increase in discharge and the high pre-event runoff (values in brackets). At spring B, the discharge volume is calculated for events 1, 2 and 3 by summation from the beginning of the rising limb to the sharp end of the peak hydrograph about 2.5 days later. The beginning and end of events could be identified clearly because discharge shows a clear dynamic response at this site. For event 2a, the summation was stopped when the discharge reached pre-event values. At this spring, the pre-event discharge is subtracted each time to take into account the antecedent discharge. Although the estimation of the event discharge volume is somewhat inexact, it enables a rough comparison of this parameter between events at each spring and for different precipitation volumes. It has to be noted that the precipitation/discharge ratios (Tab. 1) cannot be compared between the two springs, as it was not possible to delineate spring-catchments precisely, and because of differences in the calculation of the discharge volumes. The differences in the

Table 1: Precipitation and discharge volumes for the investigated storm events at both springs.

Spring A	Event 1	Event 2	Event 2a	Event 3
Total Precipitation [mm]	63.3	45.8	20.0	24.4
Discharge [m³]	389.9	309.9	-*	(111.3)*
Precipitation/Discharge [mm/m³]	0.16	0.15	-*	(0.22)*
Spring B				
Total Precipitation [mm]	62.3	49.6	12.8	49.4
Discharge [m³]	99.5	75.3	5.3	64.4
Precipitation/Discharge [mm/m³]	0.63	0.66	2.42	0.77

*Refer to text for details.

precipitation/discharge ratios suggest that there are thresholds at both springs, with a significant acceleration of runoff generation once this threshold is exceeded. Small storm events, such as event 2a, show almost no reaction in the hydrograph at spring *A* and a significantly lower discharge volume at spring B. A comparison of discharge volumes between events 2a and 3 at spring A is rather difficult because of the high pre-event runoff, which masks the discharge response during storm events.

The hydrochemical variations are in good agreement with the discharge behaviour. At spring A, the concentration of dissolved silica remains fairly constant throughout the events (Fig. 2). Only towards the end of the recession, a slight decrease of the silica concentrations is visible. In contrast, at spring B there is a typical decrease in the silica concentrations, synchronous with the peak discharge. During the investigated storm events, the silica concentration dropped to a minimum of about 70 % of the pre-event concentration, but reached the pre-event concentration again at the end of the hydrograph recession. Only during event 2a, no significant change in concentration could be detected. The concentrations of the main ions show more or less the same type of reaction. At spring A, the concentrations stay quite constant during the events, whereas at spring B a decrease in concentrations to a minimum of about 50 % of the pre-event concentration was observed. This decrease at spring B highlights the significant contribution of at least one additional storm runoff component, which is hydrochemically different. The hardly visible response at spring A indicates a

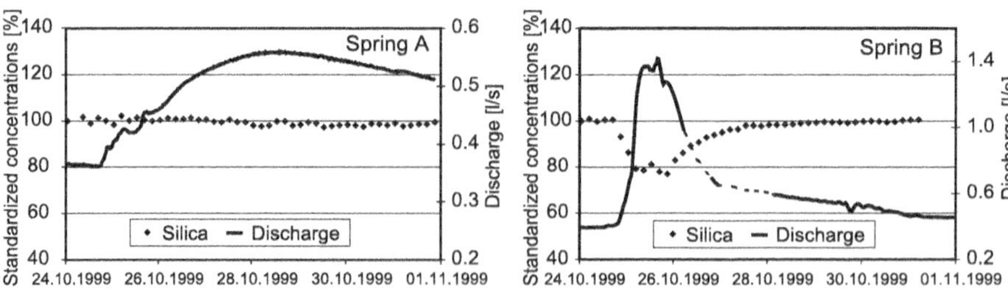

Figure 2: Observed dissolved silica concentrations, set to 100 % at the beginning of the event.

negligible contribution of storm runoff.

The runoff behaviour is also reflected in the isotopic composition of the spring water (Fig. 3). Spring A shows only small and non-systematic variations in ^2H composition. Spring B, in contrast, shows a trend towards a ^2H signature of a third water component, different from the event and pre-event signatures. This response is similar during each of the three investigated events. Because of the high δ–values of ^2H, it can be assumed that this component reflects water from the debris and drift cover, which is recharged during the summer, several months prior to the events.

To quantify the contribution to spring discharge of the different runoff components, hydrograph separations using dissolved silica and ^2H were calculated. The theory of hydrograph separations using natural tracers is discussed, for instance, by Sklash and Farvolden (1979) and Buttle (1994). At spring A, a two-component hydrograph separation using dissolved silica was carried out, accounting for the fact that only a small range of variations in hydrochemistry could be found. Assuming that the two components are direct runoff (rain water) and groundwater (hypothesis one), the fraction of the direct runoff, causing the small decrease in silica concentration, was about 3 % during the observed events (Fig. 4a). The small proportion of direct runoff could reach the spring during the event by flowing along preferential pathways (i.e. root channels, earthworm channels, etc.). The rest of the spring water was delivered from the shallow and deep groundwater. However, a distinction between those two components was not possible. A two-component hydrograph separation can also be calculated assuming a variable contribution of deep and shallow groundwater

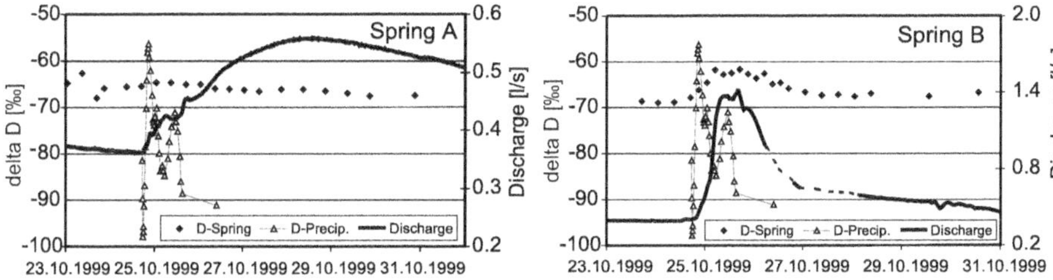

Figure 3: Observed ^2H (D) signature of spring water and precipitation at both springs.

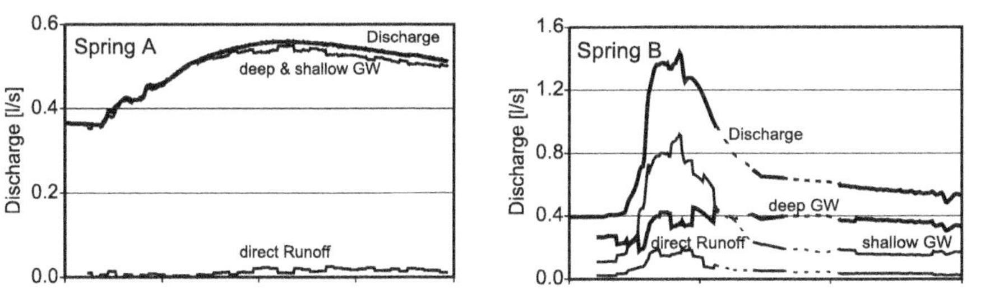

Figure 4: Hydrograph separations using dissolved silica (spring *A*), and both dissolved silica and ^2H (spring *B*).

(hypothesis two), with an increase of the latter during the event. However, the silica concentrations of these two groundwater components did not have an unique fingerprint in the observed data. Thus, an estimation of the fraction of each component is uncertain. Considering this and the small range of variations in silica concentrations, spring A shows only a slight change in the fraction of the runoff components for both hypotheses.

At spring B, a two-component hydrograph separation was not feasible, as the ^{2}H composition indicated a contribution of a third component (Fig. 3b). Thus, a three-component separation, using silica and ^{2}H, was calculated. The three components are direct runoff, with very little silica and the ^{2}H composition of rainwater, as well as the shallow and deep groundwater. The ^{2}H and silica concentrations for the deep and shallow groundwater were determined using an approach similar to the end member mixing analysis (EMMA) according to Christophersen et al. (1990), because it was not possible to measure the concentrations directly. Shallow groundwater already contributed a small proportion of base flow prior to the events, and became the major component during the peak of the event (Fig. 4b). During the three investigated events, the fraction of the direct runoff component was about 10%, whereas the deep and shallow groundwater made up approximately 40% and 50%, respectively (Fig. 4b).

3.4 Discussion and interpretation

These results show that runoff generation in the catchments of the two springs differs more clearly than would be expected from the similar location of the springs at the toe of two steep hillslopes. It is important to note that similar hydrochemical and hydraulic results have been observed for two additional events (cf. Fig. 1) that were not discussed in further detail in this paper. The reasons for the different hydrochemical and hydrological responses can be attributed to the land use and the structure of the debris and drift cover on the hillslopes above the springs. From the measured data, it cannot be determined with certainty which of the two potential factors is the most important one. However, it is likely that the root zone of the conifer forest with its many macropores, together with the stony and very permeable debris cover, enables a quick percolation and delivery of water to spring B. In contrast, the grass covered hillslope of spring A, with a slower percolation, accounts for the delayed reaction of the hydrograph and the minor contribution of direct runoff to spring discharge. Here, flood runoff was produced almost exclusively by soil and groundwater displacement (translatory flow). The fact that a highly dynamic shallow groundwater component (including contributions from soil water) is dominant during larger events at spring B, but cannot be observed at spring A, must be caused by the different permeability of the drift cover. Again, soil and groundwater displacement effects must play a key role here, as the hydrochemistry shows that the discharge water was stored in the catchment prior to the event. The fact that only minor contributions of direct runoff could be detected at spring B demonstrates an extensive mixing of event water with water stored in the soil and drift cover. Similar observations have been made, for instance, by McDonnell (1990). The runoff generation itself seems to be accelerated by increasing precipitation volumes and intensities (Table 1). Only after exceeding a certain threshold, a significant rise in the discharge volume is detectable at both springs. This might be due to the extension of partly saturated zones (perched watertables) into macroporous zones (certain soil layers) as well as the increased hydraulic connectivity of distributed subsurface reservoirs. To identify source areas and to understand the displacement effect of these subsurface storm flow components more precisely, further geophysical investigations on the structure of the drift covers are needed. In particular, combined geophysical measurements and tracer tests are required to understand the triggering process responsible for the dynamic contribution of shallow groundwater in further detail.

4. Conclusions and required future research activities in consideration of global change

As shown for the two investigated hillslope/spring systems, the processes and hydrological responses in mountainous landscapes can be very diverse, even on relatively similar hillslopes. This spatial heterogeneity of hillslope processes appears to be closely related to highly variable soil structure overlain by land use and vegetation patterns. Future changes in hydro-climatic input (e.g. rainfall, temperature and snow melt), or land use and vegetation cover, will have a significant influence on the recharge of springs and, consequently, on the discharge and composition of runoff components and their hydrochemical composition. This shows the vulnerability of springs, which are often used for the local water supply in mountainous regions (Leibundgut 1998).

Concerning the hydrochemistry of rivers and lakes, the riparian zone in the valley bottoms can also play a crucial role. Runoff coming from the hillslopes has to pass through this zone and either mixes with the stored riparian water (e.g. Burns et al. 2001) or its water chemistry is significantly modified in the riparian zone by bio-geochemical processes (e.g. Hooper 2001). The riparian zone and near-stream wetlands should therefore be highly protected as important buffer zone for potentially polluted (e.g. acidic) water from the hillslopes.

There is a range of additional global change impacts on hydrological processes besides water quantity and quality issues on hillslopes. For instance, the impact of global change on the future frequency, magnitude and character of the delivery of nutrients and fine-to-coarse sediments from hillslopes in mountainous areas is still not very well understood. In general, in order to decode the complex hillslope system with its non-linear processes and numerous feedback mechanisms, multi-disciplinary approaches that combine hydrology, ecology and other earth science disciplines are required.

This suggests a number of new and necessary research avenues that can be implemented at well-equipped test sites. Multi-technique and nested catchment studies are required to understand processes at different scales. However, not only case studies are needed, which focus on the peculiarity of specific sites or only investigate the implications of one aspect of global change, e.g. an altered rainfall regime. More integrated approaches are required to address these issues in a holistic way. In addition, much work has to be done to synthesise and generalise the findings from case studies in different environments. This is a prerequisite for the assessment of global change implications at larger scales. Last but not least, the authors would like to stress the fact that the investigation of global change needs long and reliable hydrometrical and hydrochemical records. This becomes more and more difficult in the face of the worldwide continuous decline in monitoring networks.

5. Acknowledgement

The study was supported by the German Research Foundation (Deutsche Forschungsgemeinschaft, Bonn) grant no. Le 698/8-3. The first author is a member of the working group Hydrology 2020 of the IAHS (International Association of Hydrological Sciences) and his participation is supported by the German national committee of the IHP/OHP. Parts of this paper were developed while contributing to this working group. Therefore, the support is gratefully acknowledged.

6. References

Anderson, S. P, Dietrich, W. E., Montgomery, D. R., Torres, R., Conrad, M. E., and Loague, K. (1997). Subsurface flow paths in a steep, unchanneled catchment. *Water Resource Research* **33**, 2637-2654.

Bonell, M. (1998). Selected challenges in runoff generation research in forests from the hillslope to headwater drainage basin scale. *Journal of American Water Resources Association* **34**, 765-785.

Burns, D. A., McDonnell, J. J., Hooper, R. B., Peters, N. E., Freer, J. E., Kendall, C., and Beven, K. (2001). Quantifying contributions to storm runoff through end-member mixing analysis and hydrologic measurements at the Panola Mountain Research Watershed (Georgia/USA). *Hydrological Processes* **15**, 1903-1924.

Buttle, J. M. (1994): Isotope hydrograph separations and rapid delivery of pre-event water from drainage basins. *Progress in Physical Geography* **18**, 16-41.

Christophersen, N. C., Neal, C., Hooper, R. P., Vogt, R. D., and Andersen, S. (1990). Modelling streamwater chemistry as a mixture of soilwater end-members: A step towards second-generation acidification models. *Journal of Hydrology* **116**, 307-320.

Doerr, S. H., Ferreira, A. J. D., Walsh, R. P. D., Shakesby, R. A., Leighton-Boyce, G., and Coelho, C. O. A. (2002). Soil water repellency as a potential parameter in rainfall-runoff modelling: Experimental evidence at point to catchment scales from Portugal. *Hydrological Processes* (in press).

Deutscher Verband für Wasserwirtschaft und Kulturbau (DVWK) (1999). Einflüsse land- und forstwirtschaftlicher Maßnahmen auf den Hochwasserabfluss: Wissensstand, Skalenprobleme, Modellansätze. *DVWK Materialien* **7**.

Flury, M., Flühler, H., Jury, W. A., and Leuenberger, J. (1994). Susceptibility of soils to preferential flow of water: A field study. *Water Resources Research* **30**, 1945-1954.

Hoeg, S., Uhlenbrook, S., and Leibundgut, Ch. (2000). Hydrograph separation in a mountainous catchment: Combining hydrochemical and isotopic tracers. *Hydrological Processes* **14**, 1199-1216.

Hooper, R. P. (2001). Applying the scientific method to small catchment studies: A review of the Panola Mountain experience. *Hydrological Processes* **15**, 2039-2050.

Horton, R. E. (1933). The role of infiltration in the hydrological cycle. *Transaction American Geophysical Union* **14**, 446-460.

Institut für Hydrologie (IHF) (2002). Determination of runoff generation processes and process-oriented catchment modelling. Report for the grant of the German research foundation (DGF), report of the Institute of Hydrology, University of Freiburg, No. 112, Freiburg.

Jones, J. (2000). Hydrological processes and peak discharge response to forest removal, regrowth, and roads in 10 small experimental basins, western Cascades, Oregon. *Water Resource Research* **41**, 2621-

2642.

Lange J., Greenbaum, N., Husary, S., Ghanem, M., Leibundgut, C., and Schick, A. P. (2003). Runoff generation from successive simulated rainfalls on a rocky, semi-arid, Mediterranean hillslope. *Hydrological Processes* **17**, 279-296.

Leibundgut, Ch. (1998). Tracer-based assessment of vulnerability in mountainous headwaters. IAHS Publication No. 248, p. 317.

McDonnell, J. J. (1990). A rationale for old water discharge through macropores in a steep, humid catchment. *Water Resource Research* **26**, 2821–2332.

Mehlhorn, J., Armbruster, F., Uhlenbrook, S., and Leibundgut, Ch. (1998). Determination of the geomorphological instantaneous unit hydrograph using tracer experiments in a headwater basin. *IAHS Publication* **248**, 327-336.

Mosley, M. P. (1982). Subsurface flow velocities through selected forest soils, south island, New Zealand. *Journal of Hydrology* **55**, 65-92.

Rehfuess, K. E. (1990). Waldböden: Entwicklung, Eigenschaften und Nutzung. Pareys Studientexte 29. Paul Parey, Hamburg.

Sherlock, M., Chapell, N., and McDonnell, J. J. (2000). The effects of experimental uncertainty on the calculation of hillslope flow paths. *Hydrological Processes* **14**, 2457-2472.

Sklash, M. G., and Farvolden, R. N. (1979). The role of groundwater in storm runoff. *Journal of Hydrology* **43**, 45-65.

Tilch, N., Uhlenbrook, S., and Leibundgut, Ch. (2002). Regionalisierungsverfahren zur Ausweisung von Hydrotopen in von periglazialem Hangschutt geprägten Gebieten. Grundwasser, Heft 4, 206-216.

Uhlenbrook, S., and Leibundgut, Ch. (1997a). Investigation of preferential flow in the unsaturated zone using artificial tracer. *In* "Tracer Hydrology." 7th International Symposium on Water Tracing (A. Kranjc, Ed.). pp. 181-188. A. A. Balkema, Rotterdam.

Uhlenbrook, S., and Leibundgut, Ch. (1997b). Abflussbildung bei Hochwasser. *Wasser und Boden* **49**, 13-22.

Uhlenbrook, S. (1999). Untersuchung und Modellierung der Abflussbildung in einem mesoskaligen Einzugsgebiet. Freiburger Schriften zur Hydrologie 10, Universität Freiburg.

Uhlenbrook, S., Frey M., Leibundgut, Ch., and Maloszewski, P. (2002). Residence time based hydrograph separations in a meso-scale mountainous basin at event and seasonal time scales. *Water Resource Research* **38**, 1-14.

Yair, A. (1990). Runoff generation in a sandy area, the Nizzana sands western Negev, Israel. *Earth Surface Processes and Landforms* **15**, 596-609.

Identifying Space-time Patterns of Runoff Generation: A Case Study from the Löhnersbach Catchment, Austrian Alps

Robert Kirnbauer[1*], Günter Blöschl[1], Peter Haas[1], Gabriele Müller[2], and Bruno Merz[3]

[1]*Institute of Hydraulics, Hydrology and Water Resources Management, Vienna University of Technology, Karlsplatz 13/223, A-1040 Vienna, Austria*
[2]*Federal Ministry of Agriculture, Forestry, Environment and Water Management, Hydrographical Central Office, Marxergasse 2, A-1030 Vienna, Austria*
[3]*GeoForschungsZentrum Potsdam, Telegrafenberg, D-14473 Potsdam, Germany*
**phone +43 1 58801 22320, fax +43 1 58801 22399, e-mail kirnbauer@hydro.tuwien.ac.at*

Keywords: Austrian Alps Löhnersbach, Microcatchments, Process study, Runoff coefficient, Runoff generation, Saturation areas

1. Introduction

Runoff generation is a result of the interplay of a range of processes, the relative magnitudes of which vary, among other things, with climate, catchment properties, and catchment scale. The variability of runoff generation processes within a mountain catchment and the variability from event to event is one particularly intriguing aspect. A better understanding of these spatio-temporal patterns of runoff generation is critical for obtaining realistic model simulations of events, such as extreme floods, and of run-off behaviour associated with changes in environmental and land use conditions. Estimating runoff generation is very difficult as it involves a high degree of extrapolation. Difficulties in accurately assessing runoff in mountains have been highlighted by local-scale field experiments (e.g. Scherrer 1997), observations in experimental basins (e.g. Anderson et al. 1997; Kirnbauer and Haas 1998; Torres et al. 1998; Müller and Peschke 2000; Uchida et al. 2001), and modelling studies (e.g.

U. M. Huber et al. (eds.), Global Change and Mountain Regions, 309-320.

Moore and Grayson 1991) that emphasize the spatially highly heterogeneous nature of runoff. Also, different runoff processes may dominate at different spatial scales (see e.g. Blöschl 1996; Uhlenbrook and Leibundgut 1997). Although it is possible to estimate runoff for yet unobserved situations with hydrological simulation models, the reliability of such estimates is notoriously poor, particularly when moving from the plot scale or small catchment scale to medium sized catchments (DFG 1995). There is still a gap between the understanding of runoff generation processes at the plot scale and process-based hydrological modelling at the catchment scale. Problems include excessive model complexity as a consequence of the great number of processes and, consequently, the great number of model parameters, the need for calibration, and the insufficient spatial resolution of input data (Grayson and Blöschl 2000).

One means of addressing the complexity and variety of processes is to isolate dominant processes (Gutknecht 1997). This approach is based on the hypothesis that, under certain conditions, one single mechanism (e.g. either infiltration excess overland flow, saturation overland flow, or subsurface stormflow) dominates the runoff behaviour, and other mechanisms are less important. The task then is to identify the conditions under which a given area is dominated by a certain mechanism and to describe the interactions between dominant processes and their underlying controls. Several studies have contributed to the identification of dominant mechanisms and their interactions (e.g. Dunne 1983; Anderson and Burt 1990; Peschke et al. 1999) but, so far, little work has been done in alpine terrain where the spatial heterogeneities can be much greater than in more level terrain.

The purpose of this paper is to present space-time patterns of runoff generation as inferred from runoff observations in the 16 km^2 alpine Löhnersbach catchment in the Austrian Alps. The assessment of spatial and temporal patterns of runoff is restricted by the resolution of our observational network (i.e. gauged 1 ha subcatchments) and the frequency of measurements (i.e. analysis of individual runoff events). However, further research into identifying higher resolution space-time patterns in the Löhnersbach catchment is underway.

2. Field site and instrumentation

The Löhnersbach research catchment is situated near the Saalbach skiing resort in the Salzburg region, Austria (47°20′N, 12°40′E). The research basin is a 16 km^2 alpine catchment with elevations ranging from 1100 to 2200 m asl. Slope angles range from almost 0° in the upper areas to more than 45° on the side slopes of the steep gullies. The vegetation includes spruce forests, alpine pasture and areas dominated by Rhododendron shrub. The main stream (the Löhnersbach) divides the catchment into a northwestern part and a southeastern part, which are hydrologically very dissimilar. The northwestern part, the orographic left-hand side of the valley, is vegetated by Alpine pasture. This area is exposed to the sun and not very steep. It exhibits numerous permanently saturated areas, which are fed by permanent springs. Also, the stream network density (Fig. 1) in this area is very high. In contrast, the southeastern part of the catchment is vegetated by spruce forests and the slopes are steeper. It has significantly fewer saturated areas and a lower stream network density. These differences have important ramifications for the spatial patterns of runoff generation in the catchment. It is interesting to note that these striking hydrological differences are in stark contrast to an apparently uniform geology and pedology.

The field instrumentation includes four recording rain gauges and five recording stream gauges. During field campaigns in summer, runoff has been measured daily with the salt dilution method at five additional sites within the catchment (Fig. 1). These measurements provide daily runoff data from a total of ten subcatchments, which allow some inferences on the spatial patterns of runoff generation within the catchment. Additional field observations include a visual assessment of the activity of springs and the expansion of saturated areas during storms. Much of the information on the hydrology of the catchment is of a qualitative nature and stems from an intimate knowledge of the catchment, which was acquired during numerous field campaigns in the past decade. Hydrochemical analyses are underway in this catchment and will be reported in future publications.

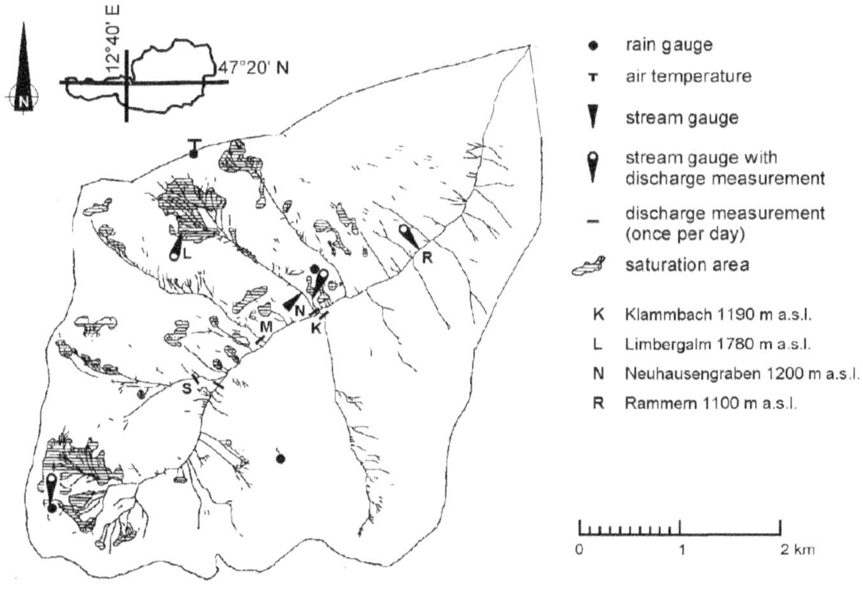

Figure 1: Drainage network, saturation areas (shaded areas) and observational network in the Löhnersbach catchment, Austria. The stream gauges used here are: R = Rammern (16 km²), K = Klammbach (2.5 km²), M = Marxtengraben (1.5 km²), S = Schusterbauergraben (1.4 km²), N = Neuhausengraben/Mündung (1.3 km²), and L = Limbergalm (0.072 km²). The main drainage line is the Löhnersbach, which flows from the southwest to the northeast.

3. Event types at various scales

Based on an understanding of what constitutes typical runoff events in the various subcatchments of the Löhnersbach, the following examples have been selected to illustrate important differences from event to event within a given subcatchment as well as differences in event types as a function of catchment scale.

Figure 2a shows an example of a hydrograph from a microcatchment (Limbergalm, see Fig. 1; topographic catchment area 7.2 ha) in the northwestern part of the Löhnersbach catchment. The shape of the hydrograph is bimodal. The first peak is flashy with response times of far less than an hour. The secondary peak is smooth indicating a delayed response with response times of about one to three days. This type of event can, on average, be observed once a year. Such events are associated with wet antecedent conditions, relatively low rainfall intensities (between 1 and 2.5 mm/15 min) and relatively large amounts of precipitation (more than 40 mm). It should be noted that part of the microcatchment (0.12 ha) is a saturated area fed by permanent springs. The different characteristics of the two peaks can be interpreted in terms of the runoff generation mechanisms in this microcatchment. The first, flashy peak is associated with direct surficial runoff from the saturated area. The second, smooth peak is due to subsurface stormflow where rainfall infiltrates in the upper portion of the microcatchment, moves along fast subsurface flow paths and discharges from the springs at the upper border of the saturated area. Apparently, a combination of long duration rainfall and wet antecedent conditions can activate subsurface flow paths, which then significantly contribute to stormflow. This interpretation is consistent with field experience and with tracer experiments (Tilch et al. 2003) in the microcatchment. It is interesting that this bimodal runoff response is quite typical of low intensity - long duration events in this and other Löhnersbach microcatchments that are partly covered by saturated areas. The differences in response times for the

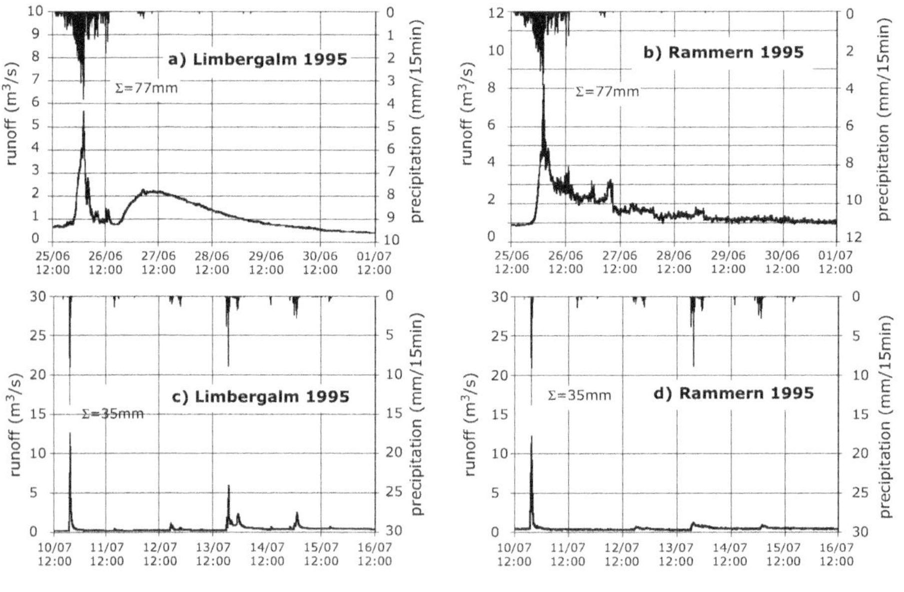

Figure 2: Examples of runoff event types. (a, b) are long duration - low intensity storms, producing a bimodal hydrograph at Limbergalm (a) and a hydrograph with a linear recession at Rammern (b). (c, d) are short duration - high intensity storms, producing unimodal flashy hydrographs both at Limbergalm (c) and Rammern (d). The topographic catchment areas of Limbergalm and Rammern are 0.072 and 16 km^2, respectively.

two types of flow paths are large enough to cause a bimodal event hydrograph. The characteristics of a number of additional events with a bimodal runoff response are given in Table 1.

At larger catchment scales, such as the 16 km^2 Löhnersbach catchment at Rammern (Fig. 1), the same storm event produces a quite different hydrograph (Fig. 2b). The large catchment also shows both a rapid runoff response within a few hours and a delayed response. However, unlike in the Limbergalm microcatchment, the delayed response is reflected in a nearly linear hydrograph recession (Fig. 2b) over two or three days rather than in a secondary peak. The interpretation of this type of response is that runoff from the 16 km^2 catchment is a combined result of runoff from microcatchments such as the one in Figure 2a. These microcatchments are likely to have a spectrum of secondary (subsurface) response times, ranging from several hours to a few days. The combined secondary response then results in a flat recession rather than a secondary peak. Hence, this hydrograph shape is characteristic of low intensity - long duration events at the scale of the entire Löhnersbach catchment (16 km^2).

A different type of runoff event is shown in Figure 2c. Although this hydrograph comes from the same microcatchment (Limbergalm) as the one shown in Figure 2a, its shape is very different. The runoff response is flashy with a single rather than a bimodal peak. This event type can be observed several times a year during thunderstorms and is characterized by relatively high rainfall intensities (between 10 and 25 mm/15min) and smaller precipitation totals (usually below 35 mm). It is clear that this flashy response is associated with direct surficial runoff from the saturated area. Apparently, short duration rainfall events with relatively low precipitation totals

Table 1: Examples of the two event types for the Limbergalm microcatchment (0.072 km^2). Autumn low flows are on the order of 0.2 l/s. The bimodal events are associated with low rainfall intensities, long rainfall durations, and a tendency for relatively large antecedent flows. The opposite is true of the flashy single peaked events. Estimated runoff volumes of the quick and delayed responses; backcalculated contributing area (assuming a runoff coefficient of unity). Recession time constants estimated from the recessions of the quick and delayed responses.

		event precip. (mm)	anteced. discharge (l/s)	runoff volume (m^3)		estim. contrib. area (m^2)		Recession time (hrs)	
Event type	Date			quick	delayed	quick	delayed	quick	delayed
Bimodal	25.6.95	77	0.8	54	194	701	2519	0.25	26
Bimodal	10.7.96	59	0.4	53	83	898	1407	0.25	107
Bimodal	7.7.97	37	0.6	47	123	1270	3324	0.33	126
Bimodal	19.7.97	60	0.8	46	82	766	1367	0.42	168
Bimodal	20.7.99	68	0.6	120	169	1764	2485	0.25	428
Flashy single peak	10.7.95	35	0.2	36	---	1028	---	0.33	---
Flashy single peak	27.7.95	44	0.2	20	---	454	---	0.25	---
Flashy single peak	28.7.96	13	0.1	8	---	615	---	0.17	---
Flashy single peak	30.7.18	13	0.5	14	---	1076	---	0.17	---
Flashy single peak	18.8.98	16	0.2	11	---	687	---	0.17	---
Flashy single peak	17.8.00	20	0.5	20.3	---	1015	---	0.17	---

are not able to activate subsurface flow paths, although the surface runoff response is quite substantial. Baseflow prior to this event type is usually low, and dry antecedent moisture conditions may contribute to the observed response. Hence, a single-peaked flashy runoff response appears to be typical for short duration - high intensity rainfall events in this and other microcatchments that are partly covered by saturated areas (Table 1).

At larger catchment scales, such as the 16 km^2 Löhnersbach catchment at Rammern (Fig. 1), the same short-duration, high-intensity event produces a similar hydrograph as in the microcatchments (Fig. 2d): There is one single flashy response with response times of a few hours or less. The interpretation of this type of hydrograph is that runoff from the 16 km^2 catchment is, again, a combined result of runoff from microcatchments, such as the one in Figure 2c. However, the timing of the responses is very similar at different catchment scales because there is little variation in the travel times of surface runoff. Also, it is likely that this type of intense rainstorm is spatially more limited. Hence, only part of the Löhnersbach catchment may have been affected, resulting in one single sharp runoff peak in the hydrograph from the 16 km^2 catchment.

Response characteristics of both event types in the Limbergalm microcatchment are summarized in Table 1. For most bimodal events, rainfall was of a low intensity - long duration type while, for the flashy unimodal event type, response intensities were generally high and durations short. The total runoff volumes for the first peak in the bimodal event type were significantly larger than the volumes of the flashy events (Table 1). It is interesting to back-calculate the effective area that contributed to the runoff peaks by assuming a runoff coefficient of unity and dividing runoff volume by rainfall depth. These effective contributing areas are on the order of 1000 m^2 for both the first peak in the bimodal event type and for the flashy events. In contrast, the effective contributing areas of the secondary peak, i.e. the delayed response, are on the order of 2000 m^2 or more. This suggests that, as the event progresses, increasingly distant flow paths are connected to the catchment outlet.

The recession constants for both event types, shown in Table 1, indicate that there is indeed a distinct contrast between the response characteristics of the two peaks in the bimodal event type. The first, flashy peaks exhibit significantly shorter response times than the second peaks of the bimodal type runoff hydrographs.

4. Spatial patterns of runoff generation

As mentioned above, the Löhnersbach catchment is hydrologically quite heterogeneous. The northwestern part of the catchment exhibits numerous permanently saturated areas, while the southeastern part does not. Field experience in the Löhnersbach catchment indicates that these differences have important implications for runoff generation. Some of these differences will be demonstrated quantitatively below.

Figure 3 shows results from individual synchronous discharge measurements at four gauges (Klammbach, Marxtengraben, Schusterbauergraben, and Neuhausengraben/Mündung). Specifically, the ratio of runoff at these four gauges relative to runoff from synchronous measurements at Rammern (i.e. the 16 km^2 Löhnersbach catchment, see Fig. 1) has been calculated and plotted against runoff at Rammern. Figure 3 thus represents the runoff contributions of the subcatchments as a function of the streamflow from the entire catchment. The four subcatchments fall into two distinct groups. The first group is the Klammbach, which comes from the southeastern part of the Löhnersbach catchment. The second group consists of the Marxtengraben, Schusterbauergraben and Neuhausengraben/Mündung, which all come from the northwestern part of the Löhnersbach catchment. Runoff contribution appears to be closely linked to flow conditions (low vs high flow). The catchment area of Klammbach covers about 15% of the Löhnersbach catchment but, during low flow conditions, the Klammbach contributes a relatively large percentage (around 20-25%) to the Löhnersbach runoff. As runoff increases, the relative contribution of the Klammbach decreases to about 15% or less. In contrast, the catchment area of Neuhausengraben/Mündung is about 8% of that of the Löhnersbach catchment but, during low flow conditions, the Neuhausengraben/Mündung contributes a much smaller percentage (less than 5%) to the Löhnersbach runoff. As runoff increases, the relative contribution of the Neuhausengraben/Mündung increases to about 10% and, occasionally, it is more than 13%. Marxtengraben and Schusterbauergraben, belonging to the same group of northwestern catchments, exhibit a very similar behaviour, although the increase is less pronounced. It is clear that the differences

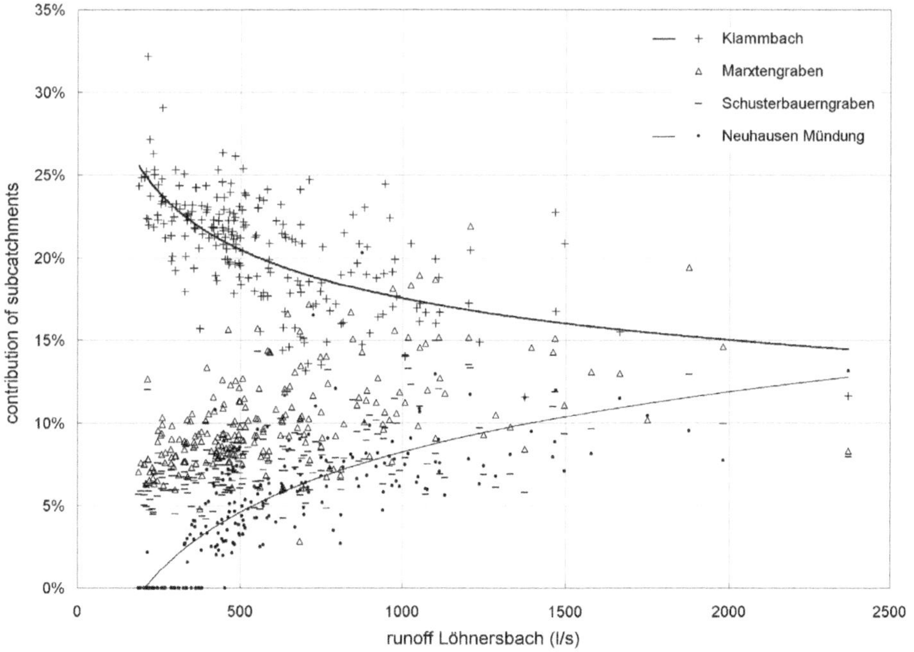

Figure 3: Runoff contribution from subcatchments of the Löhnersbach is shown relative to the runoff from the entire 16 km^2 Löhnersbach catchment (synchronous measurements). This plot illustrates the distinct spatial differences in runoff generation in the Löhnersbach catchment.

in runoff between these two groups of catchments are caused by differences in the importance of surface and subsurface stormflow generation mechanisms. Lower transmissivities and storage capacities are reflected in the presence of saturated areas in the northwestern part of the catchment. During dry conditions and long interstorm periods, the subsurface runoff contribution in the northwestern area decreases, and subsurface pathways tend to be inactive. Water penetrates more deeply into the soil layer and/or is lost to evapotranspiration. During heavy rainfall following wet antecedent conditions, both surface and subsurface flow paths become active (see Fig. 2a), which increases the runoff contribution drastically. In contrast, in the southeastern subcatchments, such as the Klammbach, where these threshold mechanisms seem to be largely absent, infiltration and groundwater recharge during rainfall and snowmelt periods, as well as groundwater storage capacities, are greater. Thus, the groundwater release (baseflow) is more pronounced and stable, even during dry periods and low flow conditions. This explains why the Klammbach contributions to total streamflow in the Löhnersbach increase with decreasing flows.

For a more quantitative assessment of the differences in runoff generation behaviour of the southeastern and northwestern subcatchments of the Löhnersbach, an attempt has been made to estimate runoff of the Klammbach (2.5 km^2) on the basis of runoff observations at Rammern (16 km^2). This estimation, in essence, amounts to downscaling the runoff of the entire 16 km^2 catchment to the scale of a microcatchment (15% of the gauged catchment). Two alternative approaches were tested: In the first approach, it was assumed that the specific runoff within the entire 16 km^2 catchment (Löhnersbach at Rammern) is spatially uniform. This assumption implies that runoff at Klammbach can be calculated as a fixed ratio of runoff at Rammern, the factor of proportionality being the ratio of the subcatchment area relative to the entire 16 km^2 catchment. This is a common assumption for estimating runoff from subcatchments when no additional data are available. In the second approach, the specific runoff generation characteristics of the Klammbach catchment, as discussed above, were taken into account. This was done by first performing a nonlinear regression analysis between simultaneous discharge observations at Klammbach and the Löhnersbach at Rammern. Specifically, the following equation was fitted:

$$Q_K(t) = a \cdot Q_R(t+dt) \cdot \exp(-b \cdot Q_R(t+dt)) \qquad (1)$$

where $Q_K(t)$ is the Klammbach runoff at time t, dt is a time lag of 30 minutes, and $Q_R(t+dt)$ is the Löhnersbach runoff at Rammern at time $t+dt$. From data of the years 1993 and 1994, the parameters were estimated as a=0.24 and b=0.0003s/l. Equation (1) was then used to estimate runoff of the Klammbach in a predictive mode. In Figure 4, the results of this downscaling approach are shown for a period in the summer of 1997. As can be seen from this figure, the assumption of specific runoff being uniform in the entire catchment leads to a significant underestimation of the Klammbach runoff, especially during low flow conditions. The approach that accounts for the specific runoff generation characteristics of the Klammbach yields significantly better runoff estimates.

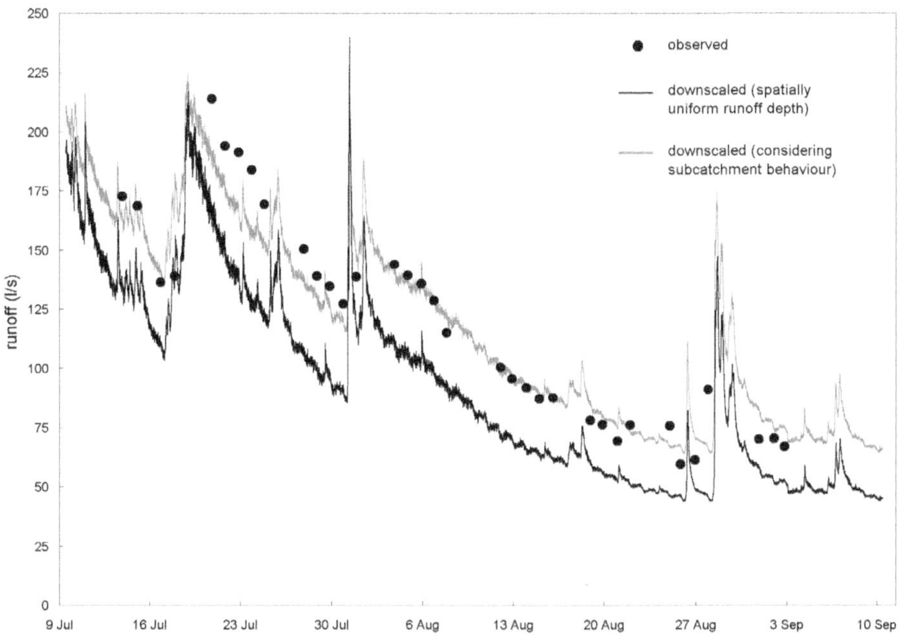

Figure 4: Runoff from the Klammbach subcatchment (2.5 km^2) downscaled from runoff measurements at Rammern (16 km^2) with two different assumptions: a) uniform runoff depth in the whole catchment (black line) and b) distinct spatial differences in runoff generation (see Equation (1) and Fig.3) taken into account. Points are runoff measurements.

5. Conclusions and future research directions

Our investigations confirm the hypothesis that, under certain conditions and in characteristic areas (e.g. saturated areas), one single mechanism dominates the runoff behaviour. Such typical runoff reactions can primarily be measured in small catchments or subcatchments where these characteristic areas take up a relatively large proportion of the total area and their runoff behaviour therefore dominates the catchment flow. Runoff observations in a small subcatchment of the highly instrumented Löhnersbach catchment, for example, display two event types: The bimodal event type consists of a flashy peak, due to saturation overland flow, and a delayed peak, due to subsurface stormflow. This event type occurs in areas that are partly covered by permanently saturated zones, provided that the storm is of a long duration - low intensity type, because these are the conditions under which subsurface flow paths are activated. The second event type is a unimodal, flashy event, which may occur in similar catchment areas but is associated with the dominance of surface flow during short duration - high intensity storms. In each of these event types, one or two runoff mechanisms dominate while the contribution from other mechanisms is insignificant. The flashy unimodal type and the initial flashy peak of the bimodal type can be observed on both the subcatchment and the total catchment scale. The second, delayed peak of the bimodal type cannot be identified on the total catchment scale because subsurface runoff contributions of different travel times are superimposed, thus forming the typical shape of the recession curve hydrograph. These findings support the existence of typical runoff events and confirm the hypothesis that it is in fact possible to isolate dominating processes in the context of runoff generation (Gutknecht 1997).

Our investigations in the Löhnersbach catchment emphasize the heterogeneity of runoff generation processes in mountainous terrain. Runoff generation was found to be very different in the northwestern and southeastern parts of the Löhnersbach catchment. During low flow conditions, the relative proportion of runoff from the northwestern area is below average, whereas the southeastern area (e.g. the Klammbach) contributes above average. However, this trend is reversed for high flow conditions and large events. These differences can be interpreted in terms of the runoff generation mechanisms that are associated with subsurface stormflow and overland flow from saturated areas and are consistent with field experience in the Löhnersbach catchment. The importance of these differences in runoff responses in the two sub-catchments has been demonstrated by estimating runoff from the Klammbach subcatchment (southeastern part of the Löhnersbach catchment) based on downscaling of runoff data from the entire Löhnersbach catchment. By taking into account the typical runoff generation characteristics of the Klammbach subcatchments, runoff estimates are significantly better than those made with the assumption of spatially uniform runoff generation.

Our results have a number of important implications for hydrological modelling. It is clear that runoff generation in small catchments is highly non-linear, because threshold effects may cause the hydrograph shape to change from event to event. However, typical runoff events do exist. It is likely that these characteristic types, identified in small and very small catchments, are also applicable in medium-sized catchments. The identification of dominant runoff processes is particularly appealing when one wishes to retain the process basis of hydrological models and, at the same time, find a strategy for coping with the complexity of runoff generation processes.

While the spatial patterns of runoff generation in the Löhnersbach catchment can be interpreted well in terms of dominant mechanisms, it is clear that the identification of these mechanisms is contingent on the existence of high-resolution field data and detailed qualitative field observations. It is much more difficult to interpret the heterogeneous spatial patterns of runoff generation in mountain watersheds if only coarse-resolution catchment attributes from regional databases are available. While mapable properties, such as stream network density and the presence of saturated zones, have been found to be valuable indicators of spatial differences in runoff generation, these types of data are usually not available in regional databases. Rather, regional databases tend to include catchment descriptors such as topographic attributes, land use, geology and pedology. These descriptors do not vary much between the two areas of the Löhnersbach catchment (Merz et al. 2000) and, hence, cannot explain the striking differences in runoff generation between the two spatial units. This finding is important for process-based approaches to modelling runoff generation in medium-sized catchments because it appears that a combination of standard GIS (Geographical Information System) layers is unlikely to capture space-time differences in runoff generation, such as those found in the Löhnersbach catchment. In larger catchments that encompass different environments, however, soil/vegetation/topography indices may be a more important control on runoff than in the Alpine catchment of the Löhnersbach. Here, a more promising strategy may be to use indices that are more closely related to field-observed runoff generation dynamics, such as the saturation areas in this study. Another possibility is an extension of the FLAB (FLächen gleicher ABflussbildung, areas of equal runoff generation processes) expert system suggested by Peschke et al. (1999), in which the dominant runoff generation mechanisms are estimated on the basis of sub-catchment scale environmental characteristics (e.g. soils, vegetation, topography) and other catchment characteristics. New rules are being developed for this expert system, making use of catchment characteristics that can be easily identified by visual inspection in the field or by the analysis of aerial photographs (e.g. bare rock surfaces, vegetation free slopes and saturation zones). Of particular importance is an expert knowledge about catchment behaviour, including qualitative observations, which were used in this study and were found to be very helpful. Quantitative information on the dynamics of subsurface flow can be assessed by irrigation and tracer experiments as performed in summer 2002 in the Löhnersbach catchment. Modelling approaches for medium-sized catchments can then be based on the concept of typical events and dominant runoff mechanisms, which may facilitate the simplification of process descriptions while retaining their most important dynamic properties.

6. Acknowledgements

Funding from the Austrian Academy of Sciences (project numbers HÖ1 and HÖ18), the Bund-Bundesländerkooperation (project number SA 26/99), and the German Science Foundation (DFG, project number Me 1844/1-1) is gratefully acknowledged.

7. References

Anderson, M. G., and Burt, T. P. (1990). "Process studies in hillslope hydrology." John Wiley & Sons, Chichester.

Anderson, S. P., Dietrich, W. E., Montgomery, D. R., Torres, R., Conrad, M. E., and Loague, K. (1997) Subsurface flowpaths in a steep, unchanneled catchment. *Water Resources Research* **33**, 2637-2653.

Blöschl, G. (1996). „Scale and scaling in hydrology," Wiener Mitteilungen, Band 132. Habilitationsschrift. Institut für Hydraulik, Gewässerkunde und Wasserwirtschaft, Technische Universität Wien.

DFG (1995). Hochwasser in Deutschland unter Aspekten globaler Veränderungen. *In* „Bericht über das DFG-Rundgespräch am 9.10.1995 in Potsdam." Potsdam-Institut für Klimafolgenforschung, Potsdam.

Dunne, T. (1983). Relation of field studies and modelling in the prediction of storm runoff. *Journal of Hydrology* **65**, 25-48.

Grayson, R. B., and Blöschl, G., Eds. (2000). "Spatial patterns in catchment hydrology: Observations and modelling." Cambridge University Press, Cambridge.

Gutknecht, D. (1997). Vielfältigkeit - Zum Umgang mit einem wichtigen Aspekt hydrologischer Prozesse. *In* „Wasserbau - Visionen für das nächste Jahrtausend, Festschrift zum 60.Geburtstag von Prof. Scheuerlein." (R. Friedrich, Ed.), pp. 183-197. D.&V. Thaur, Innsbruck.

Kirnbauer, R., and Haas, P. (1998). Observations on runoff generation mechanisms in small Alpine catchments. *In* "Hydrology, water resources and ecology in headwaters." (K. Kovar, U. Tappeiner, N. E. Peters, and R. G. Craig, Eds.), pp. 239-247. Proceedings of the HeadWater'98 Conference, Meran, Italy, Apr. 1998*). IAHS Publication* **248**.

Merz, R., Piock-Ellena, U., Blöschl, G., and Kirnbauer, R. (2000). „Skalierungsprobleme bei der Regionalisierung von Hochwässern. Endbericht an die Österreichische Akademie der Wissenschaften, HÖ 18." Institut für Hydraulik, Gewässerkunde und Wasserwirtschaft, Technische Universität Wien, Oktober 2000.

Moore, I. D., and Grayson, R. B. (1991). Terrain based prediction of runoff with vector elevation data. *Water Resoures Research* **27**, 1177-1191.

Müller, G., and Peschke, G. (2000). Hydrologische Prozessuntersuchungen auf der Basis adäquater Messnetze. *Österreichische Wasser- und Abfallwirtschaft* **52**, 94-104.

Peschke, G., Etzenberg, C., Müller, G., Töpfer, J. and Zimmermann, S. (1999). „Das wissensbasierte System FLAB: Ein Instrument zur rechnergestützten Bestimmung von Landschaftseinheiten mit gleicher Abflußbildung." IHI-Schriften H.10. Internationales Hochschulinstitut Zittau, Zittau.

Scherrer, S. (1997). „Abflußbildung bei Starkniederschlägen. Identifikation von Abflußprozessen mittels künstlicher Niederschläge." Mitteilung der Versuchsanstalt für Wasserbau, Hydrologie und Glaziologie, ETH Zürich, Heft 147.

Tilch, N., Uhlenbrook, S., Didszun, J., Leibundgut, Ch., Zillgens, B., Kirnbauer, R., and Merz, B. (2003). Entschlüsselung von Abflussbildungsprozessen mit Hilfe tracerhydrologischer Ansätze in einem alpinen Einzugsgebiet. *Österreichische Wasser- und Abfallwirtschaft* **55**, 9-17.

Torres, R., Dietrich, W. E., Montgomery, D. R., Anderson, S. P., and Loague, K. (1998). Unsaturated zone processes and the hydrologic response of a steep, unchanneled catchment. *Water Resources Research* **34**, 1865-1879.

Uchida, T., Kosugi, K., and Mizuyama, T. (2001). Effects of pipeflow on hydrological process and its relation to landslide: A review of pipeflow studies in forested headwater catchments. *Hydrological Processes* **15**, 2151-2174.

Uhlenbrook, S., and Leibundgut, C. (1997). Abflußbildung bei Hochwasser in verschiedenen Raumskalen. *Wasser und Boden* **49**, 13-22.

Soil Erosion and Runoff Generation Related to Land Use Changes in the Pyrenees

José M. García-Ruiz*, Teodoro Lasanta, Blas Valero, Carlos Martí, Santiago Beguería, Juan I. López-Moreno, David Regüés, and Noemí Lana-Renault
Instituto Pirenaico de Ecología, CSIC, Zaragoza, Spain
**phone 34-976-716026, fax 34-976-716019, e-mail humberto@ipe.csic.es*

Keywords: Experimental catchments, Experimental plots, Land use changes, Runoff, Soil erosion, Spanish Pyrenees

1. Introduction

Many scientific papers and books demonstrate the direct and indirect effects of human activities on the intensification of soil erosion processes and changes in both sediment and runoff sources (Ives and Messerli 1989). It is well known that deforestation and hillslope farming cause distinct changes in soil properties and infiltration rates, which ultimately affect soil erosion processes and the hydrological cycle at a basin and hillslope scale (Goudie 1986).

The effects of changes in land use and plant cover are especially pronounced in mountain areas, since they are high-energy environments, where sediment transfer from the hillslopes to the channels is greatly facilitated. In addition, mountains are worldwide the most important regions for the generation of runoff, and their geographic characteristics control the river regimes and water resources in the lowlands. In climates with strongly seasonal precipitation and in semi-arid climates, mountains can be considered as "islands" of humidity that support high population densities and supply water for irrigation and urban consumption in downstream areas. This is the reason why land use changes in mountains affect not only sustainability of soils and water quality in the mountains themselves but also affect the livelihood and welfare of many people in the lowlands.

U. M. Huber et al. (eds.), Global Change and Mountain Regions, 321-330.

In this paper a synthesis of the most representative results from the Pyrenees are presented as a case study to illustrate changes in soil erosion and water resources of mountain areas that are controlled by changing land use practices.

2. Data and methods

Since 1990 the Department of Soil Erosion and Land Use Changes in the Pyrenean Institute of Ecology (Spanish National Research Council, CSIC) has studied the hydrological and geomorphic consequences of different human activities in mountain areas from a multidisciplinary point of view. Research has focused especially on the Central Spanish Pyrenees, where extensive land use changes have occurred during the last 50 years.

2.1 Plot-scale investigations

The Aísa Valley Experimental Station has set up a number of erosion plots under different traditional and modern land uses to assess the effects of agricultural changes on runoff generation and soil erosion at a local scale (García-Ruiz et al. 1995). Traditionally, many hillslopes in this region were used for arable farming. Today, many of the fields are in various stages of abandonment. Our investigations include the following land uses:

- Cereal (barley)
- Fallow land (alternating every two years with cereal)
- Shifting agriculture (barley)
- Abandoned cereal (now covered by dense herbaceous communities, after 5 years of abandonment)
- Abandoned shifting agriculture (with herbs now covering 60 per cent of the soil after 4 years of abandonment)
- Burnt plot (previously dense shrub cover, which already re-established two years after the 1991 fire)
- Meadow and
- Dense shrub cover (representing the evolution of most of the hillslopes 30 years after farmland abandonment).

The experimental set-up consists of eight closed, 10 x 3 m plots, one in each land use type. Sediment yields are measured with a Gerlach trap in the lower section of each plot. A set of tipping buckets and a pluviometer, connected to data loggers, are used to record runoff and precipitation, respectively. Sediment concentration is estimated by storing the runoff in containers, which are emptied after each rainstorm event. All the plots are located in a field with a 25% slope.

2.1 Basin-scale investigations

Also, in the Arnás and San Salvador experimental catchments (285 and 136 ha in size, respectively) basin-scale information on discharge and sediment transport has been collected from two different environments: Arnás is a catchment that was previously cultivated with cereals and abandoned 40 years ago, and San Salvador is a forest area. At the outlet of both catchments a flume has been built across the river channel and an ultra-sound sensor has been installed to measure the water height from which discharge can be calculated. A turbidimeter allows us to estimate suspended sediment load and a sediment trap serves to measure the bedload. A weather station has been installed near the gauging station.

In addition, shallow landslides have been studied to identify the factors that explain their spatial distribution and elucidate their role in sediment delivery. Remote sensing and Geographical Information Systems (González et al. 1995) have been used to assess sediment sources at a larger scale (basins of more than 1,000 km^2). Information is also available on reservoir siltation. The effects of changes in land cover and climate on water resources throughout the Central Spanish Pyrenees have also been studied (Beguería et al. 2003).

3. The historical evidence

Geomorphic evidence reveals the consequences of plant cover changes in the Central Pyrenees in pre-historical and historical times. Human impact on vegetation and geomorphic processes has a long history in this region. For example, a paleosoil with charcoal and burnt branches intercalates between two sediment units in stratified scree in the Bentué Valley (Pre-Pyrenees). The soil has been dated to 3340±70 ^{14}C years BP (García-Ruiz and Valero-Garcés 1998) and represents a period of wildfires that are probably associated with a known expansion of pastures for domestic livestock.

Sediment cores from high mountain lakes show two periods of more intense erosion in the Pyrenees: one around 3950 ^{14}C years BP, and the other around the 12th century (Montserrat 1992). The former was relatively short, whereas the second represents a sudden and very significant lowering of the upper forest belt, in order to enlarge the pasture area. This increase of pastureland was related to the expansion of the Christian Pyrenean kingdoms towards the Ebro Depression and the consequent flourishing of transhumance systems. At present, the upper forest limit in many Pyrenean valleys reaches only 1600-1700 m, whereas the natural forest limit is at about 2200 m. The current subalpine landscape shows clear indications of the negative consequences of such a deforestation process: shallow translational landslides have displaced the soil from the steepest slopes, and dense rill networks occupy large areas on planar slopes (García-Ruiz et al. 1990; García-Ruiz and Valero-Garcés 1998). Similar phenomena occurred in other European mountains in the past, especially the Alps (Hollermann 1985). Also, Ballantyne (1994) recognized an intensification of solifluction in Scotland after the 16th century, due to grazing and the reduction of plant cover.

The occurrence of some large debris flows in the Pyrenees has also been attributed to deforestation and farming on steep slopes. For example, the St. Adrián de Sasave Monastery, built at the end of the 9th century, was buried by two debris flows at the beginning of the 12th century and in the second half of the 18th century (García-Ruiz and Valero-Garcés 1998). At present, the 3.2 km^2 basin is completely covered by a dense pine forest and seems to be very stable. Under present land use, large sediment yields are uncommon in the basin. However, the landscape was very different in the Middle Ages, when cultivated fields occupied most of the hillslopes below 1600 m.

A close relationship between historical farming and sediment yield can be deduced from the modern landscape, especially in the Mediterranean mountains. Many abandoned fields have lost most of their original soil cover, and surface soils tend to be stony (García-Ruiz and Puigdefábregas 1982). During the population peak, between the 18th century and the first half of the 20th century, the Pyrenean alluvial plains were characterized by bare river bars and very instable braided channels, suggesting a supply of large volumes of coarse sediment from the hillsides. Gómez-Villar (1996) demonstrated that many of the alluvial fans in the Pyrenees and the Iberian Range may, at least in part, have a human-induced origin. The results from the Arnás experimental catchment confirm that during the period of traditional exploitation (until the middle of the 20th century) almost the entire catchment was geomorphologically active and delivered high quantities of sediment. On a larger scale, the growth of the Ebro River delta since Roman times has been attributed to deforestation and hillslope farming within the entire basin. The period of maximum delta aggradation between the 15th and 19th centuries coincides with the period of increased and widespread cereal farming in the Ebro Basin, in many cases under shifting agriculture systems.

4. Hydromorphological consequences of modern land use changes

Since the last decades of the 20th century, most of the mountain areas of the world have been subject to intense and rapid land use changes. In the case of mountains in developing countries, a strong contrast can be observed between some areas, where population is still increasing, as in the Himalayas and the African mountains (Ives and Messerli 1989; Messerli and Hurni 1990; Ries 1995), and other areas where population is decreasing due to emigration toward the cities, as in parts of the Andes (Harden 1996; Inbar and Llerena 2000). This contrast is expressed as an increasing pressure on the land in the first case, and a partial abandonment of cultivated fields in the second case.

In some developed countries, such as Spain, emigration of more than 50% of the rural population to urban-industrial centers has led to a decrease in population pressure in mountain regions and farmland abandonment is a prominent process. Farm abandonment is related to factors such as the difficulties in using modern machinery on steep slopes and to new relationships between highlands and lowlands, including the construction of large reservoirs for hydropower and irrigation, and the expansion of tourism activities (Lasanta 1989). The spatial consequences of depopulation are an extensification of land uses in most of the mountainous terrain, and an intensification in the valley bottoms (García-Ruiz and Lasanta 1993).

The Aísa Valley Experimental Station and the Arnás and San Salvador experimental catchments have been used to address the environmental effects of extensification in the Spanish Pyrenees. In addition, other scientific approaches have been used to identify sediment sources and to analyse the evolution of runoff and sediment yield in the last few decades.

4.1 Plot-scale evidence

The results from the Aísa Valley Experimental Station show that runoff coefficients and soil erosion rates are clearly higher in the cultivated plots, especially in the plot with shifting agriculture, followed by the plot in fallow and the cereal plot (Fig. 1). The lowest values are recorded from the plot with dense shrub cover, the meadow plot and the burnt plot. However, the latter was subject to high erosion rates during the first six months after the fire and erosion rates decreased rapidly when dense shrub cover reestablished. The abandoned plots, with a dense herbaceous cover, show intermediate

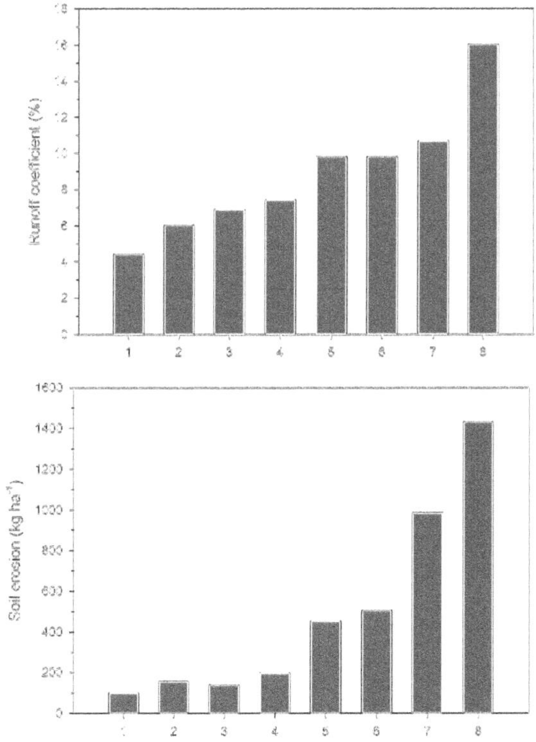

Figure 1: Runoff and soil erosion under different land uses in the Aísa Valley Experimental Station, Central Spanish Pyrenees. 1: Dense shrub cover; 2: Meadow; 3: Burnt shrub (1991); 4: Abandoned field; 5: Abandoned field after shifting agriculture; 6: Cereal field (barley); 7: Plot in fallow; 8: Shifting agriculture (barley).

runoff coefficients and soil erosion rates. These results confirm the importance of areas with traditional arable farming practices as sediment sources in mountain regions, and the progressive decrease in soil erosion after land abandonment, especially when plant recolonisation reaches the stage of a dense shrub cover (García-Ruiz et al. 1995). The experiment also demonstrates the importance of dense shrub and grassland cover for soil conservation and the moderation of surface runoff on hillslopes. In addition, our investigations explain landscape degradation in many mountain areas of intermediate elevation in the Pyrenees, especially those affected by shifting agriculture and frequent human-induced fires. The results also suggest that a substitution of dense shrub by meadows would not cause a significant increase in soil erosion, while at the same time reducing wildfire hazard and increasing fodder availability for livestock (Lasanta and García-Ruiz 1998).

4.2 Basin-scale evidence

A comparison between a deforested, extensively farmed and grazed catchment (Arnás) and a densely forested catchment (San Salvador) gives consistent information about the role of forest cover in controlling the frequency and intensity of floods as well as suspended sediment and bed load transport. The hydrograph in Figure 2 shows that the Arnás catchment reacts to almost every rainstorm event (Arnáez et al. 1999), although the response has a very high variability. One small rainfall event in a 24 hour interval is enough to produce a sudden increase in the hydrograph. This runoff response is likely due to the presence of eroded areas close to the main channel, which are inherited from past farming activities and overgrazing. The absence of a dense vegetation cover in these eroded areas likely decreases infiltration and increases the contribution of the rapidly responding surface flow relative to the more slowly responding runoff components (subsurface stormflow, baseflow). In the San Salvador catchment, the hydrograph shows a very moderate response to rainfall events, especially at the beginning of the autumn, and a long recession curve after each event. Such differences are attributed to increased interception and infiltration, and decreased evapotranspiration in the forested catchment, which leads to decreased peak flows and increased base flows. Suspended sediment transport shows even greater differences between the two catchments (Lorente et al. 2000). No bedload transport in response to rainfall events has been recorded in San Salvador, whereas this process occurs between 4 and 8 times per year in Arnás, with a high variability in total volume.

At a regional scale, the use of Remote Sensing, in combination with hydrological models, allows the identification of both sediment sources and the main factors that explain erosion history in large basins. The maps obtained show that sediment sources are closely tied to certain lithologies (especially marls) and to areas where human activities have been more intense (García-Ruiz and Puigdefábregas 1982; González et al. 1995). The location of shallow landslides that evolve into debris flows is also linked to areas with a long farming history, as well as deforested and frequently burnt areas (Lorente et al. 2002).

In summary, new trends in soil erosion and runoff generation have been identified in the Central Pyrenees in the last few years. These changes are associated with farmland abandonment and plant recolonisation and reforestation of abandoned fields. For instance, Beguería et al. (2003) have demonstrated decreasing streamflow in the Central Spanish Pyrenees independent of climatic oscillations between 1945 and 1995. The difference between the expected trends in discharge (based on climate) and the actual discharge implies that a non-climatic, time-dependent factor has determined the decrease in discharge. This negative trend in discharge of Pyrenean rivers coincides with major changes in land management, such as farmland abandonment, drastical reduction of human-induced fires and reforestation. In subalpine grasslands, a decrease in intensive livestock management in the past 30 years has allowed the establishment of tall herb communities and a slow colonisation by scrubs and trees (Tappeiner and Cernusca 1993). Similarly, a decrease in sediment yield and delivery during the last few decades has also been observed in some large reservoirs, in spite of an increase in the number and intensity of rare floods (Valero-Garcés et al. 1999; López-Moreno et al. 2003).

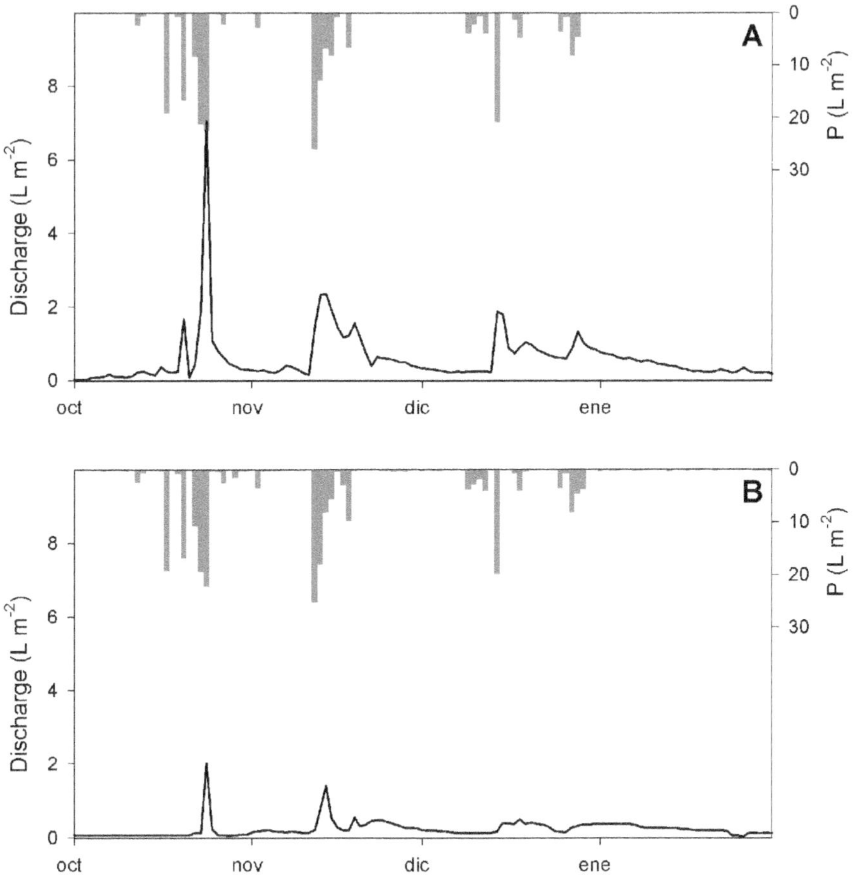

Figure 2: Precipitation and discharge in (A) Arnás and (B) San Salvador experimental catchments between October 1999 and January 2000.

5. Water and sediment: A global change perspective

Changes in plant cover and land uses cause strong changes in the hydrological cycle and, ultimately, in soil erosion, sediment delivery, runoff generation and the characteristics of streamflow (i.e. the frequency and intensity of floods). At the same time, changes in climate parameters also lead to hydrological and geomorphic changes. Probably, one of the most important challenges for the future is to discriminate between the impacts of both factors (Beguería et al. 2003). The effectiveness of correction measures will strongly depend on a clear understanding of underlying processes and their driving mechanisms.

The studies carried out in the Central Spanish Pyrenees, as in other mountains of the world, confirm the extreme importance of working at several scales and of integrating the results across the scales:

- Experimental plots are needed to obtain detailed information on sediment yield and runoff generation under different land uses during rainstorm events. In addition, more detailed information is required to better understand soil processes in relation to differences in plant cover.
- Experimental catchments give a larger-scale perspective and allow the quantification of basin-scale discharge and sediment delivery. Relationships between sediment sources, peak flow and sediment transport can be established during rainfall events. Linking hydrological processes from the plot to the catchment scale is the most important step for hydrological modeling and for forecasting the direction of changes.
- Finally, the study of large basins provides information on streamflow and sediment transport trends, and can help to address the relationships between these two parameters and climatic change. This large-scale perspective is needed to analyze the effects in the middle and lower river reaches, and the interactions between highlands and lowlands. Both water resources management policies and soil conservation measures have to take into account information on the scale of experimental plots and catchments. Monitoring reservoir sedimentation provides an excellent opportunity to assess the short-term temporal variability of sediment transport, especially during high flows, and the trends related to plant cover changes in a large basin. A longer temporal perspective (thousands of years) can be obtained from lake sediment records. These field data are fundamental for the validation of hydrological models that are used to assess the hydrological consequences of environmental change and to define future land and watershed management strategies.

Mountain areas are affected by intense and rapid human and biophysical changes. The survival of mountain populations and the preservation of unique and pristine landscapes depend on our ability to forecast changes and, especially, the intensity of hydromorphological responses. Soil productivity and the quantity and quality of water resources are strongly dependent on plant cover characteristics. Our main role as scientists is to rank the importance of the different factors, to identify the changes, to establish relationships between processes and driving mechanisms, and to predict and model the responses.

6. Acknowledgements

Funding for this research was provided by the following projects: "Water resources management in a changing environment: The impact of sediment in sustainability" (WARMICE, ENV4-CT98-0789), funded by the European Community, and "Assessment of sediment and runoff sources in relation to land use changes" (HIDROESCALA, REN2000-1789-C04-01/GLO), "Hydrological processes in semi-natural Mediterranean areas" (PROHISEM, REN2001-2268-C02-01/HID), "Hydrologic and erosive processes in Pyrenean catchments related to land use changes and climate variability" (REN2003-08678/HID) all funded by CICYT, Spanish Ministry of Science and Technology.

7. References

Arnáez, J., Martí-Bono, C., Beguería, S., Lorente, A., Errea, M. P., and García-Ruiz, J. M. (1999). Factores en la generación de crecidas en una cuenca de campos abandonados, Pirineo Central español. *Cuadernos de Investigación Geográfica* **25**, 7-24.

Ballantyne, C. K. (1994). Holocene mass movement on Scottish mountains: Dating, distribution and implications for environmental change. *In* "Solifluction and climatic variations in the Holocene." (B. Frenzel, Ed.), pp. 71-86. Gustav Fisher, Stuttgart.

Beguería, S., López-Moreno, J. I., Lorente, A., Seeger, M., and García-Ruiz, J. M. (2003). Assessing the effect of climate oscillations and land use changes on streamflow in the Central Spanish Pyrenees. *Ambio* **32**, 283-286.

García-Ruiz, J. M., and Lasanta, T. (1993). Land use conflicts as a result of land use change in the Central Spanish Pyrenees: A review. *Mountain Research and Development* **13**, 295-304.

García-Ruiz, J. M., and Puigdefábregas, J. (1982). Formas de erosión en el flysch eoceno surpirenaico. *Cuadernos de Investigación Geográfica* **8**, 85-128.

García-Ruiz, J. M., and Valero-Garcés, B. (1998). Historical geomorphic processes and human activities in the Central Spanish Pyrenees. *Mountain Research and Development* **18**, 309-320.

García-Ruiz, J. M., Alvera, B., Del Barrio, G., and Puigdefábregas, J. (1990). Geomorphic processes above timberline in the Spanish Pyrenees. *Mountain Research and Development* **10**, 201-214.

García-Ruiz, J. M., Lasanta, T., Ortigosa, L., Ruiz-Flaño, P., Martí, C., and González, C. (1995). Sediment yield under different land uses in the Spanish Pyrenees. *Mountain Research and Development* **15**, 229-240.

Gómez-Villar, A. (1996). "Conos aluviales en pequeñas cuencas torrenciales de montaña." Geoforma Ediciones, Logroño.

González, C., Ortigosa, L., Martí, C., and García-Ruiz, J. M. (1995). Use of a Geographical Information System to study the spatial organization of geomorphic processes in mountain areas. *Mountain Research and Development* **15**, 141-149.

Goudie, A. (1986). "The human impact on the natural environment." Blackwell, Oxford.

Harden, C. (1996). Interrelationships between land abandonment and land degradation: A case from the Ecuadorian Andes. *Mountain Research and Development* **16**, 274-280.

Hollermann, P. (1985). The periglacial belt of mid-latitude mountains from a geoecological point of view. *Erdkunde* **39**, 259-270.

Inbar, M., and Llerena, C. A. (2000). Erosion processes in high mountain agricultural terraces in Peru. *Mountain Research and Development* **20**, 72-79.

Ives, J. D., and Messerli, B. (1989). "The Himalayan dilemma: Reconciling development and conservation." Routledge, London.

Lasanta, T. (1989). "Evolución reciente de la agricultura de montaña: el Pirineo aragonés." Geoforma Ediciones, Logroño.

Lasanta, T., and García-Ruiz, J. M. (1998). La gestión de los usos del suelo como estrategia para mejorar la producción y la calidad del agua. Resultados experimentales en el Pirineo Central español. *Cuadernos de Investigación Geográfica* **24**, 39-57.

López-Moreno, J. I., Beguería, S., Valero-Garcés, B., and García-Ruiz, J. M. (2003). Intensidad de avenidas y aterramiento de embalses en el Pirineo Central español. *Eria* (in press).

Lorente, A., Martí, C., Beguería, S., Arnáez, J., and García-Ruiz, J. M. (2000). La exportación de sedimento en suspensión en una cuenca de campos abandonados, Pirineo Central español. *Cuaternario y Geomorfología* **14**, 21-34.

Lorente, A., García-Ruiz, J. M., Beguería, S., and Arnáez, J. (2002). Factors explaining the spatial distribution of hillslope debris flows. A case study in the Flysch Sector of the Central Spanish Pyrenees. *Mountain Research and Development* **22**, 32-39.

Messerli, B., and Hurni, H., Eds. (1990). African mountains and highlands: Problems and perspectives. African Mountains Association, Marceline, Missouri.

Montserrat, J. (1992). "Evolución glaciar y postglaciar del clima y la vegetación en la vertiente sur del Pirineo: Estudio palinológico." Instituto Pirenaico de Ecología, Zaragoza.

Ries, J. B. (1995). Does soil erosion in the high mountain region of the eastern Nepalese Himalayas affect the plains? *Physics and Chemistry of the Earth* **20**, 251-270.

Tappeiner, U., and Cernusca, A. (1993). Alpine meadows and pastures after abandonment. *Pirineos* **141-142**, 85-96.

Valero-Garcés. B., Navas, A., Machín, J., and Walling, D. (1999). Sediment sources and siltation in mountain reservoirs: A case study from the Central Spanish Pyrenees. *Geomorphology* **28**, 23-41.

The Role of Riparian Zones in Steep Mountain Watersheds

Brian L. McGlynn
Department of Land Resources and Environmental Sciences, Montana State University, 334 Leon Johnson Hall. PO Box 173120, Bozeman, MT 59717-3120, USA
phone +1 (406) 994-7690, fax +1 (406) 994-3933, email bmcglynn@montana.edu

Keywords: Hillslope, Hydrograph separation, Hydrology, Riparian, Runoff, Watershed

1. Introduction

The riparian zone encompasses the strip of land between the stream channel and the hillslope and is sometimes referred to as the valley floor, near-stream zone (Cirmo and McDonnell 1997), floodplain (Bates et al. 2000), or buffer zone (Lowrance et al. 1985). Riparian zones have been differentiated from upslope zones by unique hydrology, topography, vegetation, and soils (Hill 1996). Characteristics such as anoxic zones, gleyed soils, distinct soil color, high organic content, breaks in slope, and near-surface water tables often distinguish riparian zones from adjacent hillslopes. Because of their location, riparian zones have significant potential to regulate the movement of water and elements in surface and subsurface runoff that flows from upslope areas to the stream (Hill 1996).

In shallow soil mountain watersheds with steep slopes and poorly permeable bedrock, hillslope and riparian zone flow dynamics are predominantly controlled by topography. While the flushing of riparian zones by hillslope runoff is a first-order control on potential chemical transformation (Hill 1990) and hillslope water expression in stream flow (Hooper et al. 1997; McGlynn et al. 1999; Burns et al. 2001; McGlynn and McDonnell 2003a,b), little is known about the ratio of the hillslope inputs relative to the local riparian zone storage. Significant uncertainties also exist about the role riparian zones play in regulating water and element movement from

U. M. Huber et al. (eds.), Global Change and Mountain Regions, 331-342.

uplands to streams, despite much work on individual hillslope and riparian cross sections. However, there is increasing awareness that dynamic relationships exist between riparian zones and hillslopes, which need to be better understood (e.g. Hooper et al. 1997; Becker, this volume).

These issues are at the forefront of hydrological process studies in small watersheds (McDonnell and Tanaka 2001; Uhlenbrook et al. 2003), the development and application of hydrological and hydrochemical models, and the assessment of the potential impact of climate change scenarios on mountain hydrology. They are addressed in this paper with emphasis on the Maimai watersheds in New Zealand (South Island, west coast), and reference to Panola Mountain, Georgia, and Sleepers River, Vermont. Together, these watersheds provide an optimal comparison for in-depth analysis of the relative roles of hillslopes, riparian zones, and watershed organisation in controlling streamwater quantity, quality, and age. The following questions are discussed in this paper:

- What are the relative proportions of hillslope and riparian zone sources in watershed runoff?
- How does the riparian zone modulate hillslope inputs to the stream?
- How does the distribution of hillslopes and riparian zones affect riparian buffering in headwater watersheds? How does the hydrological function of riparian zones vary from the narrow headwater riparian zones to wider valley bottom floodplains?
- Finally, the hydrological role of riparian zones is assessed in a larger context by carrying out a watershed inter-comparison between the Maimai watershed in New Zealand and two watersheds in the US.

2. Maimai case study

2.1 Research site description

The Maimai research watersheds are a set of highly responsive, steep, wet watersheds located along the axis of a valley that forms the headwaters of the Grey River on the west coast of the South Island of New Zealand (Fig. 1). Annual precipitation is 2600 mm and runoff ratios are 64% annually, with 39% as quickflow. Soil depths are shallow, with an average of 0.6 m, overlying Old Man Gravels, a poorly permeable early Pleistocene well-cemented conglomorate. The geology and soil depths are relatively uniform across Maimai watersheds and the topography is steep and highly dissected. Hillslopes are short (<300 m), steep (34 degrees), have local relief of 100 to 150 m, are composed of regular spurs and linear hollows, and are consistent across watershed scales. Only riparian areas scale with increasing watershed size. Riparian zones are narrow in the headwaters and increase in width with watershed area to a maximum valley bottom riparian floodplain width at the 280 ha watershed scale (Fig. 2). The median upstream riparian width in the 280 ha Maimai experimental watershed is 6 m (typical of ~3 ha watersheds), the mean is 20 m (typical of 15–30 ha watersheds), and the maximum is 163 m (floodplain riparian zone at the watershed outlet) (McGlynn and Seibert 2003). Maimai has a long history of hillslope hydrological research in well-characterized sub-5 ha research watersheds. For detailed site characterization and a review of previous research related to the evolution of a detailed perceptual model of hillslope runoff generation in the Maimai watersheds, please see McGlynn et al. (2002).

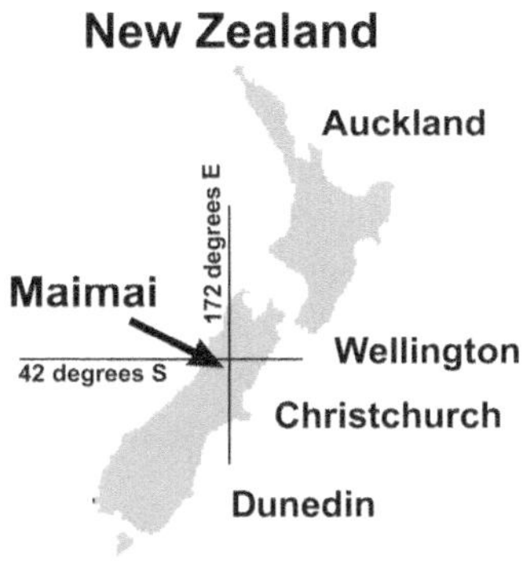

Figure 1: Research site location.

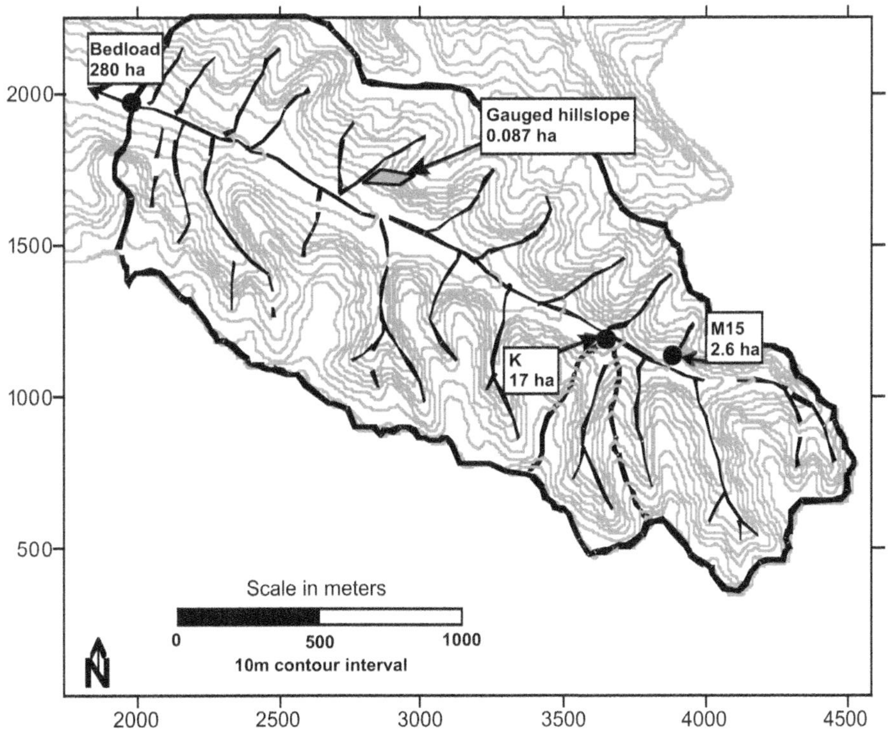

Figure 2: The Maimai nested research watersheds.

2.2 Riparian and hillslope sources of headwater watershed runoff

McGlynn and McDonnell (2003a) sought to determine the dominant controls on watershed runoff by disaggregating the watershed into discrete landscape units. These are often referred to as hydrotopes (e.g. Becker, this volume), Hydrological Response Units (HRUs) (Wigmosta et al. 1994; Leavesley and Stannard 1995), hydrogeomorphic units (Winter and Woo 1990), or dynamic contributing areas (Beven and Freer 2001), in which hydrological processes are assumed to be relatively "uniform" due to somewhat homogeneous environmental conditions. Combinations of techniques, utilizing extensive survey, hydrometric, isotopic, and solute data are necessary to obtain an unequivocal understanding and quantification of both spatial and temporal runoff sources, and thus runoff generation at the scale of dominant hydrotopes in headwater watersheds (especially hillslope and riparian zones). Therefore, hydrological processes and parameters were investigated at this scale. The following definitions for source water in time and space are used in this paper:

- Temporal:

  Event water: Rainfall associated with the current storm event

  Pre-event water: Resident watershed water prior to storm event

- Spatial:

  Hillslope water: Water originating from upland and hillslope zones (event and pre-event)

  Riparian water: Water originating from riparian zones (event and pre-event).

We measured the area-normalized hillslope runoff at a trenched and gauged hillslope (see Fig. 2), and extrapolated it to the entire hillslope area in the adjacent 2.6 ha watershed to determine the hillslope proportion of total watershed runoff (McGlynn and McDonnell 2003a). This approach assumes that the gauged hillslope is characteristic of all the hillslopes in the 2.6 ha watershed and that it adequately represents the response of similar hillslopes in the Maimai watershed. Results from landscape analyses and hydrological investigations in multiple headwater watersheds demonstrated that the gauged hillslope is comparable to other hillslopes in the Maimai headwater watersheds (McGlynn and McDonnell 2003a).

We compared the results of flow-based hydrograph separations with three-component mass balance hydrograph separations using ^{18}O and silica to separate runoff into event rainwater, pre-event hillslope water, and pre-event riparian zone water. Combining the flow-based and three-component mass balance separations, we determined the spatial sources of both old water and new rainwater runoff. Our watershed-scale observations were constrained by runoff quantity and tracer dynamics at the trenched and gauged hillslope, and with distributed instrumentation including wells, piezometers, lysimeters, and water content probe nests in hillslope and riparian zone positions.

We found that hillslope runoff comprised 2-16% of total watershed storm runoff during a small 27 mm event and 47-55% during a larger 70 mm event. However, less than 4% of the new water, collected at the watershed outlet, originated from the hillslopes during each event. We found that in the 27 mm rain event, 84-97% of total storm runoff was generated in the riparian zone. In a larger 70 mm event, riparian water dominated total flow early in the event, although the hillslope became the main contributor once hillslope runoff was initiated (four hours later). Despite the large proportion of subsurface hillslope runoff in total storm runoff during the second larger event, riparian and channel zones accounted for 96% of the new water measured at the watershed outlet. In other words, the hillslope contribution to storm runoff was primarily old water, whereas the riparian zone contributed both old and new water, which was due to the expansion and contraction of surface saturated zones near the stream channel and direct precipitation onto near-stream saturated areas (variable source area dynamics).

In the large event, proportions of riparian and hillslope runoff were similar; however, riparian water contribution was greater on the rising limb whereas hillslope water contribution was greater on the falling limb. We compared the flow-based hydrograph separation with the tracer based three-component hydrograph separation and demonstrated comparable results (McGlynn and McDonnell 2003a). During periods of hillslope runoff, the buffering potential of the riparian zone was a function of both the riparian reservoir volume and the magnitude of hillslope throughflow to the riparian zone. This indicates that the buffering potential of riparian zones is a function of the ratio of riparian volume to hillslope throughflow.

Our multi-method approach allowed the determination of the spatial sources of new and old water throughout the monitored events and the evaluation of the roles of dominant landscape units in streamflow generation. While it has been conceptualized for some time that the relative runoff contributions of hillslope and riparian zones vary throughout an event, this is the first study to quantify these changes with hydrometric, isotopic and solute data. Understanding the relationships between dominant landscape units (hillslopes and riparian zones) and runoff dynamics is the first step in the development of a conceptual model of watershed runoff at Maimai.

2.3 The relative distributions of riparian and hillslope contributing area

Plot-scale riparian research and the links between plot- and watershed-scale behaviour has been largely qualitative and speculative thus far. In addition, investigations in one small headwater watershed limit our ability to scale and transfer the results described above to the larger Maimai valley and other watersheds. We addressed the need for objective mapping and quantification of dominant landscape units (hillslopes and riparian zones), the simple evaluation of riparian zones, and the distribution of hillslope and riparian zones throughout the landscape through extensive surveying and the analysis of digital elevation models (DEMs) (McGlynn and Seibert 2003).

Topography can often be used as a surrogate for hydrological processes in steep wet watersheds with shallow soils. This has been highlighted by many studies (Beven and Kirkby 1979; Beven and Freer 2001), especially since digital elevation models became widely available. We developed an objective technique for watershed characterization that allows us to apportion landscapes into their riparian and hillslope components, assess the modulating (buffer) potential of riparian zones, and quantify hillslope-riparian-stream-watershed connections (McGlynn and Seibert 2003). Specifically, we report on new techniques to determine bulk watershed, riparian area fractions of watersheds and distributed measures of local riparian-to-hillslope ratios along the channel network.

The Maimai landscape is relatively simple; existing maps and surveys indicate that, as in many other mountain watersheds, flat valley bottom areas are relatively large along the main axis of the watershed and decrease toward the headwaters. However, techniques to quantify the abundance and distribution of riparian areas have been previously unavailable. Accumulated area maps can show us where hollows and streams are located by quantifying the upslope area that contributes runoff into each cell in a digital elevation model, highlighting convergence and divergence in the topography. Slope maps can show us the distribution of flat valley bottom areas and steep hillslopes. Integrated, topographic index calculations (Beven and Kirkby 1979) highlight well-drained and poorly drained areas of the watershed, using the combined information from accumulated area and slope maps. However, index calculations do not tell us much about the functions of the riparian zones. Also, they do not allow objective mapping of dominant landscape units or solve the uncertainties associated with hillslope-riparian-stream connections and processes.

The new technique outlined by McGlynn and Seibert (2003) reveals relationships that were previously masked. The pieces that form each step in this analysis are simple. Essentially, riparian and hillslope zones are mapped and quantified based on DEM analysis. The spatial organization of hillslope zones relative to riparian zones in the watershed (topology) is then used to evaluate the capacity of the riparian zones, associated with each stream reach, to buffer hillslope inputs that drain into the same stream reach.

We computed the accumulated watershed area along the channel network and found that 35% of the 280 ha watershed area is located in sub-watersheds smaller than 1 ha, 60% in <4 ha sub-watersheds, and 85% in <20 ha sub-watersheds (McGlynn and Seibert 2003). Hence, most area is accumulated in the smallest headwater watersheds and enters the main stem of the river via defined tributary junctions. Relatively little area enters the stream network through wider valley bottom riparian zones along the main stem (~15%).

We also evaluated the relative distributions of hillslope and riparian areas (local riparian-to-hillslope area ratios). Headwater watersheds often have smaller riparian zone percentages than larger watersheds due to widening of valley bottom zones with increasing watershed size. We found a clear dissimilarity between where most upland area is accumulated and where most riparian area is accumulated. 75% of the total hillslope area is located in >13 ha watersheds. In contrast, 75% of the riparian area is located in watersheds >14 ha. In other words, 75% of the hillslope area is associated with 25% of the riparian zone area and 25% of the hillslope area is associated with 75% of the riparian zone area. This is an important observation since the capacity of the riparian zone to modulate (buffer) hillslope inputs depends on the connection of hillslopes to riparian zones (Devito et al. 1996), and the inherent biogeochemical function of the riparian zone is influenced by its position in the landscape (Hill 2000).

The majority (60%) of the 280 ha watershed area is comprised of sub-5 ha watersheds where riparian-to-hillslope area ratios are small (typically 0.01 to 0.12). When the total riparian area of the 280 ha watershed is divided by the total hillslope area, the ratio is larger (0.14). However, this number is misleading because the local setting controls riparian zone function (Devito et al. 1996; Hill 1996; 2000). The distributed riparian-to-hillslope area ratio has a median of 0.057 and is strongly skewed toward small riparian-to-hillslope ratios that are found predominantly in <5 ha sub-watersheds. The ratio between riparian and hillslope area can be interpreted as a buffer-capacity index. For example, when ratios are small, the riparian zone has a lower capacity to buffer hillslope runoff. Hence, more hillslope water will contribute to stream flow during a storm event. However, it is important to consider additional variables, such as event magnitude, duration, and frequency, as well as antecedent conditions, when examining riparian zone function during storm events. Furthermore, variations in soil depth might be of importance in more heterogeneous watersheds (Devito et al. 1996), where the volume rather than the area of the riparian and hillslope zones may be the primary control on buffering capacity.

At Maimai, we found that headwater riparian areas were narrower, but more tightly connected to direct hillslope inputs than larger floodplain riparian zones along higher order streams draining larger watersheds. The implication is that watershed buffering potential is determined in headwater riparian zones. Larger watershed-scale valley bottoms were disconnected from where the bulk of hillslope inputs originate. The implication is that wider valley bottom floodplains have low potential to buffer hillslope runoff. When combined, the maps of hillslope and riparian area inputs to the stream network provided a distributed riparian-to-hillslope area measure and a distributed estimation of potential riparian buffer capacities.

3. Watershed inter-comparison: The hydrological role of riparian zones

Watershed inter-comparison is valuable for testing understanding gained in one watershed in other physiographic and climatic settings. Without inter-comparison, we risk focussing on the idiosyncrasies of one site, rather than the processes and controls operating across sites. Based on the research presented in the preceding sections (McGlynn and McDonnell 2003; McGlynn and Seibert 2003; McGlynn and McDonnell 2003a,b; McGlynn and Seibert 2003) and relationships suggested by Hooper et al. (1997), McGlynn et al. (1999), and Burns et al. (2001), we hypothesized that the degree of hillslope water expression in storm flow is a function of both riparian-to-hillslope reservoir ratios and landscape organization. To further test this assumption, we quantified the contributions of hillslopes and riparian zones to storm flow using chemical and isotopic techniques across three diverse (17 to 40 ha) headwater watersheds: a highly responsive, steep, wet watershed (Maimai, New Zealand), a moderately steep, snowmelt-dominated watershed (Sleepers River, Vermont), and a highly seasonal, low-relief watershed (Panola Mountain, Georgia). We monitored watershed runoff, internal hydrological response, and isotopic and solute dynamics for discrete riparian and hillslope zones within each watershed, including hillslope trenches at Maimai and Panola. We applied landscape analysis techniques to gain insight into watershed structure. Specifically, we examined where hillslope area and riparian area were accumulated and assessed the relative contributions of source waters from riparian zones, hillslope zones, and new water, based on three-component hydrograph separations utilizing ^{18}O and natural solute tracers (Fig. 3).

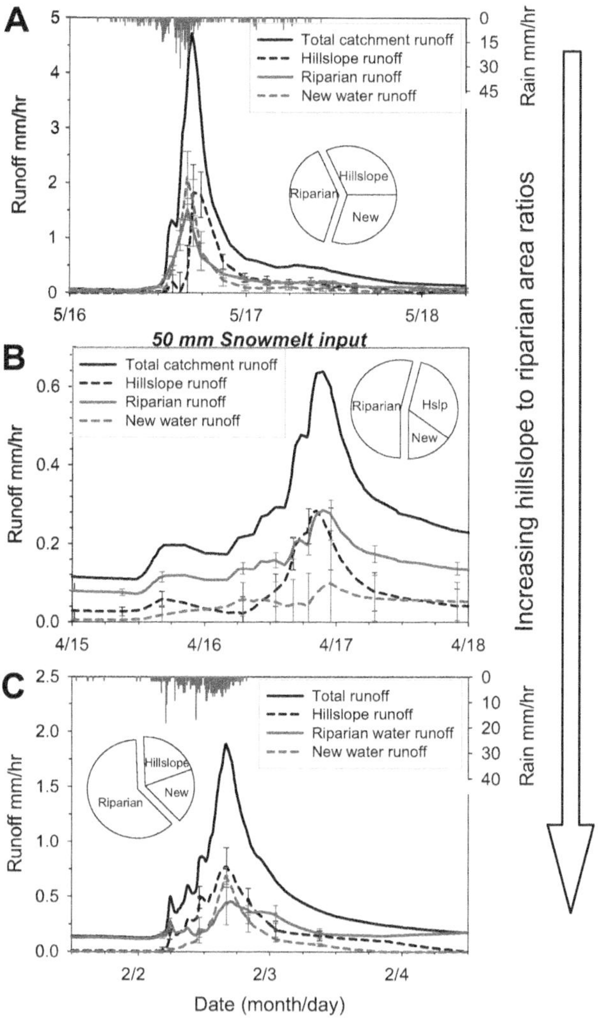

Figure 3: The contribution of different runoff types to storm flow in three watersheds with increasing ratios of riparian-to-hillslope area. Riparian proportions of storm flow increase with increasing riparian-to-hillslope area ratios. A) Maimai: 65 mm rain event with a watershed runoff ratio of 0.5. B) Sleepers River: 50 mm snowmelt event with a runoff ratio of 0.5. C) Panola Mountain: 62 mm rain event with a runoff ratio of 0.54.

We found that:

(1) Hillslope proportions of total storm flow runoff were a function of riparian-to-hillslope area ratios in all watersheds. Riparian proportions of watershed storm runoff increased with increasing riparian-to-hillslope area ratios (Fig. 3 and 4).

(2) Greater variability in hillslope contributions along the stream network (focused flow in some locations and little flow in others) suggested decreased buffering potential of riparian zones while consistently smaller hillslope influx along the network suggested more diffuse hillslope contribution and greater buffering potential. Riparian areas are better able to buffer small diffuse hillslope runoff into the riparian zone across the whole stream network rather than large inputs to a few stream reaches.

(3) The relative contributions of hillslope and riparian zones to watershed runoff were partially a function of landscape organization. Local area entering the stream channels was most variable at Maimai, with Panola and Sleepers showing an order of magnitude less variability. Median local area entering the channel network (a measure of the distribution of hillslope runoff inputs) was 0.33 ha at Maimai, and 0.016 and 0.013 ha at Sleepers and Panola, respectively. At Maimai, greater variability and a higher median area suggested a more dissected landscape, resulting in focused flow into and through the riparian zone. At Sleepers and Panola, less variability and lower median area suggested more diffuse hillslope inputs and less concentrated flow from the hillslopes into the riparian zone.

(4) The effectiveness of a given riparian buffer for runoff and biogeochemistry depended partially upon reservoir volumes, local riparian-to-hillslope area ratios, and local hillslope runoff rates.

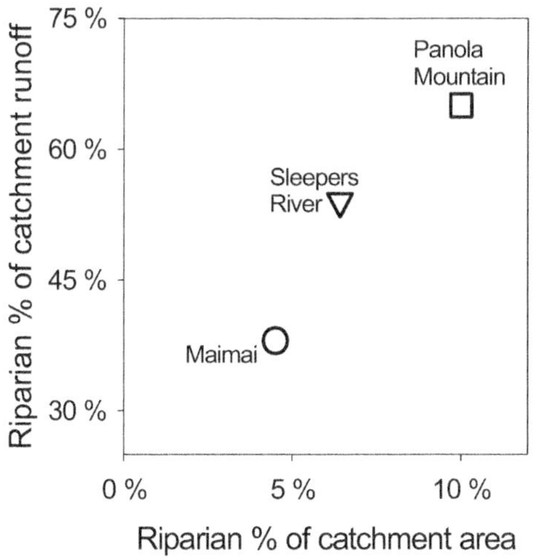

Figure 4: Percent of watershed runoff originating in the riparian zone versus percent of watershed area mapped as riparian in the Maimai, Sleeper River, and Panola Mountain watersheds. Results suggest a positive relationship between riparian zone extent and the amount of riparian source water in storm runoff.

In addition, these research results have provided first insights into riparian buffering of hillslope runoff on the watershed and landscape scales, which is still poorly understood. These new approaches and the insight gained from watershed comparison provide a way to scale hydrological processes from hillslope-riparian plots to small watersheds and eventually to greater portions of the landscape to assess the buffering potential of riparian zones and its effect on water quantity and quality. This is particularly important in the context of assessing the impact of alternative climatic regimes on the sources and buffering of watershed runoff.

4. Priorities for the future

The relative roles of hillslopes, riparian zones, and hyporheic zones (the sub-stream mixing zone for water of both sub-surface and stream channel origin) in streamflow generation and composition remain only partially understood. Historically, hydrological research has been compartmentalized. Many small watershed hydrological studies have focused on hillslopes and related flow processes. Concurrently, much riparian research has focused on nutrient transformation in the near-stream zone, while hyporheic zone research has focused on hyporheic stream connections and mixing. Few projects have attempted to integrate research on hillslope, riparian, and hyporheic processes within a watershed- or landscape-scale context to quantify first-order controls on storm flow and baseflow generation, the partitioning of old and new water, the impact on water residence times, and subsequent control on stream chemistry. Preliminary data analysis suggests that there is quantifiable interplay between riparian and hillslope zone source waters that varies across watershed scales and with storm event characteristics. The research summarized here has hypothesized that there is a shift from riparian-dominated (hydrodynamics and solutes) watershed outflow response to hillslope-dominated (hydrodynamics and solutes) watershed response during storm events and that the timing and degree of this transition is a function of event characteristics and landscape organization (e.g. the spatial distribution of riparian-to-hillslope reservoir ratios). However, a framework for the assessment of watershed hydrology in areas without the significant infrastructure that is present at long-term research watersheds remains elusive.

Hydrological reviews have identified scale as a major unresolved problem in the hydrological sciences. Hydrological process research has historically been examined at one spatial scale and then extrapolated to larger and smaller scales, or transferred to other watersheds, with little appreciation of the physical, spatial, and temporal scaling of dominant hydrological processes. To gain a realistic understanding of hydrological processes, it is, however, critical to examine the role of riparian zones in diverse watersheds across environmental gradients, including climate and topography, and across a continuum of watershed scales. These approaches are vital to progress toward a better understanding of the complex relationship between climate and watershed response.

Advances in hydrological modelling must be grounded in physical observations and empirically based relationships. Watershed hydrochemical models, for instance, often rely on solute concentrations sampled at the watershed outlet for inferences about internal processes in the entire watershed. If the model assumes that the watershed is behaving in a homogeneous manner and that the entire basin is contributing flow to the stream, it is often not representing reality. A model that takes into account outflow quantity or quality without adequately representing internal watershed dynamics does not advance our understanding and provides only limited utility. Therefore, valid conceptual models that account for the heterogeneity of mountain watersheds and are based on measurable relationships and quantitative observations are vital. These types of models are particularly important for assessing potential changes in the quantity and quality of runoff in mountain watersheds under different climate change scenarios.

5. Acknowledgements

Thanks to John Payne, Jagath Ekanayake, and Breck Bowden of LandCare Research, NZ, for logistical, technical, and collaborative support, and to Robert S. McGlynn and Kendall Watkins for invaluable assistance in the field. Thanks also to Jeff McDonnell, Rick Hooper, Carol Kendall, Jamie Shanley, and Nic Hjerdt for their contributions to this work. This work was made possible by NSF grant EAR-0196381 and the 2001 AGU Horton Research Grant awarded to BLM. This manuscript has been assigned Journal Series No. 2002-70, Montana Agricultural Experiment Station, Montana State University-Bozeman.

6. References

Bates, P. D., Stewart, M. D., Desitter, A., Anderson, M. G., Renaud, J. P., and Smith, J. A. (2000). Numerical simulation of floodplain hydrology. *Water Resources Research* **36**, 2517-2529.

Beven, K. J., and Kirkby, M. J. (1979). A physically based, variable contributing area model of basin hydrology. *Hydrological Sciences Bulletin* **24**, 43-69.

Beven, K., and Freer, J. (2001). A dynamic TOPMODEL. *Hydrological Processes* **15**, 1993-2011.

Burns, D. A., McDonnell, J. J., Hooper, R. P., Peters, N. E., Freer, J. E., Kendall, C., and Beven, K. (2001). Quantifying contributions to storm runoff through end-member mixing analysis and hydrologic measurements at the Panola Mountain Research Watershed (Georgia, USA). *Hydrological Processes* **15**,1903-1924.

Cirmo, C. P., and McDonnell, J. J. (1997). Linking the hydrologic and biogeochemical controls of nitrogen transport in near-stream zones of temperate-forested catchments: A review. *Journal of Hydrology* **199**, 88-120.

Devito, K. J., Hill, A. R., and Roulet, N. (1996). Groundwater-surface water interactions in headwater forested wetlands of the Canadian Shield. *Journal of Hydrology* **181**, 127-147.

Hill, A. R. (1990). Groundwater cation concentrations in the riparian zone of a forested headwater stream. *Hydrological Processes* **4**, 121-130.

Hill, A. R. (1996). Nitrate removal in stream riparian zones. *Journal of Environmental Quality* **25**, 743-755.

Hill, A. R. (2000). Stream chemistry and riparian zones. *In* "Streams and ground waters." (J. B. Jones, and P. J. Mulholland, Eds.), pp. 83-110. Academic Press, San Diego, CA.

Hooper, R., Aulenbach, B., Burns, D., McDonnell, J. J., Freer, J., Kendall, C., and Beven, K. (1997). Riparian control of streamwater chemistry: Implications for hydrochemical basin models. *IAHS Redbook* **248**, 451-458.

Leavesley, G. H., and Stannard, L. G. (1995). The precipitation-runoff modeling system - PRMS. *In* "Computer models of watershed hydrology." (V. P. Singh, Ed.), pp. 281-310. Water Resources

Publications, Fort Collins, CO.

Lowrance, R. R., Leonard, R., and Sheriden, J. (1985). Managing riparian ecosystem to control nonpoint source pollution. *Journal of Soil and Water Conservation* **40**, 87-91.

McDonnell, J. J., and Tanaka, T. (2001). On the future of forest hydrology and biogeochemistry. *Hydrological Processes* **15**, 2053-2055.

McGlynn, B. L., McDonnell, J .J., Shanley, J. B., and Kendall, C. (1999). Riparian zone flowpath dynamics during snowmelt in a small headwater catchment. *Journal of Hydrology* **222**, 75-92.

McGlynn, B. L., McDonnell, J. J., and Brammer, D. D. (2002). A review of an evolving perceptual model of hillslope flowpaths at the Maimai catchments, New Zealand. *Journal of Hydrology* **257**, 1-26.

McGlynn, B. L., McDonnell, J. J., Seibert, J., and Kendall, C. (2003). The effects of catchment scale and landscape organization on streamflow generation. *Water Resources Research* (in review).

McGlynn, B. L., and McDonnell, J. J. (2003a). The role of discrete landscape units in controlling catchment dissolved organic carbon dynamics. *Water Resources Research* **39**, 1090 (doi:10.1029/2002WR001525).

McGlynn, B. L., and McDonnell, J. J. (2003b). Quantifying the relative contributions of riparian and hillslope zones to catchment runoff and composition. *Water Resources Research* **39** (XXXX, doi: 10.1029/2003WR002091, in press).

McGlynn, B. L., and Seibert, J. (2003). Distributed assessment of contributing area and riparian buffering along stream networks. *Water Resources Research* **39**, 1082 (doi:10.1029/2002WR001521).

Uhlenbrook, S., and Leibundgut, C. (2002). Process-oriented catchment modelling and multiple-response validation. *Hydrological Process* **16**, 423-440.

Uhlenbrook, S., McDonnell, J. J., and Leibundgut, C. (2003). Preface: Runoff generation implications for river basin modelling. *Hydrological Processes* **17**, 197-198.

Wigmosta, M. S., Vail, L. W., and Lettenmaier, D. P. (1994). A distributed hydrology-vegetation model for complex terrain. *Water Resources Research* **30**, 1665–1679.

Winter, T. C., and Woo, M. K. (1990). Hydrology of lakes and wetlands. *In* "Surface water hydrology: The geology of North America." (M. G. Wolman, and H. C. Riggs, Eds.), pp. 159-187. Geological Society of America, Boulder, CO.

Wondzell, S. M. (1994). "Flux of groundwater and nitrogen through the floodplain of a fourth-order stream." PhD dissertation, Oregon State University, Corvallis.

The Use of Hydrological Models for the Simulation of Climate Change Impacts on Mountain Hydrology

Joachim Gurtz*, Herbert Lang, Mark Verbunt, and Massimiliano Zappa
Institute for Atmospheric and Climate Science, Swiss Federal Institute of Technology (ETH) Zurich, Switzerland
**phone +41-1-635-5228, fax +41-1-362-5197, e-mail joachim.gurtz@iac.umnw.ethz.ch*

Keywords: Hydrological modelling, Runoff, Snow cover, Swiss basins, Water balance

1. Introduction

According to the Second and Third Assessment Reports of the Intergovernmental Panel on Climate Change (IPCC 1996; 2001) the increase in mean surface air temperature of the northern hemisphere was larger in the 20th century than in any other period of the last 1000 years. The decade 1990-1999 was the warmest of this time period. It is also believed that this increase in air temperature will be accompanied by intensification of the global hydrological cycle and, in the same chain of cause and effect, by enhanced evaporation and precipitation (Schär and Frei, this volume). However, the scientific community needs to gain a better understanding of the biosphere-atmosphere system before being confident on the predictions of hydrological processes in a future climate (Frei et al. 2000; Ohmura and Wild 2002).

The assessment of the hydrological impacts of climate change is particularly challenging in the mountains (Viviroli et al. 2003). Mountainous environments are particularly sensitive hydrological systems with significant responses to climatic variations (Gurtz et al. 1997). Mountain river basins are characterized by extremely variable morphology and topography and by strong variations in vegetation and soil characteristics. The climatic elements, including snow coverage, show large variability in time and space. All these features are strongly related to altitude.

The high degree of variability needs to be adequately considered when modelling

U. M. Huber et al. (eds.), Global Change and Mountain Regions, 343-354.

hydrologic processes. The following issues require particular attention: 1) the selection of adequate model components and the definition of their structure and interactions (Gurtz et al. 2003a); 2) the spatial resolution at which modelling should be undertaken (Zappa and Gurtz 2002) and 3) the temporal resolution of the model-based simulation of hydrological processes.

For our investigations of the hydrological impacts of climate change in the Alps detailed simulation experiments were carried out with two spatially distributed hydrological models: a) the Water flow and balance Simulation Model (WaSiM-ETH) (Schulla 1997) and b) the Precipitation-Runoff-Evapotranspiration-HRU model (PREVAH) (Gurtz et al. 1999). WaSiM-ETH is a grid oriented physically based hydrological catchment model. The PREVAH model is more conceptual and based on hydrological response units (HRU) or hydrotopes (Gurtz et al. 1999). Recent work showed that both models are well suited to simulate hydrological processes in mountainous catchments (Jasper 2001; Klok et al. 2001; Verbunt et al. 2003; Gurtz et al. 2003a,b; Zappa et al. 2003). The objective of this paper is to describe the specific features of hydrological modelling in mountainous regions and the application of these models for investigating the hydrological impacts of climate variations. These investigations are essential for current and future water management purposes and for the solution of environmental problems.

2. Distributed hydrological modelling of mountainous basins

2.1 Modelling of hydrological processes

In order to determine the spatial and temporal distribution of model parameters, detailed physiographic information is required. In this context, Geographic Information Systems (GIS) that allow for processing digital terrain models (DTM) and digital landuse and soil maps are invaluable. The development and application of hydrological model components in complex alpine landscapes has been greatly facilitated by the improved availability, reliability, precision and resolution of meteorological and hydrological networks, as well as the availability of automatic data collection and transfer systems and regionalized hydrological parameter sets.

The runoff generation in mountainous catchments depends, according to the elevation and morphology of a basin, on glacier melt, snowmelt, rainfall and evapotranspiration (Gurtz et al. 2003b). The spatial and temporal superposition of these processes governs the range of the observed discharge regimes (Aschwanden and Weingartner 1985). Both WaSiM-ETH and PREVAH consist of a series of modules (or sub-models) that simulate the different components of the hydrological cycle, such as snow accumulation, snow and glacier melt, evapotranspiration, interception, percolation, groundwater recharge, soil moisture and groundwater storage, runoff concentration and flood routing (Gurtz et al. 2003a).

The use of suitable spatially distributed sub-models for the estimation of snow accumulation, snowmelt and icemelt is of high importance in mountainous regions (Zappa et al. 2003). The use of energy balance based snowmelt modules is frequently limited by the availability of adequate observations of meteorological forcing. For this reason most snow and glacier melt modules are based on temperature-index-based approaches, such as the positive-degree-day method. This approach has proved to provide quite reliable results (Ohmura 2001). However, in periods of variable weather conditions and for simulation experiments at hourly time steps it is recommended to include incoming solar radiation, vapour pressure and wind speed as additional variables for the estimation of the melting processes. The radiation-extended temperature-index-based approach proposed by Hock (1999) allowed for improvements in the estimation of snowmelt and icemelt (Klok et al. 2001; Verbunt et al. 2003). This approach includes an additional term in the classical positive-degree-day method to account for the internal daily fluctuation of potential incoming direct radiation.

Particular attention has to be paid to the calculation of evapotranspiration (Menzel 1997). The Penman-Monteith equation (Monteith 1965) has been shown to be an effective approach for evaporation calculations in mountainous catchments (Gurtz et al. 1999). Air temperature and the amount of net radiation should be site-corrected, as determined by the local elevation, aspect, slope angle, albedo, and land surface characteristics (Gurtz et al. 2003a).

The interception module is strongly related to the modelling of evaporation (Menzel 1997). The interception sub-model has to consider variations in interception storage as a function of the vegetation type. The surface depression storage capacity of rocky soils and urban areas is also defined for the estimation of interception and direct evaporation from non-vegetated surfaces.

The runoff-generation in mountainous catchments is characterized by a temporally and spatially highly variable contribution of different runoff components to the total runoff. Thus, different runoff components with different drainage paths and temporal delays can be distinguished (for details see Becker, this volume). The surface runoff component in prealpine basins is caused by both storm rainfall and large volumes of snowmelt. In high alpine basins, surface runoff is an important component in summer when the hydrological response is governed by snow and icemelt. However, the more or less delayed interflow was found to be the most important runoff component in mountainous regions (Gurtz et al. 2003a). Its dynamics mainly depend on soil characteristics (layering, differences in hydraulic conductivity and storage capacity, slope and moisture). The relative portion of interflow to total runoff is reduced with increasing glacierization and increasing exposure of rock surfaces within a catchment. Baseflow is the slow and stable runoff component. In mountainous catchments, the relative contribution of baseflow to total flow is highest in winter. However, baseflow also depends on the characteristics of groundwater storages and the amount of groundwater recharge. The absolute volume of baseflow is highest in summer, after the end of the snowmelt season (Verbunt et al. 2003). PREVAH and WaSiM-ETH includes two different model concepts to simulate the dynamic generation of runoff components. PREVAH uses a series of linear reservoirs. WaSiM-ETH computes runoff-generation using the discrete Richards equation. Two case studies showing an intercomparison between these two approaches for the computation of runoff-generation are discussed in Gurtz et al. (2003a,b).

2.2 Spatial and temporal discretization

Adequate spatial discretization is essential for obtaining a reliable representation of the small-scale variability of processes that govern both the water fluxes between soil, vegetation and atmosphere and the runoff generation. Inadequate discretization may lead to inaccurate parameterization of the spatial variability of key hydrological processes and may affect the quality of the hydrograph and soil moisture simulation. Two different approaches are generally used for the spatial discretization of a catchment: the grid based approach (WaSiM-ETH model) and the HRU or hydrotope approach (PREVAH model). Grid-based models can be more easily coupled with atmospheric models and allow for the explicit determination of the water flows from cell to cell. HRU-based models allow for the discretization of a catchment with an internally dynamical spatial resolution. In other words, the HRU size is smaller where the ensemble of soil, land surface and topography characteristics shows higher spatial variability (Zappa 2002). At equal spatial resolution (e.g. 100x100 m^2) HRU-based models require much less CPU-resources than grid-based models. HRU-based models might therefore be used at higher spatial resolution than grid-based models or for the simulation of large catchments with less CPU-time. As an example: the WaSiM-ETH application for the Dischmabach catchment (43.3 km^2) at 500x500 m^2 resolution required 3 times more CPU-time than the application of PREVAH, which was based on the aggregation of 4330 grid cells at 100x100 m^2 resolution into 334 HRUs. In this experiment, no significant difference in the quality of the runoff hydrograph was found between the two models (Gurtz et al. 2003a).

The sensitivity of the hydrograph simulation to the spatial resolution of physiographic input data was the research focus of a special study in both a pre-alpine catchment (Murg at the hydrometric gauge of Frauenfeld) and a high-alpine catchment (Dischmabach at the gauging station Kriegsmatten) (Zappa and Gurtz 2002). The impact of spatial resolution on the quality of hydrograph simulations was assessed at different resolutions (100x100 m^2 to 5x5 km^2). The overall quality of hydrograph simulations decreased gradually with decreasing resolution. Once resolution becomes coarser than 1x1 km^2, the quality of hydrograph simulations decreases rapidly.

In the pre-alpine Murg catchment, the decrease in simulation quality was very sensitive to the aggregation of land-use information. In other words, a realistic representation of landuse appears to be essential for accurately modelling runoff in the pre-alpine catchment. In case of the high-alpine Dischmabach catchment, a decrease in the quality of hydrograph simulations is closely coupled to the snowmelt season and to the resolution of the topographical information. Hence, in these types of catchments a realistic representation of snowmelt and topography is essential for accurate runoff modelling. In summary, the critical resolution for detailed spatially explicit hydrological simulations in alpine and sub-alpine catchments between 20 and 2000 km^2 is between 500x500 m^2 and 1x1 km^2 (Zappa 2002).

The necessary temporal resolution for runoff models strongly depends on the aim of the investigations. The dynamic hydrological response of mountainous catchments (e.g. storm events, flood peaks) can only be captured when the models are run at time steps on the order of one hour. For other applications, such as water balance investigations, a temporal resolution of one day may be sufficient.

2.3 Model calibration and verification

The evaluation of distributed hydrological models is limited by the availability of adequate data sets. The discharge at the catchment outlet is often the only available variable for the assessment of the model quality. However, the discharge represents an integral of the catchment response and therefore carries limited information on spatial variability of hydrological processes within the catchment. The quality of the model results has to be assessed by comparing model outputs with different observed variables (multiple-response verification). Available field observations for model evaluation are: soil moisture time series, lysimeter data (evapotranspiration, percolation, storage changes), latent heat flux, groundwater levels, snow depth and snow water equivalent, discharge and remotely sensed snow cover patterns (Gurtz et al. 2003a; Zappa 2002).

2.4 Meteorological data

For the modelling of hydrological processes up to six meteorological input variables are employed: precipitation, air temperature, global radiation, relative sunshine duration, wind speed and relative air humidity or water vapour pressure. Our investigations with distributed hydrological models in Swiss pre-alpine and alpine catchments are based on meteorological observations at weather stations and on rain gauge networks maintained by MeteoSwiss. The meteorological variables have to be interpolated. The procedure adopted for the spatial interpolation of observed meteorological information is mainly based on altitude-dependent regressions and/or inverse distance weighting (Schulla 1997; Jasper 2001).

The required meteorological information can also be obtained from simulations with climate models at regional scales (RCMs) (Kleinn 2002), numerical weather prediction models (NWPs) (Jasper et al. 2002) and general circulation models (GCMs) (Zappa 2002). In these cases, the co-ordinates of the grid elements of the GCM, RCM and NWP domains in the neighbourhood of the investigated area are considered as co-ordinates of virtual surface meteorological stations that are providing the required input for driving the hydrological model. The grid resolution in hydrological models is higher than in meteorological models, a downscaling of the simulated meteorological data to the grid of the hydrological grid model is necessary. For this purposes either bilinear interpolation (Jasper 2001) or a newly developed statistical downscaling (Kleinn 2002) might be used.

3. Climate change and mountain hydrology

3.1 Some specific remarks

One of the most spectacular hydrological impacts of the last 100 years of climate change is the strong glacier retreat in almost all mountain regions of the earth (see Haeberli, Dyrurgerov, this volume). A number of alpine glacier mass balance and energy balance studies emphasize the significance not only of temperature and net radiation but also of winter and summer snowfalls. Summer snowfalls strongly increase the albedo of the glacier surface with a strong effect on net radiation conditions and hence on melting rates and mass balance.

Hydrologically, glacier retreat indicates a negative storage change in these river catchments, which causes corresponding increases in the flow of glacier-fed streams. However, with ongoing glacier retreat a point is approached at which these rivers will change their trend and show a decrease in flow and regime changes from a glacial to a more snowmelt-dependent type with the time of maximum runoff occurring earlier in summer.

Snow cover variations are generally thought to be strongly related to air temperature (monthly, seasonal or even annual averages). However, the joint effects of precipitation and temperature during frequent but short precipitation events are critical for the percentage of snow in total precipitation and hence snow accumulation in the different elevation zones of mountain regions. Also, under mid-latitude climatic conditions a great portion of the total snow accumulation occurs during few major storm events. Hence, the air temperature during precipitation events and within the precipitation area become important input variables for a reliable simulation of future snow cover and glacier conditions. Unfortunately, such detailed scenario data are not yet available.

Figure 1 illustrates snow cover conditions at the end of the winter season (April 1st) in the Wägital basin, northern Swiss Alps, over the last 60 years. This is one of the longest time series of snow water equivalent. It is remarkable that, despite a positive temperature trend during the last 60 years, no clear decreasing trend in snow cover can be detected in the different altitudinal zones of this basin. This is in contrast to what we expected. The series of below-average snowpack for the period 1989-2002 could be the beginning of the expected trend. However this signal is still quite weak and could also represent one of the typical fluctuations that are often observed over a number of years in such long-term records. A similar situation is observed for the Claridenfirn (Swiss Glarner Alps), where at an altitude of 2900 m asl (period 1915-2001) the winter snow accumulation does not show any significant change (Glaciological Commission of the Swiss Academy of Science 2002). On the other hand, the net accumulation, as measured each year at the same point, is showing a weak trend towards decreasing amounts during the last 20 years, in agreement with the negative mass balances of glaciers in the European Alps.

Fliri (1992) has published a comprehensive overview of snow conditions in the Tirolean Alps (Austria) for the period 1895-1991, from which it is also not possible to discern a clear secular trend in winter snow conditions. At present, it is still poorly understood what drives variations of winter snow conditions in the Alps. However, studies on the synoptic-scale atmospheric forcing mechanisms of winter snow cover in the European Alps are underway. These will also include the role of the North Atlantic Oscillation (NAO) (Lifland 2003).

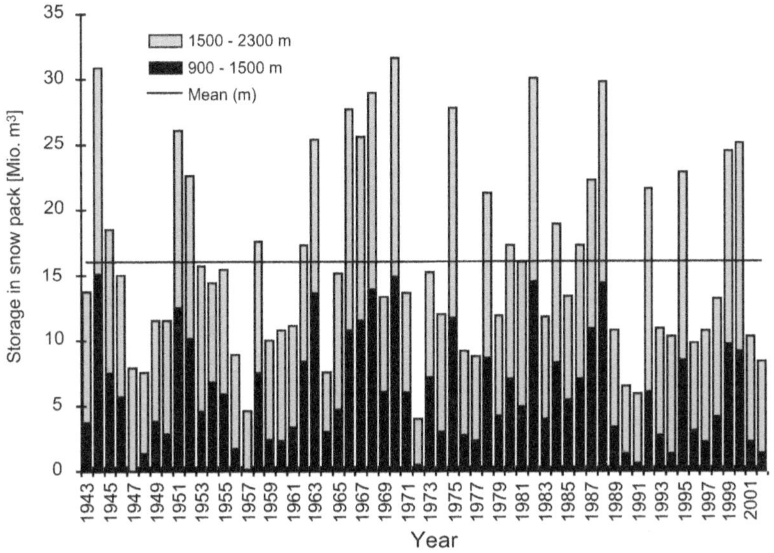

Figure 1: Changes in snowcover: Measurements of water storage volume in snow pack in the Wägital, Swiss Alps, for two altitudinal transects between 900 and 1500 m asl and between 1500 and 2300 m asl for the period 1943-2002 (Meteodat and IACETH 2002). All measurements were made on April 1st.

3.2 Examples of hydrological scenarios in a changing climate

Based on climatic scenarios for this century from various GCMs, the hydrological models described above were used to generate the corresponding hydrological scenarios (Gurtz et al. 1997; Schulla 1997; Kleinn 2002). Figure 2 presents some results of the scenario computations of hydrological climate change impacts for the pre-alpine subcatchment Frauenfeld/Murg (208.8 km^2 and 390-1031 m asl) and the alpine subcatchment Stein-Iltishag/Thur (81.3 km^2 and 850-2503 m asl). Both are situated in the Thur-River basin in the northeastern part of Switzerland. In these case studies (Grabs 1997), hydrological scenarios for the year 2050 (UKHI 2050) were simulated, using mean monthly changes between a control run and a scenario run of the model developed by the Climate Research Unit of the University of East Anglia. Model performance was evaluated with a control run for the time period 1981-1995 (Hulme et al. 1994). Scientists in different countries adopted these scenarios in an EU-project on climate change impact in the Rhine river basin (Grabs 1997).

The simulations predict distinct changes in precipitation seasonality in the study region, with higher winter and lower summer precipitation. Also, model predictions indicate an increase of temperature, wind speed and global radiation over the whole year (Grabs 1997). Further, climate change is expected to increase evapotranspiration and decrease runoff generation from snowmelt. These changes are connected with increased amounts of liquid precipitation and an increase in snow line altitude, with largest relative changes occurring in alpine catchments. In addition, model simulations show increased seasonal and month-to-month variability of runoff in both pre-alpine and alpine subcatchments. In Figure 2a, 2b and 2c the changes of seasonal runoff regimes for high discharge, medium discharge and low discharge are presented. With the exception of some winter months, there is a distinct decrease of high, mean and low discharge in the pre-alpine catchment. In the alpine catchment, model scenarios suggest a strong increase of high, mean and low discharge in winter and spring and a clear decrease in summer and autumn. These changes result in a shift to more rainfed runoff regimes in the alpine zone. For example, less summer precipitation in combination with higher evapotranspiration rates causes decreasing soil moisture and groundwater recharge, and thus decreased low flows. The statistical analysis of the

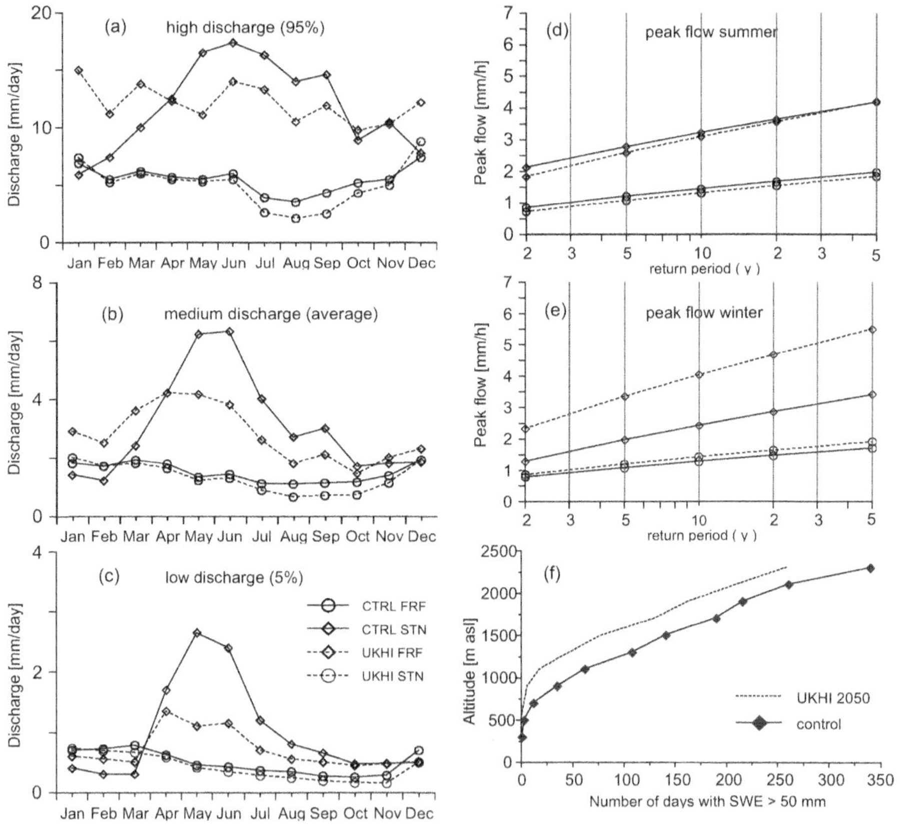

Figure 2: Left: Daily discharges for the pre-alpine Frauenfeld (FRF, circles) and the alpine Stein (STN, diamonds) catchment for high discharges (95% - in 95% of the time discharge is equal or less than these values), medium discharges and low discharges (5% - discharge exceeds these values in 95% of the time) under present climate (control run, CTRL continuous line) and under climate scenario conditions (UKHI 2050, dashed line). Right: Results of peak flow statistical analysis for Frauenfeld (FRF) and Stein (STN): summer peak flows (d) and winter peak flows (e). A return period of 10 years means that a particular peak flow is expected to occur once during a ten-year period. (f) Altitude dependence of annual average snow duration exceeding 50 mm snow water equivalent (SWE) under current climate and in a climate change scenario (UKHI 2050).

peak flows for the two catchments, shown in Figure 2d and 2e, indicates that winter peak flows will increase more strongly in the alpine than in the pre-alpine region for all return periods. Higher winter precipitation in combination with a displacement of the 0°C isotherm to higher elevations will cause substantial rises in winter peak flows in the future. In contrast, decreased precipitation and soil moisture in summer lead to a decrease in summer floods. Summer peak flows are lower for all return periods (see figure captions for further explanations) in the pre-alpine catchment, while peak flows in the alpine catchment decrease only for the shorter return periods. The opposite applies for low flows.

Model predictions also show a distinct impact of climate change on snow-cover. The mean number of days per year with snow coverage exceeding 50 mm of snow water equivalent in the Thur-River basin is reduced by about 30 days in the altitudinal range between 1000 and 2500m asl (Figure 2f).

Figure 3 gives insights into the spatial-temporal behaviour of both water balance (precipitation P minus evapotranspiration ET minus runoff R, as average between 1981 and 1996) and runoff generation for the entire Thur basin under present and changed climate conditions. Climate change scenarios are based on IPCC scenarios (IPCC-A, for details see Gurtz et al. 1997). The strongest changes occur from March to July in the altitudinal range between 1500 and 2500 m asl. In the future, discharge regimes that are typical of the Swiss plain might only be found at higher elevation ranges. The elevation range of snowmelt dependent regimes will rise from 1500 to 1800 m asl. Hence, in the Alps, the elevation ranges above 1500 m asl appears to be most sensitive to the hydrological impacts of climate change with corresponding impacts on the socio-economic conditions in these mountain regions. Several Swiss winter resorts below 1500 m asl might be forced to give up snow related tourism or to increase the use of artificial snow to ensure a sufficient snow cover throughout the winter. Many questions on the water management in alpine landscapes may arise. A sustainable compromise has to be found to optimise the use of the available water resources for hydropower production, agriculture, industry, supply of freshwater, and production of artificial snow.

3.3 Directions of future research

The impacts of climate change on mountain environments may consist of strongly differentiated hydrological responses in time and space. For a thorough analysis of climate change impacts on the hydrological cycle further investigations are needed, especially in mountainous regions:

Future developments in the application of high-resolution regional climate models (RCMs) will be critical for improving our predictions of climate change impacts on mountain hydrology. GCM simulations could be used to force RCMs. These regional climate models can provide meteorological input data for hydrological models, which is an important prerequisite for the simulation of climate change impacts on hydrological processes.

In future hydrological studies, changes in synoptic climatology should be taken into account (Kleinn 2002). The development of fully coupled climate-hydrological models that consider feedback mechanisms is important for the improvement of model simulations. Nevertheless, off-line (no feedback from the hydrological model to the atmospheric model) coupling of atmospheric and hydrologic models is important for further development of the research cooperation between climatologists and hydrologists in the future.

Better tools for an effective calibration and evaluation of macroscale applications of hydrological models have to be developed. For example, the different aspects of event-based precipitation and their effects on regional snow cover formation deserve particular attention. Also, it is critical to gain a better understanding of the spatial distribution of extreme precipitation events causing extreme floods and of their potential regional variations in the future (Frei et al. 1998; 2000; Schär et al. 1998; Schär and Frei, this volume). These research efforts should also take into account changes in land cover and glacier extent, including the resulting effects on catchment hydrology.

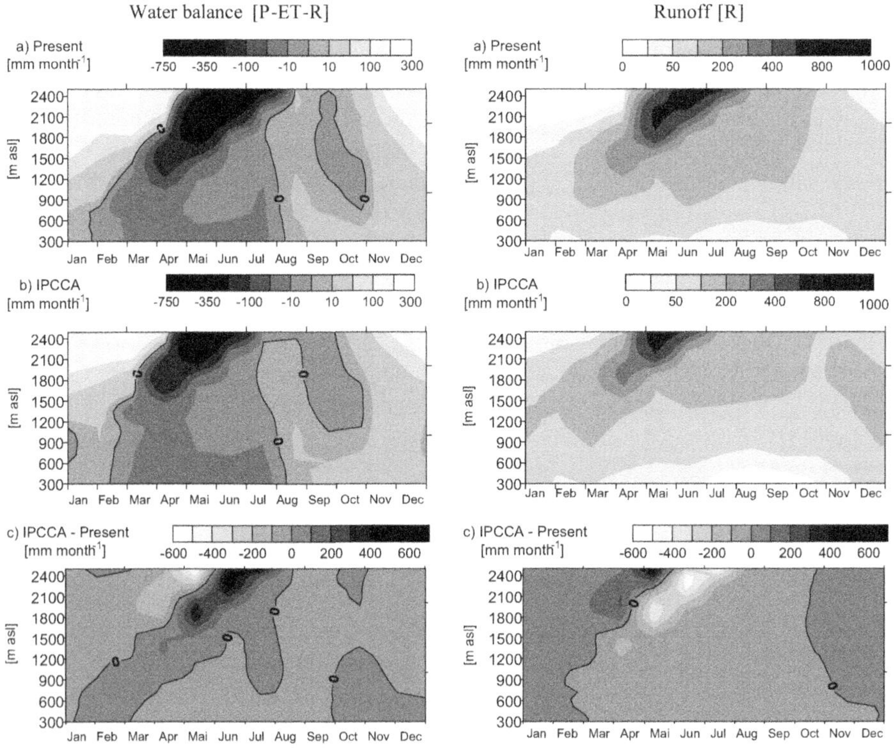

Figure 3: Monthly averages of water balance (P-ET-R) and runoff generation R, depending on the altitude, in the whole Thur-River basin under a) present, and b) changed climate conditions (IPCCA). c) Changes in water balance and runoff (IPCCA - Present), shown as the difference between the values under present and changed climate conditions. Both the control and the climate change experiments were run over a period of 16 years. A positive water balance means that water resources are stored in the basin. A negative water balance means that previously stored water resources are released. The long-term storage change is neglected.

The concentration of precipitation and freshwater resources in mountains (Viviroli et al. 2003) warrants an increased focus on mountain hydrology in future global change research. Because of the critical role of water in the climate system, and its existential significance as a basis of live and a threat to civilisations, research in the field of climate and water deserves high priority.

4. References

Aschwanden, H., and Weingartner, R. (1985). „Die Abflussregimes der Schweiz." Publikation Gewässerkunde 65, Bern.

Fliri, F. (1992). "Der Schnee in Nord- und Ost-Tirol. 1895-1991." Bd. 1, Universitaetsverlag Innsbruck, Innsbruck.

Frei, C., Schär, C., Lüthi, D., and Davies, H. C. (1998). Heavy precipitation processes in a warmer climate. *Geophysical Research Letters* **25**, 1431-1434.

Frei, C., Davies, H. C., Gurtz, J., and Schär, C. (2000). Climate dynamics and extreme precipitation and flood events in Central Europe. *Integrated Assessment* **1**, 281-299.

Glaciological Commission of the Swiss Academy of Science (2002). „The Swiss glaciers." Glaciological Report 121 and 122. VAW, Federal Institute of Technology, ETH Zurich.

Grabs, W., Ed. (1997). "Impact of climate change on hydrological regimes and water resources management in the Rhine basin." International Comission for the Hydrology of the Rhine Basin (CHR). CHR-Report No. I-16.

Gurtz, J., Baltensweiler, A., Lang, H., Menzel, L., and Schulla, J. (1997). "Auswirkungen von klimatischen Variationen auf Wasserhaushalt und Abfluss im Flussgebiet des Rheins." Abschlussbericht Nationales Forschungsprogramm 31: Klimaänderungen und Naturkatastrophen. vdf-Hochschulverlag ETH Zurich, Zurich.

Gurtz, J., Baltensweiler, A., and Lang, H. (1999). Spatially distributed hydrotope-based modelling of evapotranspiration and runoff in mountainous basins. *Hydrological Processes* **13**, 2751-2768.

Gurtz, J., Zappa, M., Jasper, K., Lang, H., Verbunt, M., Badoux, A., and Vitvar T. (2003a). A comparative study in modelling runoff and its components in two mountainous catchments. *Hydrological Processes* **17**, 297-311.

Gurtz, J., Zappa, M., Verbunt, M., and Jasper, K. (2003b). Advanced applications of distributed hydrological models in mountainous catchments. *In* "Climate and hydrology in mountain areas." (C. De Jong, D. Collins, and R. Ranzi, Eds.). John Wiley & Sons, Chichester, UK (submitted).

Hock, R. (1999). Distributed temperature-index ice- and snowmelt model including potential direct solar radiation. *Journal of Glaciology* **45**, 101-111.

Hulme, M., Conway, D., Brown, O., and Barrow, E. (1994). "A 1961-1990 baseline climatology and future climate change scenarios for Great Brittain and Europe. Part III: Climate change scenarios for Great Britain and Europe". Climatic Research Unit, University of East Anglia, Norwich.

IPCC (1996). "Climate change 1995. Second assessment report of the Intergovernmental Panel on Climate Change: Impacts, adaptions and mitigation of change." WMO/UNEP, Cambrigde University Press.

IPCC (2001). "Climate change 2001. Third assessment report of the Intergovernmental Panel on Climate Change: Impacts, adaptions and mitigation of climate change." WMO/UNEP, Cambrigde University Press.

Jasper, K. (2001). "Hydrological modelling of alpine river catchments using output variables from atmospheric models." Dissertation No. 14'385, ETH Zurich, Institute for Atmospheric and Climate Science.

Jasper, K., Gurtz, J., and Lang, H. (2002). Advanced flood forecasting in Alpine watersheds by coupling meteorological observations and forecasts with a distributed hydrological model. *Journal of Hydrology* **267**, 38-50.

Kleinn, J. (2002). "Climate change and runoff statistics in the Rhine basin: A process study with a coupled climate-runoff model." Dissertation No. 14663, ETH Zurich, Institute for Atmospheric and Climate Science.

Klok, L., Jasper, K., Roelofsma, K.P., Badoux, A., and Gurtz, J. (2001). The application of complex distributed hydrological models to a heavily glacerized Alpine river catchment. *Hydrological Sciences*

Journal **46**, 553-570.

Lifland, J. (2003). The North Atlantic Oscillation: Climatic significance and environmental impact. *EOS* **84**, 73.

Menzel, L. (1997). "Modellierung der Evapotranspiration im System Boden-Pflanze-Atmosphäre." Zürcher Geographische Schriften 67, Geographical Institute ETH Zurich.

Meteodat and IACETH (2002). "Joint longterm observation project on Wägital snow conditions." Annual Report, Meteodat GmbH Zurich and Institute for Atmospheric and Climate Science ETH Zurich.

Monteith, J. L. (1965). Evaporation and environment. *In* "The state and movement of water in living organism," Proceedings of the 19th Symposium, Society of Experimental Biology, Cambridge University Press, London, 205-234.

Ohmura, A., and Wild, M. (2002). Is the hydrological cycle accelerating? *Science* **298**, 1345-1346.

Ohmura, A. (2001). Physical basis for the temperature-based melt-index method. *Journal of Applied Meteorology* **40**, 753-761.

Schär, C., Davies, T. D., Frei, C., Wanner, H., Widmann, M., Wild, M., and Davies, H. C. (1998). Current alpine climate. *In* "Views from the Alps: Regional perspectives on climate change." (P. Cebon, U. Dahinden, H. C. Davies, D. M. Imboden, and C. Jäger, C., Eds.), pp. 21-72. MIT Press, Boston.

Schulla, J. (1997). "Hydrologische Modellierung von Flussgebieten zur Abschätzung der Folgen von Klimaänderungen." Zürcher Geographische Schriften 69, Geographical Institute ETH Zurich.

Verbunt, M., Gurtz, J., Jasper, K., Lang, H., Warmerdam, P., and Zappa, M. (2003). The hydrological role of snow and glaciers in alpine river basins and their distributed modelling. *Journal of Hydrology* (accepted).

Viviroli, D., Weingartner, R., and Messerli, B. (2003). Assessing the hydrological significance of the World's mountains. *Mountain Research and Development* **23**, 32-40.

Zappa, M., and Gurtz, J. (2002). The spatial resolution of physiographic data as sensitive variable for distributed hydrological simulations in prealpine and alpine catchments. *In* "Water resources and environment research, Proceedings of ICWRER 2002." Band 28, Volume I, 101-105.

Zappa, M. (2002). "Multiple-response verification of a distributed hydrological model at different spatial scales." Dissertation No. 14895, ETH Zurich, Institute for Atmospheric and Climate Science.

Zappa, M., Pos, F., Strasser, U., Warmerdam, P., and Gurtz, J. (2003). Seasonal water balance of an alpine catchment as evaluated by different methods for spatially distributed snow melt modelling. *Nordic Hydrology* **34** (in print).

Effects of Climate Variability and Change on Mountain Water Resources in the Western U.S.

L. Ruby Leung
Pacific Northwest National Laboratory, P.O. Box 999, Richland, WA 99352, USA
phone 1-509-372-6182, fax 1-509-372-6168, e-mail ruby.leung@pnl.gov

Keywords: Climate variability and change, Regional climate modeling, Water resources impacts, Western U.S.

1. Introduction

The western U.S. derives its water resources predominantly from cold season precipitation and storage in snowpack along the narrow Cascades and Sierra ranges, and the Rocky Mountains. Hydroclimate is modulated by the diverse orographic features across the region. Precipitation and runoff generally peak during winter and spring respectively, whereas water demand is highest during the summer. Such phase differences between water supply and demand create a necessity for water management, which is reflected by major developments of reservoirs and dams that regulate irrigation, hydropower production, and flood control during the past 50 years. Because water resources have been essential to the economic development and environmental well being of the western states, it raises concerns when recent studies suggest that global warming may exert significant impacts on snowpack and streamflow, which may seriously affect water resources in the western U.S. in the 21st century (e.g. Leung and Ghan 1999; Leung and Wigmosta 1999; Mile et al. 2000; Leung et al. 2003a; Miller and Kim 2000).

To understand how climate change may affect mountain water resources, we have taken the approach of an "end-to-end" assessment where simulations of current and future climate, produced by global climate models (GCMs), are downscaled using regional climate models (RCMs). These models then provide regional atmospheric

U. M. Huber et al. (eds.), Global Change and Mountain Regions, 355-364.

conditions for assessing climate change impacts on water resources using hydrologic models (e.g. Leung and Wigmosta 1999; Miller and Kim 2000) and water management models (e.g. Hamlet and Lettenmaier 1999; Payne et al. 2002). This suite of models guides us from a comprehensive and global view of the effects of greenhouse warming on the atmosphere-ocean-land system to regional climate change, to hydrologic response in river basins and watersheds, and finally to reservoir management. Water management models convert hydrologic response to impacts on water management objectives and enable the evaluation of adaptation strategies through modifications to existing reservoir operating rules.

Because of the large mismatch between the spatial scales resolvable by GCMs (200-400 km) and that needed to accurately represent hydrologic processes in river basins (less than 10 km), scaling is an important issue in transitioning between models in the end-to-end assessment approach. The use of RCMs, known as dynamical downscaling, has been essential in providing regionally specific climate information for impact assessment. Leung and Ghan (1995; 1998) developed a novel approach to modeling subgrid-scale orographic precipitation to improve the effectiveness of dynamical downscaling. Their method is particularly useful for linking global or regional climate models with hydrologic models (Leung et al. 1996; Leung and Wigmosta 1999) to simulate hydrologic processes in mountainous regions.

Although downscaling is an important step in the end-to-end assessment approach, it requires large computational resources to produce regional climate information by the dynamical downscaling method. It is important to evaluate whether downscaling indeed provides improved accuracy or spatial specificity that are critical in the end-to-end assessment (Leung et al. 2003b). To partly address this issue, and to illustrate the sensitivity of mountain water resources to climate variations and change, this overview summarizes recent findings based on our work in the western U.S. As discussed below, much can be learned about climate sensitivity and human-induced climate change through our understanding of natural climate variations and their effects.

2. Climate variations and effects on streamflow

The cold season climate of the western U.S. is strongly affected by variations in the atmosphere-ocean system known as the El-Niño-Southern Oscillation (ENSO), which has a time scale of 2-7 years. During the warm phase of ENSO (El Niño), large-scale atmospheric circulation favors warm-dry winter/spring conditions in the Pacific Northwest and cool-wet weather in the Southwest (e.g. Cayan 1996). The opposite is true during the cool phase of the ENSO (La Niña) cycle. The dynamical downscaling of the end-to-end assessment approach can be tested to determine how well these large-scale climate variations and their regional effects (e.g. Cayan et al. 1999; Redmond and Koch 1991) can be simulated.

Leung et al. (2003c, d) performed a 20-year regional climate simulation of the western U.S. driven by realistic large-scale atmospheric circulation and sea surface temperatures from the National Centers for Environmental Prediction (NCEP) and National Center for Atmospheric Research (NCAR) global reanalysis data set for

1981-2000. In the NCEP/NCAR reanalysis dataset (Kistler et al. 2001), observations were assimilated with a data assimilation system that includes the NCEP GCM and an analysis scheme to provide gridded atmospheric fields that are dynamically consistent with the observations. The regional climate model was based on the Penn State/NCAR Mesoscale Model (MM5) (Grell et al. 1993). Figure 1a and b shows the El Niño precipitation anomaly, which was calculated based on the NCEP/NCAR reanalyses and our regional simulations. Precipitation anomaly is defined as the difference between the mean conditions of the six El Niño years (1983, 1987, 1988, 1992, 1995, and 1998) and the long-term average (1981-2000). The NCEP/NCAR reanalysis data capture the large-scale dry conditions in the Northwest and the wet conditions in California during El Niño years. Downscaling produces more complex regional structures, including the large wet anomaly in coastal northern California and along the Sierra, and the wet-dry-wet anomaly pattern in the Olympic Mountains, along the Cascades, and on the east side of the Cascades in the Pacific Northwest.

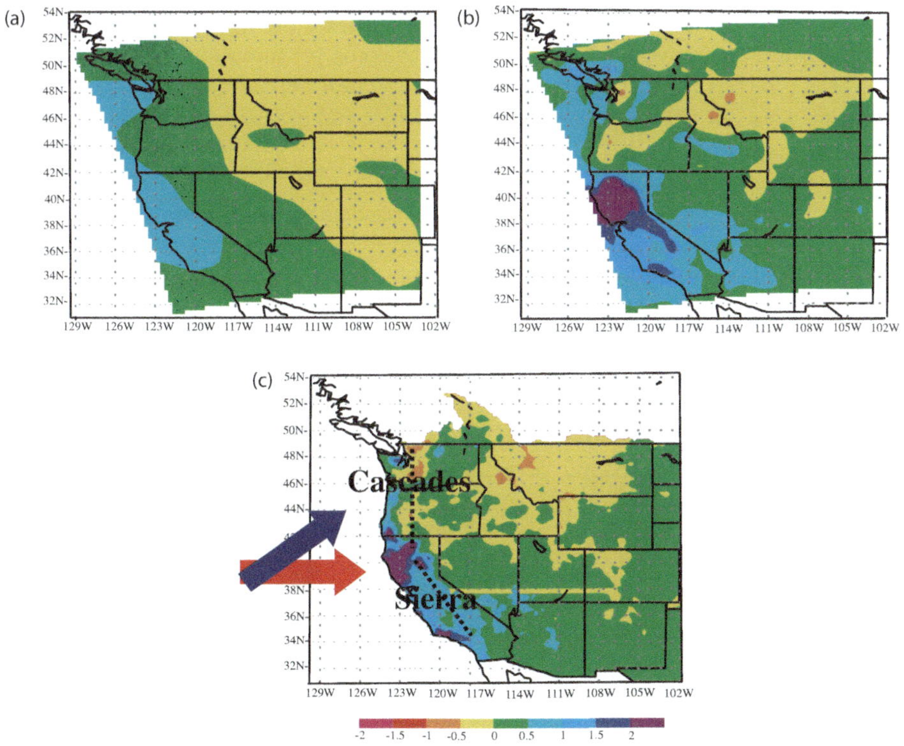

Figure 1: El Niño precipitation anomalies based on (a) the NCEP/NCAR reanalysis data set, (b) the RCM simulation, and (c) meteorological observations for January-February-March of 1981-2000. Color contours are in 0.5 mm/day intervals. Figure 1c also shows a schematic of the large-scale circulation patterns (arrows) and the orientation of the Cascades and Sierra ranges (dotted lines). The dominant winds during El Niño years are southwesterly, shown by the blue arrow, and are associated with a larger amount of atmospheric moisture. During normal years, westerly winds (red arrow) are more dominant, which typify drier conditions.

A comparison of the regional simulation (Fig. 1b) with meteorological observations (Fig. 1c) shows that the simulated regional patterns are consistent with observations. Figure 1c shows a schematic of the atmospheric flow patterns in the Pacific Northwest that explains the spatial distribution of regional El Niño precipitation anomalies. During El Niño years, a higher frequency of flows from the southwesterly direction occurs in response to changes in the location and intensity of the Aleutian Low over the North Pacific Ocean. More abundant atmospheric moisture typically accompanies the southwesterly flows, which originate near the tropical Pacific Ocean. This leads to widespread wetter conditions in the western U.S. However, on the scale of individual mountain ranges, orographic forcing can lead to more heterogeneous precipitation patterns. Orographic forcing is maximized for flows that are perpendicular to a mountain range. Consequently, in the North-South oriented Cascades range, westerly flows favor stronger orographic precipitation. During El Niño years, westerly flow is less frequent and southwesterly flow becomes dominant. Hence, the incoming air mass no longer intercepts the Cascades at a perpendicular angle and precipitation is therefore reduced along the windward (or west) side of the mountain range. In contrast, a shift towards southwesterly flows is favorable for precipitation on the Olympic Mountain, which is circular and does not possess a preferred flow axis, and on the east side of the Cascades because of increased moisture and reduced orographic blocking associated with southwesterly flows. The RCM correctly simulated the wet-dry-wet anomaly pattern associated with the complex terrain in the Pacific Northwest.

In California, both the increased moisture associated with southwesterly flows and the change in dominant wind direction from westerly to southwesterly during El Niño years are favorable for increased orographic precipitation along the northwest-southeast oriented Sierra Range. Hence, much larger wet anomalies are found in northern California during El Niño years. The RCM also correctly simulated a larger wet anomaly in the region, although the magnitude of this anomaly is too large owing to a general wet bias simulated in the Sierra Range.

Figure 2 shows the observed streamflows in the Columbia River Basin (CRB) and the Sacramento-San Joaquin River Basin (SSJ) during El Niño years compared to the long-term average over the last 45 years. These basins occupy large areas in southwestern Canada, the U.S. Pacific Northwest, and California and are a major source of water for irrigation and hydropower generation. Temperature and precipitation changes associated with ENSO strongly affect snowpack and streamflow in these river basins. In the CRB, temperature anomalies of 0.5-1.5°C and precipitation anomalies between –0.5 and 0.5 mm/day during the cold season of El Niño years cause a decrease in the magnitude of peak flows during early summer. In the SSJ, warmer temperature and increased precipitation during El Niño winters contribute to more runoff during both winter and early spring. Further, Leung et al. (2003d) found streamflow anomalies in smaller river basins that are consistent with the regional precipitation anomalies, such as the wet-dry-wet pattern in the Northwest. Hence, climate variations during ENSO events left a clear signature in the streamflow of mountain river basins in this region.

This analysis confirms that the regional El Niño precipitation anomalies, simulated by the RCM for the western U.S., are realistic. More importantly, our model results show that the RCM reproduces the observed climate sensitivity to perturbations in the large-scale atmospheric and ocean conditions. The RCM simulation closely mirrors the regional-scale precipitation anomalies that result from interactions between changes in large-scale circulation and regional topography during El Niño years. These regional-scale precipitation and temperature anomalies are important for defining streamflow anomalies in river basins. This suggests that downscaling is likely very important in developing realistic regional climate information for assessing climate change impacts on water resources in the mountainous western U.S.

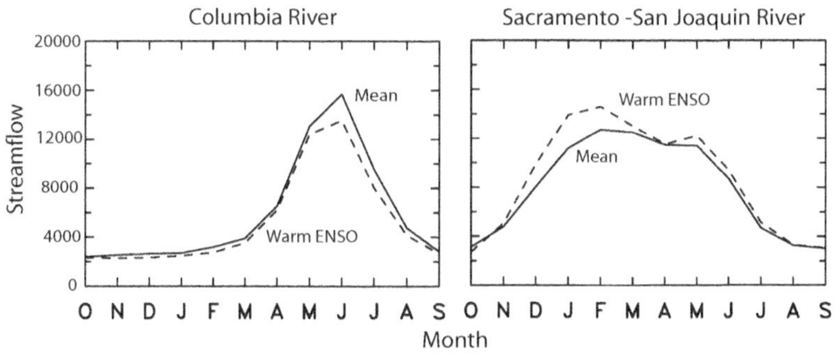

Figure 2: Observed streamflow (in m^3/s) at the Columbia River (left panel) and Sacramento-San Joaquin River (right panel): The average streamflow over the warm ENSO (El Niño) years (1952, 1958, 1959, 1964, 1966, 1969, 1970, 1973, 1977, 1978, 1980, 1983, 1987, 1988, 1992, 1995) is compared with the long-term mean over the 1950-1995 interval.

3. Climate change and hydrologic impacts

Using the end-to-end assessment approach described above, Leung et al. (2003a) carried out a numerical experiment that simulates potential climate changes in the western U.S. with global and regional climate models. To estimate uncertainty, their study used an ensemble modeling technique to capture climate variability. Three ensemble coupled ocean-atmosphere GCM simulations were performed based on the National Center for Atmospheric Research and Department of Energy (NCAR/DOE) Parallel Climate Model (PCM) (Washington et al. 2000). The simulations were initialized with ocean conditions of 1995 that were produced by data assimilation using an ocean circulation model. Different initial atmospheric conditions were used in the simulations to produce an ensemble of possible future climate, following the same emission scenario from 1995 through 2100. Changes in atmospheric greenhouse gases and aerosol concentrations followed the business-as-usual scenario (Dai et al. 2001; 2002). Hence, the concentration of CO_2 is roughly 505 ppmv in 2050 compared to 360 ppmv in the 1990s. The RCM, based on the Penn State/NCAR MM5, was used to downscale each GCM simulation. The model was run for 20 years under present climate conditions (control climate) and for 20 years under mid-century climate conditions (future climate).

Comparing the future climate with the control, temperature will have increased by 1-2°C in the western U.S. by mid-century (2040-2060). Similar to previous findings (e.g. Giorgi et al. 1997; Leung and Ghan 1999), stronger warmings were found at higher altitudes as a result of snow-albedo feedback effects; that is, reduced snowpack in the warmer climate induces further warming by reflecting less solar radiation at the surface. Precipitation changes are small and highly variable both temporally and spatially. Because of warmer temperatures, climate change effects are expressed mainly in the reduction of snowpack along the coastal mountains, such as the Cascades and Sierra. The warming reduces snowpack by over 50% in the maritime mountains (Fig. 3). In contrast, in the Rocky Mountains, snowpack change is governed by temperature as well as precipitation changes. A warming of 1-2°C in the continental climate regime is not sufficient to significantly reduce snowpack. Rather, increased snowpack reflects the small increase in precipitation during winter in the future climate.

As changes in temperature and precipitation affect snowpack in the mountains, they also alter the amount and timing of runoff in river basins of the western U.S. In the CRB, current streamflow peaks during April, which corresponds to peak snowmelt rather than precipitation, which is at a maximum in November – January. By mid-century, warming increases rainfall over snowfall. At the same time, warming also reduces snowpack. Overall, the frequency of rain-on-snow events, which are the leading cause of flooding in the basin, is increased by 5-10% during winter and spring. Hence, our results show a higher likelihood of wintertime flooding and reduced runoff in the summer, when water is in highest demand.

The SSJ basin is located further south in the subtropical zone. Streamflow peaks in February and March, which corresponds to the timing of maximum precipitation. There is a small secondary peak in streamflow during April that corresponds to snowmelt runoff. With an increase in temperature and precipitation, peak runoff in February increases because more precipitation (and a higher percentage in the form of rain) contributes directly to runoff. With the reduction in snowpack along the Sierra (see Fig. 3) and warmer temperature, which promotes earlier snowmelt, streamflow between March and May will be reduced in the future.

Such changes in the amount and timing of streamflow will affect water resources in the western U.S. Based on the global and regional climate change scenarios produced by the PCM and RCM discussed here, Payne et al. (2002) and VanRheenen et al. (2002) investigated how climate change may affect water resources in the CRB and SSJ. These studies used a macroscale hydrologic model driven by both the control and future climate scenarios of the PCM and RCM to simulate streamflows in the river basins. The streamflow simulations were then utilized by the water resources system models to simulate performance of the water system with regard to hydropower production and environmental functions. In the CRB, changes in streamflow lead to increased competition for regulated storage in the reservoir system between hydropower production and environmental flow targets for fisheries habitat protection and enhancement. Adapting reservoir management (e.g. through changes in flood evacuation and refill schedules, changes in the seasonality of energy demand in a warmer climate, and increasing storage allocation to meet environmental flow targets) to the altered streamflows can mitigate most of the negative impacts of climate change on fish, but at the expense of severe hydropower losses. In the SSJ, changes in

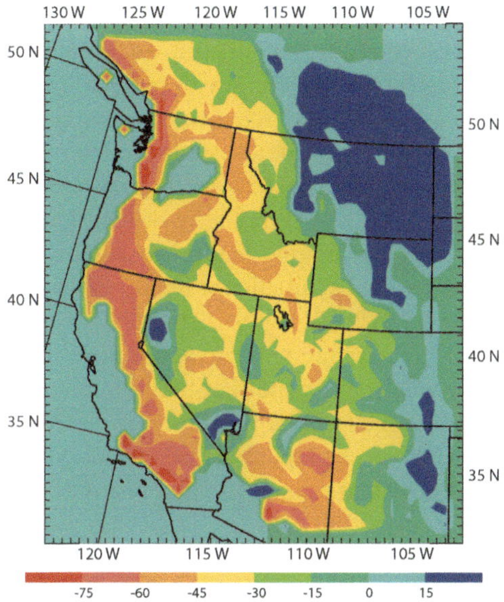

Figure 3: Percentage change in annual mean snowpack comparing results based on regional climate simulations for the mid-century (2040-2060) with the control simulation of present conditions. Color contours are in 15% intervals. Snowpack is generally reduced by over 60% along coastal mountain ranges. The change in snowpack in areas at an elevation below 600 m (snow free zone) is set to 0.

streamflow severely degrade system performance such that it is nearly impossible to maintain current system performance even with adaptive management.

To put these results in perspective, the climate change scenarios described here are based on climate projections by the PCM that shows relatively low climate sensitivity compared to other GCMs (Barnett et al. 2001). Indeed, the PCM simulated a warming of 1-2°C in the western U.S., which is small compared to results reported by other models (e.g. IPCC 2001). Yet the impacts of climate change on mountain water resources are enough to warrant concern and justify further investigations. It is likely that similar studies using climate projections based on GCMs with higher climate sensitivity may reveal even larger climate change impacts on the hydrologic and water management systems.

4. Priorities for future research

Our studies on the effects of climate variability and change suggest that hydrologic conditions in the mountainous western U.S. are very sensitive to variations or changes in temperature and precipitation. Leung et al. (2003a) showed that, at higher spatial resolution, the regional model simulated the spatial distribution of precipitation, temperature, and snowpack much more realistically than the global climate model. In the climate change scenario, differences between the spatial distribution of temperature and precipitation signals produced by the regional and global climate models were sufficiently large to induce major differences between the hydrologic responses simulated by the two models.

As computing resources continue to grow at an accelerating pace, there are increasing efforts in high-resolution modeling with regional or global climate models and hydrologic models. An important issue that needs to be addressed is: at what spatial resolution are climate and hydrologic processes adequately represented in numerical models? In mountainous regions, such as the western U.S., this depends partly on the spatial scale of the terrain features, the complexity of the models (including whether subgrid land surface heterogeneity and physical processes are treated in the models), and the spatial resolution of the data available for model inputs and evaluation. Leung and Qian (2003) and Leung et al. (2003e) show that higher spatial resolution typically improves the simulation of temperature and snow, but does not necessarily improve the simulation of precipitation. They found that precipitation is overly amplified on the windward side of mountains as model resolution increases. It appears that each model, together with the set of model input parameters and verification criteria, may yield optimal results within a certain range of spatial resolution.

As physics parameterizations and treatments of subgrid processes in climate models improve, the dependence of climate simulations on model resolution may be reduced. For example, Leung and Ghan (1995; 1998) found that the simulations of precipitation and snow in regions of complex terrain were more accurate when the RCM was applied at 90 km resolution, with the subgrid parameterization of orographic precipitation, than simulations performed at 30 km resolution but without the subgrid parameterization. Furthermore, with the subgrid parameterization, simulations become less dependent on spatial resolution. The issue of spatial resolution will remain important as high performance computing resources are becoming much more accessible.

Another important issue that needs to be addressed is the reliability of the end-to-end assessment approach in estimating climate change scenarios and hydrologic impacts. Clearly, the numerical models used in each step of the assessment introduce some bias and uncertainty because of inherent variability of the modeled systems and errors in representing the physical processes. In addition, bias and uncertainty can propagate nonlinearly through the chain of models to introduce more uncertainty into the final assessment.

One approach to characterize the uncertainty in the modeled climate change scenarios is the use of the ensemble simulation technique. This technique includes simulations based on multiple models, which are initialized by using multiple atmospheric states to capture the uncertainty associated with both climate variability and model formulation. Based on the large number of climate simulations, a reliability measure can be developed to quantify the uncertainty associated with the climate change scenarios. An example is the "reliability ensemble averaging" (REA) method developed by Giorgi and Mearns (2002). The REA method defines reliability by assessing both model performance based on the ability of climate models to reproduce present-day climate and convergence of simulated climate change across models. The modeling work of different research groups should be better coordinated to ensure similar approaches for estimating reliability of climate change scenarios, which can be very useful in quantifying uncertainty in climate change impacts.

Lastly, hydrologic impacts are usually not directly estimated from the soil moisture or runoff simulated by climate models, but from hydrologic models that are driven by global or regional climate change simulations. This one-way coupling of climate and hydrologic models is preferred because neither global nor regional climate models can normally be applied at spatial resolutions that are sufficient to accurately simulate hydrologic processes. Furthermore, in one-way coupling, bias in the simulated temperature and/or precipitation can be removed to yield more realistic hydrologic simulations in the control climate. However, in mountain regions, the lack of feedback between climate and hydrologic processes in one-way coupling can introduce errors in the assessment. For example, the enhanced warming at high altitude cannot be correctly simulated by the one-way coupled climate and hydrologic models because the positive temperature feedback resulting from changes in snow cover simulated by the hydrologic model is not taken into account in this modeling approach. The climate model only accounts for snow feedback effects based on snow cover simulated by the climate model, which can be very different from that simulated by the hydrologic model because of differences in spatial resolution and model formulation. In the future, it is likely that fully coupled climate-hydrologic models can be applied at high spatial resolution to study climate change impacts. This will eliminate issues related to feedback effects, but possible large biases associated with surface temperature and precipitation in the climate simulations can still limit our investigations of hydrologic response. Much has been and will be learned by coordinating efforts of the kind described in this summary to test the full end-to-end assessment approach.

5. References

Barnett, T. P., Pierce, D. W., and Schnur, R. (2001). Detection of anthropogenic climate change in the world's oceans. *Science* **292**, 270-274.

Cayan, D. R. (1996). Interannual climate variability and snowpack in the western United States. *Journal of Climate* **9**, 928-948.

Cayan, D .R., Redmond, K. T., and Riddle, L. G. (1999). ENSO and hydrologic extremes in the western United States. *Journal of Climate* **12**, 2881-2893.

Dai, A., Wigley, T. M. L., Boville, B. A., Kiehl, J. T., and Buja, L. E. (2001). Climates of the 20th and 21st centuries simulated by the NCAR climate system model. *Journal of Climate* **14**, 485-519.

Dai, A., Washington, W. M., and Meehl, G. A. (2002). The ACPI climate change simulations. *Climatic Change* (submitted).

Giorgi, F., Hurrell, J. W., Marinucci, M. R., and Beniston, M. (1997). Elevation dependency of the surface climate signal: A model study. *Journal of Climate* **10**, 288-296.

Giorgi, F., and Mearns, L. O. (2002). Calculation of average, uncertainty range, and reliability of regional climate changes from AOGCM simulations via the "Reliability Ensemble Averaging" (REA) method. *Journal of Climate* **15**, 1141-1158.

Grell, G., Dudhia, J., and Stauffer, D. R. (1993). "A description of the fifth generation Penn State/NCAR Mesoscale Model (MM5)." NCAR Technical Note, NCAR/TN-398+IA, National Center for Atmospheric Research, Boulder, CO.

Hamlet, A. F., and Lettenmaier, D. P. (1999). Effects of climate change on hydrology and water resources in the Columbia River basin. *Journal of the American Water Resources Association* **35**, 1597-1623.

IPCC (2001). "Climate change 2001: The scientific basis." Cambridge University Press, Cambridge.

Kistler, R., Kalnay, E., Collins, W., Saha, S., White, G., Woollen, J., Chelliah, M., Ebisuzaki, W., Kanamitsu, M., Kousky, V., van den Dool, H., Jenne, R., and Fiorino, M. (2001). The NCEP-NCAR 50-year reanalyses: Monthly means CD-ROM and documentation. *Bulletin of the American Meteorological Society* **82**, 247-268.

Leung, L. R., and Ghan, S. J. (1995). A subgrid parameterization of orographic precipitation. *Theoretical and Applied Climatolology* **52**, 95-118.

Leung, L. R., Wigmosta, M. S. Ghan, S. J., Epstein, D. J., and Vail, L. W. (1996). Application of a subgrid orographic precipitation/surface hydrology scheme to a mountain watershed. *Journal of Geophysical Research* **101**, 12803-12818.

Leung, L. R., and Ghan, S. J. (1998). Parameterizing subgrid orographic precipitation and surface cover in climate models. *Monthly Weather Review* **126**, 3271-3291.

Leung, L. R., and Ghan, S. (1999). Pacific Northwest climate sensitivity simulated by a regional climate model driven by a GCM. Part II: 2xCO_2 simulations. *Journal of Climate* **12**, 2031-2053.

Leung, L. R., and Wigmosta, M. S. (1999). Potential climate change impacts on mountain watersheds in the Pacific Northwest. *Journal of the American Water Resources Association* **35**, 1463-1471.

Leung, L. R., and Qian, Y. (2003). The sensitivity of precipitation and snowpack simulations to spatial resolution via nesting in regions of complex terrain. *Journal of Hydrometeorology* (accepted).

Leung, L. R., Qian, Y., Bian, X., Washington, W. M., Han, J., and Roads, J. O. (2003a). Mid-century ensemble regional climate change scenarios for the Western United States. *Climatic Change* (accepted).

Leung, L. R., Mearns, L. O., Giorgi, F., and Wilby, R. (2003b). Workshop on regional climate research: Needs and opportunities. *Bulletin of the American Meteorological Society* **84**, 89-95.

Leung, L. R., Qian, Y., and Bian, X. (2003c). Hydroclimate of the western United States based on observations and regional climate simulation of 1981-2000. Part I: Seasonal statistics. *Journal of Climate* (in press).

Leung, L. R., Qian, Y., and Bian, X., and Hunt, A. (2003d). Hydroclimate of the western United States based on observations and regional climate simulation of 1981-2000. Part II: Mesoscale ENSO anomalies. *Journal of Climate* (in press).

Leung, L. R., Qian, Y., Han, J., and Roads, J. O. (2003e). Intercomparison of global reanalyses and regional simulations of cold season water budgets in the Western U.S. *Journal of Hydrometeorology* (accepted).

Mile, E. L., Snover, A. K., Hamlet, A. F., Callahan, B., and Fluharty, D. (2000). Pacific Northwest regional assessment: The impacts of climate variability and climate change on the water resources of the Columbia River Basin. *Journal of the American Water Resources Association* **36**, 399-420.

Miller, N. L., and Kim, J. (2000). Climate change sensitivity analysis for two California watersheds *Journal of the American Water Resources Association* **36**, 657-661.

Payne, J. T., Wood, A. W., Hamlet, A. F., Palmer, R. N., and Lettenmaier, D. P. (2002). Mitigating the effects of climate change on the water resources of the Columbia River basin. *Climatic Change* (submitted).

Redmond, K. T., and Koch, R. W. (1991). Surface climate and streamflow variability in the western United States and their relationship to large scale circulation indices. *Water Resources Research* **27**, 2381-2399.

VanRheenen, N. T., Wood, A. W., Palmer, R. N., and Lettenmaier, D. P. (2002). Potential implications of the PCM climate change scenarios for Sacramento-San Joaquin River basin hydrology and water resources. *Climate Change* (submitted).

Washington, W. M., Weatherly, J. W., Meehl, G. A., Semtner, A. J., Bettge, T. W., Craig, A. P., Strand, W. G. Jr., Arblaster, J., Wayland, V. B., James, R., and Zhang, Y. (2000). Parallel Climate Model (PCM) control and transient simulations. *Climate Dynamics* **16**, 755-774.

Part IV: Ecological changes

Ecological processes that affect mountain ecosystems and have an influence on their ability to provide goods and services occur over a wide range of temporal and spatial scales. Accordingly, a variety of methods and approaches is required to assess Global Change impacts on ecological processes in mountain regions. The following section of this book reflects this diversity, but it also documents that considerable progress has been made over the past years both with respect to 'sectoral' research activities as well as with respect to the integration of research across scales and disciplines.

In the first two contributions, *Körner* and *Bowman* provide an overall rationale for studying ecological processes under the impact of drivers of Global Change in mountain regions. Both papers emphasize the importance of species-specific responses and the fact that numerous feedbacks exist that complicate the analysis and often make it impossible to identify simple cause-effect relationships.

In view of such complex responses, which may also differ from one region to the other, it is highly useful to develop networks that foster collaborative research on mountain issues and allow for cross-site comparisons. *Pauli et al.* document the concept, approach, and future aims of the GLORIA (GLobal Observation Research Initiative in Alpine environments) project, which focuses on documenting and understanding the widespread upward shift of mountain vegetation using a set of sophisticated observational as well as modeling procedures. *Spehn & Körner* then introduce GMBA, the Global Mountain Biodiversity Assessment; this is an observational effort that also includes research on the relationship between biodiversity and ecosystem function, with a special emphasis on slope stability.

"Global Change" is sometimes mistaken as a synonym of "Global Warming". However, it is almost certain that at least over the coming few decades changes of land-use and bio–diversity (through extinctions and invasions) will have larger effects on mountain ecosystems than changes of climate or atmospheric chemistry. *Williams & Wahren* document the effects of the various drivers of Global Change in their case study of Australian mountain vegetation, and the contribution by *Tasser et al.* deals with land-use changes and their implications for ecological processes in the European Alps.

Understanding change at the ecosystem level needs to be based on an understanding of processes at finer scales to elucidate response patterns and potential feedback mechanisms. *Harte et al.* (as well as *Körner*, *Bowman* and *Baron et al.*) focus on such process studies, using montane meadows as a case study. *Harte et al.* also address the question whether observational altitudinal gradient studies might be used as a substitute for in-situ experimental manipulations.

Due to gravitational forces, the various terrestrial and aquatic ecosystems are tightly linked in mountain regions, and ecological assessments of Global Change impacts therefore need to go beyond the consideration of terrestrial ecosystems. *Baron et al.* discuss the implications of N fertilization for both terrestrial and aquatic systems in the Rocky Mountains and the linkages between these two sub-systems. *Vinebrooke & Leavitt* focus on high mountain aquatic biota, the likely impacts of Global Changes on these systems, and their suitability as indicators.

The need for considerations across classical system boundaries is paralleled by the need to transgress spatial and temporal scales in our attempts to understand and predict Global Change impacts. In their contribution, *Schimel & Braswell* emphasize that to understand mountain system dynamics with respect to carbon relationships, we need to consider the watershed (or, as they call it, the "carbonshed") scale. Along the same line, *Shugart* provides a review of the potential of remote sensing methods for better understanding large-scale patterns and processes in mountain regions.

Many of these contributions mention the need for the integration of observational and experimental work with modeling studies. *Guisan et al.* review the state of modeling vegetation distribution above treeline, and make proposals towards long-term monitoring networks that would serve to integrate across disciplinary boundaries. *Bugmann et al.* review modeling efforts with a focus on forested systems, and discuss the problems and potentials for integration of modeling efforts across various disciplines. *Fagre* presents a case study of integrated modeling that is facing significant challenges, but that has gone a long way towards the ultimate goal of integrated assessments of Global Change in mountain regions.

Finally, *Graumlich et al.* review treeline dynamics as inferred from paleo-ecological studies and landscape ecological analyses, thus reminding ecologists who focus on the analysis of contemporary data that the link from the past across the present into the future is a very important one that should not be forgotten (see also the section on paleo changes in this book).

The Green Cover of Mountains in a Changing Environment

Christian Körner
Institute of Botany, University of Basel, Schönbeinstrasse 6, CH-4056 Basel, Switzerland
phone +41-61-267 35 10, fax +41-61-267 35 04, e-mail ch.koerner@unibas.ch

Keywords: Alpine vegetation, Biodiversity, Climatic warming, Elevated CO_2, Nitrogen deposition, Plant ecology.

1. Introduction

Slopes induce biological diversity, and nowhere else is diversity so important as on slopes. Why the first? Why the second? The inclination of a piece of land causes gravitational forces that structure the surface and climatic vectors that differentiate life conditions across those structures. The resultant multitude of microhabitats leads to a multitude of inhabitants. A major function of those inhabitants is to secure substrate against further action of gravity. Sloping terrain is only as stable as the workforce keeping it in place. It is this endless battle between the force of gravity and biological safeguards against its consequences, which governs mountain biota. If the substrate is gone, so too are most of the plants and animals.

About 8% of the vegetated surface of the globe produces biomass on mountain slopes at seasonal mean temperatures between 3°C and 12°C. One third of this area commonly called the alpine (also afro-alpine or Andean) is confined to high mountain terrain above elevations climatically suitable for tree growth. The other two thirds are covered by montane forests or their substitute vegetation of low stature, largely pasture and shrubland, which result from human land use and are driven by fire and grazing regimes. Since atmospheric pressure drops by approximately 10% per kilometer of altitude, most of this upland vegetation operates at 20-40% reduced

U. M. Huber et al. (eds.), Global Change and Mountain Regions, 367-375.

partial pressure of carbon dioxide and oxygen and increased gas diffusivity. In addition, cold temperatures and often poorly developed soils limit nutrient availability (Körner 1999).

It is plausible to assume that climate warming, increasing CO_2 concentration and enhanced soluble nitrogen deposition will exert major changes in these ecosystems. But will they? How would the consequences of these global changes compare with those of land use? Evidence accumulated over the past years suggests that there is no common biotic response to any of these environmental drivers but rather a series of context-driven responses, with each of the three atmospheric changes exerting different effects on different plant species and in different vegetation (e.g. Körner et al. 1997; Molau and Alatalo 1998). Similar findings have emerged concerning the response of vegetation in low-elevation Arctic settings (Press et al. 1998).

2. Life in the cold when it gets warmer

Most biological mountain research, including my own, began with the exploration of the presumed thermal limitations of life at high elevation. One of the big surprises, at least for me, was that what may appear to be limiting to humans is just right for species that evolved under these life conditions. Rising temperature can induce upward migration of early-responding, early-succession pioneer species and may disadvantage slower-responding, late-successional species or populations trapped on low mountain tops (Grabherr et al. 1994). Late successional species may not respond or respond so slowly that vegetation may change little, as was documented for very old clonal sedge heath in the Alps (Steinger et al. 1996). By mapping DNA in leaves across a few meters of alpine heathland, individual clones of *Carex curvula* could be identified. The maximum distance of ramets of one genet (clone, same DNA) and the rate of annual horizontal spread led to the conclusion that these clones have been occupying the very same spot for more than 1000 years, i.e. during medieval warming as well as during the “Little Ice Age”.

Due to the interdependence between radiation and temperature (warm, when the sun shines, cold when it doesn't), alpine and montane plant photosynthesis is hardly ever limited by suboptimal temperatures (Körner 1999). The way low temperature can affect plant growth is through its influence on plant development and season length. However, plants in cold climates possess a number of safety measures to mitigate the effects of cold temperatures. Most importantly, development and growth are under genetic control. First, alpine plants are “designed” to be small and to utilize the warm boundary layer near the ground (Körner 1999). Second, both alpine and montane species utilize photoperiod signals to start and, more consistently, to terminate meristematic activity. True high altitude specialists cannot be “tricked” by a warm spell. These insights have substantial implications for predicting the future of mountain biota in a warmer climate.

I believe that impacts other than the pure temperature effects associated with global warming are likely to be more significant in many regions. Most importantly, the consequences of any change in snowfall regimes and snow duration would by far exceed direct temperature effects. A more moisture-laden atmosphere, due to warmer oceans, may in fact revert the consequences of warming-only effects into a negative trend caused by enhanced snowpack (Stone 1992; Williams et al. 1998). Such a trend became apparent during the past ca. 10 years in the Swiss Alps, where heavy late winter snowfall delayed the beginning of the season. In one year (1999), a section of alpine grassland was not released from snow for the whole summer, though this was one of the warmest years on record.

It is well known that topography and exposure may outweigh direct altitudinal effects above the alpine treeline, so that climatic differences between north- and south-facing slopes at a given elevation may exceed climatic differences otherwise found across 300-500 m of elevation on similar exposures. Consequently, a most likely initial response of the vegetation is a more-or-less horizontal spatial re-allocation of communities, rather than an upward shift of vegetation belts. Recent research of our group showed that such thermal contrasts are largely absent in closed forests. A very detailed GIS analysis across the whole of the Swiss Alps revealed no difference in treeline elevation with slope exposure (Paulsen and Körner 2001), which can be explained by the shading of the ground by trees and, thus, the elimination of differences in ground warming on south- versus north-facing slopes. This hypothesis has now been confirmed by field measurements of root zone temperatures.

The recent warming of 1-2 K in the Swiss Alps has caused a growth stimulation in treeline trees (Paulsen et al. 2000), but I have not seen any advancement of sapling fronts. A dense ground cover may often limit seedling establishment, and it may take centuries for the treeline to move upslope (Ives and Hansen-Bristow 1983). However, as an early response to warming, gap filling can be observed within the treeline ecotone in many places. This process is currently under way in the Alps and the Ural (Nagy et al. 2003). In part, this may also be related to reduced grazing pressure. In the long run, a warmer summer climate should cause an advance of the treeline upslope, because the growing season temperature is the single most important determinant of treeline position worldwide, irrespective of season length (Körner 1998).

Most alpine plants terminate growing-season activity in response to photoperiodic (day length) controls, irrespective of actual temperature. For half of the 25 alpine species tested, this is also true for the beginning of growth in early summer (Körner 1999; Keller and Körner 2003). Such photoperiodic constraints do not only diminish the potential benefit of a longer thermal growing season, they may even exert an additional burden due to respiratory losses during snow free periods that are not used for assimilation and growth. It is well known that some high elevation taxa are unable to restrain their respiratory metabolism when exposed to warmer temperatures (e.g. saxifrages; Larigauderie and Körner 1995). Given the slow rate of reproduction and the predominantly clonal growth of alpine plants, the rate of selection for better-adjusted genotypes (Crawford et al. 1995) may not be able to keep up with the speed of warming.

3. Is mountain vegetation carbon saturated?

The direct "fertilization" effects of elevated CO_2 on growth of upland vegetation are most likely much smaller than we expected when we started with *in situ* CO_2

enrichment experiments in the 1990s. Our initial assumption was that, due to the reduced partial pressure of CO_2 at high elevation, carbon would be a limiting resource for alpine vegetation. However, the exposure of closed alpine heath at 2500 m altitude to four years of elevated CO_2 revealed a complete lack of responsiveness of community biomass (Fig. 1). Even a low dose of mineral fertilizer addition, which almost doubled biomass by year four, did not facilitate a positive biomass response to CO_2. Inherent slow growth seems to dampen CO_2 effects. Significant year-to-year differences in summer temperatures had also no effect on the CO_2 response. However,

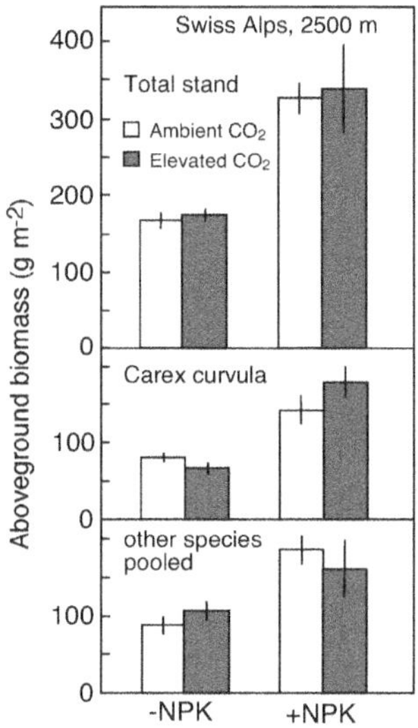

Figure 1: Four years of exposure to a 700 ppm CO_2-world have not induced a change in plant biomass in alpine grassland 2500 m above sea level and 300 m above the local treeline. The addition of only 40 kg of soluble nitrogen fertilizer (as much as is contained in lowland rainfall in Central Europe) has, however, almost doubled biomass, but still did not facilitate a response to elevated CO_2 (Körner et al. 1997).

there were small differential responses between species, which could, in the long run, accumulate into a significant biodiversity effect and this in itself could alter ecosystem properties such as productivity.

Whether or not montane forests or trees in the treeline ecotone will respond to elevated CO_2 is still an open question. I initially predicted that high elevation plant growth, including that of trees, is sink- rather than source-limited (Körner 1998) and the current mobile carbon charging of treeline trees indeed suggests no carbon limitation (Hoch et al. 2002; Hoch and Körner 2003). In other words, we found an increasing abundance of non-structural carbohydrates, such as starch and storage lipids, the closer we sampled to the altitudinal tree limit. We interpreted this as an indication of sufficient photosynthetic activity, but a limited investment into structural growth. However, initial shoot growth responses of treeline trees in a free air CO_2 enrichment (FACE) experiment in the Swiss Alps revealed a clear stimulation in the first year (Hättenschwiler et al. 2002). Similar initial responses have been seen in warm temperate pines. However, these growth differences disappeared after three years of exposure (Oren et al. 2001) despite vigorous growth of both control and test plots, which should have ensured the detection of potential growth differentiation. No significant growth stimulation by CO_2 enrichment alone has been noted in spruce in montane model communities in Switzerland (Hättenschwiler et al. 2002) and in tall trees in enclosures in northern Sweden over a three year multifactorial *in situ* experiment (B. Sigurdsson and S. Linder, pers. com.). It seems that the more a species is adapted to life in the cold and to slow growth, the less likely will carbon be a limiting resource.

4. Nitrogen, a limiting resource

Perhaps the most significant atmospheric change in the mountains of highly populated regions, such as the Alps and certain mountain ranges in North America, is enhanced N-deposition. Bowman and co-workers (1994; 2001) have documented dramatic effects of nutrient addition on alpine vegetation (see Bowman, this volume). The effect shown in Figure 1 resulted from an addition of nitrogen, with similar amounts to what is currently found in rainfall at low altitudes in Central Europe (40 kg $ha^{-1}a^{-1}$). Enhanced soluble nutrient input favours early successional, relatively fast growing species. Figure 2 illustrates that inherently slow growing species, such as cushion plants or low-stature shrubs, are rapidly overgrown by grasses, such as *Poa alpina* (Heer and Körner 2002), when nutrients are added. Hence, for those slow growing species nitrogen addition is a killer. Such biodiversity effects cause major shifts in ecosystem properties. For example, "soft-leaved", fast growing species are often more sensitive to mechanical forces on slopes, are more intensively grazed and provide better habitats for certain rodents, whose activity may transform habitats. On the other hand, such species may also regenerate more rapidly after disturbance.

Given our current knowledge, my subjective ranking of the likely impact of atmospheric changes, discussed above, is (with decreasing importance) (1) nitrogen deposition (in certain areas only), (2) indirect effects of warming via enhanced precipitation and snowpack, (3) direct effects of warming, (4) direct effects of

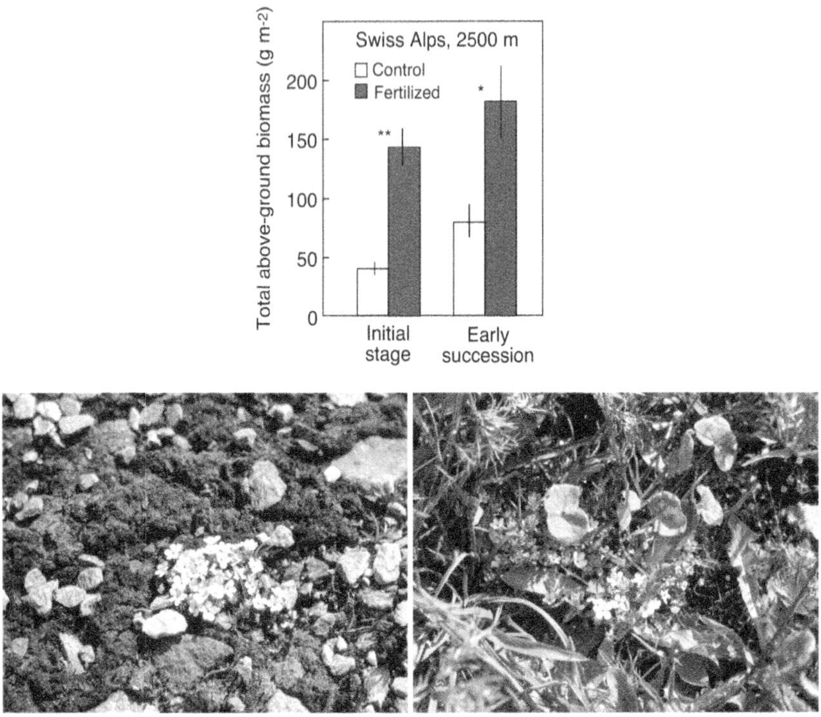

Figure 2: A yes or no test of the sensitivity of glacier forefield vegetation to massive nutrient addition (100-200 kg N $ha^{-1}a^{-1}$ over three years) revealed strong responsiveness despite a season length of only a few weeks. Slow growing cushion plants (left, *Androsace alpina*) become rapidly overgrown by vigorous grasses like *Poa alpina* (right), illustrating a complete change of community structure (Heer and Körner 2002). What is positive for growth in one species may kill others.

elevated CO_2. However, this ranking may change, depending on the time scale considered and the actual trends that become apparent in these four components of global change. Differential responses of species will play a key role in the translation of plant responses into ecosystem responses. In most parts of the world, the effects of land use by far exceed the influences of any of these atmospheric drivers of global change, a subject of a recent article (Körner 2000b). For a new global assessment of mountain biodiversity and its change consult Körner and Spehn (2002).

5. Research perspectives

Species matter. Since environmental influences affect individuals, populations and species, rather than ecosystems as such, we need to understand differential responses at the organismic level. Although we have good and plausible reasons to believe that organismic diversity is the key for understanding ecosystem functioning, the actual knowledge about the functional significance of biodiversity is very meager. Most of what we know comes from highly artificial, planted grassland communities at low elevation. So far, most of the forecasted ecosystem changes rest on the assumption of concerted changes of the individual players rather than the more realistic perspective of individualistic responses to environmental change.

It seems that the *influences of altered organism assemblages* due to environmental changes have not received the attention they deserve. It makes little sense to search for carbon sequestration or its change at the plot or landscape level, and not account for the presence or absence of certain plant and animal species. For example, the introduction or removal of a single type of grazer can lead to carbon sequestration scenarios that run counter to model predictions. It therefore seems logical to implement biodiversity aspects in all research plans, which aim at landscape level predictions. Very unfortunately, functional groups of species do not work in most cases (e.g. Molau and Alatalo 1998). Projects addressing the question of the functional significance of diverse biota on steep mountain slopes need to take into account the importance of individual species and deserve top priority. We have hardly any data to support the hypothesis that high biodiversity ensures slope stability, as plausible as we may think this "insurance hypothesis" is. The Global Mountain Biodiversity Assessment (GMBA) of DIVERSITAS has declared this as one of the focal areas for the near future (www.unibas.ch/gmba/; Körner and Spehn 2002).

Caution with warming tests. As an experimentalist, I am warning against warming experiments. There are several reasons for this. First, it is near to impossible not to confound the test with other physical changes, most importantly those of the water potential in air and soil. Second, a step change in temperature is likely to create a suite of unknown reactions, some homoeostatic for short periods others accelerating responses far beyond steady state. This may also apply to other treatments such as CO_2 enrichment and N addition, but with these factors we are very limited in alternatives. Third, we do not need such tests, because, of all global changes, temperature is the one best manipulated by nature itself. The "treatments" are found in our landscapes and cost nothing, and the responses we are likely to find are in a quasi steady state. Nature's thermal gradients are also not free from confounding gradients of factors other than temperature, but to a far lesser extent than is heating by ground cables and radiators. A north versus south side of a hill or a 300 m difference in elevation may be perfect surrogates, if carefully selected for similarity in soils and precipitation. One may even transplant large units of undisturbed soil with intact vegetation across such thermal gradients as has been done in a few instances. The differences and trends we may find in vegetation composition and ecosystem processes provide realistic prospects of responses to warming. The terrestrial ecosystem research community has left the potential of this option largely unexplored and has instead moved - trapped in its own tradition - straight into heating peat. Personally, I think these are not very instructive models and we, the experimenters, should be more open to insights developed in comparative observational non-invasive science. We will gain a lot of realism, be it for the price of more noisy data (Körner 2001).

In situ, whenever possible. When we decouple organisms from their matrix of physical, chemical and organismic links, as they developed over many millennia, we are unlikely to gain realistic response characteristics - except perhaps for ruderals or pioneer vegetation. For instance, the net outcome of CO_2 enrichment tests to a large extent reflects the boundary conditions applied by experimenters (Körner 2000a). To me the most eye-opening experience in recent years was the opposite response (positive versus negative!) of beech to four years of CO_2 enrichment, depending on whether trees grew on calcareous or siliceous soils under otherwise identical conditions (Spinnler et al. 2002). I am not aware of any other test with trees on more than one soil type - the limitations of these tests are obvious.

In summary, I believe that the most realistic insights into alpine plant responses to different components of atmospheric change will come from the study of natural environmental gradients. Except for CO_2 effects, these "*experiments by nature*" offer answers that are far closer to what we might see in the future, and often cost much less. Although I have a great interest in the effects of atmospheric changes on alpine biota, there is no doubt that these *effects will be outweighed by land use impacts* in many parts of the world. Given the extensification of land use we see in many mountain regions of the developed countries (which is also associated with severe problems), the dramatic intensification of land use in the mountains of the developing countries is often overlooked and deserves the highest research priority (see www.unibas.ch/gmba for a workshop on this topic in La Paz, 20-23 August 2003).

6. References

Bowman, W. D. (1994). Accumulation and use of nitrogen and phosphorus following fertilization in two alpine tundra communities. *Oikos* **70**, 261-270.

Bowman, W. D., and Seastedt, T. R., Eds. (2001). "Structure and function of an alpine ecosystem - Niwot Ridge, Colorado." Oxford University Press, Oxford.

Crawford, R. M. M., Chapman H. M., and Smith L. C. (1995). Adaptation to variation in growing season length in arctic populations of *Saxifraga oppositifolia* L. *Botanical Journal of Scotland* **41**, 177-192.

Grabherr, G., and Pauli, M. G. H. (1994). Climate effects on mountain plants. *Nature* **369**, 448.

Hättenschwiler, S., Handa, T., Egli, L., Asshoff, R., Ammann, W., and Körner, Ch. (2002). Atmospheric CO_2 enrichment of alpine treeline conifers. *New Phytologist* **156**, 363-375.

Heer, C., and Körner, Ch. (2002). High elevation pioneer plants are sensitive to mineral nutrient addition. *Basic and Applied Ecology* **3**, 39-47.

Hoch, G., Popp, M., and Körner, Ch. (2002). Altitudinal increase of mobile carbon pools in *Pinus cembra* suggest sink limitation of growth at the Swiss treeline. *Oikos* (in press).

Hoch, G., and Körner, Ch. (2003). The carbon charging of pines at the climatic treeline: A global comparison. *Oecologia* (in press).

Ives, J. D., and Hansen-Bristow, K. J. (1983). Stability and instability of natural and modified upper timberline landscapes in the Colorado Rocky Mountains, USA. *Mountain Research and Development* **3**, 149-155.

Keller, F., and Körner, Ch. (2003). The role of photoperiodism in alpine plant development. *Arctic, Antarctic and Alpine Research* (in press).

Körner, Ch. (1998). A re-assessment of high elevation treeline positions and their explanation. *Oecologia* **115**, 445-459.

Körner, Ch. (1999). "Alpine plant life." Springer, New York.

Körner, Ch. (2000a). Biosphere responses to CO_2 enrichment. *Ecological Applications* **10**, 1590-1619.

Körner, Ch. (2000b). The alpine life zone under global change. *Gayana Botànica* **57**, 1-17.

Körner, Ch., Diemer, M., Schäppi, B., Niklaus, P., and Arnone, J. (1997). The responses of alpine grassland to four seasons of CO_2 enrichment: A synthesis. *Acta Oecologica* **18**, 165-175.

Körner, Ch. (2001). Experimental plant ecology: Some lessons from global change research. *In* "Ecology: Achievement and challenge." (M. C. Press, N. J. Huntly, and S. Levin, Eds.), pp 227-247. Blackwell Science.

Körner, Ch., and Spehn, E., Eds. (2002). "Mountain biodiversity: A global assessment." Parthenon,

London.

Larigauderie, A., Körner, Ch. (1995). Acclimation of leaf dark respiration to temperature in alpine and lowland plant species. *Annals of Botany* **76**, 245-252.

Molau, U., and Alatalo, J. M. (1998). Responses of subarctic-alpine plant communities to simulated environmental change: Biodiversity of bryophytes, lichens, and vascular plants. *Ambio* **27**, 322-329.

Nagy, L., Grabherr, G., Thompson, D., and Körner, Ch. (2003). "Alpine biodiversity across Europe." Ecological Studies. Springer, New York (in press).

Oren, R., Ellsworth, D. S., Johnsen, K. H., Phillips, N., Ewers, B. E., Maier, C., Schäfer, K. V. R., McCarthy, H., Hendrey, G., McNulty, S. G., and Katul, G. G. (2001). Soil fertility limits carbon sequestration by forest ecosystems in a CO_2-enriched atmosphere. *Nature* **411**, 469-472.

Paulsen J., Weber, U. M., and Körner, Ch. (2000). Tree growth near treeline: Abrupt or gradual reduction with altitude? *Arctic, Antarctic, and Alpine Research* **32**, 14-20.

Paulsen, J., and Körner, Ch. (2001). GIS-analysis of tree-line elevation in the Swiss Alps suggest no exposure effect. *Journal of Vegetation Science* **12**, 817-824.

Press, M. C., Callaghan, T. V., and Lee, J. A. (1998). How will European Arctic ecosystems respond to projected global environmental change? *Ambio* **27**, 306-311.

Spinnler, D., Egli, P., and Körnse, Ch. (2002). Four-year growth dynamics of beech-spruce model ecosystems under CO_2 enrichment on two different forest soils. *Trees* **16**, 423-436.

Steinger, Th., Körner, Ch., and Schmid, B. (1996). Long-term persistence in a changing climate: DNA analysis suggests very old ages of clones of alpine *Carex curvula*. *Oecologia* **105**, 94-99.

Stone, P. B., Ed. (1992). "The state of the world's mountains." Zed Books, London.

Williams, M. W., Brooks, P. D., and Seastedt, T. (1998). Nitrogen and carbon soil dynamics in response to climate change in a high-elevation ecosystem in the Rocky Mountains, USA. *Arctic and Alpine Research* **30**, 26-30.

The Response of Alpine Plants to Environmental Change: Feedbacks to Ecosystem Function

William D. Bowman
Mountain Research Station, Institute of Arctic and Alpine Research and Environmental, Population and Organismic Biology, University of Colorado, Boulder CO, 80309-0334, USA
phone 1-303-492-2557, fax 1-303-492-8699, email bowman@spot.colorado.edu

Keywords: Alpine, Biotic change, Nitrogen deposition, Plant influences on nutrient cycling, Rocky Mountains.

1. Alpine species composition changes in response to environmental change

Alpine ecosystems occur on all continents, and potentially serve as sensitive indicators of biotic response to environmental change. Because environmental change associated with resource extraction and development is minimal in most alpine areas, biotic changes in the alpine are reflective of "indirect" anthropogenic environmental effects, including changes in climate, atmospheric chemistry, and transmission of ultraviolet radiation. Plant species respond differentially to these environmental changes, related in part to their ability to alter growth rates as resource supply changes and to changes in biotic interactions with neighbors (Theodose and Bowman 1995; Callaway et al. 2002). Thus, changes in plant species composition are likely to herald environmental change in the alpine. Floristic changes have been noted in some alpine areas, potentially associated with climate change (Grabherr et al. 1994), atmospheric pollution (Rusek 1992), and increased N deposition (Korb and Ranker 2001; see Baron et al., this volume for aquatic biotic responses to N deposition).

U. M. Huber et al. (eds.), Global Change and Mountain Regions, 377-382.

2. Changes in alpine species composition influence ecosystem function

Changes in alpine plant species composition associated with environmental change will have important feedbacks to ecosystem function, perhaps as great or greater than the direct environmental change effects. Although climatic conditions (low temperatures, drought) can impose important constraints on rates of primary production and nutrient cycling, plant species control on these processes is also important. For example, preformation of buds in slow growing dominant species, primarily forbs and sedges, may delay vegetative responses to growing season climatic conditions for several years (Walker et al. 1994; Diggle 1997). Shifts in species composition from forbs or sedges to grasses would therefore result in more rapid responses and greater sensitivity of alpine vegetation to climatic variation, due to the greater flexibility in new meristem production of the grasses. As a result, interannual variation in primary production may increase as responsive species increase in abundance.

3. Plant feedbacks influencing ecosystem response to N deposition

In addition to influencing temporal variation in production, plant species can significantly influence spatial variation in nutrient cycling and soil fertility in the alpine (Fig. 1; Steltzer and Bowman 1998; Onipchenko et al. 2001), and thus changes in plant species composition could result in significant changes in nutrient cycling (Bowman and Steltzer 1998). Because plants of fertile environments tend to enhance nutrient cycling relative to species of more infertile environments (Van der Krift and Berendse 2001), changes in species composition may have long-term effects on nutrient cycling regimes in the alpine. For example, increased nitrogen deposition may increase the cover of fast-growing species at the expense of slow-growing species tolerant of nutrient-poor conditions, leading to a eutrophying effect due both

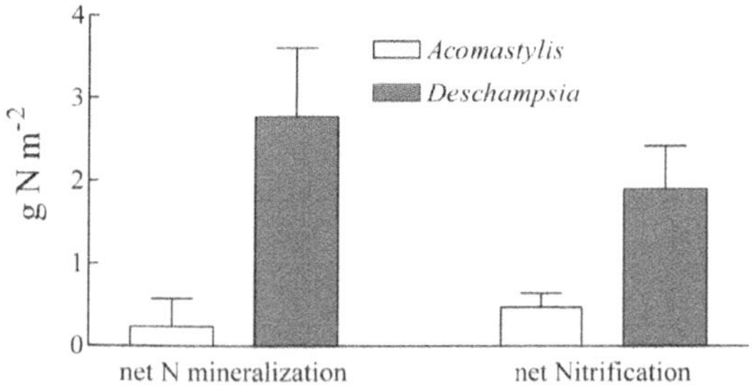

Figure 1: In situ rates of net N mineralization and nitrification for soil dominated by 2 co-dominant Rocky Mountain moist meadow alpine species, *Deschampsia caespitosa* and *Acomastylis rossii*. Measurements are from 7 replicate communities, and were made during the growing season (June-August). There were no differences in soil microclimate between soils of the 2 species (from Steltzer and Bowman 1998).

to increased external nitrogen inputs as well as increased rates of nitrogen cycling (Fig. 2). Thus, even if rates of nitrogen deposition decrease, rates of soil nitrogen cycling may remain elevated due to a change in plant species composition. This may have important consequences for nitrogen fluxes to aquatic ecosystems and the potential for buffering atmospheric inputs (Strengbom et al. 2001).

Results from several fertilization experiments on Niwot Ridge, Colorado support the hypothetical model shown in Figure 1. Grasses and some forb and sedge species are most responsive to increased supply of nutrients, at the expense of the cover of the dominant sedge species. Higher rates of root turnover, and lower C:N ratios of the litter of the responsive species tend to increase rates of net N mineralization relative to the dominant species. In addition, many of the dominant forb species contain relatively high concentrations of secondary compounds (phenolics, terpenes), which can influence soil biogeochemistry. Ongoing research is examining the influence of these compounds on microbial activity in soils to determine whether they have inhibitory or stimulatory effects, and how these effects in turn influence nutrient supply to neighboring plants. The operating hypothesis is that secondary compounds may provide some relatively slow growing forb species an effective mechanism to compete with neighbors. The secondary compounds mediate competition by limiting access of neighbors to nutrients by either supplying microbes with carbon and increasing nutrient immobilization, or by inhibiting microbial turnover of soil organic matter. The initial results of this work indicate that microbial N immobilization is higher in soils dominated by species with high concentrations of total phenolics, and that growth of neighbors is lower in the presence of litter of these species. The

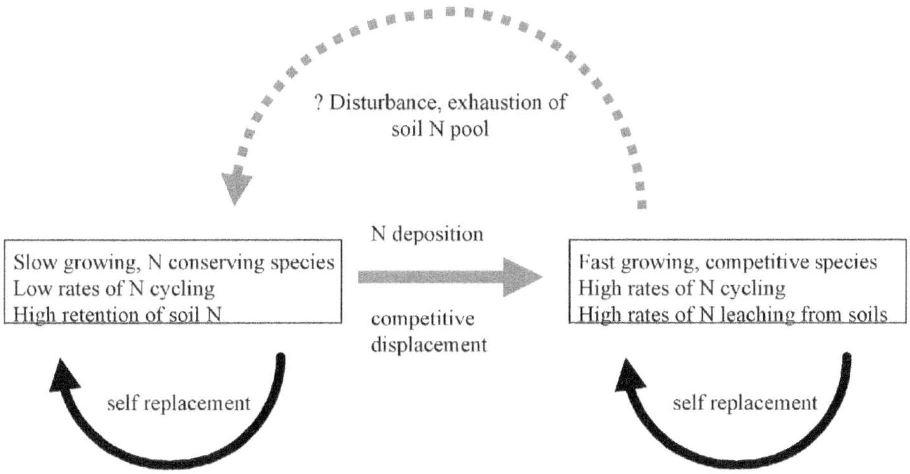

Figure 2: Conceptual model showing the generalized pattern of plant species self-replacement in communities with low or high N availabilities. In the face of higher N deposition, N conserving species of nutrient-poor environments would be replaced by more competitive species, with biotic characteristics promoting higher N cycling in soils. The new state would be preserved, even in the absence of additional N inputs. A return to the initial state would require some disturbance or exhaustion of soil N reserves, allowing species adapted to low N conditions to reestablish (from Bowman and Steltzer 1998).

presence of relatively high phenolic concentrations in the tissues of slow growing alpine species appears to be widespread in floras of the Rockies, Alps, and Tatras, indicating plant secondary compounds may be important influences on N retention, nutrient cycling, and competitive interactions in several alpine areas.

In order to more accurately estimate the influence of atmospheric N deposition in the alpine, a long-term experiment is ongoing, involving relatively low level N inputs (2, 4, and 6 g inorganic N m^{-2} yr^{-1}) into an alpine dry meadow on Niwot Ridge. This experiment will determine 1) what are the important sinks for inorganic N inputs, including soil, plant, and microbial components; 2) what level of N inputs result in saturation of the system; and 3) will vegetation composition change in response to low levels of N input? Preliminary data indicate that dry meadow alpine communities may saturate at levels around 2 g inorganic N m^{-2} yr^{-1} (Fig. 3), consistent with modeling (Baron et al. 1994) and other empirical work using a whole-catchment approach (Williams and Tonneson 2000). Although plant production and subsequently N uptake increase with relatively low inputs of N, not all of the soil solution N is taken up. Pulse-chase labeling using ^{15}N is being done in these experimental plots in conjunction with Keri Holland and Alan Townsend at University of Colorado to determine the important sinks of inorganic N, and how they change with long-term (5+ years) N fertilization. Plants and soil organic matter appear to be the largest long-term sink for increased N inputs. Microbial use of different soil carbon fractions appears to change in response to N fertilization of alpine soils (Neff et al. 2002), indicating changes in C cycling and soil stabilization may occur in response to increasing N deposition.

Fertilization experiments along with climate and N deposition monitoring in the Western Tatra Mountains in Slovakia began in 2002, in conjunction with Lubos Halada, Juraj Hresko, and Zdeno Kostka of the Slovak Academy of Sciences and Peter Fleisher of the High Tatras National Park. The goals of this research are to determine

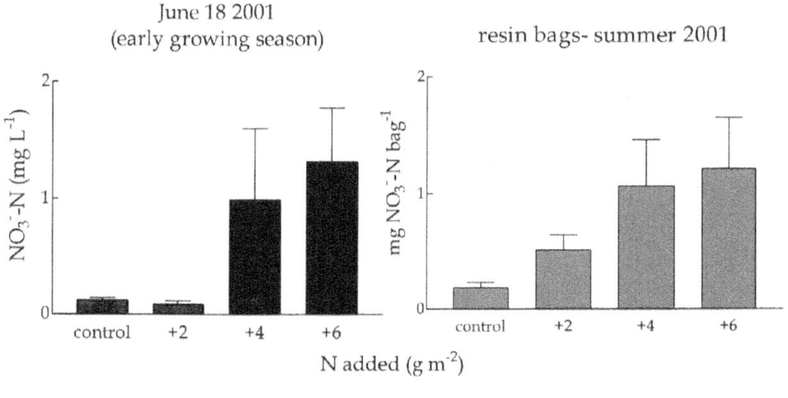

Figure 3: Estimates of soil solution concentrations of NO_3^- - N in alpine dry meadow plots subjected to control (=ambient), 2, 4, and 6 g N (NH_4NO_3) using soil microlysimeters (left) and resin bags (right). Soil microlysimeters had much higher yields in the N addition plots later in the growing season (late July), indicating that plant uptake of N early in the growing season is an important sink for inorganic N in these soils. Integrated yields of inorganic N in soil solution over the course of the growing season indicates that these soils become "N saturated" at relatively low inputs of inorganic N.

the similarities and differences in N deposition dynamics between the more highly impacted Western Tatras and the Front Range of the Colorado Rocky Mountains, which are receiving increasing levels of N depositional inputs.

4. Future research needs

Experiments are needed to determine the potential changes in species composition that will accompany specific environmental changes, and the importance of biotic interactions in determining these changes. Although climatic conditions are important in alpine ecosystems, there may be a generalized set of "responsive" species that will increase in abundance as resource levels increase due to warming, enhanced precipitation, or increased nitrogen deposition. Whether there are indicator species that are specific to particular environmental stresses is not known. More information is needed to verify that these responsive species will alter ecosystem properties in the manner outlined above. In particular, more basic research is needed on the influence of plant functional types on alpine ecosystem processes.

A better understanding of the coupling between terrestrial and aquatic ecosystems is also needed to predict the impact of biotic changes in the alpine on lake and streamwater quality (Baron et al., this volume). This includes 1) determination of the sink strength of vegetation, microbial biomass, and soils for inorganic and organic nitrogen, 2) the amount of exchange of ground water with surface waters in streams and lakes, and 3) the thresholds of nutrient inputs or terrestrial biotic change at which fluxes of elements from terrestrial to aquatic systems will increase. Manipulative experiments, particularly those using tracers, are an important component of this research effort. These experiments need to be coupled with spatially explicit process-level models to provide a predictive understanding of the effect of environmental change on the interaction between alpine terrestrial and aquatic ecosystems.

5. References

Baron, J. S., Ojima, D. S., Holland, E. A., and Parton, W. J. (1994). Analysis of nitrogen saturation potential in Rocky Mountain tundra and forest: Implications for aquatic systems. *Biogeochemistry* **28**, 1-31.

Bowman, W. D., and Steltzer, H. (1998). Positive feedbacks to anthropogenic nitrogen deposition in Rocky Mountain alpine tundra. *Ambio* **27**, 514-517.

Callaway, R. M., Brooker, R. W., Choler, P., Kikvidze, Z., Lortie, C. J., Michalet, R., Paolini, L., Pugnaire, F. L., Newingham, B., Aschehoug, E. T., Armas, C., Kikodze, D., and Cook, B. J. (2002). Positive interactions among alpine plants increase with stress. *Nature* **417**, 844-848.

Diggle, P. K. (1997). Extreme preformation in alpine *Polygonum viviparum*: An architectural and developmental analysis. *American Journal of Botany* **84**, 154-169.

Grabherr, G., Gottfried, M., and Pauli, H. (1994). Climate effects on mountain plants. *Nature* **369**, 448-448.

Korb J. E., and Ranker, T. A. (2000). Changes in stand composition and structure between 1981 and 1996 in four Front Range plant communities in Colorado. *Plant Ecology* **157**, 1-11.

Neff, J. C., Townsend, A. R., Gleixner, G., Lehman, S. J., Turnball, J., and Bowman, W. D. (2002). Variable effects of nitrogen additions on the stability and turnover of soil carbon. *Nature* **419**, 915-917.

Onipchenko, V. G., Makarov, M. I., and van der Maarel, E. (2001). Influence of alpine plants on soil nutrient concentrations in a monoculture experiment. *Folia Geobotanica* **36**, 225-241.

Rusek, J. (1992). Air pollution-mediated changes in alpine ecosystems and ecotones. *Ecological*

Applications **3**, 409-416.
Steltzer, H., and Bowman, W. D. (1998). Differential influence of plant species on soil N transformations within moist meadow alpine tundra. *Ecosystems* **1**, 464-474.
Strengbom, J., Nordin, A., Näsholm, T., and Ericson, L. (2001). Slow recovery of boreal forest ecosystem following decreased nitrogen input. *Functional Ecology* **15**, 451-457.
Theodose, T. A., and Bowman, W. D. (1997). The influence of interspecific competition on the distribution of an alpine graminoid: Evidence for the importance of plant competition in an extreme environment. *Oikos* **79**, 101-114.
Van der Krift, T. A. J., and Berendse, F. (2001). The effect of plant species on soil nitrogen mineralization. *Journal of Ecology* **89**, 555-561.
Walker, M. D., Webber, P. J., Arnold, E. A., and Ebert-May, D. (1994). Effects of interannual climate variation on aboveground phytomass in alpine vegetation. *Ecology* **75**, 393-408.
Williams, M. W., and Tonnessen, K. A. (2000). Critical loads for inorganic nitrogen deposition in the Colorado Front Range, USA. *Ecological Applications* **10**, 1648-1665.

Ecological Climate Impact Research in High Mountain Environments: GLORIA (Global Observation Research Initiative in Alpine Environments) – its Roots, Purpose and Long-term Perspectives

Harald Pauli*, Michael Gottfried, Daniela Hohenwallner, Karl Reiter, and Georg Grabherr
[1]*Department of Conservation Biology, Vegetation and Landscape Ecology, Institute of Ecology and Conservation Biology, University of Vienna, Althanstrasse 14, A-1090 Vienna, Austria*
**phone +43-1-4277-54383, fax +43-1-4277-9542, e-mail pauli@pflaphy.pph.univie.ac.at*

Keywords: Alpine vegetation, Biodiversity, Climate change, Global *in-situ* observation network, Long-term monitoring, Mountain summits.

1. Introduction

High mountain ecosystems are sensitive to climate change (Box 1). Historical records of the flora on high summits in the Alps provide an important baseline against which climate-induced effects on high mountain ecosystems can be assessed. Reinvestigations of these old "monitoring summits" have shown that mountain plants have migrated upwards during the 20th century. An increase of atmospheric temperatures since the late 19th century is the most likely cause of this upward shift (Gottfried et al. 1994; Grabherr et al. 1994; 1995; 2001a; Pauli et al. 1996; 2001a). This "summit study" underlined the importance of long-term monitoring for assessing climate change effects on mountain ecosystems and initiated the establishment of extensive monitoring networks in mountain environments.

Such a large-scale long-term monitoring network – the Global Observation

U. M. Huber et al. (eds.), Global Change and Mountain Regions, 383-391.

Box 1: High mountains: An environment of global priority for ecological climate impact research

- The high mountain biome is unique in occuring at all latitudes, thus providing an invaluable potential for tracing atmospheric influences on the geo-biosphere on a global scale.
- High mountain ecosystems are sensitive to climate change, because they are controlled by low temperature. With increasing altitude, climate-related ecological factors become dominant. Therefore, the effects of climate change may be more pronounced compared to ecosystems of lower altitude.
- Many high mountain environments comprise natural wilderness areas, where direct human influences do not mask the signals of climate change.
- A key aspect of mountains is the presence of narrow ecotones, where changes are significant over short distances (threshold phenomena). This may lead to rapid recognition of changes if boundaries shift; it also allows gradient studies within small areas.
- Mountain regions are often "hot spots" of organismic diversity, caused among other reasons by the compression of thermal life zones (Barthlott et al. 1996; Grabherr et al. 2000a). A high degree of endemism is characteristic for the upper belts of many mountain systems. The risk of biodiversity losses owing to climate warming is high, because cryophilic high mountain species would have nowhere to migrate to.

Research Initiative in Alpine Environments (GLORIA) – has been initiated at the turn of the millennium. GLORIA uses alpine ecosystems because they are distributed over all continents and all major life zones on Earth. GLORIA takes advantage of the sensitivity of alpine plants, which are ideal indicators of the ecological implications of climate change. Its purpose is to assess risks of biodiversity losses and the vulnerability of high mountain ecosystems to climate change pressures. *In-situ* observations on the species level appear to be crucial for this purpose, because plant communities will not respond to climate warming as a whole, but single species will respond in different ways consistent with their unique range of ecological tolerances (Ammann 1995; Grabherr et al. 1995; Gottfried et al. 1999a; 2002b). What is too warm for one species within a given plant community may still be appropriate for another, or where one species may respond by migration, another may have restricted possibilities to move to new habitats. Such differential movement of species could result in a disruption of the connectedness among many species in current ecosystems (Root et al. 2003), and may be accompanied by significant biodiversity losses and by changes in ecosystem functioning. Körner (2002) pointed out that biological richness insures against "system failure", and further, he argued that intact vegetation provides safety, particularly in mountain environments, where slopes are only as stable and safe as the integrity and stability of their vegetation. Functional redundancy among the species of an ecosystem may not play an important role for long periods of time, but drastic changes of the ecological constraints, e.g. owing to climate warming, can cause a species to become the life-preserver that sustains ecosystem functioning on fragile mountain slopes.

2. GLORIA – the Global Observation Research Initiative in Alpine Environments

At the same time that the Mountain Research Initiative (MRI) emerged from a long awareness building process (cf. Becker and Bugmann 1997; 2001), an Austrian IGBP/GCTE research initiative developed a concept for a worldwide *in-situ* observation network for alpine environments as a contribution to the MRI. This was the starting point of GLORIA. The concept was first presented at an international conference on global change in mountains in Oxford in December 1997 (Pauli et al. 1999a). Two approaches, the Single-Mountain and the Multi-Summit approach, were suggested in this concept. The first approach is based on one mountain per target region and focuses on vegetation studies on various spatial scales, including monitoring, experimental, and modelling components. Such extensive studies depend on master stations with existing research capacities and infrastructure (see the case study on Mount Schrankogel below). The second approach is based on several summit areas per target region, which span a range of altitudes, and focuses on the fundamental climatic gradients behind biodiversity patterns (Box 2 and Fig. 1).

In 1998 and 1999, a sampling design for the Multi-Summit approach was developed and tested in two climatically different mountain regions, the NE-Alps in Austria and the Sierra Nevada in Spain (Pauli et al. 2003b). Comparability, simplicity and economy were the main considerations in designing the Multi-Summit approach for an effective large-scale monitoring network. The low-instrument and low-cost approach, together with the short time required in the field, makes the method workable even in expedition conditions. Thus, a large number of sites in all major mountain systems of the world, including remote areas, are possible (Grabherr et al. 2000b; 2001b; Pauli et al. 2001b; Gottfried et al. 2002a).

GLORIA has attracted more than 100 experts from all continents who expressed their particular interest to participate. In 2000, the first international GLORIA-workshop was held at the inauguration conference of the Global Mountain

Box 2: GLORIA's *Multi-Summit approach*

Four summits of different elevation, from the treeline ecotone upwards, are used as monitoring sites in each *GLORIA target region*, to represent an altitudinal gradient (Fig.1). The standardised sampling design consists of different plot types arranged around each summit, to record species cover and frequency on different scales (Pauli et al. 2003a). Continuous measurements of the soil temperature are conducted to compare temperature and snow regimes.

Mountain summits are considered to be highly comparable reference units for long-term climate impact monitoring, because they:
- include all compass directions within a small area;
- are not exposed to shading effects from neighbouring land features;
- comprise a range of characteristic plant communities representative for a given altitude;
- are not particularly prone to severe disturbances, such as debris falls and avalanches;
- may function as traps for upward-migrating species, due to the absence of escape routes;
- are prominent landmarks, which can be readily relocated for re-investigations.

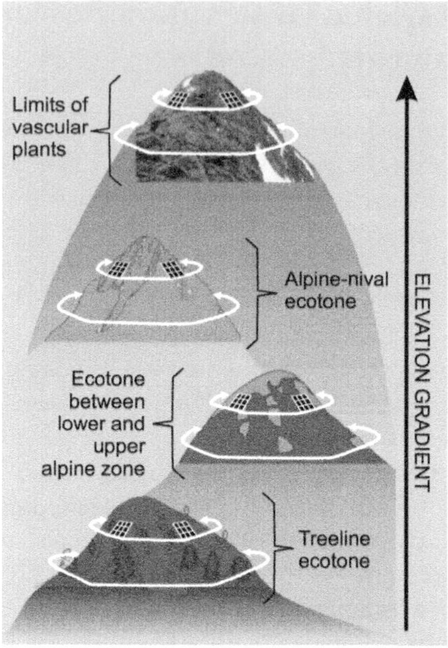

Figure 1: The *Multi-Summit approach*. Four summits of different altitude, preferably at ecotones, represent the elevation gradient of a particular target region. Each summit site is divided into an upper and a lower summit area, indicated by the white lines. Both areas are further subdivided into N, E, S, and W sections. In addition, 3m x 3m clusters are established in the four main compass directions around the summit.

Biodiversity Assessment (GMBA, a new network of DIVERSITAS; Körner and Spehn 2002). In January 2001, GLORIA-Europe, a 5th RTD-framework project of the EU, was initiated with 22 partner groups from 13 European countries. This project is a contribution to the Global Terrestrial Observing System (GTOS) and to the MRI as a pilot study towards the worldwide implementation of GLORIA (Grabherr et al. 2001c; 2002). The successfully operating project has established a total of 72 long-term observation sites in 18 target regions that are distributed over the major mountain systems of Europe. All summit sites were established according to the standardised guidelines of the *Multi-Summit approach* described in the GLORIA field manual (Pauli et al. 2003a), which can be obtained from the GLORIA-web site (www.gloria.ac.at).

All field data were assembled in a central database, maintained at the GLORIA server in Vienna. A total of 991 different vascular plant taxa (species and subspecies) were recorded. Species richness showed remarkable differences among the 18 target regions and was highest in the Alps (Southern Alps, 198 taxa; NE-Alps, 174 taxa), surprisingly low in Mediterranean mountains (e.g. Sierra Nevada, 79 taxa; Lefka Ori, Crete, 70 taxa) and lowest in the Scottish Cairngorms with only 14 taxa. Within the observed summit areas, the average species richness was different among the main compass directions. It was highest on the eastern exposure, followed by the southern, western, and northern aspects. Detailed analyses of this data set are currently being carried out by comparing the patterns of species richness, species cover, and soil temperature along elevation gradients and among the main compass directions.

The Spanish Sierra Nevada may serve as an example region to underscore the high risk of biodiversity losses. In this mountain range, the proportion of locally distributed endemic plants increases with altitude up to the highest peaks. The unique, mostly endemic high mountain flora of this small mountain range occupies only the narrow upper elevation belt. Climate change seriously threatens the survival of this characteristic low-temperature controlled flora, which exists nowhere else on the planet, because these plants would have nowhere to migrate to in increasingly warmer climates (Pauli et al. 2003b). The Sierra Nevada is not an exception. A similar critical situation can be expected in many mountain systems, such as the Australian Snowy Mountains, the North African Atlas Mountains, and the remote mountains of Hokkaido/Japan, among others.

3. A case study: Mount Schrankogel

The research site Mount Schrankogel (3497 m) was established in 1994 and lies in the Central Eastern Alps of Tyrol, Austria. It represents a master station or master site, where the *Single-Mountain approach* of GLORIA is conducted. The study area includes the entire southern slope system of the mountain from the alpine zone to the nival zone. The research site is used to develop strategies for vegetation recording and monitoring, temperature measuring, and species and vegetation modelling in an extreme high-mountain environment. Research focuses on the ecological impacts of climate warming and will support the interpretation of summit data of GLORIA's *Multi-Summit approach.*

A monitoring network within the alpine-nival ecotone (i.e. between 2900 and 3400 m asl), arranged along several transects with a total of a 1000 permanent plots, was established. Reinvestigations of the permanent plots are planned at decadal intervals. Based on these plots and on a high-resolution digital elevation model, a spatially explicit distribution model of plant species was developed (Gottfried et al. 1998; 1999b). The vegetation analysis (Pauli et al. 1999b) and the modelled vegetation patterns (Gottfried et al. 1998) showed that current species distributions are differentiated by micro-climate and highly dependent on the relief and the topographically determined patterns of disturbance and snow cover. Modelled climate warming scenarios indicate that alpine grassland species (alpine species) will shift upwards and that typical species of the alpine-nival ecotone (nival species) will suffer area losses (Gottfried et al. 1999a).

Temperature measurements at 32 sites within the alpine-nival ecotone showed different climatic constraints for alpine and nival species (Gottfried et al. 2002b). Alpine species were restricted to sites with higher temperatures and earlier snowmelt, while nival species preferred habitats with lower temperature regimes and long-lasting snow cover. Climatic constraints are both direct and indirect. Alpine species appear to be limited by the direct effects of temperature and snow cover, which restrict growth processes due to their effects on plant metabolism and the length of the growing season. In contrast, the key factor for nival species seems to be inter-species competition, which is currently low under the more extreme climatic conditions of the nival zone. Hence, climate mainly has an indirect effect on nival species. In a warmer climate with decreased snow cover, increased competition from alpine plants could lead to the extinction of cryophilic nival plants and summits would become nival "species traps" (Grabherr et al. 1995; Gottfried et al. 2002b).

Besides the above investigations, the vegetation of Mount Schrankogel's alpine zone was mapped (Abrate 1998; Dullinger 1998). The current research focus concentrates on the influence of grazing mammals on upper alpine/subnival vegetation patterns and seed dispersal, which may interfere with climate change effects (Ertl et al. 2002; Hülber et al. 2002a). Other research foci include an assessment of the environmental control mechanisms behind the reproductive phenology of high mountain plants (Hülber et al. 2002b; Keller and Körner 2003) and the sensitivity of bryophytes as indicators of climate change impacts (Hohenwallner et al. 2002). Monitoring methods for bryophytes are currently tested for an application within the *Multi-Summit approach.*

4. Aims and future research directions

GLORIA is a young initiative that aims for long-term operation. Long-term *in-situ* monitoring, based on a network of standardised observation sites, is surprisingly new in high mountain ecology. In contrast, glaciologists have started with international monitoring more than a hundred years ago and, in some parts of the world, continuous meteorological time-series from high mountain observatories date back to the 19th century. Ongoing and increasing climate warming has accelerated a broad awareness building process in the scientific community during the 1990s. This led to reconsiderations concerning the role of long-term observations as a crucial instrument for assessing the ecological implications of climate change (Grabherr et al. 2001b; 2002).

GLORIA emerged from this process and is actually the first successful attempt to establish a large-scale observation network for mountain biodiversity. It has already been demonstrated that such a network is not just an urgently required future research task, but also a feasible and cost-saving option in the long term. This was exemplified by the recent implementation of GLORIA on the European scale.

4.1 The global implementation

The major expansion of the project to a global scale and the long-term integration of the network is a forthcoming effort. This important next milestone appears to be attainable in the near future, thanks to the encouraging engagement of ecologists across the globe and the close co-operation with institutions and organisations of other research disciplines that deal with global change issues. In addition to the GLORIA-Europe project, new GLORIA field sites were established in Switzerland, Italy, New Zealand, and in the Peruvian Andes, and initial fieldwork has already been carried out in the US-Rocky Mountains.

The collaboration with MRI, GTOS, GMBA and the Centre for Mountain Studies at Perth College/Scotland will help to facilitate the integration of GLORIA into the wider scope of global change research that includes physical, biological, and social dimensions. The involvement of powerful environmental and socio-economic NGOs, such as the Worldwide Fund for Nature (WWF) and the Commission Internationale pour la Protection des Alpes (CIPRA) will ensure the flow of information to both the public and to policymakers. GLORIA also intends to take advantage of the 6th EU framework programme.

4.2 Short-term and long-term perspectives

By establishing a long-term "early warning system" for the impacts of climate change with a network of alpine sites, the GLORIA programme provides an important investment for future generations. Not only do the first regional data sets provide an extensive basis for future monitoring at decadal intervals, but when standardised within a global network, these data sets will provide a valuable means for addressing regional differentiation of climate change processes. Standardised global data sets are also invaluable for large-scale comparisons of global change effects on alpine environments. Further, cross-continental and global comparisons on the state of mountain biodiversity will enable the development of data-based model scenarios and first risk assessments for potential biodiversity losses. The state of the Earth's diverse high mountain biota can only be assessed by direct observations. GLORIA is looking forward to providing still lacking hard facts, which are critical for forthcoming political decisions.

5. Acknowledgements

We wish to thank all persons who have supported the GLORIA initiative and who have contributed to the discussion process. We acknowledge the efforts of all colleagues who have established GLORIA field sites and want to encourage all who have expressed their intention to start with the fieldwork. The GLORIA-Europe project is funded by the European Commission within the 5th RTD-Framework Programme. The observation network is further supported by the Austrian Federal Ministry of Education, Science and Culture. The Mount Schrankogel master site is financed by the Austrian Academy of Sciences from the national IGBP and UNESCO-MAB budgets.

6. References

Abrate, S. (1998). "Vegetationskarte des Schrankogel, Stubaier Alpen." Unpublished M.S. thesis, Universität Wien, Vienna.

Ammann, B. (1995). Paleorecords of plant diversity in the Alps. *In* "Arctic and alpine biodiversity: Patterns, causes and ecosystem consequences." (F. S. Chapin III, and C. Körner, Eds.), pp. 136-149. Ecological Studies. Springer, Berlin.

Barthlott, W., Lauer, W., and Placke, A. (1996). Global distribution of species diversity in vascular plants: Towards a world map of phytodiversity. *Erdkunde* **50**, 317-327.

Becker, A., and Bugmann, H. (1997). "Predicting global change impacts on mountain hydrology and ecology: Integrated catchment hydrology/altitudinal gradient studies." IGBP Report 43, Stockholm.

Becker, A., and Bugmann, H. (2001). "Global change and mountain regions. The Mountain Research

Initiative." IGBP Report 49, Stockholm.
Dullinger, S. (1998). "Vegetation des Schrankogel, Stubaier Alpen." Unpublished M.S. thesis, Universität Wien, Vienna.
Ertl, S., Hülber, K., Reiter, K., and Grabherr, G. (2002). Einfluss von Weidevieh und Wild auf die Ausbreitung alpiner Gefäßpflanzen. *In* "Bericht über das 10. Österreichische Botanikertreffen." pp. 7-10. Bundesanstalt für alpenländische Landwirtschaft, Gumpenstein, Irdning.
Gottfried, M., Pauli, H., and Grabherr, G. (1994). Die Alpen im "Treibhaus": Nachweise für das erwärmungsbedingte Höhersteigen der alpinen und nivalen Vegetation. *Jahrbuch des Vereins zum Schutz der Bergwelt* **59,** 13-27.
Gottfried, M., Pauli, H., and Grabherr, G. (1998). Prediction of vegetation patterns at the limits of plant life: A new view of the alpine-nival ecotone. *Arctic and Alpine Research* **30,** 207-221.
Gottfried, M., Pauli, H., Reiter, K., and Grabherr, G. (1999a). A fine-scaled predictive model for changes in species distribution patterns of high mountain plants induced by climate warming. *Diversity and Distributions* **5,** 241-251.
Gottfried, M., Pauli, H., Reiter, K., and Grabherr, G. (1999b). The Austrian research initiative: Global change effects at the low-temperature limits of plant life. *In* "Global change in the mountains." (M. F. Price, T. H. Mather, and E. C. Robertson, Eds.), pp. 54-56. Parthenon, New York.
Gottfried, M., Pauli, H., Hohenwallner, D., Reiter, K., and Grabherr, G. (2002a). GLORIA - the Global Observation Research Initiative in Alpine Environments: Wo stehen wir ? *Petermanns Geographische Mitteilungen* **146,** 69-71.
Gottfried, M., Pauli, H., Reiter, K., and Grabherr, G. (2002b). Potential effects of climate change on alpine and nival plants in the Alps. *In* "Mountain biodiversity - A global assessment." (C. Körner, and E. M. Spehn, Eds.), pp. 213-223. Parthenon, New York.
Grabherr, G., Gottfried, M., and Pauli, H. (1994). Climate effects on mountain plants. *Nature* **369**, 448.
Grabherr, G., Gottfried, M., Gruber, A., and Pauli, H. (1995). Patterns and current changes in alpine plant diversity. *In* "Arctic and alpine biodiversity: Patterns, causes and ecosystem consequences." (F. S. Chapin III, and C. Körner, Eds.), pp. 167-181. Ecological Studies. Springer, Berlin.
Grabherr, G., Gottfried, M., and Pauli, H. (2000a). Hochgebirge als "hot spots" der Biodiversität – dargestellt am Beispiel der Phytodiversität. *Berichte der Reinhold-Tüxen-Gesellschaft* **12**, 101-112.
Grabherr, G., Gottfried, M., and Pauli, H. (2000b). GLORIA: A global observation research initiative in alpine environments. *Mountain Research and Development* **20**, 190-191.
Grabherr, G., Gottfried, M., and Pauli, H. (2001a). Long-term monitoring of mountain peaks in the Alps. *In* "Biomonitoring: General and applied aspects on regional and global scales." (C. A. Burga, and A. Kratochwil, Eds.), pp. 153-177. Tasks for Vegetation Science. Kluwer, Dordrecht.
Grabherr, G., Gottfried, M., and Pauli, H. (2001b). High mountain environment as indicator of global change. *In* "Global change and protected areas." (G. Visconti, M. Beniston, E. D. Iannorelli, and D. Barba, Eds.), pp. 331-345. Kluwer, Dordrecht.
Grabherr, G., Gottfried, M., Hohenwallner, D., Pauli, H., and Reiter, K. (2001c). GLORIA-Europe: Report on the kickoff meeting, 25-29 April, Vienna. *Mountain Research and Development* **21**, 294-295.
Grabherr, G., Gottfried, M., and Pauli, H. (2002). Ökologische Effekte an den Grenzen des Lebens. Dossier: Klima. *Spektrum der Wissenschaft*, 84-89.
Hohenwallner, D., Zechmeister, H., and Grabherr, G. (2002). Bryophyten und ihre Eignung als Indikatoren für den Klimawandel im Hochgebirge - erste Ergebnisse. *In* "Bericht über das 10. Österreichische Botanikertreffen." pp. 19-21. Bundesanstalt für alpenländische Landwirtschaft, Gumpenstein, Irdning.
Hülber, K., Ertl, S., Reiter, K., Gottfried, M., and Grabherr, G. (2002a). Effekte von Weidetieren am alpin/nivalen Ökoton. *In* "Bericht über das 10. Österreichische Botanikertreffen." pp. 119-120. Bundesanstalt für alpenländische Landwirtschaft, Gumpenstein, Irdning.
Hülber, K., Gottfried, M., Pauli, H., and Grabherr, G. (2002b). Phänologie ausgewählter Arten am alpin/nivalen Ökoton der Zentralalpen. *In* "Bericht über das 10. Österreichische Botanikertreffen." pp. 23-25. Bundesanstalt für alpenländische Landwirtschaft, Gumpenstein, Irdning.
Keller, F., and Körner, C. (2003). The role of photoperiodism in alpine plant development. *Arctic, Antartic and Alpine research* (in press).
Körner, C. (2002). Mountain biodiversity, its causes and function: An overview. *In* "Mountain biodiversity: A global assessment." (C. Körner, and E. M. Spehn, Eds.), pp. 3-20. Parthenon, New York.
Körner, C., and Spehn, E. M., Eds. (2002). "Mountain biodiversity: A global assessment." Parthenon, New

York.
Pauli, H., Gottfried, M., and Grabherr, G. (1996). Effects of climate change on mountain ecosystems - Upward shifting of alpine plants. *World Resource Review* **8**, 382-390.
Pauli, H., Gottfried, M., and Grabherr, G. (1999a). A global indicator network for climate change effects on the vegetation in high mountain ecosystems - Proposals from an Austrian IGBP/GCTE-research initiative. *In* "Global change in the mountains." (M. F. Price, T. H. Mather, and E. C. Robertson, Eds.), pp. 25-28. Parthenon, New York.
Pauli, H., Gottfried, M., and Grabherr, G. (1999b). Vascular plant distribution patterns at the low-temperature limits of plant life: The alpine-nival ecotone of Mount Schrankogel (Tyrol, Austria). *Phytocoenologia* **29**, 297-325.
Pauli, H., Gottfried, M., and Grabherr, G. (2001a). High summits of the Alps in a changing climate. The oldest observation series on high mountain plant diversity in Europe. *In* "Fingerprints of climate change: Adapted behaviour and shifting species ranges." (G.-R. Walther, C. A. Burga, and P. J. Edwards, Eds.), pp. 139-149. Kluwer, New York.
Pauli, H., Gottfried, M., Reiter, K., and Grabherr, G. (2001b). High mountain summits as sensitive indicators of climate change effects on vegetation patterns: The "Multi Summit-Approach" of GLORIA (Global Observation Research Initiative in Alpine Environments). *In* "Global change and protected areas." (G. Visconti, M. Beniston, E. D. Iannorelli, and D. Barba, Eds.), pp. 45-51. Kluwer, Dordrecht.
Pauli, H., Gottfried, M., Hohenwallner, D., Reiter, K., and Grabherr, G. (2003a). The GLORIA field manual - Multi-Summit approach. 4th version (in prep.).
Pauli, H., Gottfried, M., Dirnböck, T., Dullinger, S., and Grabherr, G. (2003b). Assessing the long-term dynamics of endemic plants at summit habitats. *In* "Alpine biodiversity in Europe - A Europe-wide assessment of biological richness and change." (L. Nagy, G. Grabherr, C. Körner, and D. B. A. Thompson, Eds.), pp. 195-207. Ecological Studies. Springer, Heidelberg.
Root, T. L., Price, J. T., Hall, K. R., Schneider, S. H., Rosenzweig, C., and Pounds, A. (2003). Fingerprints of global warming on wild animals and plants. *Nature* **421**, 57-60.

A Global Assessment of Mountain Biodiversity and its Function

Eva M. Spehn* and Christian Körner
Global Mountain Biodiversity Assessment, Institute of Botany, University of Basel, CH-4056 Basel, Switzerland
**phone +41-61-267 35 11, fax +41-61-267 35 04, e-mail gmba@unibas.ch*

Keywords: Biological inventory, DIVERSITAS, Global Mountain Biodiversity Assessment, Insurance hypothesis, Organismic diversity, Slope stability.

1. Introduction

The montane and alpine regions of the world cover about 10% of the terrestrial area, a life zone ca. 1000 m above and below the climatic treelines in temperate and tropical latitudes, including some of the biologically richest ecosystems. The alpine life zone above the climatic treeline hosts a vast biological richness, exceeding that of many low elevation biota and covers 3% of the global terrestrial land area (Körner 1995). The overall global vascular plant species richness of the alpine life zone alone was estimated to be around 10,000 species, 4% of the global number of higher plant species. No such estimates exist for animals but based on flowering plants, high elevation biota are, as a general rule, richer in species than might be expected from the land area they cover.

Within the alpine zone, the total plant species diversity of a given region commonly declines by about 40 species of vascular plants per 100 m of elevation (Fig. 1). The upper montane forest, its substitute pastureland, and the often fragmented treeline ecotone also host a wealth of organismic diversity, often exceeding that in the alpine life zone.

U. M. Huber et al. (eds.), Global Change and Mountain Regions, 393-400.

1.1 Causes of high biological diversity in mountains

The causes of this high biological diversity at high altitude are manifold. Mountain terrain is commonly highly fragmented and topographically diverse and this high geodiversity is strongly related to biological diversity, as it reflects the multitude of life conditions in a given area. Patterns of snow distribution reinforce landscape diversity by influencing soils, length of growing season, and microclimate. Selective microenvironments, for example habitats with insufficient or excessive snow cover, are characterized by specialist communities of organisms that may exist in close proximity to one another. In the European Alps, communities with moderate snow cover are richer in species than strongly exposed communities or snowbed sites (Virtanen et al. 2002). High plant diversity in mountains may be attributed in part to the small size of alpine species. Alpine plants are on average one tenth of the size of their closest lowland relatives (Körner 1999), which increases the likelihood of a diverse suite of taxa occurring in a small area. Another important cause of high biological richness in mountains is a moderate disturbance regime. Disturbance can either be related to the dynamic state of the physical environment, which keeps plant communities at an early successional stage or by domestic livestock and/or natural grazing. Although alpine species are usually long-lived, strongly reliant on reproduction by vegetative growth, and often geographically isolated, their genetic diversity within populations is usually surprisingly high due to effective genetic and breeding systems (Körner 1999; Till-Bottraud and Gaudeul 2002).

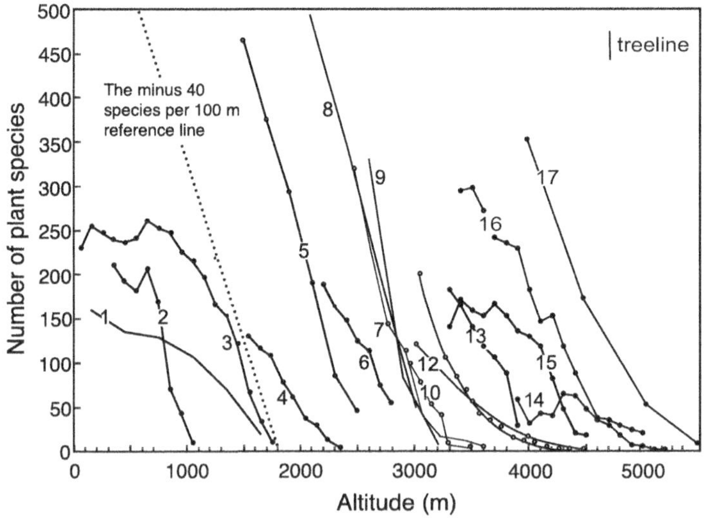

Figure 1: Examples of the elevational reduction in plant species diversity in different mountain ranges. (1) E-, NE-Greenland; (2) Clova, Scotland; (3) Aurland, S-Norway; (4) Jotunheimen, S-Norway; (5) Tatra Alps; (6) Olympus, Greece; (7) Swiss Alps; (8) Bernina, Swiss Alps; (9) West Alps; (10) Bernina, Swiss Alps; (11) Oetztaler Alps, Tyrol; (12) Montafon, Alps; (13) Oytagh, Karakorum; (14) K2-North, Karakorum; (15) Batura, Karakorum; (16) Nanga Parbat, Karakorum; (17) Hindukush. For references see Körner (2002). Figure reproduced with kind permission of the Parthenon Publishing Group.

1.2 Biodiversity provides insurance

Mountain biodiversity is perhaps the best indicator of the integrity of mountain ecosystems. Traffic routes, settlements, streams and lakes, but also water reservoirs depend on the integrity of upslope systems and mountain hydrology is strongly influenced by the type of vegetation and its stability. Diverse mountain forests offer sustainable physical barriers that provide protection from natural hazards and therefore prevent major ecosystem disturbances by mechanical forces (e.g. by avalanches). Ecosystem integrity on steep mountain slopes and in high elevation landscapes is in general a question of soil stability, which in turn depends on plant cover and rooting patterns (Körner 2002). A highly structured, diverse ground cover with different root systems is probably the best insurance for the maintenance of intact slopes that are subject to gravitational force. The many and varied manifestations of alpine environments require different mechanical solutions. Alpine vegetation must cope with physical disturbances such as the impact of heavy rainfall or hail on the ground surface, the disruptive force of surface runoff, mass movements of surficial material on unstable slopes, and snow gliding, and at the same time maintain exploitation of deep substrate moisture, and resist the effects of trampling and grazing by large herbivores. A multitude of plant structures can and do provide a maintenance function when operating in a concerted manner. Yet, natural diseases, divergent life cycles, and varying sensitivity to stress and disturbance may eliminate different players, at least periodically. The more morpho-types of plants co-occur the less likely will extreme events lead to vegetation failure and soil erosion (insurance hypothesis). Although intuitively plausible, this is a research field that is poorly supported by data and the insurance hypothesis is therefore a prime topic in the research promotion agenda of the Global Mountain Biodiversity Assessment.

2. Impacts of global change on mountain biodiversity

2.1 Impacts of climate change on mountain biodiversity

Global warming will reduce available land area for cold adapted organisms and therefore, it will be a threat to mountain plant species richness, especially in isolated ranges where high alpine plants are often restricted to small summits (Wohlgemuth 2002). The uppermost vegetation zones or the uppermost ecotones often host species, which are only uncommonly found in the zones below and are therefore distinguished as nival from the other alpine species. As a consequence of earlier snowmelt and/or climatic warming, typical alpine species may migrate into the nival niche, exerting competitive pressure on the nival flora and potentially leading to biodiversity losses (Gottfried et al. 2002).

With predicted higher temperatures, longer summers with greater incidence of drought are expected in many mountain regions. Recent drought events led to significant losses in the diversity of plant communities in the Cumbres Calchaquies of NE Argentina (Central Andes) (Halloy 2002). In the Cape Fold Mountains in

South Africa, increased incidence of fires is threatening the fynbos (macchia-like) plant communities in many moist highland and marginal arid localities. Substantial plant species replacements are possible (McDonald et al. 2002). Increasing summer temperatures, predicted for this century, are also likely to cause a loss of species in the Australian mountains (cf. Williams et al., this volume), and a stochastic loss of species from surviving alpine islands, related to increased isolation and decreased area, are expected (Kirkpatrick 2002). The responses of mammals and birds to a 30% reduction in snow cover in the Snowy Mountains of Australia over the last 45 years already caused a higher abundance of feral mammals in alpine/high subalpine areas, along with the prolonged winter presence of browsing macropods (Green and Pickering 2002). The predicted impacts of global warming on snow cover will result in a significant change in distribution of animal communities both spatially and temporally.

2.1 Impacts of land use changes on mountain biodiversity

Although global climatic changes can dramatically affect the distribution of plant species in the alpine zone, these changes will most likely be superseded by heavy anthropogenic impacts, such as overgrazing and inappropriate land management in the short term. Of all global change impacts on mountain biodiversity, land use is the most important factor. It is encouraging that traditional upland grazing systems and land management has contributed to the establishment of rich biota in a sustainable way. One example is the Andringitra Massif in south-central Madagascar, which is known as a biodiversity hotspot and is characterized by a high degree of local endemism (Bloesch et al. 2002). Traditional land-use appears to be the key to the preservation of this hotspot of mountain biodiversity, as it replaces former natural drivers of biodiversity, such as fire and (extinct) large herbivores. This traditional knowledge is currently threatened by population pressure and poverty combined with the disappearance of traditional land-use methods and may be lost (Körner 2002). Thus, different approaches to alleviate increasing human pressure on mountain ecosystems and its consequences on biodiversity are needed.

Sarmiento et al. (2002) demonstrated that the traditional long fallow system in the paramo, located in the upper belt of the Northern Andes, reduces the local biodiversity and generates a low economic income for the Andean farmers. Currently, high land use pressure is pushing up the agricultural frontier into the pristine paramo, representing a risk to plant diversity conservation. The paramo is characterised by a very diverse flora with many endemics and particular adaptations to these cool tropical environments.

The best theoretical alternative to retain or enhance both the local biodiversity and the economic profit is to conserve large areas of natural vegetation and to manage the remaining land with a sustainable, but intensive system. In the East African mountains, the natural vegetation has vanished except for a few patches. In regions such as those, human population growth and survival needs exceed land carrying capacity, and biodiversity protection becomes a low priority. A four-year test with different grazing regimes, including exclosures, revealed that high stocking rate pasturing is not necessarily detrimental to species richness and ground cover (Mohamed-Saleem and Woldu 2002). There is also evidence that a certain level of forest use in tropical montane forests in Bolivia is compatible with the conservation of endemic plant taxa (Kessler 2002), where endemism reaches a maximum in moderately anthropogenically disturbed forests at about 3500 m a.s.l.

Medicinal plants are one of the most valuable resources at high altitudes. For example, a survey of the available literature reveals that about 2500 species from the Indian subcontinent are used for local medicinal purposes or commerce/trade, involving the pharmaceutical industry (Purohit 2002). 1748 of these species are from the Indian Himalayan region and 44% of these plants are from the sub-alpine and alpine zones. These are also the species with high economic returns. As this resource is more and more exploited, production- and/or processing-based strategies need to be developed to ensure the sustainable use of medicinal mountain plants.

3. The Global Mountain Biodiversity Assessment

The Global Mountain Biodiversity Assessment (GMBA), initiated by the Swiss Academy of Sciences in 1999, is a global research network dealing with biological richness, its function and change at the cool high elevation limit of the biosphere. GMBA is part of DIVERSITAS (Paris), an international global change research programme on biodiversity sciences.

The understanding of biological diversity in mountains requires a three-dimensional global approach. Firstly, a horizontal, biogeographic dimension with a zonal emphasis is necessary on a global scale (e.g. major mountain regions at different latitudes - from the tropics to the poles). Mountains provide an excellent opportunity for such a global network, as they exist in every climatic zone. Secondly, a vertical, bioclimatologic dimension is required on a regional scale, focusing on the alpine and montane zone (i.e. above and below climatic treeline) and on elevational transects along mountain slopes. Thirdly, a temporal dimension can explain how past environmental changes have shaped current diversity and provide potential analogues for future predictions of global change impacts.

The main goal of GMBA is to document the great biological richness of the mountains of the world and its change induced by both direct and indirect human influences ("global change"), to synthesize existing knowledge and to initiate new research activities with an emphasis on large-scale comparisons. These include cross- and intercontinental comparisons of the upper montane zone, the treeline ecotone and the alpine regions, as well as elevational transects. Another task is to shape a corporate identity, which will help to increase the political visibility of mountain biodiversity issues, and to create a global scientific community involved in mountain biodiversity research, in order to induce transfer of knowledge and cooperation between globally scattered mountain researchers. Most importantly, GMBA wants to investigate the human influence on natural and cultural landscapes in the mountains with the task to preserve mountain biodiversity and to encourage sustainable development of rural areas. To meet this objective, GMBA initiated a project on "High mountain biodiversity and sustainable land use in the tropics/subtropics" with two thematic workshops in Africa and the Andes.

4. Future research needs

Biodiversity research is often seen as an inventory effort, and we certainly need more and better (i.e. in terms of comparability) documents of what the biological richness of regional mountain biota is. Our current mountain biodiversity database has large gaps, with some groups of organisms missing completely for some regions. Therefore, rapid improvement of a mountain biodiversity database is an area of prime engagement of GMBA. Given different levels of development worldwide, a more even distribution of research efforts is needed to arrive at a more balanced understanding of biodiversity and conservation needs. Since we have neither the resources nor the time for a complete biological inventory of all mountain biota across the globe, groups of key stone organisms and taxonomic ratios between groups are promising tools in biodiversity assessments. Taxonomic ratios are based on the assumption that diversity within certain groups of organisms is closely linked with diversity in other groups. In addition, inventories of organismic taxa do not require the visitation of every square km of mountain landscape. 90% of the taxa and measures of overall biotic richness often can be retrieved in sample areas of 10-20 km^2 (or less) within a given biogeographic zone. A promising tool for up-scaling local inventories is remote sensing (satellite) data, offering new avenues of documenting habitat and community diversity over large areas (Braun et al. 2002).

Science needs to provide facts not only on mountain biological diversity itself, but also on its functional significance for mountain ecosystems. This scientific evidence is seen as an addition to, rather than a substitute for, other values of biodiversity, such as the general ethical, aesthetical and economic value. In other words:

1. Both present and future inventories of biological richness need to be analysed by quantitative methods and for functional significance.
2. We need empirical evidence for the insurance hypothesis, because sustained integrity of ecosystems provides the strongest scientific justification for the protection of biodiversity. The insurance hypothesis proposes that high biodiversity buffers the effects of environmental changes on ecosystem processes because different species respond differently to these changes, leading to functional compensations among species. In other words, less affected species can take over the ecosystem function of strongly affected species (e.g. soil protection against erosion).
3. Ecosystem services, such as productivity of upland pastures or erosion control, need to be demonstrated and quantified, which requires experiments.
4. Research needs to explore climate change and management scenarios, which serve both the sustained integrity of diverse mountain biota and human needs.

The current scientific basis on which the benefits of diversity and the potential drawbacks of change can be assessed is rather limited. Much of the current debate is based on observational, plausibility-oriented and theory-based reasoning. Future assessments of mountain biodiversity need to develop a deeper and more functionally oriented search for answers, one of the major tasks of the Global Mountain Biodiversity Assessment (GMBA) and its international networking activity.

5. References

Braun, G., Mutke, J., Reder, A., and Barthlott, W. (2002). Biotope patterns, phytodiversity and forestline in the Andes, based on GIS and remote sensing data. *In* "Mountain biodiversity: A global assessment." (Körner, C., and Spehn, E. M., Eds.), pp. 75-89. Parthenon, London.

Bloesch, U., Bosshard, A., Schachenmann, P., Rabetaliana Schachenmann, H., and Klötzli, F. (2002). Biodiversity of the subalpine forest/grassland ecotone of the Andringitra Massif, Madagascar. *In* "Mountain biodiversity: A global assessment." (Körner, C., and Spehn, E. M., Eds.), pp. 165-176. Parthenon, London.

Gottfried, M., Pauli H., Reiter, K., and Grabherr, G. (2002). Potential effects of climate change on alpine and nival plants in the Alps. *In* "Mountain biodiversity: A global assessment." (Körner C., and Spehn E. M., Eds.), pp 215-226. Parthenon, London.

Green, K., and Pickering, C. (2002). A scenario for mammal and bird diversity in the Snowy Mountains of Australia in relation to climate change. *In* "Mountain biodiversity: A global assessment." (Körner, C., and Spehn, E. M., Eds.), pp. 241-250. Parthenon, London.

Halloy, S. R. P. (2002). Variations in community structure and growth rates of high Andean plants with climatic fluctuations. *In* "Mountain biodiversity: A global assessment." (Körner, C., and Spehn, E. M., Eds.), pp 227-240. Parthenon, London.

Kessler, M. (2002). Plant species richness and endemism of upper montane forests and timberline habitats in the Bolivian Andes. *In* "Mountain biodiversity: A global assessment." (Körner, C., and Spehn, E. M., Eds.), pp. 59-74. Parthenon, London.

Kirkpatrick, J. B. (2002) Factors influencing the spatial restriction of vascular plant species in the alpine archipelagos of Australia. *In* "Mountain biodiversity: A global assessment." (Körner, C., and Spehn, E. M., Eds.), pp. 155-164. Parthenon, London.

Körner, C. (1995). Alpine plant diversity: A global survey and functional interpretations. *In* "Arctic and alpine biodiversity: Patterns, causes and ecosystem consequences." (Chapin, F.S. III, and Körner, C., Eds.), pp. 45-62. Ecological Studies 113, Springer, Berlin.

Körner, C. (1999). "Alpine plant life." Springer, Berlin.

Körner, C. (2002). Mountain biodiversity, its causes and function: An overview. *In* "Mountain biodiversity: A global assessment." (Körner, C., and Spehn, E. M., Eds.), pp. 3-20. Parthenon, London.

McDonald, D., Midgley, C. F., and Powrie, L. (2002). Scenarios of plant diversity in South African mountain ranges to climate change. *In* "Mountain biodiversity: A global assessment." (Körner, C., and Spehn, E. M., Eds.), pp. 263-268. Parthenon, London.

Mohamed-Saleem, M. A., and Woldu, Z. (2002). Land use and biodiversity in upland pastures in Ethiopia. *In* "Mountain biodiversity: A global assessment." (Körner, C., and Spehn, E. M., Eds.), pp. 279-284. Parthenon, London.

Purohit, A. N. (2002). Biodiversity in mountain medicinal plants and possible impacts of climatic change. *In* "Mountain biodiversity: A global assessment." (Körner, C., and Spehn, E. M., Eds.), pp. 269-276. Parthenon, London.

Sarmiento, L., Smith, J. K., and Monasterio, M. (2002). Balancing conservation of biodiversity and economic profit in the high Venezuelan Andes: Is fallow agriculture an alternative? *In* "Mountain biodiversity: A global assessment." (Körner, C., and Spehn, E. M., Eds.), pp. 285-296. Parthenon, London.

Till-Bottraud, I., and Gaudeul, M. (2002). Intraspecific genetic diversity in alpine plants. *In* "Mountain biodiversity: A global assessment." (Körner, C., and Spehn, E. M., Eds.), pp. 23-34. Parthenon,

London.
Virtanen, R., Birnböck, T., Dullinger, S., Pauli, H., Staudinger, M., and Grabherr, G. (2002). Multi-scale patterns in plant species richness of European high mountain vegetation. *In* "Mountain biodiversity: A global assessment." (Körner, C., and Spehn, E. M., Eds.), pp. 91-102. Parthenon, London.
Wohlgemuth, T. (2002). Environmental determinants of vascular plant species richness in the Swiss alpine zone. *In* "Mountain biodiversity: A global assessment." (Körner, C., and Spehn, E. M., Eds.), pp. 103-116. Parthenon, London.

Potential Impacts of Global Change on Vegetation in Australian Alpine Landscapes: Climate Change, Landuse, Vegetation Dynamics and Biodiversity Conservation

Richard J. Williams[1*] and Carl-Henrik Wahren[2]
[1]*CSIRO Sustainable Ecosystems, PMB 44, Winnellie NT 0821, Australia*
[2]*Boreal Ecology Cooperative Research Unit, PO Box 756780, University of Alaska Fairbanks, Fairbanks Alaska 99775, USA. Current address: School of Agriculture, La Trobe University, Bundoora, Victoria, 3083, Australia*
**phone +61 8 89448426, fax +61 8 89448444, email dick.williams@csiro.au*

Keywords: Australia, Disturbance, Grazing, ITEX, National park, Open top chamber.

1. Introduction

The alpine and subalpine regions of south-eastern mainland Australia are small and restricted, covering an area of only approximately 11,000 km^2 in a continent of 7.7 million km^2 (Williams and Costin 1994; Costin et al. 1999; Williams et al. 2003). The most extensive are the Kosciuszko plateau in New South Wales (NSW), the Bogong High Plains in Victoria, the Central Plateau in Tasmania and the mountains of south-west Tasmania (Kirkpatrick 1994; 1997; Williams and Costin 1994; Costin et al. 1999). These areas are of prime importance as catchments for the supply of high quality water to adjacent lowlands; for hydroelectricity generation; for recreation in both summer (e.g. walking, horse riding) and winter (mainly skiing); and for nature conservation. In Victoria and Tasmania, the high country is also used for the summer grazing of domestic cattle. Because of their unique combination of geomorphic, biotic and land-use characteristics, and despite their limited distribution, Australia's high mountain regions are of national and international significance (Kirkpatrick 1994). In

U. M. Huber et al. (eds.), Global Change and Mountain Regions, 401-408.

recognition of this, most of the Australian Alps are designated National Park.

2. Australian high mountain environments

2.1 Landscape features

Over the past 50 years, there has been considerable scientific effort directed towards understanding the distribution and dynamics of Australian alpine landscapes (Costin et al. 1999). Compared with many other mountain systems of the world, the zone above the climatic treeline (alpine life zone) in the Australian high country is comparatively narrow (Costin et al. 1999). Hence, the impact of climate warming on Australian alpine (and high subalpine) ecosystems could be greater than elsewhere, for the simple reason that this biolclimatic zone may retract substantially, or even disappear (Williams et al. 2003). The rate and direction of climate-induced change in ecosystems depend to a large degree on the state of the ecosystem. For the Australian Alps, this relationship between state and ecosystem responses to climate means that landuse-climate interactions will be an important determinant of the magnitude and direction of landscape changes. Predicting these changes is a substantial challenge, because, although the long-term vegetation dynamics are reasonably well-understood for the major plant communities (see below), there has been little systematic investigation of the potential impacts of climate change on the vegetation of the Australian Alps. Yet, establishing strong causal links between climate and vegetation responses is necessary to both increase our understanding of how Australian alpine ecosystems function and allow sound predictions of how these systems are likely to respond to environmental change.

A major characteristic of the Australian mountains is their well-developed mantle of soil. The most widespread soil type, occurring on most well-drained terrain, is alpine humus soil (Williams and Costin 1994). This is a friable, organic loam, and may be up to one metre deep. Organic soils, peats in particular, occur in sites subject to water-logging. All alpine soils in Australia are relatively low in clay, high in organic matter, acidic (pH 4-5) and nutrient poor. They are susceptible to erosion by water, frost and wind, particularly where vegetation has been removed or the soil surface disturbed (Williams and Costin 1994; Kirkpatrick 1997; Costin et al. 1999).

Australian alpine vegetation consists of about ten major structural formations (Williams and Costin 1994; Kirpatrick and Bridle 1998; Williams et al. 2003), which may occur both above and below the climatic treeline (Fig. 1). These include heathlands, grasslands, herbfields, fens, bogs and specialized formations such as feldmarks and bolster heaths. The vegetation dynamics of these plant communities are well understood, largely because the Australian Alps have been the focus of numerous long-term (25-50 year) studies that have been designed to elucidate ecological processes underlying observed patterns and changes in the vegetation (Wimbush and Costin 1979a, b, c; Williams and Ashton 1987; Wahren et al. 1994; 1999; 2001a, b, c, Kirkpatrick 1997).

Figure 1: Typical high mountain landscapes in south-eastern mainland Australia. Top: Wetland complex in high subalpine valley (1770 m) near Mt Cope, Victoria; bottom: Heathlands, grasslands and herbfields above treeline near Mt Nelse (1880 m), Victoria.

2.2 Disturbance regimes

The disturbance regimes within Australian alpine vegetation have been carefully studied. Australian alpine vegetation is subject to natural, small-scale disturbance, caused by the wind, frost heave, water flow, drought, and native herbivores (Williams 1990a). Fire is rare (Wahren et al. 2001c). This small-scale disturbance regime is a fundamental feature of Australian alpine vegetation, and results in localised patches, such as plant litter, canopy gaps, or patches of bare ground, wherein different life forms and species may regenerate. Such small-scale disturbances are important for the long-term dynamics and maintenance of species diversity of all alpine plant communities (Wimbush and Costin 1979 a, b, c; Wahren et al. 1994; 1999; 2001b). Despite recurrent natural or endogenous disturbances, most Australian alpine plant communities usually have a complete cover of vegetation, with little bare ground (<2%).

3. Potential impacts of global change on Australian alpine vegetation

Global change, both climatic and land use, will bring numerous threats to biodiversity and ecosystem function in the Australian alpine environment via its effects on ecological processes such as the small-scale natural disturbance regimes (Williams 1990a; Williams et al. 2003). These threats include grazing by domestic livestock, introduction of exotic species, slope development in ski resorts, growing recreational pressures and global climate change. These different impacts will be discussed in turn below.

3.1 Grazing by domestic livestock

Long-term experimental studies have shown clearly that grazing by domestic livestock causes declines in the abundance of plant species palatable to cattle, and increases in the amount of bare ground (Wimbush and Costin 1979a, b, c; Wahren et al. 1994). Wetlands and snow patch herbfields are extremely sensitive to trampling (Wimbush and Costin 1983; Wahren et al. 1999). Trampling may initiate stream entrenchment, followed by a dramatic increase in the velocity of water, and thus erosion potential (Costin et al. 1999). In wetlands, peat and *Sphagnum,* the major hummock-building moss, may be removed from wetlands, degrading both their high water-holding capacity, and biodiversity conservation values. In snow patch herbfields, where slopes are steep and soils remain moist until well into the growing season, cattle grazing and trampling can increase the amount of bare ground from less than 2% to more than 50% (Wahren et al. 2001a).

3.2 Exotic species

There are about 70 exotic weeds in the Australian alpine vegetation – about 15% of the flora (Mallen 1986). Most weeds are concentrated in ski villages and other highly disturbed areas such as roadsides. About 50 occur in the natural vegetation away from ski resorts, but only a handful are widespread. These include sheep sorrell (*Rumex acetocsella*), cat's ear (*Hypochoeris radicata*), dandelion (*Taraxacum officinale*) and white clover (*Trifolium repens*). Common sources of weeds are garden escapes (especially from gardens in ski villages), rubbish dumps, and species used prior to the 1980s in revegetation efforts. Introduced animals include the Rabbit, Hare, Fox, Feral Dog, Feral Cat, House Mouse, Feral Horse and Sambar deer. Introduced birds are less prominent, but include the Common Starling (around ski villages and construction sites), the Common Skylark (occasionally present on low alpine herbfields) and the Common Blackbird.

3.3 Tourism, skiing and hydro-electric developments

Development of tourist facilities, ski resorts and hydro-electric schemes will continue to result in substantial modification of the alpine landscape, such as tree removal; shrub slashing; and removal of large volumes of soil to construct buildings, roads, dams and aqueducts. Through such disturbances, whole plant communities, such as bogs or snow patch vegetation, may completely disappear from some areas. Such alpine vegetation cannot be replaced. Tourist developments are often located within National Parks, or near Park boundaries. In all cases, expansion of these industries will threaten alpine landscapes, via the process of incremental loss of whole patches of alpine vegetation, and severe modification of others (Williams et al. 2003).

3.4 Climate change

Of the potential threats to landscape integrity in the Australian Alps, it is global climate change, especially rising temperatures, that is capable of having the most severe and long-term impacts. At the same time, it is the least understood factor, and will be the most difficult to manage for. Various scenarios are painted for the Australian Alps, but the common ones are rising temperatures (1-2°C by the middle of the 21st century), reduced winter snow cover, and increased summer rainfall (Williams and Costin 1994; Hughes 2003).

Because of their potential to cause extinctions in the alpine biota, there are three main issues relating to the effects of climate warming on alpine landscapes that concern researchers and land managers in the Australian alps. These are (1) the limited capacity of alpine species to migrate; (2) the unknown capacity of alpine species to adapt; and (3) the potential for alpine vegetation to be invaded by species from lower altitudes, particularly trees and shrubs from the subalpine and montane zones, as these migrate to higher altitudes in response to warming.

We know little about the first two. In general, species respond to changing environments by migration, adaptation, or some combination of the two, and have done so through evolutionary time. However, the specialised biota of the alpine landscapes in Australia cannot rely on migration as a response to climate change, because these species have virtually nowhere else to go. Species may adapt, but the manner by which they may do so, the rate of adaptation, the extent of genetic variation within populations, and the potential for extinctions, are all but unknown for Australian mountain species.

With respect to the third concern, given that the growth of woody species appears to be dependent on summer temperatures in alpine environments, a combination of higher summer temperatures and reduced winter snow pack are likely to increase the establishment, survival and growth of shrubs (Williams 1990b). Establishment of woody vegetation, however, is dependent on the availability of patches of bare ground for establishment of the juveniles (Williams and Ashton 1987). Such patches are not provided by climate change alone, and, given the almost complete cover of vegetation under endogenous disturbance regimes in Australian Alpine environments, the potential for shrub expansion into grasslands and snow-patch herbfields (Wahren et al. 2001a) will depend mostly on the availability of bare patches and therefore on local disturbance regimes. Given the increases in bare ground that accompanies cattle grazing, where this practice continues, the amount of bare ground and potential for shrub invasion will be considerably greater than elsewhere.

If summer rainfall increases with climate change, then wetland vegetation will assume an even greater role in regulating both the quality and rate of catchment discharge. To achieve this, however, the wetlands must be in optimal condition, with a minimum of entrenched drainage channels. Given that most wetlands are currently degraded or in the process of recovering from grazing-induced degradation (Wahren et al. 1999; 2001b) then the removal of grazing from the whole of the high country must occur in order to achieve optimal conditions for the maintenance of viable wetlands.

3.5 Systematic studies of climate change in the Australian Alps: A potential approach for the future

Within the alpine zone, we need to know how climate change will affect the vegetation of both well-drained (supporting grassland, heathland and herbfield) and poorly-drained (the bogs and their associated peat soils) sites. Therefore, in order to construct a knowledge framework within which to manage for climate change in the Australian Alps additional information will be required concerning the dynamics of plant species at the boundaries between the montane forests, the subalpine woodlands and the treeless alpine vegetation, as well as the dynamics of vegetation within the alpine zone.

Although the vegetation dynamics of Australian alpine communities are relatively well understood there has been no systematic ecological research focusing on the potential impact of climate warming on the Australian Alps. There is therefore a need for both species and community level research. Initial studies need to focus on individual plant species because global change, particularly climate change (temperature, precipitation) acts not, in the first instance, on communities or ecosystems, but on species and processes, such as soil nutrient cycling and rate of mineralization. Community and landscape responses follow as a consequence of the differential responses of populations of constituent species. Focusing on specific species in studies of ecological response to climate change may therefore yield more timely results and provide simpler indicators of climate change and biotic response than community-based studies.

One approach that could easily be used in Australia to study the effects of climate change on alpine vegetation is that developed by the International Tundra Experiment: ITEX (Molau and Molgaard 1996). The ITEX program, which commenced in 1990, and currently has sites in several countries, particularly in North America and Europe, has as its main objectives:

a. to quantify the change in the environment brought about by experimental warming;
b. to understand the potential of tundra plant populations to climate warming, either through acclimation or adaptation;
c. to partition the effects of global warming on key morphological, phenological, and physiological traits into environmental and genetic components.

The core ITEX experiment involves the use of small open-topped chambers (OTC) or greenhouses, which passively warm ambient air temperatures within chambers by 1.5-3°C. Responses will concentrate on the timing of key phenological events (e.g. dates for first leaf, flower bud, flower opening), quantitative measures of whole plant morphology (e.g. leaf length, internode length, inflorescence number), canopy height and species cover (Molau and Molgaard 1996).

A complementary approach is to assess biological variability along environmental gradients, in conjunction with small-scale experiments (ca. $10m^2$ scale; e.g. see Harte, this volume). This is a commonly used technique in global change research, especially the Global Change and Terrestrial Ecosystems (GCTE) Program, which is based on the concept of substituting space for time (Koch et al. 1995). In the alpine context, such a design along an altitudinal gradient supposes that by studying environmental variability along the gradient, the potential future conditions (under predicted global warming scenarios) at upper altitudes will be present at the current lower altitudes. Used in combination with ITEX-type experiments, gradient studies are powerful tools for studying the potential impacts of global climate change on landscapes.

4. Conclusion

Climate change poses clear threats to ecosystem function and biodiversity of the Australian Alps. Other manifestations of global change include ski resort development, weed encroachment, and continued livestock grazing. Threats to biodiversity from these uses will continue to grow, even in the unlikely event of little or no changes to climate. Climate change will interact with these exogenous agents of disturbance, but the magnitude and direction of responses are as yet unknown and difficult to predict for Australian alpine environments. Given the rarity of the Australian alpine environment, a substantial investment in research in these areas is warranted. The ITEX approach offers enormous potential in this area, and the Australian Alps offer an excellent opportunity to test the manipulative, experimental components of global change research.

5. References

Costin, A. B., Gray, M., Totterdell, C. J., and Wimbush, D. J. (1999). "Kosciuzsko Alpine Flora." CSIRO, Melbourne.

Hughes, L. (2003). Climate change and Australia: Trends, scenarios and impacts. *Austral Ecology* (in press).

Kirkpatrick, J. B. (1994). "The international significance of the natural values of the Australian Alps." A Report to the Australian Alps Liaison Committee. Australian Alps Liaison Committee, Canberra.

Kirkpatrick, J. B. (1997). "Alpine Tasmania: An illustrated guide to the flora and vegetation." Oxford University Press, Melbourne.

Kirkpatrick J. B., and Bridle, K. (1998). Environment and floristics of ten Australian alpine vegetation formations. *Australian Journal of Botany* **47**, 1-21.

Koch, G. W., Vitousek, P. M., Steffen, W. L., and Walker, B. H. (1995). Terrestrial transects for global change research. *Vegetation* **121**, 53-65.

Mallen, J. (1986). Introduced vascular plants in the high altitude and high latitude areas of Australia, with particular reference to the Kosciusko area, New South Wales. *In* "Flora and fauna of alpine Australasia:

Ages and origins." (B. A. Barlow, Ed.), pp 249-258. CSIRO, Melbourne.

Molau, U., and Molgaarad, P., Eds. (1996). "ITEX Manual." Danish Polar Centre, Copenhagen.

Wahren, C.-H. A., Papst, W. A., and Williams, R. J. (1994). Long-term vegetation change in relation to cattle grazing in subalpine grassland and heathland on the Bogong High Plains: An analysis of vegetation records from 1945 to 1994. *Australian Journal of Botany* **42**, 607-639.

Wahren, C. H., Williams, R. J., and Papst, W. A. (1999). Alpine and subalpine wetland vegetation on the Bogong High Plains, south-eastern Australia. *Australian Journal of Botany* **47**, 165-188.

Wahren, C. H., Williams, R .J., and Papst, W. A. (2001a). Alpine and subalpine snow patch vegetation on the Bogong High Plains, SE Australia. *Journal of Vegetation Science* **12**, 779-790.

Wahren, C. H., Williams, R. J., and Papst, W. A. (2001b). Vegetation change and ecological processes in alpine and subalpine Sphagnum bogs of the Bogong High Plains, Victoria, Australia. *Arctic, Antarctic, and Alpine Research* **33**, 357-68.

Wahren, C. H., Williams, R. J., and Papst, W. A. (2001c). Early post-fire regeneration in subalpine heathland and grassland in the Victorian Alpine National Park, south-eastern Australia. *Austral Ecology* **26**, 670-679.

Williams, R. J. (1990a). Cattle grazing within sub-alpine heathland and grassland communities on the Bogong High Plains: Disturbance, regeneration, and the shrub-grass balance. *Proceedings of the Ecological Society of Australia* **16**, 255-265.

Williams, R. J. (1990b). Growth of sub-alpine shrubs and snowgrass following a rare occurrence of frost and drought in south-eastern Australia. *Arctic and Alpine Research* **22**, 412-422.

Williams, R. J., and Ashton, D. H. (1987). Effects of disturbance and grazing by cattle on the dynamics of heathland and grassland communities on the Bogong High Plains, Victoria. *Australian Journal of Botany* **35**, 413-431.

Williams, R. J., and Costin, A. B. (1994). Alpine and subalpine vegetation. *In* "Australian vegetation." (R. H. Groves, Ed.), pp. 467-500. Cambridge University Press, Melbourne.

Williams, R. J., Mansergh, I. M., Wahren, C.-H. A., Rosengren, N. J., and Papst, W. A. (2003). Alpine landscapes. *In* "Ecology: An Australian perspective." (P. M. Attiwill, and B. Wilson, Eds.), pp. 352-369. Oxford University Press, Oxford.

Wimbush, D. J., and Costin, A. B. (1979a). Trends in vegetation at Kosciusko. I. Grazing trials in the subalpine zone, 1957-1971. *Australian Journal of Botany* **27**, 741-87.

Wimbush, D. J., and Costin, A. B. (1979b). Trends in vegetation at Kosciusko. II. Subalpine range transects, 1959-1978. *Australian Journal of Botany* **27**, 789-831.

Wimbush, D. J., and Costin, A. B. (1979c). Trends in vegetation at Kosciusko. III. Alpine range transects, 1959-1978. *Australian Journal of Botany* **27**, 833-871.

Wimbush, D. J., and Costin, A. B. (1983). Trends in drainage characteristics in the subalpine zone at Kosciusko. *Proceedings of the Ecological Society of Australia* **12**, 143-154.

Ecological Effects of Land-use Changes in the European Alps

Erich Tasser[1*], Ulrike Tappeiner[1,2], and Alexander Cernusca[2]
[1] *European Academy of Bozen, Dursusallee 1, I-39100 Bozen, Italy*
[2] *Institute of Botany, University of Innsbruck, Sternwartestr. 15, A-6020 Innsbruck, Austria*
**phone +39-0471-055311, fax +39-0471-055399, e-mail erich.tasser@eurac.edu*

Keywords: Biodiversity, Bio-geochemical cycles, Comparative measurements, Hydrology, Natural hazards, Transect across the Alps.

1. Introduction

In many mountain regions, there have been dramatic changes in agricultural land use in recent decades. In some cases, these are related to changes in technology, such as the increased use of machine harvesting of hay or a switch from one breed of grazing animals to another. In other cases, the trend has been to abandon agriculture on less productive and least accessible land (Lambin et al. 1999). In the European Alps, for example, 16% of all farm holdings were abandoned within ten years (1980-1990). In addition, almost 70% of the farms that are still in operation today are run only as a secondary source of income. With regard to the land use issue, this means that an average of about 20% of the agricultural land of the Alps has been abandoned, and in some areas as much as 70% (Tappeiner et al. in press). In contrast, farming in the better agricultural locations is being intensified. Hence, land-use changes are considered to be a major driving force behind changes in landscape patterns, ecosystem function and dynamics in Europe (MacDonald et al. 2000).

Management-induced changes in ecosystem structure and function include changes in biodiversity, biogeochemical cycles and hydrological processes (Cernusca et al. 1999a). Changes in land-use can also affect the transport of sensible and latent heat, CO_2, nutrients, and pollutants by altering the exchange processes between

U. M. Huber et al. (eds.), Global Change and Mountain Regions, 409-420.

ecosystems and the lower layers of the atmosphere. Thus, feedback effects between land-use changes and regional climate changes can be expected (Tappeiner and Bayfield 2003). In mountain regions, which can be considered as hydrologically and geomorphically "high energy" environments, changes in ecosystem processes could lead to changes in rates of erosion, magnitude of floods, snow gliding, and avalanches (Tasser et al. 2003).

To quantify farm management induced land-use changes and their ecological consequences in European mountain ranges, several integrated research projects have been carried out over the last several years. Among others, these include the two EU-FP4 projects ECOMONT (Cernusca et al. 1999b) and SUSTALP (Tappeiner et al. 2003) and the INTERREG II-Program INTEGRALP (Tasser et al. 2001). In addition, over a Europe-wide transect, the carbon cycling in mountain landscapes is under investigation in the EU-FP5 Project CARBOMONT (http://botany.uibk.ac.at/forschung/forschungsprojekte/carbomont/).

2. Data and methods

The results presented in this article are based primarily on studies conducted within the framework of ECOMONT and INTEGRALP along a south-north transect that crosses the Alps from Monte Bondone (Trentino, Italy), through the Passeier Valley (South Tyrol, Italy), to the Stubai and Lech Valleys (North Tyrol, Austria) (Fig. 1). The study areas differ from each other in the configuration of their environmental variables, such as geology, climate, and slope aspect, but also in their distinct agricultural history. This is because they all belong to different agrarian regions of the Alps. From an agricultural point of view, the Lech Valley and the study area in Trentino (Monte Bondone) must be classified as "in decline". Characteristics of these regions are a high rate of farm closures and land abandonment, an increasing proportion of older farm managers and a high percentage of very small farms. The Passeier Valley, in contrast, is typical for many areas in the eastern Alps, with a specialization in raising livestock, an economical integration of tourism, and therefore, a relatively high proportion of part-time farming. This is coupled with a relatively low rate of farm abandonment. The Stubai Valley is a region that is characterized by intensive tourism with, in the context of the Alps, below-average employment in agriculture and a slightly above-average percentage of part-time farmers. However, the increase in part-time farming by almost 20% in the last decade clearly demonstrates that this is a fairly recent development in this region. In all study areas, meadows and pastures have been maintained for over 500 years. During the past decades, however, management of many grasslands has been discontinued. Today, there are pastures and meadows with different degrees of use, alongside areas that have lain abandoned for 3-60 years. This mosaic of different sites has allowed us to use a comparative approach in order to assess the ecological effects of different land use intensities in this region. The investigations took place at two different levels: the ecosystem and the landscape level.

Comparative measurements at the ecosystem level were carried out in different Alpine ecosystems in all project areas. Data were collected on vegetation (including canopy structure and ecophysiology), the budgets of energy, carbon, and water, as well as soil properties (including erosion rates). The investigation of the interface between vegetation, soils and the atmosphere, with the main emphasis on bio-geochemical cycles, has been of particular interest. Further, botanical and zoological investigations were carried out to gain insight into biodiversity at the ecosystem level.

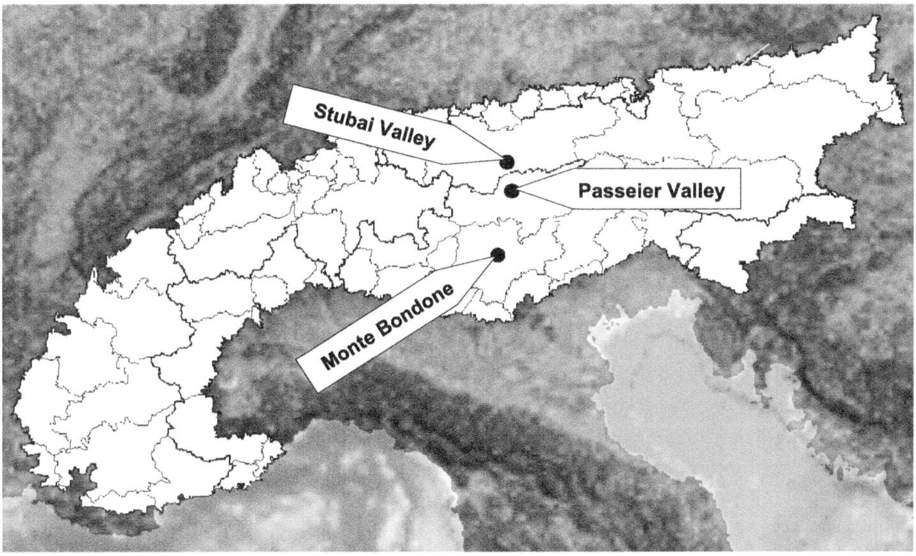

Figure 1: Research sites of ECOMONT and INTEGRALP in the Alps.

To extend our knowledge to larger spatial and temporal scales, we have also been investigating ecological trends related to land use changes on the landscape level. For these studies, we have concentrated on the Passeier Valley, where an area of about 270 km^2 has been investigated. Examples of key research issues include changes in vegetation, landscape diversity, and soil water balance at the patch scale and at the catchment scale. In addition, investigations focused on the potential implications of land-use changes for natural hazards, including analyses of the effects of trampling, erosion (hill slope erosion, landslides), and snow gliding.

The plot-level measurements and the spatial database then served as input for various analyses and models (Cernusca et al. 1999b). We modeled gas exchange from the leaf level, to single plants, and ultimately to the landscape level. At the landscape level, GIS models, in combination with multivariate statistical analysis, were successfully applied to predict the effects of land use changes on different landscape parameters. The models show the most promising results for the prediction of vegetation patterns, soil depths and meteorological phenomena, and the estimation of potential risks in relation to natural and/or human factors. Plot-level results were scaled up to entire study areas.

3. Recent results

Our results clearly indicate that, depending on the type and intensity of land management, not only characteristic vegetation communities have developed, but also related ecosystem structure and functions have been influenced and changed. The most important changes will be discussed in the following sections.

3.1 Land-use changes and biodiversity

Our results from the Passeier Valley support the well-known finding reported in the literature that forested areas have increased over the last 150 years at the cost of pasture and meadows. Inconveniently situated areas, such as high Alpine hay meadows and steep slopes, have been abandoned and are reverting to unmanaged forests. In easily accessible and more productive areas, the landscape has been cleared and has become more homogeneous. Prior to this intensification of agriculture, tilled fields were often alternated with meadows and interrupted by woods and hedgerows. Today, however, fruit and vegetable cultivation dominates the picture in lower elevation areas, and hay meadows are common at higher elevations. Depending on the type and intensity of management, characteristic vegetation communities have developed. Simulation experiments, for example, have shown that only 55% of the present vegetation distribution in the study area is accounted for by natural local factors (e.g. climate and topography). When factors of human influence are added to the model, this leads to a considerably higher prediction accuracy of more than 80 % (Tappeiner et al. 1998; Tasser and Tappeiner 2002). These findings emphasize the strong influence of human impact on the vegetation pattern.

The shifts in vegetation, related to changing management strategies, are often accompanied by changes in biodiversity (Olsson et al. 2000). As our results show, both intensification and abandonment lower plant biodiversity relative to traditional land use patterns (Tasser and Tappeiner 2002) (Fig. 2a). With primary (Orthoptera) and secondary (Carabidae) consumers, however, the intensification of hay meadows does not lead to a reduction of diversity and species richness. In fact, quite the opposite is true. These hay meadows belong to the land-use types with the highest species richness in primary and secondary consumers (Guido and Gianelle 2001). Through abandonment and natural reforestation, however, the diversity diminishes markedly. Unlike vegetation composition, the composition of the orthopteran assemblages is not determined by the nutrient conditions, but is mainly related to the canopy structure resulting from land-use practices. In summary, then, we see that traditional uses result in the highest degree of flora biodiversity, at the ecosystem as well as the landscape level. For the fauna, on the other hand, the interrelations are more complex.

3.2 Land-use changes and bio-geochemical cycles

According to our results, land use changes affect bio-geochemical cycles in many ways (Fig. 3).

Consequences of intensification: Fertilizing, usually with animal dung, improves

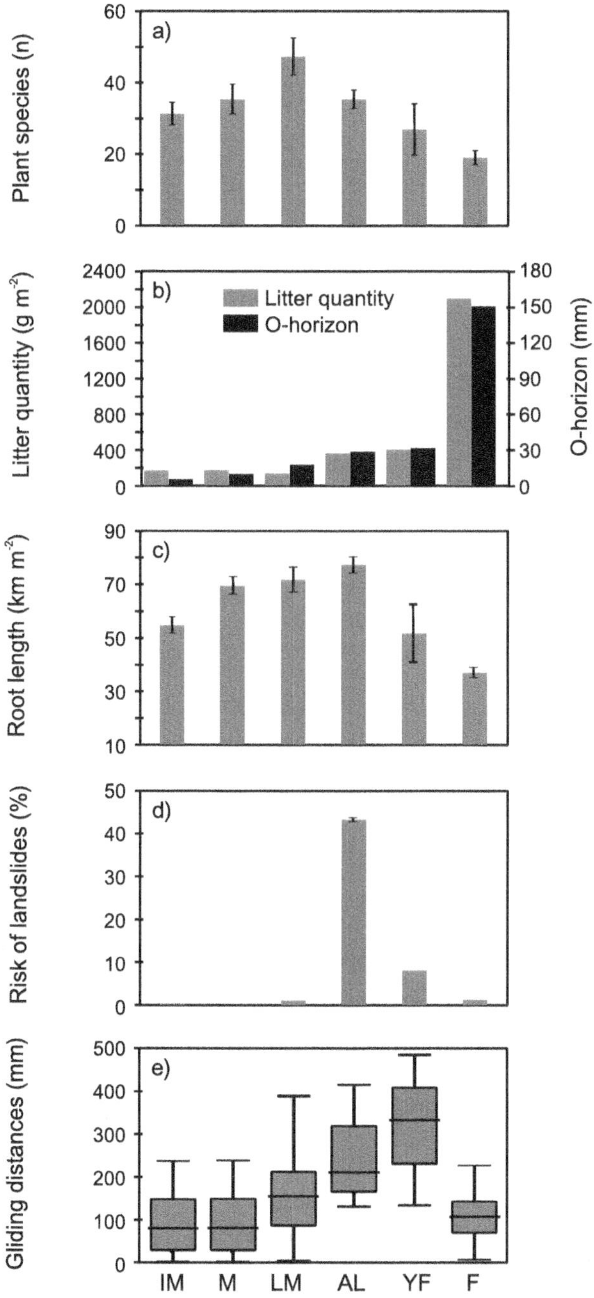

Figure 2: Changes in a) the average number of plant species; b) litter quantity and O-horizon thickness; c) root length per m² soil surface; d) risk of landslides in topsoil; and e) snow gliding distance, in different land-use and succession types. IM = intensively used hay meadow; M = hay meadow, cut every year; LM = lightly used hay meadow, cut every 1-2 years; AL = abandoned land; YF = young forest; F = forest.

nutrient supply to the plants and brings about the intensification of hay meadow cultures. Plant-available mineral nitrogen is strongly related to net nitrogen mineralization (NNM), because mineralization is the key process in making organic nitrogen available to plants (Zeller et al. 2000; 2001). Our investigation of NNM has shown a significant increase with land use intensification, which results in a higher release of plant available nitrogen (N_{min}). In addition, hay harvesting reduces carbon storage in the ecosystem. Therefore, the results of intensification are a reduced accumulation of carbon and an increase in decomposition, which together lead to a reduction of both the C/N ratio and the total amount of N in the soil (Gamper and Tasser 2002).

Especially the amount and type of plant-available nitrogen (N_{min}, sum of NO_3 and NH_4) significantly affect plant species composition (Olff et al. 1994; Tasser and Tappeiner 2002). A better nutrient supply allows nitrophilous species to immigrate. These species have higher photosynthetic capacity, mass-based leaf nitrogen contents, and photosynthetic nitrogen use efficiencies, at both leaf and whole plant levels. This indicates enhanced nutrient cycling on managed meadows as compared to abandoned areas and the ability of the dominant species to more efficiently incorporate nitrogen for increased potential CO_2 uptake (Cernusca et al. 1999b). The canopy grows higher because of the increased productivity of species that are already present and the immigration of nitrophilous and naturally taller growing meadow species. Light levels are reduced by the taller canopy structure so that the low-growing summer species gradually decline. Furthermore, mowing suppresses all of the more slowly developing species (e.g. dwarf shrubs) that have not sufficiently matured before

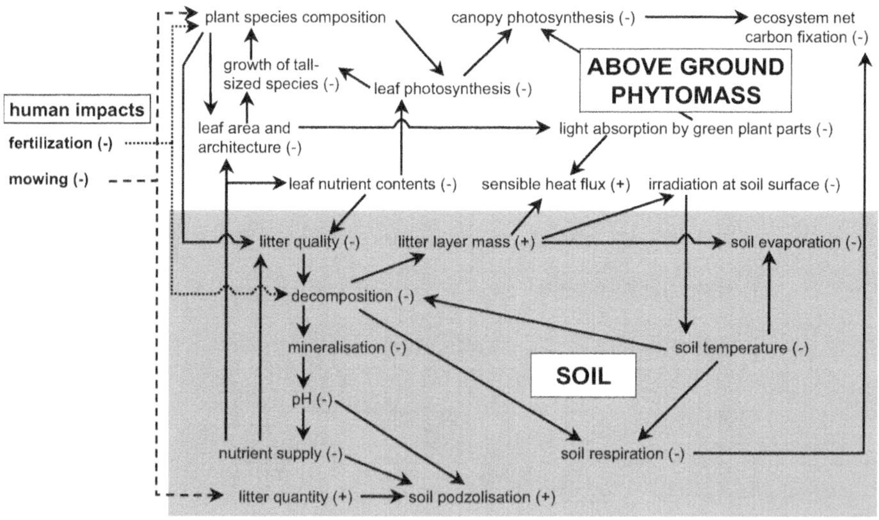

Figure 3: Effects of abandonment of mountain meadows and pastures on several major ecosystem processes (after Cernusca et al. 1999a). Arrows indicate relationships between the processes. (+) and (-) indicate positive and negative effects of abandonment on the processes, as compared to managed meadows and pastures.

mowing. In summary, from a global point of view, intensification of hay meadows leads to a higher nutrient release, reduced nutrient pools, and a higher CO_2 uptake at the net ecosystem level.

Consequences of abandonment: The absence of mowing also leads to strong changes in the ecosystem. Primarily, it causes a sudden rise in the yearly accumulation of litter. Accumulation of dead plant material within the canopy after abandonment has the following consequences for exchange processes (Tappeiner and Cernusca 1998): (1) Abandonment leads to increasing accumulation of dead plant material within the canopy. As a consequence a larger part of the absorbed radiation energy is converted into sensible heat, and therefore evapotranspiration is reduced. (2) A reduction in the use of photosynthetically active radiation by assimilatory plant components occurs, and (3) canopy photosynthesis and water use efficiency decrease.

In the long term, the increase in litter build-up and decomposition results in changes in soil chemistry. The increase in humic acids produced by the decomposition process leads to acidification of the soil, to shifts in the ion ratios, and to an onset of soil podzolization. As a result, vegetation composition changes. Altered nutrient conditions and the cessation of mowing allow acidophilous species, such as dwarf shrubs, to immigrate. As a consequence, the accumulating litter is composed of plant material that takes longer to decompose (twigs, dwarf shrub leaves) (Gamper and Tasser 2002). The results of these changes become apparent in a shift in the microbial biomass and the NNM. The fungal proportion of the microbial biomass increases strongly (Zeller et al. 2001), which leads to a reduction in the NNM and, ultimately, significantly decreases decomposition rates.

Thus, abandonment leads to a decrease in total soil respiration, coupled with decreases in litter decomposition and nutrient availability. Due to a high proportion of dwarf shrubs, abandoned areas are characterised by a high concentration of biomass in the upper soil layers (Fig. 2b). This special structure causes distinct shading effects on the soil surface, and lower soil temperatures and soil heat fluxes, which result in direct changes in the chemical properties of the A-horizon. Thus, there is a positive correlation between the duration of land abandonment and the organic carbon content of the soil, whereas the nitrogen content correlates negatively. Although the supply of plant available nutrients decreases, the nutrient pool increases because of the sharp increase in the thickness of the O-horizon.

3.3 Land-use changes and hydrology

The soils of the studied land use types all proved to have only small differences in soil physical properties (Cernusca et al. 1999b). The clearly higher organic horizons of abandoned areas and forests can store more water than those of intensively managed hay meadows, while the opposite is true for the mineral soil horizons. In the uppermost part of the mineral soil horizons, both large pores and infiltration rates increase from intensively managed hay meadows to forests. The reasons are changes in litter composition and amount, an increase of root biomass and consequent changes in humus content. On the other hand, field capacity decreases with an increase of large pores. Similar soil types under intensively managed hay meadows therefore show higher water storage capacity and more plant-available water than under reduced management. As a whole, these differences in soil physical properties do not significantly affect the soil water balance. Distinct differences are, however, apparent in the evapotranspiration rates of the studied land use types: Forests and hay meadows transpire significantly more water into the atmosphere than, e.g. pastures and fallow fields. Evapotranspiration rates range from 38–58% of precipitation across all land-use types. The remaining water infiltrates the soils or, to a smaller extent, runs off the surface to eventually join the waterways (Tasser et al. 2001). The consequences of changes in evapotranspiration rates become evident in area water balance models of the smallest watersheds: The total reforestation of formerly cultivated areas would result in a reduction in run-off of 7 to 52% (Bou-Vinals unpubl.), and changes in forest cover would therefore have significant effects on the risk of flooding.

3.4 Land-use changes and natural hazards

Our investigations have shown that the vulnerability of alpine ecosystems to landslides, hillslope erosion and snow gliding processes can increase with both the reduction and intensification of agricultural use (Newesely et al. 2000; Tasser et al. 2003) (Fig. 2d, e). When taking into account the effects of exposure and inclination, lightly managed meadows and pastures are significantly less prone to landslides and hillslope erosion than intensively managed meadows and abandoned grasslands. However, it is not the land-use activities themselves that lead to changes in risks for landslides and hillslope erosion, but rather the direct or indirect effects on vegetation and soil properties. It is a well-established fact that root growth prevents landslides, and our studies confirm that especially the impact of cultivation on floristic biodiversity and vegetation cover, and thereby on root diversity and density, is a significant factor. The landslide risk decreases with increasing root density, whereby the highest degree of root density is achieved under low-intensity management (Tasser et al. 2001) (Fig. 2c). Vegetation cover is the determining factor for hillslope erosion. The lower-density vegetation cover of intensively used meadows and pastures provides less protection against soil loss than low-intensity management.

Besides snow depth and slope angle, land use also plays a decisive role in snow gliding and the formation of avalanches. The management of meadows and pastures leads to a reduction in the rate of gliding, whereas land abandonment leads to a distinct increase in the short term (Newesely et al. 2000). The rate of snow gliding is thereby closely related to canopy characteristics. Fallow fields are predominantly covered by long-stemmed grasses and low-growing dwarf shrub communities, which promote snow gliding. However, in the long term, when taller-growing dwarf shrub communities and young trees take over, the risk decreases strongly, as these serve as obstacles for snow gliding.

4. Perspectives for future research

A number of perspectives for future research have emerged among the different topics of our research. Besides the further development and application of new scientific methods, it has become evident that the complexity and sensitivity of mountain areas require continued concerted research efforts, which must be guided by the needs and demands of sustainability at the ecosystem, landscape, and global scales. The consequences of developments in agriculture and forestry are of special significance because of the large land areas involved. Of course, there are other forms of land use, for instance in the tourism, energy, and sports sectors, which need to be taken into consideration in future research projects (see Tappeiner and Bayfield 2003). In the following, we will concentrate on research topics that, from our point of view, are especially important, concerning land-use changes in agriculture and forestry at the levels of ecosystems, landscapes and regions.

The consequences of changes in agricultural and forestry land use at the ecosystem level have probably been most thoroughly investigated. Over the last few years, data have been gathered from a wide variety of national and international projects from all over the world, which concentrated on single cause-effect relationships. These investigations focused on the effects of different types of land-use on, for example, vegetation, bio-chemical processes, or water balance. Of course there are still open questions at the ecosystem level that we feel should be given more attention in the future:

- *Gas exchange:* Management-caused changes in the vegetation composition lead to changes in the canopy structure and the related gas exchange with the atmosphere. Future research on this topic should focus on the linkages between the spatial heterogeneity of ecosystems and its effects on gas exchange processes. In addition, the combined effects of land use changes and global warming on the uptake of CO_2 should be investigated for different ecosystems.
- *Water balance:* Depending on the type of land use, more or less water is stored in the ecosystem, or released into the atmosphere, which, in turn, can influence the flow characteristics of streams and the development of flood events. The effects of land-use changes on water balance will, in our opinion, become an important field of research, especially because superimposed climate change impacts will add further complexity to the chain of effects (see Joint Water Project of the Global Environmental Change Programmes IHDP, WCRP, IGBP and Diversitas; http://www.jointwaterproject.net/)
- *Soils:* Our studies support the results of earlier studies which indicate that a change in land use does not only affect the vegetation cover, but also initiates chemical and physical changes in the soil with effects on soil water budgets and erosion rates. Such changes may occur much faster than generally assumed, which greatly increases the need for research on soil water balance and erosion risk.
- *Applicability of our results in a global context:* Until now, our research projects have only concentrated on the mountain ranges of Europe. It is not yet known if these results are applicable to mountains of other climate zones, such as tropical or subtropical mountain ranges. Future projects should address this question.

Despite the high standard of instruments and field methods available today, research on the environmental consequences of land-use change at the *landscape level* have often not gone beyond generalized statements. The combination of field investigations at the ecosystem level and modeling at the landscape level, using tools such as GIS, allow us to analyze spatial and temporal patterns and to up-scale processes from the plot level to landscapes. From our point of view, it is of special importance for future research activities that results obtained at the ecosystem level can be scaled up to the landscape level, to enable the clarification of the following key questions concerning the ecological effects of large-scale land-use changes: (1) What consequences do large-scale land-use changes have on biodiversity? (2) What are the consequences of large-scale land-use changes on water resources and melting of permafrost? (3) Would there be an increase in erosion and avalanche risk if large areas of alpine meadows were to be abandoned? The results of up-scaling are, of course, only of scientific relevance if they can be corroborated by ground-truthing at the landscape level. For this, measurements of aggregate factors, such as the total runoff and gas exchange of a given area, or remote sensing data are of critical importance (see Cernusca et al. 1999b). The determination of factors, relevant for such a large-scale investigation, should prove to be an interesting study in itself.

Only recently, validated geo-statistical approaches have begun to offer an additional possibility for developing future scenarios on the effects of land use change at the landscape level (Tappeiner et al. 1998). These approaches are particularly valuable because they enable a separate evaluation of natural and anthropogenic impacts on ecosystems and both the impact of land use and future climate change can be taken into consideration. We see the development, validation, and application of such models as a research direction that will be of particular importance in the future.

The landscape level will be an important starting point for interdisciplinary research approaches. First links between ecological and socio-economical models should be initiated at this level. For instance, different agricultural management models (e.g. dairy farming, cattle breeding, vineyards, orchards, abandonment) could be modified to include ecological consequences of land-use change at the landscape level (Tappeiner et al. 2003). This would allow the analysis of interactions between social, economical and ecological components of a unit of manageable size, i.e. that of a single farm holding.

An interdisciplinary approach on the landscape level offers the basis for investigating exchange processes on *regional levels* and enables the aggregation of data on multiple scales. An important part of research activities at the regional level is not only to determine which socio-economical conditions lead to which ecological consequences, but also to recognize the triggering factors for past, present, and future changes in agriculture. The disciplines of natural, social, economical and political sciences must find common links, such as those initiated with the project SUSTALP. The goal should be to develop, by means of this interdisciplinary approach, a system of coupled land use/land cover, ecological, hydrological, and atmospheric models, which also include the triggering political, economical, and social factors of land-use changes. The potential effects of environmental changes (e.g. climate change) should also be taken into consideration. Such coupled models are necessary for different spatial and temporal domains, i.e. from ecosystem to landscape and even regional to global scales, and for timescales from days and weeks to centuries. Moreover, where consequences are cumulative but non-linear, and their impacts are deferred only until critical thresholds are transgressed, the longer-term perspective of paleo-research may be invaluable (Becker and Bugmann 2001). Such a modeling tool, with the help of scenario techniques, would enable the analysis of possible future land use/land cover scenarios with their hydrological and ecological implications. Scenarios that could be explored include the effects of demographic changes in mountain populations, new agrarian techniques or new political circumstances (like the current changes of the Common Agricultural Policy of the EU), and environmental changes.

In summary, our experience in investigating land-use changes and their ecological effects in the European Alps suggests that combining experimental field measurements with dynamic, process-based, and spatially oriented modeling can result in broadening narrow disciplinary and spatial-temporal boundaries. The information this provides, however, represents only an important first step. Making it accessible to the users will be the critical next step. Especially in decision-making processes, it is necessary to develop scenarios that integrate socio-economic models. This requires a dynamic, process-oriented modeling approach – from the ecosystem to the regional level – with a view towards sustainable development.

5. References

Becker, A., and Bugmann, H. (2001). Global change and mountain regions. The Mountain Research Initiative. *IGBP Report* **49**.

Cernusca, A., Bahn, M., Bayfield, N., Chemini, C., Fillat, F., Graber, W., Rosset, M., Siegwolf, R., Tappeiner, U., and Tenhunen, J. (1999a). "ECOMONT: Ecological effects of land-use changes on terrestrial mountain ecosystems." Unpublished Final report. University of Innsbruck, Innsbruck.

Cernusca, A., Tappeiner, U., and Bayfield, N. (1999b). "Land-use changes in European mountain ecosystems (ECOMONT): Concept and results." Blackwell, Berlin.

Gamper, S., and Tasser, E. (2002). Soil development depending on land use and vegetation changes in sub-alpine areas. *In* "Interdisciplinary mountain research." (R. Bottarin, and U. Tappeiner, Eds.), pp. 180-191. Blackwell, Berlin.

Guido, M., and Gianelle, D. (2001). Distribution patterns of four Orthoptera species in relation to microhabitat heterogeneity in an ecotonal area. *Acta Oecologica* **22**, 175-185.

Lambin, E. F., Baulies, X., Bockstael, N., Fischer, G., Krug, T., Leemans, R., Moran, E. F., Rindfuss, R. R., Sato, Y., Skole, D., Turner, B. L., and Vogel, C. (1999). Land-use and land-cover change (LUCC): Implementation strategy. A core project of the International Geosphere-Biosphere Programme and the International Human Dimensions Programme on Global Environmental Change. *IGBP Report* **48**.

MacDonald, D., Crabtree, J. R., Wiesinger, G., Dax, T., Stamou, N., Fleury, P., Gutierrez Lazpita, J., and Gibon, A. (2000). Agricultural abandonment in mountain areas of Europe: Environmental consequences and policy response. *Journal of Environmental Management* **59**, 47-69.

Newesely, Ch., Tasser, E., Spadinger, P., and Cernusca, A. (2000). Effects of land-use changes on snow gliding processes in alpine ecosystems. *Basic and Applied Ecology* **1**, 61-67.

Olff, H., Berendse, F., and De Visser, W. (1994). Changes in mineralization, tissue nutrient concentrations and biomass compartmentation after cessation of fertilizer application to mown grassland. *Journal of Ecology* **82**, 611-620.

Olsson, E. G. A., Austrheim, G., and Grenne, S. N. (2000). Landscape change patterns in mountains, land use and environmental diversity, Mid-Norway 1960–1993. *Landscape Ecology* **15**, 155-170.

Tappeiner, U., and Cernusca, A. (1998). Effects of land-use changes in the Alps on exchange processes (CO_2, H_2O) in grassland ecosystems. *In* "Hydrology, water resources and ecology in headwaters." (K. Kovar, U. Tappeiner, N. E. Peters, and R. G. Craig, Eds.), pp. 131-138. *IAHS Publication* **248**.

Tappeiner, U., Tasser, E., and Tappeiner, G. (1998). Modelling vegetation pattern using natural and anthropogenic influence factors: Preliminary experience with a GIS based model applied to an Alpine area. *Ecological Modelling* **113**, 225-237.

Tappeiner, U., and Bayfield, N. (2003). Management of mountainous areas. *In* "UNESCO Encyclopedia of Life Support Systems (EOLSS)." (in press).

Tappeiner, U., Tappeiner, G., Hilbert, A., and Mattanovich, E. (2003). "SUSTALP. Evaluation of EU-Instruments: Their contribution to a sustainable agriculture and environment in the Alps." Blackwell, Berlin (in press).

Tasser, E., Tappeiner, U., and Cernusca, A. (2001). "Südtirols Almen im Wandel." Europäische Akademie Bozen 28, Bozen.

Tasser, E., and Tappeiner, U. (2002). Impact of land use changes on mountain vegetation. *Applied Vegetation Science* **5**, 173-184.

Tasser, E., Mader, M., and Tappeiner, U. (2003). Effects of land use in alpine grasslands on the probability of landslides. *Basic and Applied Ecology* **4**, 271-280.

Zeller, V., Bahn, M., Aichner, M., and Tappeiner, U. (2000). Impact of land-use changes on nitrogen mineralization in subalpine grasslands in the Southern Alps. *Biology and Fertility of Soils* **31**, 441-448.

Zeller, V., Bardgett, R. D., and Tappeiner, U. (2001). Site management effects on soil microbial properties of subalpine meadows: A study of land abandonment along a north-south gradient in the European Alps. *Soil Biology and Biochemistry* **33**, 639-649.

Climate Interactions in Montane Meadow Ecosystems

John Harte
Energy and Resources Group, University of California, Berkeley, CA 94720 USA
phone 1-510-642-8553, fax 1-510-642-1085, e-mail jharte@socrates.berkeley.edu

Keywords: Climate-ecosystem feedback, Colorado Rockies ecosystem response to climate change, Gradient studies, Warming manipulation.

1. Introduction

Climate change can alter ecosystems and thereby trigger feedback effects that can either enhance or retard the climate change (Lashof et al. 1997). Such feedbacks are especially likely in montane and high-latitude ecosystems where soils are carbon-rich (Whittaker 1975; Schlesinger 1997), ecotones are prevalent as a result of topographic variability, vegetation is sensitive to climatic variables such as snowmelt date and length of growing season (Körner 1992; Harte and Shaw 1995; Goulden et al. 1998), and climate change is expected to be large due to snow-albedo feedback (Groisman et al. 1994). Predicting the chronology and magnitude of such feedbacks is a major challenge in ecology today, as well as an important issue both for global climate change science and policy and, locally, for people whose livelihood is dependent upon montane climatic and ecological regimes.

To investigate montane climate-ecosystem interactions, we are conducting three types of field studies in subalpine meadow habitat. Central to the research is a climate manipulation experiment that uses overhead electric heaters to warm five 30m^2 Rocky Mountain meadow plots (matched with five control plots) by an amount anticipated from global warming models during the middle of this century. The site, called the "warming meadow" is located at 2920 m, 38°53'N, 107°02'W on the western slope of the Colorado Rockies in Gunnison Co. CO, USA. In 1988 we designed the

U. M. Huber et al. (eds.), Global Change and Mountain Regions, 421-427.

experimental facility and in 1990 began collecting data. Since then we have been routinely monitoring effects of the manipulated climate change on soil microclimate, carbon and nitrogen fluxes and pool sizes, and plant growth, flowering success, physiological vigor, phenology, and species diversity.

To extend our knowledge to larger spatial and temporal scales, we are also investigating ecological trends in thirty $16m^2$ meadow plots that lie along a natural elevational and climate gradient within the same drainage (The Upper East River Valley) as the warming meadow. The experimental manipulation provides insights into causal mechanisms governing short-term responses to climate change, while gradient studies help elucidate longer-term phenomena; together they are contributing to the construction of an increasingly unified understanding of ecosystem-climate interactions across a range of space and time scales.

The climate manipulation experiment has allowed us to identify the characteristics of plants whose growth and reproductive success are either severely depressed, unaffected, or enhanced by climate warming. To determine the functional role of those species likely to be eliminated under a warmer climate, we have recently begun a third type of long-term field investigation in meadow plots located near the warming meadow: a double manipulation experiment in which removal of climate-sensitive species is crossed with a nitrogen addition. This experiment will lead to an improved understanding of how an anticipated local loss of species will affect ecosystem resilience to an anthropogenic stress in montane meadow habitat.

This combined use of experimental climate manipulations and observations along climate gradients, coupled with mathematical modeling, provides a novel integrated approach to understanding spatial scaling and feedback in climate-ecosystem interactions. It will contribute to our capacity to develop the next generation of global change models, which need to incorporate ecological as well as physical mechanisms.

2. Recent results

The direct physical effect of the climate manipulation has been to lengthen the snow-free growing season by approximately two weeks at each end, and, during the growing season, to increase soil temperatures at 12 cm depth by approximately 1.5°C and decrease soil moisture by about 15% on a gravimetric basis (Harte et al. 1995). The ecological responses to these climatic changes have included a decrease in annual production, physiological vigor, and flowering success of various species of forbs (Harte and Shaw 1995; Loik and Harte 1996; 1997; de Valpine and Harte 2000; Loik et al. 2000), and an increase in the annual production and physiological vigor of the dominant shrub in the plots, *Artemisia tridentata*, a sagebrush (Shaw et al. 2000; Perfors et al. 2003) (Fig. 1, 2). The forbs that were most sensitive to heating were shallow-rooted species, such as *Erigeron speciosus*, the dominant forb in the plots.

We also observed an increase in the rate of net nitrogen mineralization in the heated plots (Shaw and Harte 2001a) and a 15% decrease in soil organic matter in the top 10 cm of the heated plots (Saleska et al. 1999; 2002) (Fig. 3.). The loss of soil organic matter resulted not from an altered rate of decomposition but rather from the negative net effect of heating on vegetation productivity, which caused a decrease in litter input to soil. We have constructed a mathematical model (Saleska et al. 2002) of carbon cycling in montane meadows that explains the observed decrease in soil carbon in the heated plots and at the same time successfully predicts soil carbon levels along our elevational gradient.

Two ecologically-mediated feedbacks could potentially result from these responses to warming. The loss of soil organic matter was not balanced by an increase in living

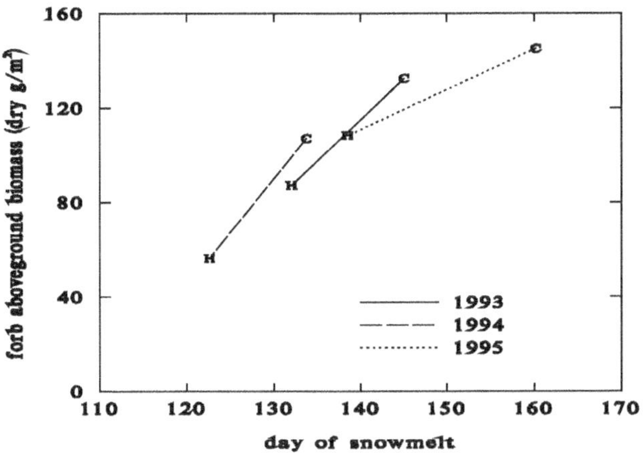

Figure 1: Dependence of treatment-averaged forb aboveground biomass on treatment-averaged date of snowmelt (H = heated, C = control); note that effect of interannual variability in meltdate on forb production is qualitatively consistent with effect of treatment, though weaker in magnitude in control plots.

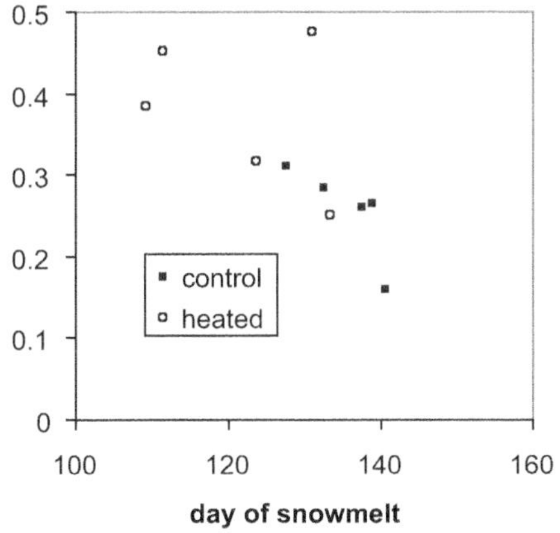

Figure 2: Age-detrended growth rate of *A. tridentata* in heated and control plots versus annually averaged snowmelt date for 1993-2000. With the exception of one outlier, meltdate explains both within and between treatment growth rate (R^2_{adj} = 0.82; treatment significant at p = 0.04 with all data). Explanation of age-detrending procedure and other methods in Perfors et al. (2003).

biomass and thus there was a net loss of total ecosystem carbon. If our results could be extrapolated to large spatial scales, the corresponding increase in atmospheric carbon dioxide would add significantly to the rate of increase of this greenhouse gas, thereby augmenting global warming. The shift in vegetation community structure leads to another, more localized, feedback because the solar reflection coefficients (albedos) of forbs and sagebrush differ significantly. Sagebrush has a lower albedo and thus the increasing dominance of this species under a warmer climate would lower regional albedo and augment the warming.

Currently sagebrush is found on the western slope of the Colorado Rockies at elevations up to 3100 m, but only in scattered patches as the high elevation limit is approached. We anticipate that future climate warming will result in the increasing dominance of sagebrush in montane meadows of the western U.S. as these scattered patches expand.

Although the warming manipulation has resulted in a loss of soil organic matter, we hypothesize that soil organic matter and total ecosystem carbon in the heated plots will recover back toward, or even above, control plot levels during the next decade. This hypothesis is motivated by the fact that shifts in plant community composition can alter both the quantity of litter input to soils and the quality of that litter and the soil organic matter that the litter produces. The observed decrease in soil organic matter resulted from the decline in the quantity of annual litter input, but at the same time the quality of that input is decreasing because sagebrush litter is more recalcitrant than is forb litter (Shaw and Harte 2001b; Saleska et al. 2002). The mathematical model developed to understand the already observed effects of warming on carbon cycling predicts a recovery of the soil carbon level. Continuing research at our sites will provide further tests of the model predictions. We note that if the hypothesized recovery and overshoot of soil carbon is correct, then the sign of the feedback will shift from positive to negative.

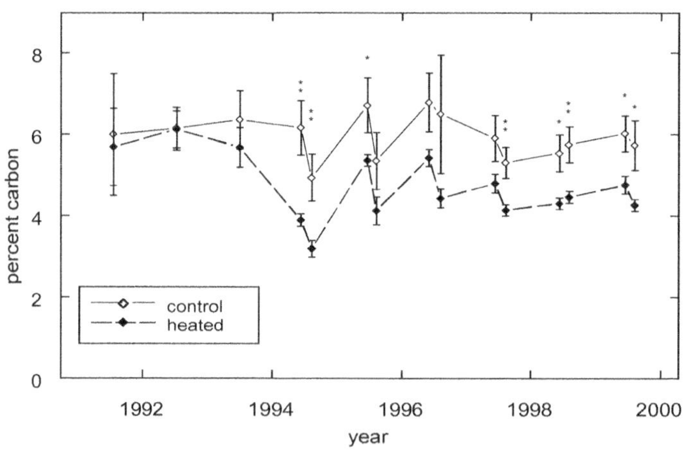

Figure 3: Soil carbon levels in control and heated plots (means ± 1 standard error; n = 5 per treatment in 1991-93, 20 per treatment in 1994-99). * $p<.05$, ** $p<.01$; two-sample t-test (Saleska et al. 2002).

In addition to these effects of heating on vegetation community composition and on soil organic matter, we have observed negative spatial correlations across control plots between plant species richness, on the one hand, and length of growing season, net nitrogen mineralization rate, and extent of sagebrush dominance, on the other. Figure 4 provides a conceptual model for the multiple inter-related influences on plant species diversity. Because our climate manipulation resulted in an increase in all three of these correlates to plant species richness, we hypothesize an eventual decline in plant species richness under a warmer climate. This is motivated by the fact that every correlational connection in Figure 4 suggests such a decline will occur.

We emphasize that, to date, no species have disappeared in any of the warming meadow plots –and thus we have no direct empirical evidence for a future warming-induced decline in forb species richness. Testing this hypothesis, and developing a better understanding of the ecological consequences of declining plant species richness in montane meadows are important priorities for our future research.

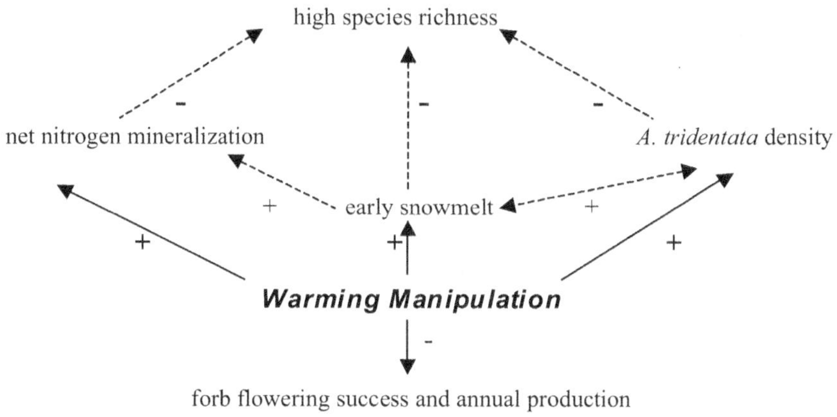

Figure 4: Observed relationships between forb species richness and other ecological parameters. Causal relationships, deduced from the manipulation experiment, are indicated with solid arrows, while purely observational correlations across a range of spatial scales are indicated by dashed arrows.

3. A priority for the future

Experimental climate manipulation in montane meadow habitat at spatial scales on the order of 10 m^2 and over time scales of a decade clearly indicates that even a relatively small amount of climate warming can induce statistically and ecologically significant changes in vegetation communities and biogeochemical processes. Moreover, these ecological responses to warming could result in sizeable feedback effects that might potentially further alter the climate of montane regions. Of the utmost priority now is to develop the capacity to extend the insights obtained from such experimental manipulations to spatial scales on the order of landscapes and whole biomes and to temporal scales on the order of a century.

Using a combination of climate manipulation experiments, observational studies along climate gradients, and a variety of other targeted ecological approaches (plant species removals, biogeochemical manipulations, laboratory soil incubations, mathematical models) we have made some progress in quantifying the generality of our findings from the plot-scale manipulations. Moreover, this methodologically pluralistic approach has allowed us to unravel some of the mechanisms governing climate-ecosystem interactions; it is only by understanding mechanisms that we will, with confidence, be able to scale up our findings in space and time.

Central to the issue of extrapolating in space and time is testing the "space-for-time" assumption that many ecologists implicitly assume to be true but rarely test. The hypothesis states that ecological changes observed along a spatial (in this case also climate) gradient can be directly used to predict how ecosystems at a particular point in space along the gradient will respond to a change in climate over time. Clearly if this assumption were valid, then the enormous effort, use of personnel, and expense of conducting manipulation experiments would be unnecessary and simpler observational work along climate gradients would allow us to predict responses to climate change at large spatial scales.

For certain ecological response variables we have found the assumption to be surprisingly valid. For example, comparing the response of plant phenology to a. climate manipulation, b. variation in climate along an elevational gradient, and c. interannual variation in climate, we have found that for each of these sources of climate variability, the timing of snowmelt explains nearly 100% of the variance in the date of plant flowering in montane meadows for a wide range of species, including both early and late blooming ones. Moreover, the dependence of flowering time on each of these sources of variability is identical, which means that observations along gradients predict response to manipulation (Dunne et al. 2003).

In contrast, when the space-for-time assumption is naively applied to other response variables, such as soil organic matter, the assumption fails dramatically. However, a combination of laboratory soil incubations at a variety of moisture and temperature conditions, foliage and litter chemical analyses, litter bag experiments, and vegetation censusing along a gradient have permitted us to develop, calibrate, and validate a mathematical model (DWP, or "Decomposition Weighted Productivity") that predicts soil organic matter both along gradients and in response to manipulations (Saleska et al. 2002). With the aid of this relatively simple and robust model, the ecological factors that will govern future short- and long-term changes in soil organic matter have been identified and tested.

In summary, our experiences investigating climate-ecosystem dynamics in montane meadows of the Western U.S. suggest that developing the ability to reliably forecast future ecological conditions under a changing climate will require a combination of experimental, observational, and mathematical approaches. More specifically, the goal should be the establishment of validated mathematical models that simultaneously predict short-term ecological responses to plot-scale manipulations and long-term established patterns along large spatial gradients. Grand and complex global models are probably going to be of less value in this endeavor than are targeted, relatively simple models that lack large numbers of difficult-to-measure parameters, and therefore more readily permit validation.

4. References

de Valpine, P., and Harte, J. (2001). Plant responses to experimental warming in a montane meadow. *Ecology* **82**, 637-648.

Dunne, J. A., Taylor, K., and Harte, J. (2003). Subalpine meadow flowering phenology responses to climate change: Integrating experimental and gradient approaches. *Ecological Monographs* **73**, 69-86.

Goulden, M. L., Wolfsy, S. C., Harden, J. W., Trumbore, S. E., Crill, P. M., Gower, S. T., Fries, T., Daube, B. C., Fan S. M., Sutton, D. J., Bazzaz, A., and Munger, J. W. (1998). Sensitivity of boreal forest carbon balance to soil thaw. *Science* **279**, 214-217.

Groisman, P. Y., Karl, T. R., and Knight, R. W. (1994). Observed impact of snow cover on the heat-balance and the rise of continental spring temperatures. *Science* **263**, 198-200.

Harte, J., and Shaw, R. (1995). Shifting dominance within a montane vegetation community-results of a climate-warming experiment. *Science* **267**, 876-880.

Harte, J., Torn, M. S., Chang, F., Feifarek, B., Kinzig, A. P., Shaw, R., and Shen, K. (1995). Global warming and soil microclimate: Results from a meadow warming experiment. *Ecological Applications* **5**, 132-150.

Körner, C. (1992). Response of alpine vegetation to global climate change. *In* "Greenhouse impact on cold-climate ecosystems and landscapes." (M. Boer, and E. Koster, Eds.), pp. 85-96. Catena supplement 22, Catena Verlag, Cremlingen-Destedt, Germany.

Lashof, D. A., DeAngelo, B. J., Saleska, S. R., and Harte, J. (1997). Terrestrial ecosystem feedbacks to global climate change. *Annual Review of Energy and the Environment* **22**, 75-118.

Loik, M. E., and Harte, J. (1997). Changes in water relations for leaves exposed to a climate warming manipulation in the Rocky Mountains of Colorado. *Environmental and Experimental Botany* **37**, 115-123.

Loik, M., and Harte, J. (1996). High temperature tolerance for *Artemisia tridentata* and *Potentilla gracilis* under a climate change manipulation. *Oecologia* **108**, 224-231.

Loik, M., Redar, S., and Harte J. (2000). Photosynthetic responses to a climate-warming manipulation for contrasting meadow species in the Rocky Mountains, Colorado, USA. *Functional Ecology* **14**, 166-175.

Perfors, T., Harte, J., and Alter, S. (2003). Enhanced growth of sagebrush (*Artemisia tridentata*) in response to manipulated ecosystem warming. *Global Change Biology* (in press).

Saleska, S. R., Harte, J., and Torn, M. S. (1999). The effect of experimental ecosystem warming on CO_2 fluxes in a montane meadow. *Global Change Biology* **5**, 125-141.

Saleska, S., Shaw, M., Fischer, M., Dunne, J., Holman, M., Still, C., and Harte, J. (2002). Plant community composition mediates both large transient decline and predicted long-term recovery of soil carbon under climate warming. *Global Biogeochemical Cycles* **16**, 1055 (doi: 10.1029/2001GB001573).

Schlesinger, W. H. (1997). "Biogeochemistry: An analysis of global change." Academic Press, San Diego.

Shaw, M. R., Loik, M. E., and Harte, J. (2000). Gas exchange and water relations of two Rocky Mountain shrub species exposed to a climate change manipulation. *Plant Ecology* **146**,197-206.

Shaw, M. R., and Harte, J. (2001a). Response of nitrogen cycling to simulated climate change: Differential responses along a subalpine ecotone. *Global Change Biology* **7**, 193-210.

Shaw, M. R., and Harte, J. (2001b). Control of litter decomposition in a subalpine meadow-sagebrush steppe ecotone under climate change. *Ecological Applications* **11**, 1206-1223.

Whittaker, R. H. (1975). "Communities and ecosystems." Macmillan, New York.

High Elevation Ecosystem Responses to Atmospheric Deposition of Nitrogen in the Colorado Rocky Mountains, USA

Jill S. Baron[1,2*], Koren R. Nydick[1,6], Heather M. Rueth[3], Brenda Moraska Lafrançois[4,7], and Alexander P. Wolfe[5]
[1]*Natural Resource Ecology Laboratory, Colorado State University, Fort Collins CO 80523 USA*
[2]*United States Geological Survey*
[3]*Ecosystem Center, Marine Biological Laboratory, Woods Hole MA 02543 USA*
[4]*Department of Fishery and Wildlife Biology, Colorado State University, Fort Collins CO 80523 USA*
[5]*Department of Earth and Atmospheric Sciences, University of Alberta, 1-13 ESB, Edmonton, AB T6G 2E3, Canada*
[6]*Aquatic, Watershed and Earth Resources Department, Utah State University, Logan UT 84322 USA*
[7]*St. Croix Watershed Research Station, National Park Service, 16910 152nd St. North, Marine on St. Croix, MN 55047 USA*
**phone 01-1-970-491-1968, fax 01-1-970-491-1965, email jill@nrel.colostate.edu*

Keywords: Acidification, Eutrophication, Forests, Lakes, Nitrogen, Rocky Mountains.

1. Introduction

The rapid rise in human populations and technological advances since 1850 have caused changes in several global scale biogeochemical cycles, including the global nitrogen cycle. The Haber-Bosch process to convert atmospheric nitrogen gas (N_2) to ammonia (NH_3) is now almost universally used to fertilize food crops. The production of nitrogen oxides (NO_x) from combustion for industrial purposes, energy production, and transportation is the other large source of reactive nitrogen to the atmosphere.

U. M. Huber et al. (eds.), Global Change and Mountain Regions, 429-436.

Combined, these two human alterations have added approximately 140 Tg N yr^{-1} to the global reactive N pool, a value that now exceeds natural source contributions of about 100 Tg N yr^{-1} (Galloway and Cowling 2002).

Commensurate with the rise in emissions has been an increase in nitrogen (N) deposition in some parts of the world, with ecological consequences. These consequences include shifts in terrestrial plant assemblages and aquatic algal communities, and lake and stream acidification (Vitousek et al. 1997; Aber et al. 1998). We have been exploring the response of high elevation U.S. Rocky Mountain ecosystems to nitrogen deposition for a number of years. We have taken advantage of a natural meteorological barrier at the mountain crest to compare alpine lakes and subalpine forests and soils on the east side, where N deposition ranges from 3-6 kg N yr^{-1}, with similar systems on the west side, where N deposition is lower. Although the amount of N is low compared with other regions of the world receiving chronically high N deposition loads, high mountain ecosystems are sensitive to change due to harsh climate, large expanses of exposed bedrock and shallow soils, very low vegetation biomass, dilute waters, and a snowmelt-driven hydrology (Baron et al. 2000a). Mountain environments are generally considered to be oligotrophic, so even the slight introduction of a limiting nutrient can lead to marked change. Our conceptual model of response is that vegetation will respond to additional N with increased productivity and, possibly, changes in species composition (Bowman, this volume). When N saturation occurs excess N will begin to influence soil cation exchange and lake fertility. Ultimately soils will become depleted in base cations, leading to lake acidity. Both lake eutrophication and acidification will affect species composition (Fig. 1).

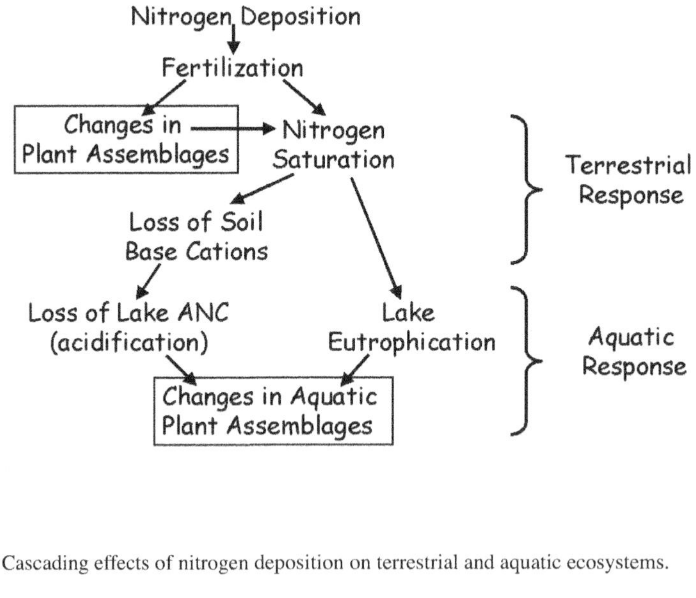

Figure 1: Cascading effects of nitrogen deposition on terrestrial and aquatic ecosystems.

The Colorado Front Range is the easternmost front of the Rocky Mountains in Colorado. It is part of the South Platte River basin, where the Great Plains are home to more than two million people and extensive crop and animal agriculture. Paleolimnological reconstructions of nitrogen deposition proxies (diatom species assemblages) suggest increases began around 1950, commensurate with post-war immigration to Colorado and increased agricultural activity (Baron et al. 2000a; Wolfe et al. 2001). Air masses carrying aerosols and gases rise into the mountains against the prevailing winds primarily in spring and summer months (Parrish et al. 1990; Williams et al. 1996). At the highest elevations air masses curve back to the east as they become entrained in prevailing westerly winds, creating the circumstance of greater N deposition east of the mountain crest (Baron et al. 2000a; Williams and Tonnessen 2000). Our study sites were located between 3000 and 4000 m elevation on public lands that have not been logged or otherwise disturbed. Subalpine forests are composed of 300-700 year old *Picea engelmanii* and *Abies lasiocarpa.* There are hundreds of lakes in the crystalline bedrock; they are clear, low ionic strength cirques that are ice-covered from November to June.

Our investigations in the Colorado Front Range have looked for ecological and biogeochemical changes in otherwise undisturbed forest and lake environments. We have examined forest and soil biogeochemical properties in regions of low and higher N deposition (Rueth and Baron 2002). We have also conducted fertilization experiments in high- and low-N deposition regions to determine the responsiveness of forests with differing nutrient status (Rueth et al. 2003). In aquatic ecosystems we have compared lake nutrient status across our natural meteorological barrier, and explored changes that occur with N additions using both high-resolution paleolimnological records and experimental additions of N to enclosures in lakes of high and low N status (Wolfe et al. 2001; Lafrançois 2002; Nydick 2002).

2. Summary of research findings

2.1 Terrestrial findings

Forest stands showed significant differences in soil and foliar chemistry, and microbial activity depending on their location east or west of the mountain barrier (Rueth and Baron 2002). Six forest stands east of the mountain crest received greater N deposition than five forest stands located west of the mountain crest, but the stands were otherwise similar in exposure (northeast-facing stands), elevation (between 3000 and 3500 m), and mean January and July temperatures. West sites were drier than east sites (860 vs. 1040 mm yearly precipitation), but as the difference was in amount of precipitation that fell as snow, soil moisture remained high through July at all sites. East (high N) sites had greater foliar N and lower C:N ratios than west sites. East sites also had lower lignin:N ratios, and higher potential net mineralization rates. When C:N ratios dropped below 29, as they did in east-side organic horizon soils, mineralization rates increased linearly, an indication of greater N availability (Table 1).

When a fertilization experiment was conducted in two old-growth coniferous

Table 1: Foliar and soil chemistry, and soil microbial responses from six high N and five low N deposition forest stands of the Colorado Front Range. Values are means (std. dev.), and an asterisk (*) denotes significance at 0.05.

	High N (East Side)	Low N (West Side)
Foliar chemistry		
% N	1.14 (0.1)	0.99 (0.1)*
C:N	45.6 (4.2)	52.1 (6.6)*
N:Mg	11.9 (1.7)	9.66 (1.6)*
Soil Characteristics		
Organic soil %N	1.39 (0.2)	1.08 (0.2)*
Organic soil %C	38.8 (4.8)	34.2 (5.9)
Organic soil C:N	25.9 (2.7)	32.5 (5.0)*
Lignin:N	22.2 (3.1)	28.3 (5.7)
Microbial mineralization rate (μg N g^{-1} d^{-1})	3.42 (2.7)	0.69 (1.0)*
Microbial nitrification rate (μg N g^{-1} d^{-1})	0.57 (1.5)	0.06 (0.3)
Response to fertilization		
Foliar %N	no change	significant increase
Organic soil %N	no change	significant increase
Inorganic soil %N	significant increase	no change
Mineralization rate	significant increase	no change

forests, one east and one west of the mountain barrier, the responses differed by location. East side soils with greater total soil N (991 kg ha^{-1}) and low C:N ratios (C:N of 24) showed significant increases in microbial mineralization rates and inorganic soil N over neighboring controls, but foliar N and organic layer soil N remained unchanged. In contrast, fertilization of west side stands (soil C:N ratio of 36, total soil N pool of 605 kg ha^{-1}) showed no change to microbial mineralization rates, but significant increases in foliar and organic soil percent N. The difference in the size of the soil organic N pool and C:N ratio between east and west sites is attributed to N deposition, and these characteristics control the responsiveness of coniferous forests and soils. Additional N inputs to the east (high N) site will enhance N mineralization rates and leaching losses. The west site is still N-limited, and additional N from fertilization is used to enhance biomass. We predict continued fertilization will narrow the C:N ratio to a point where increased biogeochemical N cycling and fluxes will be detected (Rueth et al. 2003).

2.2 Aquatic findings

The mean NO_3 concentration from 30 east side lakes (10.5 μmol L^{-1}) was significantly higher than from 14 west side lakes (6.6 μmol L^{-1}, P=0.02 after concentrations were adjusted to account for seasonal variability; see Baron et al.

2000 for details). There was no pattern of concentration for conductivity with lake location east or west (8.8 μS cm^{-2} compared with 10.3 μS cm^{-2}, p=0.12), but the Acid Neutralizing Capacity (ANC) was significantly lower in east side lakes, with a mean of 49.0 μeq L^{-1} compared with the west side mean of 72.2 μeq L^{-1} (p=0.01 for adjusted values).

Effects of elevated NO_3 on lake biota and nutrient cycling were studied with surveys and micro- and meso-scale experiments. As shown in Figure 1, there is a potential for N to induce eutrophication in alpine lakes, but only if N is a limiting nutrient. Experimental N additions did not cause eutrophication responses in a high NO_3 lake on the east side, but productivity was enhanced with the addition of phosphorus (P), another often-limiting nutrient. N additions to low NO_3 lakes enhanced phytoplankton productivity and biomass, and caused shifts in species dominance from chrysophytes toward chlorophytes, cyanophytes, and diatoms (Fig. 2). A survey of 15 mountain lakes with a range of NO_3 concentrations confirmed that chrysophytes were associated with low NO_3 lakes, while cyanophytes and chlorophytes were more prevalent in the higher NO_3 lakes (Lafrançois 2002). Species

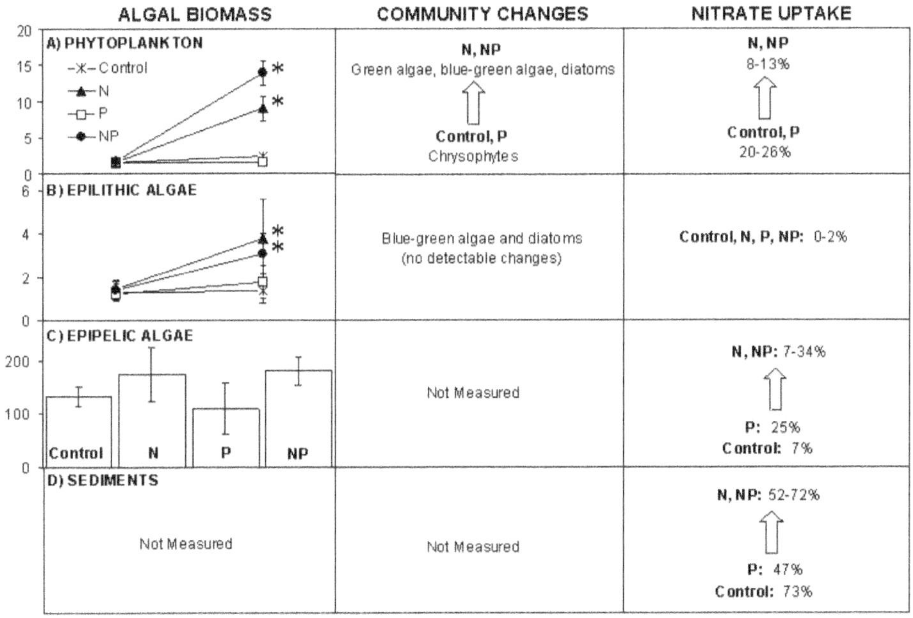

Figure 2: Algal biomass, community changes, and nitrate uptake in a low N lake for A) phytoplankton, B) eplithic algae, C) epipelic algae, and D) sediments in experimental mesocosms. Responses were measured over two weeks and were combined for three different experiments conducted over two summers. Units for algal biomass are mg m^{-3} (phytoplankton) or mg m^{-2} (epilithic and epipelic algae) chlorophyll a. Experimental treatments with added nitrate (N), added phosphate (P), and nitrate plus phosphate (NP). Asterisks on the algal biomass graphs indicate a significant difference from the control at $\alpha \leq 0.05$. Nitrate uptake was determined with ^{15}N isotope. Control and P treatments received only trace applications (<10% of ambient nitrate) of ^{15}N.

richness generally declined with eutrophication (Nydick 2002). When experiments were expanded to look at the response of mountain biota to simultaneous fertilization and acidification in high and low N lakes, phytoplankton species composition was altered and biomass again increased in the low N lakes, but the high N lake responded only to P additions. Combined additions of nutrients and acid, however, altered phytoplankton composition in both lakes more dramatically than from nutrient addition alone, favoring chlorophyte taxa in the low N lake, and chlorophytes and the dinoflagellate Gymnodinium in the high N lake. Changes in algal biomass are clearly related more to nutrient additions than acidification, and while species composition changed with fertilization, even stronger changes were observed with acidification (Lafrançois 2002).

The experimental results confirm a shift in species toward mesotrophic diatoms that has been observed in lake sediment records with the addition of N. Before 1900, diatom assemblages in two high N east side lakes were dominated by typical oligotrophic alpine flora, including *Aulacoseira* spp., *Fragilaria pinnata*, *F. construens* var. *venter*, and *Achnanthes* spp. Between 1950 and 1970 the diatom composition shifted to the mesotrophic taxa *Asterionella formosa* and *Fragilaria crotonensis* and these are now the dominant diatom taxa in many high N lakes (Wolfe et al. 2001). These are species that have been associated with disturbance and slight nutrient enrichment in other studies of European and North American lakes (Andersen et al. 1995; Hall et al. 1999). While sediment records from three low N lakes also show a slight shift away from oligotrophic conditions, the amplitude of change in the low N lakes is less than in the high N lakes (Wolfe, pers. comm.). This suggests that either our mountain crest barrier is not impermeable, or that regional enrichment is occurring, with locations closer to large sources exhibiting greater symptoms.

While the addition of NO_3 affected phytoplankton species and biomass, we found it was not an important factor in determining benthic invertebrate species composition in a wide variety of lakes. Benthic invertebrates were more responsive to elevational gradients and the presence or absence of fish (Lafrançois et al. 2002; Nydick 2002). Zooplankton were somewhat responsive to nutrient additions, but more work needs to be done to determine the effect of N additions on food webs.

Benthic algal responses to nutrients were more variable and differed by substrate. While algae on sediments did not respond to N amendments, algal biomass on clay tiles sometimes responded. However, a ^{15}N isotope tracer experiment revealed that the benthos dominated biogeochemical cycles, and especially NO_3 uptake. Retention of added NO_3 and concurrent biological alkalinity generation were similar between lakes, showing that benthic microbial activity and not nutrient limitation of phytoplankton determined ecosystem NO_3 uptake capacity. N additions generated more than 200 μeq L^{-1} of alkalinity in the low N lake (Nydick 2002).

3. Future research directions

Among the many interesting conclusions that have come from our work, there are two that stand out in their importance for future research and even future policy-making directions.

The first is that high mountain environments are very responsive to slight changes in N availability. Even our fertilization experiments yielded rapid response in spite of low levels of inputs. Forest stands were fertilized with 25 kg ha^{-1} yr^{-1}, while lake nutrient additions increased enclosure concentrations by only 1000 μg L^{-1}.

a. This has profound implications for biodiversity and trophic status of remote mountain ecosystems particularly in the northern hemisphere, since N emissions are increasing sharply. Do some species gain and others lose when N no longer acts as a limiting nutrient? What are the large-scale implications for ecosystems?
b. It appears microbial populations control N cycling in both terrestrial and shallow aquatic environments, but we know very little about these organisms and their response to chronic N additions. How are fungi and bacteria affected by increasing N availability? Is there a threshold of N availability above which N microbial activity is impaired? What are the large-scale implications for ecosystems?
c. Soils and lake sediments hold very large stores of carbon on Earth. Recent work has suggested that N amendments to alpine tundra influences mobilization and turnover of recalcitrant soil carbon in alpine tundra (Neff et al. 2002). The implications are that N stimulates microbial ability to mine carbon. If this is occurring commonly in regions receiving chronic inputs of nitrogen there are ramifications to carbon cycling that have barely begun to be explored. What are the large-scale implications for ecosystems? For regional and global carbon cycles?

The second conclusion is that dramatic changes occur when there are multiple stressors that are different than fertilization effects from N additions alone.

a. Experimental results with P and acid additions confirm this in our aquatic studies. Further research is warranted as mountain ecosystems become ever more inhabited, used, and exposed to an increasing array of exotic chemicals and species introductions.
b. Simulations with climate scenarios also suggest terrestrial nutrient cycling in mountain ecosystems will differ under warmer climate regimes (Baron et al. 2000b). Warmer temperatures could lead to more snowmelt during winter and earlier snowmelt in the spring, with potential for greater soil N retention and reduced N leakage into aquatic ecosystems. Currently soils are warm enough under winter snowpack to support microbial activity, but very dry; addition of meltwater may stimulate soil microbes to take up nitrogen. Earlier snowmelt in spring will enhance vegetation growth and nitrogen uptake during the time of year when inorganic N is most plentiful. Even though enhanced water stress during summer might reduce plant N uptake ability later in the growing season, it is the spring snowmelt season when N is most abundant, so increased N uptake and storage during this season may affect N losses to aquatic systems. The interactions of climate and N cycling need further exploration and experimentation.

Once-isolated mountain ecosystems are increasingly subject to multiple influences, either via atmospheric deposition, climate change, or direct human activities such as non-native species introductions. Our challenge is to integrate across disciplines to understand the interactive effects, and convey this understanding to policy-makers so the knowledge can be used to manage mountain resources, and the forces that lead to their change, more effectively.

4. References

Aber, J., McDowell, W., Nadelhoffer, K., Magill, A., Berntson, G., Kamakea M., McNulty, S., Currie, W., Rustad, L., and Fernandez, I. (1998). Nitrogen saturation in temperate forest ecosystems: Hypothesis revisited. *BioScience* **48**, 921-934.

Anderson, N. J., Renberg, I., and Segerstrom, U. (1995). Diatom production responses to the development of early agriculture in a boreal forest catchment (Kassjön, northern Sweden). *Journal of Ecology* **83**, 809-822.

Baron, J. S., Rueth, H. M., Wolfe, A. P, Nydick, K. R., Allstott, E. J., Minear, J. T., and Moraska, B. (2000a). Ecosystem responses to nitrogen deposition in the Colorado front range. *Ecosystems* **3**, 352-368.

Baron, J. S., Hartman, M. D., Band, L. E., and Lammers, R. B. (2000b). Sensitivity of a high-elevation Rocky Mountain watershed to altered climate and CO_2. *Water Resources Research* **36**, 89-99.

Galloway, J. N., and Cowling, E. B. (2002). Reactive nitrogen and the world: 200 years of change. *Ambio* **31**, 64-71.

Hall, R. I., Leavitt, P. R, Quinlan, R., Dixit, A. S., and Smol, J. P. (1999). Effects of agriculture, urbanization, and climate on water quality in the northern Great Plains. *Limnology and Oceanography* **44**, 739-756.

Lafrançois, B. M. (2002)."Algal and invertebrate responses to atmospheric nitrogen deposition in Rocky Mountain lakes." Ph.D. dissertation, Colorado State University, Fort Collins, Colorado.

Lafrançois, B. M., Carlisle, D. M., Nydick, K. R., Johnson, B. M., and Baron, J. S. (2002). Environmental characteristics and benthic invertebrates in Colorado mountains lakes. *Western North American Naturalist* (in press).

Neff, J. C., Townsend, A. R., Gleixner, G., Lehman, S. J., Turnbull, J., and Bowman, W. D. (2002). Variable effects of nitrogen additions on the stability and turnover of soil carbon. *Nature* **419**, 915-917.

Nydick, K. R. (2002). "Mountain lake responses to elevated nitrogen deposition." Ph. D. dissertation, Colorado State University, Fort Collins. Colorado.

Parrish, D. D., Hahn, C. H., Fahey, D. W., Williams, E. J., Bollinger, M. J., Hübler, G., Buhr, M. P., Murphy, P. C., Trainer, M., Hsie, E. Y., Liu, S. C., and Fehsenfeld, F. C. (1990). Systematic variations in the concentration of NO_x (NO plus NO_2) at Niwot Ridge, Colorado. *Journal of Geophysical Research* **95**, 1817-1836.

Rueth, H. M., and Baron, J. S. (2002). Differences in Englemann spruce forest biogeochemistry east and west of the Continental Divide in Colorado, USA. *Ecosystems* **5**, 45-57.

Rueth, H. M., Baron, J. S., and Allstott, E. J. (2003). Responses of old-growth Engelmann spruce forests to nitrogen fertilization. *Ecological Applications* (in press).

Vitousek, P. M., Aber, J. D., Howarth, R. W., Likens, G. E., Matson, P. A., Schindler, D. W., Schlesinger, W. H., and Tilman, D. G. (1997). Human alteration of the global nitrogen cycle: Sources and consequences. *Ecological Applications* **7**, 737-750.

Williams, M. W., Baron, J. S., Caine, N., Sommerfeld, R., and Sanford, R. (1996). Nitrogen saturation in the Rocky Mountains. *Environmental Science and Technology* **30**, 640-646.

Williams, M. W., and Tonnessen, K. A. (2000). Critical loads for inorganic nitrogen deposition in the Colorado Front Range, USA. *Ecological Applications* **10**, 1648-1665.

Wolfe, A. P., Baron, J. S., and Cornett, R. J. (2001). Unprecedented changes in alpine ecosystems related to anthropogenic nitrogen deposition. *Journal of Paleolimnology* **25**, 1-7.

Mountain Lakes as Indicators of the Cumulative Impacts of Ultraviolet Radiation and other Environmental Stressors

Rolf D. Vinebrooke* and Peter R. Leavitt
Department of Biology, University of Regina, Regina, Saskatchewan, Canada S4S 0A2
**phone 01 306 585 4267, fax 01 306 337 2410, e-mail rolf.vinebrooke@uregina.ca*

Keywords: Climate change, Dissolved organic matter, Ecotonal sensitivity hypothesis, Multiple stressors, Paleoecology, Treeline lakes, Ultraviolet radiation.

1. Introduction

High elevation lake ecosystems are regarded as potentially sensitive indicators of global change because of their cold and dilute abiotic environment, low biodiversity, poor functional redundancy, and relative lack of local human perturbations (Skjelkvåle and Wright 1998; Sommaruga 2001; Battarbee et al. 2002; Psenner et al. 2002). Mountain lakes located near treeline are expected to be the most responsive to long-term impacts of stratospheric ozone depletion and increased flux of solar ultraviolet-B radiation (UV-B; 290-320 nm), climatic warming, and other stressors because of sharp transitions in control processes (Fig. 1) associated with vegetation development and snowpack albedo (Vinebrooke and Leavitt 1998; 1999a; Fyke and Flato 1999). As detailed below, increased flux of solar UV-B and global warming may be already interacting to restructure food webs and biogeochemical cycles in many mountain lakes (Leavitt et al. 1997; Sommaruga-Wögrath et al. 1997).

Ultraviolet radiation (UV; 290-400 nm) can be a key environmental factor affecting certain mountain lakes (Fig. 1). For example, solar irradiance of DNA-damaging UV-B increases by up to 20% per 1000 m of lake elevation (Blumthaler et al. 1992). Similarly, underwater exposure to UV can be high because concentrations

U. M. Huber et al. (eds.), Global Change and Mountain Regions, 437-447.

of UV-attenuating dissolved organic matter (DOM) are among the lowest in the world, and because populations of suspended algae with sunscreen pigments are only sparsely developed in the surface waters of clear mountain lakes (Laurion et al. 2000). In particular, pronounced changes in underwater attenuation of UV occur near tree line, where coniferous vegetation and associated acidic soils favour the development of chromophoric DOM (CDOM), the main regulator of UV penetration in mountain lakes (Sommaruga et al. 1999; Laurion et al. 2000). Because UV attenuation increases exponentially with CDOM concentration, minor variations in CDOM levels can substantially alter biotic exposure to UV in clear mountain lakes that contain less than 2 mg of dissolved organic carbon per liter (DOC l^{-1}) (Vinebrooke and Leavitt 1998; Laurion et al. 2000). In contrast, we expect that the modern impacts of increased UV-B flux on biota are less pronounced in very clear (< 1 mg DOC l^{-1}) alpine lakes because of the potential for acclimation and adaptation resulting from long-term exposure to intense solar irradiance.

This chapter first reviews the direct and indirect causes and effects of UV impacts on biota and ecological processes within mountain lakes, and then considers the evidence for the role of UV in regulating the structure and function of these ecosystems. We conclude by identifying the potential cumulative impacts of UV and other environmental stressors, including global warming, acidic deposition, fisheries management, and land use practices. It is our view that future environmental change will involve synergistic and antagonistic interactions between global and local stressors, which will determine the net effect of UV on sensitive mountain lake ecosystems situated near treeline.

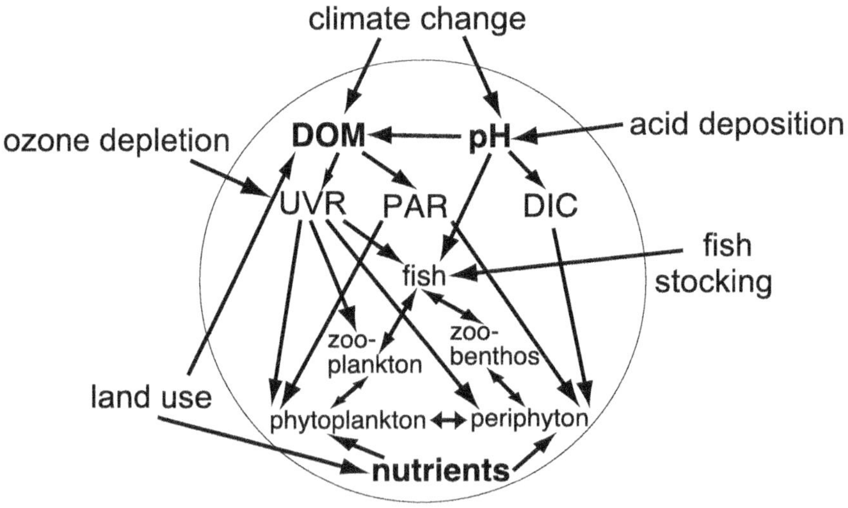

Figure 1: Principle pathways and mechanisms of ultraviolet radiation (UV) impacts on mountain lakes. UV penetration is regulated by external controls of environmental change (outside circle) that moderate the abundance of master chemical variables (bold) including UV-absorbing dissolved organic matter (DOM). PAR = photosynthetically active radiation, DIC = dissolved inorganic carbon. Only control pathways that regulate UV impacts or pathways of ecosystem change are included.

2. Modern impacts of UV on mountain lakes

2.1 Spatial and temporal patterns

Surveys of lakes along altitudinal gradients have identified strong correlations between metrics of UV exposure and aquatic community parameters (Sommaruga et al. 1999; Vinebrooke and Leavitt 1999b). Biotic exposure to UV varies with season, atmospheric and meteorological conditions (e.g. ozone, aerosols, clouds), lake chemistry and morphometry (DOM, pH, depth, refugia), and organism behaviour. In general, changes in CDOM content explain most of the variance in UV transparency of mountain lakes during the ice-free season (Laurion et al. 2000), although turbid inputs can also substantially affect the underwater light environment in glacial-fed mountain lakes. In general, phytoplankton abundance and community composition can be influenced by herbivore biomass, and the presence of photo-protective pigments, or depth refuges from UV, whereas the abundance of rock-associated (epilithic) algae is positively correlated with DOM and negatively correlated with lake elevation (Vinebrooke and Leavitt 1999b; Laurion et al. 2002). Although invertebrate distribution is not strongly related to DOM content or UV exposure, animals are known to exhibit a wide variance in the degree of photo-protective pigmentation by mycosporine-like amino acids (MAA), carotenoids (Tartarotti et al. 2001) and melanin (Hessen 1996). Recent experiments suggest that production of these compounds in response to UV exposure involves both metabolic and ecological costs (Hessen 1996; Hansson 2000). Unfortunately, clear identification of landscape patterns of invertebrate response to UV may be obscured because heavily pigmented invertebrates may be subject to selective predation from visually-feeding fish (Hansson 2000).

2.2 Mechanisms of UV effects

Most short-term laboratory and field experiments demonstrate that exposure to intense UV, especially UV-B, damages cell structures, decreases metabolic efficiency, and reduces individual growth or survival (Karentz et al. 1994). However, the lethality of UV depends also on differences in cellular structure and organism size, sunscreen pigment content, repair mechanism efficiency, and avoidance behaviours (Sommaruga et al. 1999). Consequently, analyses of *in situ* populations reveal that the degree of sensitivity to UV varies substantially among viruses, bacteria, algae, and zooplankton (Fig. 2; Cabrera et al. 1997; Halac et al. 1997; Vinebrooke and Leavitt 1998; 1999a). Similarly, field experiments indicate that aquatic communities in clear mountain lakes can exhibit differential sensitivities to UV depending on their taxonomic composition, habitat type (e.g. water column, rocks, sediments), and trophic position (Fig. 2; Vinebrooke and Leavitt 1998; 1999a). As a result, extrapolation from small-scale laboratory and field experiments to larger spatial and temporal scales has proven to be a challenge (Sommaruga 2001).

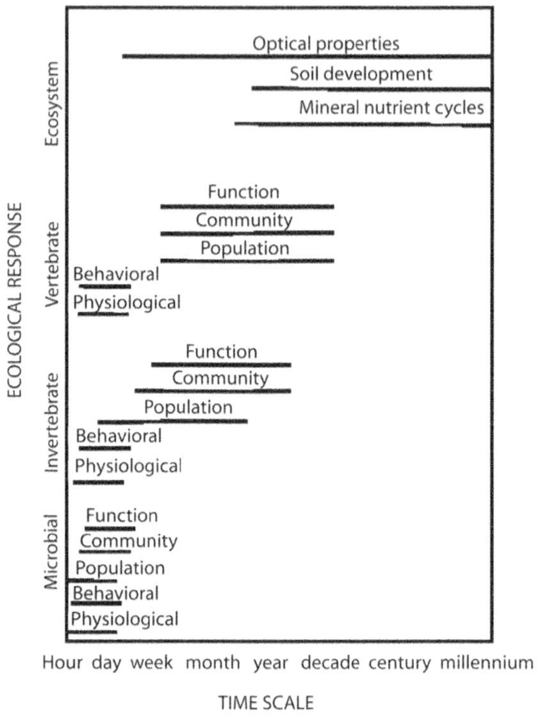

Figure 2: Time scales of ecological responses to the cumulative impact of increased UV-B as mediated by other environmental factors in mountain lakes. Each bar represents the expected period of maximum impact by UV-B on a particular ecosystem component. In general, short-term effects of changes in UV-exposure primarily occur at both lower trophic and cellular levels, while long-term impacts are most pronounced at higher trophic and ecosystem levels.

2.3 Field evidence of UV effects on mountain lake ecosystems

To date, few *in situ* studies have demonstrated detrimental effects of UV on biological communities and related ecological processes (e.g. production, grazing rates) in mountain lakes. Synthesis of early literature suggests that primary producers and zooplankton may be less sensitive to UV than are viruses, bacteria, and heterotrophic protists (Sommaruga 2001), although only conflicting and limited evidence exists for evaluating the relative impacts of UV on primary and secondary production. For instance, short-term studies (Vinebrooke and Leavitt 1996; Sommaruga et al. 1997; Villafane et al. 1999) tend to show more pronounced suppression of primary producers and bacteria than do long-term experiments in alpine lakes (Halac et al. 1997; Vinebrooke and Leavitt 1998; 1999a; Sommaruga et al. 1999). Such discrepancies may be attributable to the relative lack of both physical refugia and potential for compensatory species responses to UV exposure during short-term, small-scale experiments. Similarly, mobile invertebrate species appear less impacted by UV in field trials, suggesting that experimental scale may be a critical factor determining the net response of communities to UV exposure (Vinebrooke and Leavitt 1999a). Regardless of the mechanisms involved, comparison of results from long-term (>1 month) experiments reveals a wide variety of community responses to UV and several apparently contradictory patterns (Sommaruga 2001).

The importance of physical refuges in moderating UV damage is clearly demonstrated in field experiments that compare UV impacts among aquatic communities within the shallow waters of alpine lakes (Vinebrooke and Leavitt 1996; 1998; 1999a). These studies reveal that UV suppresses eukaryotic algae attached to exposed hard surfaces (epilithon) while promoting growth of algae that inhabit soft sediments (epipelon). In contrast, alpine phytoplankton often consist of small cyanobacterial species with high photo-protective and repair capabilities, which are typically unaffected by UV (e.g. Halac et al. 1997; Vinebrooke and Leavitt 1999a). Differential responses of algae to UV appear to arise because sessile epilithic species lack a physical refuge, while mobile epipelic taxa and planktonic cyanobacteria can either actively avoid UV within the lake sediments (Vinebrooke and Leavitt 1999a) or produce sunscreen pigments (Leavitt et al. 1997). Interestingly, the effects of UV on the epilithon may decline over the years as a function of community development (Vinebrooke and Leavitt 1996; 1998; 1999a), suggesting that minor variations in UV have little impact on mature benthic communities in clear alpine lakes. For instance, persistent mats of cyanobacteria are common in shallow, translucent mountain lakes and ponds (Vinebrooke and Leavitt 1999b), in part because exposure to UV initiates production of photoprotective pigments (Leavitt et al. 1997).

2.4 Interactions with other environmental factors

Experimental evidence of interactive effects of UV, CDOM and inorganic nutrients on algal community structure and growth demonstrates that the biological impacts of UV depend upon chemical conditions within clear mountain lakes (Vinebrooke and Leavitt 1998; Xenopoulos et al. 2002). Therefore, chemical changes and their effects on photo-sensitivity and UV-attenuation may have more pronounced impacts on primary producers than does increased flux of solar ultraviolet-B radiation in clear mountain lakes. For example, nutrient availability may determine the extent of UV-damage in biota because a higher nutritional status should enhance the potential for photoprotection and repair.

Synthesis of apparently contradictory data from field surveys and experiments suggests that UV impacts on lakes may be strongest for sites located at tree line, or which otherwise experience rapid changes in CDOM derived from terrestrial sources. On the one hand, in situ experiments reveal few prolonged effects of exposure to modern UV flux on primary or secondary production in either highly transparent or CDOM-rich lakes (Sommaruga 2001). On the other hand, both surveys and manipulative field trials show that the biomass of epilithic algae increases in response to elevated CDOM levels because of the photo-protective and nutritive qualities of CDOM (Vinebrooke and Leavitt 1998; 1999b). Together, these patterns suggest that CDOM acts as a master control of UV impacts on biota (Fig. 1), particularly in lakes at treeline where CDOM influx may vary substantially because of short-

(e.g. droughts; Leavitt et al. 1997) or long-term climatic variability (e.g. reflected in treeline migration; Pienitz and Vincent 2000). In contrast, CDOM levels are consistently low at elevations above treeline (Sommaruga et al. 1999), leading to the establishment of populations that are highly adapted to elevated UV exposure (benthic algae, planktonic cyanobacteria, pigmented invertebrates). In many montane lakes at low elevations, CDOM levels are sufficiently high to attenuate damaging irradiance and restrict detrimental impacts of UV to biota that inhabit shallow-water habitats or which become trapped in surface waters exhibiting temporary thermal stratification (Xenopoulos et al. 2000). Hence, the treeline is expected to represent an ecotone where mountain lake ecosystems exhibit critical threshold responses to climate change and UV exposure. The evaluation of this *ecotonal sensitivity hypothesis* is difficult, but could be conducted by manipulating hydrologic budgets of lakes near treeline to vary CDOM influx and exposure to UV.

3. Long-term impacts of UV on lake ecosystems

According to our hypothesis, the most substantial variations in UV exposure and ecosystem damage should occur on time scales that reflect changes in the factors regulating inputs of UV-absorbing CDOM into mountain lakes situated near treeline (Fig. 1). Although individual biotic responses to UV usually occur in less than 1 year, variations in UV flux (ozone depletion), and development or loss of terrestrial CDOM sources (climate change) occur over a broader range of time (Fig. 2). Further, because factor interactions can often introduce substantial time lags (see below), we feel that the principle UV impacts will be evident only on time scales that extend from decades to centuries.

Paleoecological reconstructions of past levels of CDOM and UV exposure confirm that past variance in underwater irradiance regimes is greater than that arising from the past century of anthropogenic environmental change (ozone depletion, acid deposition, land use) (reviewed in Leavitt et al. 2002). Further, these studies suggest that most aspects of ecosystem structure and function co-vary with past changes in CDOM and UV. In particular, lake production may decline by orders of magnitude during high irradiance events. In general, these retrospective studies use changes in sediment biogeochemistry (organic matter content) or fossil assemblages (diatoms, sunscreen pigments) to reconstruct past levels of CDOM. When coupled with modern lake-surveys that relate CDOM concentration to optical properties of water, fossil-inferred CDOM content can be used to estimate past penetration of UV within the water column. The impacts of past changes in UV exposure on lake biota can then be evaluated by comparing changes in UV penetration with variations in the abundance and composition of sedimentary remains derived from algae (diatom microfossils, pigments), invertebrates (carapaces, mandibles, resting eggs), and, potentially, fish (scales, bones, or inferred from prey fossils).

Comparisons of fossil reconstructions from tree-line lakes in mountain and sub-polar regions reveal that algal production can decline up to 10-fold in response to reductions in CDOM arising from climate change (Leavitt et al. 1997; Pienitz and Vincent 2000; Leavitt et al. 2002). Variability in UV exposure and algal abundance at decadal time scales appears to arise from changes in temperature or precipitation (e.g. droughts) that reduce the transportation of CDOM from terrestrial sources to lakes (Leavitt et al. 1997). For lakes at the tree-line ecotone, small changes in CDOM content (1-3 mg DOC l^{-1}) can lead to exponential increases in UV penetration and loss of most deep-water refuges. At the century scale, development of coniferous forests and soils due to climatic warming can increase CDOM content by >10-fold, leading to similarly large increases in algal abundance if CDOM is delivered to the lake (Pienitz and Vincent 2000; Leavitt et al. 2002). Thus, because of the tight linkage between climatic change and CDOM supply in the past, we propose that most future impacts of UV on mountain lakes will arise because of climatic variability and its interactions with mechanisms regulating CDOM content and optical properties (Fig. 1).

4. Cumulative impacts of increasing UV-B and global change on mountain lakes

There is a growing awareness in the scientific community that the cumulative impacts of global change on aquatic ecosystems will result from interactions among environmental stressors, rather than their individual effects (Schindler 2001). As shown elsewhere in this volume, mountain environments are impacted by several natural and anthropogenic stressors, including climate change, atmospheric pollution, ozone depletion, stocking of fish, and land use (urbanization, agriculture, forestry). In turn, these external sources of environmental stress impact three *master control variables* (DOM, pH, and mineral nutrient content; bold in Fig. 1) that directly and indirectly regulate physical and biological processes in mountain lakes.

The vulnerability of mountain lake ecosystems to UV-B and global change is expected to depend on the relatedness of community tolerances to UV-B and other environmental stressors. If community tolerance to UV-B is negatively correlated with tolerance to another stressor (e.g. warming events), then loss of species and their functional performance is expected to be greater than if tolerances are unrelated. In other words, negatively correlated tolerances should result in multiple stressors having a synergistic impact on communities. Alternatively, communities that display positively correlated tolerances should be less vulnerable to the cumulative effect of UV-B and other stressors. For example, communities that are already stressed by extreme conditions in mountain lakes may be co-tolerant of certain novel stressors arising from global change. Natural stressors that induce community tolerance to anthropogenic stressors would therefore result in an antagonistic interaction in which exposure to one stressor minimizes the impact of the other. Below we use this conceptual framework to forecast the impacts of multiple stressors on climatically sensitive mountain lakes. Consistent with the theme of this chapter, we will focus mainly on future direct and indirect impacts of UV-B.

The impacts of increased UV-B flux on mountain lake ecosystems resulting from stratospheric ozone depletion are likely mediated by interactions with other environmental stressors. For example, global warming could reduce CDOM concentrations in shallow lakes located near tree line, resulting in greater exposure to UV-B (Leavitt et al. 1997) and increased water temperatures (Schindler 2001). In this scenario, warmer water temperatures may alleviate the adverse biological impacts of UV-B by enhancing photorepair processes (Williamson et al. 2002). Acid deposition can also increase exposure to UV-B by removing CDOM from the water column (Donahue et al. 1998). However, acidification may amplify the impacts of UV-B on the productive capacity of mountain lakes because increased acidity suppresses UV-tolerant cyanobacteria (Vinebrooke et al. 2003). Climate warming may mitigate the synergistic impact of acidification and UV-B on biota by enhancing weathering and organic mineralization rates in acid-sensitive alpine catchments (e.g. Sommaruga-Wögrath et al. 1997). Interestingly, extensive stocking of mountain lakes with exotic zooplanktivorous fish is known to lead to selective predation of pigmented invertebrates (Donald et al. 2001; Schindler and Parker 2002), thereby suppressing key mechanisms of invertebrate photo-protection from UV (Hansson 2000). Here, UV-B and introduction of sportfish might exert a synergistic impact on biodiversity and ecosystem processes in clear, naturally fishless mountain lakes. Finally, the degree of ecological damage to mountain lakes by UV may depend on whether land-use practices alter influx of CDOM and mineral nutrients. For example, agriculture, forestry and urbanization can increase inputs of mineral nutrients (N, P) and CDOM to lakes, which can reduce the adverse biological impacts of UV (Vinebrooke and Leavitt 1998). However, regardless of which regulatory pathway predominates, it seems likely that the impacts of future environmental change will be expressed through influences on master variables such as CDOM, rather than directly through UV-B.

The impacts of global stressors are difficult to forecast at century time scales, although paleoecological reconstructions may provide a model basis for prediction. For example, climate warming over several centuries is known to favour development of coniferous forests and soils, the principle supply of CDOM to most mountain lakes (Pienitz and Vincent 2000). Under these conditions, biological production of forested lakes may increase by up to 10-fold relative to tree-less lakes, especially once thresholds of CDOM (~1-2 mg DOC l^{-1}) are surpassed and deepwater refugia from UV are established (Leavitt et al. 2002). Climate warming may lead to upward migration of treeline in fewer than 50 years if the magnitude of warming is sufficient, and aquatic communities can respond with lags of less than a decade (Ammann et al. 2000). Unfortunately, forecasts of the impacts of other external variables are much less certain, mainly because of insufficient information on the potential magnitude of future variability.

5. Conclusion

The current state of knowledge of the impacts of increasing UV-B on mountain lakes indicates that they are context-dependent (Fig. 1 and 2). First, UV-B effects are expected to depend on lake elevation and landscape position. Our *ecotonal sensitivity hypothesis* predicts that UV-B impacts are greatest for lakes located near tree line, where small changes in environmental factors controlling influx or removal of CDOM greatly alter potential exposure to UV. Second, future impacts of increasing UV-B on mountain lakes will likely depend on interactions with other environmental stressors (Fig. 1). In particular, coordinated research activities are urgently needed to identify synergistic interactions between UV-B and other global stressors because these represent worst-case scenarios for the loss of aquatic biodiversity and impairment of ecosystem function. We also recommend that the emphasis of future studies should be on the use of ecologically realistic and large-scale approaches to evaluate the long-term impacts of changes in UV-B on whole communities and ecosystem processes in mountain lakes (cf. Fig. 2).

The hypothesized sensitivity of mountain treeline lakes to environmental change suggests that the ecosystem services that they provide to human society are also vulnerable to the cumulative impacts of global environmental stresses. For example, mountain lakes often serve as sources of high-quality drinking water. The cumulative impact of UV and other stressors on microbial communities has the potential to reduce the water-quality of mountain-fed reservoirs by promoting nuisance species, such as odour-causing algae (Watson et al. 2001). Mountain lakes also often provide habitat for coldwater native and introduced sportfish species. Climate-driven increases in UV exposure and decreases in food availability can severely impair the early developmental stages of fish (Williamson et al. 1999), resulting in a decline in the productive capacity of harvestable fish populations. Therefore, future research needs to both examine the impacts of global change on mountain treeline lakes, and develop appropriate adaptive management strategies.

6. Acknowledgements

This research was supported by NSERC Discovery Grants to RDV and PRL.

7. References

Ammann, B., Birks, H. J. B., Brooks, S. J., Eicher, U., von Grafenstein, U., Hofmann, W., Lemhadl, G., Schwander, J., Tobolski, K., and Wick, L. (2000). Quantification of biotic responses to rapid climate changes around the Younger Dryas: A synthesis. *Palaeogeography, Palaeoclimatology, Palaeoecology* **159**, 313-347.

Battarbee, R. W., Thompson, R., Catalan, J., Grytnes, J.-A., and Birks, H. J. B. (2002). Climate variability and ecosytem dynamics of remote alpine and arctic lakes: The MOLAR project. *Journal of Paleolimnology* **28**, 1-6.

Blumthaler, M., Ambach, W., and Rehwald, W. (1992). Solar UV-A and UV-B radiation flux at two alpine stations at different altitudes. *Theoretical and Applied Climatology* **46**, 39-44.

Cabrera, S., Lopez, M., and Tartarotti, B. (1997). Phytoplankton and zooplankton response to ultraviolet radiation in a high-altitude Andean lake: Short- versus long-term effects. *Journal of Plankton Research* **19**, 1565-1582.

Donahue, W. F., Schindler, D. W., Page, S. J., and Stainton, M. P. (1998). Acid-induced changes in DOC quality in an experimental whole-lake manipulation. *Environmental Science and Technology* **32**, 2954-2960.

Donald, D. B., Vinebrooke, R. D., Anderson, R. S., Syrgiannis, J., and Graham, M. D. (2001). Recovery of zooplankton asssemblages in mountain lakes from the effects of introduced sport fish. *Canadian Journal of Fisheries and Aquatic Sciences* **58**, 1822-1830.

Fyke, J. C., and Flato, G. M. (1999). Enhanced climate change and its detection over the Rocky Mountains. *Journal of Climate* **12**, 230-243.

Halac, S., Felip, M., Camarero, L., Sommaruga-Wögarth, S., Psenner, R., Catalan, J., and Sommaruga, R. (1997). An in situ enclosure experiment to test the solar UV-B impact on microplankton in a high-altitude mountain lake. I. Lack of effect on phytoplankton species composition and growth. *Journal of Plankton Research* **19**, 1671-1686.

Hansson, L.-A. (2000). Induced pigmentation in zooplankton: A trade-off between threats from predation and ultraviolet radiation. *Proceedings of the Royal Society of London B* **267**, 2327-2332.

Hessen, D. O. (1996). Competitive trade-off strategies in Arctic Daphnia linked to melanism and UV-B stress. *Polar Biology* **16**, 573-579.

Karentz, D., Bothwell, M. L., Coffin, R. B., Hanson, A., Herndl, G. J., Kilham, S. S., Lesser, M. P., Lindell, M., Moeller, R. E., Morris, D. P., Neale, P. J., Sanders, R. W., Weiler, C. S., and Wetzel, R. G. (1994). Impact of UV-B radiation on pelagic freshwater ecosystems: Report of working group on bacteria and phytoplankton. *Ergebnisse der Limnologie* **43**, 31-69.

Laurion, I., Ventura, M., Catalan, J., Psenner, R., and Sommaruga, R. (2000). Attenuation of ultraviolet radiation in mountain lakes: Factors controlling the among- and within-lake variability. *Limnology and Oceanography* **45**, 1274-1288.

Leavitt, P. R., Vinebrooke, R. D., Donald, D. B., Smol, J. P., and Schindler, D. W. (1997). Past ultraviolet radiation environments in lakes derived from fossil pigments. *Nature* **388**, 457-459.

Leavitt, P. R., Hodgson, D. A., and Pienitz, R. (2002). Past UV environments and impacts on lakes. *In* "UV effects in aquatic organisms and ecosystems." (E. W. Helbling, and H. Zagarese, Eds.). Vol 2 in Comprehensive Series in Photosciences, Royal Society of Chemistry (in press).

Pienitz, R., and Vincent, W. F. (2000). Effect of climate change relative to ozone depletion on UV exposure in subarctic lakes. *Nature* **404**, 484-487.

Psenner, R., Rosseland, B. O., and Sommaruga, R. (2002). Preface. High mountain lakes and streams: Indicators of a changing world. *Water, Air, and Soil Pollution: Focus* **2**, 1-4.

Schindler, D. W. (2001). The cumulative effects of climate warming and other human stresses on Canadian freshwaters in the new millennium. *Canadian Journal of Fisheries and Aquatic Sciences* **49**, 1431-1438.

Schindler, D. W., and Parker, B. R. (2002). Biological pollutants: Alien fish in mountain lakes. *Water, Air, and Soil Pollution: Focus* **2**, 379-397.

Skjelkvåle, B. L., and Wright, R. F. (1998). Mountain lakes: Sensitivity to acid deposition and global climate change. *Ambio* **27**, 280-286.

Sommaruga, R. (2001). The role of solar UV radiation in the ecology of alpine lakes. *Journal of Photochemistry and Photobiology B: Biology* **62**, 35-62.

Sommaruga, R., Psenner, R., Schafferer, E., Konig, K. A., and Sommaruga-Wögrath, S. (1999). Dissolved organic carbon concentration and phytoplankton biomass in high-mountain lakes of the Austrian Alps: Potential effect of climatic warming on UV underwater radiation. *Arctic, Antarctic, and Alpine Research* **31**, 247-253.

Sommaruga-Wögrath, S, Koinig, K. A., Schmidt, R., Sommaruga, R., Tessadri, R., and Psenner, R. (1997). Temperature effects on the acidity of remote alpine lakes. *Nature* **387**, 64-67.

Tartarotti, B., Laurion, I., and Sommaruga, R. (2001). Large variability in the concentration of mycosporine-like amino acids among zooplankton from lakes located across an altitude gradient. *Limnology and Oceanography* **46**, 1546-1552.

Villafane, V. E., Andrade, M., Lairana, F., Zaratti, F., and Helbing, E. W. (1999). Inhibition of phytoplankton photosynthesis by solar ultraviolet radiation: Studies in Lake Titicaca, Bolivia. *Freshwater Biology* **42**, 215-224.

Vinebrooke, R. D., and Leavitt, P. R. (1996). Effects of ultraviolet radiation on periphyton in an alpine lake. *Limnology and Oceanography* **41**, 1035-1040.

Vinebrooke, R. D., and Leavitt, P. R. (1998). Direct and interactive effects of allochthonous dissolved organic matter, inorganic nutrients and ultraviolet radiation on an alpine littoral food web. *Limnology and Oceangraphy* **43**, 1065-1081.

Vinebrooke, R. D., and Leavitt, P. R. (1999a). Differential responses of littoral communities to ultraviolet radiation in an alpine lake. *Ecology* **80**, 223-237.

Vinebrooke, R. D., and Leavitt, P. R. (1999b). Phytobenthos and phytoplankton as potential indicators of climate change in mountain lakes and ponds: A HPLC-based pigment approach. *Journal of the North American Benthological Society* **18**, 15-33.

Vinebrooke, R. D., Schindler, D. W., Turner, M. A., Findlay, D. L., Paterson, M., and Mills, K. H. (2003).

Trophic dependence of ecosystem resistance and species compensation in experimentally acidified Lake 302S (Canada). *Ecosystems* **6** (in press).

Watson, S. B., Satchwill, T., Dixon, E., and McCauley, E. (2001). Under-ice blooms and source-water odour in a nutrient-poor reservoir: Biological, ecological, and applied perspectives. *Freshwater Biology* **46**, 1553-1567.

Williamson, C. E., Grad, G., De Lange, H. J., Gilroy, S., and Karapelou, D. M. (2002). Temperature-dependent ultraviolet radiation responses in zooplankton: Implications for climate change. *Limnology and Oceanography* **47**, 1844-1848.

Williamson, C. E., Hargreaves, B. R., Orr, P. S., and Lovera, P. A. (1999). Does UV play a role in predation and zooplankton community structure in acidified lakes? *Limnology and Oceanography* **44**, 774-783.

Xenopoulos, M. A., Prairie, Y. T., and Bird, D. F. (2000). Influence of ultraviolet-b radiation, stratospheric ozone variability, and thermal stratification on the phytoplankton biomass dynamics in a mesohumic lake. *Canadian Journal of Fisheries and Aquatic Sciences* **57**, 600-609.

Xenopoulos, M. A., Frost, P. C., and Elser, J. J. (2002). Joint effects of UV radiation and phosphorus supply on algal growth rate and elemental composition. *Ecology* **83**, 423-435.

The Role of Mid-latitude Mountains in the Carbon Cycle: Global Perspective and a Western US Case Study

David Schimel[1*] and B. H. Braswell[2]
[1]*National Center for Atmospheric Research, 1850 Table Mesa Drive, Boulder CO 80305, USA*
[2]*Morse Hall, University of New Hampshire, 39 College Road, Durham, NH 03824-3525, USA*
**phone 01-1-303 497 1610, fax 01-1-303 497 1695 or 492 8699, email schimel@ucar.edu*

Keywords: Carbon, Ecosystem models, Eddy covariance, GPP, Remote sensing, Water

1. Introduction

The International Geosphere Biosphere Program report on mountain ecosystems stresses the potential role of mountainous regions in the Earth's geophysical cycles (Becker and Bugmann 2001). However, mountain environments have rarely been addressed specifically in studies of terrestrial carbon dynamics. Although it was first suggested that the US carbon sink was localized in eastern US forests (Fan et al. 1998), more recent studies that partition the US sink into specific regions suggest that a significant fraction is located in the western US (Schimel et al. 2000; Pacala et al. 2001; Schimel et al. 2002). As increasing development puts pressure on arable lands in North America and Temperate Asia, forests and other high carbon storage ecosystems are increasingly relegated to mountain landscapes. Inspection of recent land cover databases (e.g. IGBP or DeFries et al. 2000) shows clearly that in Temperate North America, Europe and China, a large fraction of forested landscapes is found in major and minor mountain ranges. Figure 1 shows an index of carbon uptake in forests based on forest cover from satellite observations (Defries et al. 2000) and growing season length (with longer growing seasons indicating a higher carbon uptake potential). Growing season lengths are scaled to eddy covariance estimates of carbon uptake

U. M. Huber et al. (eds.), Global Change and Mountain Regions, 449-456.

per growing season day (Falge et al. 2002). Since the majority of current terrestrial sinks are found in the Northern Hemisphere mid-latitudes, montane forests have the potential to contribute significantly to current carbon sinks. We conducted a pilot study using model results and satellite imagery from the Western US to carry out a preliminary assessment of the role of mountains in forest carbon uptake (Schimel et al. 2000).

Most western US mountain regions are covered by forests, which are, for the most part, recovering from historical harvesting and experiencing active fire suppression (Tilman et al. 2000). While we currently perceive western landscapes as pristine natural systems, the Rockies, Sierra, and Cascades were essentially deforested between 1860 and the beginning of the 20th century, during the mining, railroad building and settlement period. Currently, the fraction of old growth forest remaining in the west is variously estimated as 5-15%, however this number must be interpreted with caution. In some regions, high elevation old-growth forests of limited current economic value are excluded. In other cases, relatively young forest stands are not included in the category of old growth forests even in systems that are characterized by short natural disturbance cycles. As well, in some areas where fires were actively suppressed throughout the 20th century, particularly adjacent to urban areas, lower treelines are decreasing in elevation, and forests are moving into relatively productive regions. While the proportion of total forestland is probably higher than oft-quoted numbers (taking into account regional variations in reporting of old growth and high-elevation forests), it remains that a very considerable fraction of the more productive forest lands have experienced some degree of historical disturbance and is in the process of regeneration.

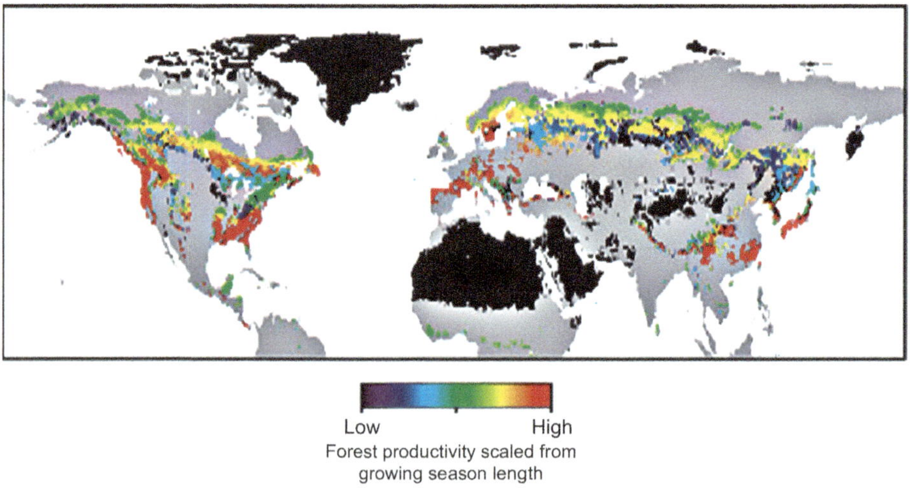

Figure 1: Regions of high carbon storage in the Northern Hemisphere, indicating that in Europe, North America and China, most of the mid-latitude "hot spots" are in mountain ranges. Exceptions to this are found in the boreal (which contributes relatively little globally to carbon uptake) and in parts of the southeastern US.

Fire suppression is another factor resulting in carbon accumulation, as fire suppression tends to increase carbon accumulation in the forest understorey, soils and dead plant material. Since the initial harvesting of western forests, fire suppression has been actively, and for the most part successfully, implemented. The annual area burned in the 1990s only amounted to approximately 15% of the presettlement burn (Tilman et al. 2000). Fire suppression favors unhealthy forest stands with many small trees, and frequent pest and pathogen outbreaks. While these ecosystems are unstable in the long term, they contain large quantities of carbon stored in litter, dead wood and the forest understorey, increasing short-term carbon sequestration. The "permanence" of this storage is thus debatable.

Wildfire suppression in regrowing stands is thought to have a significant effect on carbon sequestration in Western US forests. Pacala et al. (2001) estimate significant sinks as a result of fire suppression in western pine forests. Schimel et al. (2000) also showed significant sinks in mountain forest biomes. The fire regime prior to the European settlement of the Western US was dominated by frequent small-scale low-intensity fires and sporadic large-scale stand-replacing fires (Pyne 1982; Veblen and Lorenz 1991). These fires occurred mainly in the dry seasons and during drought conditions and were due to lightning strikes and land management practices of the indigenous peoples. As a result of frequent surface fires, forests were broadly maintained in healthy, but relatively low carbon states (Tilman et al. 2000). Current fire suppression efforts have reduced the annual area burned to 10-15% of the presettlement levels (Pyne 1982; Tilman et al. 2000). Incorporation of fire suppression activity into ecosystem simulation models suggests massive consequences for carbon storage in western landscapes (Houghton et al. 1999; Pacala et al. 2000). In the most comprehensive effort to date to reconstruct national carbon budgets, Pacala et al. (2000) concluded that about 25% of US carbon uptake (0.12 Gt C year) could be due to fire suppression in western coniferous forests.

Fire suppression is not the only mechanism responsible for carbon uptake in mountain environments. In the US and Europe, as recreational, watershed and other non-consumptive land uses have increased in the mountains, forest harvest and pasture maintenance practices have decreased. In Europe, the abandonment of high pastures, formerly used for livestock husbandry, is allowing significant expansion of high- and mid-elevation forests, creating significant carbon sinks (Cernusca et al. 1998). In temperate Asia, where large populations impose a high demand for agricultural land, remaining forests are largely in mountainous areas.

Thus, the carbon budget of the western US and the health of western forests is best understood in terms of:

- Climate, with favorable water balance permitting increased productivity and therefore carbon storage at higher elevations;
- Historical land use, with past forest disturbance setting the stage for widespread forest regrowth, especially in the more productive areas;
- Current land use, with fire suppression favoring high carbon storage but at-risk ecosystems.

Quantifying carbon sequestration in the mountains will require extensive model-data integration, using measurements to calibrate and constrain models, and models to interpolate observations. Foreseeable measurement approaches will allow the development of accurately calibrated models and algorithms for extrapolating carbon fluxes over complex terrain. Below, we illustrate the potential of models and remote sensing, as well as the unique data requirements for operating in complex terrain.

2. Modeling and remote sensing of mountain biogeochemistry

We used the Biome-BGC and century biogeochemistry models, to explore potential carbon uptake patterns in the US (Fig. 2). The results are drawn from the Vegetation and Ecosystem Modeling and Analysis Project (VEMAP) (Schimel et al. 2000). These model experiments were driven by historical climate, reconstructed from 1895 to 1993 using over 8000 long and short-term weather stations, as well as about 700 high elevation stations from the SNOTEL (SNOw TELemetry) network. Short-term stations were linked geostatistically to long-term stations to create a complete pseudo-network of 98-year long records, which were then gridded to obtain a spatially distributed climate record. In the gridding procedure, temperature and precipitation were statistically corrected for elevation, aspect and mountain valley inversions. Thus, the unique features of mountain climate, as well as large-scale temporal variations were all included in the data set. This extensive data adjustment is required because most extant gridded data bases used in comprehensive models do not account well for effects of elevation and orography on microclimate. If significant carbon stores are, as we suggest, located in mountain landscapes, then accurate climate drivers are required for model-based estimates of mountain climate and carbon storage.

The ecosystem models were "spun up" and then run from 1895-1993. The models also included the effect of increasing atmospheric CO_2. Vegetation definitions were fixed and based on reconstructed actual vegetation. Agriculture was treated explicitly, using USDA county-level information (Schimel et al. 2000) and 18 crop-management combinations that were simulated using the Century model. Century agricultural results were blended on an area coverage basis into both the Century and Biome BGC results. The VEMAP results have been independently compared to observations for validation, and agree reasonably well with data (Schimel et al. 1997; Jenkins et al. 2000).

Our results show that 70% of the Western US carbon sink occurs at elevations above 750 m (Fig. 2), an elevation range dominated by hilly or mountainous topography (50-85% complex terrain; Fig. 2). This comprises 20-40% of total uptake for the lower 48 states. The pattern is striking in the semi-arid western US, in which most low-elevation ecosystems are dry and dominated by biomes with low carbon density (Fig. 2); foci of high carbon uptake are found in the Sierra Nevada and Rocky Mountains. Figure 2 shows the results for the 1980s, a time when much of the Sierra Nevada and Cascade mountains were affected by drought, and so carbon uptake in the productive Eastern Sierra and Pacific Northwest may have been lower than normal.

The VEMAP results are probably at least qualitatively correct. However, these simulations were run without detailed disturbance and management regimes. The

Western mountains have been intensively managed over the past century. Despite the impression of vast wilderness areas, most of the mid- and low-elevation montane forests have been logged or otherwise disturbed, beginning in the pioneer mining era with intensive harvesting and burning during the railroad era. After the early period of forest utilization for construction and fuelwood, industrial harvesting began and continues to this day (Veblen and Lorenz 1991). In the VEMAP simulation illustrated, only a very simple land use and natural disturbance history was applied, consistent with broad-scale statistics but likely wrong in many details. Model results thus do not take into account all of the factors acting to modify spatial and temporal patterns of carbon exchange in the Western US.

Remote sensing provides wall-to-wall coverage of ecosystems but does not provide direct estimates of Net Ecosystem Exchange (NEE), a measure of the net exchange of carbon between ecosystems and the atmosphere. However, a new operational satellite product provides regular estimates of fractional intercepted photosynthetically active radiation (FPAR) from the MODerate Resolution Imaging Spectroradiometer instrument on NASA's TERRA and AQUA spacecraft. FPAR has been shown to be

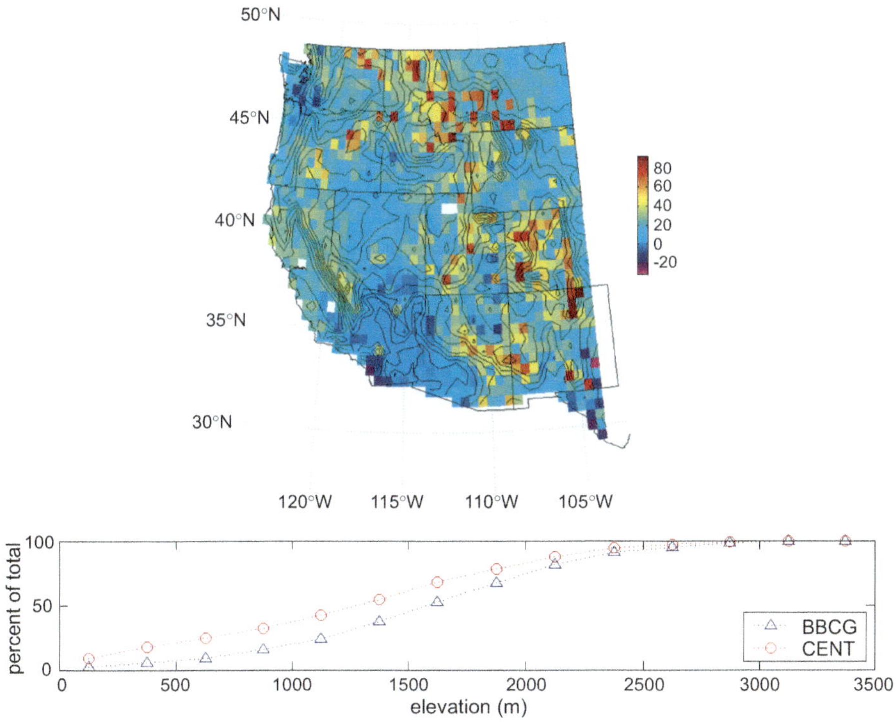

Figure 2: Upper: Mapped net ecosystem exchange in the western US over topographic contours (1980-1989 simulations; Schimel et al. 2000), based on the Century Ecosystem model. Highest fluxes can be observed in the mountains. Fluxes in the Pacific Northwest and California are depressed for this decade because of drought. Lower: Lines show results from Century and Biome-BGC models and agree on the basic carbon distribution with elevation (upper line graph) with 75% of carbon storage in complex high elevation topography.

highly correlated with Gross Primary Productivity (GPP), the gross flux of carbon into the biosphere via photosynthesis, and Net Primary Productivity (NPP), the balance of carbon uptake and respiration in vegetation, and provides an additional check on our model simulations. The GPP image from MODIS is produced using retrieved FPAR, estimate of radiation, temperature and precipitation and other ancillary data and shows a pattern (Fig. 3) clearly corresponding to the model results in Figure 1. The satellite-based GPP supports the argument that most highly productive and high carbon storage potential systems in the Western US are in montane forests and complex topography. High rates of GPP can support significant amounts of carbon storage, although much of the annual GPP is respired or burned in each year.

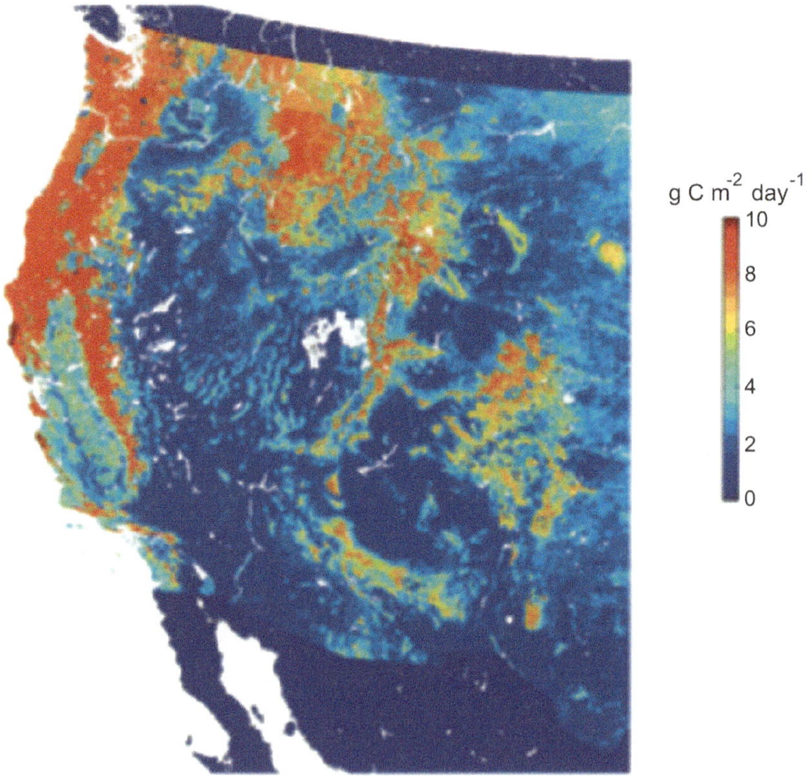

Figure 3: Mapped Gross Primary Productivity (GPP) for the US, from the MODIS instrument aboard the NASA TERRA satellite for peak growing season 2001.

3. Conclusions and future research directions

- Mountains are important contributors to carbon uptake in the western US, as a result of their unique climate within a semi-arid to arid region, and because of historical harvesting and present-day fire management.

- Any changes to fire regime or fire management practices are likely to have widespread impacts on forest ecosystem function, affecting carbon storage and tightly linked water resources.
- While fire suppression has probably led to a larger carbon sink in the Western US, this sink may not be stable. Increased fuel accumulation due to fire suppression contributes to increased fire risk in drought years and carbon sinks could rapidly turn into carbon sources. Improved carbon management must consider both the amount of carbon stored and the stability of that storage as climate and fire regimes evolve.
- Carbon storage in mountain forests of the western US is largely fed by snowpack moisture, and improved understanding of snowpack and high-elevation water dynamics will be important for forecasting the future of these mountain forests (cf. Körner, this volume).
- The pattern of forest regrowth in mountainous areas is pervasive in the Northern Hemisphere and amply supported anecdotally and by satellite observations. However, the land use histories and current drivers vary considerably. Future research must evaluate both common factors and regional differences and provide improved methodology for quantitative measurements of forest regrowth in montane regions.

4. References

Becker, A., and Bugmann, H., Eds. (2001). Global change and mountain regions: The Mountain Research Initiative. IGBP Report No. 49. Stockholm, Sweden.

Belward, A. S., Estes, J. E., and Kline, K. D. (1999). The IGBP-DIS 1-km land-cover data set DISCover: A project overview. *Photogrammetric Engineering and Remote Sensing* **65**, 1013-1020.

Cernusca, A., Bahn, M., Chemini, C., Graber, W., Siegwolf, R., Tappeiner, U., and Tenhunen, J. (1998). ECOMONT: A combined approach of field measurements and process-based modelling for assessing effects of land-use changes in mountain landscapes. *Ecological Modelling* **113**, 167-178.

DeFries, R. S, Hansen, M. C., Townshend, J. R. G., Janetos, A. C., and Loveland, T. R. (2000). A new global 1-km dataset of percentage tree cover derived from remote sensing. *Global Change Biology* **6**, 247-254.

Falge, E., Baldocchi, D., Tenhunen, J., Aubinet, M., Bakwin, P., Berbigier, P., Bernhofer, C., Burba, G., Clement, R., Davis, K. J., Elbers, J. A., Goldstein, A. H., Grelle, A., Granier, A., Guomundsson, J., Hollinger, D., Kowalski, A. S., Katul, G., Law, B. E., Malhi, Y., Meyers, T., Monson, R. K., Munger, J. W., Oechel, W., Paw, K. T., Pilegaard, K., Rannik, U., Rebmann, C., Suyker, A., Valentini, R., Wilson, K., and Wofsy, S. (2002). Seasonality of ecosystem respiration and gross primary production as derived from FLUXNET measurements. *Agricultural and Forest Meteorology* **113**, 53-74.

Fan, S., Gloor, M., Mahlman, J., Pacala, S., Sarmiento, J., Takahashi, T., and Tans, P. (1998). A large terrestrial carbon sink in North America implied by atmospheric and oceanic carbon dioxide data and models. *Science* **282**, 442-446.

Gurney, K. R., Law, R. M., Denning, A. S., Rayner, P. J., Baker, D., Bousquet, P., Bruhwiler, L., Chen, Y. H., Ciais, P., Fan, S., Fung, I. Y., Gloor, M., Heimann, M., Higuchi, K., John, J., Maki, T., Maksyutov, S., Masarie, K., Peylin, P., Prather, M., Pak, B. C., Randerson, J., Sarmiento, J., Taguchi, S., Takahashi, T., and Yuen, C. W. (2002). Towards robust regional estimates of CO_2 sources and sinks using atmospheric transport models. *Nature* **415**, 626-630.

Jenkins, J. C., Birdsey, R. A., and Pan, Y. (2001). Biomass and NPP estimation for the mid-Atlantic region (USA) using plot-level forest inventory data. *Ecological Applications* **11**, 1174-1193.

Pacala, S. W., Hurtt, G. C., Baker, D., Peylin, P., Houghton, R. A., Birdsey, R. A., Heath, L., Sundquist, E. T., Stallard, R. F., Ciais, P., Moorcroft, P., Caspersen, J. P., Shevliakova, E., Moore, B., Kohlmaier, G.,

Holland, E., Gloor, M., Harmon, M. E., Fan, S. M., Sarmiento, J. L., Goodale, C. L., Schimel, D., and Field, C. B. (2001). Consistent land- and atmosphere-based US carbon sink estimates. *Science* **292**, 2316-2320.

Pyne, S. (1982). "Fire in America: A cultural history of wildland and rural fire." University of Washington Press, Washington.

Schimel, D. S., Emanuel, W., Rizzo, B., Smith, T., Woodward, F. I., Fisher, H., Kittel, T. G. F., McKeown, R., Painter, T., Rosenbloom, N., Ojima, D. S., Parton, W. J., Kicklighter, D. W., McGuire, A. D., Melillo, J. M., Pan, Y., Haxeltine, A., Prentice, C., Sitch, S., Hibbard, K., Nemani, R., Pierce, L., Running, S., Borchers, J., Chaney, J., Neilson, R., and Braswell, B. H. (1997). Continental scale variability in ecosystem processes: Models, data, and the role of disturbance. *Ecological Monographs* **67**, 251-271.

Schimel, D., Melillo, J., Tian, H. Q., McGuire, A. D., Kicklighter, D., Kittel, T., Rosenbloom, N., Running, S., Thornton, P., Ojima, D., Parton, W., Kelly, R., Sykes, M., Neilson, R., and Rizzo, B. (2000). Contribution of increasing CO_2 and climate to carbon storage by ecosystems in the United States. *Science* **287**, 2004-2006.

Schimel, D. S., Kittel, T. G. F., Running, S., Monson, R., Turnipseed, A., and Anderson, D. (2002). Carbon sequestration studied in Western US mountains. *EOS* **83**, 445-449.

Tilman, D., Reich, P., Phillips, H., Menton, M., Patel, A., Vos, E., Peterson, D., and Knops, J. (2000). Fire suppression and ecosystem carbon storage. *Ecology* **81**, 2680-2685.

Turnipseed, A. A., Blanken, P. D., Anderson, D. E., and Monson, R. K. (2002). Energy budget above a high-elevation subalpine forest in complex topography. *Agricultural and Forest Meteorology* **110**, 177-201.

Veblen, T. T., and Lorenz, D. C. (1991). "The Colorado Front Range: A century of ecological change." University of Utah Press, Salt Lake City.

Wofsy, S. C., and Harriss, R., Eds. (2002). "The North American Carbon Program." UCAR, Boulder CO.

Remote Sensing Detection of High Elevation Vegetation Change

Herman H. Shugart
Department of Environmental Sciences, University of Virginia, Charlottesville, Virginia 22901, USA
phone 1-924 7642, fax 1-924 4761, email hhs@virginia.edu

Keywords: Africa, Carbon dioxide, Climate change, Mt Kenya, Tropical mountains, Vegetation zones.

1. Introduction

A striking change associated with modern human society has been the increase in atmospheric CO_2 due to the increased burning of fossil fuels (coal, petroleum, natural gas) since the industrial revolution (Sarmiento and Siegenthaler 1992; Sarmiento and Bender 1994). One potential consequence of this atmospheric change is the so-called "greenhouse effect", a global climatic warming induced by elevated atmospheric CO_2 modifying the atmosphere's opacity to infra-red radiation. Because CO_2 is an essential component of plant photosynthesis, an increase in ambient CO_2 levels immediately leads to the question as to whether these changes might be altering plant function globally and might also be changing vegetation patterns.

Depending on the location, dynamic global vegetation models or DGVMs (Cramer et al. 2001) can simulate the synergistic and antagonistic interactions between recent climate change (drought, temperature increase) and CO_2 fertilization. Clearly, the observation of direct CO_2 effects (photosynthesis and water-use efficiency changes) and indirect effects (changes caused by a changed climate caused by changed CO_2-levels) on natural vegetation would be a valuable monitoring objective in terrestrial ecosystems. Of course, the direct and indirect effects of CO_2 on vegetation in many regions are confounded by the history of land-use and disturbances.

Monitoring of high-altitude vegetation in remote (or even inaccessible regions) is

U. M. Huber et al. (eds.), Global Change and Mountain Regions, 457-465.

not a panacea to understanding global and local change of terrestrial vegetation, but there are unique features that high-altitude vegetation offers. Many mountaintops are inaccessible and, while their vegetation is potentially fragile, they can be relatively protected from large-scale land use changes. Many such locations are in mountain parks or protected areas. Hence, in these ecosystems, the direct and indirect effects of CO_2 can be more easily separated from land use impact. However, in many cases, the long-term instrumentation of mountaintops with climate monitoring stations is scanty, which makes it difficult to assess the impact of climate changes on high-altitude vegetation. This relative disadvantage may be offset by the possibility of a high sensitivity of mountain vegetation to the direct and indirect effects of CO_2.

In this paper, we explore the capabilities of remote sensing to detect change in tropical mountain vegetation. It will be argued in this chapter that the detection of vegetation change on tall equatorial mountains would be especially valuable for the detection of potential CO_2 fertilization effects. In these locations, vegetation reaches very high altitudes and, thus, grows under very low partial pressures of CO_2. One outcome of this study is a demonstration of the capability to design a satellite-based world-wide monitoring system for montane vegetation change. We show that these monitoring systems can detect recent changes in tropical mountain vegetation. These changes are consistent with increased plant performance from increasing levels of CO_2 and/or climate change. Future field studies are critical for determining whether or not CO_2 fertilization has been the primary driver behind the observed changes.

2. Why might high-elevation vegetation serve as a potential indicator of direct CO_2 effects on plants?

Those of us who have walked or climbed at high altitudes immediately recognize the limiting effects of lower partial pressures of oxygen. The analogous low partial pressures of CO_2 at high elevations lead one to expect a similar effect of altitude on plant CO_2 uptake and photosynthesis. If high elevation vegetation is limited by CO_2, one further might expect it to be more responsive to atmospheric CO_2 increases than vegetation in lowlands. However, logical this may seem, this is not a topic without debate. Some of this debate stems from the difficulty in applying experimental controls to observational data and the possibility that other factors (notably climate change) may also account for some of the observed changes in vegetation. Nevertheless, several lines of evidence and deduction imply a potential response to elevated CO_2 at high elevations.

Several studies (e.g. Woodward and Bazzaz 1988; Woodward and Kelly 1995) that have investigated changes in stomatal density of plants along altitudinal gradients show an increase in stomatal density with decreasing partial pressure of CO_2 at higher elevations. The increase in the number of stomata implies a potential increase of stomatal conductance of plants and, hence a greater rate of water loss and CO_2 gain by the leaf. Woodward (1987) provided an important example on how observations at high altitudes can be relevant to global studies of environmental change. Inspired by studies of stomatal number variation in plants grown at different altitudes (and the associated different partial pressures of CO_2), Woodward found that the numbers of stomata per unit area on plant leaves collected over the past 200 years from the Midlands of England have varied inversely in relation to the increase in CO_2 concentration since the Industrial Revolution (around 1860). Woodward was able to duplicate the changes in stomatal index, observed in archival plant specimens, by assessing the stomatal density of modern plants that were grown under altered partial pressures of CO_2 in experimental conditions (also see Woodward and Bazzaz 1988). There are species differences in the response to CO_2 changes. For example, Woodward and Kelly (1995) found that 74% of 100 plant species reduced stomatal index with increased CO_2, but some species showed no significant responses and a few increased their stomatal index.

Changes in stomatal density could potentially modify the stomatal resistance to fluxes of H_2O and CO_2 and thus the plant water-use efficiency (the ratio of carbon fixed by photosynthesis to water lost by transpiration). One would expect leaves with more stomata of a given aperture to have less resistance (greater conductance) to the inward diffusion of CO_2 and outward diffusion of water. Stomatal conduction is the inverse of stomatal resistance.

Carbon isotope composition in plant species also shows significant temporal and altitudinal trends. CO_2 with the ^{13}C isotope of carbon moves at a lower rate into the leaf than CO_2 with the ^{12}C isotope at a given stomatal resistance. The greater the stomatal resistance (or the lower the stomatal conductance), the greater is the discrimination against $^{13}CO_2$. The ratio of $^{13}C/^{12}C$ isotopes in the Woodward (1987) herbarium specimens that were examined for stomatal density have decreased systematically since the Industrial Revolution - consistent with a lower stomatal resistance in plants grown in the lower CO_2 pre-industrial revolution atmosphere (Woodward 1993). Körner et al. (1988) found that the $^{13}C/^{12}C$ ratio in leaves of over 100 plant species (or ecotypes) from several different mountain ranges show increasing ^{13}C concentrations with elevation. This is consistent with the higher stomatal indices and associated higher conductance with altitude. This increased ^{13}C concentration of high elevation plants growing under lower CO_2 partial pressures (due to altitude) is consistent with Woodward's (1993) observations of higher ^{13}C content in plant specimens grown in the historically lower CO_2 partial pressures of the pre-Industrial Revolution atmosphere.

At the whole-plant level, LaMarche et al. (1984) also noted that the lower partial pressures of CO_2 at high elevation should reduce the gradient between the external atmosphere and the interior of the leaf and should therefore lead to a reduced rate of CO_2 diffusion through the leaf stomata (see also Tranquillini 1979). They felt that plants at higher elevations are more likely to be limited by CO_2 and hence to respond to increased CO_2 levels with an increase in primary production. Pointing out possible complicating factors as well as theoretical considerations, Cooper (1986) and Gale (1986) both criticized this view. Subsequently, they were rejoined by LaMarche et al. (1986) who cited the experimental work of Mooney et al. (1964; 1966). Mooney and his colleagues had found that photosynthesis rates show a clear decrease with increasing elevation. Subsequently, Körner and Diemer (1987) inspected 112 pairs of plant species in field experiments at elevations of 600 and 2600m in the Austrian Alps. Alpine plants appeared to profit more from elevated CO_2 levels than lowland plants of the same species (Körner, this volume).

Several observations of tree growth on mountains seem consistent with a direct effect of the recent anthropogenic increases in atmospheric CO_2. LaMarche et al. (1984) found an increase in tree ring-widths between the years 1840 and 1970 in bristlecone pines (*Pinus longaeva* and *P. aristata*) at high elevation sites from New Mexico, Colorado and California. Growth trends showed a monotonic increase and a dampened sensitivity in the response to year-to-year fluctuations in climate. They interpreted this growth increase as the first documented growth enhancement of natural vegetation from the increase in atmospheric CO_2. They also reported increased growth rings for another high-elevation species, limber pine (*P. flexilis*) in central Nevada, which showed a similar pattern with a continued or even accelerated growth rate increase in the 1970s. Further corroboration came from additional bristlecone pine data from two upper tree line sites located in the White Mountains of Eastern California (LaMarche et al. 1984). However, in Nevada, Graumlich (1991) measured tree ring growth indices in other species of subalpine coniferous trees (foxtail pine, *P. balfouriana*; lodgepole pine, *P. murrayana*; western juniper, *Juniperus occidentalis*), but failed to find such a trend in three of five locations.

Street-Perrott et al. (1997) inspected ratios of carbon isotopes in sediments of African mountain lakes on Mount Kenya that cover the last 20,000 years. The $\delta^{13}C$ values from high elevation lakes during the last glacial maximum indicate a severe carbon limitation of tropical mountain plants at high altitudes. This led them to the conclusion that tropical tree lines were significantly influenced by carbon limitation rather than temperature changes in the past.

3. Mountain vegetation as a potential monitor of CO_2 changes in the atmosphere

If high-elevation vegetation is more likely to be carbon-limited, as the studies discussed in the section above imply, the direct effects of increased CO_2 on vegetation should be more pronounced at higher elevations than in lowland vegetation. High-elevation sites may therefore provide excellent locations for monitoring the effects of carbon fertilization. This should be most observable in the tallest equatorial mountains where vegetation grows at very high elevations and where the ambient natural CO_2 partial pressures are particularly low. However, tall equatorial mountains are often remote, and a lack of roads makes it difficult to access their upper reaches with heavy scientific equipment. It is therefore sensible to inspect these locations using remote sensing techniques as a logical precursor to more expensive and difficult long-term ground studies.

Earlier studies (e.g. Körner and Diemer 1994) indicate that long-term monitoring of changes in high-elevation vegetation could be a useful means of detecting vegetation response to changes in atmospheric CO_2. Because of other possible environmental changes with elevation, such as land use and climate change, global long-term monitoring networks of vegetation changes are required and these need to be augmented by field studies at key sites. Monitoring studies should focus on tropical mountain environments where vegetation is likely particularly carbon-limited and should therefore be the best indicator for the direct effects of carbon fertilization. Unfortunately, equatorial tall mountains are often cloud covered, particularly at their summits, and are thus poor targets for remote sensing applications. Nonetheless, several recent studies have demonstrated the utility of remote sensing to detect vegetation changes associated with land clearing or disturbances in tropical mountains (e.g. Shao et al. 1996; Cushman and Wallin 2000; 2002; Liu et al. 2002). Local vegetation surveys in mountains indicate relatively slow rates of spatial changes in vegetation in response to environmental change (Franklin et al. 1971; Sprugel 1976; Marr 1977; Kullman 1986; Slatyer 1990; Noble 1993). In order to detect these changes, relatively high-resolution remote-sensing technology is required.

4. Measuring vegetation change on high-elevation equatorial mountains

My colleagues and I have inspected US reconnaissance satellite imagery to assess changes in high-elevation vegetation in equatorial regions around the world (Shugart et al. 2001). While these investigations were initially undertaken to determine if there has been measurable vegetation response to increased ambient atmospheric CO_2, global climatic trends may explain or contribute to observed vegetation changes. We selected symmetrical mountains close to the equator and avoided sites on or near active volcanoes with the goal of determining whether high-resolution satellite imagery could be used to monitor vegetation change. The results of two case studies from Mt. Kenya (Kenya) and Mount Kilimanjaro (Tanzania) are summarized in this paper. With the cooperation of the Kenyan Forest Health Management Centre, over 120 ground-located control points were used on Mt. Kenya to obtain a current vegetation map (Fig. 1) from Landsat TM data. This map is generally indicative of how remote sensing information can be used to rapidly produce detailed vegetation maps for remote mountain regions at the 25m resolution of the TM satellite.

The principal difference between Mounts Kenya and Kilimanjaro is that a slightly drier climate at Mt. Kilimanjaro results in the absence of a bamboo zone that occurs at elevations between 2200-3100m on Mt. Kenya. Vegetation change was detected at all sites surveyed. The changes were of two types: (1) changes in vegetation type (or zone) apparently due to species migration; and (2) a transition of sparse vegetation cover to more dense vegetation cover, either with or without a shift in plant species. These changes are consistent with a hypothesized increase in plant vigor due either to an increase in ambient CO_2 levels in the atmosphere and/or to a simultaneous increase in temperature at all three sites.

On Mt. Kenya, the most striking change was observed along the boundary between heathland and *Hagenia-Hypericum* forest over a 9 year time period (January 1984 to March 1993). Based on interpretation of the imagery and a field visit, it was determined that this area of vegetation change has a canopy that is similar to the thin ribbon of *Hagenia-Hypericum* forest that constitutes the interface between the bamboo and heathland vegetation zones. The same vegetation change was detected

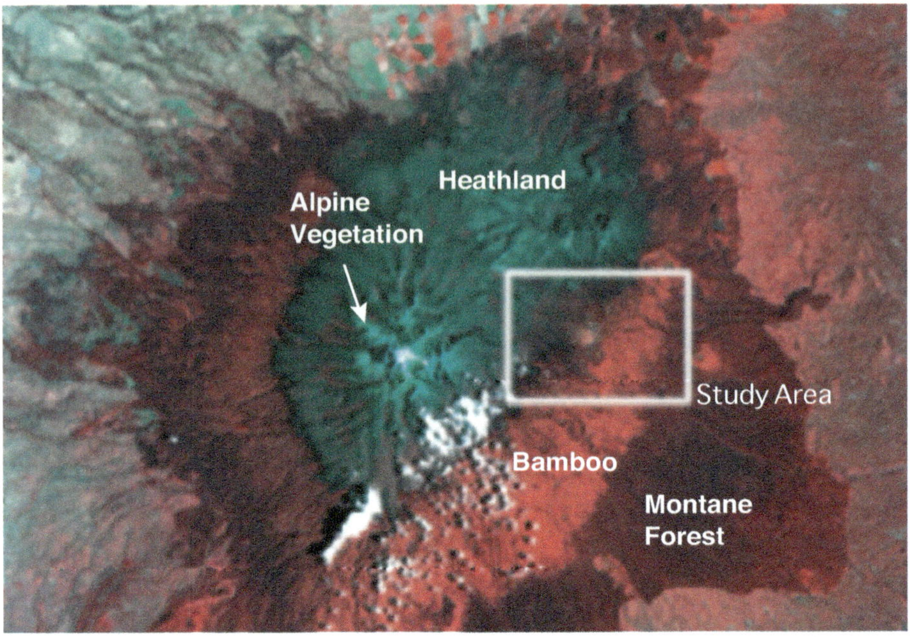

Figure 1: Landsat TM false color composite, from 30 January 1995 (red=band4, green=band3, blue=band2). This satellite image shows the vegetation zones of Mt. Kenya. The study area is indicated by a rectangle. Vegetation zones are from: H. Rehder, M. E. Beck, B. Kokwaro, and J. O. Kokwaro (1988). *Phytocoenologia* **16,** 433-436; Figure from Shugart et al. (2001). Montane Rain Forest occurring from 1500-2900 m altitude with Meru Oak (*Vitex keniensis*), East African Camphorwood (*Ocotea usambarensis*), East African Yellow-wood (*Podocarpus falcatus* and *P. latifolius*) and Parasol Tree (*Polyscias kikuyuensis*).

a. Bamboo (*Arundinaria alpina*) zone at 2200-3100 m altitude. This zone tends to have distinct boundaries, but may also be found mixed with species from the montane zone. This species is not found on the drier Mt. Kilimanjaro.

b. *Hagenia-Hypericum* Forest zone found on wet soils and distributed as a thin ribbon between heathland and either bamboo or montane rain forest at altitudes of 2700-3200 m. This vegetation typically has a distinctive boundary. It is also occasionally found in pockets within the heathland zone. Typical species include: East African Redwood (*Hagenia abyssinica*), *Hypericum keniensis*, and African Pencil Cedar (*Juniperus procera*).

c. Heathland zone at an altitude of 2900-4000 m. The zone has two distinct communities: Ericaceous bush and open moorland. Ericaceous bush occurs throughout the altitude range of the heathland. Typical vegetation includes small trees, shrubs and scrubby vegetation, including *Nidorella arborea*, *Protea kilimandscharica*, *Philippia keniensis*, *Erica arborea*, groundsel (*Dendrosenecio brassica*), tree-groundsel (*Dendrosenecio johnsonii batticomei*), and *Lobelia keniensis*, along with occasional grassy tussocks. Open moorland occurs only in the wetter areas over an altitude range of 3100-3900 m and contains mostly tussock grasses (e.g. *Festuca pilgeri*) with sedges (such as *Carex monostachya*) and herbaceous plants (*Alchemilla johnstonii*).

d. Alpine zone found at 3900-4500 m. The boundary between the heathland and the alpine zone is not distinct. The species living in the two zones are similar and have overlapping altitudinal ranges. The transition from heathland to alpine zone is a gradual decrease in the density of the vegetation. The species that is strikingly unique in the alpine zone is the giant tree groundsel (*Dendrosenecio keniodendron*), which grows in a scattered fashion throughout the zone. In some parts of this zone, *D. keniodendron* forms open groves called *Dendrosenecio* woodlands.

e. Nival zone, which covers the altitude range from about 4500 m to the summit (5900 m for Mt. Kilimanjaro and 5200 m for Mt. Kenya).

and mapped using civilian satellite images, Landsat MSS imagery from 24 January 1976 and Landsat TM imagery from January 1995. From the analysis of the imagery, the estimated area of increase in the *Hagenia-Hypericum* vegetation type at this site over the time period from 1984 to 1993 is 1.28 km^2 or about 1% of the 135 km^2 study area.

Results from the Mt. Kilimanjaro sites were consistent with the Mt. Kenya results. For this analysis, images were digitized and precisely georeferenced to 1:50,000 scale maps obtained from the United Kingdom Ordnance Survey (Shugart et al. 2001). The digitized images were entered into a geographic information system (GIS), and differences between images were identified using image brightness (reflectivity) information, rather than the manual interpretation used for the Mt. Kenya analysis. The most pronounced increase in vegetation cover again occurred in the transition area between the heathland zone and the montane rain forest. Overall, the study site on the south side of Kilimanjaro showed a 17% increase in vegetation cover in 9 years whereas the study site on the north side showed a 22% increase.

On Mt. Kenya and Mt. Kilimanjaro, over a roughly ten-year period of investigation, species movement/vegetation change was detected at all of the sites reported. On Mt. Kenya, movements of the boundary between the heathland and *Hagenia-Hypericum* forest amounted to very substantial shifts in vegetation zones over a decade. On Mt. Kilimanjaro, vegetation changes were most striking around the Mawenzi peak where significant vegetation change (filling in of bare ground and sparsely vegetated areas) was seen over approximately 20% of the area. These results may be consequences of temperature increases at these elevations, recovery of the vegetation from disturbance, or changes in management policies in the two mountain parks. Certainly, the results are indicative of increased vegetation cover. The observed patterns of plant growth are also consistent with what one might expect from an increase in atmospheric CO_2, if low partial pressures of CO_2 are limiting plant performance at high elevations. *In situ* observations of other environmental conditions and an assessment of land use history are needed to eliminate the possibility that other changes in these environments may be producing these same effects and to assess the impact of synergistic or antagonistic interactions between different environmental changes. The observed changes in equatorial, high-altitude vegetation suggest a need to investigate the causes of such changes with additional studies of plant performance.

5. Implications for future studies

What is clearly illustrated in these studies is the capability of an archival network, using remote sensing imagery augmented by field studies, to collect primary data on vegetation change on high equatorial mountain peaks where direct effects of CO_2 enrichment of the atmosphere could be most pronounced. The studies reported here demonstrate that a high-elevation archival network should be able to detect both changes in ecotones and apparent increases in vegetation cover within a vegetation zone.

The development of commercially available high resolution (≈1m) satellite imagery with the IKONOS satellite and the cloud penetrating capabilities of synthetic aperture RADAR on the Japanese JERS-1 satellites create a potential to initiate an archive of vegetation changes in a variety of locations. This chapter argues that one such archive should be developed for high mountains, particularly the tall tropical mountains. There is a considerable body of circumstantial evidence that such locations may be early indicators of the direct effects of CO_2 on vegetation, but this evidence will be debatable until environmental monitoring and process studies are in place in these same locations. Such a monitoring network would be a relatively inexpensive detection system for wide-spread global change.

6. References

Cooper, C. F. (1986). Technical comments: Carbon dioxide enhancement of tree growth at high elevations. *Science* **231**, 859.

Cramer, W., Bondeau, A., Woodward, F. I., Prentice, I. C., Betts, R. A., Brovkin, V., Cox, P. M., Fisher, V., Foley, J. A., Friend, A. D., Kucharik, C., Lomas, M. R., Ramankutty, N., Sitch, S., Smith, B., White, A., and Young-Molling, C. (2001). Global response of terrestrial ecosystem structure and function to CO_2 and climate change: Results from six dynamic global vegetation models. *Global Change Biology* **7**, 357-373.

Cushman, S. A., and Wallin, D. O. (2000). Rates and patterns of landscape change in the Central Sikhote-alin Mountains, Russian Far East. *Landscape Ecology* **15**, 643-659.

Cushman, S. A., and Wallin, D. O. (2002). Separating the effects of environmental, spatial and disturbance factors on forest community structure in the Russian Far East. *Forest Ecology and Management* **168**, 201-215.

Franklin, J. F., Moir, W. H., Douglas, G. W., and Wiberg, C. (1971). Invasion of subalpine meadows by trees in the Cascade Range, Washington and Oregon. *Arctic and Alpine Research* **3**, 215-224.

Gale, J. (1986). Technical comments: Carbon dioxide enhancement of tree growth at high elevations. *Science* **231**, 859-860.

Graumlich, L. J. (1991). Subalpine tree growth, climate and increasing CO_2: An assessment of recent growth trends. *Ecology* **72**, 1-11.

Körner, C., and Diemer, M. (1987). *In situ* photosynthetic responses to light, temperature and carbon dioxide. *Functional Ecology* **1**, 179-194.

Körner, C., and Diemer, M. (1994). Evidence that plants from high altitudes retain their greater photosynthesis efficiency under elevated CO_2. Functional Ecology **8**, 58-68.

Körner, C., Farquhar, G. D., and Roksandic, Z. (1988). Carbon isotope discrimination in plants from high altitudes. *Oecologia* **74**, 623-632.

Kullman, L. (1986). Recent tree-limit history of *Picea abies* in the southern Sweden Scandes. *Canadian Journal of Forest Research* **16**, 761-771.

LaMarche, V. C., Jr., Graybill, D. A., Fritts, H. C., and Rose M. R. (1984). Increasing atmospheric carbon dioxide: Tree ring evidence for growth enhancement in natural vegetation. *Science* **225**, 1019-1021.

LaMarche, V. C., Jr., Graybill, D. A., Fritts, H. C., and Rose, M. R. (1986). Technical comments: Carbon dioxide enhancement of tree growth at high elevations. *Science* **231**, 860.

Liu, Q. J., Takamura, T., Takeuchi, N., and Shao, G. (2002). Mapping of boreal vegetation of a temperate mountain in China by multitemporal Landsat TM imagery. *International Journal of Remote Sensing* **23**, 3385-3405.

Marr, J. W. (1977). The development and movement of tree islands near the upper limit of tree growth in the southern Rocky Mountains. *Ecology* **58**, 1159-1164.

Mooney, H. A., Strain, B. R., and West, M. (1964). Photosynthetic efficiency at reduced carbon dioxide tensions. *Ecology* **47**, 490-491.

Mooney, H. A., Wright, R. D., and Strain, B. R. (1966). The gas exchange capacity of plants in relation to vegetation zonation in the White Mountains of California. *American Midlland Naturalist* **72**, 281-297.

Noble, I. R. (1993). A model of responses of ecotones to climate change. *Ecological Applications* **3**, 395-403.

Sarmiento, J. L., and Siegenthaler, U. (1992). New production and the global carbon cycle. *In* "Primary productivity and the biogeochemical cycles of the sea." (P. G. Falkowski, and A. D. Woodhead, Eds.), pp. 317-332. Plenum Press, New York.

Sarmiento, J. L., and Bender, M. (1994). Carbon biogeochemistry and climate change. *Photosynthesis Research* **39**, 209-234.

Shao, G., Zhao, G., Zhao, S., Shugart, H. H., Wang, S., and Schaller, J. (1996). Forest cover types derived from Landsat TM imagery for Changbai Mountain Area of China. *Canadian Journal of Forest Research* **26**, 206-216.

Shugart, H. H., French, N. H. F., Kasischke, E. S., Slawski, J. J., Dull, C. W., Shuchman, R. J., and Mwangi, J. (2001). Detection of vegetation change using reconnaissance imagery. *Global Change Biology* **7**, 247-252.

Slayter, R. O. (1990). Alpine and valley bottom treelines. *In* "The scientific significance of the Australian Alps." (R. Good, Ed.), pp.15-26. Australian Alps National Parks Liaison Committee and Australian Academy of Science, Canberra.

Sprugel, D. G. (1976). Dynamic structure of wave-regenerated *Abies balsamea* forests in northeastern United States. *Journal of Ecology* **64**, 889-911.

Street-Perrott, F. A., Huang, Y., Perrott, R. A., Eglinton, G., Barker, P., Khelifa, L. B., Harkness, D. D., and Olago, D. O. (1997). Impact of lower atmospheric carbon dioxide on tropical mountain ecosystems. *Science* **278**, 1422-1426.

Tranquillini, W. (1979). "Physiological ecology of the Alpine timberline." Springer, New York.

Woodward, F. I. (1987). Stomatal numbers are sensitive to increasing CO_2 from pre-industrial levels. *Nature* **327**, 617-618.

Woodward, F. I. (1993). Plant responses to past concentrations of CO_2. *Vegetatio* **104/105**, 145-155.

Woodward, F. I., and Bazzaz, F. A. (1988). The response of stomatal density to CO_2 partial-pressure. *Journal of Experimental Ecology* **39**,1771-1781.

Woodward, F. I., and Kelly, C. K. (1995). The influence of CO_2 concentration on stomatal density. *New Phytologist* **131**, 311-327.

Monitoring Networks for Testing Model-Based Scenarios of Climate Change Impact on Mountain Plant Distribution

Antoine Guisan[1*] and Jean-Paul Theurillat[2]
[1]Department of Ecology and Evolution, University of Lausanne, CH-1015 Lausanne, Switzerland
[2]Centre Alpien de Phytogéographie (CAP), Fondation J.-M. Aubert, CH-1938 Champex, Switzerland
**phone +41-21-692-42-54, fax +41-21-692-42-65, e-mail antoine.guisan@ie-bsg.unil.ch*

Keywords: Climate change, Distributional changes, Monitoring network, Mountain flora, Permanent plots, Phenological change, Predictive modelling, Swiss Alps.

1. Introduction

In recent years, predictive modelling of plant species' distribution has been shown to be a powerful method for obtaining preliminary assessments of potential ecological impact of rapid climatic change (e.g. Brzeziecki et al. 1995; Kienast et al. 1996; Saetersdal and Birks 1997; Iverson and Prasad 1998; Lischke et al. 1998; Gottfried et al. 1999; Guisan and Theurillat 2000; 2001; Bakkenes et al. 2002). Such models give static results: they reveal where suitable species' habitats might be located in a climatically changed future, but they do not explicitly consider all the processes leading to the predicted changes. A basic assumption behind their application is thus to consider present and future distributions of species to be in equilibrium, or at least in pseudo-equilibrium, with their environment (Guisan and Theurillat 2000). Although this assumption obviously does not hold in all ecological situations, scenarios obtained from these models nevertheless constitute an interesting spatially-explicit and quantitative basis for discussing how climate change might impact plant distribution. Examples of such discussions are provided in the next section.

U. M. Huber et al. (eds.), Global Change and Mountain Regions, 467-476.

The wedding of this discipline with others, such as dynamic vegetation modelling (e.g. gap models), phenological modelling, geogenetics (spatial population genetics) and vegetation monitoring, offers great promise for the future. Monitoring and phenological investigations provide a valuable basis for predictive modelling in the context of climate change. More and more empirical evidence is accumulating each year on the impact of climate change on organisms, such as upslope migration of species along elevation gradients (Grabherr et al. 1994).

However, these distributional shifts are somewhat "late" signals that biological impact is occurring. A much earlier signal is the one provided by phenological measurements, obtained by monitoring small but significant changes in the biological rhythms of populations of organisms. The recent literature contains many such examples, some of which are presented below.

In this paper, we intend to show that modelling exercises should not be disconnected from field reality. First, we present the use of predictive distribution models used in plant ecology to generate impact scenarios. We then review some data on the impact of climate change on plant phenology, and the potential of phenological modelling. Finally, we illustrate how specific vegetation monitoring designs, which include phenological measurements, can be set up as part of any modelling exercise. In the long term, such data sets will be invaluable for testing the different scenarios of future species distribution changes that are predicted by present ecological models.

2. Developing scenarios of climate change impact on plant distribution

A variety of statistical models are currently used to simulate the spatial distribution of plant species (see Guisan and Zimmermann 2000 for a review), among which generalized linear models (GLM) have become very popular in ecology in recent years (see Guisan et al. 2002). They all rely on establishing a quantitative relationship between the occurrence of a taxon (and possibly other measures of its fitness, such as vitality and abundance) and a set of variables that characterize its natural environment. Hence, most useful are those models that include climatic predictors, since these usually have a more direct effect on plant survival than, for instance, topographic ones (Guisan and Zimmermann 2000).

In Switzerland, models have been developed for alpine and subalpine plant communities (Binz and Wildi 1988; Fischer 1990; Zimmermann and Kienast 1999), forest communities (e.g. Brzeziecki et al. 1993; 1995; Kienast 1993; Kienast et al. 1995; 1996), individual plant species (Guisan et al. 1998; Guisan and Theurillat 2000; 2001) as well as plant species assemblages and biodiversity reconstructed by superimposing maps of individual species' predictions (see Guisan and Theurillat 2000).

These models result in spatial predictions indicating locations of the most suitable (or unsuitable) habitats for the species or communities in question. Changing the climatic parameters in the models provides different climate change impact scenarios (e.g. Saetersdal and Birks 1997; Iverson and Prasad 1998; Gottfried et al. 1999;

Guisan and Theurillat 2001; Bakkenes et al. 2002). The various results show the future location of suitable habitats as a consequence of a prescribed climate change, but also the location of currently suitable habitats that may become unsuitable after climate change (Fig. 1), for instance, due to exclusion by more competitive invaders. Sensitivity studies can be set up by considering various intensities of climate change (e.g. different increases in temperature) in the model and by visualizing the consequent changes at the level of plant distribution. To our knowledge, very few models have been published that consider alpine plant species at a local resolution in a mountain

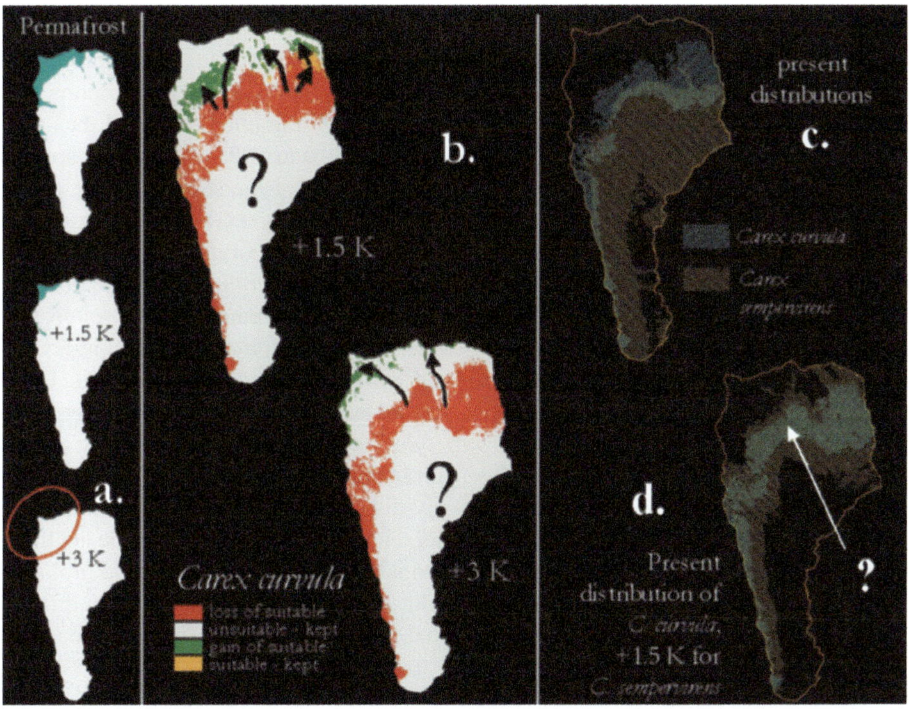

Figure 1: Impact of two climate change scenarios (+1.5 K and + 3 K) on the distribution of the alpine sedge *Carex curvula*. The topography of the area is shown in Figure 2. a) Disappearance of permafrost at the highest elevations for the two scenarios (model by Keller 1992). It will be necessary to assess whether areas with recently melted permafrost will be stable and how plants will be able to colonize them. b) Simulated responses of *C. curvula* to the two climate change scenarios. Figure legend was abbreviated and means the following: *loss of suitable habitat, unsuitable habitat unchanged, gain of suitable habitat, suitable habitat unchanged.* Some questions that arise are: Will migration towards higher elevations be possible for this species? Or, are there some natural or man-made barriers that might prevent such migration? How long can the species remain in its current location? c) Map showing the present predicted distributions of *C. curvula* and *C. sempervirens*. The narrow overlap between both distributions is shown in light blue; one hypothesis here is that the latter sedge limits the occurrence of the former at lower elevation through competition (Guisan et al. 1998). d) Map showing the present predicted distribution of *C. curvula* and the future distribution of *C. sempervirens* after a + 1.5 K warming. Questions that arise are: What would happen after a +1.5 K warming? Would *C. curvula* be outcompeted by *C. sempervirens* and other species from lower elevations or would its dense underground root system prevent the establishment of other plants? (see Guisan et al. 1998).

environment (e.g. Gottfried et al. 1999).

To fill this gap, such a study was conducted on 62 alpine and subalpine plant species in a small study area of the Swiss Alps (see Guisan and Theurillat 2001). Various intensities of climate warming (+1.5K, +3K, +4.5K) were simulated by a transfer-function approach taking into consideration the future locations of suitable species habitats after equilibrium has been hypothetically recovered. Results show that a strong impact could be expected on the distribution of several species (Table 1; Guisan and Theurillat 2001). Similar results were obtained on a European scale (Huntley et al. 1995; Bakkenes et al. 2002) and elsewhere (e.g. Iverson and Prasad 1998).

One would expect strict alpine species to be at higher risk of local extinction than species also distributed at lower elevations, as the latter have a wider tolerance to those ecological factors co-varying with elevation (e.g. climatic ones), and thus they run a lower risk of local extinction (Theurillat and Guisan 2001). Subalpine species would also be globally at lower risk since the total area of the subalpine belt would be only slightly reduced by moderate warming, as shown by a GIS-based analysis of the Swiss Alps (Theurillat and Guisan 2001). Our results support this hypothesis, with most of the alpine species showing a decrease in the extent of their suitable habitats.

Except for one poorly modeled species, scenarios for alpine species are tenable. Indeed, when considering the whole of Switzerland, some physiographic predictors, such as slope angle, are unequally distributed with elevation (Theurillat et al. 1998; Theurillat and Guisan 2001; Körner 2001). As elevation increases, fewer gentle slopes are available, causing a decrease in the variety of habitats. If climate warming affects these high elevation areas, some species from lower elevation would become able to extend their range upwards and competitively displace some native alpine and nival species. As possible escape routes to higher elevations are not available to these species (these areas are already covering the tops of mountains), they would probably suffer most from any warming, be it only slight. As a result, they might be confined to small vegetation patches at the periphery of habitats colonizable by plants, or become locally extinct.

Scenarios for subalpine and lower elevation species deserve further research, since their full elevation range was not taken into consideration in this study (Guisan and

Table 1: Overall change in predicted plant species occurrence, relative to the present potential distribution, resulting from three warming scenarios. The first four categories represent possible species responses (column total is always 62). The last row is a complementary view of the categories "Occurrence decreases" and "Extinct in the study area". It provides the number of species in each scenario for which the range of potential habitat would decrease by more than 90%.

	+ 1.5 K	+ 3 K	+ 4.5 K
Occurrence increases	18 (29.0%)	14 (22.6%)	12 (19.4%)
No change	5 (8.1%)	5 (8.1%)	5 (8.1%)
Occurrence decreases	38 (61.9%)	42 (67.7%)	42 (67.7%)
Extinct in the study area	1 (1.6%)	1 (1.6%)	3 (4.8%)
Percent decrease > 90%	2 (3.2%)	11 (17.7%)	24 (38.7%)

Theurillat 2001). This can result in truncated response curves for several subalpine species at their lower elevation range, and introduce a bias in the predictions. The University of Lausanne is currently carrying out studies on the full elevation gradient in the Swiss Prealps of Canton de Vaud (from about 400 m to 3000m asl).

By providing a visual prognosis of the reduction or expansion of species habitats (Fig. 1), these climate change impact scenarios can help to identify the most dramatic trends that can be expected, such as the disappearance of all suitable habitats for a species. In this case, a species might well be able to persist in certain locations, less affected by climate change. However, its survival in the long-term would be seriously jeopardized, especially in the case of small, fragmented populations, as reproduction might, for instance, be affected (Guisan and Theurillat 2001).

An interesting case study is provided by the application of two warming scenarios (+1.5 K and + 3 K) to the model of *Carex curvula*, a dominant sedge in the alpine belt. In a previous study, Wagner and Reichegger (1997) suggested that moderate warming would favour upward colonization of this important sedge in the eastern Alps. To test this hypothesis, we first applied a model predicting the partial (+1.5 K scenario) and then total (+ 3 K) disappearance of permafrost and permanent snow from the area (Fig. 1a; based on a model by F. Keller 1992; see Guisan et al. 1998). To assess the impact of these warming scenarios on the distribution of *C. curvula*, a model was adjusted for this species, from which maps were drawn that compare the potential species distribution under present and future climates (Fig. 1b).

We know that *C. curvula* is a slow-growing, clonal species, showing a C-S strategy (competitive and stress-tolerant; according to Grime 1977). Moreover, its growth is highly dependent upon the availability of sufficient light, and it is mainly distributed on siliceous substrate in the alpine belt, on flat or slightly sloping terrain, avoiding steep slopes affected by periglacial phenomena (Guisan et al. 1998). Based on these characteristics, will it be possible for this species to migrate towards higher elevation? An important question is whether or not soils will have the ability to develop there, once the permanent snow or permafrost will be totally removed, even though many alpine species are able to colonize bare soils. Furthermore, some natural barriers (e.g. steep slopes) might prevent its migration, as discussed by Rupp et al. (2001) for tree species in the Arctic. And finally, suitable habitats may simply not be available at higher elevations (e.g. no gentle slopes), and the species would have to resist competitors to maintain viable populations in its present location, where it should not need to adapt or acclimatize since it can already grow in milder climate at a lower elevation.

Another important component of the discussion is the consideration of possible biotic interactions since, for example, the lower distribution limit of *C. curvula* might be determined by the dominance of *C. sempervirens* (Guisan et al. 1998), as suggested by the present distribution of both species (Fig. 1c). When overlapping the present distribution of *C. curvula* with that predicted for *C. sempervirens* in a +1.5 K warming scenario (Fig. 1d), it is apparent that *C. curvula* will face increasing competition from *C. sempervirens* and other species from lower elevations, such as some *Ericaceae*. Its dense root system might help *C. curvula* to resist invasion and prevent the establishment of any dominant invader (see Guisan et al. 1998). This is a point of great interest since some individuals of *C. curvula* might be as much as 2000 years old (Steinger et al. 1996; Körner this volume), which would mean that they might have already survived several climate changes in the past, and thus resisted several invasions. A recent study suggests, however, that alpine conditions have prevailed at the investigated site throughout the Holocene (Carnelli 2002). Nevertheless, these questions regarding individual species will need verification in the future.

At a higher organizational level, Guisan and Theurillat (2000) assessed the possible impacts on species *richness* and *composition*, two collective properties of individual species' distribution (Austin 2002). Results showed that, under present climatic conditions, similar predictions of species' richness (SR) could be obtained (i) by overlapping all individual species predictions (cumulative model) or (ii) by fitting a model directly to the species counts. However, the two approaches provided different results when applied under changed climate conditions. The direct SR model predicted little or no shift upward in elevation, but rather a less directional spreading of species richness hot spots at middle elevation. In contrast, the cumulative SR model predicted a shift that would parallel the individual shifts in species ranges on which it is based. The latter might appear more realistic since most paleoecological studies suggest that, in the past twenty thousand years, species responded individualistically to change in climate (Davis 1989; Huntley 1991). Still, some particular sites might currently have a higher potential for hosting a rich flora (e.g. due to a particular substrate) and, although species composition may change at these locations under future climatic conditions, it is likely that species richness will persist. Further investigation would be needed here. Species composition was also affected by climate change in our simulations (Guisan and Theurillat 2000), which additionally supports the view of individualistic responses of species to climate change.

3. Impacts on plant phenology

Several phenological shifts have been reported recently (e.g. Menzel and Fabian 1999; Hughes 2000; Peñuelas and Filella 2001; Fitter and Fitter 2002; Walther et al. 2002; see also Theurillat and Guisan 2001) that show good correlations with regional climate change. This widespread correspondence between phenological and climatic events supports the strong relationships that have been observed between air temperature and phenological stages of shrub and tree species in the Swiss Alps (Theurillat and Schlüssel 2000). Observed shifts are mainly reported for leaf unfolding and first spring flowering, but also for leaf colour change or abscission in autumn. In their review Peñuelas and Filella (2001) mention an earlier onset of biological spring (i.e. leaf unfolding) in Europe of about eight days for the three decades 1969-1998. Similar changes are reported elsewhere, depending on the type of climate. In addition, the same authors report precocity of the time of first flowering for many species, starting for instance a week earlier on average in Hungary for the locust tree (*Robinia pseudoaccacia*), although observations were made over a much longer period (1851-1994). Very similar shifts in flowering time are cited for Europe by Walther et al. (2002) and for America by Chuine et al. (2000). Walther et al. (2002) document a shift of about 1.4-3.1 days earlier per decade for the past 30 to 48 years, which also adds up (on average) to nearly a week over three decades (exactly 6.75 days if taking an average of 2.25 days/decade). This change is the result of an increase in mean air temperature of about 0.5-0.6 K during the same period. For comparison, an upward shift in elevation of 100 m results in a decrease of 0.55 K and a shortening of the growing season by nine days over the year, six in spring and three in autumn (Theurillat and Guisan 2001).

A more detailed study by Fitter and Fitter (2002) of 385 British plant species further suggests that the greatest changes in the last fifty years occurred during the last decade (1991-2000), with a mean advance in the onset of flowering of 4.5 days. Furthermore, these ground observations are confirmed by remote sensing studies (although on a larger scale), which suggest an advance of the growing season of eight days between 1982 and 1990 (these estimates are based on the normalized-difference vegetation index, NDVI, Myneni et al. 1997).

Phenological changes thus appear to have started before the first shifts in species' distribution were noticed. Furthermore, phenology was shown to be an important determinant of species' distribution (Chuine and Beaubien 2001) and should therefore be considered when assessing the consequences of global warming on plant distribution. Several phenological models have recently been developed in the context of climate change (e.g. Chuine et al. 2000; Theurillat and Schlüssel 2000; Chuine and Beaubien 2001). They provide a powerful tool that is complementary to the modelled impact scenarios on plant distribution discussed in Section 2. In particular, combining them with spatial or dynamic models would seem to be a promising approach.

4. Toward an integrated strategy for predicting and detecting climate change signals in plants

Predictive distribution models combined with phenological models thus represent key tools to generate hypotheses on the way climate change might affect alpine plants in the future. They should be of immediate use for assisting environmental managers to take decisions, hopefully based on some kind of precautionary principle. However, *testing* these model-based scenarios and related hypotheses will be another extremely important task to be conducted in the relatively short-term (five to ten years), although it seems to have received little attention in recent years. It represents a major opportunity to improve our understanding of plant distribution. What design will we use in the near future to evaluate those scenarios derived from present models? If we look at current monitoring networks there are very few available. Grabherr et al. (1994) showed that some mountain summits with good historical vegetation records could be used for such assessments. However, these sites only provide a narrow view of vegetation change at the very upper end of the elevation gradient. Furthermore, such a design might not suffice to track all types and rates of plant migration that could result from climate warming. This is because, realistically, not all elevations might undergo an equivalent temperature change (i.e. the magnitude of change might depend on elevation). National monitoring networks have been developed in several countries. For example, Switzerland has set up a national Biodiversity Monitoring project, but its grid design is too coarse to allow efficient tracking of upslope species migration (i.e. not enough points along mountain slopes). An efficient design in this respect could be that of Grabherr and co-workers (see Pauli et al. this volume), who distributed successive rectangular transects, composed of several 1 m² plots, along a relatively short elevation gradient (2900 and 3400 m asl) reaching the top of the Schrankogel in the Austrian Alps (see Gottfried et al. 1999).

However, to our knowledge, none of these designs includes both permanent plot mapping and phenological measurements, which means that, for some species, ongoing changes might not be registered. What is needed is to set up adequately designed strategies to accurately monitor the changes in phenological events as well as upward migrations of plants. In this respect, mountains represent natural laboratories of exceptional value, because they concentrate wide environmental gradients (e.g. vegetation) over very short distances (Körner 2001). Such a design has been set up in a small study area of the Swiss Alps between 1993 and 1999 (Fig. 2).

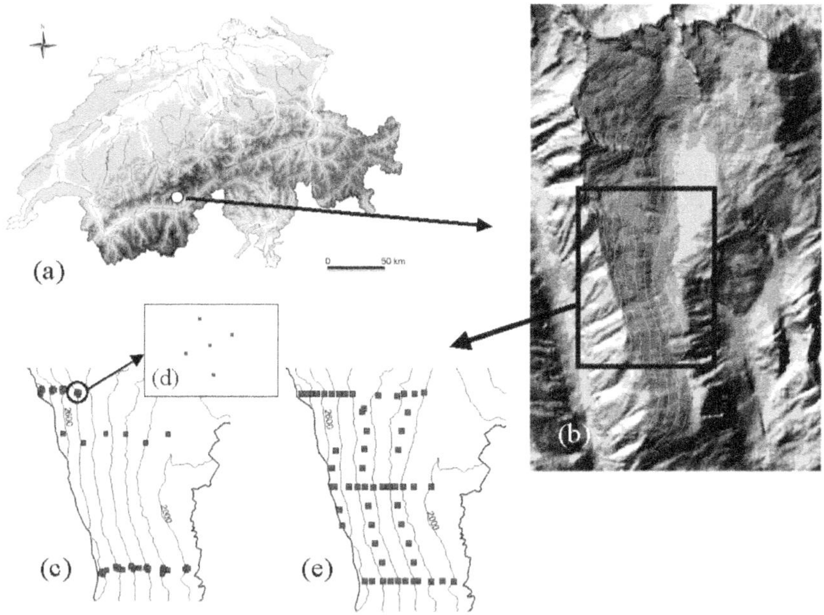

Figure 2: Network of permanent plots (incl. vegetation mapping) and phenological plots set up within the framework of the ECOCLINE Project (Theurillat et al., Schlüssel 1999). Plots are distributed along an elevation gradient in the Belalp study area (Valais, Swiss Alps) ranging from 1900 m to 3400 m. A similar design was also set up in another study area of the Swiss Alps (Champex, Valais). (a) Location of the study area; (b) Location of the transect within the study area; (c) Spatial design of the permanent plots (PER) within the transect; at least two permanent plots are located within the same elevation class; in each plot, precise vegetation mapping was done and the list of species and their abundance was reported; each plot has been visited four times, in 1993, 1995, 1997 and 1999; (d) several PER plots (replicates) were set up at some locations; (e) Spatial design of the phenological plots (PHE) within the transect; 14 key species were surveyed at regular intervals in 1997/1998 (and partly 1999) in each PHE plot. Six species are chamaephytes (dwarf shrubs, like *Vaccinium myrtillus* and *Rhododendron ferrugineum,* and one dwarf willow) and eight are hemicryptophytes (like *Anthoxanthum odoratum*). None of the species were present in all plots.

Two levels of monitoring have been considered here:

1) *The population level*, through measuring phenological stages for a few key species and their relationship to air temperature and other climatic parameters; the plots and surveyed individuals were marked in the field.
2) *The species composition and diversity level*, through accurate itemizing and mapping of permanent plots (vegetation and species mapping in each plot), using advanced differential geopositioning systems (DGPS) to locate plot boundaries (centimeter accuracy).

Plots are distributed approximately every 100 m, which corresponds to a temperature change of about a half-degree (mean lapse rate of 0.55°C/100 m) and were visited at regular intervals during a six years period.

Such a scheme should prove sensitive enough to detect most ongoing species migration upslope while also being able to capture early phenological signals. It should further allow for the testing of current model predictions and hypotheses in a relatively short time.

5. Acknowledgments

To Harald Bugmann for his useful comments on a first draft. To Julie Warrillow and Mel Reasoner for the linguistic revision.

6. References

Austin, M. P. (2002). Spatial prediction of species distribution: An interface between ecological theory and statistical modelling. *Ecological Modelling* **157**, 101-118.

Bakkenes, M., Alkemade, J. R. M., Ihle, F., Leemans, R., and Latour, J. B. (2002). Assessing the effects of forecasted climate change on the diversity and distribution of European higher plants for 2050. *Global Change Biology* **8**, 390-407.

Binz, H. R., and Wildi, O. (1988). "Das Simulationsmodell MaB-Davos." Schlussbericht Schweiz. MaB-Programm Nr. 33, Bundesamt für Umweltschutz, Bern.

Brzeziecki, B., Kienast, F., and Wildi, O. (1995). Modeling potential impacts of climate change on the spatial distribution of zonal forest communities in Switzerland. *Journal of Vegetation Science* **6**, 257-258.

Carnelli, A. L. (2002). "Evolution of the vegetation at the subalpine-alpine ecocline during the Holocene: Comparative study in the Aletsch region, Val d'Arpette and Furkapass (Valais, Switzerland)." Unpublished PhD Thesis, University of Geneva.

Chuine, I., Cambon G., and Comtois, P. (2000). Scaling phenology from the local to the regional level: Advances from species-specific phenological models. *Global Change Biology* **6**, 943-952.

Chuine, I., and Beaubien, E. G. (2001). Phenology is a major determinant of tree species range. *Ecology Letters* **4**, 500-510.

Fischer, H. S. (1990). Simulating the distribution of plant communities in an alpine landscape. *Coenoses* **5**, 37-43.

Fitter, A. H., and Fitter, R. S. R. (2002). Rapid change in flowering time in British plants. *Science* **296**, 1689-1691.

Gottfried, M., Pauli, H., Reitter, K., and Grabherr, G. (1999). A fine-scaled predictive model for changes in species distribution patterns of high mountain plants induced by climate warming. *Diversity and Distributions* **5**, 241-251.

Grabherr, G., Gottfried, M., and Pauli, H. (1994). Climate effects on mountain plants. *Nature* **369**, 448.

Grime, J. P. (1977). Evidence for the existence of three life history strategies in plants and its relevance to

ecological and evolutionary theory. *American Naturalist* **111**, 1169-1194.
Guisan, A., Theurillat, J.-P., and Kienast, F. (1998). Predicting the potential distribution of plant species in an alpine environment. *Journal of Vegetation Science* **9**, 65-74.
Guisan, A., and Theurillat, J.-P. (2000). Equilibrium modelling of alpine plant distribution and climate change: How far can we go ? *Phytocoenologia* **30**, 353-384.
Guisan, A., and Zimmermann, N. E. (2000). Predictive habitat distribution models in ecology. *Ecological Modelling* **135**, 147-186.
Guisan, A., and Theurillat, J.-P. (2001). Assessing alpine plant vulnerability to climate change: A modeling perspective. *Integrated Assessment* **1**, 307-320.
Guisan, A., Edwards, T. C., and Hastie, T. (2002). Generalized linear and generalized additive models in studies of species distribution: Setting the scene. *Ecological Modelling* **157**, 89-100.
Hughes, L. (2000). Biological consequences of global warming: Is the signal already apparent? *TREE* **15**, 56-61.
Huntley, B. (1991). How plant respond to climate change: Migration rates, individualism and the consequences for plant communities. *Annals of Botany* **67**, 15-22.
Huntley, B., Berry, P. M., Cramer, W., and McDonald, A. P. (1995). Modelling present and potential future ranges of some European higher plants using climate response surfaces. *Journal of Biogeography* **22**, 967-1001.
Iverson, L. R., and Prasad, A. M. (1998). Predicting abundance of 80 tree species following climate change in the eastern United States. *Ecological Monographs* **68**, 465-485.
Keller, F. (1992). Automated mapping of mountain permafrost using the program PERMAKART within the geographical information system ARC/INFO. *Permafrost and Periglacial Processes* **3**, 133-138.
Kienast, F., Wildi, O., and Brzeziecki, B. (1996). Potential impacts of climate change on species richness in mountain forests - An ecological risk assessment. *Biological Conservation* **83**, 291-305.
Körner, C. (2001). Why are there global gradient in species richness? Mountains might hold the answer. *TREE* **15**, 513-514.
Lischke, H., Guisan, A., Fischlin, A., and Bugmann, H. (1998). Vegetation response to climate change in the Alps: Modeling studies. *In* "Views from the Alps: Regional perspectives on climate change." (P. Cebon, U. Dahinden, H. C. Davies, D. Imboden, and C. C. Jaeger, Eds.), pp. 309-350. MIT Press, Cambridge, Massachusetts.
Menzel, A., and Fabian, P. (1999). Growing season extended in Europe. *Nature* **397**, 659.
Myneni, R. B., Keeling, C. D., Tucker, C. J., Asrar, G., and Nemani, R. R. (1997). Increased plant growth in the northern high latitudes from 1981 to 1991. *Nature* **386**, 698-702.
Peñuelas, J., and Filella, I. (2001). Responses to a warming world. *Science* **294**, 793-795.
Rupp, T. S., Chapin III, F. S., and Starfield, A. M. (2001). Modelling the influence of topographic barriers on treeline advance at the forest-tundra ecotone in Northwestern Canada. *Climatic Change* **48**, 399-416.
Saetersdal, M., and Birks, H. J. B. (1997). A comparative ecological study of Norwegian mountain plants in relation to possible future climatic change. *Journal of Biogeography* **24**, 127-152.
Schlüssel, A. (1999). Phàéologie, diversité et structure de la végétation dans l'écocline subalpin-alpin. Ph.D. thesis, University of Geneva.
Steinger, T., Körner, C., and Schmid, B. (1996). Long-term persistance in a changing climate: DNA analysis suggests very old ages of clones of alpine *Carex curvula*. *Oecologia* **105**, 94-99.
Theurillat, J.-P., Felber, F., Geissler, P., Gobat, J.-M., Fierz, M., Fischlin, A., Küpfer, P., Schlüssel, A., Velutti, C., and Zhao, G.-F. (1998). Sensitivity of plant and soils ecosystems of the Alps to climate change. *In* "Views from the Alps: Regional perspectives on climate change." (P. Cebon, U. Dahinden, H. C. Davies, D. Imboden, and C. C. Jaeger, Eds.), pp. 225-308. MIT Press, Cambridge, Massachusetts.
Theurillat, J.-P., and Schlüssel, A. (2000). Phenology and distribution strategy of key plant species within the subalpine-alpine ecocline in the Valaisian Alps (Switzerland). *Phytocoenologia* **30**, 439-456.
Theurillat, J. P., and Guisan, A. (2001). Potential impact of climate change on vegetation in the European Alps: A review. *Climatic Change* **50**, 77-109. See also the Erratum **53** (2002), 529-530.
Wagner, J., and Reichegger, B. (1997). Phenology and seed development of the alpine sedges *Carex curvula* and *Carex firma* in response to contrasting topoclimates. *Arctic and Alpine Research* **29**, 291-299.
Walther, G.-R., Post, E., Convey, P., Menzel, A., Parmesan, C., Beebee, T. J. C., Fromentin, J.-M., Hoegh-Guldberg, O., and Bairlein, F. (2002). Ecological responses to recent climate change. *Nature* **416**, 389-395.

Projecting the Impacts of Climate Change on Mountain Forests and Landscapes

Harald Bugmann*, Bärbel Zierl, and Sabine Schumacher
Mountain Forest Ecology, Department of Forest Sciences, Swiss Federal Institute of Technology Zurich, CH-8092 Zurich, Switzerland
**phone+41-1-632 3239, fax+41-1-632 1146, e-mail bugmann@fowi.ethz.ch*

Keywords: Catchment hydrology, Ecological modeling, European Alps, Forest succession, Integrated modeling, Natural disturbances.

1. Introduction

Mountain forests fulfil a multitude of functions, including the provision of timber, fuelwood, edible and medicinal plants, the storage of carbon, the purification of air and water, the regulation and reduction of peak streamflow, the protection from natural hazards, and the contribution to the aesthetic beauty of the landscape. The importance of these functions varies greatly from one mountain region to the other, but in some way, forested landscapes and their fate under a changing climate are important for the capability of mountain regions to provide many of the goods and services that humanity depends on.

Empirical evidence as well as modeling studies suggest that mountain vegetation could be particularly sensitive to anthropogenic climate change (e.g. Grabherr et al. 1994; Bugmann 1997; Theurillat and Guisan 2001). Forest ecosystems are dominated by long-lived organisms that are able to withstand adverse environmental conditions for some time. Thus, in spite of their general sensitivity to climatic parameters, we may expect little change in these ecosystems over time periods of several years. This makes experimental approaches for studying climate change impacts on forests and forested landscapes difficult, and underlines the importance of synthetic tools to study these phenomena. Quantitative, mathematical models are among the tools that are

U. M. Huber et al. (eds.), Global Change and Mountain Regions, 477-487.

suitable for projecting the long-term effects of environmental changes on mountain forests.

In mountain ecosystems, several factors are of crucial importance that may be nearly negligible elsewhere. Examples include (1) the role of slope angle and aspect for determining energy balance, (2) the significance of lateral (downhill) water flows because of the steep slopes, (3) the role of diverse root systems for slope stabilization, (4) high topographic heterogeneity, which induces strong biotic heterogeneity, and (5) the importance of vegetation, particularly forests, on steep slopes for protecting ecosystems and human infrastructure from natural hazards such as avalanches and rockfall. In addition, the tight coupling of the various processes in mountain regions makes it necessary to consider a range of ecological and hydrological processes simultaneously if we are to be able to project the impacts of climate change on these systems.

Historically, research into the impacts of climate change on mountain landscapes has taken place largely in a sectoral manner. For example, many forest ecologists examined the impacts on forest structure and composition using gap models (e.g. Bugmann et al. 2001); biogeochemists investigated possible changes in carbon, water and nitrogen exchange between forests and the atmosphere using physiologically based "big leaf" models (e.g. Keyser et al. 2000); landscape ecologists looked at the impacts of a changing disturbance regime on vegetation patterns at the landscape scale (e.g. He et al. 2002); and hydrologists examined changes in streamflow and related variables, including the risk of flooding (e.g. Pruski and Nearing 2002). Studies such as these are certainly timely and welcome because they highlight the potential for climate-induced changes in ecosystem goods and services in mountains.

However, most of these studies did not take into account the tight linkages between the various processes mentioned above, which are particularly pronounced in mountain areas. For example, gap modeling applications typically have neglected the effects of lateral water flows; hydrologists have tended to disregard the effects of vegetation changes on the flow regime by assuming a constant vegetation type across time; and landscape ecologists have focused on the effects of a changed disturbance regime, while de-emphasizing the importance of successional processes.

In the following section, we give an overview of recent research results obtained in our group that deal with the impacts of climate change on forests and landscapes in mountain regions; some of these studies continue to be sectoral, whereas others have begun to address the linkages between the sectors and processes mentioned above. In the last section, we try to pull together the insights gained so far to derive an outlook regarding future research directions.

2. Case studies

2.1 Forest structure, composition and biomass at the patch scale

Possible changes in the structure, composition and biomass of mountain forests have been a focus of our research over the past years (e.g. Bugmann 1996; 1997; Bugmann and Solomon 2000) because such changes could have significant impacts on the protective function, the carbon storage, and the aesthetic value of mountain forests. We have adapted, improved and refined forest gap models (Bugmann 2001), a widespread class of forest succession models that had originally been developed by Botkin et al. (1972).

Forest gap models simulate the establishment, growth and mortality of individual trees on small patches of land (often $^1/_{12}$ ha) as a function of the species' autecological requirements and the extrinsic and intrinsic conditions of the stand. To obtain forest development at larger spatial scales, the successional patterns of patches from many simulation runs are averaged. This concept is supported by many plant succession studies, which show that a forest ecosystem may be described by the average growth dynamics of a multitude of patches with different successional ages. The model FORCLIM (the acronym stands for *For*ests in a changing *Clim*ate), which is used here, consists of three independent submodels, representing the abiotic environment (FORCLIM-E), tree population dynamics (FORCLIM-P), and soil organic matter turnover (FORCLIM-S). The FORCLIM model was derived from an earlier gap model with the particular aim of a reliable representation of climatic influences on tree population dynamics.

We used this model to study the impacts of a range of climate scenarios for the year 2100 on forest properties at sites along an altitudinal gradient in the European Alps, including steady-state considerations as well as an evaluation of the dynamic properties of the simulated forest (cf. Bugmann 1997; 1999).

Forests on a north-facing and a south-facing slope at the site Bever in the Upper Engadine Valley, which is characterized today by a fairly continental climate, differ dramatically in their simulated species composition under current conditions (Fig. 1 top). In the simulation, the south-facing slope is characterized by a mixture of European larch (*Larix decidua*) and pine, with a strong dominance by Swiss stone pine (*Pinus cembra*) and some Scots pine (*Pinus sylvestris*). The north-facing slope, however, is simulated to be covered by forests that are dominated by Norway spruce (*Picea abies*). These simulation results correspond well to descriptions of the natural vegetation of this area, as discussed by Bugmann (1999).

Notably, these forests differ in their response to a regionalized scenario of climate change (cf. Gyalistras et al. 1994) that was applied across 100 years (mimicking the period 2000-2100; Fig. 1 top), in spite of their close spatial proximity. On the north-facing slope, the dominance of Norway spruce observed under current conditions continues under the future climate, while maple (*Acer* spp.), elm (*Ulmus* spp.) and other deciduous species are of minor importance only. On the south-facing slope, however, the stone pine forest is replaced by a mixed forest where Norway spruce is the most abundant species but contributes less than half of the total biomass (Fig. 1 top). Thus, under this climate scenario, future forests are simulated to be more similar to each other on opposing slopes than under current conditions, but considerable differences continue to exist many centuries into the future.

In addition, we wanted to know whether it might be possible to maintain the current species composition at two high-elevation sites, where tourism is the most important source of income today (Davos and Bever, Switzerland). In the face of environmental change, it is conceivable that attempts would be undertaken to mitigate

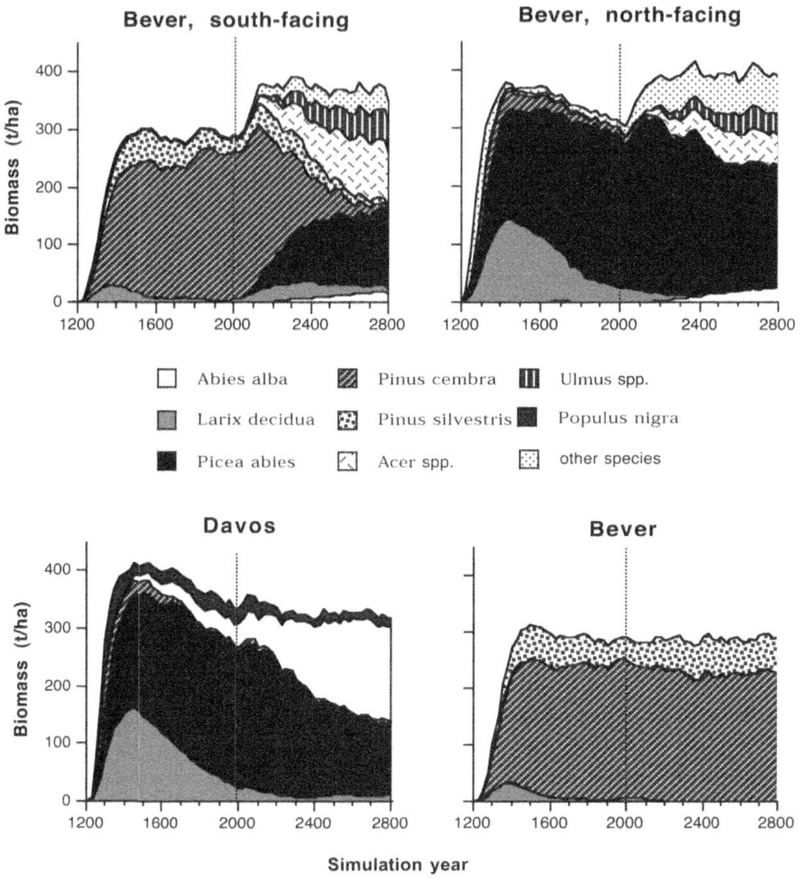

Figure 1: Impacts of climate change on forest composition and biomass in the European Alps. The scenarios of climate change are from Gyalistras et al. (1994) as extrapolated to the year 2100 by Bugmann (1997); they involve an increase of monthly mean temperature between 1.5 and 4.2°C and a precipitation increase between 0.9 and 3.8 cm/month, depending on season and site. Top: Response of forests on north- and south-facing slopes near Bever, Upper Engadine Valley under unmanaged conditions (Bugmann, unpublished). Bottom: Simulation results at two subalpine sites under the assumption that those tree species immigrating under a new climate would be eliminated by management in an attempt to maintain the current species composition.

climate-induced changes in the forest cover. This could be implemented, for example, by eliminating tree species that are immigrating but are less desirable from the point of view of landscape aesthetics, or by planting tree species that are characteristic of the current species assemblage but are unable to regenerate naturally in the future (e.g. because of a lack of chilling). We assessed the success of such management regimes through a series of simulations. These showed that any planting efforts would be irrelevant under the range of climate scenarios considered because according to the model it is not the regeneration niche that will determine the success or failure of the current species under a future climate, but rather competitive effects as the trees become larger. The "elimination" scenario, however, would be fairly successful in maintaining current forest composition far into the future (cf. Fig. 1 bottom), except for the conifer *Abies alba*, which would become quite abundant at Davos. Note that such a management practice would be quite difficult and costly to implement across large areas, and these artificially maintained forest stands would be likely to become more susceptible to disturbance agents, including pests and pathogens, which are not included in the current version of the ForClim model.

2.2 Disturbance dynamics and vegetation patterns at the landscape scale

Looking at patch scale dynamics (see above) is important for a number of reasons, but real mountain landscapes are strongly shaped by natural (e.g. windstorms, wildfires, avalanches) and anthropogenic (e.g. management) disturbances, and such large-scale spatial phenomena cannot be modeled at the patch scale. Changes of disturbance regimes are expected for the 21st century as a consequence of global change, particularly changes of climate and land use. However, the interactions between climate, disturbance dynamics and vegetation properties are not well researched. We started looking into these issues using temperate mountain regions in the European Alps as case studies. First, the sensitivity of the fire regime to altered environmental conditions, and to spatial heterogeneity of fuel availability is addressed. Secondly, the sensitivity of forest landscape patterns to the impacts of changes in fire regimes and climate is investigated. Other drivers of landscape dynamics, such as windthrows and avalanches, are also considered via simple parameterizations.

Landscape-level ecological simulation models are important tools for improving our understanding of the ecological consequences of such changes. We based our work on LANDIS, a modeling framework of landscape dynamics (He and Mladenoff 1999). LANDIS is a spatially explicit, raster-based, and stochastic model developed to simulate and analyze the interactions between forest plant behavior, site conditions and disturbance regimes such as fire, windthrow, and forest management. Vegetation succession, however, is implemented in a fairly simple manner, reflecting the assumption that the disturbance regime is of overriding importance for shaping landscape dynamics. In "weakly" disturbed landscapes such as those of the European Alps of the 20th century, we found that the succession submodel of LANDIS had to be reformulated. We therefore developed a succession submodel that (1) better represents the species' relative competitive ability and (2) includes their responses to changes in the physical environment in more detail. With this approach we obtained a landscape-scale simulation model that incorporates a semi-mechanistic submodel of tree-tree interactions, a feature that is unique to our approach and allows us to study the transition from weakly disturbed to strongly disturbed landscapes. To derive relationships between the parameters that characterize the fire regime and climatic variables, an extensive analysis of fire data was conducted. To expand the bioclimatological space beyond European conditions, data from the (more fire-prone) western US were used.

As a first step, we used the new LANDIS version to examine the response of forests in the Dischma valley (Switzerland) to changes in climate that are directly tied to a change in the fire regime. The Dischma valley is located in the eastern part of the Swiss Alps in the transition zone between the wet northern Alps and the dry central Alps. It is a 43.3 km^2, north-northwest facing catchment. Elevation ranges from 1668 to 3146 m asl. The current vegetation pattern is influenced by the harsh climate, steep slopes, avalanches, wind disturbances and anthropogenic impacts. Fire plays no significant role in the present landscape.

The LANDIS model was initialized with forest characteristics derived from measured data that correspond to the current situation in the valley. Notably, the alpine treeline is significantly lower (ca. 2000-2100 m asl) today than under natural conditions because of historic and present cattle and sheep grazing as well as harvesting activities. Simulations were run for 500 years into the future under (1) current climatic conditions, assuming no forest management and no grazing, thus allowing the forest to expand into what is currently alpine tundra; and (2) a simple climate change scenario (year-round increase of temperature by 2°C and reduction of precipitation by 20%) that also includes a moderate fire regime. In all simulations, avalanche tracks were not available for spontaneous reforestation, i.e. we assumed a regular recurrence of avalanches, which thus prevents the establishment and persistence of large trees.

The simulation results for the current climate (Fig. 2, left panels) show that the dominance of *Picea abies* at the landscape scale (below 1900 m asl) would continue, which is reasonable considering that the species is thought to be the climax tree in this area. A considerable fraction of the forested area at lower elevations in the valley would be dominated by the early successional *Larix decidua*, as a consequence of windthrow disturbance. *Pinus cembra*, which is currently just a minor component of the forests at higher altitudes, would start to dominate the zone just below treeline. Treeline itself would shift upwards by roughly 150-200 m, in the absence of livestock grazing. Biomass is around 200-300 t/ha at the lower elevations in the valley and decreases rapidly between 1850 to 2100 m asl to very low values above 2100 m asl. All these features are characteristic of elevational gradients in high mountain catchments of the central European Alps, and the modified LANDIS version is capable of capturing them adequately.

The simulation results under the climate change scenario (Fig. 2, right panels) suggest that a strong increase of treeline elevation (up to ca. 2500 m asl) would occur, with a concomitant upward shift of the various forest types along the elevational gradient. Notably, these simulations suggest no significant "individualistic" behavior of the tree species under this scenario of climate change. Spatial heterogeneity would increase in the zone dominated by *P. abies*, due to the more frequent occurrence of disturbances, including wildfires. Note that species that are currently confined to the montane zone were not able to immigrate in this simulation because they are not included in the present model version. Further investigations will be necessary to evaluate the interactions of forest composition with the disturbance regime; the modified LANDIS model is an important tool towards achieving this goal.

2.3 Mountain ecosystem services at the catchment scale

As mentioned above, ecological and hydrological processes are tightly coupled

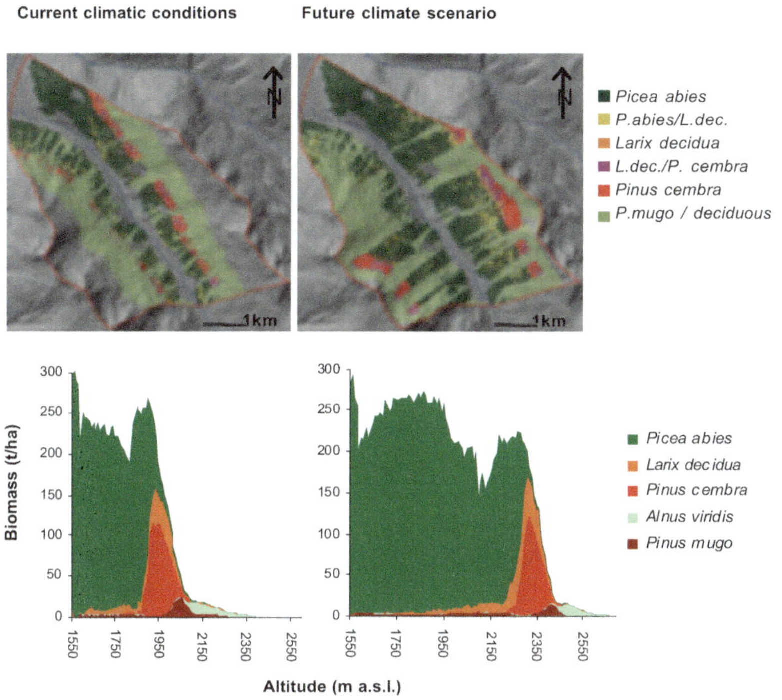

Figure 2: Impacts of climate change on landscape patterns in the Dischma valley, based on simulation results with the LANDIS model. Top: Landscape composition under current conditions (left) and under a simple scenario of a future climate (right). Bottom: Biomass distribution by altitude, excluding avalanche tracks, under current climatic conditions (left) and the future climate scenario (right).

in mountain regions. The integration of biogeochemistry and catchment hydrology in mountain regions requires simulating both the vertical fluxes between soil, vegetation, and atmosphere, and the lateral fluxes between adjacent vegetation patches. Band et al. (1993) and White and Running (1994) combined distributed flow modeling based on TOPMODEL with an ecophysiological canopy model based on BIOME-BGC and a climate interpolation scheme based on MTCLIM to build the regional hydroecological simulation system RHESSys. We build upon this tool in a research project that focuses on global change effects on mountain ecosystem goods and services, which is part of the EU project "Advanced Terrestrial Ecosystem Analysis and Modeling" (ATEAM, see http://www.pik-potsdam.de/ateam).

RHESSys is a spatially distributed daily time step model, designed to solve the coupled cycles of water, carbon, and nitrogen in mountain catchments. It has been widely used to simulate hydroecological processes in various mountain catchments in North America (Fagre et al. 1997; Hartmann et al. 1999; Baron et al. 2000; Tague and Band 2001).

We use RHESSys to analyze the impact of climate and land use change on river discharge and carbon storage in mountain catchments in Europe. As a first case study, we chose the Dischma catchment (cf. description in section 2.2). Within a few kilometers of the catchment, there are two meteorological stations maintained by MeteoSwiss. At the lower end of the catchment, river discharge is measured.

To perform a preliminary evaluation of the climatic sensitivity of hydrological and ecological processes in the Dischma catchment, we conducted simulations under current climate as well as under scenarios of a changed climate. Current climatic conditions are based on daily weather observations from 1981 to 2000. To generate two simple climate change scenarios, the data for the current climate were altered by either adding 2°C to minimum and maximum daily temperatures, or by decreasing daily precipitation by 10% throughout the year. Using these 20-year climate scenarios, river discharge and carbon storage were simulated with the model RHESSys.

The simulation results suggest a complex pattern of responses in carbon storage to shifts in temperature and precipitation (Fig. 3, top). Depending on the location within the catchment, rising temperature can lead to either an increase or a decrease of ecosystem carbon storage. In particular, water availability appears to control the responses of photosynthesis and carbon allocation to rising temperatures.

Decreased precipitation in the climate change scenario implies an increase in incoming solar radiation through decreased cloudiness. Consequently, there are two contrasting effects on photosynthesis. On the one hand, the precipitation decrease

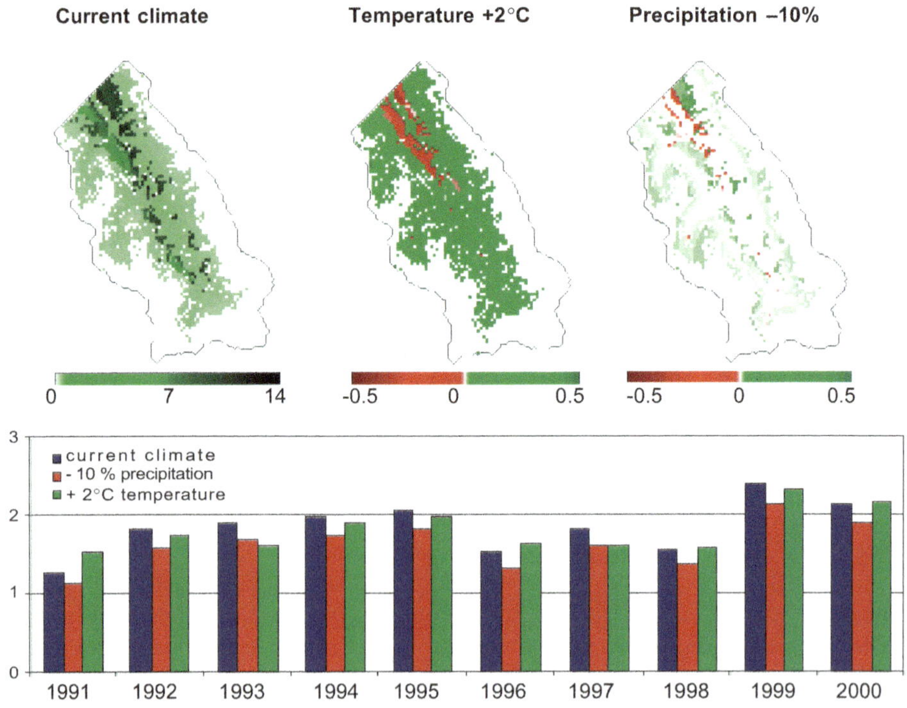

Figure 3: Impacts of climate change on carbon storage and streamflow in the Dischma valley. Top: Total carbon storage [kg/m^2] under current climate, and changes in carbon storage [kg/m^2] under two simple scenarios of climate change. Bottom: Response of average annual river discharge [m^3/s] in the Dischma under the same climate scenarios

leads to reduced soil water availability, which in turn decreases photosynthesis through stomatal closure. On the other hand, the increase of solar radiation brings about an increase of photosynthate production. The balance of the two effects determines the overall response of carbon storage, which obviously varies considerably within the catchment (Fig. 3, top).

As expected, decreasing precipitation diminishes river discharge, but it has no effect on the seasonality of discharge. In contrast, the timing of hydrological responses is highly sensitive to temperature changes. On average, annual river discharge decreases with rising temperatures as evapotranspiration increases (Fig. 3, bottom). However, an analysis of the seasonal effects of a temperature increase (data not shown) suggests strong seasonal differences: winter discharge increases, whereas summer discharge decreases with rising temperatures.

These seasonal differences illustrate the control of temperature over snow accumulation and snow melt, which in turn affects river discharge. Under a warmer climate, more precipitation during the winter season falls as rain, and the snow cover melts earlier. This water flows directly to the river, increases winter river discharge, and is unavailable to the ecosystems in the catchment, because the vegetation is dormant during this time of the year. It is well known that in the present climate, snow cover has an important function with respect to storing and buffering winter precipitation peaks, and releasing the water during the first half of the growing season. Our scenario study suggests that this function might be lost under a warmer climate.

3. Future research directions

A fairly sophisticated set of tools is available today for projecting the impacts of climate change on mountain forests and landscapes, resulting from research efforts that typically have focused on one single spatial (and often also temporal) scale, and most of these efforts have proceeded in a sectoral manner. As a consequence, we are not in a position to simultaneously evaluate the effects of (1) climatic changes on long-term successional processes; (2) changes in the disturbance regime and the associated vegetation feedbacks, and (3) the link between lateral flows of water and either the disturbance regime or successional processes. However, an integrated evaluation of these processes is required for mountain regions because of the tight coupling between them. Thus, how do we achieve this integration?

At first sight, one might perceive the best strategy to be the full coupling of existing models such as ForClim, LANDIS, and RHESSys. According to this strategy, one would try to combine the biogeochemical routines of the BIOME-BGC model, which is a part of RHESSys, with the description of tree population dynamics from ForClim and the representation of disturbance dynamics from LANDIS. We believe that there are two major reasons why such an endeavor would be exceedingly difficult, if not futile:

First (and more pragmatically), the coupling of complex ecosystem models is prone to great conceptual and technical difficulties. The various models differ in their underlying assumptions, in their representation of ecosystem components, and in the process formulations. It is not obvious that the input-output linkages between the different models could always be satisfied. Rather, we would expect that the seemingly simple enterprise of "linking" models would bring about the need to re-evaluate, adjust and revise model formulations (not just output-input relationships), which thus would lead to a modeling effort of its own.

Second (and more fundamentally), each of these models represents a fairly complicated and supposedly realistic representation of some aspects of ecosystem dynamics. Each model is characterized by several hundred parameters, and the coupling would most likely lead to an "integrated" model of greatest complexity whose dynamics are hard to understand, difficult to track to the underlying processes, and not amenable to a rigorous analysis of sensitivity and uncertainty. Thus, we surmise that such an "integrated" model would be of little value as a research tool, let alone as an assessment or management tool.

Therefore, we advocate a different route for achieving a more integrated model of ecological and hydrological processes in mountain regions. In the case of the LANDIS model (cf. section 2.2), we have attempted to replace a very simple representation of successional processes by a more detailed description that was developed by *abstracting* from the patch-level processes that are incorporated in forest gap models (such as ForClim, section 2.1). Here, *abstracting* refers to the process of deriving simplified, more aggregated descriptions of ecological phenomena for incorporation into coarser-scale models. Note that this is not the same as coupling ForClim to LANDIS. We believe that a similar approach could be taken towards integrating successional processes and disturbance dynamics into hydro-ecological simulators such as RHESSys, i.e. by abstracting crucial information from more detailed ecological models, and incorporating "scaled" process formulations into RHESSys, instead of attempting to couple the full models with each other. We still have a long way to go towards this goal, but we think this approach is promising and should be pursued further.

Finally, we would like to emphasize that a truly "integrated" approach to a model-based assessment of global change effects in mountain catchments (or mountain regions) would have to go far beyond the natural sciences, on which this contribution focused. Including land cover and land use dynamics together with socio-economic aspects (and perhaps even atmospheric dynamics) in a single modeling framework is an entirely different challenge. As in the case of natural sciences, the models for many sectoral considerations are available. Even more so than within the natural sciences, we are convinced that a truly integrated assessment of global change in mountain regions cannot be achieved simply by coupling existing models – a significant amount of research efforts will be required to arrive at solutions for this challenge.

4. References

Band, L. E. , Patterson, P., Nemani, R., and Running, S. W. (1993). Forest ecosystem processes at the watershed scale: Incorporating hill slope hydrology. *Agricultural and Forest Meterology* **63**, 93-126.

Baron, J. S., Hartmann, M. D., Band, L. E., and Lammers, R. B. (2000). Sensitivity of a high-elevation Rocky mountain watershed to altered climate and CO_2. *Water Resources Research* **36**, 89-99.

Botkin, D. B., Janak, J. F., and Wallis, J. R. (1972). Some ecological consequences of a computer model of forest growth. *Journal of Ecology* **60**, 849-872.

Bugmann, H. (1996). A simplified forest model to study species composition along climate gradients. *Ecology* **77**, 2055-2074.

Bugmann, H. (1997). Sensitivity of forests in the European Alps to future climatic change. *Climate Research* **8**, 35-44.

Bugmann, H. (1999). Anthropogene Klimaveränderung, Sukzessionsprozesse und forstwirtschaftliche Optionen. *Schweizerische Zeitschrift für das Forstwesen* **150**, 275-287.

Bugmann, H. (2001). A review of forest gap models. *Climatic Change* **51**, 259-305.

Bugmann, H. K. M., and Solomon, A. M. (2000). Explaining forest biomass and species composition across multiple biogeographical regions. *Ecological Applications* **10**, 95-114.

Bugmann, H. K. M., Reynolds, J. F., and Pitelka, J. F., Eds. (2001). How much physiology is required in forest gap models for simulating long-term vegetation response to global change? *Climatic Change* **51**, 249-557.

Fagre, D. B., Comanor, P. L., White, J. D., Hauer, F. R., and Running, S. W. (1997). Watershed responses to climate change at Glacier National Park. *Journal of the American Water Resources Association* **33**, 755-765.

Grabherr, G. M., Gottfried, M., and Pauli, H. (1994). Climate effects on mountain plants. *Nature* **369**, 448.

Guisan, A., and Theurillat, J. P. (2001). Potential impact of climate change on vegetation in the European Alps: A review. *Climatic Change* **50**, 77-109.

Gyalistras, D., von Storch, H., Fischlin, A., and Beniston, M. (1994). Linking GCM generated climate scenarios to ecosystems: Case studies of statistical downscaling in the Alps. *Climate Research* **4**, 167-189.

Hartmann. M. D., Baron, J. S., Lammers, R. B., Cline, D. W., Band, L. E., Liston, G. E., and Tague, C. (1999). Simulations of snow distribution and hydrology in a mountain basin. *Water Resources Research* **35**, 1587-1603.

He, H. S., and Mladenoff, D. J. (1999). Spatially explicit and stochastic simulation of forest-landscape fire disturbance and succession. *Ecology* **80**, 81-99.

He, H. S., Mladenoff, D. J., and Gustafson, E. J. (2002). Study of landscape change under forest harvesting and climate warming-induced fire disturbance. *Forest Ecology and Management* **155**, 257-270.

Keyser, A. R., Kimball, J. S., Nemani, R. R., and Running, S. W. (2000). Simulating the effects of climate change on the carbon balance of North American high-latitude forests. *Global Change Biology* **6** (Supplement 1), 185-195.

Pruski, F. F., and Nearing, M. A. (2002). Runoff and soil-loss responses to changes in precipitation: A computer simulation study. *Journal of Soil and Water Conservation* **57**, 7-16.

Tague, C., and Band, L. (2001). Simulating the impact of road construction and forest harvesting on hydrological response. *Earth Surface Processes and Landforms* **26**, 135-151.

White, J. D., and Running, S. W. (1994). Testing scale dependent assumptions in regional ecosystem simulations. *Journal of Vegetation Science* **5**, 687-702.

Assessing Climate Change Effects on Mountain Ecosystems Using Integrated Models: A Case Study

Daniel B. Fagre[1*], Steven W. Running[2], Robert E. Keane[3], and David L. Peterson[4]
[1]US Geological Survey, West Glacier, Montana
[2]University of Montana, Missoula, Montana
[3]US Forest Service, Missoula, Montana
[4]US Forest Service, Seattle, Washington
**phone 406 888 7922, fax 406 888 7990, email dan_fagre@usgs.gov*

Keywords: Ecological modeling, Global change, National parks

1. Introduction

Mountain systems are characterized by strong environmental gradients, rugged topography and extreme spatial heterogeneity in ecosystem structure and composition. Consequently, most mountainous areas have relatively high rates of endemism and biodiversity, and function as species refugia in many areas of the world. Mountains have long been recognized as critical entities in regional climatic and hydrological dynamics but their importance as terrestrial carbon stores has only been recently underscored (Schimel et al. 2002; this volume). Mountain ecosystems, therefore, are globally important as well as unusually complex. These ecosystems challenge our ability to understand their dynamics and predict their response to climatic variability and global-scale environmental change.

To meet this challenge, mountain scientists increasingly are modeling the vast array of relationships that comprise ecosystem dynamics. Dynamic modeling can examine the interactions between land management strategies and climatic change to develop appropriate responses to future human demands on mountain systems. Modeling provides spatially and temporally explicit, quantified results that can be

U. M. Huber et al. (eds.), Global Change and Mountain Regions, 489-500.

validated in the field, thus providing feedback to our understanding of ecosystem dynamics. Modeling results, particularly maps and other visual tools, also give a concrete dimension to our understanding of the scale and magnitude of potential future changes. Modeling alerts scientists and land managers to apparently counter-intuitive outcomes of ecosystem responses to climate change or management decisions. For instance, in an early modeling exercise for northwest Montana, USA, Running and Nemani (1991) found that streamflow in a warmer future climate decreased by 30% in the Swan Range even when precipitation was increased by 10% in a particular climate change scenario. This unexpected response was due to enhanced forest growth, and increased evapotranspiration, resulting from the earlier snowmelt and extended growing season.

There is a rich legacy of models that address climate and weather, hydrology, forest growth (e.g. gap dynamics and succession), forest fires (e.g. fuel loading) and land cover change (cf. Bugmann et al., this volume). Much less common, however, are attempts to fully integrate models from various disciplines to create a robust system that adequately addresses the entire range of ecosystem dynamics. In addition, fine-resolution modeling of entire mountain ranges (i.e. regional ecosystem scale) is not as common as global or continental scale modeling or watershed/catchment scale modeling. However, this is the scale that is germane to policy decisions such as in the western US and Canada, i.e. in those areas that contain most of the mountainous terrain of North America. This paper describes our efforts to implement an integrated regional modeling approach while characterizing potential future responses of a mountain ecosystem to climate change. Our study area was Glacier National Park in northwestern Montana, USA. Glacier Park is a 4082 km^2 mountain wilderness that straddles the continental divide and contains over 150 summits of up to 3150 m elevation in the Lewis and Livingston mountain ranges.

2. Model background

The integrated modeling program for Glacier Park was built on previous efforts and developed into two related models, the Regional Hydro-Ecological Simulation System (RHESSys) and FIRE-BGC (Fire BioGeoChemical).

2.1 Modeling ecosystem patterns and processes

RHESSys is an evolving group of regional hydroecological models that are designed to simulate the coupled cycles of water, carbon, and nitrogen. This modeling system has been modified to provide appropriate input/output data to each sub-model and collectively address ecosystem patterns and processes in mountain regions (Running et al. 1989; Band et al. 1993). Several spatial databases are required to initialize and run RHESSys, including digital elevation models (i.e. topographic relief) and soil properties. Remotely sensed data from satellite platforms provide another spatial database on the distribution and density of vegetation on the ground, required for calculating biophysical processes. Data sources include AVHRR (Advanced

Very High Resolution Radiometer) at 1 km^2 resolution and Landsat TM (Thematic Mapper) at 30 m resolution. From these data, vegetation indices are calculated, such as Leaf Area Index (LAI), that estimate the potential photosynthesis. Combined with climatic and other environmental data, these indices are used by models to estimate the productivity of vegetation (for details see White et al. 1998). In RHESSys, vegetation is not represented as community types but as broad plant functional types, which differ in their carbon storage capacity and ecophysiological properties, such as photosynthesis and respiration. These vegetation estimates are overlaid on a digital elevation model (DEM) that represents the park's topography in three dimensions at up to 10 m resolution, and a topographic soils index (TSI) that combines existing soil data and estimates of water movement through the terrain (Band et al. 1993). The result is a static model-based description of the current state of Glacier Park.

From this starting point, the simulation is made dynamic by supplying daily meteorological variables to the model. These are generated for different aspects and elevations of the mountain landscape by taking daily data from a meteorological station in the valley bottom and using MT-CLIM (a Mountain Climate simulator) (Hungerford et al. 1989) to estimate maximum and minimum temperatures, relative humidity, short-wave radiation and other climate variables. Based on information about slope, aspect, elevation, soil moisture holding capacity, daily estimated weather and estimated initial vegetation biomass, spatially-explicit calculations of daily tree growth can be made. At the heart of RHESSys is the biogeochemical model FOREST-BGC (Running and Gower 1991) that estimates ecosystem attributes, such as gross primary productivity, at daily time-steps. The simulation results of FOREST-BGC are mapped across the mountain landscape on an annual basis. FOREST-BGC interacts with a hydrological routing model to estimate daily soil moisture and stream discharge of water that is not used in evapotranspiration. Thus, the results from this integrated modeling system for a specific watershed include net primary productivity as well as net ecosystem carbon exchange, aboveground and soil carbon pools, carbon-to-nitrogen ratios and daily to annual hydrologic discharge.

2.2 Modeling ecosystem structure and fire disturbance

FIRE-BGC is a closely related integrated model that emphasizes the structural components of the mountain forest ecosystem and includes forest fires as a key disturbance factor. This model simulates forests in a similar way to how they are perceived by people (forest structure and composition) rather than as estimates of photosynthetic activity, which is not directly observable. The model does not yet include hydrologic routing and discharge estimates. FIRE-BGC shares FOREST-BGC as a common component with RHESSys for estimating nutrient cycling and other ecosystem processes but merges it with the forest gap model FIRESUM (Keane et al. 1996). This sub-model allows FIRE-BGC to estimate rates and trajectories of post-disturbance succession and provides important compositional and structural attributes, such as tree species dominance, stand age, and coarse woody debris. These latter attributes are critical for estimating fuel loads, which greatly affect the potential occurrence of large, stand-replacing wildfires, and for simulating fire intensity and size, using companion models such as FARSITE (Fire Area Simulator) (Finney and Ryan 1995). FIRE-BGC is a multi-scale model, working from individual trees and aggregating up through plots and tree stands to landscapes. A multi-scale approach is necessary because some ecosystem processes, such as fire and seed dispersal, occur only at the landscape level, whereas other processes, such as organic matter accumulation and tree establishment, can be modeled at the stand level. FIRE-BGC is tree-species specific and includes competitive relations between trees that alter the responses to the broad environmental drivers included in RHESSys. Thus, for a mountain ecosystem such as Glacier Park, FIRE-BGC can map mosaics of forest stands of different ages and composition, resulting from both climatic variability and forest fires, and can provide details of tree stand dynamics, such as depth of duff and litter on the forest floor. FIRE-BGC also interfaces well with other forest science research and models so that different forestry management scenarios can be examined. For example, FIRE-BGC and a smoke management model could be paired to estimate the long-term effects of different fire suppression policies on regional air quality.

2.3 Model validation

To determine whether these models provide an accurate picture of the dynamics of this mountain ecosystem, we gathered field data on key ecosystem processes to compare with the estimates from the models. Many of these field studies have been generating information for 10 years, providing a spatially-extensive dataset for other mountain research endeavors and further model validation. We focused on two watersheds for initial model development and field validation. Lake McDonald watershed is approximately 462 km^2, is located west of the continental divide, and receives predominantly maritime climatic influences. St. Mary watershed is similar in size, abuts the Lake McDonald watershed east of the continental divide, and has stronger continental climatic influences. Both watersheds are relatively pristine and undeveloped, have extensive conifer forests, receive most of their annual precipitation as snow, and contain remnant glaciers.

Results of field validation have been previously reported (Fagre et al. 1997; White et al. 1998) but are summarized below. The agreement between model results and field observations is clearly scale-dependent. For example, for RHESSys results, good agreement with field data was obtained for the distribution of snow water equivalent (Fagre et al. 1997; White et al. 1998) at hillslope and watershed scales (e.g. r^2=0.95) but less so for point or plot scales (e.g. r^2 = 0.78). A major constraint to fine-resolution snow estimates was the inadequacy of satellite-based LAI estimates at appropriate scales. LAI values were derived from 30 m pixels but considerable variation in forest canopy density can occur within that pixel. The forest canopy, in turn, influences the snow water equivalent because it intercepts and sublimates falling snow. Thus, the model estimates of snow distribution are limited by LAI estimates. Another issue was the variability in snow distribution, due to micro-topographic relief that was not included in the models. For watershed-scale simulations, however, the snow estimates proved sufficient. For both watersheds, hydrologic discharge simulations tracked daily discharge data closely, except for storm events where some over-prediction was noted

(White et al. 1998). At Glacier National Park, only 4 of 84 watersheds have as much as 3% of their area covered by glacial ice and 18 watersheds have only 1%. Nonetheless, in watersheds with remnant glaciers, observed discharge values during late summer were higher than model simulations, which underscored both the contributions of glacial meltwater to streamflow and the need to include this source in future models of mountain hydrology in the region. Additionally, modeled daily estimates of stream temperatures throughout the watershed closely matched daily measurements from 7 monitored streams (Fagre et al. 1997).

Carbon budget estimates for the watersheds indicated close agreement with observed values for soil CO_2 effluxes and productivity for both low and high elevation forests that cover 75% of the watersheds. For grassland sites, modeled soil CO_2 effluxes and productivity were higher than observed because of difficulties in LAI estimation for grasses. Forest stem production estimates agreed with observed values when aggregated by hillslope areas > 10 ha.

White et al. (1998) concluded that RHESSys generated reasonable estimates of ecosystem processes and attributes for these watersheds. These estimates included net primary productivity, evapotranspiration, available nitrogen and other major forest processes driving ecosystem change. Some estimates of ecosystem attributes, such as net primary productivity, were much less sensitive to scale than hydrologic discharge (White and Running 1994).

Results from FIRE-BGC simulations were also compared to field data, collected from 110 circular 0.4 ha forest plots in the Lake McDonald and St. Mary watersheds. Ecological characteristics of all plant communities were assessed by choosing plots with representative combinations of slope, aspect and elevation. An additional 98 ground-truth plots, distributed across both watersheds, were used for validation of satellite imagery classifications. A 44-year climate record was used to drive a FIRE-BGC simulation of tree growth and predicted tree ring widths. These ring widths for 44 years compared well with those taken from actual trees and suggested that FIRE-BGC was capturing annual variation in growth responses adequately (Keane et al. 1997). However, young tree growth was over-predicted while large tree growth was under-predicted for shade intolerant trees, underscoring the need to improve carbon allocation routines for a greater diversity of stand conditions.

In summary, both models were able to make reasonable estimates of most ecosystem attributes and processes for which we obtained field observations. Validating these models by comparing results to observed data provided confidence that the models were accounting for most major ecosystem processes at Glacier National Park.

3. Results of modeling future climate scenarios

After having established the models' capacities to simulate the Glacier Park mountain ecosystem and its responses to current climatic variability, we applied various climate scenarios to estimate potential future conditions of the park. This predictive, or forecasting, ability provides managers with a valuable tool for assessing a range of climate change scenarios and ecosystem responses for Glacier National Park but does not actually predict the future. Rather, RHESSys and FIRE-BGC translate possible future climate change into spatially-explicit effects on the park landscape with a reasonable degree of confidence.

3.1 Direct vegetation responses to climate change

One scenario of climate change was based on an evaluation of four general circulation models and several downscaling approaches to provide a "most likely" climate change scenario for the Glacier Park area (Ferguson 1997). This downscaling effort is especially critical for mountain environments with strong climatic gradients that are not captured by general circulation models. This scenario projects a 30% annual precipitation increase and a 0.5°C annual temperature increase by 2050. Based on this climate change, the FIRE-BGC model projected shifts in the distribution and dominance of tree species, including a reduction in subalpine fir (*Abies lasiocarpa*), as treelines rise, and a significant expansion of Engelmann spruce (*Picea engelmannii*) at the expense of lodgepole pine (*Pinus contorta var. latifolia*). Because this climate change scenario did not lead to greater fire frequency, these vegetation changes appear to be due to increased precipitation and less snowpack persistence. Using another climate change scenario, RHESSys tested the effects of an extremely variable climate but without long-term increases in temperature or precipitation. After 120 years, long-term net primary productivity of conifers in Glacier Park decreased by 4% on the western side of the continental divide and by 13% on the eastern side (White et al. 1998). The eastern side currently has a more continental climate that is drier and inherently more variable. Thus, the additional climatic variability simulated in this scenario stressed vegetation to a greater degree. According to RHESSys, broad-leaved shrubs and alpine vegetation increased by 2-7%, perhaps because periods of low snowfall increased establishment rates into areas where current snowpack prohibits establishment. Grass net primary productivity at the forest-grassland ecotone decreased because severe drought conditions became more frequent under this more variable scenario. In fact, the lower treeline (the forest-grassland ecotone) rises under this scenario due to water stress, reducing the amount of forest cover in the St. Mary watershed. These RHESSys results do not account for any increased fire frequency due to the periodically drier environment and vegetation on the eastern side. Only the plant physiological responses, such as water and nitrogen stress, are estimated. Undoubtedly, some vegetation shifts would be accelerated by altered fire frequencies, producing even greater changes in this mountain ecosystem.

Ecosystem models can also be used to elucidate changes in limiting factors that drive plant interactions and ecosystem function. For example, in both models water and nitrogen indices are calculated as part of the carbon allocation process, and these indices integrate information about water stress, nutrient availability, and the potential ratio of shoot-to-root growth. Under current conditions, growth is limited by nitrogen availability for conifer forests and shrubs, but grasses are water limited. However, under the extremely variable climate scenario, some limitations change. For instance, alpine vegetation is water-limited under the current climate but becomes nitrogen limited under the variable climate scenario. The relative nitrogen limitation for conifer forests decreases and grasses at lower treeline become much more water-limited. Partly these limitation shifts are due to resource availability as competition for resources changes but physiological requirements also change in some instances. Nitrogen sequestration increases with woody biomass accumulation. This, in turn, may decrease nitrogen cycling and therefore increase nitrogen limitation in alpine vegetation. These shifts in limitations, as simulated by the ecosystem models, can provide ecologists with more specific constraints to predict how individual species will fare under changing environmental conditions and how biodiversity patterns will change for different vegetation types. For example, as atmospheric nitrogen deposition increases in alpine environments, the nitrogen-limited alpine vegetation under the variable climate scenario will respond differently than the water-limited alpine vegetation under the current climate.

3.2 The response of fire disturbance to climate change

Wildfire is the primary disturbance process in northern Rocky Mountain forests and greatly influences carbon cycles in mountain ecosystems. Under future climate scenarios, FIRE-BGC clearly indicates that the resulting more productive forest landscapes will be exposed to more frequent and severe fires than the same landscapes experienced historically, even with the predicted increase in annual precipitation (Keane et al. 1997). Fuel loads will increase more quickly under generally wetter, warmer conditions but inherent climatic variability will still ensure the occasional drought and hot temperatures that will lead to more intense and extensive wildfires. This change in the frequency and severity of fires would nearly double smoke emissions in the future, jeopardizing the pristine air quality that the Glacier Park area currently enjoys and posing a management challenge for park managers who need to restore historic fire frequencies. Without restoring historic fire frequency, during which frequent, low-intensity fire kept fuel loads from building up, the risk of catastrophically intense fires that can consume most of the biomass is increased and could limit any vegetation growth in the affected area. Because fire frequency has been altered by humans throughout the northern Rocky Mountains for the past century, fuel loads have built up to levels that could lead to larger and more intense fires than might have been experienced in the past. Keane et al. (1997) examined the interplay of different fire management policies coupled with different climate scenarios. In the absence of fire suppression, after 250 years of simulation, fires burned over 55% and 67% of the Glacier Park landscape under a current and future climate scenario, respectively. The resulting landscapes were more productive and diverse than the landscapes that developed during the simulations without fire. These latter landscapes became marginally productive and tended to respire more of the carbon dioxide fixed by photosynthesis than did the communities in which fire played a major role. Keane et al. (1997) found that the fire-maintained early successional communities create overall landscapes that release less carbon to the atmosphere than landscapes without fire. This is true under both current and future climate scenarios and even when carbon emissions during fires are considered. There are implications of long-term fire suppression for global carbon balance if other landscapes behave similarly to those simulated in this study.

4. Modeling at larger scales

4.1 The CLIMET project

Much of the integrated modeling program described here had been focused on two watersheds in Glacier National Park of approximately 500 km^2 each. This is the scale at which many policy decisions regarding natural resource protection are made. We felt the need to test these models at larger scales of mountain national parks (approx. 5000 km^2) and mountain parks within their mountain-dominated regions (approx. 20,000 km^2) to examine scale issues. Another need was to integrate landscape fragmentation and other human disturbance into model results and to test model performance in other mountain systems. Accordingly, the CLIMET (**C**limate **L**andscape **I**nteractions – **M**ountain **E**cosystem **T**ransect) project was developed. This project investigates a transect of three distinct mountain bioregions, with large mountain national parks as core research sites, from the Pacific Coast to the Rocky Mountains. The national parks are Glacier in the northern US Rocky Mountains, North Cascades in the Cascade Mountains, and Olympic in the Olympic Mountains near the Pacific Coast. All are large, wilderness-dominated parks near the United States-Canada border. Each park encompasses mountains with similar topographic relief, numerous glaciers and expansive conifer forests; each is characterized by high winter snowfall, acts as the headwaters for its region and contains relatively intact floral and faunal assemblages.

These parks represent a transect of climatic influences, with dominant air masses providing Olympic with a maritime climate, North Cascades with a transitional climate and Glacier with a more continental climate. Olympic has the greatest landscape fragmentation outside its borders; Glacier has the least. Precipitation varies dramatically between westside and eastside locations within each park. For example, precipitation in the Olympic Mountains ranges from >600 cm/yr on Mt. Olympus to only 40 cm/yr in the northeastern rainshadow. This contrast in precipitation over relatively small distances has a profound impact on microclimate, vegetation distribution and disturbance regimes. It allows us to model and compare climatically distinct paired watersheds within each park and scale up to the bioregion and CLIMET scales.

4.2 First results

Three years of field data have been collected in the paired watersheds from each park for modeling. In addition, the DAYMET climate model (Thornton et al. 1997) was applied to the CLIMET transect. DAYMET takes existing climate data from meteorological networks, interpolates daily temperature and precipitation between existing stations, and extrapolates these parameters across topographic features and network gaps, using seasonally adjusted lapse rates. It also estimates relative humidity and radiation; both key requirements for ecosystem models. The end result is spatially-explicit climate data at 1 km^2 resolution for the past 18 years that will be a major template upon which ecosystem models are run. The DAYMET product suggests that the CLIMET transect is climatically more diverse than other areas in the US (Fig. 1), providing a good test of mountain ecosystem responses to regional climatic variation.

Temporal climate variability for CLIMET was also examined and our results suggest that multi-decadal patterns have been significant influences and should be accounted for in the future scenarios used to drive models. The Pacific Decadal Oscillation (PDO) is a broad-scale, recurring pattern of ocean-atmospheric variability (Mantua et al. 1997) that affects mountain snowpacks even on the eastern edge of the CLIMET area (Selkowitz et al. 2002). A principal components analysis of five paleoproxy reconstructions of the PDO suggests that the PDO has been a robust feature of North Pacific climate variability since 1840 and has probably been operating at least since 1600 (Gedalof et al. 2002). However, it has not been uniformly coherent and seems to have been stronger in the 20th century, based on intercorrelation between the proxies. Nonetheless, the PDO has influenced regional tree growth and other natural resources to varying degrees for a long time and the environmental effects of decadal-scale climate variability need to be taken into account in future integrated model scenarios.

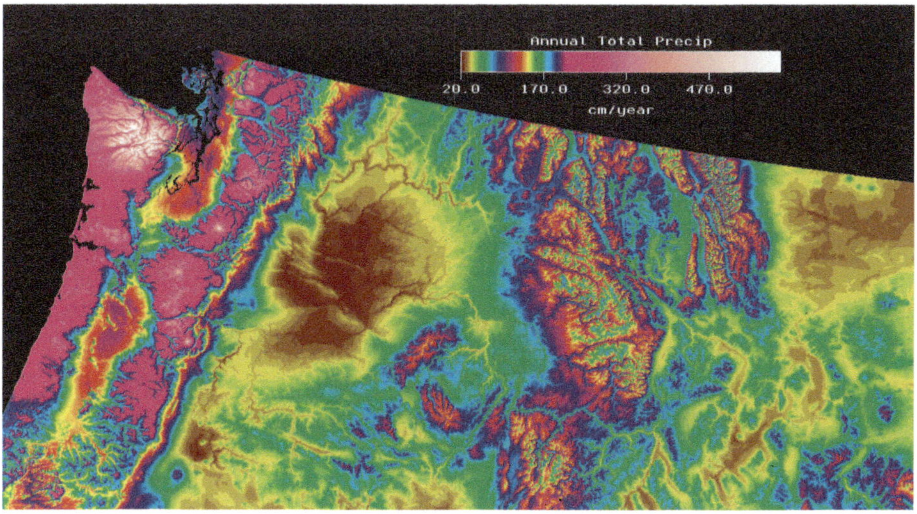

Figure 1: Distribution of annual precipitation from 1980-1997 across the Climate Landscape Interactions – Mountain Ecosystem Transect (CLIMET) program described in text. The western US contains remarkable climatic and ecologic diversity, ranging from temperate rainforests to deserts and including three major mountain systems. The upper edge of the figure corresponds to the 49°N parallel separating the US and Canada. The Olympic Peninsula, surrounded by the Pacific Ocean, is on the upper left. The Rocky Mountains of Idaho and Montana (US) are in the center right and the Great Plains are on the far right.

The first results from integrated modeling of the CLIMET transect are based on DAYMET climatic data and BIOME-BGC, a model with elements from RHESSys adapted for regional scales (Kang et al. 2002). Leaf Area Index (LAI) values from field sites in the Olympic and North Cascades mountains of CLIMET show close correspondence with potential LAI estimated by BIOME-BGC (Fig. 2). Net Primary Productivity (NPP) was estimated and mapped across this topographically complex region for current conditions (Fig. 3). The spatial distribution of NPP appears to be reasonable, based on comparisons to vegetation cover maps. NPP data from field validation sites in two of the mountain ranges indicate that, at a regional scale, the model is able to estimate a key ecosystem attribute and will prove as useful as it did at the watershed scale. Future work will focus on analyzing spatial patterns of NPP that relate to land use history, the relative role of protected areas, and responses to potential future climates.

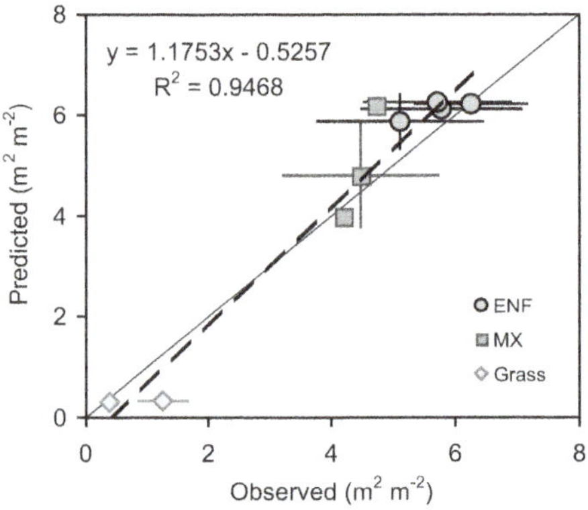

Figure 2: Relationship between field measured Leaf Area Index (LAI) and potential LAI from an integrated ecosystem model for the CLIMET transect. LAI is a proxy measure of vegetation productivity. Field sites were four watersheds in the Olympic and North Cascades mountains of western Washington State located in the northwestern US. ENF = evergreen needle forest, MX= mixed deciduous forest, Grass = several types of grasslands (from Kang et al. 2002).

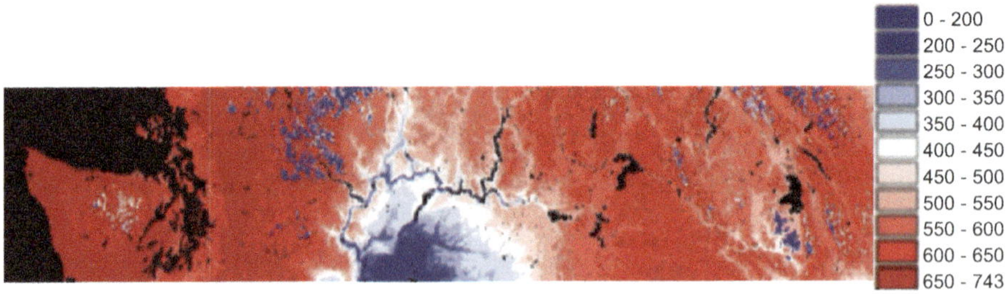

Figure 3: Map of potential Net Primary Productivity (NPP) (gC m^{-2} y^{-1}) from an integrated ecosystem model for the CLIMET transect extending from the Pacific Ocean to the Rocky Mountains in the northwestern US. NPP calculated for 1980-1997 (from Kang et al. 2002).

5. Modeling challenges in mountain areas

Modeling is a valuable adjunct to mountain research programs, sharpening our ecological understanding of phenomena at broader scales and allowing us to view multiple outcomes of our actions that would not be possible in the real world. The Mountain Research Initiative (Becker and Bugmann 2001) explicitly recognizes this value and promotes modeling as a key tool in mountain investigations and management. Numerous other modeling efforts are underway in mountain regions, as modeling becomes a more common tool in many scientific investigations. In the US, similar integrated modeling approaches at ecosystem scales have been applied to climate and landscape change in the Colorado Rocky Mountains (Baron et al. 2000). Other modeling approaches also may provide similar results as the mechanistic models. A geo-spatial model applied to Glacier Park suggests similar vegetation responses (Hall and Fagre 2003) to those described in this paper and, when applied to glacier mass balance, predicts that even the largest glaciers will be gone by 2030 under a scenario of current warming rates. The flexibility of integrated models is evident in the downscaling of RHESSys/FIREBGC components to the alpine treeline ecotone (Cairns 1994) and the upscaling to the CLIMET project. Schimel et al. (2002) describe the integrated modeling of carbon flux for all of the western US mountain areas in support of a national assessment on carbon sequestration. Thus, modeling clearly has moved beyond its development as a science tool and, increasingly, is becoming part of the policy arena.

However, a number of challenges remain. One is to develop integrated models that are interactively multi-scale. Management problems exist, and decisions often have to be made, at specific spatial and temporal scales that may not coincide with the scale of an integrated model's projections. We propose that management decisions can be greatly improved by examining other scales than those prescribed by law or policy. Encouraging multi-scale perspectives in policymaking will become increasingly important as global interactions increase with human population densities. Another challenge is to more fully integrate the drivers for human-caused landscape change and this will require better integration with socioeconomic models. While introducing even greater complexity into models that are already complex, achieving such a goal will not only improve our collective management of mountain resources but will also convincingly demonstrate the importance of mountains for people.

6. References

Band, L. E., Patterson, P., Nemani, R. R., and Running, S. W. (1993). Forest ecosystem processes at the watershed scale: Incorporating hillslope hydrology. *Agricultural and Forest Meteorology* **63**, 93-126.

Baron, J. S., Hartman, M. D., Band, L. E., and Lammers, R. B. (2000). Sensitivity of a high-elevation Rocky Mountain watershed to altered climate and CO_2. *Water Resources Research* **36**, 89-99.

Becker, A., and Bugmann, H., Eds. (2001). "Global change and mountain regions: The Mountain Research Initiative." IGBP Report 49, Stockholm.

Cairns, D. M. (1994). Development of a physiologically mechanistic model for use at the alpine treeline ecotone. *Physical Geography* **15**, 104-124.

Fagre, D. B., Comanor, P. L., White, J. D., Hauer, F. R., and Running S. W. (1997). Watershed responses to climate change at Glacier National Park. *Journal of the American Water Resources Association* **33**, 755-765.

Ferguson, S. A. (1997). A climate-change scenario for the Columbia River Basin. *In* "Interior Columbia Basin Ecosystem Management Project: Scientific Assesssment." (T. M. Quigley, Ed.), 9p. Research Paper PNW-RP-499. U.S. Department of Agriculture, Forest Service, Pacific Northwest Research Station, Portland, OR.

Finney, M. A., and Ryan, K. C. (1995). Use of the *Farsite* Fire Growth Model for fire prediction in US National Parks. *In* "Proceedings of The International Emergency Management and Engineering Conference (TIEMEC'95)." (J. D. Sullivan, J. L. Wybo, and L. Buisson, Eds.), pp. 183-189.

Gedalof, Z., Mantua, N. J., and Peterson, D. L. (2002). A multi-century perspective of variability in the Pacific Decadal Oscillation: New insights from tree rings and coral. *Geophysical Research Letters* **29**, 2204 (doi:10.1029/2002GL015824).

Hall, M. P., and Fagre, D. B. (2003). Modeled climate-induced glacier change in Glacier National Park, 1850-2100. *BioScience* **53**, 131-140.

Hungerford, R. D., Nemani, R. R., Running, S. W., and Coughlan, J. C. (1989). "MTCLIM: A mountain microclimate simulation model, INT-414." Intermountain Research Station, Ogden, UT.

Kang, S., Kimball, J. S., Running, S. W., Michaelis, A., and Zhao, M. (2002). Comparisons of MODIS productivity and potential productivity in the Pacific Northwest and BOREAS area. American Geophysical Union Annual Meeting, San Francisco, California, US, December 5-10, 2002.

Keane, R. E., Morgan, P., and Running, S. W. (1996). "FIRE-BGC: A mechanistic ecological process model for simulating fire succession on coniferous forest landscapes of the Northern Rocky Mountains." INT-RP-484, US Department of Agriculture, Forest Service, Intermountain Research Station, Ogden, UT.

Keane, R. E., Hardy, C. C., Ryan, K. C., and Finney, M. A. (1997). Simulating effects of fire on gaseous emissions and atmospheric carbon fluxes from coniferous forest landscapes. *World Resource Review* **9**, 177-205.

Keane, R. E., Morgan, P., and White, J. D. (1999). Temporal patterns of ecosystem processes on simulated landscapes in Glacier National Park, Montana, USA. *Landscape Ecology* **14**, 311-329.

Mantua, N. J., Hare, S. R., Zhang, Y., Wallace, J. M., and Francis, R. C. (1997). A Pacific interdecadal climate oscillation with impacts on salmon production. *Bulletin of the American Meteorological Society* **78**, 1069-1079.

Running, S. W., and Nemani, R. R. (1991). Regional hydrologic and carbon balance responses of forests resulting from potential climate change. *Climatic Change* **19**, 349-368.

Running, S. W., Nemani, R. R., Peterson, D. L., Band, L. E., Potts, D. F., Pierce, L. L., and Spanner, M. A. (1989). Mapping regional forest evapotranspiration and photosynthesis by coupling satellite data with ecosystem simulation. *Ecology* **70**, 1090-1101.

Running, S. W., and Gower, S. T. (1991). Forest-BGC, a general model of forest ecosystem processes for regional applications: II. Dynamic carbon allocation and nitrogen budgets. *Tree Physiology* **9**, 147-160.

Schimel, D., Kittel, T., Running, S., Monson, R., Turpinseed, A., and Anderson, D. (2002). Carbon sequestration studied in western US mountains. *EOS Transactions* **83**, 445-446.

Selkowitz, D. J., Fagre, D. B., and Reardon, B. A. (2002). Interannual variations in snowpack in the crown of the continent ecosystem. *Hydrological Processes* **16**, 3651-3665.

Thornton, P. E., Running, S. W., and White, M. A. (1997). Generating surfaces of daily meteorological variables over large regions of complex terrain. *Journal of Hydrology* **190**, 214-251.

White, J. D., and Running, S. W. (1994). Testing scale dependent assumptions in regional ecosystem simulations. *Journal of Vegetation Science* **5**, 687-702.

White, J. D., Running, S. W., Thornton, P. E., Keane, R. E., Ryan, K. C., Fagre, D. B., and Key, C. H. (1998). Assessing simulated ecosystem processes for climate variability research at Glacier National Park, USA. *Ecological Applications* **8**, 805-823.

Detecting Global Change at Alpine Treeline: Coupling Paleoecology with Contemporary Studies

Lisa J. Graumlich*, Lindsey A. Waggoner, and Andrew G. Bunn
Big Sky Institute, Montana State University, Bozeman MT 59715, USA
**phone 01-406-994-5320, fax 01-406-994-5122, email lisa@montana.edu*

Keywords: Alpine treeline, Landscape ecology, Paleoecology, Scale, Tree rings.

1. Introduction

Mountain ecosystems provide unique opportunities to detect and understand global change impacts due to their strong altitudinal gradients coupled with the presence of parks and biosphere reserves in many mountain areas where direct human impacts are minimal (Graumlich 2000; Becker and Bugmann 2001). Alpine treeline, the distinctive boundary between forest and tundra on high mountains, has been a particularly important focus of this research. While alpine treeline can appear to be a simple ecotone and thus a ready indicator of changing temperatures, a rich history of research has revealed that the dynamics of treeline are complex. Issues of lags and inertia, as well as multiple drivers across diverse scales abound. In this essay, we outline key issues for understanding alpine treeline in the context of global climate change. While our views strongly reflect research that has been done in the temperate zone, particularly in North America, the questions underlying the interpretation of treeline are applicable to many biotic indicators of climate change.

Detecting and understanding change in alpine treeline has been a fertile avenue for inquiry in global change research for several reasons. Given the importance of temperature in controlling the elevation of alpine treeline, this ecotone is likely to be one of the first forested ecosystems to register climate impacts as changes in structure and function. Detection and attribution of contemporary climate impacts at alpine treeline is aided by the context provided by paleoecological records. Precise dates of

U. M. Huber et al. (eds.), Global Change and Mountain Regions, 501-508.

the incursion of woody seedlings into tundra can be obtained through tree-ring based demographic studies. Past treeline movements can be inferred by tree-ring dating of dead wood. Remnant wood above current treeline provides firm evidence of previous ecotone fluctuations, which place contemporary changes into the context of long-term variability. Finally, by integrating precisely dated demographic studies with observational and proxy climate data, we can disentangle the role of different climate factors, extreme events, and lags in the overall impact of climate on ecotone dynamics (Lloyd 1997; Lloyd and Fastie 2003).

Research to date has largely focused on using paleoecological records to document *where* and *when* climatic variation alters high elevation forests (Graumlich 1994; Rochefort et al. 1994; Kullman 1998). Change is ubiquitous, although not synchronous, at time scales of decades to centuries. More recently, researchers have been seeking to build on this understanding of the patterns of change by addressing *how* climatic variability affects ecological processes at treeline. One objective of such efforts is to enhance our ability to forecast ecosystem change (*sensu* Clark et al. 2001). Below we describe efforts to link pattern to process at treeline. In particular, we emphasize strategies that meld the high temporal resolution approaches of paleoecology with tools from landscape ecology that exploit spatial patterns to explore dynamics.

2. Recent results

2.1 Climatic controls on treeline – a scale issue

A starting point for any discussion of changes at alpine treeline is the nature of factors that control the current position of the ecotone. Findings from recent research on treeline position mirror results from other fields in ecology in emphasizing the importance of scale in detecting pattern and sorting out driving factors as well as feedback processes (Urban et al. 1987; Levin 1992). There is a general consensus that the position of alpine treeline is ultimately controlled by climate but proximately subject to a variety of biotic and abiotic processes at finer spatial scales (Körner 1998). At the subcontinental scale, treeline position is best predicted by seasonal mean temperature. At the regional scale, evidence from mid-latitudes indicates that temperature and precipitation interact strongly to produce treeline position (Lloyd and Graumlich 1997). At the scale of the organism itself, treeline is predicted by microtopography (Bunn, unpublished data). Finally, biotic factors can interact with the physical features of the landscape to produce a more complex pattern than might be expected on the basis of physical factors alone (e.g. Lescop-Sinclair and Payette 1995; Weisberg and Baker 1995; Lloyd et al. 2003). An important implication of these issues of scale is that changes in treeline position are an important, although not simple, indicator of climate change (Noble 1993; Malanson 2001).

Two readily observable features of alpine treeline that can be used to infer trajectories of change are seedling establishment and the presence of remnant wood above current treeline. The former portends future increases in the elevation of treeline while the latter indicates where and when treeline was higher in the past.

Seedling establishment at and above current treeline is widely reported at mid and high latitudes (Table 1; also see summaries in Graumlich 1994; Lloyd et al. 2003) and is often interpreted as evidence for global warming. The entries in Table 1 include unpublished observations from researchers who work at treeline in the western United States and reflects the informal consensus that treeline is moving upslope in most areas. However, this interpretation is complicated by observations from the same researchers who commonly report the presence of remnant wood from the last millennium up to 100 m above current treeline in the western United States (Table 1) and elsewhere (e.g. Kullman and Kjällgren 2000). A conclusion from the data presented in Table 1 and other published reports in European, South American, Australian and Asian mountain ranges is that there is potential for utilizing evidence of current and past treeline fluctuations to increase our understanding of ecotone dynamics.

Table 1: Alpine treeline sites in the western United States offer ample evidence for contemporary and past treeline fluctuations. The information presented here is intended to complement existing summaries of records indicating change at treeline such as Graumlich (1994) and Rochefort et al. (1994). In developing this summary, we contacted researchers with active field programs at treeline and queried them regarding evidence for treeline change. The number of researchers reporting remnant wood above current treeline indicates that this resource has not been fully exploited for its value in interpreting past and present treeline dynamics.

Location	Lat/Long	Elevation (m)	Seedling recruitment	Remnant wood above treeline	Notes	Citation
Scenic Point, Glacier National Park, MT	48.29°N 113.19°W	2300	no	yes	not dated	Pederson, G. (2002 pers. comm.)
Grass Mountain, Beartooth Plateau, MT	45.16°N 109.53°W	3000	yes	yes	advance from AD 1150-1550 and ~1850-present, retreat AD 1600-1850	Graumlich (unpubl.)
Lemhi Range, ID	44.62°N 113.62°W	2800-3000	yes	yes	seedling establishment AD 1890-1960, deadwood dates to ~400BP	Winter (1984)
Mt. Washburn, Yellowstone National Park, WY	43.70°N 110.51°W	2900	yes	yes	advance AD 950-1250 and AD 1840-present, retreat AD 1300-1800	Graumlich (unpubl.)
Wyoming Peak, WY	42.60°N 110.62°W	3400	no data	yes	not dated	Griggs (1946)
Sangre de Cristos, CO	37.96°N 105.59W°	3500	no data	yes	not dated	Woodhouse (2002 pers. comm.)
White Mountains, CA	37.58°N 118.26°W	3500	yes	yes	retreat AD 1100, 1550, seedlings AD 1900-present	LaMarche (1973)
Southern Sierra Nevada, CA	36.47°N 118.23°W	3300-3500	yes	yes	advance AD 300-200, 800-900; retreat AD 1100-1200, 1400-1500 and 1700-1900	Lloyd and Graumlich (1997)
San Francisco Mountain, AZ	35.25°N 111.65W°	3500	no data	yes	not dated, scattered locations	Salzer, M. (2002 pers. comm.)
Front Range, CO	40°N 106° W	3500	no data	yes	not dated; scattered locations	Woodhouse (2002 pers. comm.)

Interpreting the degree to which the present patterns of seedling establishment at and above treeline represents a response to unprecedented warming during the last millennium (Mann et al. 1998) requires that we understand the causes of treeline fluctuations in the relatively recent past. In this regard, the annual or near annual resolution that can be obtained from tree-ring dating of remnant wood is critical to inferring causation when these data are combined with similarly resolved estimates of past climate. The strength of such an approach is illustrated in the work of Lloyd and Graumlich (1997) in the Sierra Nevada. They reconstructed a 3500 yr history of the fluctuations of treeline and tree abundance by dating live trees and remnant wood at and above treeline. By combining these results with dendroclimatic reconstructions of past climate, they concluded that decade- to century-scale patterns of treeline fluctuations may be best understood as a regional response to the interaction of growing season temperature and drought. These conclusions parallel the model-based interpretations by Bugmann and Pfister (2000). Specifically, Lloyd and Graumlich (1997) discovered that treeline forests expanded upslope during a warm and wet period early in the last millennium. Subsequently, treeline forest contracted during two severe multi-decadal droughts (Table 1). Given that severe, multi-decadal droughts have been identified as regular components of the climate of western North America (Woodhouse and Overpeck 1998), severe drought and concomitant changes in snow depth and soil moisture may play a key role in the interpretation of alpine treeline demography elsewhere.

2.2 The importance of site characteristics for tree recruitment

If drought is a factor in understanding treeline fluctuations on time scales of decades to centuries in the western United States, how can we exploit site characteristics to identify those treeline sites where the response can be distinguished as solely a temperature response? Recent research by Bunn and colleagues (in review) investigated the demographic structure of treeline sites with different thermal and soil moisture regimes in the Sierra Nevada. Using proxies for solar radiation and soil moisture derived from an elevation model, Bunn and colleagues mapped relative values of "brightness" and "wetness" over a 25 km^2 area. Within the mapped area, they identified sites falling into the highest or lowest 25th percentile for combinations of brightness and wetness (e.g. areas that were exceptionally xeric or mesic and exceptionally bright or dark). They then inventoried forest structure (i.e. basal area and seedling density) and age-class structure (by extracting increment cores) in 90 x 90 m plots in a random subset of the sites. Results depicting rates of stem recruitment into two stands at treeline and representing two of the four possible site categories are depicted in Figure 1. At the bright and wet sites, there is a dramatic increase in tree recruitment, especially from 1950 onward. At the dark and dry sites, recruitment rates are more or less steady over the past 600 years with some indication of a slowing in recruitment in the past 100 years. Further work by Bunn et al. (in review) indicates that this is a robust result: forest age class structure can be predicted by physical setting with bright and wet sites having the greatest proportion of the stems recruited in the last 100 years. These contemporary results are coherent with the paleoecological studies of Lloyd and Graumlich, which show that persistent episodes of high precipitation are critical in allowing forest expansion to take place under warming conditions. In related analyses, Bunn found that seedling (i.e. individuals less than 2.5 cm diameter at breast height) numbers and age distribution could not be predicted by the brightness and wetness indices. In essence, seedlings were abundant under all microsite conditions. The work by Bunn and colleagues clearly indicates that it is necessary to go beyond seedling presence or absence to detect directional change in the alpine treeline ecotone and that seedling recruitment into sapling and larger size classes is a more suitable indicator of change. This work also indicates that sites where change can be detected will be those sites where other environmental factors, especially soil moisture, substrate, and nutrients, are not limiting (Butler et al. 1994; Kupfer and Cairns 1996).

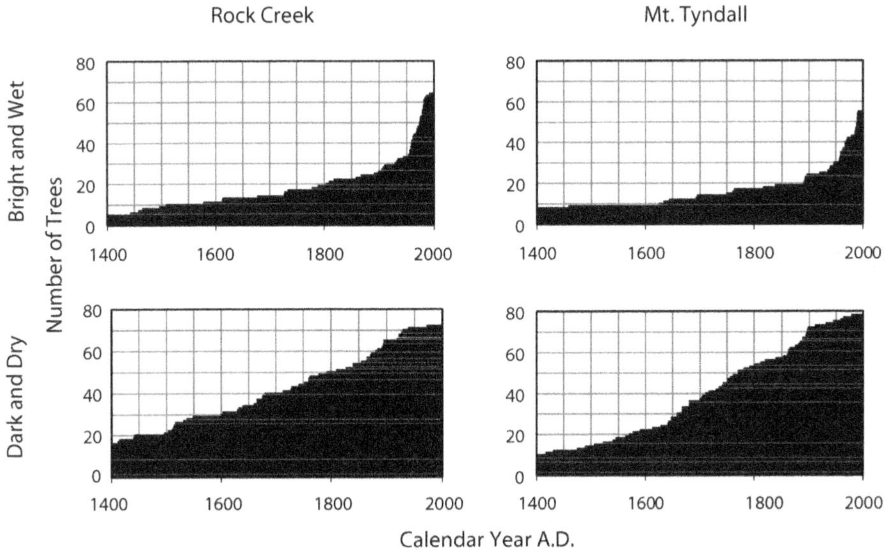

Figure 1: Four age class charts are shown for 90 x 90 m plots from different physical settings (rows) at two different sites (columns) in Sequoia National Park, USA. Rock Creek (36° 25' N, 118° 19' W) and Mt. Tyndall (36° 39' N, 118° 20' W) are both treeline sites near or at the alpine treeline (elev. 3300 m) dominated by foxtail pine (*Pinus balfouriana* Grev. and Balf.). Bright and wet vs. dark and dry sites were identified based on indices for solar radiation receipt and topographic convergence as computed with a 30 m digital elevation model. Age class data were obtained from cores extracted from all stems greater than 2.5 cm diameter at breast height in each plot.

Elements of this approach are proving to be useful in interpreting the significance of remnant wood above current treeline. At sites in the Sierra Nevada and in the Greater Yellowstone Area, we have found hundreds of dateable pieces of remnant wood above treeline (Fig. 2). Our research group is systematically mapping dead wood at and above treeline at several sites in the Sierra Nevada and Absaroka Mountains (Montana) for the purpose of understanding when and where treeline fluctuations occurred in the past millennium. One objective of this project is to understand where the evidence in form of remnant wood is consistently and reliably preserved. After accounting for any biases due to differential preservation, we will assess the relationship between site characteristics of slopes where treeline has changed in the past millennium and slopes where we currently see seedling and sapling establishment above treeline. Conventionally, the first-order interpretation of the presence of deadwood above treeline is that recent changes at treeline are still within the envelope of variability of the last millennium. This simple interpretation belies the complexity of drivers and modes of response as summarized in this essay. In other words, different combinations of temperature, precipitation and site characteristics may produce similar changes in treeline elevation. The presence of deadwood above current treeline indicates that at specific periods during the past millennium and in specific topographic contexts treeline expanded and subsequently retreated. Higher-than-present treeline during the last millennium is likely related to periods with a combination of warm and wet conditions rather than simply warmer-than-present climate (Lloyd and Graumlich 1997). By combining classic high temporal resolution paleoecological analyses with a more nuanced analysis of the role of site factors (e.g. slope, topography, environmental heterogeneity), we seek to understand how synergistic and persistent combinations of temperature and precipitation array themselves on complex topography in such a way

Figure 2: The Mt. Tyndall (36° 39' N, 118° 20' W) area of Sequoia National Park, USA, is shown in a 4 m multi-spectral satellite image from the IKONOS sensor. It has been draped over an elevation model to provide a three dimensional view. Alpine treeline occurs at about 3300 m and can be detected as an intricate boundary between dark (tree) pixels and bright (granite substrate) pixels. Moist vegetation (e.g. meadows) appears as red in the false color image. The yellow dots represent the locations of remnant wood found above current treeline located by field reconnaissance and recorded with a Geographic Positioning System (GPS). The spatial pattern of remnant wood indicates that specific slope positions are more apt to record treeline fluctuations, or to preserve remnant wood, than others.

as to enhance the recruitment of seedlings into adult age classes. We explicitly assume that factors that limit tree establishment do not change uniformly over space and time (Gavin et al. 2001). More simply, our goal is to move beyond detecting climatic change using alpine treeline towards a predictive and mechanistic understanding.

3. Future directions

We see the melding of paleoecological approaches with spatial analysis tools, including remote sensing, as the key to refining the role of alpine treeline as an indicator of global change. Concrete products from such an approach would be maps of change potential for subalpine forests along major mountain chains of the world. These maps would identify sites best suited as indicators of a specific aspect of global change (e.g. change in growing season degree days). We envision the development of a network of strategically placed sites where *in situ* and remotely sensed data can be gathered to monitor and gauge the effects of change that would complement other observing systems (e.g. glacial monitoring network).

Beyond its role as an indicator of change, the alpine treeline ecotone has long fascinated researchers as a landscape where pattern and process are intimately linked. Changes from tundra or meadow vegetation to a forested landscape will entrain a host of feedbacks, involving snow, albedo, hydrology and biogeochemistry. Efforts to monitor and predict the trajectory of change at treeline will have direct relevance to such questions as the role of snow-albedo feedbacks in altering the pace of warming at high elevations as well as to questions of carbon sequestration and release. Of critical importance in all these studies is the continued effort to integrate our understanding of high-elevation landscapes as systems in which pattern and process interact at a variety of temporal and spatial scales.

4. References

Becker, A., and Bugmann, H., Eds. (2001). Global change and mountain regions: The Mountain Research Initiative. *In* "Implementation Plan," IGBP Report 49 / IHDP Report 13 / GTOS Report 28, Stockholm.

Bugmann, J., and Pfister, C. (2000). Impacts of interannual climate variability on past and future forest composition. *Regional Environmental Change* **1**, 112-125.

Butler, D. R., Malanson, G. P., and Cairns, D. M. (1994). Stability of alpine treeline in Glacier National Park, Montana, U.S.A. *Phytocoenologia* **22**, 485-500.

Clark, J. S., Carpenter, S. R., Barber, M., Collins, S., Dobson, A., Foley, J. A., Lodge, D. M., Pascual, M., Pielke, R., Pizer, W., Pringle, C., Reid, C. W., Rose, K. A., Sala, O., Schlesinger, W. H., Wall, D., and Wear, D. (2001). Ecological forecasting: An emerging imperative. *Science* **293**, 657-660.

Gavin, D. G., McLachlan, J. S., Brubaker, L. B., and Young, K. A. (2001). Postglacial history of subalpine forests, Olympic Peninsula, Washinton, USA. *The Holocene* **11**, 177-188.

Graumlich, L. J. (1994). Long-term vegetation change in mountain environments. *In* "Mountain environments in changing climates." (M. Beniston, Ed.), pp. 32-45. Routledge, London.

Graumlich, L. J. (2000). Global change and wilderness areas: Disentangling natural and anthropogenic changes. *In* "Proceedings Wilderness Science in a Time of Change." Ogden, Utah, U.S. Department of Agriculture, Forest Service, Rocky Mountain Research Center.

Griggs, R. F. (1946). The timberlines of Northern America and their interpretation. *Ecology* **27**, 275-289.

Körner, C. (1998). A re-assessment of high elevation treeline positions and their explanation. *Oecologia*

115, 445-459.
Kullman, L. (1998). Tree-limits and montane forests in the Swedish Scandes: Sensitive biomonitors of climate change and variability. *Ambio* **27**, 312-321.
Kullman, L., and Kjällgren, L. (2000). A coherent postglacial tree-limit chronology (*Pinus sylvestris* L.) for the Swedish Scandes: Aspects of paleoclimate and "recent warming," based on megafossil evidence. *Arctic, Antarctic, and Alpine Research* **32**, 419-428.
Kupfer, J. A., and Cairns, D. M. (1996). The suitability of montane ecotones as indicators of global climatic change. *Progress in Physical Geography* **20**, 253-272.
LaMarche, V.C. (1973). Holocene climatic variations inferred from treeline fluctuations in the White Mountains, California. *Quaternary Research* **3**, 632-660.
Lescop-Sinclair, K., and Payette, S. (1995). Recent advance of the arctic treeline along the eastern coast of Hudson Bay. *Journal of Ecology* **83**, 929-936.
Levin, S. A. (1992). The problem of pattern and scale in ecology: The Robert H. MacArthur award lecture. *Ecology* **73**, 1943-1967.
Lloyd, A. H. (1997). Response of treeline populations of foxtail pine (*Pinus balfouriana*) to climate variation over the last 1000 years. *Canadian Journal of Forest Research* **27**, 936-942.
Lloyd, A. H., and Graumlich, L. J. (1997). Holocene dynamics of treeline forests in the Sierra Nevada. *Ecology* **78**, 1199-1210.
Lloyd, A. H., and Fastie, C. L. (2003). Recent changes in treeline forest distribution and structure in interior Alaska. *Ecoscience* (in press).
Lloyd, A. H., Rupp, T. S., Fastie, C. L., and Starfield, A. M. (2003). Patterns and dynamics of treeline advance on the Seward Peninsula, Alaska. *Journal of Geophysical Research* (in press).
Malanson, G. P. (2001). Complex responses to global change at alpine treeline. *Physical Geography* **22**, 333-342.
Mann, M. E., Bradley, R. S., and Hughes, M. K. (1998). Global-scale temperature patterns and climate forcing over the past six centuries. *Nature* **392**, 779-787.
Noble, I. R. (1993). A model of the responses of ecotones to climate change. *Ecological Applications* **3**, 396-403.
Rochefort, R. M., Little, R. L., Woodward, A., and Peterson, D. L. (1994). Changes in sub-alpine tree distribution in western North America: A review of climatic and other causal factors. *The Holocene* **4**, 89-100.
Urban, D. L., O'Neill, R. V., and Shugart, H. H. (1987). Landscape ecology. *Bioscience* **37**, 119-127.
Weisberg, P. J., and Baker, W. L. (1995). Spatial variation in tree seedling and krummholz growth in the forest-tundra ecotone of Rocky Mountain National Park, Colorado, USA. *Arctic and Alpine Research* **27**, 116-129.
Winter, M. H. (1984). "Altitudinal fluctuations of upper treeline at two sites in the Lemhi Range, ID." Unpublished M.S. thesis, University of Kansas, Lawrence.
Woodhouse, C. A., and Overpeck, J. T. (1998). 2000 Years of drought variability in the central United States. *Bulletin of the American Meteorological Society* **79**, 2693-2714.

Part V: Human Dimensions

Global change is a force of nature. Since the middle of the 20th century, however, humans became a significant and sometimes dominating environmental force with tremendous implications for the Earth's system. As such, the human system is not only subject to global change but constitutes one of the main driving forces behind global change. The human dimension of global change therefore includes economic and social aspects apart from the bio-physical aspects.

Mountain people that live in marginal and remote areas are most vulnerable to the unprecedented impacts of global change. However, also a large part of the urban population in the lowlands that is dependent on many of the goods and services provided by mountain areas is highly vulnerable. Consequently, the ultimate objective of global change research is not the generation of scientific data and knowledge, but rather the improvement of the quality of life. The following examples of integrated global change research place people and their activities in the center and successfully link disciplines from natural and social sciences.

Beniston's chapter that analyzes the risks associated with climatic change in mountain regions highlights the system's vital functions for land and people living outside their natural boundaries. the connection between uplands and lowlands can become a threat when climate change triggers risks that cross the natural border. *Price* shares this insight in the context of forests in sustainable mountain development. He illustrates how forestry science and practice has shifted from an emphasis on wood production to an emphasis on their multiple functions, products, and roles for upland and lowland people. The new understanding of an interconnected upland-lowland system calls for more strategic research and consistent scientific information. In a similar vein, *Lebel* analyses the interface between forest governance and forest resources in Northern Thailand. He demonstrates that institutions both cause and respond to environmental change. Therefore, better understanding of such cross-scale interactions is crucial for sound decision-making. Another example for the inter-linkages between institutions and upland ecosystems is provided by *Jianchu & Wilke*'s article on Yunnan, China. They show how state simplifications shape the political discourse on the management of mountain systems, often resulting in inappropriate interventions and policies instead of being more supportive of local traditions, ensuring economic benefits for mountain people, and involving them in decision-making processes. The need to acknowledge local people's values and views is also highlighted by *Ramakrishnan*. He points out that social, cultural and spiritual activities are not confined to the community but reach far into the natural system. The author outlines two possibilities of how modern knowledge could complement traditional ecological knowledge to ensure sustainable land use in mountain environments.

Looking beyond the boundary is also proposed by *Hansen & DeFreis* in their contribution on land use intensification around nature reserves in mountains. Based on the observation that human activities outside the boundaries of nature reserves alter the ecosystem inside and *vice versa*, the authors suggest spanning the true system boundaries far outside the designated boundaries. *McCracken & Huband* stress the point that conservation does not necessarily imply the exclusion of human activities from a certain area. In contrary, particular farming systems, such as mountain livestock systems, are important in maintaining nature conservation value.

Jodha's essay on economic globalization and its repercussions from fragile mountains and communities in the Himalayas, adds a new perspective. Apart from various sources of risk, such as the declining resilience of traditional mountain systems, he identifies potential opportunities created by globalization. Both, risks and opportunities, are the result of the mountain-specific conditions, such as fragility, inaccessibility, diversity, and marginality.

In the article on research partnerships for mitigating global change syndromes in mountain areas, *Hurni et al.* take stock of the knowledge available on the impact of global change and moves forward towards future research. The three selected case studies demonstrate that the mitigation of syndromes of global change, i.e. clusters of environmental, social and economic problems, requires transdisciplinary research and concerted actions of multiple actors and all stakeholders concerned. The Mount Kenya example given by *MacMillan & Liniger* supports the view that future research should be integrated, long-term and multi-level across the region. Their article on monitoring and modeling for the sustainable management of water resources stresses the fact that the availability of sound scientific data obtained from monitoring and modeling is a precondition to the resolution of water conflicts between upland and lowland communities. *Schreier*'s essay on challenges in mountain watershed management analyses internal and external pressures on local water resources emphasizing the need of integrated watershed management, which was, so far, hampered by the poor state of knowledge on global change impacts on hydrological processes. Giving examples of legal and economic instruments and social organization models of how upland people can receive compensation for their protection and maintenance services, *Koch-Weser* successfully bridges the gap between science and mountain people's reality.

All these articles demonstrate that the challenge of global change can only be met if scientists join hands with researchers from different disciplines, local stakeholder groups, politicians and decision-makers.

The Risks Associated with Climatic Change in Mountain Regions

Martin Beniston
Department of Geosciences, University of Fribourg, Pérolles, CH-1700 Fribourg, Switzerland
phone +41-26-3009011, fax +41-26-3009746, e-mail martin.beniston@unifr.ch

Keywords: Biodiversity, Climatic change, Climate modeling, Extreme events, Risk, Water resources.

1. Introduction

The Earth's environment is continuously subjected to various stresses through natural processes and human interference. With the rapid industrialization and population growth that the 20th century has experienced worldwide, however, humankind has added a new dimension of stress to the global environment in general, and mountain regions in particular. In some instances, environmental degradation is inevitable because of the basic requirements of human populations, particularly where those are growing rapidly; in other cases, environmental damage is a direct result of mismanagement and over-exploitation of natural resources (Beniston 2000). The sensitivity of a given mountain region to changes in environmental conditions depends largely upon the climatic, geological and biological features of the region considered. Changes in these controlling factors, particularly through direct human interference or indirect effects such as climatic change, may have significant impacts upon numerous mountain environments.

2. Modeling changes in climate and climatic extremes in mountain regions

The current spatial resolution of General Circulation Models (GCM) is generally

U. M. Huber et al. (eds.), Global Change and Mountain Regions, 511-519.

too crude to adequately represent the orographic detail of most mountain regions and therefore the complexities of regional climates. On the other hand, most climatic impacts research requires information at fine spatial definition, where the regional details of topography or land-cover are important determinants in the response of natural and managed systems to change. Since the mid-1990s, the scaling problem related to complex orography has been addressed through regional modeling techniques, pioneered by Giorgi and Mearns (1991), and through statistical-dynamical downscaling techniques (e.g. Zorita and von Storch, 1999).

In this context, it is recognized that many natural and managed systems are far more sensitive to climatic extremes than to changes in mean climate. In the Alps, a warmer and perhaps more extreme climate is likely to result in increasingly frequent hazards, related to the reduction of slope stability (e.g. through permafrost degradation and loss of vegetation cover following shifts in ecosystem distribution). An increase in the frequency of intense precipitation events will enhance the severity of natural hazards, such as floods, slope erosion, and avalanches. Research on the impacts of extreme climatic events thus needs to focus on cryospheric, hydrologic and biospheric issues both at high elevations, where the sensitivity of natural systems to change is high (e.g. Keller et al. 2000), and at lower elevations. Extreme climatic events are often accompanied by significant losses resulting from damage to infrastructure and to forests and crops (as witnessed during the storms of end-December 1999 where the total costs across Europe exceeded US$20 billion). Hence, the quantification of future trends in weather extremes in a changing global climate, and their translation into economic losses, are of long-term strategic interest to mountain countries such as Switzerland.

In the context of a Swiss network research program (NCCR-Climate), the Department of Geography of the University of Fribourg is currently investigating extreme climate events such as wind-storms, high precipitation, and the persistence of heat or drought (Goyette et al. 2001; 2003; Jungo et al. 2002). Modeling climatic extremes addresses more than a simple set of simulations in which the differences in temperature and precipitation between a particular baseline and some point in the future are quantified. At the spatial scales relevant to mountain environments, there are numerous interactive systems that need to be taken into account, in particular land-use patterns, snow and ice, water resources, and ecosystems. Decadal-scale variability of the climate system that can modulate the intensity and frequency of certain extremes also need to be considered when attempting to simulate shifts in extreme events. For example, changes in the extremes of temperature and moisture in the Alpine region can be particularly sensitive to the behavior of the North Atlantic Oscillation, as shown by Jungo and Beniston (2001) and Beniston and Jungo (2002).

In order to understand the fundamental mechanisms underlying regional extremes in wind and temperature, their variability and persistence, and their shifts in a changing global climate in the 21st century, investigations at the University of Fribourg make use of the Canadian Regional Climate Model CRCM-2 (e.g. Laprise et al. 1998). The numerical modeling of climatic processes at the regional scale enables the identification of causal mechanisms and, in addition, allows a certain degree of predictability. The regional climate model (RCM) procedures currently applied to the simulation of wind storms over Western Europe and their impacts on the Alps involve a numerical "cascade" process, by which the model successively increases its spatial definition (resolution) in order to focus upon regions where there may be high sensitivity of infrastructure, forests, and slope stability. In the work by Goyette et al. (2001), for example, a low-resolution (60 km grid mesh) captures the essence of the synoptic weather systems that generate local or regional extremes, and then "feeds" a 5-km version of the model to ameliorate the quality of regional-scale simulations. Finally, in order to reproduce and document areas particularly affected by wind damage, or related to extended heat and drought stress, a 1-km resolution is applied to the region of interest. Figure 1 illustrates the CRCM-2 cascade process, zooming in progressively from the Europe-Atlantic scale to the scale of a large mountain valley.

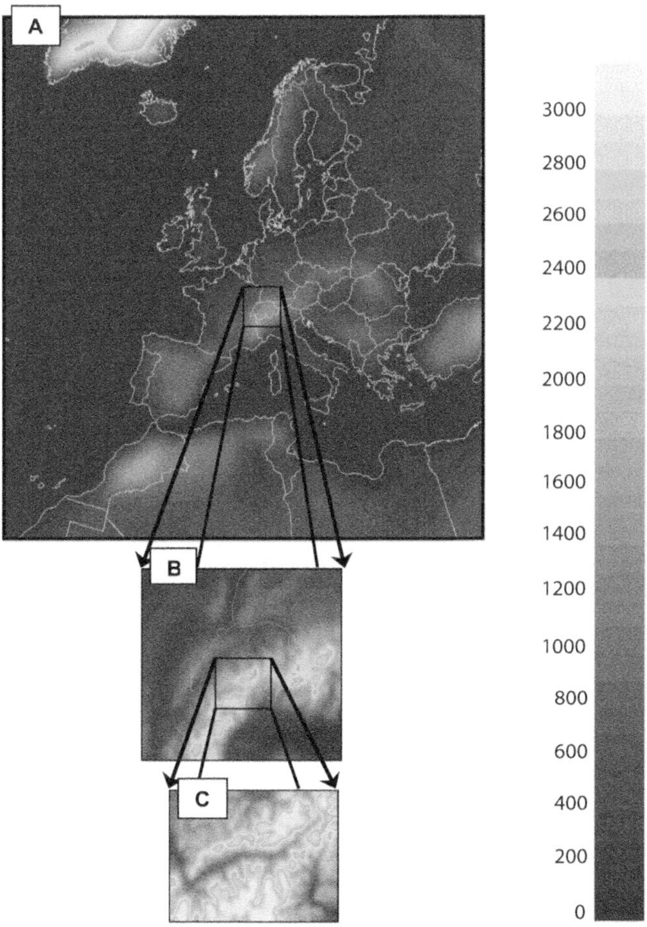

Figure 1: Illustration of the regional climate model "self-nesting" scheme that allows the investigation of climatic processes at increasingly finer scales (after Goyette et al. 2001). Domain A uses a 60-km grid, domain B a 5-km grid, and domain C a 1-km grid. This latter resolution enables the impacts of severe climatic events to be quantified. The legend on the right is terrain height in meters.

In such types of climatic change investigations, risk analysis is a necessary part of the approach, because it is necessary to quantify potential increases in damage in order to prepare for adequate response strategies by stakeholders, essentially local authorities. Risk and cost-benefit analyses help to determine whether the probable increases in damage costs warrant investments for protection infrastructure or other measures. On the basis of regional climate scenarios, and a quantification of the potential impacts on the natural environment, new strategies essentially linked to sustainability can be tracked and publicized. Such strategies should be promoted as they constitute a way of mitigating the consequences of global change on a wide range of economic, political and social sectors in mountain regions.

3. Risks associated with climatic change in mountain regions

A warming climate will enhance the hydrological cycle, implying higher rates of evaporation, and a greater proportion of liquid precipitation compared to solid precipitation; these physical mechanisms, associated with potential changes in precipitation amount and seasonality, will affect soil moisture, groundwater reserves, and the frequency of flood or drought episodes. Though water is present in ample quantity at the Earth's surface, the supply of water is limited and governed by the renewal processes associated with the global hydrological cycle. With the expansion of human settlements and the growth of industrial activities, water has been increasingly used for the assimilation and discharge of wastes. This resource has been taken for granted, and only in the past few decades has increasing water shortage and declining water quality from pollution drawn attention to the inherent fragility and scarcity of water. This has led to concerns about water availability to meet the requirements of the 21st century.

Snow and ice are, for many mountain ranges, a key component of the hydrological cycle, and the seasonal character and amount of runoff is closely linked to cryospheric processes. A changing climate is likely to lead to shifts in seasonal snow pack; glacier melt will influence discharge rates and timing in the rivers that originate in mountains. In most temperate mountain regions, the snow-pack is close to its melting point, so that it responds rapidly to apparently minor changes in temperature. As warming progresses in the future, regions where snowfall is the current norm will increasingly experience precipitation in the form of rain. For every °C increase in temperature, the snowline will rise by about 150 m.

Because of the sensitivity of mountain glaciers to temperature and precipitation, the behavior of glaciers provides some of the clearest evidence of atmospheric warming over the past decades (Haeberli and Beniston 1998). The volume of ice in a glacier, and correspondingly its surface area, thickness, and length, is determined by the balance between inputs (accumulation of snow and ice) and outputs (melting and calving). As climate changes, the balance between inputs and outputs may be altered, resulting in a change in thickness and the advance or retreat of the glacier. Temperature, precipitation, humidity, wind speed, and other factors such as slope and the reflectivity of the glacier surface all affect the balance between inputs and outputs (Fitzharris et al. 1996).

There is widespread evidence that glaciers are retreating in many mountain areas of the world. Since 1850, the glaciers of the European Alps have lost about 30 to 40% of their surface area and about half of their volume (Haeberli and Beniston, 1998). Similarly, glaciers in the Southern Alps of New Zealand have lost 25% of their area over the last 100 years, and glaciers in several regions of central Asia have been retreating since the 1950s (Fitzharris et al. 1996; Meier 1998). Glacial retreat is also prevalent in the higher elevations of the tropics. Glaciers on Mt. Kenya and Kilimanjaro have lost over 60% of their area in the last century (Hastenrath and Greischar 1997), and accelerated retreat has been reported for the tropical Andes (Thompson et al. 2000).

Empirical and energy-balance models indicate that 30 – 50% of existing mountain glacier mass could disappear by the year 2100 if global warming scenarios in the range of 2-4°C indeed occur (Fitzharris et al. 1996; Haeberli and Beniston 1998). The smaller the glacier, the faster it will respond to changes in climate. As a result, many glaciers in temperate mountain regions would lose most of their mass within decades.

Because mountains are the source region for over 50% of the globe's rivers, the impacts of climatic change on hydrology are likely to have significant repercussions not only in the mountains themselves but also in populated lowland regions that depend on mountain water resources for domestic, agricultural, energy and industrial purposes. Water resources for populated lowland regions are influenced by mountain climates and vegetation; snow feeds into the hydrological basins and acts as a control on the timing of water runoff in the spring and summer months. Hydrological systems are also controlled by soil moisture, which largely determines the distribution of ecosystems, groundwater recharge, and runoff; the latter two factors sustain river flow and can lead to floods.

Significant shifts in climatic conditions will also have an effect on social and economic systems in many regions through changes in demand, supply and water quality. In regions which are currently sensitive to water stress (arid and semi-arid mountain regions), any shortfalls in water supply will enhance competition for water use for a wide range of economic, social, and environmental applications. In the future, such competition will be sharpened as a result of larger populations, which will lead to heightened demand for irrigation and perhaps also industrialization, at the expense of drinking water (Beniston 2002).

Because of increasing population, the additional demand will be accompanied by a sharp decline in water availability per capita. A consumption of 1,000 m^3 of water per year, per capita is considered a standard for "well-being" in the industrialized world. Projections of annual water availability per capita within the next 20 years, however, show a declining trend in many parts of the world, including those that are considered to have ample water resources (Shiklomanov 2001). Figure 2 illustrates the changes that are projected between current and future water availability. This is a reflection of the combined influence of environmental change (e.g. modified precipitation patterns in a changing climate) and socio-economic trends (e.g. sustained population growth in many of these countries, implying a reduction in per capita availability even if water amounts remain the same as today). In many instances, the impact of population

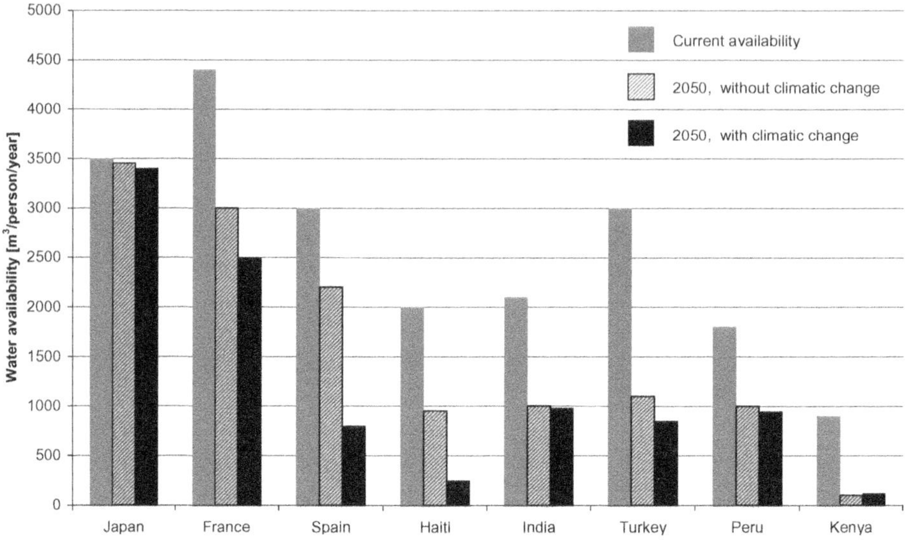

Figure 2: Water availability under current climatic conditions; in 2050 without shifts in climate; and in 2050 with climatic change (after Shiklomanov 2001).

growth may in fact be larger than that of climatic change itself, the latter being merely an exacerbating factor (IPCC 2001).

Geomorphological hazards are part of a larger group of natural hazards, including floods, soil or water quality risks due to agricultural technologies, desertification, sudden weather changes and wildfires. Climate change could alter the magnitude and/or frequency of a wide range of geomorphologic processes. Increases in extreme precipitation events, associated with snowmelt, could increase the frequency and severity of floods. In the mountain regions of Asia affected by the Monsoon, for example, loss of revenue in agriculture and damage to infrastructure related to Monsoon storms represent a high proportion of annual loss, as reported by Carver and Nakarmi (1995). Increased severity and frequency of storms during the Monsoon period in the future may lead to higher rates of erosion and more frequency of flooding in the Himalayas. Such extreme events would affect erosion, discharge and sedimentation rates. Furthermore, sediments deposited in large quantities on agricultural lands, irrigation canals and streams could lead to disruptions of agricultural production.

Erosion by rivers at the base of slopes increases slope steepness and hence the potential for increased hazards. Slow movement of landslides and gradual creep of rock and soil down-slope can be as destructive as fast slides but are less likely to cause loss of life. Debris flows, on the other hand, are rapid flows of water laden with rocks and sediments. Material on steep slopes that becomes saturated with water after prolonged intense rain or abrupt snow melt may develop a debris flow or mud flow (Dikau et al. 1997) resulting in destruction to infrastructure, forests and ecosystems. Loss of life can be high because of the unexpected and rapid nature of debris-flow events.

A further hazard leading to decreased slope stability is related to changes in the cryosphere at high elevations. The retreat of glaciers, and the corresponding changes in landscape could be one of the more immediately perceptible signals of climate change (Haeberli and Beniston 1998). Because the recolonization of vegetation on sparse and fragile mountain soils is slow, deglaciated morainic deposits can remain unprotected against erosion for decades to centuries. On slopes whose angles are steeper than 25-30°, stability problems can arise in newly exposed areas or those where permafrost thawing becomes significant. With the melting of the present permafrost zones at high mountain elevations, rock and mud-slide events can be expected to increase in number and, possibly, in severity. This will certainly have a number of economic consequences for mountain communities, where loss of life and the damage costs to constructions are certain to rise in proportion to the number of landslide events. In many mountainous regions, tourist resorts such as those in the Alps and the Rocky Mountains, or large urban areas close to mountains and areas characterized by steep slopes (e.g. suburbs of South American Andean cities, Hong Kong, or Los Angeles) have spread into high-risk areas, thus enhancing their exposure to landslide hazards.

Fire is an element that is of particular importance in many ecological systems; it is destructive in numerous circumstances, but also plays a valuable role in the recycling of organic material and the regeneration of vegetation. Changing climatic conditions are likely to modify the frequency and intensity of fire outbreaks, but there are other factors that need to be considered as well. For example, changes in fire-management practices and forest dieback in response to external stress factors can impact fuel characteristics and can lead to increased fire risk (King and Neilson 1992).

With climatic change as projected by the IPCC (2001), prolonged periods of summer drought would transform areas already sensitive to fire into regions of sustained fire hazard. The coastal ranges of California, the Blue Mountains of New South Wales (Australia), Mt. Kenya, and mountains on the fringes of the Mediterranean Sea, already subject to frequent fire episodes, would be severely affected. In addition, many of these regions are located close to major population centers, and thus considerable damage to infrastructure and disturbances to economic activities at the boundaries of many large urban areas might be expected. Cities such as Los Angeles and the San Francisco Bay Area in California, Sydney in Australia, coastal resorts close to the mountains in Spain, Italy, and southern France could become more vulnerable in the future as fire hazards increase in response to climatic change and urban centers expand in response to population pressures. Fires could also break out in regions that are currently relatively unaffected, as critical climatic, environmental and biological thresholds for fire outbreaks are exceeded (e.g. Johnson 1992).

4. Conclusions

Mountains are unique features of the Earth system in terms of their scenery, their climates, their ecosystems; they provide key resources for human activities well beyond their natural boundaries; and they harbor extremely diverse cultures in both the developing and the industrialized world. The protection of mountain environments against the adverse effects of economic development should be a priority for both today's generation and the generations to come (Beniston 2000).

In facing up to environmental change, human beings are going to have to think in terms of decades and centuries. Many of the impacts of these profound changes may not become unambiguously apparent for several generations. Many of the policies and decisions related to pollution abatement, climatic change, deforestation or desertification would provide opportunities and challenges for the private and public sectors. A carefully selected set of national and international responses aimed at mitigation, adaptation and improvement of knowledge can reduce the risks posed by environmental change to water resources and natural hazards.

5. References

Beniston, M. (2000). "Environmental change in mountains and uplands." Arnold Publishers, London, and Oxford University Press, New York.

Beniston, M. (2002). Climatic change. Implications for the hydrological cycle and for water management. *In* "Advances in global change research." (M. Beniston, Ed.). Kluwer, Dordrecht.

Beniston, M., and Jungo, P. (2002). Shifts in the distributions of pressure, temperature and moisture in the alpine region in response to the behavior of the North Atlantic Oscillation. *Theoretical and Applied Climatology* **71**, 29-42.

Carver, M., and Nakarmi, G. (1995). The effect of surface conditions on soil erosion and stream suspended sediments. *In* "Challenges in resource dynamics in Nepal: Processes, trends and dynamics in middle mountain watersheds." Proceedings of an ICIMOD Workshop (International Center for Integrated Mountain Development), (H. Schreier, P. B. Shah, and S. Brown, Eds.), pp. 155-162. Kathmandu, Nepal.

Dikau, R., Gärtner, H., Holl, B., Kienholz, H., Manni, P., and Zimmermann, M. (1997). Untersuchungen zur Murgangaktivität in Mattertal, Wallis, Schweiz. *In* "Proceedings of the Interpraevent Conference, Garmisch-Partenkirchen," pp. 397-408.

Fitzharris, B. B., Allison, I., Braithwaite, R. J., Brown, J., Foehn, P., Haeberli, W., Higuchi, K., Kotlyakov, V. M., Prowse, T. D., Rinaldi, C. A., Wadhams, P., Woo, M. K., and Youyu Xie (1996). The Cryosphere: Changes and their impacts. *In* "Second Assessment Report of the Intergovernmental Panel on Climate Change (IPCC)." Chapter 5, pp. 241-265. Cambridge University Press.

Giorgi, F., and Mearns, L. O. (1991). Approaches to the simulation of regional climate change: A review. *Reviews of Geophysics* **29**, 191-216.

Goyette, S., Beniston, M., Jungo, P., Caya, D., and Laprise, R. (2001). Numerical investigation of an extreme storm with the Canadian Regional Climate Model: The case study of windstorm Vivian, Switzerland, February 27, 1990. *Climate Dynamics* **18**, 145-168.

Goyette, S., Brasseur, O., and Beniston, M. (2003). Application of a new wind gust parameterisation: Multi-scale case studies performed with the Canadian RCM. *Journal of Geophysical Research* (in press).

Haeberli, W., and Beniston, M. (1998). Climate change and its impacts on glaciers and permafrost in the Alps. *Ambio* **27**, 258-265.

Hastenrath, S., and Greischar, L. (1997). Glacier recession on Kilimanjaro, East Africa, 1912-89. *Journal of Glaciology* **43**, 455-459.

IPCC (2001). "Climate change. The IPCC Third Assessment Report." Cambridge University Press, Cambridge and New York. Volumes I (The Scientific Basis), II (Impacts, Adaptation, and Vulnerability) and III (Mitigation).

Johnson, E. A. (1992). "Fire and vegetation dynamics. Studies from the North American boreal forest." Cambridge University Press, Cambridge.

Jungo, P., and Beniston, M. (2001). Changes in the anomalies of extreme temperature anomalies in the 20th century at Swiss climatological stations located at different latitudes and altitudes. *Theoretical and Applied Climatology* **69**, 1-12.

Jungo, P., Goyette, S., and Beniston, M. (2002). Daily wind gust speed probabilities over Switzerland according to three types of synoptic circulation. *International Journal of Climatology* **22**, 485-499.

Keller, F., Kienast, F., and Beniston, M. (2000). Evidence of the response of vegetation to environmental change at high elevation sites in the Swiss Alps. *Regional Environmental Change* **2**, 70-77.

King, G. A., and Neilson, R. P. (1992). The transient response of vegetation to climate change: A potential source of CO_2 to the atmosphere. *Water, Air and Soil Pollution* **64**, 365-383.

Laprise, R, Caya, D., Giguère, M., Bergeron, G., Côté, H., Blanchet, J.-P., Boer, G. J., and McFarlane, N. A. (1998). Climate of Western Canada under current and enhanced greenhouse gas concentration as simulated by the Canadian Regional Climate Model. *Atmosphere and Oceans* **36**, 119-167.

Meier, M. (1998). Land ice on Earth: A beginning of a global synthesis. *In* "Unpublished transcript of the 1998 Walter B. Langbein Memorial Lecture," American Geophysical Union Spring Meeting, Boston, MA, 26 May 1998.

Shiklomanov, I. A., Ed. (2001). World water resources at the beginning of the 21st century. UNESCO Publications, Paris.

Thompson, L. G., Mosley-Thompson, E., and Henderson, K. A. (2000). Ice core paleoclimate records in tropical South America since the Last Glacial Maximum. *Journal of Quaternary Science* **15**, 377-394.

Zorita, E., and von Storch, H. (1999). The analog method - a simple statistical downscaling technique: Comparison with more complicated methods. *Journal of Climate* **12**, 2474-2489.

Forests in Sustainable Mountain Development

Martin F. Price
Centre for Mountain Studies, Perth College, UHI Millennium Institute, Crieff Road, GB-Perth PH1 2NX, UK
phone +44-1738-877217, fax +44-1738-877018, e-mail martin.price@perth.uhi.ac.uk

Keywords: Forests, Mountains, Policy, Research, Sustainable development

1. Introduction

In 2000, the Task Force on Forests in Sustainable Mountain Development of the International Union of Forestry Research Organizations (IUFRO) published a state-of-knowledge report (Price and Butt 2000). The terms of reference for the Task Force recognized the need for such a report, deriving from four linked trends:

- a widespread shift in the science and practice of forestry, from an emphasis on the production of wood towards integrated management recognizing that forests serve multiple functions and produce a wide range of goods;
- changing expectations regarding the roles of mountain forests among populations around the world, in an increasingly urbanized global society;
- rapid rates of change, both perceived and measured, in the cover and uses of forests and adjacent ecosystems in mountain regions around the world;
- the growing recognition of the global importance of mountain ecosystems and their inhabitants.

The report summarises existing information and needs for new research on issues of global significance, and recognises that sound scientific information is essential for developing policies. It was prepared during 1999 by 124 social and natural scientists and practitioners from around the world, who provided 90 contributions, ranging from global overviews to local case studies on the very wide range of topics relating to the intersections of forests and sustainable development in mountain areas. This paper is

U. M. Huber et al. (eds.), Global Change and Mountain Regions, 521-529.

largely based on these contributions.

2. The global extent of mountain forests

The first issue to address when preparing the report was to define the extent of the world's mountain forests. This was not a simple task, as there are many definitions of both mountains and forests. Consequently, the first map of the world's mountains, objectively defined according to rules relating to altitude, slope, and relative relief, was produced (Kapos et al. 2000). The basic data comprised a global database recording the average altitude of every square kilometre of the Earth's surface. The resulting map shows that mountains cover 24% (35,813,437 km^2) of the Earth's land surface. This map was then overlain on an existing global database of forests. Putting these two databases together shows that 28% (5,179,248 km^2) of the world's forests are mountain forests. Despite the precision of this number, further accuracy is needed and much work remains in order to refine definitions, especially at regional and national scales. Nevertheless, these statistics show that mountain forests cover an important proportion of our planet.

3. Global change and mountain forests

Mountain forests are subject to many forces of change, both natural and man-made, and often acting together. Mountain ecosystems are particularly dynamic in both space and time. Major disturbances, such as fire, wind, and avalanches, may occur only infrequently, but have significant influences on mountain forests (Peterson et al. 2000). Increasingly, the occurrence of such "natural" disturbances, or "hazards", is influenced by human activities, both locally and at broader scales. This is true both in developing countries, where forest cover is typically declining – the greatest rate of forest loss in any biome is in tropical upland forests (1.1% per year: FAO 1993) – and throughout most of the temperate zone, where the area and/or density of mountain forests is generally stable or increasing through both spontaneous reforestation and planting (Piussi 2000). Research conducted across the world shows that these processes of change are remarkably complex. In developing countries, the concept of "more people, fewer trees" is increasingly questioned (Scherr and Templeton 2000). Population density is clearly one factor, but others include property rights, forest regulations, land management institutions, access infrastructure, and timber harvesting policies. These processes are long-term, so that comparable studies focusing on land and forest productivity, environmental quality, market evolution, the impacts of technological and institutional innovation, and participatory management – and their interactions over time – are needed to understand past and current change and develop scenarios for possible future paths of change.

To manage these processes of change, and especially to prioritise the targeting of scarce resources, considerable new research is also needed on biophysical themes, such as the dynamics of ecological processes, the selection and establishment of forest species, and the impacts of changing forest cover and composition on water flows and erosion. Such research by natural scientists needs to be linked to, and integrated with, research by social scientists on the interactions of the key economic, political and demographic factors. Such interdisciplinary (i.e. natural with social science) research is becoming ever more important as mountain forests are increasingly affected by large-scale forces, both economic and atmospheric. Two major atmospheric forces, which may interact with and affect mountain forests in unpredictable ways, are regional air pollution and climate change (Beniston 2000; Innes 2000). As climates change, decision-makers at all levels – from mountain villagers to government officials and representatives of global forestry companies – will have to decide which species to plant and nurture where and for what purposes. In some cases, they will also have to agree which areas to leave to change according to whatever forces become most influential. Appropriate levels of intervention and appropriate location-specific criteria, indicators, and guidelines for monitoring will have to be defined.

This initial discussion shows clearly that change in mountain forests is not just an issue for those living in the mountains and concerned with forestry. At a global scale, perhaps the greatest value of mountain forests is that they occupy much of the upper watersheds of the rivers, which supply at least half of humanity with fresh water. The management of these forests affects the timing, quantity, and quality of water flowing downstream. In humid temperate areas, up to half of the water flowing to the lowlands comes from the mountains; in semi-arid and arid areas, up to 90-95% (Liniger and Weingartner 2000). These figures are persuasive, yet there are still no good global overviews of the quantitative contributions of mountain regions to regional water budgets, or of the roles of forests compared to other types of land use or vegetation in influencing these contributions. Hydrological research over the last decades has typically focussed on small watersheds, or parts of larger watersheds, and each study has tended to use different methodologies, making comparison difficult. A critical need with regard to the linkages between mountain forests and water flows is to bring together many types of information, preferably in spatial databases or geographic information systems (GIS). These can be invaluable tools for allowing diverse interests – including scientists, foresters, engineers, local communities, government employees, and political representatives – to jointly consider various water resource scenarios for the future (Schreier 2000). This requires not only further research to supply the necessary data, but also the commitment of governments to provide full access to relevant information, which has not always been the case.

4. Sustainable development and mountain forests

4.1 Values and knowledge

Reflecting the diverse range of stakeholders concerned with mountain forests, it is increasingly recognised around the world that "multifunctionality" must be a keyword for planning and managing these forests (Buttoud 2000). This concept, deriving particularly from the forestry profession, recognises that mountain forests have diverse values to many different groups of people. For instance, recent research in Switzerland has shown that, while foresters regard the protective functions of mountain forests as most important, members of the public perceive that they are most valuable for recreation and nature (Zimmermann and Schmithüsen 2000). Research on public perceptions and attitudes could generate an understanding of different value systems and complement and challenge the knowledge and judgement of experts and politicians. This is essential for informed and equitable decision-making. In addition, it must be recognised that western "objective" science is not the only source of knowledge. Traditional ecological knowledge can be complementary; training and education, both from traditional knowledge-holders to scientifically trained individuals and *vice versa*, are necessary (Thomson et al. 2000). Both types of knowledge should be important inputs to environmental education in mountain areas (for indigenous people and visitors) and further afield to raise awareness of the diverse values of mountain forests to different groups. In this respect, the tourism industry should be a particularly important actor, given that mountains are primary tourism and recreation destinations, second only to coastal areas in the global tourism market (Godde et al. 2000).

The indigenous knowledge of mountain people is particularly important with regard to their diverse uses of both plants and animals (Ramakrishan, this volume). Mountain forests are global "hotspots" of biological diversity for various reasons, including evolution and migration of species over geological time, isolation, contrasting conditions on different slopes and at different altitudes, and diverse microhabitats. However, scientific knowledge of the biological diversity of mountain forests varies greatly, both from one area to another, and also with regard to different groups of flora and fauna. Greater emphasis needs to be given to inventory and taxonomic description, as well as understanding and prioritising human uses (Grabherr 2000). As much of this knowledge is the intellectual property of local people, it is essential that scientists work in partnership with them – and that they derive appropriate benefits from such research.

4.2 Forest products and systems

One aspect of the remarkable diversity of mountain forests is the many plants and animals, which are, or can be, used to produce non-timber forest products (NTFPs). These are of great value to mountain people for subsistence and sale, especially in hard times (e.g. crop failure, unemployment) (Arnold and Ruiz Pérez 2000; Shrestha and Pokharel 2000). While this is particularly true in developing countries, NTFPs – such as mushrooms and herbs - can also contribute significantly to mountain economies in industrialised countries. However, as the value of these traditional products grows in wider economies, pressures towards excessive production or harvesting tend to increase, and a growing proportion of profits is often taken by the wealthiest mountain villagers, middlemen, or outsiders. Thus, apart from an emphasis on maintaining indigenous knowledge, collaboration between people and groups who traditionally have not collaborated is increasingly needed. Development agencies, NGOs, and governments have key roles to play in this. They should also be involved in research and action to ensure that use and management rights are protected or established, and processing and marketing facilities developed. The aim should be that a greater proportion of the value added remains with mountain people while, at the same time, environmental protection is fostered and biological diversity is maintained.

Many similar issues relate to agroforestry in mountain areas. Agroforestry systems have similar biophysical structures to natural forests, but with a greater density of species valued for local use or sale. Again, a key issue is how to establish and maintain agroforestry systems that provide the best distribution of economic, societal, and ecological benefits to both local people and others depending on mountain landscapes. Major areas of research, which need to be continued, are soil erosion control, soil fertility, improved fallows, biomass transfer, and the selection and breeding of species (especially indigenous) which provide key benefits (Atta-Krah and Tang 2000). While many of these species provide valuable NTFPs, others are important sources of wood for many purposes. The provision of fuel wood is particularly important, given that this is the major fuel source for most mountain people (Schweizer and Preiser 1997). However, although many case studies exist on the production and demand of wood fuels in mountain areas and a regional overview has been put together for the Hindu Kush-Himalaya (Rijal 2000), reliable national and regional data compilations are generally lacking. In this context, interdisciplinary research is needed on production, demand, land tenure, appropriate species, and sustainable energy consumption.

Until very recently, mountain forestry in much of the world focussed primarily on wood harvesting. However, the "traditional" concept of sustained-yield forestry, developed in the lowland forests of Europe two centuries ago and applied widely in mountain forests around the world, has shown to be inappropriate for these multifunctional forests. This concept presumed that all forest functions could be achieved through the annual removal of a sustained yield of timber. However, it was based on three assumptions that often do not hold in mountain forests – if they ever did: 1) all areas of the forest can be managed; 2) there is a sustained demand for wood; and 3) there is a sufficient workforce (Price 1990). The concept of multifunctionality, mentioned above, is a direct response to the recognition that sustained-yield forestry is generally inappropriate for mountain forests. Nevertheless, timber harvesting remains an important activity in these forests (Heinimann 2000). There is a continued need to compare means of harvesting and extracting timber, using not only economic, but also environmental and societal criteria, in order to minimise negative impacts and maximise benefits. The information resulting from such comparisons is of value not only for planning and management, but also for training the forestry workforce.

4.3 Economics and benefits

Many decisions regarding the management of mountain forests are based on economic data. Economic datasets that specifically address mountain forests are available for Austria, Germany and Switzerland (Sekot 2000) but are rare for other countries. The available data show that forestry in the Alps has higher costs of access and logging, and lower levels of profitability and value added per hectare, than lowland forestry. For appropriate decision-making regarding the future of mountain forests, economic data need to be disaggregated at least to the level of mountain/non-mountain. Also, mountain forestry needs to be better established in reporting systems.

In addition, common definitions and protocols are needed for data collection, linked to research on appropriate multipliers (Gregersen 2000; Sekot 2000). The provision of well-justified, comparable economic data is critical for defining levels of equitable compensation of mountain people, forest owners and enterprises by the downstream beneficiaries of mountain forest management (Zingari 2000). Further research is needed to evaluate the "downstream" benefits from mountain forest management. While it may be possible to value some of these benefits in economic terms, many - such as protection against natural hazards, protection of watersheds, and maintenance of landscape diversity and aesthetics - can only be expressed in non-market terms. Yet, although such research is at an early stage, and the full suite of quantitative data is not yet available, mechanisms for compensation exist (Koch-Weser, this volume). They recognise the essential services provided by mountain forests to downstream populations and, with regard to the conservation of biological diversity and the sequestration of carbon, the global community.

4.4 Institutions

These issues lead into the question of the appropriate institutions for planning and implementing the management of mountain forests. Around the world, the number of stakeholders concerned with these forests is growing, and various studies suggest that power relationships are more important than the actual participation of stakeholders in decision-making (Dubois 2000). It is worth noting that institutions for the cooperative management of mountain forests have very long histories in many cultures around the world (Kissling-Näf 2000), and various new models are being developed (Joshi 2000). As mountain forests are increasingly influenced by regional and global forces, these experiences should be of great value for shaping and reshaping institutions for the future. Critical analysis and inter- and intra-regional comparisons are highly desirable. The same holds for research and action on legislation and policies relating to mountain forests, given recent shifts towards proactive, multifunctional, and cross-sectoral measures based on monitoring and incentives (Schmithüsen and Zimmermann 2000). However, successes and failures of policy implementation need to be evaluated in order to develop and implement more effective policies, which meet the goals of the diverse stakeholders in mountain forests.

5. Looking ahead: Collaboration, integration, and communication

The state of the art in research in mountain forests is that we have many snapshots, but not enough consistent data sets, especially with regard to economic data for both market and non-market goods and services. Analyses or syntheses, which are sufficiently broad in both space and time and bring together insights from a sufficiently diverse range of perspectives, are lacking. Such broad analyses and syntheses are of ever greater importance in a period of economic, political, and ecological change; they are necessary to provide the foundations of integrated policies which recognise the key roles of mountain forests for many sectors of society.

While analyses and syntheses are important, so is fundamental research; many needs have been highlighted throughout this paper. In mountain forests around the world, scientists from a wide range of natural and social scientific, as well as technical, disciplines are undertaking research. Yet this is too infrequently multi- or interdisciplinary, despite increasing recognition of the added value of collaborative work for informing policy-making and implementation. The planning and implementation of future research should be more strategic, and should also recognize that mountain people, although often poorly trained in the "scientific method", can complement western science with local expertise (Branney and Hobley 2000). To bring together and integrate these different worldviews, new technologies such as GIS can be of great value. The outcomes of research also need to be communicated more effectively, using all appropriate media (Pandey 2000) to all those concerned, in mountain villages, forestry companies, government agencies, academic institutions, non-governmental organisations, parliaments, regional policy-making bodies, and global organisations.

6. References

Arnold, J. E. M., and Pérez, M. R. (2000). Income from non-timber forest products. *In* "Forests in sustainable mountain development: A state of knowledge report for 2000." (M. F. Price and N. Butt, Eds.), pp. 300-306. CAB International, Wallingford.

Atta-Krah, K., and Tang Ya (2000). Agroforestry in highlands and mountain areas. *In* "Forests in sustainable mountain development: A state of knowledge report for 2000." (M. F. Price, and N. Butt, Eds.), pp. 270-284. CAB International, Wallingford.

Beniston, M. (2000). Climate oscillations and extremes. *In* "Forests in sustainable mountain development: A state of knowledge report for 2000" (M. F. Price, and N. Butt, Eds.), pp. 70-76. CAB International, Wallingford.

Branney, P., and Hobley, M. (2000). Participatory research – is this research? *In* "Forests in sustainable mountain development: A state of knowledge report for 2000." (M. F. Price, and N. Butt, Eds.), pp. 479-486. CAB International, Wallingford.

Buttoud, G. (2000). Approaches to multifunctionality in mountain forests. *In* "Forests in sustainable mountain development: A state of knowledge report for 2000." (M. F. Price, and N. Butt, Eds.), pp. 187-194. CAB International, Wallingford.

Dubois, O. (2000). Institutions for the collaborative management of mountain forests. *In* "Forests in sustainable mountain development: A state of knowledge report for 2000." (M. F. Price, and N. Butt, Eds.), pp. 443-449. CAB International, Wallingford.

FAO (1993). "Forest resources assessment 1990 – Tropical countries." FAO Forestry Paper 112, Food and Agriculture Organization of the United Nations, Rome.

Godde, P., Price, M. F., and Zimmermann, F. M., Eds. (2000). "Tourism and development in mountain regions." CAB International, Wallingford.

Grabherr, G. (2000). Biodiversity of mountain forests. *In* "Forests in sustainable mountain development: A state of knowledge report for 2000." (M. F. Price, and N. Butt, Eds.), pp. 28-38. CAB International, Wallingford.

Gregersen, H. M. (2000). Income from mountain timber and wood products. *In* "Forests in sustainable mountain development: A state of knowledge report for 2000." (M. F. Price, and N. Butt, Eds.), pp. 234-239. CAB International, Wallingford.

Heinimann, H. R. (2000). Forest operations under mountainous conditions. *In* "Forests in sustainable mountain development: A state of knowledge report for 2000." (M. F. Price, and N. Butt, Eds.), pp. 224-230. CAB International, Wallingford.

Innes, J. (2000). Responses of mountain forests to environmental change: Forest decline, air pollution and other anthropogenic and natural factors. *In* "Forests in sustainable mountain development: A state of knowledge report for 2000." (M. F. Price, and N. Butt, Eds.), pp. 76-81. CAB International, Wallingford.

Joshi, A.L. (2000). Leasehold forestry, joint forest management and community forestry as appropriate programmes for mountain development. *In* "Forests in sustainable mountain development: A state of knowledge report for 2000." (M. F. Price, and N. Butt, Eds.), pp. 452-459. CAB International, Wallingford.

Kapos, V., Rhind, J., Edwards, M., Price, M. F., and Ravilious, C. (2000). Developing a map of the world's mountain forests. *In* "Forests in sustainable mountain development: A state of knowledge report for 2000" (M. F. Price, and N. Butt, Eds.), pp. 4-9. CAB International, Wallingford.

Kissling-Näf, I. (2000). Forests as common property in the Swiss Alps. *In* "Forests in sustainable mountain development: A state of knowledge report for 2000." (M. F. Price and N. Butt, Eds.), pp. 459-465. CAB International, Wallingford.

Liniger, H. and Weingartner, R. (2000). Mountain forests and their role in providing freshwater resources. *In* "Forests in sustainable mountain development: A state of knowledge report for 2000." (M. F. Price, and N. Butt, Eds.), pp. 370-380. CAB International, Wallingford.

Pandey, S. (2000). Information and communication systems for sustainable mountain forestry: A brief guide to available mechanisms and resources. *In* "Forests in sustainable mountain development: A state of knowledge report for 2000." (M. F. Price, and N. Butt, Eds.), pp. 508-520. CAB International, Wallingford.

Peterson, D. L., Prichard, S. L., and McKenzie, D. (2000). Disturbance in mountain forests. *In* "Forests in sustainable mountain development: A state of knowledge report for 2000." (M. F. Price, and N. Butt, Eds.), pp. 51-58. CAB International, Wallingford.

Piussi, P. (2000). Expansion of European mountain forests. *In* "Forests in sustainable mountain development: A state of knowledge report for 2000." (M. F. Price and N. Butt, Eds.), pp. 19-25. CAB International, Wallingford.

Price, M. F. (1990). "Mountain forests as common-property resources: Management policies and their outcomes in the Colorado Rockies and the Swiss Alps." Forstwissenschaftliche Beiträge 9, Professur Forstpolitik und Forstökonomie, ETH Zürich.

Price, M. F., and Butt, N., Eds. (2000). "Forests in sustainable mountain development: A state of knowledge report for 2000." CABI International, Wallingford.

Rijal, K. (2000). Energy from the Hindu Kush – Himalayan mountain forests. *In* "Forests in sustainable mountain development: A state of knowledge report for 2000." (M. F. Price, and N. Butt, Eds.), pp. 249-255. CAB International, Wallingford.

Scherr, S. J. and Templeton, S. R. (2000). Impacts of population increase and economic change on the mountain forests of developing countries. *In* "Forests in sustainable mountain development: A state of knowledge report for 2000." (M. F. Price and N. Butt, Eds.), pp. 90-97. CAB International, Wallingford.

Schmithüsen, F. and Zimmerman, W. (2000). The role of forest and environmental legislation in sustainable land-use practices. *In* "Forests in sustainable mountain development: A state of knowledge report for 2000" (M. F. Price, and N. Butt, Eds.), pp. 401-410. CAB International, Wallingford.

Schreier, H. (2000). Research, planning, and implementation of watershed management. *In* "Forests in sustainable mountain development: A state of knowledge report for 2000." (M. F. Price and N. Butt, Eds.), pp. 380-389. CAB International, Wallingford.

Schweizer, P., and Preiser, K. (1997). Energy resources for remote highland areas. *In* "Mountains of the World: A global priority." (B. Messerli, and J. D. Ives, Eds.), pp. 157-170. Parthenon, New York.

Sekot, W. (2000). Income from timber: The economics of mountain forestry in Central Europe. *In* "Forests in sustainable mountain development: A state of knowledge report for 2000." (M. F. Price, and N. Butt, Eds.), pp. 239-247. CAB International, Wallingford.

Shrestha, T. B., and Pokharel, S. (2000). The potential of medicinal and aromatic plants for sustainable mountain development in Nepal. *In* "Forests in sustainable mountain development: A state of knowledge report for 2000." (M. F. Price, and N. Butt, Eds.), pp. 312-318. CAB International, Wallingford.

Thomson, A. J., Jimmie, M. N., Turner, N. J., and Mitchell, D. (2000). Traditional knowledge, western science and environmental ethics in forest management. *In* "Forests in sustainable mountain development: A state of knowledge report for 2000." (M. F. Price, and N. Butt, Eds.), pp. 181-186.

CAB International, Wallingford.

Zimmerman, W., and Schmithüsen, F. (2000). The importance of empirical research on public perceptions and attitudes towards forests for participatory policy development. *In* "Forests in sustainable mountain development: A state of knowledge report for 2000." (M. F. Price, and N. Butt, Eds.), pp. 176-181. CAB International, Wallingford.

Zingari, P. C. (2000). Sustainably balancing downstream and upstream benefits in European mountain forest communities. *In* "Forests in sustainable mountain development: A state of knowledge report for 2000." (M. F. Price, and N. Butt, Eds.), pp. 155-160. CAB International, Wallingford.

Institutional Dynamics and Interplay: Critical Processes for Forest Governance and Sustainability in the Mountain Regions of Northern Thailand

Louis Lebel
Unit for Social and Environmental Research, Faculty of Social Sciences, Chiang Mai University, Chiang Mai 50200, Thailand
phone/fax 66-53-854-347, email llebel@loxinfo.co.th or louis@sea-user.org

Keywords: Forest governance, Institutions, Livelihoods, Political economy, Thailand

1. Introduction

The main argument of this paper is that changes in the formal and informal institutions that govern natural resources in mountain regions of northern Thailand have been critical for environmental changes, livelihoods and sustainability. Over the past decade, there have been new insights from interdisciplinary research on how societies interact with environmental changes in mountain regions. These have underlined the importance of institutions as both causes and responses to environmental change, and how institutions themselves arise from the way environmental and sustainability problems are constructed. In this chapter, these more general findings will be illustrated primarily through examples from recent and ongoing research in the mountain region of Northern Thailand. Taken together, these various studies challenge long-held beliefs about what constitutes problems in environmental change and sustainability, underline the need for a better understanding of cross-scale interactions, and point the way towards a more open and accountable science in support of sustainability.

U. M. Huber et al. (eds.), Global Change and Mountain Regions, 531-540.

2. Institutional causes of and responses to environmental change

The policy of successive (Siam) Thai governments since around 1900 has been to assume greater control over decision-making and management of forestlands, and to find ways to limit access to land, timber, and other forest products (Table 1, Pragtong and Thomas 1990). The limited capacity and resources of the bureaucracy, however, have allowed many peripheral areas to develop relatively undisturbed until well into the second half of the past century. The nationalization of forest resources (Vandergeest 1996) has been accompanied by a shift towards the use of western scientific systems of knowledge about how to log teak forests, establish pine plantations, and more recently, conserve wildlife and habitat in protected areas. As a result, traditional ecological knowledge and property rights systems have been re-bundled or dismissed as irrelevant in the state's drive towards modernization. Formal property rights for land and timber have over the last century been transformed several times by the state as a result of changing policies towards national security, logging, narcotics control, macro-economic and rural development policies, and conservation (Ganjanapan 2000; Contreras 2003). Most of these changes in laws and regulations were applied to the whole nation by the central bureaucracy, despite very different forest conditions in different regions.

Table 1: Summary of some of the major historical institutional changes directly related to forest management in Thailand.

Year	Institutional Changes
1896	Establishment of the Royal Forest Department
1900-1910	Several regulations regarding management of teak concessions, for example, minimum girths, cutting cycles and block sizes
1913	Forest Conservation Law – first state attempts to control non-teak forest products
1938	Protection and Reservation Forests Act – started process of mapping out different forest uses – implementation initially very slow.
1941	Forest Act – replaces Forest Conservation Law of 1913. Further revisions in 1948 and 1951.
1947	Establishment of Forest Industry Organization, a state logging enterprise
1960-61	Wildlife Conservation and Protection Act and National Park Act – began process of demarcation of protected areas.
1964	National Forest Reserves Act – facilitated commercial exploitation by reducing need for local community consultation with decision-making left largely with Royal Forest Department.
1985	National Forest Policy – adopted FAO goals of 40% forest cover for nation in both conservation and production forests.
1989	Logging ban and revoking of timber concessions
1997	New Thai Constitution and 8th National Economic and Social Development Plan – promoted idea of decentralization.
2003	Community Forestry Bill – that would allow use and management decisions to be made by community rather than state organizations: multiple versions debated for over a decade – highly restricted version may come into force.

The north has maintained relatively more areas of forest than other regions of Thailand, in part, as a result of its complex topography. As elsewhere in Thailand, virtually all lowland plains and larger inter-montane valleys have been cleared of native vegetation for agriculture, plantations, and urban development. Some of the irrigation systems for rice around Chiang Mai town, for example, are centuries old. Forest cover persists in smaller upland catchment areas, often as part of traditional fallow-based land-use systems, and in more remote and steep terrain, much of it now inside the boundaries of national parks. Access to forestland is an important part of the traditional livelihoods of the numerous ethnic groups, which are numerically dominant in the upland areas, both as part of their fallow-based rotational systems, as well as for timber and non-timber forest products. The mountainous landscape of northern Thailand is, therefore, important both for conservation of biodiversity and the livelihoods of farmers (Rerkasem et al. 2002; Santasombat 2003).

Over the past 30 years, wars in neighbouring countries have had a major impact on immigration into the highland areas, increasing pressure on land and water resources, as well as reinforcing negative attitudes of the Thai state towards the uplands (Forsyth 1999; Vandergeest 2003). Public debate over the magnitude and consequences of forest loss in northern Thailand has been intense, in part because of perceived threats to the economically and symbolically (*rice bowl* of Thailand) important irrigated agriculture in the lowlands around Chiang Mai, and further downstream, the central plains around Bangkok (Laungaramsri 2002). The result has been battles and political gridlocks over legally recognizing rights to citizenship, community forests, agricultural land and villages in upland watersheds, many of which are now within the boundaries of national parks. Negotiating resolutions in these conflicts has been made more complex by differences in language, culture and land management systems among the ethnic minorities and Thai as well as significant in-migration in some border areas arising from armed conflicts and poor economic conditions in neighbouring countries.

3. Rules on paper, rules in use

Institutions are "systems of rules, decision-making procedures, and programs that give rise to social practices, assign roles to the participants in these practices, and guide interactions among the occupants of the relevant roles" (Young et al. 1999); they include both rules on paper and rules in use. Formal institutions, such as government laws and regulations that are enforced by police, soldiers or inspectors, are the most obvious type of institution. However, appearances can be deceptive. In Thailand, many of the laws concerning forest protection have proven impossible to implement or easy to circumvent with the right connections. The logging ban has not applied, for example, to some senior forestry or military officials, resulting in periodic scandals in the press. Likewise, villagers in remote areas have often been able to make compromises and deals with local government officials over clearing land for agriculture and forest access, where state laws would make such activity illegal. Flexibility in local institutional arrangements has both positive and negative implications for social justice and sustainability.

Moreover, in most mountain areas there were probably earlier institutions, both formal and informal, governing access and use of forest and forest-derived lands (Tan-kim-yong 1997; Poulsen et al. 2001). Thus, apart from formal rules, there is a whole range of informal institutions that are critical for understanding the causes of, as well as vulnerabilities to, environmental changes. The capacity for local institutions to adapt to new technologies, larger and more mobile populations, as well as direct threats to human security, appears to vary greatly from place-to-place, with examples of both success and failure to manage local resources sustainably under more traditional and modern contexts.

4. Institutional interplay

The system of forest governance in Northern Thailand has changed substantially in the past hundred years. New institutions have been introduced at local, state and international levels, while many local institutions have been abolished or significantly transformed. As the number and complexity of overlapping institutions that deal with different aspects of forest governance increases, the success or effectiveness of a particular institution increasingly depends not only on its own characteristics but also on how it interacts with other institutions, or the institutional interplay (Young 2002). Interactions can be characterized as vertical (across levels of governance) or horizontal (on the same level of governmental organization).

Over the past century, different branches of the Thai government developed their own policies in the key area of land tenure and settlement. This resulted in strong horizontal interplay between different systems of rules and the implementing organizations (Lebel, in preparation a). Among the many bureaucratic players the Royal Forest Department, the Department of Land Development and the Ministry of Interior have been key. The history of conflicts over rights to land for settlement and agriculture, to timber and non-timber products, and to water and watershed services has been intertwined.

Vertical interplay is also a relatively modern phenomenon, at least for the inhabitants of the more mountainous region. Mountain people were, at least partly and probably intentionally, insulated from the civilization building projects of various competing kingdoms in the lowlands (Scott 1998). Interplay of state and local institutions, has been a major process influencing the management of forestlands, as noted before, primarily through the submission and replacement of local institutions. Interplay has been highly asymmetric starting from centralized state decisions, operational guidelines and goals.

Several new institutions at the international level have emerged over the past two decades. The "International Forest Regime", however, remains fragmented and largely ineffective. The 1992 Earth Summit produced two soft law instruments, "Agenda 21", chapter 11 of which focuses on deforestation, and the non-legally binding "Forest Principles" statement. It also resulted in two conventions, the "Convention on Biological Diversity" and the "Framework Convention on Climate Change" that refer to forests. A decade of intergovernmental dialogues since then, however, has been unable to establish clear rules or standards, as well as coordination mechanisms between institutions, or to provide a regular forum for dialogue or conflict resolution where issues of forest management have been concerned. As elsewhere in Asia, domestic factors, especially corporate interests in timber harvesting and then plantations, have been a crucial factor in the state responses to the "International Forest Regime" (England 1996; Dauvergne 2001).

5. Sustainable livelihoods

Ethnic minority communities in the uplands of northern Thailand include many of the poorest in the nation but, overall, economic and health indicators suggest wellbeing is improving. The consequences of institutional and environmental changes for livelihoods are not easy to summarize as they are confounded by the many other social and economic changes affecting the uplands. The enforcement of restrictions on access and use of forestland and products, as well as tenure insecurity, have undoubtedly been an important challenge to livelihoods. On the other hand, improvements in road and communication infrastructure have made access to commodity, labour, and credit markets much easier.

A key livelihood strategy has been the diversification of income sources, which in turn can both remove or increase dependencies on forest ecosystem goods and services (Lebel et al., in preparation b). In some locations, there is a strong competitive advantage for earning income from tourism, for instance, through providing elephant rides and bamboo-rafting experiences. Here, there can be strong incentives for maintaining a forest-like setting. Other places, because of their proximity to good water sources, roads and market channels, may expand and intensify the cultivation of higher value temperate crops (e.g. cut-flowers, lychees, and stone-fruit), which grow better in the cooler upland climate.

The manipulation of watershed functions is a deliberate traditional practice, for example, through choice of areas for clearing and forest preservation in Karen villages (Tan-Kim-Yong 1997). Whether the customary institutions that govern these practices will persist, or be replaced by new ones over the layout of sprinkler irrigation systems and the diversion of upstream water, remains to be seen. What is clear from research is that upland farmers often show a remarkable capacity to adapt their land-use systems, natural resource institutions and culture to a wide variety of challenges and opportunities (Rerkasem and Rerkasem 1995; Thong-Ngam et al. 1995; Battersbury and Forsyth 1999).

To what extent these capacities will be effective in reducing vulnerability to particular aspects of future global environmental changes in the mountain regions of northern Thailand has been little studied. We note, with concern, that total consumption of water for agriculture, forestry, human settlements and industry has grown rapidly, and in many places, now often approaches the total potential supply with current technologies. Changes in rainfall patterns under climate change could greatly exacerbate these problems creating intense competition and conflict over water resources. The point remains that the detailed structure of water- and land-rights, and the process by which they were arrived at, will probably continue to have important consequences for the vulnerability of different places, sectors and people to environmental changes arising and driven by processes at various scales.

6. Governance and knowledge

The consequences of the current trends in livelihood activities for forest ecosystem goods and services are uncertain. Changes in forest conditions are caused not only by a variety of actors but also by interactions between institutions that change incentives for these actors at multiple scales. The way a number of nascent political and agricultural market institutions unfold will matter greatly.

For instance, in the more mountainous districts of Chiang Mai province, the larger ethnic minorities with citizenship rights are now taking their places in local government (Tambon Administrative Councils). Although these government bodies do not yet have jurisdiction over critical forestlands in their area (which remains with the Royal Forest Department), increasing political power could change some of these arrangements at least locally. While issues of ethnic identity (cf. Vandergeest 2003) may grow less important with economic and cultural exchange, the tension between upland and lowland water, land and forest user groups will probably intensify.

Future governance systems should aim to retain a certain amount of the flexibility that is inherent where state capacity to implement is weak. Institutional arrangements need to be able to respond to improving as well as deteriorating forest conditions. They also need to be sensitive to the wider social, economic and political contexts of transformation (Fig. 1). The key policy issue is therefore how to foster resource management institutions that promote resilience of both the ecological and social systems to a suite of stresses, challenges and potential surprises, including but not restricted to those from global environmental change. The way water- and land-rights are constructed and are allowed to evolve is critical to whether capacities to cope and adapt will be fostered or suppressed in the groups most at risk. Multi-stakeholder processes appear crucial.

In Thailand, governance is no longer seen as the sole responsibility of the state. Local communities have contributed to forest governance in the past and should continue to do so in the future. The poor record of the Thai state in managing forest resources strongly argues for a high level of local participation in decision-making, monitoring and the formulation of rules. An open public policy process that contains mechanisms to achieve objectives at different scales is still lacking. For these reasons, one of the most important areas of future research will be on institutional interplay and the potential for new and rebuilt cross-scale institutions. An effective governance system should be sensitive to the need and plight of the most vulnerable parts of the population, often found among those with the least capacity to influence the political process.

At the same time, high quality research on the impacts of land-use and climate change on the goods and services obtained from mountain landscapes is also needed to help clarify polarized debates. State agencies and non-governmental organizations in Thailand have commonly justified their watershed management and land tenure policies based on extrapolations of scientific findings from small-scale and single land-use studies to the complex landscapes of northern Thailand, the entire Chao Phraya basin and even the greater Mekong basin region. Current research on landscape hydrology, erosion and temporal and spatial rainfall variability suggests that such extrapolations across scales are misleading (Forsyth 1996; Schmidt-Vogt 1998; Thomas et al. 2003).

Throughout the region there is a need to better harness research-based knowledge in support of transitions to sustainability. This is not just an issue of appropriate technology choice and refinement, but also one of coming up with institutional arrangements that encourage sustainable practices and social equality.

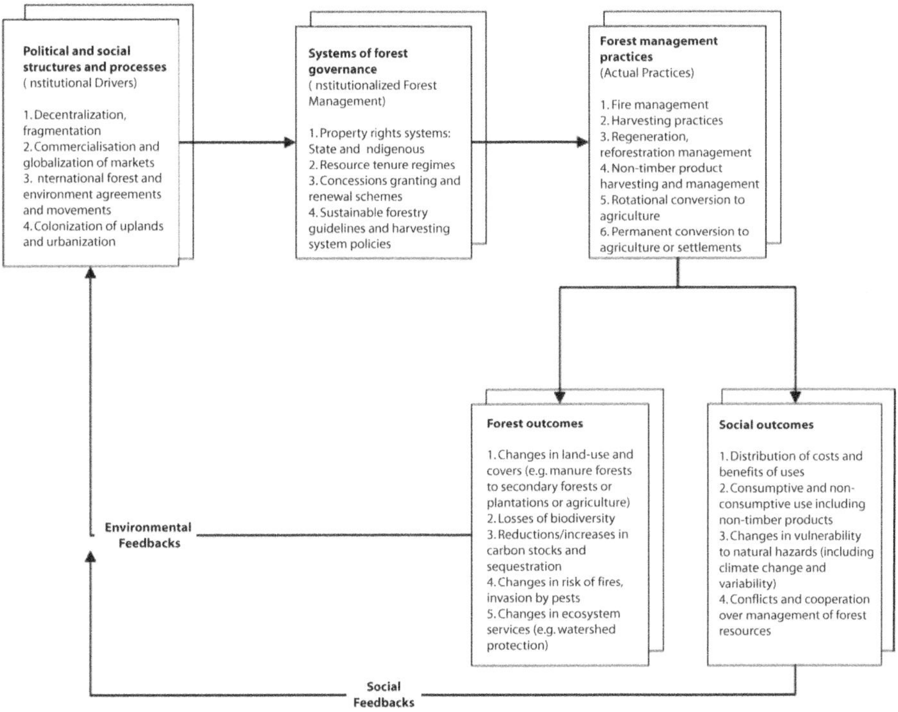

Figure 1: Systems of forest governance and actual practices modify the influences of the political and social structures and processes, which ultimately drive changes in forest land-use and conditions. Changes in forest conditions and the social outcomes of forest management and land-uses influence the institutional drivers of future change in a system that feeds back on itself (adapted from Contreras et al. 2001).

7. Cross-scale sustainability puzzles

The consideration of place and scale is central to the analysis of most sustainability issues. Both ecological and social processes vary with scale, and cross-scale interactions, such as institutional interplay, are among the main sources of complexity. Scale, however, is not politically neutral. The selection of scale may intentionally or unintentionally privilege certain actors or groups. The adoption of a particular scale in science, institution building, or in a policy, limits the types of problems that can be addressed, the modes of explanations that are allowed, and which generalizations are likely to be used in analysis.

Appeals to scale can be an argument that empowers state institutions. Most states view indigenous knowledge and institutions as local in scope, relevance and power, whereas the rules and knowledge of the state are seen as bigger in scale and hence more important. On the other hand, the source of many problems associated with the management of ecosystem services may result exactly from the centralization and uniformity in bureaucratic operations that hinder local participation, adaptation and learning. Ecosystem services are strongly dependent on scale. For example, forests provide carbon storage and biodiversity (as public goods) and timber for a house (as an individual or shared private good). For some cultures, they may also provide a wide diversity of foods and subsistence-use products, whereas other cultures may not even be aware of most of these products or their uses. Scale is thus a critical issue for governance of resources, and especially so in mountain regions where upland-lowland linkages involve different users, interests and histories of institutional development. A multi-scale perspective will often be needed in studies of environmental change in mountain regions.

Comparative, historical and experimental research is needed in the upland areas of Asia on how interactions between institutions and ecological changes play out at various scales. New research should add a cross-scale perspective to the already substantial body of knowledge about local institutions. This will greatly help communities and governments to better understand and react to the consequences of globalization through measures such as market integration, sharing and dissemination of knowledge, new technologies, and institutional innovations. A bottom-up as well as a top-down approach is needed because it is far from clear that design principles, derived from the analysis of an institution's performance at one scale, are transferable to other scales (cf. Berkes 2002; Young 2002). Furthermore, the theoretical aspects of these two growing areas of research need to be integrated better. This could be achieved with models that consider how different sets of rules interact and how actors at various scales either develop trust and cooperation or dissolve into factions and become engaged in conflicts that end in stalemate. Focused research on successful, as well as failed, institutional interventions is required, and these interventions need to be assessed against a comprehensive set of scale-sensitive indicators of sustainability. Some key questions to be addressed are:

1. Under what conditions and for what types of institutions does cross-scale interaction, or institutional interplay, result in better use of knowledge in sustainable management practices and just forms of governance?
2. What are the prospects of re-designing or establishing new institutions or cross-boundary organizations to help bridge gaps across scales? What form should these take?
3. What are the consequences of initial scale choices in problem definition for institutional design and performance?
4. How can institutional interplay and other forms of cross-scale interactions in socioeconomic systems empower the poor and (politically) marginalized?

Finally, a new focus is needed on building institutions that can learn. If sustainability in the uplands is considered as a strategy for maintaining adaptive capacity and sources of innovation, rather than developing the perfect crop or forest management system, then learning must be a large part of institutional designs. The fit between institutions and their ecosystems will never be perfect, but the possibility of co-evolution through education and adaptation should at least be attempted. This focus intersects closely with issues of scale and current patterns of governance. We need larger-scale frameworks and programmes that enable, rather than hinder, local adaptation. A key question is:

5. What types of organizations and institutions, and what forms of interplay among them, enhance the likelihood that environmental changes can be detected or foreseen, and then foster appropriate investments in adaptive reactions?

Many of the most pressing and challenging sustainability puzzles arise from interactions across scales of social and ecological organization. Empirical and integrative research on natural resource management in mountain regions could make an important contribution to the wider theories and models about sustainability and cross-scale interactions.

7. References

Batterbury, S., and Forsyth, T. (1999). Fighting back: Human adaptation in marginal environments. *Environment* **41**, 6-11, 25-30.

Berkes, F. 2002. Cross-scale institutional linkages: Perspectives from the bottom up. *In* "The dilemma of the commons." (E. Ostrom, N. Dorsak, P. C. Stern, S. Stonich, and E. U. Weber, Eds.), pp. 293-321. National Academy Press, Washington DC.

Contreras, A. P., Lebel, L., and Pasong, S. (2001). "The political economy of tropical and boreal forests." IDGEC Scoping Report 3, Dartmouth, USA.

Contreras, A. P. (2003). "The Kingdom and the Republic: Forest governance and political transformation in Thailand and the Philippines." Ateneo de Manila University Press, Manila.

Dauvergne, P. (2001). "Loggers and degradation in the Asia-Pacific: Corporations and environmental management." Cambridge University Press, Cambridge.

England, P. (1996). UNCED and the implementation of forest policy in Thailand. *In* "Seeing forests for trees: Environment and environmentalism in Thailand." (P. Hirsch, Ed.), pp. 53-71. Silkworm Books, Chiang Mai.

Forsyth, T. (1996). Science, myth and knowledge: Testing Himalayan environmental degradation in Thailand. *Geoforum* **27**, 375-292.

Forsyth, T. (1999). Questioning the impacts of shifting cultivation. *Watershed* **5**, 23-29.

Ganjanapan, A. (2000). "Local control of land and forest: Cultural dimensions of resource management in Northern Thailand." Regional Centre for Social Science and Sustainable Development, Chiang Mai, Thailand.

Laungaramsri, P. (2002). "Redefining nature: Karen ecological knowledge and the challenge to the modern conservation paradigm." Earthworm Books, Chennai, India.

Lebel, L. (in preparation a). "Institutional dynamics and interplay and the governace of forests in Northern Thailand." Unpublished USER Working Paper. Unit for Social and Environmental Research, Chiang Mai University.

Lebel L, Garden P, Myint C, Khrutmuang S. (in preparation b). Biodiversity and Sustainable Livelihoods in the Uplands ofNorthern Thailand. Unpublished USER Working Paper. Unit for Social and Environmental Research: Chiang Mai University.

Poulsen, E., Skov, F., Lakanvichian, S., Thanisawanyangkura, S., Borgtoft, H., and Hoiris, O. (2001). "Forest in culture - Culture in forest: Perspectives from Northern Thailand." Research Centre on Forest and People in Thailand, Denmark.

Pragtong, K., and Thomas, D. E. (1990). Evolving management systems in Thailand. *In* "Keepers of the forest: Land management alternatives in Southeast Asia." (M. Poffenberger, Ed.), pp. 167-186. Ateneo de Manila University Press, Manila.

Rerkasem, K., and Rerkasem, B. (1995). Montane mainland South-east Asia: Agroecosystems in transition. *Global Environmental Change* **5**, 313-322.

Rerkasem, K., Yimyam, N., Korsamphan, C., Thong-Ngam, C., and Rerkasem, B. (2002). Agrodiversity lessons in mountain land management. *Mountain Research and Development* **22**, 4-9.

Schmidt-Vogt, D. (1998). Defining degradation: The impacts of swidden on forests in northern Thailand. *Mountain Research and Development* **18**, 135-149.

Santasombat, Y. (2003). "Biodiversity, local knowledge and sustainable development." Regional Centre for Social Science and Sustainable Development, Chiang Mai University.

Scott, J. C. (1998). "Seeing like a state: How certain schemes to improve the human condition have failed." Yale University Press, New Haven.

Tan-kim-yong, U. (1997). The Karen culture: A co-existence of two forest conservation systems. *In* "Development or domestication? Indigenous peoples of Southeast Asia." (D. McCaskill, and K. Kampe, Eds.), pp. 219-236. Silkworm Books, Chiang Mai, Thailand.

Thomas, D. E., Preechapanya, P., and Saipothong, P. (2003). "Landscape agroforestry in upper tributary watersheds of Northern Thailand." ICRAF, Chiang Mai University.

Thong-Ngam, C., Shinawatra, B., Healy, S., and Trebuil, G. (1995). Farmer's resource management and decision-making in the context of changes in the Thai highlands. *In* "Proceedings of First Symposium on Montane Mainland Southeast Asia in Transition." 12-16 November, 1995, Chiang Mai University, pp. 462-487.

Vandergeest, P. (1996). Mapping nature: Territorialization of forest rights in Thailand. *Society and Natural Resources* **9**, 159-175.

Vandergeest, P. (2003). Racialization and citizenship in Thai forest politics. *Society and Natural Resources* **16**, 19-37.

Young, O. R., Agrwal, A., King, L. A., Sand, P. H., Underdal, A., and Wasson, M. (1999). "Institutional dimensions of global environmental change. Science Plan." IHDP Report 9, Bonn.

Young, O. R. (2002). Institutional interplay: The environmental consequences of cross-scale interactions. *In* "Drama of the commons." (E. Ostrom, T. Dietz, N. Dolsak, P. C. Stern, S. Stonich, and E. U. Weber Eds.), pp. 263-291. National Academy Press, Washington DC.

State Simplifications of Land-Use and Biodiversity in the Uplands of Yunnan, Eastern Himalayan Region

Jianchu Xu[1*] and Andreas Wilkes[2]
[1]*Kunming Institute of Botany, The Chinese Academy of Sciences, Heilongtan, Kunming, Yunnan, China 650204*
[2]*Department of Anthropology, University of Kent at Canterbury, United Kingdom*
[*]*phone 86-871-5212143, fax 86-871-5150226, e-mail xujianchu@mail.kib.ac.cn*

Keywords: Biodiversity conservation, China, Indigenous knowledge, State simplifications.

1. Introduction

Uplands, mountains, or highlands are both biogeographic and cultural terms that refer to mountainous areas, their biological components and agricultural practices (Sajise and Baguninon 1982). In public perception, mountain regions are often associated with geographical and socio-political peripheries, due to their often remote locations, their higher proportion of ethnic minorities, their landuse and livelihood practices, and their political status. The preamble to Chapter 13 of Agenda 21 states the importance of mountain ecosystems as follows:

"Mountains are an important source of water, energy and biological diversity. Furthermore, they are a source of such key resources as minerals, forest products and agricultural products and of recreation. As a major ecosystem representing the complex and interrelated ecology of our planet, mountain environments are essential to the survival of the global ecosystem. Mountain ecosystems are, however, rapidly changing. They are susceptible to accelerated soil erosion, landslides and rapid loss of habitat and genetic diversity. On the human side, there is widespread poverty among mountain inhabitants and loss of indigenous knowledge" (Menzies 2002).

In this statement, mountains are still primarily seen as exploitable sources of environmental goods and services that benefit the global community rather than

U. M. Huber et al. (eds.), Global Change and Mountain Regions, 541-550.

as sources of livelihoods for local mountain people. Mountain people are often not adequately integrated in decision-making processes. Political discourse on the management of upland ecosystems has often been dominated by either the government or lowland peoples, who may perceive the management practices of mountain peoples as harmful to mountain environments.

This paper examines the recent dynamics of resource governance in the uplands of Yunnan. Particular attention is given to how state policies and interventions have been based on "state simplifications" (Scott 1998) - generalizations that do not necessarily fit the specific circumstances of a certain mountain region. This paper highlights the importance of unique adaptations of mountain people and their livelihood systems to highly diverse mountain environments in ways that ensure their material as well as spiritual survival. Therefore, we advocate that mountain areas should be seen as a potential source of solutions, in contrast to the mainly problem-oriented discussion on mountain environments.

2. Mountain regions in Yunnan

Yunnan is located in the eastern Himalayan Region of Southwest China. The province has a wide range of elevation, from the highest peak (6740 m asl) in the alpine temperate zone to sub-tropical valleys as low as 76 m asl. The "roof" of Southeast China includes the headwaters of the Yangtze, Salween, Irrawaddy, Mekong, Red, and Pearl Rivers. Yunnan's ecosystems are therefore not only important to mountain people but also to lowland people, as well as a large number of stakeholders in the downstream regions in Burma, Laos, Vietnam, and Thailand.

Yunnan's uplands have historically been home to diverse indigenous cultures, with 25 officially recognized ethnicities counting more than 14 million people. Indigenous people have practiced complex land use systems, such as agro-pastoralism among Tibetans, shifting cultivation among the Lisu and Jinuo, terraced paddy cultivation of the Hani, hunting and gathering among the Kucong (Lahu) and Dulong, and intensive lowland paddy cultivation among Dai and Bai people. In the larger valleys, intensive rice-based agriculture can support larger settlements. These valley centers are often foci of political and economic power. Through migration and trade, they also influence the economic activities, labor force and land use of the surrounding upland communities.

3. State simplifications and state-led processes

Since 1949, the Chinese state has recognized the importance of agricultural production for economic development and has focused its governmental programs and policies on agricultural development. However, these land use policies have been developed based on lowland perspectives on land tenure, technology and taxation.

This phenomenon has been defined as "state simplifications" (Scott 1998). The key point of this concept is that states, in dealing with diverse natural and social environments, attempt to make these environments comprehensible by creating "thin simplifications". These are generalizations that ignore specific local circumstances, and therefore frequently lead to problems such as a lack of acceptance for new land use policies and detrimental environmental and social impacts of implemented land use policies. The interests of the state and farmers regarding land are often different. The mismatch of the government's land tenure policies and the farmers' land use practices is often a source of conflict in rural communities. Frequent changes in land tenure policy have exacerbated these problems in Yunnan.

Since the 1950s, the Chinese state has clearly identified itself as a developmental state. Rural policy in the 1950s focused on promoting agricultural development by bringing productive forces into the public realm. The priority agenda for Yunnan's government was to assist upland ethnic groups in making the transition from a "primitive society" to a "socialist society". To reach this goal, "work teams" were sent to live and work with mountain people, and to assist in the implementation of land reform and the establishment of collectives and communes. Along with institutional reforms, these teams introduced "improved" agricultural technologies and practices.

Following the ideological excesses and the institutional collapse of the Cultural Revolution, in 1978 the Central Government redirected its focus towards modernization with the objective of generating wealth. In practice, this led to a large expansion of the shifting cultivation area in the early 1980s, due to both poorly defined land tenure and ambiguities in the application of land reforms to areas of shifting cultivation. In 1982, the government addressed the associated deforestation problem with a new policy (*linyesanding*) that promoted the sedentarization of shifting cultivators.

In the 1980s, the focus on modernization led to the creation of several local policies and land conversion practices, such as the establishment of tea, rubber, sugarcane, and tobacco plantations. Since the late 1990s, one of the policy strategies has been to develop Yunnan into a Green Economy Province. This has involved bio-prospecting and the development of bio-resource processing enterprises. For example, there is policy support for the exploitation of medicinal and other plant resources used by Yunnan's ethnic minorities. Following the rapid development of Yunnan's tourism industry in the 1980s and 1990s, another policy strategy has promoted the identification of cultural assets, such as ethnic clothing, handicrafts, songs and dances, for packaging as marketable products.

4. State simplifications and resource policies

Land use and forestry management in China are in a state of transition. Changes include: a) a shift from forest use for subsistence needs to the generation of cash income; b) a shift from state to collective tenure, and even private ownership, as exemplified in the implementation of the Wasteland Auction policy; c) a shift from a policy emphasis on forest production to ecological functions and the conservation of biological diversity, particularly after the disastrous floods on the Yangtze River in 1998; d) a shift from centralized planning of reforestation to decentralized planning and multi-stakeholder participation; e) a shift from mono-culture to more diversified tree intercropping and agro-forestry; and f) a shift from traditional cultural values and beliefs to modern value systems.

4.1 Land use practices: The example of shifting cultivation

Shifting cultivation, or swidden farming, is a system of alternating clearance of forest and a short cultivation period with a long fallow period during which forests regenerate and soils recover. Shifting cultivation is commonly found in the humid tropical and sub-tropical areas of southwest Yunnan. Today, this agricultural practice provides food and cash income to 2 million farmers. However, this livelihood system is increasingly under threat from state agricultural and forestry policies and practices, as well as global market forces.

The state has consistently seen shifting cultivation as an unsustainable farming practice and the principal driving force behind deforestation in Yunnan. This view should be critically re-examined. As long as population pressure is not too high, composite swidden systems can be sustainable and highly diverse agroecosystems. Shifting cultivation is often only one component of composite swidden systems, which may also include homegardens, fishponds, cash crop plantations and even irrigated rice paddies in upland areas. A number of practices enhance biodiversity and forest regeneration by means of: a) protection of useful native tree species in swiddens, b) the combination of annual crops and perennial tree crops, c) selective weeding to preserve forest tree seedlings, d) planting of favored native tree species during swidden-fallow, and e) even domesticating native plants and enhancing agrobiodiversity. Furthermore, small patches of forest in swidden-fallow fields serve as centers of seed dispersal (Pei and Xu 1997). The swidden cultivator's goal is not to destroy forest but to obtain a continuous harvest of cultigens while managing the succession towards a new forest of high diversity. The management of such diversity is dependent on indigenous technologies and customary institutions, often grounded in indigenous cosmovisions.

Using aerial photographs and satellite imagery, we found that secondary vegetation covers 92% of the landscape in southern Yunnan (Fox et al. 1995; 2000; Xu et al. 1999). Our research also suggests that major land-use changes have been associated with the switch from swidden cultivation for subsistence to large-scale cash crop production, including both rice paddy and tree plantations (Long et al. 1999). Cash crop production can either result in a tree-dominated land cover (e.g. rubber, fruit trees, cardamom, or tea), or in a land cover composed of annuals (e.g. maize, sugarcane, cassava, and upland rice) in the uplands of Yunnan. In either case, biodiversity probably declines. Although a large-scale increase of plantation-style forest in non-forested areas increased total forest coverage in China from 5.2% in 1950 to 13.9% in 1995, natural forests declined to 30% of the total forest area during this interval (Zhang et al. 2000).

State forest policy has presented significant threats to shifting cultivation. A new state policy, the Natural Forest Protection Program, was implemented after 1998. Logging in designated forest areas, including secondary succession in swidden-fallow, was banned. This policy was associated with the reallocation of tenure rights, as swidden-fallow became subject to natural forest conservation regulations. The social acceptance of forest management strategies is at risk due to insecure forest land tenure, and instable development policies. Although the protection of designated forest reserves alongside agroforestry systems is important, their acceptance can only be ensured, if the livelihoods of mountain people are not threatened.

4.2 Land degradation, floods and "upland conversion"

Following a long period of forest decline that began in the late 1950s, overall forest cover in Yunnan has increased in recent decades, from 26% in 1978 to 34% in 1997 before the historical floods of the Yangtze River. This suggests an improvement in land cover conditions and associated ecosystem goods and services. Among the factors contributing to these improvements has been the promotion of collective as opposed to state forest management in the early 1980s, and of large-scale state afforestation programs. However, this increase in forest cover does not stop the loss of biodiversity, as much of the new forest areas are monoculture (particularly pine) plantations.

It is commonly believed that the deforestation associated with traditional agroforestry practices in mountain watersheds is responsible for increased flooding in downstream areas. After the flooding of the Yangtze River in 1998, for instance, a logging ban affecting local communities was announced. However, historically, the greatest damage to forest ecosystems has come from state-sponsored, large-scale logging operations. Since they involve extensive clear-cutting of forest stands, large-scale logging operations have resulted in the loss of wildlife habitats and endangered species, and an increase in soil erosion and landslides in the region (Zhang 2000). Between the 1970s and 1990s, large and essentially unregulated concessions were given to private or state enterprises. Timber extraction peaked in the mid-1980s and decreased in the 1990s due to the depletion of forest resources and increasing access problems (Fig. 1). Most of the forest clearance and erosion that is visible in Northwest Yunnan today is the result of this period of logging. Local people were, however, involved in state forestry practices. They usually received wages for their labour, while the revenue from timber sales remained in the hands of the state and private companies.

In 1999, a new state policy, called Upland Conversion Program (*tuigeng huanlin huancao*), was introduced. In order to decrease soil erosion, this policy aims to convert agricultural land on slopes steeper than 25 degrees into forest or grassland. Under this policy, local farmers receive monetary support to buy tree seedlings (ca. USD 90 per ha of converted land). In addition, farmers receive unprocessed rice (2250 kg per ha per year) for five years if land is planted in cash crop trees, and for 8 years if land is planted with trees that have a primarily ecological function.

The potential impact of upland conversion on indigenous people is great, affecting over 650,000 ha of land and one fourth of the total mountain population of Yunnan (Table 1). Although the policy is an important measure for erosion control, it poses a threat to local land use practices, land tenure, and customary rights. The threat to indigenous livelihoods is exacerbated where no provisions have been made for the gathering of non-timber forest products and fuel wood or the inclusion of agro-forestry in the upland conversion and reforestation schemes. In the long term, cash and grain compensation is thus insufficient to ensure the welfare of local people.

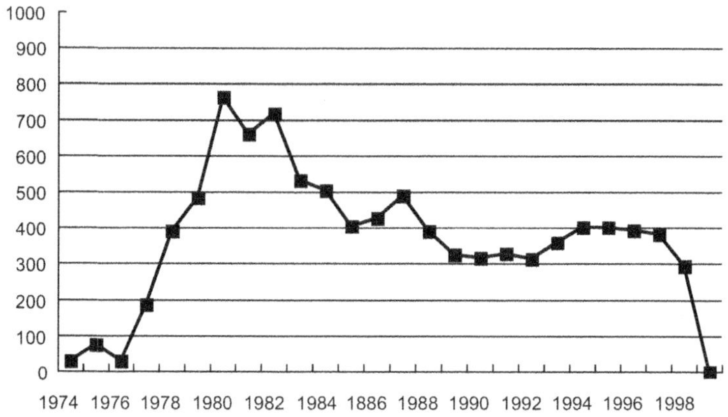

Figure 1: Timber extraction in Diqing Prefecture, NW Yunnan, 1974-1999 (in 1,000 m^3) (from: Xu and Wilkes 2003).

Table 1: Farmland within major watersheds of Yunnan (from Xu 2002).

Major watersheds	Total farm-land area (10,000ha)	Farmland area >25° (10,000ha)	Farmland area >25° in each water-shed (%)	Steep farmland area of the total area of land >25° in the province (%)	Average yield of farmland (t/ha)	Popula-tion (10,000)
Jinsha-Yangtze	215	20.2	9.4	27	2.9	341
Nanpan-Pearl	140	10.6	7.6	14.2	2.6	168
Honghe-Red	99	19.0	19.3	25.4	2.4	199
Lancang-Mekong	137	16.6	12.1	22.3	2.0	185
Nujiang-Salween	30	6.7	22.4	8.9	1.7	53
Drong-Irrawaddy	22	1.7	7.6	2.2	2.5	61
Total	643	75.6	11.6	100	2.4	1007

Note: Crops on sloping farmland are primarily corn and wheat and, to a lesser extent, potatoes and buckwheat.

4.3 Biodiversity conservation

Yunnan has about 17,000 flowering plant species (63% of China's native species), 793 bird species (64%), and 300 mammal species (51%). About half of the endangered species in China are protected in 120 nature reserves, which cover 6.5% of Yunnan's land area. Biodiversity is conventionally defined as diversity in habitats, genetics, and species (McNeely et al. 1990). Biodiversity, is not just a property of natural ecosystems, but also a product of the interaction of social and bio-physical systems (Pei and Sajise 1993) and is therefore linked to indigenous knowledge and cultural diversity. For instance, shifting cultivation may actually increase, rather than reduce, the diversity of habitats and species in the landscape (Xu et al. 1997).

From the 1980s, conservation policy in China was dominated by an approach that excluded local communities from involvement in resource utilization and management within reserves. However, biodiversity has economic, cultural and ecological functions, and can be managed through a variety of social institutions. Increasingly, conservation specialists recognize the importance of social factors in biological resource management and the protection of biodiversity. Since the mid-1990s, "co-management" approaches involving communities in the management of nature reserves have been increasingly advocated (Li 2000). It has also been recognized that sacred sites and landscapes can play an important role in biodiversity conservation (Pei 1991; Ramakrishnan 1996; this volume).

4.4 Perceptions of poverty

The way poverty is understood determines the way state resources for poverty reduction are allocated and the way poverty reduction initiatives are formulated. The poverty line has historically been based on two primary indicators: cash income and food availability. However, surveys commonly find that the vernacular understanding of poverty focuses more on access to natural resources, the resilience or vulnerability of livelihood assets, and non-material factors, such as feelings of happiness and respect, which may be related to access to power and ethnic identity.

As measured by official poverty criteria, Yunnan is one of China's poorest provinces (Li 1994). The government has made great efforts to alleviate poverty over the last 50 years. Yet poverty remains a problem in many regions. Wang (2000) describes how earlier state interventions aimed at increasing grain yields (e.g. by cultivation of permanent land plots on steep hillsides) but, because they were based on a poor understanding of local ecology, led to severe ecological destruction. More recently, adverse environmental conditions (e.g. poor natural resource base and physical remoteness) have been more prominent as criteria for identifying the poor. Hence, resettlement into more geographically favorable regions has become a widely adopted state measure to alleviate poverty. Because resettlement schemes involve the redistribution of already reclaimed land, equity issues are particularly prevalent, and there is considerable cause for concern over the social impacts of these schemes.

5. Conclusion: Creating mountain perspectives

Currently, natural resource management in the mountain regions of Yunnan is dominated by "lowland" patterns of thinking, which serve to channel the benefits derived from mountain resources to lowland economic and political centers. The negative impacts of the marginalization of upland people challenge government agencies to devise policies and establish institutions that 1) are more supportive of indigenous knowledge, culture and livelihoods, 2) ensure economic benefits for mountain people, and 3) involve upland people in decision-making processes. This can be achieved through increased decentralization of state policy formulation and program planning, and by supporting locally based knowledge innovation processes. These changes require at least three important pre-conditions.

Firstly, the impacts of traditional land use on biodiversity need to be assessed from a mountain perspective rather than through the simplifications of lowlanders, taking into account the dynamic and holistic relationship between indigenous communities and ecosystems. State simplifications, such as the view that shifting agriculture is necessarily unsustainable and destructive, need to be replaced by careful assessments that take into account the diversity of local environmental conditions, traditional land ownership and land use practices, including indigenous knowledge on these factors.

Secondly, resource tenure issues must be addressed. Although the overall area of forest cover in Yunnan has increased, tenure security has not improved. Frequent changes in resource policy have decreased the area of available arable land, and therefore directly impacted the livelihoods of indigenous mountain people. These policies also restrict the involvement of local people in making key decisions over land use and resource management. Ironically, some "community forests" are now off-limits to villagers, leaving no possibility for developing sustainable utilization approaches, and serve as forest reserves with little or no community benefit.

Access to land and natural resources, on which local people's knowledge and customs are based, needs to be ensured. Important tenure rights include not only resource ownership, but also fair arrangements for allocating management, harvesting and marketing rights. It must be recognized that local farmers and communities have preferential rights to sustainable resource use, such as the development of eco-tourism in protected forest areas. In addition, strategies should be developed that promote diversification of resource use (Xu and Wilkes 2002).

Thirdly, the decentralization of local policy formulation, planning and action processes will require capacity building and empowerment of local communities through village democratic processes such as elections. Capacity building among community members, NGOs and government staff should ensure that adequate skills are acquired at the appropriate level. New technologies and resource management processes require effective training, impact assessment, and follow-up support. Government agencies may have to first develop these skills in order to provide appropriate support for local communities. Such support should aim beyond the simple transfer of modern technologies, and also include support for local processes of innovation. With this approach, indigenous knowledge can play a role alongside external knowledge in devising solutions to contemporary problems and opportunities.

6. Acknowledgements

The field research on Land Use and Land Cover in the Mekong-Lancang Basin Project is supported by the Ford Foundation, Beijing Office, P. R. China.

7. References

Alcorn, J. (1990). Indigenous agroforestry strategies meeting farmers' needs. *In* "Alternatives to deforestation: Steps toward sustainable use of the Amazon rain forest." (A. Anderson, Ed.), pp. 141-148. Columbia University Press, New York.

Brookfield, H., and Padoch, C. (1994). Appreciating agrodiversity. *Environment* **36**, 6-45.

Fox, J., Krummel, J., Yarnasarn, S., Ekasingh, M., and Podger, N. (1995). Land use and landscape dynamics in northern Thailand: Assessing change in three upland watersheds. *Ambio* **24**, 328-334.

Fox, J., Dao M. T., Rambo, A. T., Tuyen, N. P., Le Trong C., and Leisz, S. (2000). Shifting cultivation without deforestation: A case study in the mountains of northwestern Vietnam. *BioScience* **50**, 521-528.

Li, C. (2000). Policies and implementation in the nature reserves of Yunnan Province. *In* "Links between cultures and biodiversity." (J. Xu, Ed.). Yunnan Science and Technology Press, Kunming.

Li, W. (1994). Human resources development and poverty alleviation: A study of 23 poor counties in China. *Asia-Pacific Population Journal* **9**, 3-18.

Long C., Fox, J., Xing, L., Lihong, G., Kui, C., and Jieru, W. (1999). State policies, markets, land-use practices, and common property: Fifty years of change in Yunnan, China. *Mountain Research and Development* **19**, 133-139.

McNeely, J. A., Miller, K. R., Reid, W. V., Mittermeier, R. A., and Werner, T. B. (1990). Conserving the world's biological diversity. IUCN, WRI, CI, WWF-US, and the World Bank.

Menzies, N. (2002). 'Nice view up there' - Discordant visions and unequal relations between the mountains and the lowlands." Keynote address presented at the Third Conference on Montane Mainland Southeast Asia (III MMSEA). *In* "Proceedings of MMSEA Conference, Lijiang. August 2002." (Xu J., Ed.) (in press).

Nagata, H. (1996). The effect of forest disturbance on avian community structure at two lowland forests in peninsular Malaysia. *In* "Conservation and faunal biodiversity in Malaysia." (A. Abidin, A. Hasan, and Z. Akbar, Eds.), pp. 93-101. Malaysia University, Bangi.

Pei, S. (1991). Conservation of biological diversity in temple-yards and holly hills by the Dai ethnic minorities of China. *Ethnobotany* **3**, 27-35.

Pei, S. J., and Sajise, P. E., Eds. (1993). Regional study on biodiversity concepts, frameworks and methods. *In* "Proceedings of the Southeast Asian Universities Agroecosystem Network (SUAN) and Program on Environment (ENV), East-West Center Workshop, Xishuangbanna, Yunnan Province, P.R. China, October 24-30, 1993."

Pei, S., and Xu, J. (1997). Biodiversity and sustainability in swidden agroecosystems: Problems and opportunities (a synthesis). *In* "Collected research papers on biodiversity in swidden agroecosystems in Xishuangbanna." (S. J., Pei et al., Eds.), pp. 173-177. Yunnan Education Publishing House, Kunming (in Chinese).

Ramakrishnan, P. S. (1996). Conservation of the scared species to landscape. *Nature and Resources* **32**, 11-19.

Sajise, P. E., and Baguninon, N. T. (1982). Upland development: A critical resource management issue. *In* "Proceedings of workshop on Ecological Basis for Rational Resource Utilization in the Humid Tropics of Southeast Asia, January 18-22, 1982." (Kamis Awang, Ed.). Universiti Pertanian Malaysia, Kuala Lumpur.

Scott, J. (1998). "Seeing like a State: How certain schemes to improve the human condition have failed." Yale University Press, London.

Wang, D. (2000). Heavy price to pay: A study on the relationship between social and environmental development in the Nujiang Canyon region. *In* "Links between cultures and biodiversity." (J. Xu, Ed.), pp. 622-632. Yunnan Science and Technology Press, Kunming.

Xu, J. (2002). "Policy review and institutional capacity analysis."Yunnan Environmental Development Program, Unpublished DFID Project Report.

Xu, J., Chen, S., and Pei, S. (1997). Swidden agriculture in the context of political economy. *In* "Collected research papers on biodiversity in swidden agroecosystems in Xishuangbanna." (S. J. Pei et al., Eds.), pp. 155-164. Yunnan Education Publishing House, Kunming (in Chinese).

Xu, J., Fox, J., Lu, X., Podger, N., Leisz, S., and Ai, X. (1999). Effects of swidden cultivation, population growth, and state policies on land cover in Yunnan, China. *Mountain Research and Development* **19**,

123-132.

Xu, J., and Wilkes, A. (2002). People and ecosystems in mountain landscape of NW Yunnan, SW China: Causes of biodiversity loss and ecosystem degradation. *Global Environmental Research* **6**, 103-110.

Xu, J., and Wilkes, A. (2003). Biodiversity impact analysis in Northwest Yunnan, Southwest China. *Biodiversity and Conversation* (accepted).

Zhang, P., Guofan, S., Guang, Z., Le Master, D. C., Parker, G. R., Dunning Jr., J. B., and Qinglin, L. (2000). China's forest policy for the 21st century. *Science* **288**, 2135-2136.

Mountain Biodiversity, Land Use Dynamics and Traditional Ecological Knowledge

P. S. Ramakrishnan
School of Environmental Sciences, Jawaharlal Nehru University, New Delhi 110067, India
phone 2670-4326, 2617-2438, fax 2617-2438, 2616-9962, e-mail psr@mail.jnu.ac.in

Keywords: Adaptive landscape management, Mountain landscape, Sacred groves, Sacred landscape, Sustainable forestry, Sustainable mountain agriculture, Traditional ecological knowledge

1. Introduction

Traditional mountain societies are characterized by their close interconnection with nature and natural resources. They depend upon natural resources and biodiversity for their sustainable livelihood concerns (Ramakrishnan 1992a; Ramakrishnan et al. 1994; 1996). This linkage with nature and natural resources extends beyond the economic realm; social, cultural and spiritual dimensions also play a significant role (Ramakrishnan et al. 1998). Traditional mountain societies have a holistic view of the ecosystem and the social system. This relationship with nature is based on coexistence rather than competition, which results in agricultural strategies that are adapted to the natural environment and the sustainable use of natural resources. The result of this relationship is a set of institutional arrangements that evolved towards ecological prudence. The ultimate objective is the sustainable use of natural resources through compromises between environmental risks on the one hand, and productivity concerns on the other.

Traditional Ecological Knowledge (TEK), centered on the manipulation of biodiversity, determines the land use dynamics in many tropical mountain regions. Consequently, the land use system is based on diversification rather than homogenization of the landscape. This approach has ensured sustainability in the past and has become significant in the context of global change adaptation strategies (Walker et al. 1999), because diverse landscapes may render mountain societies less

U. M. Huber et al. (eds.), Global Change and Mountain Regions, 551-561.

vulnerable to environmental changes. In the wake of modernization, the linkages of traditional mountain people with nature have often been weakened, resulting in economically marginalized mountain societies and social disruptions (Ramakrishnan 1999a,b). Due to contrasting value systems of development agents and indigenous people, the transfer-of-technology approach of recent decades often failed to enhance economic development in mountain areas. Also, this approach excluded mountain people from active participation (Ramakrishnan et al. 1994). In order to ensure the success of developmental strategies, it is crucial to place the sustainability of mountain systems in the context of TEK and people's participation.

2. What is TEK?

Traditional societies, often referred to as indigenous or tribal people, have accumulated a lot of empirical knowledge from their experience with nature and natural resources. This traditional wisdom is based on the intrinsic realization that man and nature form part of an indivisible whole, and therefore should live in partnership. This ecocentric view of traditional societies is widely reflected in their attitudes towards plants, animals, rivers, and the earth. These timeless truths of man-nature ethics are passed on in iconography and sculptural imagery (Vatsayan 1993).

Much descriptive literature has been accumulated on the food and medicinal species used by traditional societies, which in itself is important (National Academy of Sciences 1975; Hladik et al. 1993). In addition, a lot of environmental knowledge can be gleaned from investigations on traditional land management practices. Another aspect of TEK which needs to be addressed is how ecosystem processes are altered by societal perceptions and decision-making processes. In other words, how does the socio-ecological system function as an integrated whole? It is only in recent times that ecologists have started to look at the dynamics of these socio-ecological relationships, operating at varied spatial (sub-specific, species, ecosystems and landscapes) and temporal scales.

Our understanding of TEK is based on the economic, ecological and socio-cultural benefits, both tangible and intangible, that traditional societies may derive from the surrounding landscape (Ramakrishnan 2001):

a. Economic: Traditional crop varieties, lesser known plants and animals of food value, and medicinal plants harvested from the wild can provide an important economic basis for mountain societies and can buffer periods of food scarcity;
b. Socio-ecological: The way in which mountain societies conserve and manipulate biodiversity contributes towards ecosystem resilience, and strengthens the capacity of mountain people to cope with environmental change, to control soil water regimes and hydrology and to manage soil fertility through enhancement of soil biological processes (Ranakrishnan 1994);
c. Socio-cultural: Cultural, spiritual and religious belief systems of mountain people are centered on the concept of sacred species, sacred groves and sacred landscapes, which can play an important role in biodiversity conservation.

3. Managing mountain agroecosystems

Mountain societies are at various stages of socio-economic development. Mountain agroecosystems range from casually managed shifting agriculture on one extreme to modern monocultures and plantation systems with high-energy inputs on the other. A variety of sedentary agroforestry practices occupy an intermediate position along this developmental gradient (Swift et al. 1996).

3.1 TEK in traditional agroecosystems

Biodiversity-centered TEK in traditional agroecosystems is geared towards soil conservation and fertility management on steep hill slopes. In the shifting agricultural system of north-east India, for example, farmers manipulate crops and associated biodiversity at the species and sub-specific level in order to sustain soil fertility and optimize production under conditions of environmental uncertainties (Box 1). The soil management of traditional mountain societies is based on the intimate knowledge of spatial differences in soil fertility and the effects of different crops, plant residues and non-food plants on erosion control and soil fertility.

TEK is not static. For example, the slash and burn agriculture in north-east India has adapted to declining forest cover and land degradation, as well as increasing market pressures. Traditional agriculture has increasingly been replaced by rotational fallow practices and eventually by sedentary systems. These systems emphasize nitrogen-fixing legumes as part of the mixed and/or rotational farming practices (Ramakrishnan 1992a). Furthermore, large-scale deforestation by outside forces has disrupted traditional agroforestry. Traditional farming practices have often been replaced by modern agriculture that is primarily aimed towards higher productivity through energy subsidies. Loss of agrobiodiversity and the related ecological knowledge, and erosion of traditional value systems are some of the consequences. Intensification of agriculture results in practices that trigger new environmental and social problems, such as accelerated land degradation, abandonment of farmlands, and population migration to lowlands for off-farm employment opportunities, as is happening now in the Central Himalayan region in India (Ramakrishnan et al. 2000).

3.2 Developmental pathways for TEK-based agroecosystems

In the mountain regions of India, there have been many attempts to introduce modern high-input agricultural practices that are primarily aimed at a rapid increase in production levels. Over the last hundred years or so, such approaches to agricultural development have time and again been rejected by the local mountain communities, because they are not related to their traditional value systems. Balancing concerns for higher agricultural production with local value systems becomes a challenging task that requires an adjustment of agricultural development to local circumstances. Realizing that modern agriculture, based on the formal knowledge system, is only one of the possible developmental pathways, two additional models were considered appropriate for traditional mountain societies. These are: (i) the incremental pathway, and (ii) the contour pathway (Swift et al. 1996). What these pathways imply in real terms is best illustrated through some specific examples.

Modern agriculture: For decades, agricultural research and extension efforts have been steered towards modernization. At this extreme, the resulting production types stand apart from the rest of the landscape as an artificial entity. They represent an attempt to convert the natural ecosystem into a controllable set of biological and chemical elements, almost irrespective of the background ecological conditions. In this agricultural strategy, narrow production targets are pursued, and the background ecological conditions are only taken into account to provide information on production potential. Intensified crop production through external energy inputs ensures high productivity levels, though based on short-term socio-economic considerations and at costs to the environment.

Box 1: Biodiversity linked TEK and soil fertility management (from: Ramakrishnan 1992a; Ramakrishnan et al. 2002).

Traditional societies in north-eastern India grow crop species with low nutrient requirements on the nutrient-poor slopes, and species with higher requirements on the relatively fertile lower slopes.

With a shortening of the shifting agricultural cycle (about 5-6 year long fallows) farmers tend to preferentially grow nutrient-use efficient tuber and vegetable crops on the nutrient-poor soils, whereas the emphasis lies on cereals that are nutrient-demanding with longer cycles (> 10 years of fallow).

In mixed cropping systems, where species are sown soon after the first monsoon rains, farmers harvest crops sequentially as the different crops mature over a period of a few months. This practice ensures continuous crop cover and reduces competition for space and other soil resources (e.g. water and nutrients). After harvesting the edible parts of the cultivated crop, the harvested biomass is left in the field. This practice limits the export of soil nutrients and serves as a means of erosion control.

Weeded plants are put back into the system and some weeds are left *in situ* for nutrient conservation and erosion control.

Earthworms form an important component of many traditional agricultural systems. Earthworms increase plant-available nutrients and improve soil structure, thus reducing the threat of erosion. The international Tropical Soil Biology & Fertility programme therefore promotes eco-technologies that increase the density of earthworms (e.g. tillage systems that leave surface residues) and other soil organisms. A high diversity of insects/arthropods reduces pest pressure and thus the need of chemical pesticides.

Crop species and other plant species that are valued both in traditional agricultural and natural forest systems often provide important ecosystem services; for example, these key species often play a critical role in nutrient enrichment of the soil; indeed, indigenous people attribute value to these species that goes beyond human nutrition (e.g. sacred species). The use of culturally valued species for the rehabilitation of degraded agroecosystems ensures community participation.

Traditional eco-technologies, such as irrigation systems that include water harvesting (roof-top systems, village ponds) and elaborate above- or belowground canals for water distribution, enhance soil biological processes and soil fertility by providing ideal conditions for the growth of soil micro- and macro-organisms in seasonally dry monsoon climate.

The contour pathway: In contrast to modern agriculture, understanding the relationships between structure and function in natural ecosystems provides valuable guidelines on what crops might be planted, what environmental problems might be encountered, and what production goals are reasonable (Swift et al. 1996). Working with nature, rather than dominating it, this approach involves active planning with the nature of the background ecosystem fully in mind. Building upon traditional knowledge, with appropriate inputs from the formal knowledge system, forms the basis for designing agroecosystem models that are adapted to the biophysical limitations or "contours" at a given location. In a sense, this pathway attempts to integrate traditional and modern agricultural strategies. It seeks to work with the ecological forces that provide the basis for the agricultural system, while acknowledging the social, economic and cultural requirements of the farming communities. Many agroforestry systems with low and medium intensity management fall into this category. Examples are agro-pastoral systems, compound farms, and traditional systems based on the home garden concept.

There have been many attempts to design sustainable agricultural systems to meet with the specific needs of mountain societies. Sloping Agricultural Land Technology (SALT), developed in the southern Philippines, is one such system (Tacio 1993). SALT is based on the planting of annual and perennial crops in three to five bands between double rows of nitrogen fixing trees and shrubs. Crops are planted along contours for soil conservation. The objective here was to establish a stable ecosystem that minimizes soil erosion, ameliorates the chemical and physical properties of the soil and increases the food security of farmers. These objectives were realized in the initial experimental phase in the Philippines, and attempts are now being made to introduce this technology in other Asian regions (Partap and Watson 1994). However, the initial reaction to this practice has been disappointing. This lack of success is primarily linked to reasons such as uncertain land tenure systems and small and fragmented land holdings, which hinder a consistent application of this technology at larger spatial scales (Ramakrishnan 2001). In addition, the high monetary and labor costs of the initial investment and the subsequent maintenance are restrictive. Hence, the poor success of this approach is not a technology problem, but rather a socio-economic problem. Moreover, SALT is alien to many local communities, because this approach (i) is based on a generalized model developed in experimental plots, which may not be transferable to other locations, (ii) promotes certain non-native plant species (e.g. the fast growing multi-purpose legume *Lucaena leucocephala* from South America), and (iii) is built around formal rather than TEK. For these reasons, SALT has not been able to gain acceptance in the Indian subcontinent.

The incremental pathway: Many traditional agricultural systems need to initially be developed through TEK-based incremental changes, because any drastic change may not find acceptance in local communities. Thus, one may have to consider short-term compromises that may be constrained by ecological, economic, social and/or cultural factors, while not losing sight of a more ideal long-term strategy. The chief difference between the incremental and the contour pathways lies in the relative proportion of traditional vs. formal knowledge used in designing improved agricultural technologies. The incremental pathway more strongly incorporates TEK.

The most comprehensive study on the incremental pathway as a potential route for agricultural development in tropical mountain systems is a case study on the shifting agricultural system of Nagaland, north-eastern India. The conclusions of this study have broad implications for a land use system that is prevalent all over Asia, Africa and Latin America. The Nagaland initiative in north-east India (Box 2) arose from the in-depth analysis of the environmental impacts and redevelopment possibilities of shifting agriculture (Ramakrishnan 1992a). The initiative aims at a participatory development plan for the entire State of Nagaland on the basis of a decentralized village development plan. The village development boards are based on the value systems of a large variety of local ethnic groups. Agricultural strategies are based upon TEK and have emphasized forest fallow management, building upon traditional shifting agriculture. A further emphasis has been on the conservation of natural and human-managed biodiversity in this national biodiversity hot spot.

To ensure sustainable land use in tropical mountain environments, we need to increase our understanding of the organization and functioning of these human-managed ecosystems. Future agricultural development needs to be based on TEK in order to ensure sustainable livelihoods to mountain societies, while taking into consideration environmental conservation concerns. Diversified agroecosystem models, based on a value system that mountain societies are able to relate to can be more easily adapted to changing environmental conditions, which will reduce the vulnerability of mountain people to global change impacts (Swift et al. 1996; Ramakrishnan 2001).

4. Managing and rehabilitating forest ecosystems

A significant portion of natural forest ecosystems in developing countries, largely located in mountains, is degraded to secondary formations (Chokkalingam et al. 2000). In addition, natural forests in both the developed and developing world have been converted to forest plantations to varying degrees. The conversion to plantations essentially implies the replacement of species-rich natural ecosystems with species-poor monocultures. Firstly, many of these plantations require intensive management and are therefore dependent on a variety of external inputs. Thus, the long-term sustainability of such conversions, both in ecological and economic terms, often remains uncertain. The maximization of economic returns has formed an important basis of forestry practices in the developed world and, until recent times, this has also been the case in the developing world. Only recently, mixed plantation forestry practices, that have long-term sustainability without external inputs, have started to receive increasing attention. These mixed forest systems are rich in non-timber forest products. With this additional livelihood source, pressure on the remaining natural forest ecosystems has decreased. Hence, such a change in plantation forestry has been important both for forest conservation and carbon sequestration in the context of global change (Ramakrishnan 1992b).

Box 2: An Indian government initiative (NEPED and IRRR 1999) for improving the shifting cultivation system in Nagaland (based on this author's analysis).

The local government, with support from the India-Canada Environmental Facility, initiated a project to strengthen shifting agriculture (*jhum*) in Nagaland with a strong focus on TEK. Land use redevelopment was initiated through participatory extension and dissemination and the incorporation of gender issues. VDBs (Village Development Boards), constituted on the basis of the local value system, are seen as appropriate vehicles for land use development. The project involves about 1200 villages and about 200 replicated experimental plots in about 5500 ha of farmland.

Farmers have adopted tree-based intensified *jhum* systems. These systems are based on agroforestry principles and are tested by villagers in 870 villages, covering a total area of about 33,000 ha (38 ha per village x 870 villages). In these experimental plots, local adaptations and innovations of agroforestry practices, such as soil and water management, are assessed.

Edible legume cover crops are cultivated in mixed as well as pure cropping systems as part of an extended *jhum* cropping phase of three to four years to increase N-fixation and, consequently, to prolong soil fertility.

Nepalese Alder (*Alnus nepalensis*), a common tree species in the north-eastern region, is often planted and/or protected in cultivated and fallow *jhum* plots. This species fixes up to 120 kg N ha^{-1} per year. The use of Nepalese Alder is only a starting point and additional tree species have been identified for the restoration of soil fertility during the fallow phase of *jhum*.

Ten selected tree species, commonly used for construction and fuel wood, can be harvested 5-10 years after planting, and 20 additional tree species that are of value for timber have been identified and introduced into *Jhum* plots in consultation with local communities. Mixed tree plantations were shown to be superior to monocultures for fallow recovery and are recommended.

The widespread thatch grass (*Imperata cylindrica*), a perennial weedy grass that commonly grows in less fertile and dry areas all over the tropics, is controlled through dense cropping of Cassava, which is part of the *jhum* cropping system.

Agroforestry related cultivation of non-traditional crops, such as tea and oyster mushrooms, are being tested as additional possibilities.

Improving the yield from the home garden systems through the cultivation of additional vegetables is another option for cash income; similarly, multipurpose bamboo cultivation, including bamboo shoots as a food item, has also been suggested.

Biodiversity conservation is a key objective in redeveloped *jhum* systems, since biodiversity decreases with a shortening of the traditional *jhum* cycles.

Traditional means of rainwater harvesting and erosion control are incorporated into the improved *jhum* practices, where appropriate.

4.1 The role of TEK in forest ecosystem management

It is being increasingly recognized that natural forest management, and indeed rehabilitation of degraded natural systems, demand attention to ecological, socio-economic and even cultural dimensions, and the interactions between these factors. Recognizing this complexity is even more critical in developing countries, where a large proportion of the human population is directly dependant upon forests not only for timber but also for non-timber forest products (Patnaik 2002). When evaluating tropical mountain forests, it is important to move away from a mere profit-based assessment of timber value and instead move towards an appreciation of the combined ecological, socio-economic and spiritual value of forests.

Indeed, the cultural and spiritual importance of a large number of species could form an important basis for forest conservation and management. There is increasing evidence that ecological key species are often also selected by mountain societies (Ramakrishnan et al. 1998). For example, *Ficus* spp. (fig tree), which are considered sacred in many Asian and African societies, and *Quercus* spp. (oak), which are culturally valued trees in the Central Himalayan region, also play an important ecological role. These trees regenerate soil fertility by absorbing plant nutrients from the deep soil layers and by contributing them to the upper soil through high quality leaf litter. Increased soil organic matter leads to higher water holding capacity, which improves plant moisture supply during the dry season (Ramakrishnan 2001). The tree species also contribute to enhancing above- and belowground biodiversity in forest and agriculture systems. Consequently, these species play an important role in agroforestry and forestry rehabilitation (Ramakrishnan 2001). Capitalizing on the cultural and religious value of ecologically important tree species could ensure community participation in forest conservation and management.

4.2 The role of sacred landscapes in forest biodiversity conservation

The concept of sacred landscapes or sacred groves is a worldwide phenomenon (Box 3). It developed from the recognition that any modification of nature should actively maintain ecosystems in a diverse and productive state, based on locally evolved TEK. In sacred landscapes, restricted land use is permitted, whereas land use is excluded from scared groves. Sacred groves and sacred landscapes, which were not only widespread in the tropics (Ramakrishnan et al. 1998) but also existed in temperate regions at one point in ecological history (Hughes 1998), are indicative of the cultural linkage of local communities with their ecosystems. The guiding principles that regulate the use of natural resources are embedded in the formal and informal institutions of traditional mountain societies (Ramakrishnan et al. 1998). Examples are restrictions on resource use and on the timing of harvest by village committees. In some of the sacred landscapes, such as the Demajong landscape, which is sacred to the Tibetan Buddhists of the Sikkim region in the Eastern Himalayas, traditional institutional arrangements differentiate between permissible small-scale perturbations and non-permissible large-scale perturbations (Ramakrishnan et al. 1998). Thus, a recent hydroelectric project in this region met with strong resistance from local mountain communities, which lead to the abandonment of this project.

We can learn many lessons from the integrated way in which sacred landscapes are managed by traditional institutions in the developing tropics. The more recently evolved biosphere reserve concept of UNESCO, indeed a rediscovery of the sacred landscape of traditional societies (Ramakrishnan et al. 2002), is an attempt to move towards an integrated management strategy in order to conserve natural resources for sustainable land use. Sacred groves, which are completely protected ecosystems,

Box 3: Examples of sacred mountain landscapes (from Bernbaum 1997; Ramakrishnan et al. 1998).

The northern hills and plains of the Ganges river system in the Indian subcontinent are a sacred landscape for the Hindus who form the vast majority of the population in the region. The land, river tributaries, human settlements, a chain of temples dating back to antiquity, the holy mountain cities of Gangotri, Jamnotri, Kedarnath, Badrinath, Rishikesh, Haridwar, etc., and many sites in the northern plains form a set of interconnected ecosystems bound together by the sacred river.

Worshipped by the Hindus and Buddhists, and tucked away in the folds of the Himalaya, the symmetrical Mount Kailash rises above the Tibetan plateau. This mountain is the legendary Mount Meru or Sumeru, and represents the cosmic axis around which the Universe is organized. This belief has even penetrated into the belief system of the distant Balinese and Javanese of Indonesia. All the major sacred rivers of Hindu mythology originate from this region that is considered the cradle of human civilization in this part of the world. Conservation of natural resources in the Himalayan mountain region could be linked to these belief and value systems.

The Buddhist Dai (T'ai) tribe of Xishuangbanna in the Yunnan Province in southwest China has many holy hills. Nong Ban and Nong Meng form hundreds of small or large forested reserves, with human-managed ecosystems and villages interspersed throughout the region.

For many traditional societies mountains represent supernatural beings: e.g. Mount Fuji is seen as God Mysterious in Japan.

have been important for the conservation of high forest biodiversity in Asian mountain regions and could form the basis for the rehabilitation of degraded forest ecosystems.

5. Mountain landscape management: A vision for the future

There is increasing recognition across the different disciplines that mountain regions exhibit a symbiotic relationship between biophysical and social systems. In the modern context, where external pressures often determine land use changes, it is critical that this interconnectedness is recognized and that TEK is incorporated in developmental strategies. For example, in the plantation areas of the Western Ghat mountains in India, the combination of modern soil and water management technologies with traditional soil fertility management (Senapati et al. 2002) and water harvesting technologies (Agarwal and Narain 1997) can help to ensure sustainable land use in a region that is subject to both seasonally heavy monsoon rains and drought (Ramakrishnan et al. 2000).

To be successful, sustainable landscape management needs to be locally adapted. This requires small-scale agricultural development strategies that are based on community participation and are implemented through local institutions. In mountain regions and in other fragile environments, the sustainability of agroecosystems requires an easily adaptable management strategy (Ehrenfeld 1991), which is specifically designed to accommodate pronounced environmental variability and ecosystem complexity within a landscape mosaic. Coping with environmental uncertainties in the context of global change demands prudent management of natural and human-managed biodiversity for a sustainable future in mountain regions.

6. References

Agarwal, A., and Narain, S. (1997). "Dying wisdom. A state of India's environment." Centre for Science and Environment Report 4, New Delhi.

Bernbaum, E. (1997). The spiritual and cultural significance of mountains. *In* "Mountains of the world: A global priority." (B. Messerli, and J. D. Ives, Eds.), pp. 39-60. Parthenon, Carnforth, Lancs.

Chokkalingam, U., Smith, J., de Jong, W., Sabogal, C., Dotzauer, H., and Savenije, H. (2000). "Towards sustainable management and development of Tropical secondary forests in Asia: The Samarinda proposal for action." CIFOR, Bogor, Indonesia.

Ehrenfeld, D. (1991). The management of biodiversity: A conservation paradox. *In* "Ecology, economics, ethics: The broken circle." (F. H. Bormann, and S. R. Kellert, Eds.), pp. 26-39. Yale University Press, New Haven.

Hladik, C. M., Hladik, A., Linares, O. F., Pagezy, H., Semple, A., and Hadley, M. (1993). "Tropical forests, people and food: Biocultural interactions and applications to development." UNESCO-MAB Book Series 13, UNESCO, Paris, and Parthenon, Carnforth, Lancs.

Hughes, J. D. (1998). Sacred groves of the ancient mediterranean area: Early conservation of biological diversity. *In* "Conserving the sacred: For biodiversity management." (P. S. Ramakrishnan, K. G. Saxena, and U. M. Chandrashekara, Eds.), pp. 101-121. UNESCO, Oxford, and IBH Publishers, New Delhi.

National Academy of Sciences (1975). "Underexploited Tropical plants with promising economic value." National Academy of Sciences, Washington, D.C.

NEPED and IRRR (1999). "Building upon traditional agriculture in Nagaland." Nagaland Environmental Protection and Economic Development, Nagaland, India, and International Institute of Rural Reconstruction, Phillippines.

Partap, T., and Watson, H. R. (1994). "Sloping Agricultural Land Technology (SALT): A regenerative option for sustainable mountain farming." ICIMOD Occasional Paper 23, Kathmandu.

Patnaik, S. (2002). Sustainable non-timber forest products management: Challenges and opportunities. *In* "Managing traditional ecologcial knolwledge for biosphere management in South and Central Asia." (P. S. Ramakrishnan, R. K. Rai, R. P. S. Katwal, and S. Mehndiratta, Eds.), pp. 97-127. UNESCO, Oxford, and IBH Publishers, New Delhi.

Ramakrishnan, P. S. (1992a). "Shifting agriculture and sustainable development: An interdisciplinary study from North-Eastern India." UNESCO-MAB Series, Paris, Parthenon, Carnforth, Lancs (republished by Oxford University Press, New Delhi, 1993).

Ramakrishnan, P. S. (1992b). Tropical forests: Exploitation, conservation and management. *Impact of Science on Society* **42**, 149-162.

Ramakrishnan, P. S. (1994). The *jhum* agroecosystem in north-eastern India: A case study of the biological management of soils in a shifting agricultural system. *In* "The management of Tropical soil biology and fertility." (P. L. Woomer, and M. J. Swift, Eds.), pp. 189-207. Wiley-Sayce, Exeter.

Ramakrishnan, P. S. (1999a). Ecological and human dimensions of 'global change' research. *In* "Global change in the mountains." (M. Price, Ed.), pp.176-179. Parthenon, Carnforth, Lancs.

Ramakrishnan, P. S. (1999b). The impact of globalisation on agricultural systems of traditional societies. *In* "Sustainable agriculture and environment: Globalization and the impact of trade liberalisation." (A. K. Dragun, and C. Tisdell, Eds.), pp. 185-200. Edward Elgar, Cheltenham.

Ramakrishnan, P. S. (2001). "Ecology and sustainable development." National Book Trust of India, New Delhi.

Ramakrishnan, P. S., Campbell, J., Demierre, L., Gyi, A., Malhotra, K. C., Mehndiratta, S., Rai, S. N., and Sashidharan, E. M. (1994). Ecosystem rehabilitation of the rural landscape in South and Central Asia: An analysis of issues. Special Publication, UNESCO (ROSTCA), New Delhi.

Ramakrishnan, P. S., Das, A. K., and Saxena, K. G. (1996). "Conserving biodiversity for sustainable development." Indian National Science Academy, New Delhi.

Ramakrishnan, P. S., Saxena, K. G., and Chandrashekara, U. M. (1998). "Conserving the sacred: For biodiversity management." UNESCO, Oxford, and IBH Publishers, New Delhi.
Ramakrishnan, P. S., Chandrashekara, U. M., Elourd, C., Guilmoto, C. Z., Maikhuri, R. K., Rao, K. S., Sankar, S., and Saxena, K. G. (2000). "Mountain biodiversity, land use dynamics and traditional ecological knowledge." UNESCO, Oxford, and IBH Publishers, New Delhi.
Ramakrishnan, P. S., Rai, R. K., Katwal, R. P. S., and Mehndiratta, S. (2002). "Managing traditional ecologcial knolwledge for biosphere management in South and Central Asia." UNESCO, Oxford, and IBH, New Delhi.
Senapati, B. K., Naik, S., Lavelle, P., and Ramakrishnan, P. S. (2002). Earthworm-based technology application for status assessment and management of traditional agroforestry systems. *In* "Managing traditional ecological knowledge for biosphere management in South and Central Asia." (P. S. Ramakrishnan, R. K. Rai, R. P. S. Katwal, and S. Mehndiratta, Eds.), pp. 139-169. UNESCO, Oxford, and IBH, New Delhi.
Swift, M. S., Vandermeer, J., Ramakrishnan, P. S., Anderson, J. M., Ong, C. K., and Hawkins, B. A. (1996). Biodiversity and agroecosystem function. *In* "Functional roles of biodiversity: A global perspective." (H. A. Mooney et al., Eds.), pp. 261-298. J. Wiley and Sons, New York.
Tacio, H. D. (1993). Sloping agricultural land technology (SALT): A sustainable agroforestry scheme for the uplands. *Agroforestry Systems* **22**, 145-152.
Vatsayan, K. (1993). "Prakriti." Indira Gandhi National Centre for the Arts, New Delhi.
Walker, B. H., Steffen, W. L., and Langridge, J. (1999). Interactive and integrated effects of global change on terrestrial ecosystems. *In* "The terrestrial biosphere and global change: Implications for natural and managed ecosystems." (B. Walker, W. Steffen, J. Canadell, and J. Ingram, Eds.), pp. 329-375. Cambridge University Press, Cambridge.

Land Use Intensification around Nature Reserves in Mountains: Implications for Biodiversity

Andrew J. Hansen[1*] and Ruth S. DeFries[2,3]
[1]*Ecology Department, Montana State University, Bozeman, MT, 59717, USA*
[2]*2181 Lefrak Hall, Department of Geography, University of Maryland, College Park, MD, 20742, USA*
[3]*Earth System Science Interdisciplinary Center, University of Maryland, College Park, MD, 20742, USA*
**phone 406 994-6046, fax 406 994-3190, email hansen@montana.edu*

Keywords: Biodiversity, Human impacts, Land use, Nature reserves

1. Introduction

Many mountain environments are experiencing increases in population density and are undergoing rapid intensification of human land uses, such as recreation, resource extraction, agriculture, and housing. One consequence of increasing human impact is the alteration of nature reserves, many of which are in mountain regions. While nature reserves are generally well-protected within their borders (Bruner et al. 2001), evidence is mounting that many reserves are nonetheless losing native biodiversity (Newmark 1987; 1995; 1996; Woodroffe and Ginsberg 1998; Brooks et al. 1999). A contributing factor is the conversion of surrounding habitats for agriculture, logging, settlements, and other human activities (Sala et al. 2000). We suggest that human land use activities outside the boundaries of reserves both affect and are affected by nature reserves, so that the true system boundaries span far outside the designated boundaries (Fig. 1). In this review, we examine the complex interactions among socioeconomic systems, land use, biophysical factors, and biodiversity within and around reserves and point out research needs.

U. M. Huber et al. (eds.), Global Change and Mountain Regions, 563-571.

2. Land use intensification

Land area under human use has increased many-fold over the last few centuries to satisfy needs for food, fiber, and shelter. Cropland is the most extensive type of land cover conversion, though land has also been converted through logging and settlements. Although we do not have precise estimates of land use change over the past centuries, researchers estimate that cropland increased globally from 4.15 to 17.92 million km^2 from 1700 to 1990 (Ramankutty and Foley 1999), currently covering approximately 32% of the earth's land surface (Houghton 1994). Cropland expansion generally followed the history of economic development. In Europe and South and Southeast Asia, land was already extensively cleared for cropland prior to 1700. After 1700, cropland expanded most notably in North America and the former Soviet Union. Since 1850, Latin America, Africa, Australia, and South and Southeast Asia experienced exponential rates of land clearing. Globally, the area of forest and woodland decreased by 19% from 1700 to 1980, while grassland decreased by 8% (Richards 1990). During the last few decades, loss of tropical forests has accelerated (DeFries et al. 2002), while cropland in many temperate parts of the world is currently being abandoned. In addition to habitat loss from land cover conversion, land use change potentially causes increasing human pressures on nature reserves from subsistence hunting, fuelwood collection, extraction of forest products, recreational use, and other human activities.

Some mountainous areas are undergoing particularly high rates of change. For example, the Rocky Mountain West was the fastest-growing region of the United States during the 1990s (US Census). In the US and many other wealthy countries, migration to mountain areas is due to their high level of natural amenities (Rudzitis 1999). Many people are moving to locations of outstanding scenery, outdoor recreation, and access to wilderness (Rasker and Hansen 2000). In the relatively wild Greater Yellowstone Ecosystem, for example, urban areas expanded by 348% from 1975-95, and the number of rural residences increased by more than 400%

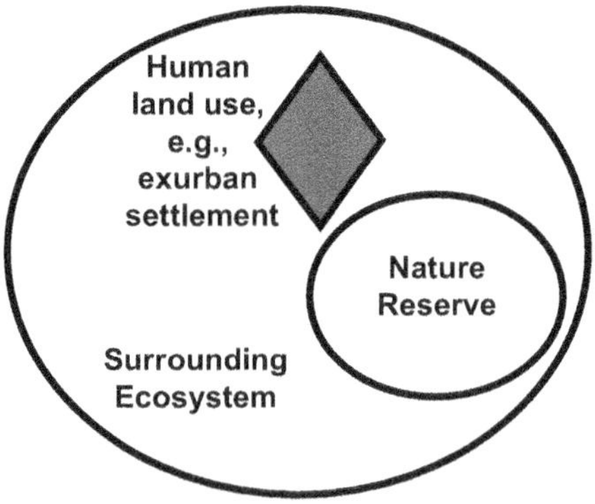

Figure 1: Conceptual diagram of nature reserves as parts of larger ecosystems. Land uses such as exurban development may alter interactions between reserves and the larger ecosystem, and degrade ecological processes and biodiversity within the reserve.

(Hansen et al. 2002). Impacts on ecosystem function and biodiversity are increasing due to expanding outdoor recreation (Knight 1999) and the intensification of rural home (Hansen et al. 2002) and road construction (Forman 2003). This is in contrast to developing countries, where land use intensification in mountain environments is driven by the need to secure food and to access increasingly scarce fuelwood, timber, forage, water, and soil resources (e.g. Newmark 1996; Liu 2001).

3. Land use change and nature reserves

Globally, ecosystems lie along a continuum from those allocated exclusively for human use to those dedicated for the maintenance of nature. National parks and similar nature reserves are afforded the highest level of protection from resource consumption by humans. This approach is founded on the assumption that ecosystems will be self-sustaining if logging, farming, human settlement, and other consumptive land uses are absent or severely restricted (Bruner et al. 2001). Consequently, nature reserves are considered to be critical refuges for native species and ecological processes.

Many nature reserves, however, are failing to meet these expectations. Climate within reserves is being altered by regional land use (Chase et al. 1999; Lawton et al. 2001) and is affected by global climate change. Ecological processes, such as natural disturbance and nutrient cycling are being disrupted (Baker 1992; Pringle 2001). Exotic diseases, weedy plants, and non-native animals are invading nature reserves (Stohlgren 1998). Perhaps most alarmingly, native species have been going extinct in protected areas, even in the time since park establishment (Newmark 1987; 1995; 1996; Woodroffe and Ginsberg 1998; Brooks et al. 1999).

Examples abound of land use around nature reserves altering biodiversity within reserves. Populations of elephant in the Mt Kilamanjaro Region of East Africa, for example, have declined dramatically, possibly due to the conversion of key seasonal habitat to croplands (Coughenour et al. 2000). Logging practices on corporate lands in Indonesia have fueled severe fires that devastated portions of adjacent protected areas (Jepson et al. 2001). In the Greater Yellowstone Ecosystem in the USA, productive low-elevation riparian habitats on private lands outside of the national parks are population source areas for bird species within the mountainous parks (Hansen and Rotella 2002). Rural home development has been rapid near these riparian habitats, favoring exotic predators (e.g. raccoons) that have decreased bird reproduction, converted bird source areas to sink areas and increased the risk of extinction for some bird species in the national parks (e.g. Fig. 2).

The vulnerability of nature reserves to regional land use change is partially influenced by the spatial extent and configuration of the coupled human-ecological system that the reserve is contained within (Hansen and Rotella 2001). Nature reserves that include a range of biophysical conditions and habitats are in general less vulnerable to the effects of land use change. The key ecosystem properties that influence the functioning and biodiversity of reserves have characteristic spatial extents (Allen and Starr 1982). Natural disturbance regimes, nutrient flows, and organism movements, for example, occur across specific portions of regional landscapes (Pringle 2001). If reserves are designated so as to encompass the full spatial domain of the underlying ecosystem, they are likely less vulnerable to changes outside of reserve boundaries. On the other hand, reserves that are spatially incomplete are especially vulnerable to change in the surrounding landscape as human activities alter the disturbance regimes, nutrient pathways, and migratory habitats that are strongly linked to the ecosystem within the reserve. In practice, most reserves do not contain the full spatial extents underlying ecosystem function. The vulnerability to land use change depends then on the location, extent, and type of change relative to key habitats or other ecosystem properties.

Nature reserves in mountain environments are particularly likely to not encompass the broader underlying ecosystem. These environments are characterized by strong altitudinal gradients in climate, soils, and primary productivity. These factors strongly influence ecological processes, such as natural disturbance and the movements and spatial scaling of organisms (Hansen and Rotella 1999). Nature reserves often include only the high-elevation portion of these gradients. This restricted gradient adversely affects the protection of organisms that move across the altitudinal range on a daily or seasonal basis or during different parts of their life history. Organisms often rely on habitats outside of the protected area, where their habitats are increasingly threatened by land use change and intensification (Hansen and Rotella 2002).

Mountain environments are also strongly influenced by the spatial patterning of disturbances. Disturbances tend to be initiated in particular landscape settings and move to other locations in the landscape. Interactions between the location where disturbance gets started (initiation zones) and locations where disturbances move to (run-out zones) influence the functioning of nature reserves (Hansen and Rotella 2001). In southwest

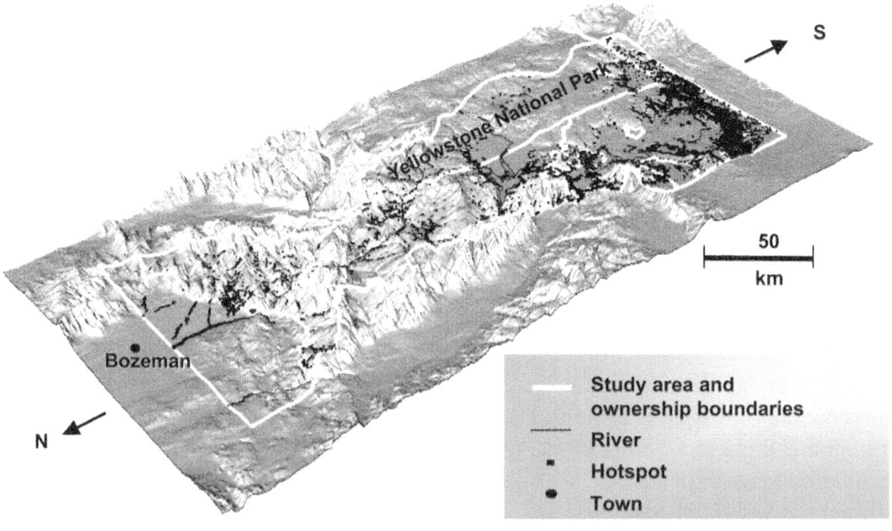

Figure 2: Distribution of bird hotspots (bird species richness and total abundance ≥60% of max.) across the northwest portion of Greater Yellowstone. Intense land use near bird hotspots has reduced bird reproduction and the viability of subpopulations in Yellowstone National Park (from Hansen and Rotella 2002).

Montana, for example, lightning strikes more frequently ignite fires in dry valley-bottom grasslands than in moister conifer forests in the uplands (Arno and Gruell 1983). These fires then spread upslope to the conifer forests. Thus, the juxtaposition of grasslands and conifer forests strongly influences the regional fire regime. Local disturbance regimes can best be maintained in nature reserves that include the disturbance initiation zones within their boundaries (Baker 1992). In the case of the example above, a reserve placed only in the upland conifers may suffer more or less frequent fire depending on the management of the valley-bottom grasslands outside of the reserve.

It is also important to include disturbance run-out zones within reserve boundaries. Run-out zones may contain unique abiotic conditions and habitat patterns important to ecological processes and organisms. For example, flood severity often increases from headwaters to large floodplains. The large scour area and bare gravel bars that form on flood plains support vegetation communities not found in other landscape settings. A reserve that does not contain this disturbance run-out zone will not include these unique riparian vegetation communities.

4. Humans as integral elements of nature reserve systems

Land use around nature reserves is often not random, but related to the presence of the reserve. Reserves are economic engines for tourism, science, and education-based industry in the surrounding region (Rasker and Hansen 2000). In both wealthy and poor countries, reserves have become magnets for migrants settling on their borders to take advantage of the economic and aesthetic opportunities provided by reserves. An attractive nature reserve draws residences and businesses to the adjacent lands and influences local land use and socioeconomic activity (Johnson and Rasker 1995). In less developed countries, impoverished peoples may relocate to reserve boundaries to legally or illegally extract forage, wood, or meat from within reserves, altering vegetation patterns and increasing wildlife mortality (Campbell and Hofer 1995). In the mountainous tropical forests of Borneo, Indonesia, for example, human settlements have formed at the edges of protected areas to allow illegal extraction of valuable tropical hardwoods at the expense of orangutan populations (Curran and Leighton 2000).

These changes in regional demographics and socioeconomics associated with the reserve may generate feedbacks that either favor or inhibit the ecological functioning of the reserve. For example, local business people may support reserve policies to protect the water quality, scenery, and/or wildlife that attract their customers. Alternatively, the feedbacks from the human component of the system may degrade the protected ecosystem. Examples include the negative effects of nature-seeking rural homeowners on biodiversity in Yellowstone National Park (Hansen et al. 2002) or of the high rates of poaching on endangered tigers and other wildlife in the Sikhote-Alin State Biosphere Reserve in the Russian Far East (Miquelle et al. 1999).

The type and magnitude of human influences on reserves may be associated with human population density, wealth, education, culture, law enforcement, political stability, proximity to global markets, and local human benefit from the reserve. Such interactions are poorly understood, however. In a survey of 93 tropical parks, Bruner et al. (2001) found that protection from habitat alteration within the reserves was positively associated with density of guards, deterrence of illegal activity, and direct compensation programs to local communities. However, park effectiveness was not associated with park enforcement capacity, accessibility, budget, or number of staff working on economic development or education. Many integrated development and conservation projects were founded on the notion that local peoples will be more likely to maintain nature reserves if they directly benefit from the reserves. Recent analyses in Africa, however, contradict this assertion (Newmark and Hough 2000). Wealthy and educated peoples may strongly support the legal protection of nature reserves, but strongly degrade native biodiversity in the reserves through land conversion on reserve boundaries and intense backcountry recreation within reserves (Hansen et al. 2002).

5. Research needs

The challenge of reserve management is to understand both the interactions between the ecological and human components of the nature reserve system and the spatial scales at which these operate (Fig. 1). Within reserve boundaries, human activities and infrastructure relating to tourism, law enforcement, resource extraction (legal and illegal), and research may strongly interact with ecological components of the reserve. Hence, reserve management has long focused on restricting human activities within the reserve to minimize negative impacts on ecological processes and organisms. Less recognized are the human and ecological interactions in the surrounding region and their influence on reserve function.

We suggest that interactions between nature reserves and surrounding regions can be better understood by considering both the spatial scaling of the underlying ecosystem components and the feedbacks between the reserve and the surrounding human community. Ecological processes and organisms are distributed in space such that interactions are strongest between specific locations and at certain spatial scales. For example, flows of nutrients are strongly linked among locations within a watershed and linkages are weaker with adjacent watersheds (Pringle 2001). Similarly, the fire regime of a given place is heavily influenced by the properties of the area upwind or downslope that influence the probability of ignition and fire spread. Populations of organisms often migrate over a defined region of the landscape to obtain temporally-shifting resources. Quantifying the spatial domain of key ecosystem processes in and around reserves is the basis for adapting regional management strategies to maintain ecosystem functioning and preserve biodiversity. In recent years, the conservation community has recognized the importance of maintaining corridors for organism dispersal between nature reserves (Noss and Cooperrider 1994). However, the spatial extent of different human-ecological systems encompassing reserves is not well studied.

Another important research issue for nature reserves in mountainous systems is ecosystem response to potential future climate change. Halpin (1997) predicted that

47-77% of biosphere reserves globally will undergo a change in ecoclimatic zone under a doubling of atmospheric CO_2. The magnitude of predicted changes varied among reserves. Those at higher latitudes and higher elevations are more likely to change than those in equatorial locations. The ability of organisms to cope with climate change is also likely to vary with reserve. Organisms have a greater ability to relocate to suitable habitats in reserves that include a wide range of biophysical conditions. Also, reserves that are connected to other reserves by semi-natural habitats, such as along mountain chains, are more likely to exchange organisms with other reserves under climate change. Local analyses are needed to determine how much change a given nature reserve is likely to experience and what management strategies might be used to cope with these changes.

Understanding feedbacks between reserves and local human communities is essential to effective reserve management. In order to develop location-specific management strategies we need to address a range of research questions for different ecosystems and socio-economic conditions. For example, does the designation of a wilderness area or national park lead to increased human population density in surrounding areas as suggested by Rudiz and Johansen (1991)? Do newcomers act to conserve reserve values or exploit them in tourist-based businesses and intense recreation? Does providing park-related jobs to local people decrease poaching in the reserve, or increase poaching because the earnings are used to expand poaching operations as was found by Ferraro and Kramer (1997)? In order to address these questions, reserves should be studied as coupled human-ecological systems with complex interactions, emergent properties, and positive and negative feedbacks.

The spatial extent of the coupled natural and human system around reserves could be quantified based on ecological and socioeconomic flows. Maps of the spatial extent of the ecosystem could be developed based on flows of water and nutrients within watershed boundaries, initiation and run-out zones of natural disturbances, and movements of key organisms. The boundaries of the socioeconomic system could be mapped based on the flow of currency, people, and natural resources between nature reserves and the surrounding region.

The human-natural system encompassing nature reserves can also be better understood through simulation modeling. Effective models would be dynamic, spatially-explicit, multi-scale, would link with land cover observations and would couple land use and biophysical processes. Such models could be developed by integrating elements from three disciplines: land use change modeling (Veldkamp and Lambin 2001); ecological modeling of population dynamics (Dunning et al. 1995); and satellite observations of land cover change (Kaufmann and Seto 2001). Multi-agent systems for simulating the coupled system are especially attractive for this topic because they allow dynamic interactions between the landscape (possibly initialized based on satellite observations) and actors (which could be specified as people, animals, economic units, ecological agents such as fire, or a variety of other agents). Such models could be used to test hypotheses about the human-natural system, identify thresholds and nonlinearities in relationships among state variables, and assess ecosystem feedbacks and vulnerability of reserves in relation to land use changes in the surrounding region.

A goal of such studies should be to help managers conceptualize the regional factors that bear upon nature reserves and better understand regional interactions. Hypothesis testing could provide managers with objective information on the linkages between reserves and regional socioeconomics. Simulation models could be used to evaluate the likely outcomes of current management strategies and strategies designed to sustain both nature reserves and local communities. The results would allow managers to compare likely outcomes of current management strategies to management objectives and to revise these strategies accordingly.

6. References

Allen, T. F. H., and Starr, T. B. (1982). "Hierarchy: Perspectives for ecological complexity." University of Chicago Press, Chicago.

Baker, W. (1992). The landscape ecology of large disturbances in the design and management of nature reserves. *Landscape Ecology* **7**, 181-194.

Brooks, T. M., Pimm, S. L., and Oyugi, J. O. (1999). Time lag between deforestation and bird extinction in tropical forest fragments. *Conservation Biology* **13**, 1140-1150.

Bruner, A. G., Gullison, R. E., Rice, R. E., and da Fonseca, G. A. B. (2001). Effectiveness of parks in protecting tropical biodiversity. *Science* **291**, 125-128.

Campbell, K., and Hofer, H. (1995). People and wildlife: Spatial dynamics and zones of interaction. *In* "Serengeti II. Dynamics." (A. R. E. Sinclair, and P. Arcese, Eds.), pp. 534-570. University of Chicago Press, Chicago.

Chase, T. N., Pielke, R. A., Kittel, T. G. F., Baron, J. S., and Stohlgern, T. J. (1999). Potential impacts on Colorado Rocky Mountain weather due to land use changes on the adjacent Great Plains. *Journal of Geophysical Research* **104**, 16673-16690.

Coughenour, M., Reid, R., and Thornton, P. (2000). "The Savanna Model." Future Harvest, Washington, D.C.

Curran, L. M., and Leighton, M. (2000). Vertebrate responses to spatiotemporal variation in seed production of mast-fruiting Dipterocarpaceae. *Ecological Monographs* **70**, 101-127.

DeFries, R., Houghton, R., Hansen, M., Field, C., Skole, S., and Townshend, J. (2002). Carbon emissions from tropical deforestation and regrowth based on satellite observations for the 1980s and 1990s. *In* "Proceedings of the National Academy of Sciences 99." pp. 14256-14261.

Dunning, J. B., Stewart, D. J., Danielson, B. J., Noon, B. R., Root, T. L., Lamberson, R. H., and Stevens, E. E. (1995). Spatially explicit population models: Current forms and future uses. *Ecological Applications* **5**, 3-11.

Ferraro, P., and Kramer, P. (1997). Projets integres de conservation et de development. (P. J. Ferraro, R. Tshombe, R. Mwinyihali, and J. A. Hart, Eds.), pp. ??-??. Wildlife Conservation Society Working Paper No. 6, New York.

Forman, R. T. T. (2003). "Road ecology: Science and solutions." Island Press, Washington, D.C.

Green, G. M., and Sussman, R. W. (1990). Deforestation history of the eastern rain forests of Madagascar from satellite images. *Science* **248**, 212-215.

Halpin, P. N. (1997). Global climate change and natural-area protection: Management responses and research directions. *Ecological Applications* **7**, 828-843.

Hansen, A. J., and Rotella, J. J. (2001). Nature reserves and land use: Implications of the "Place" Principle. *In* "Applying ecological principles to land management." (V. Dale, and R. Haeuber, Eds.), pp. 57-75. Springer, New York.

Hansen, A. J., and Rotella, J. J. (2002). Biophysical factors, land use, and species viability in and around nature reserves. *Conservation Biology* **16**, 1-12.

Hansen, A. J., Rasker, R., Maxwell, B., Rotella, J. J., Wright, A., Langner, U., Cohen, W., Lawrence, R., and Johnson, J. (2002). Ecology and socioeconomics in the New West: A case study from Greater Yellowstone. *BioScience* **52**, 151-168.

Houghton, R. A. (1994). The worldwide extent of land-use change. *BioScience* **44**, 305-313.

Jepson, P., Jarvie, J. I., MacKinon, K., and Monk, K. A. (2001). The end for Indonesia's lowland forests?

Science **292**, 859-861.
Johnson, J. D., and Rasker, R. (1995). The role of economic and quality of life values in rural business location. *Journal of Rural Studies* **11**, 405-416.
Kaufmann, R., and Seto, K. (2001). Change detection, accuracy, and bias in a sequential analysis of Landsat imagery in the Pearl River Delta, China: Econometric techniques. *Agriculture, Ecosystems, and Environment* **85**, 95-105.
Knight, R. L. (1999). Private lands: The neglected geography. *Conservation Biology* **13**, 223-224.
Lawton, R. O., Nair, R. S., Pielke, R. A. S., and Welch, R. M. (2001). Climatic impacts of tropical lowland deforestation on nearby montane cloud forests. *Science* **294**, 584-587.
Liu, J., Linderman, M., Ouyang, Z., An, L., Yang, J., and Zhang, H. (2001). Ecological degradation in protected areas: The case of Wolong Nature Reserve for giant pandas. *Science* **292**, 98-101.
Miquelle, D. G., et al.??? (1999). A habitat protection plan for the Amur tiger: Developing political and ecological criteria for a viable land–use plan. *In* "Riding the tiger: Tiger conservation in human-dominated landscapes." (J. Seidensticker, S. Christie, and P. Jackson, Eds.), pp. 273-295. Cambridge University Press, Cambridge.
Newmark, W. D. (1987). A land-bridge island perspective on mammalian extinctions in western North America parks. *Nature* **325**, 430-432.
Newmark, W. D. (1995). Extinction of mammal populations in western North America parks. *Conservation Biology* **9**, 512-526.
Newmark, W. D. (1996). Insularization of Tanzanian parks and the local extinction of large mammals. *Conservation Biology* **10**, 1549-1556.
Newmark, W. D., and Hough, J. L. (2000). Conserving wildlife in Africa: Integrated conservation and development projects and beyond. *BioScience* **50**, 585-592.
Noss, R. F., and Cooperrider, A. Y. (1994). "Savings nature's legacy: Protecting and restoring biodiversity." Island Press, Washington, D.C..
Pringle, C. (2001). Hydrologic connectivity and the management of biological reserves: A global perspective. *Ecological Applications* **11**, 981-998.
Ramankutty, N., and Foley, J. (1999). Estimating historical changes in global land cover: Croplands from 1700 to 1992. *Global Biogeochemical Cycles* **13**, 997-1027.
Rasker, R., and Hansen, A. J. (2000). Natural amenities and population growth in the Greater Yellowstone region. *Human Ecology Review* **7**, 30-40.
Richards, J. F. (1990). Land transformation. *In* "The Earth as transformed by human action." (B. L. Turner, W. C. Clark, R. W. Kates, J. F. Richards, J. T. Mathews, and W. B. Meyer, Eds.), pp. 163-178. Cambridge University Press, Cambridge.
Rudiz, G., and Johansen, H. E. (1991). How important is wilderness? Results from a United States survey. *Environmental Management* **15**, 227-233.
Rudzitis, G. (1999). Amenities increasingly draw people to the rural West. *Rural Development Perspectives* **14**, 9-13.
Sala, O. E., Chapin, F. S. I., Armesto, J. J., Berlow, E., Bloomfield, J., Dirzo, R., Huber-Sanwald, E., Huenneke, L. F., Jackson, R. B., Kinzig, A., Leemans, R., Lodge, D. M., Mooney, H. A., Oesterheld, M., Poff, N. L., Sykes, M. T., Walker, B. H., Walker, M., and Wall, D. H. (2000). Global biodiversity scenarios for the year 2100. *Science* **287**, 1770-1774.
Serneels, S., and Lambin, E. (2001). Impact of land-use changes on the wildebeest migration in the northern part of the Serengeti-Mara Ecosystem. *Journal of Bigeography* **28**, 391-408.
Stohlgren, T. J. (1998). Regional trends of biological resources-Rocky Mountains. *In* "Status and trends of the Nation's biological resources, Vol. 2." (M. J. Mac, P. A. Opler, C. E. Puckett, Haecker, and P. D. Doran, Eds.). Department of the Interior, U.S. Geological Survey, Reston, VA.
Veldkamp, A., and Lambin, E. F. (2001). Predicting land-use change. *Agriculture, Ecosystems, and Environment* **85**, 1-6.
Woodroffe, R., and Ginsberg, J. R. (1998). Edge effects and the extinction of populations inside protected areas. *Science* **280**, 2126-2128.

Nature Conservation Value of European Mountain Farming Systems

David I. McCracken* and Sally Huband
Research Division, Scottish Agricultural College, Auchincruive, Ayr KA6 5HW, United Kingdom
**phone +44-1292-525299, fax +44-1292-525333, e-mail d.mccracken@au.sac.ac.uk*

Keywords: Biodiversity, Europe, High nature value, Mountain farming systems, Pastoralism.

1. Wider context

1.1 High nature value farming systems

High nature value (HNV) farming areas are regarded as farmland where there are intimate relationships between farming practices and biodiversity and where the continuation of those farming practices is essential for the maintenance of this biodiversity value (e.g. Bignal 1998; Luick 1998; Ostermann 1998; Webb 1998; Zervas 1998). By the mid 1990s, there was a growing recognition that particular farming systems (many of them in mountainous areas) were important in maintaining nature conservation value over much of the wider European countryside, but it was also recognised that there was little information available on the range of such systems being practised across Europe. To redress some of this imbalance, the UK Joint Nature Conservation Committee and the World Wildlife Fund (WWF) funded a pilot study of nine European countries: Greece, France, Hungary, Ireland, Italy, Poland, Portugal, Spain and the United Kingdom (Beaufoy et al. 1994; Bignal et al. 1994b; Bignal and McCracken 1996a,b; 2000).

The three main aims of the study were to establish the broad characteristics of HNV farming systems within each country, to obtain an indication of where these systems were still being practised and to highlight some of the pressures they were

U. M. Huber et al. (eds.), Global Change and Mountain Regions, 573-582.

facing. Differences in data availability dictated that different approaches were taken across all nine countries, but in general this involved drawing information from a combination of sources, such as maps of specific agricultural land use or the extent of agriculturally Less Favoured Areas, agricultural statistics, the location of semi-natural habitats and/or known agricultural or ecological characteristics of specific systems.

1.1.1 Extent and distribution

The approximate distribution of low-intensity farming systems across the nine countries is indicated in Figure 1, while an estimate of the amount of such farmland in each country is provided in Table 1. HNV farmland is associated with low-intensity agricultural systems and mostly survives in upland and remote areas where there are considerable physical constraints on the development and modernisation (especially mechanisation) of agricultural practices. The southern European countries included in the study have the greatest variety and area of land under HNV farming systems, with Spain, Portugal and Greece in particular all having over 60% of their utilised agricultural area under such systems. A preliminary estimate across the nine study countries showed that there were more than 55 million hectares of land under these systems, 30 million hectares of which were associated with livestock systems alone. Spain and Portugal together contained approximately half of the HNV farmland resource occurring across these nine countries.

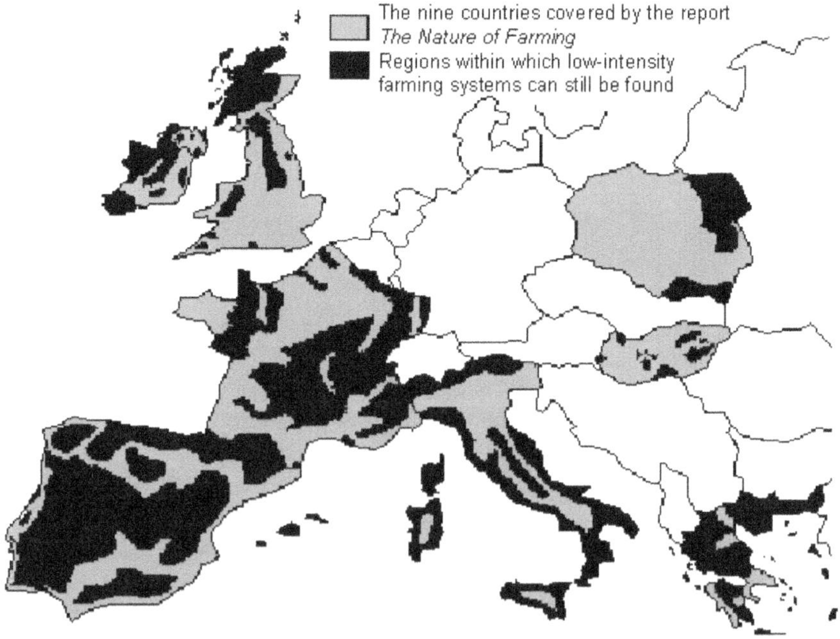

Figure 1: Principal areas within which low-intensity farming systems can be found in each of the nine European countries considered by Beaufoy et al. (1994) and reported by Bignal et al. (1994b).

Table 1: Estimated area of high nature conservation value (HNV) farmland in each of the nine European countries studied by Beaufoy et al. (1994) and reported by Bignal and McCracken (1996b; 2000).

	Land surface (ha)	Utilised agricultural area (UAA ha)	Estimated area of HNV farmland (ha)	HNV farmland as % of UAA
France	54,702,600	31,016,400	7,754,000	25
Greece	13,194,400	9,183,300	5,600,000	61
Hungary	9,303,000	6,493,500	1,500,000	23
Italy	30,122,500	22,650,000	7,100,000	31
Poland	31,267,700	19,135,800	2,735,000	14
Portugal	9,208,200	4,558,100	2,735,000	60
Republic of Ireland	7,028,300	5,650,750	2,000,000	35
Spain	50,478,200	30,589,800	25,000,000	82
United Kingdom	24,413,900	18,425,500	2,000,000	11
Total	**229,709,500**	**147,703,150**	**56,424,000**	**38**

1.1.2 Agricultural characteristics

Although the study identified a wide variety of HNV systems still being practised across these nine countries, it was possible to group these under four broad headers:

- *Livestock systems*: Low-intensity livestock systems are now mainly practised in the upland and mountainous regions of Europe and in the arid zones in the south. Such systems are very variable and range from semi-wild and largely unmanaged cattle and horses in remote regions of Spain, to dairy farms producing specialist cheeses and incorporating closely managed hay meadows in the French Jura. Low-intensity sheep systems are the most widespread livestock type covering large areas of upland, mountain or dry pasture of grassland and scrub. *Transhumance* and other seasonal movements of livestock between grazing areas are also an important characteristic of these systems in many areas of southern and eastern Europe. Across the nine countries, HNV livestock systems occur over large areas of the uplands and mountains in Scotland and northern England, western and central Ireland, southern and eastern France, northern Spain, central and northern Portugal and northern Italy. Smaller areas are found in the mountains of western Hungary and south-east Poland. Central to all these systems is a long continuity of sustainable use of large areas of grasslands, heaths and woodlands (many of which are in communal use). Virtually all the remaining high nature conservation value grasslands across Europe are associated with low-intensity livestock systems.
- *Arable systems*: Once common across all of Europe, low-intensity arable systems are now mainly confined to the Mediterranean regions, where the dryland (non-irrigated) systems in Spain, Portugal, southern Italy and Greece are particularly significant. These systems are low yielding and use fallowing (in association with grazing) to maintain soil fertility. Traditionally, 30% to 80% of this land is left uncultivated or fallow each year, with some fallows lasting for five years or more. This management creates and maintains a *pseudo-steppe* landscape of great importance for nature conservation. Although the vast majority of such systems

do not occur in mountain areas, in many instances the stubbles and fallows are closely linked with the *transhumance* systems, as they provide important foraging areas for the livestock on their movements between their winter quarters and the summer mountain pastures.

- *Permanent crop systems*: Permanent crops (such as olives, fruit and vines) are an important component of the Mediterranean lands. Much of this cultivation has been intensified in recent years and the surviving low-intensity systems are generally in the poorer areas where farming is less specialised. The majority of HNV permanent crop systems are associated with fruit orchards or olive groves where there is a low input of chemicals and where harvesting often occurs by hand. Many of these orchards and olive groves are also regularly grazed and/or used for hay production. Few of these crop systems occur in mountain areas, but they can provide an important forage resource for livestock on *transhumance* to and from mountain areas.
- *Mixed systems*: There are still areas of Europe where truly small-scale mixed systems, using little or no inorganic fertiliser or pesticides, survive. Some are virtually subsistence farming and most are in remote (often mountainous) areas where farming is often combined with other occupations. There are many examples throughout Europe and most are highly localised, such as *crofting* in northern Scotland, the *tanya* system in central Hungary, *coltura promiscua* in Italy, and *minifundia* in central and northern Portugal.

1.1.3 HNV livestock systems in mountain areas

The descriptions above are very broad, but they do serve to provide an indication of the agricultural characteristics and, to a lesser extent, the nature conservation value of HNV farming systems. Although it is recognised that further development of the approach is necessary to establish the full extent of HNV systems across Europe, it is clear that mountainous areas hold a high proportion of the remaining HNV systems and that their survival is often dependant on farming practices which form a link between mountain and lowland areas. The majority of HNV mountain livestock systems have been in existence for hundreds and, in some cases, thousands of years. The annual farming cycle practised in these systems has created and maintained ecosystems of high biodiversity, within which many plants and animals can find the resources that they require for some or all of the stages of their annual lifecycle. Indeed many mountain landscapes, habitats and wildlife communities, upon which we now place high nature value, are dependant on the continuation of low-intensity livestock farming.

One example of such a HNV mountain livestock system is the Greek transhumance system. The most common agricultural land use in Greece is low-intensity rearing of sheep and goats for meat and milk, with the livestock often herded in large mixed flocks (Bignal and McCracken 1996a). This system covers much of the mainland and is especially significant in maintaining the nature conservation value of mountainous areas. Many of the mountain pastures are of great botanical interest, and high grazing pressure for a short period each year is essential to prevent scrub encroachment and maintain floristic diversity. In addition, grazing livestock ensure that open alpineareas are replenished with the carrion that is essential for foraging by scavengers (such as Egyptian vulture, *Neophron percnopterus*, and griffon vulture, *Gyps fulvus*) and predators (such as golden eagle, *Aquila chrysaetos*, and Bonelli's eagle, *Hieraaetus pennatus*).

1.2 Ecological characteristics

Within any ecosystem, biodiversity is generally greater within areas that (a) contain a wide range of niches (e.g. different habitats, different vegetation structures), (b) are subject to medium levels of disturbance (e.g. through climatic or management factors) and (c) occur at a large enough scale to allow enough individuals to survive and maintain viable populations and to provide a variety of environmental conditions at any one time to allow sufficient choice of suitable habitats. From what is known of HNV mountain farmland areas, it is clear that these are of high nature conservation value because:

- They continue to utilise and maintain a high proportion of semi-natural vegetation. This may be largely by default in that climatic and topographic constraints limit the intensification of vegetation management. However, the outcome is a greater range of ecological niches over much of the area utilised within the pastoral system.
- These climatic and topographic constraints also generally mean that not all of the land in an area is available for utilisation by all the different land use components of the system (e.g. grazing by domestic animals, growth of fodder crops). Hence, more intensively managed semi-natural vegetation is generally found within a mix of more natural habitats.
- The constraints imposed on the vegetation by climate and topography control not only the type but, just as importantly, the timing of the management that is applied to the vegetation. Hence, the annual management is generally synchronised with the annual natural growth cycle of the vegetation and so is not imposed at a time when it would be detrimental to a wide range of the plant species involved.
- For most of the year, the nutritional value of such semi-natural vegetation is generally low, which places limits on the number of animals and the duration of grazing intervals in a given area. It also leads to a need for larger areas to be utilised. Hence, grazing pressure on any one area is generally either low or (in closely shepherded flocks) only high for a very short period, which leads to a greater heterogeneity of vegetation structures.
- The associated need to produce fodder to carry livestock through the winter and the constraints on the amount of fodder which can be grown mean that (a) there is a limit to the total number of animals that can be supported and (b) there is a need to move animals to other areas during the period of growth and harvesting of winter fodder in the summer. Both these factors markedly reduce grazing pressure on any one area of land over the course of the year. In addition, not only do the

fodder crops introduce further heterogeneity into the landscape, but many of these are also of extremely high nature conservation value in their own right.

- Grazing by livestock introduces small-scale perturbations into the vegetation, resulting in an increase in biodiversity. Hence, herd behaviour can introduce seasonal and cyclic pressures, which are virtually impossible to produce in any other way - not only through their grazing but also through their trampling, dunging, resting and ruminating in favoured places, and selecting foraging areas in relation to the seasonal availability of herbage.
- The habitats of many wildlife species are naturally unstable and it is common for populations to disappear from one area and for new ones to appear when a suitable niche becomes available elsewhere. These pastoral systems and associated farming practices are maintained at a scale and intensity which ensures sufficient area of potentially suitable habitat and allows scope for these cycles of colonisation and re-colonisation to take place.
- By the same token, these systems are much more favourable to a wider range of wildlife species (especially the larger vertebrates) because they are practised over a wider scale and therefore (a) the conditions required at any one time of year (especially by more mobile species) can be found at a wide variety of locations and (b) the different requirements by these species at different times of year are catered for, i.e. through changes in the mix of structures and habitats in any one area through the year.

The biological importance of HNV mountain systems therefore relates both to the spatial and temporal diversity that they introduce. In a spatial context, they produce a patchwork of habitats - meadows, grass pastures, crops, woodland, fallows, natural pastures (including alpine grassland, heath, moorland, marshland, bog, wood-pasture) as well as more intensively managed land around settlements and farmsteads. In a temporal context, not all land is managed in the same way at the same time; so neighbouring farms with essentially the same production systems may sow and harvest crops at different times. This produces a patchwork of the same crop at different stages of development. In a similar fashion, adjacent pasture under different ownership will be grazed in different ways (e.g. with different animals and at different stock densities) and at different times of the year. This diversity provides much more favourable conditions for plants and animals (especially invertebrates) to find areas with suitable conditions for the completion of their lifecycles (Bignal and McCracken 2000).

1.3 A vision for rural Europe

Details of many HNV farming systems and the threats facing these have been provided in many of the references cited above and in the proceedings of conferences held by the European Forum on Nature Conservation and Pastoralism (e.g. Curtis et al. 1991; Curtis and Bignal 1991; Bignal and McCracken 1992; Bignal et al. 1994a,b; McCracken et al. 1995; Poole et al. 1998; Pienkowski and Jones 1999). Many HNV mountain farming systems and their associated biodiversity value are under increasing threat (either from intensification of farming practices or from the abandonment of farming altogether). For example:

- HNV mountain systems depend on breeds of livestock that are well adapted to natural conditions and to practices such as *transhumance*. For example, Avileña cattle in central Spain can walk 20-40 km per day on the journey to their summer mountain pastures. However, the intensification of European agriculture and the drive for increased productivity has lead to the development of modern breeds that produce a lot of milk and meat at the expense of losing the characteristics that allowed traditional breeds to adapt to harsh environmental conditions. One result of this development has been the abandonment of remote pastures in many mountain areas and a loss of the biodiversity that depends on grazing impacts.
- HNV mountain systems are increasingly becoming economically and socially unviable across Europe. Farming populations in these areas are ageing. The farming systems necessary for the maintenance of the high nature value offer poor incomes and little in the way of social infrastructure to attract young people. In addition, the market for products from mountain systems is limited since, although high quality food products from HNV mountain farming systems are often highly appreciated, in practice the vast majority of consumers usually choose cheaper, intensively produced food.
- Current concerns may also mean that some production processes may need to be changed to meet hygiene regulations. However, for many small farms, the relatively low value of agricultural products does often not justify the required investments.

Taken together, these different factors put a severe threat to the continuation of these HNV farming systems. Future policies will therefore need to shift financial support structures away from intensive agricultural production towards broader socio-economic objectives, in which the maintenance of low-input, biologically diverse systems and their rural communities is paramount. At the same time, there needs to be developed a clear distinction between farming systems that are of relatively high biodiversity and those which are not.

Tubbs (1996; 1997) and the European Forum on Nature Conservation and Pastoralism have outlined such a vision for the European countryside as:

> *... one in which the greater proportion of the land surface will comprise a diversity of low-intensity farm systems together with large tracts of unenclosed land sustaining extensive pastoralism - mountain, moorland, heath, wood-pasture, uncultivated or intermittently cultivated steppe and dry grassland, and saltmarsh. This is not a vision which denies intensive production, but this may become confined to regions of naturally high productivity, in which increasingly tight management of pesticides and fertilisers is combined with technological research and development into management methods (e.g. fertiliser and machinery) which will limit their necessity. There are important places in the vision for special conservation management of places of national or international importance to nature conservation. Equally, there is a place*

for innovation involving the recreation of wilderness, but not from the extensive pastoral lands forming integral parts of low-intensity farmland.

2. Priorities for action

The majority of HNV livestock systems are now confined to mountainous areas of Europe and are a policy priority because they are fundamental to the maintenance of a wide range of habitats and species of European conservation importance. However, before appropriate policies can be put in place, it is essential to understand how such systems function, their agricultural and ecological interactions and current policy driving forces. To this end, the European Commission funded a concerted action (*PASTORAL: the agricultural, ecological and socio-economic importance of free-ranging livestock systems in Europe*) to bring together individuals from throughout Europe with a range of backgrounds and disciplines to consolidate present knowledge, identify the gaps which need to be filled and build up a network of researchers and contacts with the complementary expertise to fill those gaps.

All four meetings held to-date (in Spain, Romania, Scotland and France) each integrated field-visits with the formal discussion sessions to provide the delegates with opportunities to speak directly to shepherds, landowners and local experts. In this way, the delegates obtained a first-hand impression of what the management practices consist of, how these link to produce the environmental conditions of biodiversity value and what the key issues are affecting the continuing viability of the enterprises. These discussions also proved useful in helping to highlight some potential changes to policy, which could help improve the viability of HNV mountain farming systems.

From the discussions, it became clear that one of the major threats to HNV mountain farming systems is that there is a widespread misconception (not only among politicians, farmers and consumers but also among many ecologists and conservationists) that these farming systems are not necessary from a biodiversity perspective. There is therefore an urgent need to raise awareness among these groups of the importance of these systems, especially with regard to the linkages between the agricultural practices and the biodiversity and landscapes these produce and maintain. In this respect, it is also essential to ensure that the scale and extent of these systems still existing in the mountains across Europe is appreciated fully.

It was also evident that there is a lack of detailed knowledge of how many of these systems function, especially with regard to the intimate relationships between the timing and type of farming practices and the biodiversity value. Characterisation of these systems must therefore be given immediate priority if there is to be any chance of developing appropriate policies for their survival. To obtain the necessary understanding, it will be essential that information is collected and summarised under at least three broad headings:

1. *Production System:* How does the enterprise operate (e.g. how much land is used) and what sort of annual farming cycle is practised (e.g. when and where are the livestock located at different times of year)? What breeds are involved, what are the main products (e.g. cheese, animals for slaughter)? What are the sources

of food for the livestock at different times of year? How complex are the land ownership/rental issues? How is the division of labour split between shepherds and owners of livestock?

2. *Viability of the System:* What are the main costs (e.g. rental of land, land guarantees, labour, winter feed) and who meets these? What are the main sources of income (e.g. sale of different products, fees paid for different services) and who benefits from these (e.g. shepherds, owners of livestock, owners of pastures)? How does the income received by the shepherds and owners differ in scale? Where do the main constraints lie for the system/enterprise (e.g. financial or social factors)? What are the implications of current EU support measures for the survival of these systems?
3. *Biodiversity Value of the System:* What is the biodiversity value of each of the different components of the system? How is biodiversity linked to the different elements of the farming practices? Who is responsible for the production and maintenance of the different elements of biodiversity value? How will this biodiversity value be affected by likely changes to farming practices? What are the implications for the survival of biodiversity value under current EU support measures?

Information is therefore needed on how HNV mountain farming systems function in terms of ecology or economics. In addition, to be sustainable in the long term, an understanding is required of the socio-economic and cultural aspects associated with the different farming practices. Socio-economic factors, such as the viability of a farming operation or the willingness of farmers to continue certain practices, will often be the main driving force behind whether the long-established management practices associated with the HNV mountain farming systems will be maintained or not. Only through such a detailed focus on each farming system will it be possible to (a) really understand how and why each system is of biodiversity importance, (b) fully understand the links between all the drivers affecting each system and (c) move towards the position where we begin to understand the full implications of policy changes across Europe or how policy should be developed to reflect the actual environmental needs at the local and regional level. However, current pressures on these systems are high. Action is therefore required quickly to ensure the continued survival of Europe's HNV mountain farming systems together with the valued landscapes and biodiversity riches that they maintain.

3. Acknowledgements

This paper has been prepared with the financial assistance from the Commission of the European Communities RTD programme Quality of Life and Management of Living Resources under project reference QLRT-2000-00559. The content does not necessarily reflect the views of the Commission and in no way anticipates future Commission Policy in this area. The PASTORAL project was co-ordinated by the Scottish Agricultural College (SAC) on behalf of a partnership of SAC, The European Forum on Nature Conservation and Pastoralism, ALTERRA, Institute for European

Environmental Policy, Asociacion para el Analasis y Reforma de la Politica Agro-rural, Universidad de Autónoma de Madrid, Escola Superior Agraria de Castelo Branco and Coordination Paysanne Européenne. Further information on the project is available on the website at: http://www.sac.ac.uk/envsci/external/Pastoral/default.htm.

4. References

Beaufoy, G., Baldock, D., and Clark, J. (1994). "The nature of farming: Low-intensity farming systems in nine European countries." Institute for European Environmental Policy, London.

Bignal, E. M. (1998). Conservation of biodiversity by supporting high-nature-value farming systems. *Journal of Applied Ecology* **35**, 949-954.

Bignal, E. M., and McCracken, D. I. (1992). "Prospects for nature conservation in European pastoral farming systems". Joint Nature Conservation Committee, Peterborough.

Bignal, E. M., and McCracken, D. I. (1996a). The ecological resources of European farmland. *In* "The European environment and CAP reform: Policies and proposals for conservation." (M. Whitby, Ed.), pp.26-42. Centre for Agriculture and Biosciences International, Wallingford.

Bignal, E. M., and McCracken, D. I. (1996b). Low-intensity farming systems in the conservation of the countryside. *Journal of Applied Ecology* **33**, 413-424.

Bignal, E. M., and McCracken, D. I. (2000). The nature conservation value of European traditional farming systems. *Environmental Reviews* **8**, 149-171.

Bignal. E. M., McCracken, D. I., and Curtis, D. J., Eds. (1994a). "Nature conservation and pastoralism in Europe." Joint Nature Conservation Committee, Peterborough.

Bignal, E. M., McCracken, D. I., Pienkowski, M. W., and Branson, A. (1994b). "The nature of farming: Traditional low-intensity farming and its importance for wildlife." World Wide Fund for Nature, Brussels.

Curtis, D. J., and Bignal, E. M. (1991). "The conservation role of pastoral agriculture in Europe." Scottish Chough Study Group, Argyll.

Curtis, D. J., Bignal, E. M., and Curtis, M. A., Eds. (1991). "Birds and pastoral agriculture in Europe." Scottish Chough Study Group, Argyll and Joint Nature Conservation Committee, Peterborough.

Luick, R. (1998). Ecological and socio-economic implications for livestock-keeping systems on extensive grasslands in south-western Germany. *Journal of Applied Ecology* **35**, 979-982.

McCracken, D. I., Bignal, E. M., and Wenlock, S. E., Eds. (1995). "Farming on the edge: The nature of traditional farmland in Europe." Joint Nature Conservation Committee, Peterborough.

Ostermann, O. P. (1998). The need for management of nature conservation sites designated under Natura 2000. *Journal of Applied Ecology* **35**, 968-973.

Pienkowski, M. W., and Jones, D. G. L., Eds. (1999). "Managing high-nature-conservation-value farmland: Policies, processes and practices." European Forum on Nature Conservation and Pastoralism, Islay.

Poole, A., Petretti, F., Pienkowski, M. W., McCracken, D. I., Petretti, F., Bredy, C., and Deffeyes, C., Eds. (1998). "Mountain livestock farming and EU policy development." European Forum on Nature Conservation and Pastoralism, Islay.

Tubbs, C. (1996). Wilderness or cultural landscapes: Conflicting conservation philosophies. *British Wildlife* **7**, 290-296.

Tubbs, C. (1997). A vision for rural Europe. *British Wildlife* **9**, 79-85.

Webb, N. R. (1998). The traditional management of European heathlands. *Journal of Applied Ecology* **35**, 987-990.

Zervas, G. (1998). Quantifying and optimising grazing regimes in Greek mountain systems. *Journal of Applied Ecology* **35**, 983-986.

Economic Globalisation and its Repercussions for Fragile Mountains and Communities in the Himalayas

N. S. Jodha
Agricultural and Rural Income Diversification Programme, International Centre for Integrated Mountain Development (ICIMOD), P.O. Box 3226, Kathmandu, Nepal
phone 977-1-5525313, fax 977-1-5524509, email njodha@icimod.org.np

Keywords: Globalisation, HKH region, Mountain specificities, Potential opportunities, Risks-vulnerabilities.

1. Introduction

This essay deals with the repercussions of rapid economic globalisation for mountain environments and communities in the Hindu Kush-Himalayan (HKH) region. The subject, despite its importance, has not received systematic attention beyond the protests and debates by Non-Governmental Organisations (NGOs) and others (Roy 1997). We present the information and understanding generated by a recently concluded exploratory study on the subject supported by the MacArthur Foundation (Jodha 2002). After discussing the key features of the rapid economic globalisation in the HKH region, we examine how mountain-specific conditions (mountain specificities), such as fragility, inaccessibility, diversity, and marginality, interact with the globalisation process. We identify the risks and opportunities created by globalisation for mountain areas and communities. The prognosis that derives from our discussion is supported by emerging evidence from selected mountain areas of China, India, Nepal, Pakistan and Bangladesh (Jodha 2002).

U. M. Huber et al. (eds.), Global Change and Mountain Regions, 583-591.

2. Economic globalisation

Broadly speaking, economic globalisation is a market-driven process promoted to guide and integrate economic transactions between different countries on a global scale. Globalisation changes the patterns of international economic links and associated risks and opportunities for the participating countries and communities. Although the central element of globalisation is the key role accorded to the global market in economic transactions, what differentiates it from the historically evolving international economic relations is as follows:

a. The reinforcing role of the interconnectedness of economic transactions. Examples are: Trading of several mountain products, ranging from apples to coffee, under open general license (OGL), in India. This encourages the import of foreign products (from New Zealand, Australia, South Korea etc.), pushing local products out of the market; the governments' efforts to increasingly attract foreign direct investment (FDI), causing transfer of community natural resources (discarding customary rights etc.) to investors without substantial benefits for local economies.
b. The facilitative and speed-promoting integrative role of information technology, which reduces part of the constraints created by limited physical accessibility.
c. The power accorded to formal institutional arrangements on a global scale, such as the World Trade Organisation (WTO), which promotes global perspectives at the cost of local concerns.

Globalisation operates through different provisions of trade policies (e.g. no import duties and no export subsidies) and the de-regulation and liberalisation of economic policies and processes, designed to facilitate an internationally unrestricted flow of resources, products and services as guided by the global market forces (SAWTEE 2003). However, the present discussion is not directed towards the whole range of provisions and practices of globalisation. Instead, we focus on those that are of particular relevance to mountain areas. Specifically, we address how the changes promoted by increasingly internationally linked market processes and the marginalisation of the economic role of the state and local communities, manifest themselves in mountain areas and affect mountain people.

An emergent finding is that, despite the fact that globalisation is promoted as a means to global growth and prosperity; it also carries risks for the participants. The participants unprepared for such change are likely to encounter more risks and limited gains in the process. Mountain regions like the Himalayas and their communities are particularly vulnerable, both due to their specific bio-physical circumstances (fragile mountain environments) as well as historical processes (e.g. permanent under-investment and negative socio-economic effects of external interventions).

3. Mountain context and globalisation: A generalized picture

In mountain areas, bio-physical factors strongly shape socio-economic processes. In particular, mountain areas are characterized by a high degree of fragility, marginality, limited accessibility, diversity and specific niche resources as well as human adaptations to these conditions (Table 1). In this discussion, we refer to these inter-linked conditions generating opportunities and constraints as mountain specificities. The causes, manifestations, and implications of mountain specificities are discussed elsewhere (Jodha 1997). Mountain specificities vary between different mountain regions and are linked in several ways. Furthermore, they jointly (or individually) shape the pace and pattern of change in mountain areas and also determine the relevance and effectiveness of any interventions, including those associated with globalisation (Kreutzmann 1995).

Mountain specificities not only determine the mix of risks and opportunities due to globalisation, but strongly affect the capacities of mountain areas and communities to wisely handle them. Table 1 provides a summary of socioeconomic conditions that facilitate gains from globalisation and shows whether or not mountain specificities are conducive to these conditions. This table provides a broad framework to understand and address globalisation issues in the HKH region, and to help mountain communities and decision makers in developing strategies and options to benefit from globalisation. A conclusion from the overview provided in Table 1 is that mountain areas, due to their bio-physical constraints and limited infrastructure, are not likely to readily satisfy the conditions conducive to enhanced economic performance under globalisation. For example, the required resource use intensification or the ability to profitably utilize increased economic input is obstructed by fragility and marginality as well as limited accessibility. For instance, under agricultural development programmes, mountain communities (like other rural communities) in Himalayan countries are offered subsidised chemical fertilizers, seeds etc to increase productivity. However, marginal soils do not sufficiently respond to these inputs leading to low production-efficiency. Even where soils are more productive and responsive to new inputs the costs of acquiring and transporting inputs and for profitably marketing agricultural products are extremely high due to accessibility problems. Some of the mountain valleys with better external links are an exception to the above.

Limited accessibility also obstructs effective and equitable external links essential for enhanced economic performance through trade as well as for effective transfer of successful development strategies from other regions. Fragility not only obstructs the development of adequate infrastructure but also adds to the cost of mobility for improved economic performance. Because of the steep slopes and the high risk of natural hazards, it is difficult to construct stable roads in many mountain areas. Most of the roads are subject to landslides during the rainy season, obstructing transportation and communication for weeks or months. The construction and maintenance of mountain roads is often much more costly than in plains. Frequent earthquakes in the Himalayas further exacerbate the risk of natural hazards. The overall economic consequences are high costs for transportation and mobility.

Diversity and marginality obstruct advantages of large-scale surplus generation for reinvestment. For example, large-scale industrial agriculture is poorly suited to marginalized mountain regions that are characterized by high ecological and socio-economic diversity. Hence, due to their bio-physical constraints and limited infrastructure, mountain regions are unequipped to readily adapt to globalisation and are likely to face greater risks and limited gains, at least in the short run (Jodha 2000a; Karki 2000).

4. Emerging sources of risk

Despite obstructions and apparent limits to gains (Table 1), globalisation-related changes are manifested in the HKH region. These changes, when viewed in the context of (i) driving forces and operational mechanisms of globalisation; (ii) prevailing resource use systems; and (iii) external links (markets, technologies) suggest a range of risks and potential opportunities for mountain areas. A brief summary of exemplary issues in the HKH provide the context for understanding the impacts of globalisation on mountain environments and communities (Kreutzmann 1995; Jodha 2000b).

a. *Land-use change*. Significant changes in land use and cropping patterns in the HKH provide evidence for the incompatibilities between mountain specificities (e.g. inaccessibility, fragility, marginality) on the one hand and the driving forces and operational mechanisms of globalisation (e.g. profit-driven selectivity and resource use intensification) on the other. Uncontrolled, external demand-driven resource use intensification and over-extraction conflict with historically evolved patterns and practices, which have traditionally helped to balance intensive and extensive resource uses and are characterized by organically interlinked, diversified production systems. The observed changes carry both economic and environmental risks.

b. *Extraction of resources*. Current trajectories suggest that existing patterns of over-extraction of mountain niche resources (e.g. timber, minerals, Non-Timber Forest Products, hydro-power) will be accentuated following the increased role of the global market combined with the persistence of unequal highland-lowland economic links (Jodha 2000a). This inference is based on the following common elements of over-extractive public sector interventions in the past and the globalisation-driven initiatives of today:

   • Externally conceived, top-down, generalised initiatives (priorities, programmes, investment norms) with little concern for local circumstances and perspectives, and non-involvement of local communities.

   • Indiscriminate intensification of resource use to meet external demands, without adequate development of infrastructure and ignoring side effects on local production systems and the environment. Environmental impacts include extensive deforestation, land slides, increased surface run-off, decline in soil moisture, decreased soil fertility and biodiversity and are linked to increased socio-economic vulnerabilities.

   • Insensitivity of decision makers to the side-effects of over-extraction of niche resources. The trend is likely to accentuate both due to unaccountability of market forces for negative side effects of their ventures and weakened authority of the state to enforce any regulations that address these problems, despite protests by communities and NGOs.

Table 1: The necessary pre-conditions for ensuring economic gains from globalisation and their status in mountain areas.

Mountain features	**Conditions and processes conducive to gains from globalisation**					
	Relating to production processes			**Relating to post production processes**		
	Resource use intensification	Specialisation and economies of scale	Tradable surplus generation	Infrastructure, access to markets	Equitable external links	Human response capacities
Limited Accessibility: high costs of mobility, low dependability of external support or supplies	(-)	(-)	(-)	(-)	(-)	(-)
Fragility: high vulnerability to degradation with increased land use intensity	(-)	(-)	(-)	(-)	(-)	(-)
Marginality: limited and low earning opportunities, resource scarcities and uncertainties, cut off from the ‚mainstream' economy, social vulnerability	(-)	(-)	(-)	(-)	(-)	(-)
Diversity: temporally and spatially highly diversified products/land use patterns	(+)	(-)	(+)	(-)	(-)	(-)
Niches: potential for numerous, unique products/ land uses	(+)	(+)	(+)	(-)	(-)	(-)
Human adaptation mechanisms: traditional sustainable resource management, diversification, recycling, adjusting demands to changing supply situation	(-)	(-)	(-)	(-)	(-)	(-)

Note: (-) and (+) indicate "extremely limited" and "relatively high" degree of convergence between mountain specificities and potential opportunities gains from globalisation. The situation may differ between more accessible (commercialised) and poorly accessible areas, as illustrated by the contrasting situation of the different mountain areas in China, India and Nepal (Jodha 2002).

c. *Declining resilience.* Globalisation can increase the vulnerability of mountain communities as a result of the erosion of communities' traditional adaptation measures combined with a reduction in public sector support following the WTO-imposed economic and structural reform programmes. Elements of declining resilience and increasing vulnerability, observed in the HKH study areas, include the following:

• Traditional adaptation strategies, based on diversification, local resource regeneration, recycling, and collective sharing etc. are being abandoned due to new market-driven incentives and approaches to production and consumption. Increased emphasis on cash cropping, mono-cropping (especially through arrangements involving private input suppliers and product buyers), a shift towards high-value crops that are dependent on external inputs, and a relocation of staple food crops to more marginal and steeper slopes were observed in Hunnan (China), Himachal Pradesh (India), and parts of Nepal and Pakistan.

• A decrease in public sector support and welfare activities (including production and transport subsidies, support for environmental constraints, public distribution systems etc.) in all study areas. The withdrawal of state support has led to increased dependence on external, unfamiliar, market-determined processes with greater exposure to market risks. Even the wealthier and more progressive farmers in the study areas expressed this concern.

d. *A decline in the economic value of mountain niche products* can be observed in several mountain areas:

• The promotion of large-scale green house facilities for efficient production of traditional mountain products in plains, supported by market-driven incentives, technologies, and infrastructure. For instance, certain off-season vegetables, mushrooms, and flowers that are naturally better suited to mountain areas are now produced at lower costs and higher quality in green house complexes in plains, as observed in India and China.

• Similarly, initiation of liberal trade policies has exposed mountain niche products to external competition without alerting and preparing mountain farmers. For example, apple farmers in the hill areas of India and Nepal are currently faced with competition through imports from Australia, South Korea and other areas. Thus, some mountain niche products are being marginalised by liberal trade policies, accepted under pressure from the World Trade Organisation (SAWTEE 2003).

• While the economic value of some mountain niche products is adversely affected by market competition, other niche resources are exposed to market-driven over-extraction due to uncontrolled external demand and vanishing regulatory systems (Karki 2000). These include timber, hydropower, herbs, and minerals, as reported from different parts of India, Nepal, Pakistan, China and Bangladesh.

e. *Changed state policies*, disregarding customary laws and mountain people's access rights to community land resources, are an emerging repercussion of globalisation-driven changes in many parts of the HKH region. The loss of customary regulations for conservation and protection of resources can encourage over-extraction. Often, the state's changed approach to customary land rights in mountain areas favours the market agencies to acquire and control the right and access to natural resources (forest, water bodies, crop lands, scenic spots, mineral deposits, etc.), in the name of promoting area development. Transfers of land from local communities to international and national agencies can be observed in Tibet,

Hunnan (China), Uttaranchal Pradesh (India) and Balochistan (Pakistan) and in some more accessible valleys in Nepal.

5. Potential opportunities for mountain areas

Notwithstanding the problems discussed above, globalisation may not necessarily be a source of gloom and doom. It also creates potential for new opportunities in mountain areas, although the discourse on the subject is generally dominated by the aspect of risk rather than opportunity. Any effort highlighting the potential opportunities created by globalisation, therefore, carries the risk of being interpreted as poorly grounded. With this caution in mind, we summarise the potential globalisation benefits for mountain regions, based on our exploratory research. However, it should be noted that the realisation of most of these potential opportunities requires capacity enhancement of mountain economies and communities and conscious efforts on the part of external agencies to link mountain economies/communities to global economies as equal partners. Furthermore, the protection and sustainable use of fragile mountain environments needs to be a main global concern. Mountain areas and communities need to be adequately compensated for the environmental services they provide on a regional and global scale (see Koch-Weser, this volume), which will ensure that environmental protection becomes a viable livelihood option for marginalized mountain communities (Jodha 2000b). The potential areas of globalisation gains for mountain areas are listed below:

i. The first category of potential opportunities relates to the specific mountain products and services (e.g. medicinal herbs, flowers, organic food products, and mountain tourism) with global demand, in which these areas may have comparative or exclusive advantages (Papola 1998). Some areas such as Kunming (China), Himachal and Sikkim (India), Ilam (Nepal), North West Frontier Province (Pakistan) and Bhutan have already made some progress in these fields of comparative advantage.

ii. The increased information, awareness and capacities of mountain communities, generated through enhanced links and partnerships with external market developmental agencies, constitute another important source of opportunities. Enhanced flow of information can make mountain products and services more competitive in the global market. Also, even in the remote trans-Himalayan areas, mountain communities are profitably using information technologies for planning area allocation to specific crops; storage and long-distance trading of products and tourism related activities. The constraints of physical inaccessibility can, to a certain extent, be compensated for by information technology.

iii. Another potential opportunity is the development of niche products by improving niche-promoting infrastructure. This could be achieved through the support of more resourceful global agencies or firms that are attracted by the untapped potential of mountain areas. Local area development through the development of niche products for corporate profits is being emphasised in many areas. For example, a number of multi-national pharmaceutical or floriculture firms are helping local farmers to produce these products by providing technology support and processing facilities. In some cases, the preliminary processing is also done by local farmers. The farmers benefit by producing revenue-generating high-value products. Several mountain areas in Nepal, China, India and Pakistan already have product- and service-based partnerships with international firms. The interest of global firms in these arrangements is guided by the fact that global profitability and competitiveness cannot be sustained by the "extraction" approach of the past. Hence, a shift toward resource regeneration (including environmental protection) and the involvement of local communities is increasingly seen as essential for long-term success. However, governments have to guard against these ancillary relationships becoming exploitative.

In addition, the enhanced links with global agencies and firms that have sufficient technological and financial capital to develop mountain resources, may aid in dealing with mountain-specific constraints (e.g. poor access, isolation, fragile slopes, marginality). Examples of the development of infrastructure and support services by the private sector have already emerged in different mountain areas, especially in China, Pakistan and India. In the past, states have often argued that a lack of funds has prevented the construction of costly roads in remote mountain areas with small and scattered populations. The emerging public-private sector partnership can help in addressing this problem.

iv. Enhanced links with global agencies may also help to manage the risks emanating from different mountain specificities. Accordingly, the globalisation-driven initiatives should address incompatibilities between mountain specificities and globalisation processes. For instance, the globalised market system should advance new technologies and means for: promoting low-intensity land use with high-value products for fragile areas; road construction techniques with less damage to fragile slopes; the development of marginal areas through increased investment and appropriate land use technologies; human capacity enhancement in marginal communities with institutional and financial support; developing resource diversification approaches with high payoff using new technological and management systems; enhancing mountain niches; upgrading traditional technologies; etc. These are only some of the potential possibilities to increase globalisation benefits for mountain people through resourceful and forward looking external market agencies.

v. There is yet another set of opportunities where economic (global) and environmental concerns can be simultaneously addressed. For example, large cardamom, shed-grown coffee, various mushrooms and herbs with high export-demand perform better in well-forested areas. Where these are promoted for profit, environmental conservation will be a precondition. Also, organic farm products necessitate effective resource conservation measures. Profit- or market-driven environmental conservation is an important potential option for mountain development in the future, especially in a developing country context (Papola 1998; Rongsen 1998).

However, the realisation of the above-mentioned potential opportunities requires a more pro-active and positive role of the private sector (in association with NGOs, governments donors, and communities) in support of sustainable development of mountain areas. Notwithstanding the doubts on such changes from different quarters, this may not be an unconceivable possibility (Jodha 2000b; 2000a; 2002).

6. References

Jodha, N. S. (1997). Mountain agriculture. *In* "Mountains of the World: A global priority." (B. Messerli and J. D. Ives, Eds.). Parthenon, New York.

Jodha, N. S. (2000a). Poverty alleviation and sustainable development in mountain areas: Role of highland – lowland links in the context of rapid globalisation. *In* "Growth poverty alleviation and sustainable resource management in mountain areas of South Asia." (M. Banskota, T. S. Papola, and J. Richter, Eds.), pp. 541-570. ICIMOD, Kathmandu.

Jodha, N. S. (2000b). Globalisation and fragile mountain environments: Policy challenges and choices. *Mountain Research and Development* **20**.

Jodha, N. S. (2002). Globalisation and fragile mountains. Final narrative report of the research planning grant (submitted to the MacArthur Foundation). ICIMOD, Kathmandu.

Karki, M. D. (2000). Commercialisation of natural resources for sustainable livelihoods: The case of forest products. *In* "Growth, poverty alleviation and sustainable resource management in mountain areas of South Asia." (M. Banskota, T. S. Papola, and J. Richterm, Eds.), pp. 293-320. ICIMOD, Kathmandu.

Kreutzmann, H. (1995). Globalisation, spatial integration and their impact on sustainable development in Northern Pakistan. *Mountain Research and Development* **15**, 213-227.

Papola, T. S. (1998). High value enterprises for sustainable livelihood: Trends, experiences and policies in the Hindu Kush-Himalayan region. Paper presented at "Mountains 2000 and beyond." International Conference on Sustainable Development of Hindu Kush-Himalayan Region, Wildbad Kreuth, Germany.

Rongsen, L. (1998). Enterprises in mountain specific products in Western Sichuan, China. MEI Discussion Paper 98/7. ICIMOD, Kathmandu.

Roy, S. (1997). Globalisation, structural change and poverty: Some conceptual and policy issues. *Economic and Political Weekly* **32**, 217-235.

South Asian Watch on Trade, Economics and Environment (SAWTEE) (2003). Globalisation and mountain farmers: Tapping opportunities and mitigating threats. SAWTEE, Kathmandu.

Research Partnerships for Mitigating Syndromes of Global Change in Mountain Regions

Hans Hurni*, Hanspeter Liniger, and Urs Wiesmann
Centre for Development and Environment (CDE), Institute of Geography, University of Berne, Steigerhubelstrasse 3, CH-3008 Berne, Switzerland
**phone +41-31-631 88 22, fax +41-31-631 85 44, email hurni@giub.unibe.ch*

1. Introduction

Key problems in mountain areas and at highland-lowland interfaces are largely related to human impact in these fragile ecosystems and may be intensified by the indirect effects of human activities in surrounding lowland areas. On the positive side, mountain regions are the world's freshwater reservoirs; they are important areas for agriculture, have resources that can be exploited for mining and tourism, and exhibit great biodiversity within small areas. The combined effects of various key problems in a mountain area can lead to a so-called "mountain syndrome"; most mountain systems show key symptoms of this syndrome or have the potential for their development (NCCR North-South 2000). The syndrome concept, developed by the German Advisory Council for Global Change (WBGU 1996), provides a framework for focused research. Its basic assumption is that typical clusters of ecological, social and economic problems or symptoms can be identified in specific regions of the world, such as mountain areas. These typical problem clusters are called "syndromes of global change" and are seen by WBGU (1996) as representative, specific functional patterns of non-sustainable development. Given this assumption, the syndrome concept allows primarily for integrated, situation-specific differentiation of global change.

Mitigating the cumulative occurrence of clustered political, institutional, social, economic and environmental problems in mountain areas will require the concerted action of multiple actors and all stakeholders concerned or affected (cf. Hurni 1998).

U. M. Huber et al. (eds.), Global Change and Mountain Regions, 593-603.

Research partnerships can furnish important contributions to knowledge generation and management by facilitating the development of measures for mitigating the mountain syndrome in many parts of the world, particularly in developing and transition countries. The following sections of this paper report on three major experiences with current research partnerships at the Centre for Development and Environment (CDE) over the past 25 years. The first example relates to water management in the Mount Kenya area, the second to soil and water conservation in mountainous areas worldwide, and the third to syndrome mitigation research in different mountain contexts. These experiences increasingly demonstrate the need for transdisciplinary research, a challenge that is fully met in new research programmes directed by CDE.

2. Mitigating conflicts over scarce water resources in the Mount Kenya area

The Mount Kenya region offers a great deal of beautiful scenery and attracts tourists from all over the world. What these tourists may not see, however, is the crucial function of Mount Kenya as a water tower for its lower slopes and adjoining lowland areas. This function is becoming ever more crucial, as populations in these areas are growing at a rapid pace and new land use systems require far more water (Wiesmann et al. 2000). Aside from the freshwater requirements of urban centres and tourist resorts, increased agricultural production plays the most significant role. First, agro-pastoralist smallholders are increasingly using small-scale irrigation in order to counterbalance the high risks caused by great variability in rainfall. Although justified by the severe problems of survival they face as the largest group of inhabitants in the Upper Ewaso Ng'iro basin, their water claims are virtually unlimited. Second, water requirements for year-round export-oriented horticultural production are high, even though advanced water-saving technologies, such as drip irrigation, are used. Urban centres, agro-pastoralist settlements in the upper reaches, and horticultural enterprises locate their water intakes high up on the mountain in the tributaries of the Ewaso Ng'iro River. As a result, 60-95% of the available river water is abstracted during the dry seasons in the upper reaches of the basin; up to 90% of this abstraction is unauthorized (Gichuki et al. 1998). These developments have led to a very significant decrease in the low flow of the river in the lowlands. This implies that the river will dry up completely during drought years. Downstream populations, as well as wildlife and related tourism, are heavily affected by the virtual loss of one of their key natural resources. The great majority of the basin's inhabitants face severe difficulties earning a livelihood (Künzi et al. 2002). These difficulties are now being aggravated by the heavy decrease in the low flows of river water, and current trends indicate that problems will continue to increase in the future. Against this background, current conflicts between the major water users in the highland-lowland system of Mount Kenya are growing and could turn violent if not appropriately addressed (Wiesmann et al. 2000).

Given the ecological, socio-economic, and socio-political realities in the basin, it is obvious that there are no easy solutions to the problem of over-utilization of river water and the corresponding problems of securing a livelihood and resolving multiple water use conflicts. From a regional point of view, bottom-up approaches alone, which aim to address needs and conflicts at the grassroots level, will contribute further to problems in the basin beyond the local level. However, top-down approaches, which aim at basin-wide planning of water supply and water use regulations, will also fail, owing to problems of acceptance and unmanageable control systems that would be required for their implementation at the local level. Consequently, there must be a search for solutions that do not consist of a single approach, but combine different approaches in a concerted strategy (Kiteme et al. 1998). Such a strategy must address several levels, including relevant decision-making processes from the local to the national scale. Moreover, this multilevel strategy should not only concentrate on the problems of over-utilization of water resources, but must also address the severe problems of securing a livelihood, as well as the various conflicts directly or indirectly related to water use. In this sense, a multilevel strategy will include aspects that can be summarized by the key words "multi-scale", "multi-stakeholder", and "multi-sectoral" (Wiesmann et al. 2000).

At first glance, in view of the complexity of the dynamics and problems in the Ewaso Ng'iro basin, it seems impossible to design and implement a strategy that includes all the above aspects of a multilevel approach. However, one way out of this dilemma is to split the general strategy into single components that are agreeable to the different actor categories and stakeholders concerned, and to specify them at a level where sectoral implementation is possible. Along these lines, a long-term collaborative effort that began more than 20 years ago, involving the Universities of Nairobi and Berne, in cooperation with the Government of Kenya, has identified feasible components of a multilevel strategy. Without going into great detail, this strategy involves two separate groups of components, which are largely on the way to being implemented by a broad range of institutions at different levels.

The first group of components directly addresses problems of water use and water demand. In other words, they concentrate on sustainable and demand-oriented supplies of river water. A key criterion of these components is to maintain a low flow of at least 1.5 m^3/s in downstream parts of the basin (Wiesmann and Kiteme 1998). From the top to the bottom levels, the following components are promising in this respect:

- *At the level of the entire basin:* Encouragement of negotiations among different categories of water users on the amount of river water available to each category. Amounts are best negotiated among water user associations, as self-control mechanisms can be established and coordinated in each case by the catchment board.
- *At the district planning level:* Establishing water supply development plans to ensure that implementation funds and efforts are applied in the most needy communities, and to establish location-specific principles of supply, taking into account the allocated portions of river water and the availability of other components of the water cycle.

- *At the level of single supply systems:* Increasing water use efficiency and water storage capacities by setting conditions for assistance in rehabilitating or implementing single water supply systems. Experience has shown that this requires increased coordination and a common understanding among governmental and nongovernmental agencies.
- *At all levels:* Supporting the above components of negotiation, planning, and implementation by creating awareness and providing support for decision-making. This requires well-founded knowledge of the human and natural environments, monitoring of human impacts in the basin, and development of decision-making support tools that provide easy access to this information and help in the search for appropriate management options. It also requires clear commitment from authorities at different levels and is thus related to aspects of good governance.

The above group of strategic components addresses the supply side of the water problems in the basin. However, even if these strategic components are fully implemented, problems will persist, due to the virtually unlimited demand for water. A second group of components must therefore indirectly reduce or divert the demand for river water. Examples of such components can again be identified at different levels.

- *At the household and farming level:* Water demand for small-scale irrigation by agro-pastoral smallholders can be reduced by diminishing the risk of crop failure in semiarid areas. Experience has shown that drought-tolerant crops and water conservation technologies can reduce water demand if they fit into local farming systems and household strategies.
- *At the regional planning level:* The pressures of water demand can be decreased and the potential for innovation increased by enhancing off-farm activities that are part of agro-pastoral smallholders' household strategies. Experience has shown, for example, that credit schemes or improved infrastructure in rural centres can promote off-farm economic sectors.
- *At the national level:* There is a danger that the success of a multilevel water strategy will have little effect if further immigration into the upper parts of the basin cannot be restricted. This would require limiting further subdivision of large ranches, which in turn requires amendments to national land tenure policies. Experience has shown that such changes at the national policy level can be justified by linking them to arguments for wildlife protection and environmentally friendly tourism.

These examples of components in a multilevel water strategy illustrate that implementation of the strategy is bound to be a flexible, long-term process that involves different actor categories, stakeholders and institutions at different stages. Research that participates in and supports such a process faces the challenges of interdisciplinarity and close collaboration between researchers and concerned societies; in short, it faces the challenge of transdisciplinarity. This process will entail implementation of single components, as well as monitoring, evaluation, and modification of water use strategies, with the objective of solving the highly complex problems of water use, water demand, and related conflicts in the overall basin. It is therefore bound to require continuous social and institutional communication, as well as a willingness to modify the strategy after learning from successes and failures.

3. Soil and water management: Applying WOCAT to mountainous areas

High rainfall, steep slopes and erodible soils can induce high surface runoff, soil erosion and landslides, particularly in mountain areas. On the positive side, there have been many achievements in sustainable land use and in avoiding and combating land degradation, not only in mountain areas, but worldwide. However, most of this valuable knowledge is not well documented or easily accessible, and comparison of different experiences is difficult. As a consequence, such knowledge often remains a local, individual resource, unavailable to others working in the same environments and seeking to accomplish similar tasks.

In order to overcome this gap in knowledge documentation and exchange, an international programme known as the World Overview of Conservation Approaches and Technologies (WOCAT) was initiated in 1992 by the Centre for Development and Environment (CDE), in collaboration with numerous institutions working globally at national and international levels. WOCAT's mission is to provide tools that allow specialists in soil and water conservation (SWC) to share their valuable knowledge, that assist them in their search for appropriate SWC technologies (WOCAT 2002a) and approaches (WOCAT 2002b), and that support them in making decisions in the field and at the planning level. SWC technologies are defined as agronomic, vegetative, structural and management measures that control soil degradation and enhance productivity in the field. SWC approaches are defined as ways and means of support that help to introduce, implement, adapt and apply SWC technologies in the field.

Although WOCAT is not exclusively active in mountain areas, it is highly relevant to land use in mountain areas, as information collected over the past ten years clearly reveals. A framework for documentation and evaluation of SWC has been developed over this period. It consists of a set of three comprehensive learning tools for documenting all relevant aspects of SWC and a computer-based database system for data entry, retrieval and evaluation. All tools, results and outputs are accessible in three languages via the Internet (www.wocat.net), in the form of books and maps, or on a CD-ROM (Liniger and Schwilch 2002). The questionnaires on SWC technologies and approaches assist in the compilation of case studies and a sound evaluation of one's own SWC activities. The SWC mapping addresses the issue of where degradation problems and their treatments occur (WOCAT 2002c).

The tools and activities developed by WOCAT are incorporated and integrated into existing programmes at the international, national and sub-national levels. At the international level, they fit well with the recent international project on Land Degradation Assessment in Dryland Areas (FAO 2002), the UN Convention to Combat Desertification (CCD) and the Convention on Biodiversity (CBD). With respect to the implementation of the Framework Convention on Climate Change

(FCCC), WOCAT can help document and assess the impact of improved land management on carbon sequestration. At the national and project levels, WOCAT has been integrated into ongoing governmental, non-governmental and other development projects as part of their efforts to document and evaluate their experiences and tap knowledge for improved decision making. Additionally, WOCAT tools and results have been increasingly used in training and education for universities and in extension programmes.

Through the WOCAT network, data have so far been collected in 34 countries, comprising 271 SWC technologies and 191 SWC approaches. In addition, over 400 SWC specialists from 35 countries have been trained in the use of WOCAT tools. Although WOCAT is used worldwide, not only in mountain areas, a query of the database illustrates that 47% of the technologies are implemented on slopes greater than 16%, and 54% are applied above 1000 m altitude. Thus, 30% of the documented cases fulfil both conditions. This shows the importance of conservation measures in mountain areas, i.e. on steep land and at higher altitudes.

A detailed analysis of the data available so far shows that in some fields of knowledge, SWC specialists had great difficulties providing information. The most important information gaps relate to the impacts of SWC, be they ecological, social or economic (e.g. cost-benefit analysis, impacts on the environment, on yields, etc.). According to WOCAT, about half of the documented SWC activities had a strong applied research component. However, data analysis reveals significant gaps in essential information required for the successful implementation of SWC. There is also no appropriate documentation about the spatial extent of degradation, and especially about the extent and effectiveness of conservation and good land management practices.

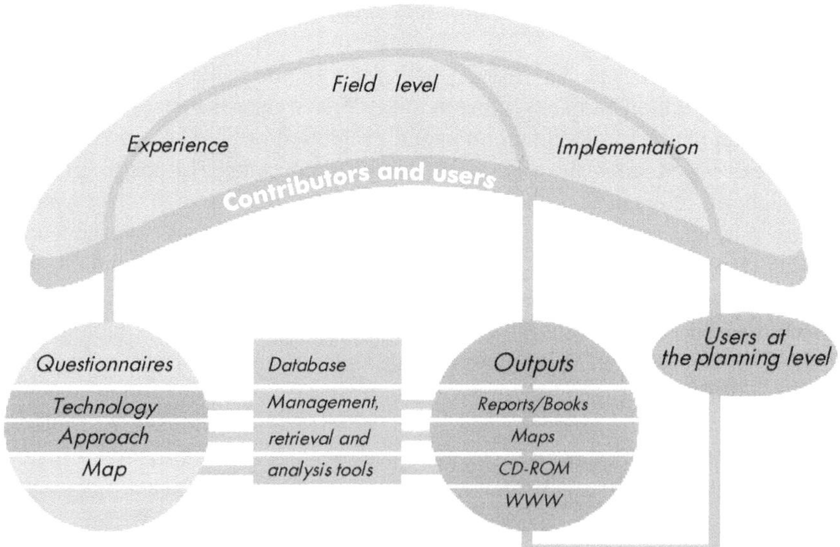

Figure 1: WOCAT tools and processes for documentation, evaluation and exchange of knowledge.

WOCAT was not primarily designed as a research programme, but it has shown that collaboration between development-oriented research and implementation is crucial for successful documentation and exchange of knowledge. It is clear that research contributes towards better understanding of degradation and improved implementation of "good" land management practices by:

- identifying important gaps/needs, e.g. the impacts of land use (ecological, social, economic);
- assessing the spatial extent of degradation and good land use practices by combining the WOCAT map tool with remote sensing, surveys, modelling, etc.;
- seeking solutions based on land users' experiences;
- contributing to up- and down-scaling between the local, regional and global levels;
- assessing the impacts of land use on natural resources and identifying key indicators and threshold values for monitoring;
- documenting agro-biodiversity.

The WOCAT initiative will thus continue to document, evaluate and exchange SWC achievements, and many of these address issues in mountain areas. A major effort will be made to promote good land management practices through the compilation and publication of a world map on SWC (Hurni and Meyer 2002). More support from applied research is needed in order to enhance understanding of the impacts of soil and water conservation in mountain areas and other ecoregions of the world.

4. Mitigation research and its application in mountain areas

Between 1999 and 2001, a Swiss consortium of institutions with experience in research partnerships developed a joint proposal known as the "National Centre of Competence in Research North-South" (NCCR North-South 2000), together with a large number of national institutions in developing and transition countries ("the South"). In a highly competitive selection process involving over 230 applicants, the NCCR North-South was approved as one of 14 national programmes in June 2001. The programme aims to carry out research on major syndromes of global and local change in developing and transition countries, and to make significant contributions to designing measures for mitigating these syndromes. Since mid-2001, the programme has been realised through research partnerships between institutions in Switzerland and countries affected by syndromes of change, as well as through co-operation with regional and international programmes pursuing similar goals. Research in the NCCR North-South programme focuses on the following three syndrome contexts: (1) urban and peri-urban areas, (2) semi-arid areas, and (3) highland-lowland areas, represented by 9 regions selected as so-called "Joint Areas of Case Study" (JACS, cf. Fig. 2).

Mountain and highland-lowland syndrome contexts are found in all but one of the JACS, the West African region. Research on the syndrome context of highland-lowland interactive areas has long concentrated on mountains alone, e.g. on mountain ecology (Ives et al. 1997; Gerrard 1990), mountain cultures (Manjari 1995), and mountain agriculture and tourism (Mountain Agenda 1999). However, the inclusion of Chapter 13 on mountain development in UNCED's Agenda 21 strengthened a research focus that incorporates interrelations between highlands and lowlands. The starting point for this stronger focus was the observation that most mountain and highland systems in the developing countries of the tropics and sub-tropics are resource-rich zones, by contrast with the situation in most developed countries (Myers et al. 2000).

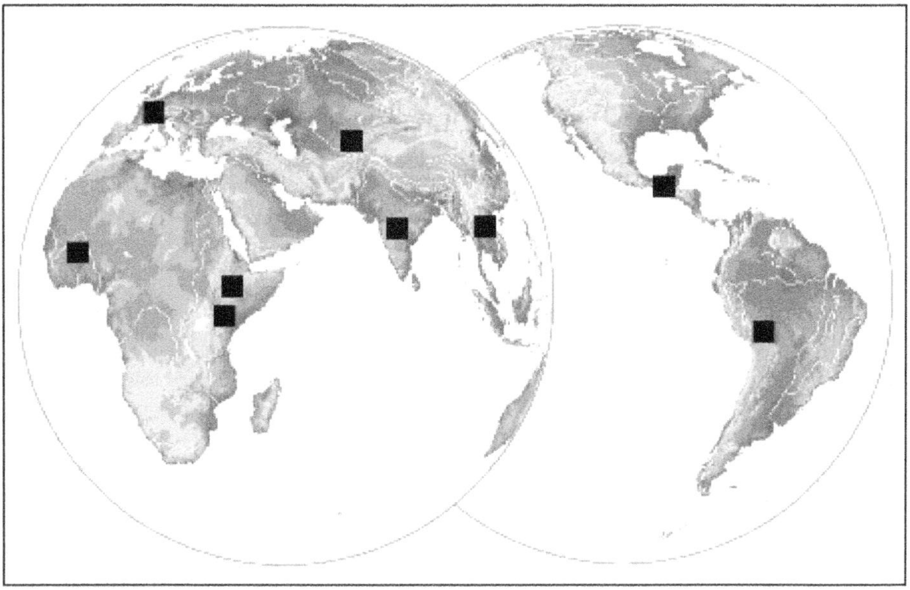

Figure 2: Focal research partnership regions (JACS: Joint Areas of Case Studies) of the Swiss National Centre of Competence in Research NCCR North-South.

The following points proved to be especially important within the stronger research focus on highland-lowland interactions:

(1) Mountains and highlands act as crucial water towers for surrounding lowlands, which are generally drier (Mountain Agenda 1998). Against this background, research is increasingly focusing on the ecological impacts of resource use changes in the uplands and their reflection in the availability of water in the lowlands (El-Swaify and Hurni 1996).

(2) The advantageous position of many mountains and highlands with respect to resources encourages a concentration of a wide range of interests among many different types of stakeholders on these resource islands (Wiesmann et al. 2000). In combination with the dependency of stakeholders in the lowlands on highland resources, this concentration of interests leads to multiple conflicts (Libiszewski et al. 1997) that must be addressed through multi-sectoral and multi-level strategies (Blaikie 1992).

This brief overview of current research highlights the fact that the syndrome contexts of highland-lowland areas are focal points of resource conflicts that result from inter-zonal ecological relations and cumulative socio-economic interests.

The overall strategy of a research approach focusing on syndrome mitigation consists of three major principles that are applied through different research and action components of the NCCR North-South.

The first principle is to focus both disciplinary and multi-disciplinary research on the study of individual core problems in individual projects, thereby relying on available theoretical and methodological competence and existing information and databases. The integration of knowledge will be gradual, first in situation-specific case study areas, and later in joint areas of case study (cf. JACS), in which several individual projects collaborate. In these joint areas of case study, the syndromes of global and local change have already been analysed, leading to a critical assessment and appraisal of the research concept and its usefulness for developing mitigation measures and strategies.

The second principle constitutes a move from explicative analysis to action-oriented research. Since the development of mitigation strategies and measures is the ultimate goal of the NCCR North-South, there is a need to transcend the "traditional" research objective of achieving better understanding and analysis of the status, dynamics and causes of core problems and syndrome contexts, in order to move towards a scientific contribution to strategies of sustainable development. This requires new concepts and methodologies based on transdisciplinarity, particularly at the stage of formulating research themes and integrating research results into societal processes – for example by adapting norms or developing strategies and measures (cf. Hurni and Ludi 2000).

The third principle calls for a move from research partnerships to societal empowerment. While research partnerships between institutions in the North and South constitute a relatively new approach, with benefits on both sides, this by itself will not suffice. The NCCR North-South ensures in particular that societal empowerment is promoted through these partnerships. This is achieved by addressing perceived needs, by allowing the participation of various stakeholders in the formulation and analysis of research projects, by valuation of research results, by integration of research-based outputs in societal processes such as policy-making, legislation and planning, and by implementing technological and procedural innovations, both in joint areas of case study and elsewhere.

Taken together, these three principles of syndrome mitigation delineate the challenges faced by research that aims to make contributions to more sustainable mountain development. The first challenge is to link profound disciplinary research with interdisciplinary approaches. The second is to interact with stakeholders to develop action-oriented findings. The third is to develop research foci in close communication with society and its needs and requirements. This triple challenge can be summed up by the term "transdisciplinarity" (Hurni and Wiesmann 2002).

The examples discussed above illustrate that the call for transdisciplinarity in development- and conservation-oriented mountain research is relevant not only in complex programmes such as the NCCR North-South. It is also valid in more narrow and focused approaches, such as the one developed by WOCAT, and in relation to particular problems in specific mountain environments, as illustrated in the case of Mount Kenya.

5. References

Blaikie, P. M. (1992). The state of land management policy, present and future. *In* "Soil conservation for survival." (Kebede Tato, and H. Hurni, Eds.). Soil and Water Conservation Society (SWCS) in Co-operation with International Soil Conservation Organisation (ISCO), Ankeny.

El-Swaify, S. A., and Hurni, H. (1996). "Transboundary effects of soil erosion and conservation in the Nile basin." Land Husbandry, Vol. 1. Oxford and IBH Publishers, New Delhi.

FAO (2002). "Land degradation assessment in drylands – LADA project." World Soil Resources Reports 97, FAO, Rome.

Gerrard, A. J. (1990). "Mountain environments: An examination of the physical geography of mountains." MIT Press, Cambridge MA.

Gichuki, F. N., Liniger, H. P., MacMillan, L., Schwilch, G., and Gikonyo, G. (1998). Scarce water: Exploring resource availability, use and improved management. *In* "Resources, actors and policies – towards sustainable regional development in the highland-lowland system of Mount Kenya." pp. 15-28. *Eastern and Southern Africa Journal* **8**, Special Number, Nairobi.

Hurni, H. (1998). A multi-level stakeholder approach to sustainable land management. Introductory keynote to the 9th ISCO Conference, Bonn. *Advances in GeoEcology* **31**, Vol. 2, 827-836.

Hurni, H., and Ludi, E. (2000). "Reconciling conservation with sustainable development. A participatory study inside and around the Simen Mountains National Park, Ethiopia." Produced with the assistance of an interdisciplinary group of contributors. Centre for Development and Environment (CDE), Berne.

Hurni, H., and Wiesmann, U. (2001). Transdisziplinäre Forschung im Entwicklungskontext: Leerformel oder Notwendigkeit? Forschungspartnerschaft mit Entwicklungsländern: Eine Herausforderung für die Geistes- und Sozialwissenschaften. Tagung der Schweizerischen Akademie der Geistes- und Sozialwissenschaften (SAGW) und der Schweizerischen Kommission für Forschungspartnerschaften mit Entwicklungsländern (KFPE), pp. 33-45. SAGW, Berne.

Hurni, H., and Meyer, K. (2002). "A world soil agenda – Discussing international actions for the sustainable use of soils." Geographica Bernensia, Centre for Development and Environment, Berne.

Ives J. D., Messerli, B., and Spiess, E. (1997). Mountains of the world: A global priority. *In* "Mountains of the world: A global priority." (B. Messerli, and J. D. Ives, Eds.), pp. 1-15. UNU and Parthenon, London.

Kiteme, B. P., Künzi, E., Mathuva, J. M., and Wiesmann, U. (1998). A highland-lowland system under transitional pressure: A spatio-temporal analysis. *Eastern and Southern Africa Geographical Journal* **8**, Special Number, 45-54.

Künzi, E., Wiesmann, U., and Maina, F. (2002). Innovation and adaptation in a new environment: Knowledge management among peasants in the upper Ewas Ng'iro river region, Kenya. *In* "Local environmental management in a north-south perspective." (M. Flury, and U. Geiser, Eds.), pp. 225-235. IOS Press, Amsterdam.

Libiszewski, S., and Baechler, G. (1997). Conflicts in mountain areas: A predicament for sustainable development. *In* "Mountains of the world: A global priority." (B. Messerli, and J. D. Ives, Eds.). UNU and Parthenon, London.

Liniger, H. P., and Schwilch, G. (2002). Better decision making based on local knowledge: WOCAT method for sustainable soil and water management. *Mountain Research and Development* **22**.

Manjari, M. (1995). Cultural diversity in the mountains: Issues of integration and marginality in sustainable development. Prepared for the Consultation on the Mountain Agenda Lima, Peru February 22-27, 1995, Committee on Women, Population and Environment. Boston University, Boston.

Mountain Agenda (1999). "Mountains of the world: Tourism and sustainable mountain development." Prepared for the Commission on Sustainable Development (CSD). Berne.

Myers, N., Mittermeier, R. A., Mittermeier, C. G., da Fonseca, G. A. B., and Kent, J. (2000). Biodiversity hotspots for conservation priorities. *Nature* **403**, 853-858.

NCCR North-South (2000). National Centre of Competence in Research North-South: Research partnerships for mitigating syndromes of global change. Proposal approved by the Swiss National Science Foundation and the Federal Council to Swiss Parliaments. CDE, Berne.

WBGU (1996). "Welt im Wandel: Herausforderung für die deutsche Wissenschaft." Wissenschaftlicher Beirat der Bundesregierung Globale Umweltveränderungen (WBGU)." Springer, Berlin-Heidelberg.

Wiesmann, U., and Kiteme, B. P. (1998). Balancing ecological sustainability and short-term needs: A regional approach to water supply planning. *Eastern and Southern Africa Geographical Journal* **8**, Special Number, 77-90.

Wiesmann, U., Gichuki, F., Kiteme, B. P., and Liniger, H. P. (2000). Mitigating conflicts over scarce water resources in the highland-lowland system of Mount Kenya. *Mountain Research and Development* **20**, 10-15.

WOCAT (2002a). "Questionnaire on SWC technologies. A framework for the evaluation of soil and water conservation (revised)." Centre for Development and Environment, Institute of Geography, University of Berne, Berne.

WOCAT (2002b). "Questionnaire on SWC approaches. A framework for the evaluation of soil and water conservation (revised)." Centre for Development and Environment, Institute of Geography, University of Berne, Berne.

WOCAT (2002c). "Questionnaire on the SWC map. A framework for the evaluation of soil and water conservation." Centre for Development and Environment, Institute of Geography, University of Berne, Berne.

Monitoring and Modelling for the Sustainable Management of Water Resources in Tropical Mountain Basins: The Mount Kenya Example

Lindsay MacMillan[1*] and Hans Peter Liniger[2]
[1]*Centre for Mountain Studies, Perth College, UHI Millennium Institute, UK*
[2]*Centre for Development and Environment, University of Bern, Switzerland*
**phone 44 (0)1738 877371, fax 44 (0)1738 877018, email lindsay.macmillan@perth.uhi.ac.uk*

Keywords: Curve number equation, Hydrological modelling, Landuse change, Model calibration, Tropical Africa

1. Introduction

The Upper Ewaso Ng'iro North river basin, which drains the north-western slopes of Mount Kenya in central Kenya, epitomises the African highland-lowland system. Extending over a vast region (15,200 km^2), it encompasses an extreme eco-climatological gradient that ranges from the glaciated peaks and indigenous forests of Mount Kenya to the semi-arid and arid land of the lowland plains (Fig. 1). The mountain forms a great natural asset in terms of water resources with plentiful rainfall (1500 mm/yr) supplying perennial rivers that radiate lifelines to the dry lowlands below. Thus, Mount Kenya is one of the major "water towers" (Liniger et al. 1998b; Liniger and Weingartner 2000) in Eastern Africa. Increasing pressures on the mountain from population increase and agricultural development have the potential to endanger this asset and cause conflict between upstream and downstream water users (Hurni et al., this volume). Rapid population growth has attained levels as high as 7-8% per annum (Kiteme et al. 1998). Migrants initially moved to the lower mountain slopes, attracted by good soils, high rainfall and proximity to rivers and transport, but latterly, forced by the pressure for land, they have settled on the dry plains, extending the migration zone into marginal areas for production (Kiteme et al. 1998; Liniger et al. 1998a).

U. M. Huber et al. (eds.), Global Change and Mountain Regions, 605-616.

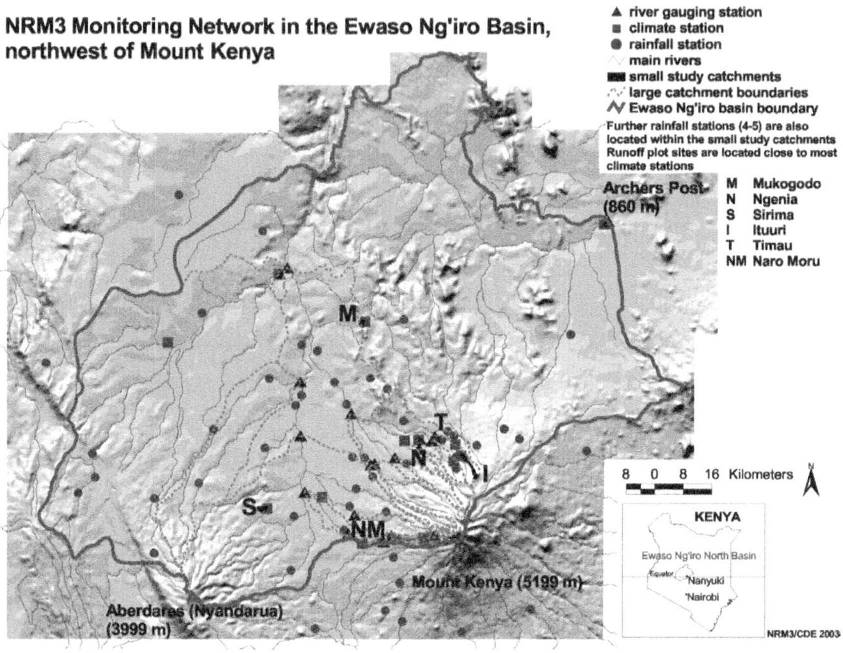

Figure 1: The hydrological monitoring network in the Ewaso Ng'iro North basin, northwest of Mount Kenya.

Landuse change is thus both historical and ongoing in the basin, with the primary changes presently being (i) intensive irrigated horticulture replacing rainfed arable farming on the semi-humid lower mountain slopes, (ii) small-scale cropland and communal over-grazing replacing the relatively dense (cover over 20% area) grass and bushland cover on the semi-arid lower mountain slopes and the highland plateau, and (iii) overgrazing of bushland savannah in the basement lowlands, resulting in grasslands with sparse vegetation cover or bare, exposed soils (Liniger et al. 1998a). Small-scale cropland is increasingly replacing natural forest on the north-western mountain slopes, despite most forest areas being protected. The natural division of the highland-lowland system is thus mirrored by settlement patterns, landuse change and accompanying water use. Heavy use in the upper parts of the basin for public water supply and agriculture reduces the flow and increases the pressure on downstream water resources for pastoralists, ranchers, conservation interests and tourism. Overall, the accumulation of changes to the natural system in the Ewaso Ng'iro basin has led to a situation in which competition for water is serious, and potential for conflict amongst users is increasing (Gichuki et al. 1998a). The urgent need to tackle the problem of water resource management within the basin by taking an integrated, multi-level, sustainable approach across the basin has been recognised (Flury et al. 1998; Hurni et al., this volume). This is no simple endeavour in any mountain basin and is commonly hampered by a lack of data and tools for objectively quantifying landuse change impacts.

This paper presents a rare example of a relatively intensive, variable scale, monitoring network in tropical highland Africa. Examples of results from this monitoring are given, which reveal the importance of vegetation cover and landuse in the generation of storm flow, groundwater recharge and decline in river flows. The calibration of a hydrological model developed for this region, together with some early results from the simulation of landuse change scenarios, are also presented.

2. Monitoring the impact of land use change on water resources

The availability of detailed climatic and hydrological data in mountain regions is in general limited, and certainly in African mountain systems, there are few databases of long-term, strategic, quality controlled monitoring (Gustard and Cole 2002). In the Mount Kenya region, a combination of governmental, academic and private monitoring activities have provided a comprehensive database of long-term climatic, hydrological, soil and landcover information at several different scales (Fig. 1). These data have been pooled together, quality controlled and stored in a single computer database and a GIS (Gichuki et al. 1998b; MacMillan et al. 1999). This electronic data compilation facilitates ease of access to information, which can be used to detect environmental changes within the highland-lowland system, such as landuse change (Niederer 2000), decreasing river flows (Gichuki et al. 1998a), and increasing water abstractions (Gikonyo, forthcoming). The monitoring programme and the research for this paper have been supported by various agencies: the Swiss Agency for Development and Cooperation (SDC), the Swiss National Science Foundation (SNSF), the Rockefeller Foundation, and the Swiss National Centre of Competence (NCCR) North-South.

Three ongoing major landuse changes have been identified: rainfed arable farming to intensive irrigated horticulture, natural grassland and bushland to small-scale cropland and to communal over-grazing (Liniger et al. 1998a). For example, a recent comparison of satellite images has quantified a 10% change from natural grass- and bushland areas to small-scale crop and agroforestry systems over the lower mountain footzone (2200 km^2) between 1984 and 1995 (Niederer 2000). A replacement of natural forest with plantation forest or small-scale cropland has also been identified, but to a lesser extent.

Hydrological monitoring at several scales and sites has revealed that vegetation cover has a key impact on water resources and their management across the basin and thus there is a need to know the consequences of the above landuse changes. For example, in the mountain forest zone with its deep soils and rainfall in excess of evapotranspiration, the main challenge is to maintain the storage of water in the soil and thus to sustain dry season river flows through delayed groundwater contributions (Liniger et al. 1998a). Soil moisture and runoff plot monitoring has shown both the importance of relatively dense vegetation cover and vegetation type on storm flow generation and groundwater recharge. On the northwest slopes of Mount Kenya at 2890 m asl, on study plots with relatively dense grassland, very little runoff was recorded, whilst between 20-45% of rainfall during heavy storms generated direct runoff from cropland (Kironchi 1999). On the southern slopes, at 2400 m asl, soil moisture measurements have shown the influence of forest and crop production on water yield in the Mount Kenya forest zone (Njeru and Liniger 1994). Cypress plantation shows greater water use compared to the natural forest, whilst least water is used by crop. Under the plantation, there was no groundwater recharge, whilst crop yielded higher groundwater contributions than natural forest.

Below the forest zone, on the lower mountain slopes and on the semi-arid basement plateau, where rainfall is lower and evaporation high, the challenge is to conserve water and minimise losses (Liniger et al. 1998b). Runoff monitoring at both the runoff plot (2x10m) (Liniger and Thomas 1998; Kironchi 1999) and small catchment (1-8 km^2) scales (MacMillan, forthcoming) has shown the paramount importance of a relatively dense vegetation cover in reducing surface runoff. Figure 2 shows rainfall and runoff (streamflow depth equivalent) from four small (1.3-8.1 km^2) representative catchments across the basin (Table 1). The effect of landcover is clearly observed, with Mukogodo (sparse grassland, 2-20% cover) generating twice the flow of Sirima (treed grassland) from half the rainfall. With well drained, sandy soils, the high runoff at Mukogodo has been attributed to the presence of sparse grassland and bare soils, which are subject to crusting and hence the promotion of overland flow. Runoff in the remaining three catchments is proportionally much lower. However, the higher flows generated at Ngenia are again attributed to crusting in an overgrazed central belt and on eroded footpaths within an otherwise cropped catchment. Sirima is a catchment that generates very low runoff, despite its poorly drained soil conditions, which might be expected to promote higher flows. The low yield is attributed to the presence of a relatively dense grass cover in large areas of the catchment, which increases infiltration. These observations at the catchment scale are supported at the point scale by infiltration measurements across the Ewaso Ng'iro basin (Liniger 1992).

At the larger catchment scale, dry season river flows of the Timau (59 km^2) and Naro Moru (63 km^2) catchments (Fig. 1), which drain Mount Kenya, show a distinct decline since the 1960s (Gichuki et al. 1998a). This decline has been attributed to an increase in irrigation abstractions of which as much as 55-80% is due to unauthorised over-abstractions, as revealed by extensive field monitoring (Gathenya 1992; Gikonyo, forthcoming). Exceptionally high flows were recorded during the El Niño year of 1998.

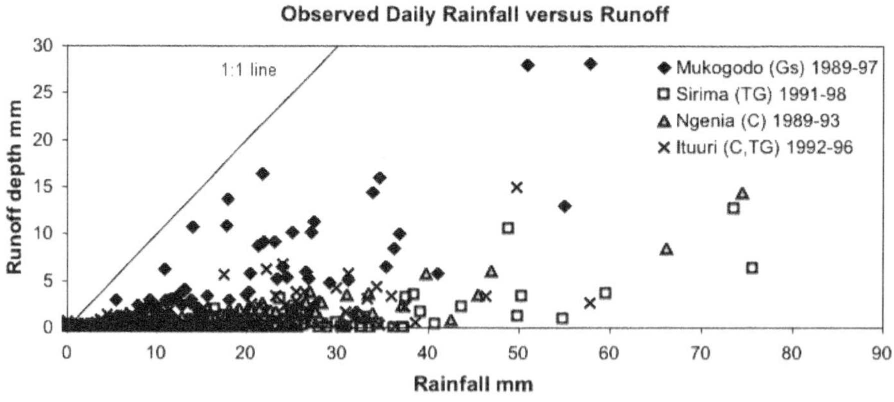

Figure 2: Observed catchment runoff under different landcovers: Mukogodo (sparse Grassland), Sirima (Treed Grassland), Ngenia (Crop), and Ituuri (Crop and Treed Grassland).

The examples given above show the evident need to continue to monitor the changing situation in a basin, where the balance between available water resources and increasing demands is fragile.

Table 1: Summary of characteristics of small study catchments.

Characteristic	Mukogodo	Sirima	Ngenia	Ituuri
Geomorphic zone	Basement plateau	Highland plateau	Lower mountain slopes	Upper mountain slopes
Eco-climatic zone	Semi-arid to arid	Semi-arid	Semi-humid to semi-arid	Humid to semi-humid
Elevation m od	1770 - 1890	1970 – 2080	2110 - 2220	2670 - 3620
Mean slope %	5.5	4.3	3.5	12.0
Area km^2	2.4	3.2	1.3	8.1
Main Landcover	Sparse grass & bare soils	Treed grass	Crop	Crop & Treed grass
Main Soils	Lixisols well drained, but prone to crusting	vertic Luvisols poorly drained, swelling/ cracking	Acrisols well drained	Andosols well drained
Data Record	1989-98	1989-98	1989-98	1992-98
Ave. annual rainfall mm	392	864	791	884
Ave. ann. pan evaporation mm	2240	1673	1484	1437
Ave. ann streamflow mm *	50	19	30	98
Max daily rainfall mm	62	134	86	103
Max 15 min rain intensity mm	66	78	94	110
Mean no. of raindays per year	62	109	106	167
Max peak streamflow m^3/s	50.52	13.26	4.87	22.9
Mean no. of flow events/yr	12	11	20	35
Max daily flow mm *	28.1	53.3	14.4	15.1
% Events with discharge < 5% of the rainfall (Q/P < 5%)	20	64	43	74

Notes: Grass (>20% plant cover), sparse Grass (2-20% plant cover), bare soils (<2% plant cover), Treed (20-50% canopy cover).
*Flow expressed as mm depth equivalent over catchment area.

3. Modelling the impact of land use change on water resources

Our monitoring results have shown that landuse changes are taking place in the Upper Ewaso Ng'iro North river basin and that they have a significant impact on the water balance. These research outcomes, coupled with the recognition of increased conflict potential amongst water users, emphasize the need for developing management tools that examine different aspects of landuse change impacts on the hydrological system (Gichuki et al. 1998a). This has provided the motivation for recent research, co-ordinated by the Universities of Nairobi and Bern, to establish three models for assessing change scenarios in the basin; a crop yield model (Njeru, forthcoming), a hydrological model (MacMillan, forthcoming) and a water allocation model (Gikonyo, forthcoming). This paper presents the results of the development of the hydrological model.

Whilst a plethora of hydrological models of varying complexity and purpose exist worldwide, the great majority of these models have been developed in industrialized countries. Consequently, they mainly focus on the hydrological processes found in temperate regions and their parameters and mathematical equations may thus be inappropriate for tropical mountain environments. In addition, funding in African mountain regions is commonly insufficient to collect data in the quantity and quality required for these models and to provide the financial and institutional commitment for computing facilities and staff training. This situation prompted the establishment of an appropriate hydrological model, which can be used to assess the impact of landuse change on the water balance of catchments in the Ewaso Ng'iro North river basin, whilst also taking into consideration the realities of data and training availability in developing countries (MacMillan, forthcoming).

3.1 Field measurements

Measured data from four small (1.3-8.1 km^2) representative catchments (three ephemeral and one perennial, Fig. 1) were used to identify the dominating hydrological factors, processes and relationships in the basin (Fig. 2, Table 1), a procedure vital to model selection, formulation and verification (Grayson et al. 1993). Figure 2 shows clear differences in daily runoff between catchments under different landcover, as discussed above. In Mukogodo, observed overland flow from areas of bare, crusted soil, caused by overgrazing, results in a very fast response to rainfall (time to peak flow typically 10 minutes) with high peak flows and high volumes (Table 1, Fig. 2). Infiltration is reduced and recovery of grass without intervention is unlikely. The presence of even a sparse grass cover (2-20%) improves infiltration. Detailed analysis of the data in Figure 2 showed that at Mukogodo, the long-term cover status of the grass proved to be a greater influence on the volume and rate of runoff than any short-term feature, such as rainfall intensity or storm duration. At Sirima, where perennial grass cover is relatively dense (>20% cover), runoff is much reduced (both peaks and volumes), and the response is considerably slower (time to peak typically >1 hr) and less variable. Occasional very high surface flows can arise due to the influence of soil swelling. At Sirima, grass cover is less variable and short-term antecedent conditions appear to have a more dominant influence on flow generation, which is predominantly sub-surface flow. At Ngenia, on the lower mountain slopes, the mixed crop cover reduces runoff, but less so than the relatively dense grass and bush cover of Sirima. In general, there is a high probability of runoff generation from rainfall events greater than 20 mm at Ngenia, whilst at Sirima a higher threshold of 30 mm is required. The faster response times observed at Ngenia (time to peak flow typically 40 minutes) are most likely due to the presence of a small area of sparse grass and bare soils (17%), which increases the surface runoff component where crusts have formed. The upper mountain catchment of Ituuri shows a mixture of crop in the lower reaches and grass on the upper slopes. The rainfall-runoff relationship in this catchment is similar to the cropped Ngenia catchment. The high rainfall at Ituuri generally infiltrates into the soils, quickly making its way to the stream channel via subsurface flow (time to peak flow typically one hour) or percolating into the deep soils and slowly issuing to the river as groundwater baseflow.

3.2 Model development and calibration

Starting with an existing hydrological model developed in Kenya (Thomas 1994) and based on a well-established methodology (USDA SCS 1985), the detailed field data from the four catchments described above were used for calibration and validation. The NRM3 model (named after the Natural Resource Monitoring, Modelling and Management (NRM3) research project) is a daily, simplified, semi-distributed model based on the concept of the Hydrological Response Unit (HRU). Parameters are obtainable from catchment measurements, rather than by model optimisation. Input data are minimal and include precipitation and evaporation pan time series, soil parameters (type, depth, available water capacity and drainage class) and landcover parameters (type, daily interception rate, Soil Conservation Service (SCS) curve number, critical soil depth, root depth, crop coefficient). The model utilises a Geographical Information System (GIS) to represent the spatial heterogeneity of the land surface characteristics in a grid-based distributed mode. Seven GIS input layers are required (catchment area, drainage network, rainfall data, evaporation pan data, soil type, landcover type and elevation). The model computes a water balance for each grid square, where the key component is the computation of daily storm runoff from net incoming daily rainfall, using the Soil Conservation Service curve number equation (USDA SCS 1985). This equation utilises a single retention parameter, the curve number, which is obtained from SCS tables according to soil type and landuse and then further adjusted for antecedent conditions in the catchment. The method was originally derived from analysis of rainfall-runoff data from small catchments, mostly in the mid-west US (USCS 1985).

In the model, the remaining infiltration then enters a three-layer store representing the unsaturated zone (which is subject to upward evapotranspiration and downward percolation), a shallow saturated zone and a deep saturated zone. Groundwater contributions to streamflow are lumped for the catchment and are determined by a simple, two-component linear function to reflect the contrast between higher groundwater contributions during and immediately after the rainy seasons and the lower contributions during the dry season. The model output is available in both time series format (daily catchment or grid cell water balance and catchment streamflow) and spatial GIS format (displaying the simulated, internal water balance across the catchment).

Key features of the model, developed specifically for the Ewaso Ng'iro, include (i) the introduction of a seven-fold, composite, landuse classification system, which takes into account the hydrological influence of vegetation and the occurrence of multi-layer and multi-type vegetation at the scale of the mapping units (Liniger et al. 1998a), (ii) the use of a distributed model to represent variability in landcover, which has been shown to influence runoff generation in the basin, (iii) the introduction of a dynamic component to reflect the influence of grass cover seasonality on runoff generation, (iv) modifications to represent the impact of cracking and crusting soils, and (v) minimal data input and computer hardware.

Although the model has been developed for application in ungauged catchments, model calibration and validation with field data was essential for transferring a methodology based on US field experiments (the curve number equation) to a new region. Based on an acceptance criterion of 70% model efficiency (Nash and Sutcliffe 1970), both daily and monthly streamflow in the three ephemeral catchments (Fig. 3) were successfully simulated (MacMillan, forthcoming). Model efficiencies ranged between 82 and 96%, with the exception of the daily runoff simulations at Mukogodo. The greater scatter observed in the Mukogodo results is most likely due to the increased sensitivity of this semi-arid catchment to processes at the event timescale, notably storm intensity, which is not represented in the model. Poorer daily model performance may also be related to the fact that intra-seasonal changes in landcover conditions are not taken into account in the model and to errors in the measurement of high flows used for calibration. In all catchments, there is a paucity of high rainfall events (> 40 mm), which although rare, often provide a great proportion of the annual flow. The poor simulation of extreme rainfall events is a key restriction to successful validation of the perennial, higher mountain catchment, Ituuri, where generally low model efficiencies (0.17-0.7) at both timescales were obtained.

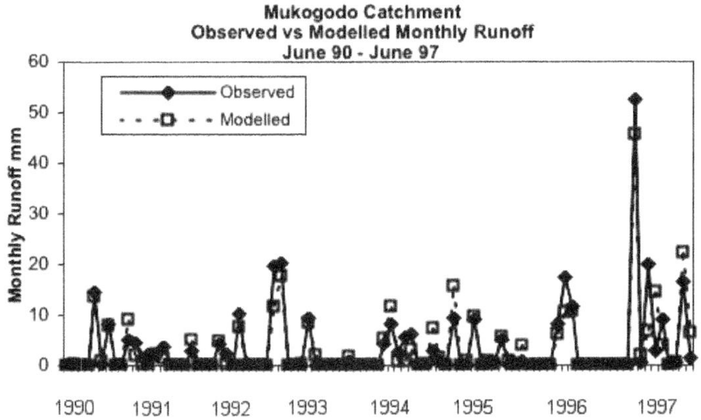

Figure 3: Observed versus modelled monthly runoff at Mukogodo.

3.3 Sensitivity analysis

Model sensitivity was tested for five selected input parameters/variables. Of the five examined, rainfall and landcover, the two key factors in the SCS curve number equation, have the largest impact on model performance at the scale of small catchments. We also tested the impact of the density and location of rainfall data on model performance. Each study catchment has 5 rain gauges (6 at Ituuri). Assuming that only one rain gauge was available, results have shown that the location of single gauges has a more significant impact on simulated catchment flows than the number of rain gauges. Whilst the five-gauge representation yields the highest model efficiencies, a well-positioned single gauge may yield similar performance. In contrast, a poorly positioned single gauge can yield unacceptably low model efficiencies at both timescales. Care is also required in the accuracy of the landcover representation. Calibration was based on an aerial photograph sourced landcover map. An alternative satellite image interpretation of landcover yielded reduced model performance, which at Mukogodo, was unacceptable. The impact of scale on model performance was examined by comparing model outputs for grid sizes, ranging from 50 m (the calibration resolution) to 500 m. The results show that a significant loss in model efficiency occurs at the 500 m grid scale. This is particularly pronounced for elongate catchments, such as Ngenia, which can lose high runoff generation areas during the rasterisation process.

3.4 Model scenario simulations

Landuse change scenarios, reflecting the prevailing changes of natural grassland to small-scale cropland and to communal over-grazing were simulated. The change is simply made by altering the pattern of pixels in the GIS landcover datafile. The model then makes the appropriate parameter changes according to landcover type depicted. Hypothetically assuming a 100% change from relatively dense grassland to small-scale cropland, the model estimates a doubling of mean annual runoff (streamflow) from the small study catchments on the lower mountain slopes. Evaporation is reduced and percolation increased. However, as the flows involved remain low, the main implications of this change may be associated with potentially increased erosion and irrigation requirements for crop production in such marginal environments. Assuming a complete restoration of fairly dense grass cover at Mukogodo would result in almost zero simulated flow generation in this semi-arid environment, whilst complete deterioration to bare soil would result in a doubling of the current mean annual simulated runoff (streamflow) (Fig. 4). Similarly at Sirima, degradation of the current relatively dense grass cover would result in a doubling or tripling of mean annual simulated runoff, assuming 100% sparse grassland and 100% bare soils conditions, respectively. The implications for erosion are likely to be severe (Liniger and Thomas 1998).

4. Future directions

Mount Kenya and its surroundings are undergoing rapid changes (population growth, land use change, land use intensification), resulting in increased competition for water resources in a region where water is scarce and conflicts over water are growing. The need to monitor the impact of these changes is clear and a great deal has been achieved in developing and maintaining a quality-controlled, monitoring network and database in the Ewaso Ng'iro basin. The key factors, which can lead to future success in managing the water resources in this environment, are continued monitoring and further development of models for predicting the impact of land use/land cover changes on water resources. The need for continued monitoring cannot be emphasised enough. Short-term records can make it difficult to detect change, given the natural variability of hydrological systems (Beven 2001) and can limit the ability to confidently predict the impact of change, as experienced on the upper mountain slopes in this study (MacMillan, forthcoming). Unfortunately, there has been a tendency to reduce funding for long-term monitoring. The Mount Kenya network and monitoring activities in small catchments in Ethiopia, Eritrea (Herweg and Stillhardt 1999), and southern Africa (Gustard and Cole 2002) are rather unique within Africa and have faced serious constraints in raising the necessary funding. In order to assess the impacts of current land use dynamics and to develop and test new hydrological methods and tools, the maintenance of the well-established, high-quality monitoring network on Mount Kenya is critical from a local as well as an international perspective.

With regard to modelling, a rigorous examination of the model performance at the small catchment scale has prepared a sound basis for future model applications at the small catchment scale and for future testing at the large catchment scale. Further testing necessary for the simulation of forested catchments and larger basins (Notter 2003) is underway. The model should be expanded to include soil erosion and sediment yields in order to predict the impact of different soil and water conservation measures, such as the ones identified through the WOCAT (World Overview of Conservation Approaches and Technologies) programme (Hurni et al., this volume). It is crucial for monitoring and modelling activities to consider mountains and their lowlands as an integrated system and to assess the impact of land use change on water resources both in the mountains and in the surrounding lowlands (Liniger 1995). In this way, monitoring and modelling play an important complementary role in making an essential contribution towards assessing the impacts of landcover change in mountain areas and in providing tools for better decision making in the field and at the different planning levels.

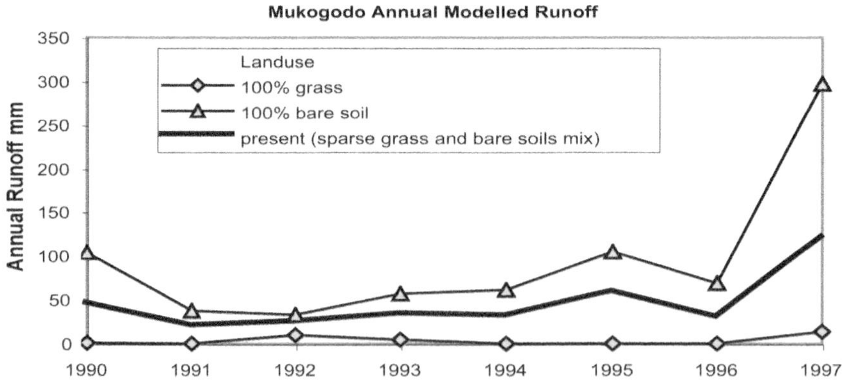

Figure 4: Modelled annual runoff (Mukogodo) under different landuse change scenarios.

5. References

Beven, K. J. (2001). Rainfall-runoff modelling: The primer. John Wiley & Sons, Chichester.

Flury, M., Mwangi, I. K., Obiero, S. V., Ndegwa, E. D., and Eggmann-Betschart, C. (1998). Stakeholders in the limelight: Principles of actor-centred resource management. *In* "Resources, actors and policies – towards sustainable regional development in the highland-lowland system of Mount Kenya." *Eastern and Southern Africa Journal* **8**, 97-105.

Gathenya, J. M. (1992). Hydrological investigations in the Mount Kenya sub-catchment of the Ewaso Ng'iro River. Geographica Bernensia A10, University of Bern.

Gichuki, F. N., Liniger, H. P., MacMillan, L., Schwilch, G., and Gikonyo, G. (1998a). Scarce water: Exploring resource availability, use and improved management. *In* "Resources, actors and policies – towards sustainable regional development in the highland-lowland system of Mount Kenya." *Eastern and Southern Africa Journal* **8**, 15-28.

Gichuki, F. N., Liniger, H. P., and Schwilch, G. (1998b). Knowledge about highland-lowland interactions: The role of a natural resource information system. *In* "Resources, actors and policies – towards sustainable regional development in the highland-lowland system of Mount Kenya." *Eastern and Southern Africa Journal* **8**, 5-14.

Gikonyo, J. (forthcoming). Water allocation and abstraction: Upper Ewaso Ng'iro river basin case study. Unpublished PhD Thesis, Department of Agricultural Engineering, University of Nairobi.

Grayson, R. B., Bloschl, G., Barling, R. D., and Moore, I. D. (1993). Process, scale and constraints to hydrological modelling in GIS. HydroGIS 93: Application of GIS in hydrology and water resources. *In* "Proceedings of the Vienna Conference, April 93." *IAHS Publication* **211**.

Gustard, A., and Cole, G. A. (2002). "FRIEND: A global perspective 1998-2002." Centre for Ecology and Hydrology, UK, Wallingford.

Herweg, K., and Stillhardt, B. (1999). "The variability of soil erosion in the highlands of Ethiopia and Eritrea. Soil Conservation Research Programme Ethiopia." Research Report 42, Centre for Development and Environment, University of Bern.

Kironchi, G. (1999). Influence of soil, climate and land use on soil water balance in the Upper Ewaso Ngiro North Basin of Kenya. Unpublished PhD thesis, Department of Soil Science, University of Nairobi.

Kiteme, B. P., Wiesmann, U., Kunzi, E., and Mathuva, J. M. (1998). A highland-lowland system under transitional pressure: A spatio-temporal analysis. *In* "Resources, actors and policies – towards sustainable regional development in the highland-lowland system of Mount Kenya." *Eastern and Southern Africa Journal* **8**, 46-53.

Liniger, H. P. (1992). Soil cover and management: Attractive water and soil conservation for the drylands in Kenya. *In* "Proceedings of the 7th International Soil Conservation Conference (ISCO), Vol. 1, Sydney, 27-30 Sep. 1992," pp. 130-139.

Liniger, H. P. (1995). "Endangered water: A global overview of degradation, conflicts and strategies for improvement." Development and Environmental Reports 12, Centre for Development and Environment, Bern.

Liniger, H. P., and Thomas, D. B. (1998). GRASS: Ground cover for the Restoration of the Arid and Semi-arid Soils. *Advances in GeoEcology* **31**, 1167-1178.

Liniger, H. P., and Weingartner, R. (2000). Mountain forests and their role in providing freshwater resources. *In* "Forests in sustainable mountain development: A state of knowledge report for 2000." (M. F. Price, and N. Butt, Eds.), pp. 370-380. CABI/IUFRO, Wallingford, UK.

Liniger, H. P., Gichuki, F. N., Kironchi, G., and Njeru, L. (1998a). Pressure on land: The search for sustainable use in a highly diverse environment. *In* "Resources, actors and policies – towards sustainable regional development in the highland-lowland system of Mount Kenya." *Eastern and Southern Africa Journal* **8**, 29-44.

Liniger, H. P., Weingartner, R. Grosjean, M., Kull, C., MacMillan, L., Messerli, B., Bisaz, A., and Lutz, U. (1998b). "Mountains of the World, Water Towers for the 21st Century: A Contribution to Global Freshwater Management. Mountain Agenda." Paul Haupt, Bern.

MacMillan, L. C., Liniger, H. P., and Gichuki, F. N. (1999). Stream flow modelling of the catchments of Mount Kenya: Data quality aspects. *In* "African mountain development in a changing World." (H. Hurni, and J. Ramamonjisoa, Eds.), pp. 193-206. African Mountains Association, African Highlands Initiative, and United Nations University, Antananarivo.

MacMillan, L. C. (forthcoming). "Hydrological model development for assessing impacts of landuse change in a tropical basin." Unpublished PhD thesis, Centre for Development and Enviroment, University of Bern, Switzerland.

Nash, J. E., and Sutcliffe, J. (1970). River flow forecasting through conceptual models part I: A discussion of principles. *Journal of Hydrology* **10**, 220-290.

Niederer, P. (2000). "Classification and multitemporal analysis of land use and land cover in the Upper Ewaso Ngiro Basin (Kenya) using satellite data and GIS." Master thesis, University of Bern, Bern.

Njeru, L. (forthcoming). "Crop water use and production in the Upper Ewaso Ng'iro Baisn." Unpublished PhD thesis, Department of Agricultural Engineering, University of Nairobi.

Njeru, L., and Liniger, H. P. (1994). „The influence of vegetation on the water resources of the Naro Moru catchment: A water balance approach." Discussion papers D17. Laikipia Research Programme, Centre for Development and Environment, University of Bern.

Notter, B. (2003). "Rainfall-runoff modelling of mesoscale catchments in the Upper Ewaso Ng'iro basin." Unpublished Diploma thesis, Department of Geography, University of Bern.

Thomas, M. K. (1994). Development of a streamflow model for rural catchments in Kenya. *Laikipia Mt. Kenya Papers* **D11**, 52.

USDA SCS (1985). Hydrology. *In* "National Engineering Handbook, Section 4." US Department of Agriculture, Washington DC.

Challenges in Mountain Watershed Management

Hans Schreier
Institute for Resources and Environment, University of British Columbia, Vancouver, B.C. Canada
phone 01-604-822 4401, fax 01-604-822 9250, e-mail star@interchange.ubc.ca

Keywords: Alternative water management, Climate change, Land use, Non-point source pollution, Scaling, Water quality, Water security.

1. Introduction

People living in mountain regions are discovering that managing watersheds for local needs is no longer a viable option. External pressure from the lowlands adds substantially to the difficulties of trying to sustain the resources used by local indigenous populations. These problems are best exemplified by addressing the hydrological and water quality issues that emerge from a combination of natural processes and the integrated effects of all land-use activities.

Internal pressures on watersheds come from relatively high population growth in many parts of the Himalayas, Andes, and African Highlands. Most mountain populations are engaged in subsistence agriculture and there is a lack of opportunity to move to the market economy. Despite significant out-migration, the population pressure remains high. To improve food security, agricultural production is intensifying and pressure on forest resources is also increasing, despite the trend to move towards community based management of resources (IDRC 1997). The impact of these land-use changes on hydrology and water quality is now a major topic of concern because integrated watershed management is still not a reality in most countries and few researchers are addressing the cumulative effects.

Of even greater concern are the external pressures that originate from lowland populations. Harvesting mountain resources to sustain urban growth and lowland activities is becoming a major challenge facing mountains. Mountains are the World's

U. M. Huber et al. (eds.), Global Change and Mountain Regions, 617-625.

watertowers and demand for mountain water has been rising sharply in recent decades. Rapid urbanization, agricultural intensification, and industrial development pose the most imminent threats to mountains. Based on the most recent prediction, the urban portion of the global population is to increase from 50% to 70% over the next 15-20 years. According to the World Bank (Pehu et al. 2002), food production has to be increased by some 50% to accommodate an extra 1.5 billion people and to improve the food security for another one billion, currently living at less than subsistence levels. Agriculture is using 70% of all freshwater resources, and irrigated agriculture, which covers 17% of the agricultural land, produces some 40% of the global food supply. Since the arable land base is shrinking, agriculture has to become more intensive and land use activities are expanding into marginal land in order to meet the food needs of the global population. At the same time, the urban diet is shifting from a grain-based staple diet to one that is dominated by meat. Global meat consumption is increasing at twice the rate of population growth and much of the increases in meat consumption will take place in developing countries. Over 50% of the meat production now takes place in confined industrial operations where the animals are stall-fed. More than 3/4 of the corn and soybean production in the developed world is used to produce meat, which is a most inefficient way to grow food since it requires enormous inputs of nutrients, energy and water. To grow a typical staple crop requires about 1000-1500 liters of water per kg of grain. In contrast, to produce one kg of chicken requires 3500-4500 liters of water, and to produce one kg of beef, depending on quality, requires some 15000–30000 liters of water per kg of meat (Gleick 2000). This shift in diet, together with the increasing demand from expanding urban centers, suggests that a water crisis will have serious effects on fragile mountain systems. Mountain watersheds will become battlegrounds over clean water to feed the appetite and lifestyle of the urban population.

It is not easy to balance external and internal pressures on water resources because mountain populations will try to increase their food security, which requires more water, and until recently they have not received many benefits for energy generation and supply of drinking water to lowland populations. Added to this is the rapid emergence of summer and winter mountain tourism, which is a water-demanding industry that puts extra pressure on mountain water resources.

The combined effects of external and internal pressures are impacting the hydrological cycle and water quality in mountains in ways that are not entirely predictable. It is challenging to isolate land-use effects from those induced by climate change, as both are changing at the same time.

2. Land-use impacts on hydrology and global climate change

Due to the high variability of weather conditions in mountains, it is difficult to isolate which proportion of discharge is impacted by land use changes and which results from climate change and variability. Such interactions can be additive, synergistic, counteractive, or complementary. It is well known that most of the current climate models are unable to discern changes in rainfall patterns and streamflow in mountains related to global warming. However, there is now sufficient evidence to suggest that seasonal decline in streamflow is apparent in the Rocky Mountains, the Himalayas, and the Andes. Most of these shortages result from over-allocation of water for irrigation, human consumption, and various demands by the lowland population for clean energy, food production, and industrial activities. What is still unclear is how many of these shortages are due to short- and long-term climatic variability.

Large differences exist in the use of water between watersheds. Storing water for those periods when supplies are insufficient is a necessity in most of the major mountain watersheds but construction of new reservoirs is becoming unpopular in many parts of the world due to environmental concerns about aquatic biodiversity, maintaining fish migration, displacing people, and unfavorable economics. According to Gleick (2000), very few new reservoirs were built in North America over the past 5 years and the number of old dams that have been demolished due to high repair costs has outstripped dam construction by a large margin. The lack of initiative for new reservoir construction is due to unfavorable economics since most of the easily accessible and geotechnically favorable sites have already been used. Upgrading existing structures makes rehabilitation costs prohibitive and the longevity of many projects have been reduced due to heavy sediment accumulation within existing reservoirs (Galay et al. 1995; Bird 2001). Displacing people is no longer an easy option in democratically governed countries. In contrast, new dam development plans are emerging in China, Nepal, India, Laos, and Vietnam, suggesting that we are entering a new phase of dam construction and water diversion in these parts of the world.

Water shortages in cities have now reached the point where cities like Los Angeles, Las Vegas and Phoenix have acquired all possible water-use rights in the Colorado River system all the way to the Colorado Rockies. In fact, negotiations are well under way to acquire more water for urban use from the agricultural irrigation sector because no new untapped resources are available in the Colorado River system. The same is true for cities like Quito, Mexico City, Lima, and La Paz, all of which are located in the mountains and are searching for new water supplies. As the costs and technical challenges for water treatment are increasing, clean freshwater resources become an attractive alternative. Entire watersheds in mountains are being earmarked as water supplies for the mega-cities. The best example is New York City (Herring 1999), which recently acquired land in a major watershed in the Catskill Mountains in order to be able to regulate the land use and guarantee safe long-term supplies for the city. In addition, they entered into a partnership with all mountain stakeholders to assure that best management practices are followed in those areas remaining under private holdings.

Water bottling companies are tapping cool clean mountain spring water for use by the urban population because we have now wrongly convinced a large proportion of the urban population that bottled water is safer than the water supplied by the municipalities. The demand for bottled water is increasing at 7% per year (Ferrier 2001) and is now a $22 billion industry annually.

Probably the largest concern is how to share water amongst different users. Many multi-stakeholder negotiation processes are now under way in an attempt to guarantee equitable water use in watersheds. The new emphasis is on maintaining sufficient water flow at critical times of the year for the survival of fish and other aquatic biota, when water uses and evaporative losses are at their highest and supplies are at their lowest. In many North American watersheds, water use license allocations exceed the supply, in particular during a dry year. If all permit holders would use their allocated amounts at the same time there would be no water flowing in many of these rivers during the summer. Efforts are now under way to buy back licenses, and to reduce rate allocations during dry years, to guarantee the survival of aquatic biota. The Yellow River is another river that shows the evidence of over-use, having no water flowing in the downstream section of the river for several months of the year. Since there has been a steady decline in hydrometric monitoring in the mountains of the world it is difficult to show trends that are primarily caused by climate change.

The importance of water supplies from the mountains in Europe has been shown by Liniger et al. (1998). They show that about 50% of the discharge of the Rhine, Rhone, and Po rivers into the North Sea and Mediterranean Sea originate from the European Alps and this percentage is significantly higher during the summer. The risk of changing the mountain water portion due to climate change is of concern due to the overwhelming evidence of rapid glacial melt (Haeberli, this volume). In British Columbia, Moore and Demuth (2001), and Moore and McKendry (1996) have shown that the hydrological response to climate change might not be linear but in the form of a step function. This suggests that there might be thresholds beyond which the hydrological systems go into another phase and in many systems these changes might only be evident as global warming proceeds.

An assessment of climatically driven changes in water supplies needs to be coupled with surface infiltration/runoff processes because these are heavily impacted by land use activities. Compacting soils and creating impervious surfaces are equally or more important than climatically induced effects because they change the infiltration and water storage capacity of the soil in dramatic ways. Research aimed at discerning these compounding effects on stream hydrology in watersheds is still very scarce. Lavkulich (1992) estimated that a 10% reduction in infiltration rates in the soils due to land use changes would exclude 4×10^{12} m^3 of water from short and long-term storage, which is more than all water discharged annually through all Canadian Rivers. It is not only soil compaction but also the loss of organic matter that has dramatic impacts on infiltration/run-off processes. Soil organic matter can absorb and hold large quantities of moisture and improves soil structure, which facilitates rainfall infiltration into the soil. Most land use activities have resulted in large losses of soil carbon and managing agricultural land and changing other land use practices has been estimated to have contributed up to 50% of the annual global CO_2 emissions between 1970 and 1990 (Kimble et al. 2002). This rate has now declined to about 20% in the latter part of the 1990's (Kimble et al. 2002), due to conservation and afforestation efforts in some parts of the world. Maintaining soil organic matter and encouraging soil carbon sequestration through improved land use should become a main concern in mountain watershed research because it improves hydrology and soil quality, and there is the added potential to obtain carbon credits.

3. Water quality issues in mountain watersheds and global climate change

Water quality is of equal concern because if we pollute the headwaters then groundwater sources, which are already under stress, will become the only safe supplies in lowlands. All land uses have some impact on water quality, and long distance atmospheric transport and changes in hydrology have already been shown to affect lakes and rivers in very remote mountain regions in Canada (Schindler et al. 1999).

Global climate change is affecting water quality by altering rainfall and evaporative processes. This will result in major shifts in land use as people adjust to climatic extremes. At the same time, declining flows have large impacts on solution chemistry and precipitation processes that can alter thermoclines in lakes and bacterial activities (Schindler 2001). Insufficient attention has been given to the generation and management of sediments. Sediments influence chemical oxygen demand, dissolved oxygen levels, general water chemistry, phosphorus and metal content in the water, and impact aquatic biota. There is now sufficient evidence to show that sediments detained behind dams contribute not only to a reduction of the life-span of dams but are also detaining large quantities of nutrients, particularly phosphorus. Studies by Stockner et al. (2000) have shown that fish productivity in the upper Columbia River, Canada, has declined significantly due to the reduction of phosphorus availability below the dam. Most of the phosphorus is adsorbed by sediments and detained behind the dam and this only became apparent some 10-15 years after the construction of dams. Recent phosphorus fertilization of lakes and streams downstream of the reservoirs has shown excellent results in stimulating fish productivity. However, fertilizing lakes and rivers is risky because poor mixing and applications of excess amounts can result in adverse water quality effects leading to eutrophication, a process that is commonly associated with excess nutrient applications (fertilizer and manure) from agriculture. New efforts are also underway in many parts of the world to enhance forest productivity by fertilizing plantation forests in mountains. This coupled with the lack of precision in nutrient management and the low costs encourage eutrophication processes.

The amount of fertilizer and manure used is changing rapidly in many mountain watersheds as farmers try to increase their productivity. Studies by Schreier et al. (1995), Brown et al. (1999; 2000), and von Wetarp (2001) have shown that within a 6-year period between 1994 and 2000, typical annual phosphorus budgets for farms in the Jhikhu Khola watershed in Nepal have gone from an annual P-deficit to a significant surplus. This is caused by increasing the annual crop rotation from two to three, introducing potatoes and tomatoes for marketing, and using higher nutrient inputs to maximize productivity. This is leading to widespread eutrophication in the streams that are used for drinking water downstream. With the introduction of potatoes and tomatoes, the use of pesticides has increased four- to tenfold and this is becoming one of the major health concerns in mountain watersheds. In a comparison between eight watersheds in the Andes and Himalayas pesticide concerns, seasonal water shortages, conflicts in water use, and health were identified as key common issues (Schreier et al. 2002). At the same time, information was shared on how best to resolve these issues in innovative ways.

Milk and meat production are agricultural activities that are of interest to mountain farmers. Overstocking of rangeland is contributing to soil erosion, and increased stall-feeding increases the pathogen risk in water supplies. Another concern raised by Harvell et al. (2002) is that climate change is altering pathogen dynamics, changing the spread of diseases, and the rate of decomposition.

The main problem with maintaining water quality in watersheds is the non-point source origin of pollutants and the cumulative effects. These problems have been addressed in a strictly sectoral manner and very few researchers have taken an integrated approach. To what extent the water quality is influenced by global climate change is difficult to isolate because land use activities are the main sources of pollution, but climate change is likely forcing land use changes, and modifications of streamflow and evaporation. How fast the human response to this is has yet to be documented.

4. Changing course in watershed management

Given the complexity and the threat imposed by global climate change, it is evident that a precautionary approach is needed in watershed management. Traditionally water resources have been managed in a sectoral manner and it is now suggested that a complete shift, and in some cases a complete reversal of management techniques, is needed if we hope to adopt and buffer the changes induced by the combination of land use and climate effects. Table 1 shows the changes that are needed to alter traditional practices towards innovative methods that have greater emphasis on protection. Research is needed to clearly document that these changes will be sufficient to minimize potential impacts due to global climate change.

Table 1: Changes needed to move from traditional to innovative approaches in watershed management.

Traditional Approach	Innovative Approach
Draining wetlands	Creating wetlands
Channelizing streams	Recreate natural channels
Building flood protection structures	Design flood concepts
Focus on point source pollution	Focus on non-point source pollution
Focus on single pollutants	Focus on interactions and cumulative effects
Creating impervious surfaces	Reducing impervious surfaces
Compacting soils	Minimizing and remediating soil compaction
End of pipe treatment	Source control
Piping stormwater into streams	Stormwater detention (ponds) & Infiltration
Building dams	Removing dams & creating bypass systems
Removing large woody debris	Adding large woody debris (to create diversity)
Sectoral water management	Integrated watershed approach for multiple use

5. Future research initiatives

Modeling efforts have to be coupled with process-based studies because mountain watersheds are highly complex, extremely variable, and sensitive to both climate change and human land use.

Our state of knowledge is particularly poor in mountain systems in the following areas:

1. *The dynamic changes in land use and surface alterations and their effects on the infiltration/runoff processes.* There are a number of remote sensing and GIS based studies that have addressed the land use dynamics issue but few have been coupled with research on how this affects soil hydrological processes (Brown et al. 2000). Adaptive management is now the buzzword on how to cope with climate change, but unless we understand how land use affects infiltration and water flow we are unable to make educated decisions on how we can counteract the added problems induced by greater climatic variability.
2. *Non-point sources of pollution and cumulative effects.* This issue dominates research activities in the lowlands but due to the intensification and expansion of land use activities in the mountains, and the potential shifts induced by climate change, this topic is now of equal importance to mountains. This issue has to be addressed in a watershed context and new research is needed to deal with diffuse sources of pollutants. Models need to be linked to GIS and process studies and sediment, nutrient and carbon budgets at a watershed scale have yet to be modeled in a way that inspires confidence in the calculations.
3. *Scaling problems.* How to scale up from mini-watersheds to watersheds and river basins is still more an art than a science. There is now enough evidence to suggest that processes are non-linear and that factors and processes change, as we move from micro- to meso- to macro-scales (Schreier et al. 2000). This type of research is particularly pertinent because we need to make linkages between upland systems and distant impacts in the lowlands. The challenge is to measure the combined changes in different land uses and how they affect water use and quality as we move to larger systems (Schreier and Brown 2002).
4. *Isolating impacts induced by climate variability.* This is probably the most challenging research issue because of the large uncertainties and poor historic databases available for water resources in mountains and because of the compounded impact of land use change. Simulation modelling, scenario development, and sensitivity testing are some of the research techniques that should be used.
5. *Improving the communication of science to decision makers.* Mountain communities, even in the most remote parts of the world, have now access to information technology tools. This provides new opportunities for sharing and comparing information and for improving communication. As was shown with the example of the "Himalayan-Andean Watershed Comparison Project", it is now possible to share experiences globally via the Internet and convert them into multi-media CD-ROMs. This allows us to identify innovative management

practices in one mountain region and then test their applicability in other mountain systems. It also allows us to present scientific information to decision makers in a more attractive and effective manner. These tools have now also been expanded to include distance education courses, offered through the Internet, which has allowed professionals in mountains access to the latest scientific knowledge by distance education and communication. We are hopeful that these tools will reduce the digital divide and allow mountain people to become active participants and greater beneficiaries of global development.

6. References

Bird, J. (2001). A global water policy arena: The world commission on dams. *In* "Globalization and water resources management: The changing value of water." Proceedings, AWRA/IWLRI University of Dundee International Specialty Conference, 7 pp.

Brown, S., Schreier, H., Shah, P. B., and Lavkulich, L. M. (1999). Soil nutrient budget modelling: An assessment of agricultural sustainability in Nepal. *Soil Use and Management* **15**, 101-108.

Brown, S., Schreier, H., and Shah, P. B. (2000). Soil phosphorus fertility degradation: A GIS based assessment. *Journal of Environmental Quality* **29**, 1152-1160.

Ferrier, C. (2001). "Bottled water: Understanding a social phenomenon." Discussion paper, World Wildlife Fund (http://www.panda.org/livingwaters/pubs.htm).

Galay, V., Okaji, T., and Nishino, K. (1995). Erosion from the Kulekhani Watershed during the July 1993 rainstorm. *In* "Challenges in mountain resource management in Nepal. Processes, trends and dynamics in Middle Mountain watershed." (H. Schreier, S. Brown, and P. B. Shah, Eds.), pp. 13-24. IDRC Ottawa, and International Centre for Integrated Mountain Development (ICIMOD), Kathmandu.

Gleick, R. H. (2000). "The World's water: 2000-2001." A biennial report on freshwater resources. Island Press, Washington, D.C.

Harvell, C. D., Mitchell, C. E., Ward, J. R., Altizer, A., Dobson, A. P., Ostfeld, R. S., and Samuel, M. D. (2002). Climate warming and disease risks for terrestrial and marine biota. *Science* **296**, 2158-2162.

Herring, J. H., Ed. (1999). Watershed management tactics in the New York City Watershed. *Water Resources Impact* **1**, 2-19.

IDRC (1997). Community-based natural resource management in Asia. Proceedings of International Development Research Centre Workshop, International Development Research Centre (IDRC). Ottawa, Canada, 161 pp.

Kimble, J. M., Lal, R., and Follett, R. F. (2002). "Agricultural practices and policies for Carbon sequestration in soil." Lewis, Baco Raton.

Lavkulich, L. M. (1992). Soil: The environmental integrator. Environmental soil science, anthropogenic chemicals and soil quality criteria. *In* "Symposium Proceedings of the Canadian Land Reclamation Association and Canadian Soil Science Society," University of Alberta, Edmonton, Alberta, pp. 1-44.

Liniger, H. P., Weingartner, R., and Grosjean, M. (1998). "Mountains of the world: Water towers for the 21st century." Report for the UN Commission on Sustainable Development, CSD session "Strategic approaches to freshwater management," Paul Haupt, Bern.

Kimble, J., Lal, R., and Follett, R. F. (2002). "Agricultural practices and policies for carbon sequestration in soil." Lewis, Boca Raton.

Moore, R. D., and McKendry, I. G. (1996). Spring snowpack anomaly patterns and winter climatic variability, British Columbia, Canada. *Water Resources Research* **32**, 623-632.

Moore, R. D., and Demuth, M. N. (2001). Mass balance and streamflow variability at Place Glacier, Canada, in relation to recent climate fluctuations. *Hydrological Processes* **15**, 3473-3486.

Pehu, E., Ed. (2001). "Science and technology in securing food for the next century – Challenges, issues, and options. Security." World Bank Report, 40 pp.

Schindler, D. W. (1999). From acid rain to toxic snow. *Ambio* **28**, 350-355.

Schindler, D. W. (2001). The cumulative effects of climate warming and human stresses on Canadian freshwaters in the new millennium. *Canadian Journal of Fisheries and Aquatic Sciences* **58**, 18-29.

Schreier, H., Shah, P. B., and Brown, S. (1995). "Challenges in mountain resource management in Nepal.

Processes, trends and dynamics in Middle Mountain watershed." IDRC Ottawa, and International Centre for Integrated Mountain Development (ICIMOD), Kathmandu, 286 pp.

Schreier, H., Brown, S., and Shah, P. B. (2000). Soil-sediment nutrient transport dynamics in a Himalayan watershed. International Symposium on the Role of Erosion and Sediment Transport in Nutrient and Contaminant Transfer. IAHS Proceedings, *International Association of Hydrological Sciences (IAHS) Publication* **263**, 1-7.

Schreier, H., and Brown, S. (2002). Scaling issues in watershed assessments. *Water Policy* **3**, 475-489.

Schreier, H., Brown, S. J., and Bestbier, R. (2002). The Himalayan-Andean watershed comparison project (Comparing 4 watersheds in the Andean Mountains with 4 in the Himalayas). International Development Research Centre (IDRC) and Institute for Resources of Environment, University of British Columbia, Canada (9 CD-ROMs) (http://www.ire.ubc.ca/himal/index.htm).

Stockner, J. G., Rydin, E., and Hyenstrand, P. (2000). Cultural origotrophication: Causes and consequences for fisheries. *Fisheries* **25**, 7-14.

von Westarp, S. (2001). Linkages between agricultural intensification, soil-fertility dynamics, and low-cost drip irrigation in the Middle Mountains of Nepal. M.Sc. Thesis, Deptartment of Soil Science, University of British Columbia.

Overcoming the Vertical Divide: Legal, Economic, and Compensation Approaches for Sustainable Management of Mountain Watersheds

Maritta R. v. Bieberstein Koch-Weser
Earth3000, Palais am Festungsgraben, D-10117 Berlin, Germany
phone ++49-30-20455995, fax ++49-30-20455997, email mkochweser@earth3000.org

Keywords: Environmental Service Payments, Mountains, Sustainable Water Management, Watersheds, Water Sources

1. Issues

1.1 The need for environmental service agreements

Sustainable water development and the mitigation of large-scale natural disasters in river basins depend in large measure on the ways in which upstream water sources and soils are protected. Environmental services provided by mountains are often only noticed when they are lost, as in the case of downstream floods caused by upstream deforestation. As half of humanity depends on fresh water that originates in mountain watersheds, protecting environmental services provided by mountain regions is highly important for global environmental security.

However, in most regions of the world, downstream people have no tradition of negotiating environmental safeguards with mountain people upstream, nor do they have legal and economic instruments and social organization models to do so.

Upstream people rarely take into account the value of environmental services provided by mountain regions when making land use decisions. This is explained by the fact that they traditionally have not received any compensation for services such as soil protection by mountain forests. In the absence of economic incentives, they are

U. M. Huber et al. (eds.), Global Change and Mountain Regions, 627-636.

often not willing to invest in soil conservation practices to protect watersheds for the benefit of their downstream neighbours.

As a result, there is a dangerous trend of accelerating erosion in headwater catchments. The world over, one can observe a lack of effective, long-term agreements on environmental maintenance and compensation between upstream and downstream communities. Environmental service agreements are now urgently needed, in the face of observable, global trends towards environmental degradation in mountain areas.

Region-specific approaches need to be developed for the valuation and contracting of upstream environmental services by downstream communities and enterprises that are dependent on reliable, high-quality water supplies and the prevention of natural disasters.

1.2 Downstream effects of environmental mismanagement

The impact of mountain ecosystem degradation through unsustainable forestry and agricultural practices can be tremendous and costly for downstream communities. Impacts of clear-cutting include the lowering of water tables and associated impacts on aquifers and wells, siltation of hydropower and irrigation reservoirs through hillside erosion, and - further downstream - less water retention for the dry season and more violent floods in the rainy season. Agricultural run-off can pollute renewable freshwater resources, and changes in water level can increase salinity or can allow arsenic and other pollutants to reach the surface. The loss of mountain forest cover accounts for increases in erosion and natural hazards, such as avalanches, landslides, and floods. Global damages on property and infrastructure can amount to tens of billions of dollars globally every year.

1.3 The plight of mountain dwellers

The protection of mountain watersheds depends on people. Mountain communities tend to be comparatively poor and isolated. In many of the poorer mountain regions, their lives leave no room for choosing the environmental high road. Instead, farmers will work any land – no matter how fragile - in their struggle for sheer short-term survival.

In many instances, traditional practices - which may have guaranteed sustainable use in past centuries - have made way for unsustainable land use patterns. Populations have outgrown the carrying capacity of the land, and farmers have moved onto increasingly more fragile, steep land for arable agriculture and livestock husbandry. In some regions, such as the Andes, former lowland populations that are entirely inexperienced in mountain farming are now being pushed into mountain regions in their quest for subsistence plots.

1.4 Fragility of mountain environments

The world's remaining mountain forests present a foremost hope against erosion. They still cover more than 9 million km^2, with almost 4 million km^2 above 1000 m, and represent 28 % of the world's closed forest area. People benefit from mountain forests in many ways. In general, forests slow the rate of surface run-off in a watershed, ensuring a certain base flow, minimizing flooding and increasing water quality.

Despite these benefits, mountain forests have been disappearing at a startling, unprecedented rate in the last decade (see Price, this volume). Deforestation tends to be driven by population growth, uncertain land tenure, inequitable land distribution, illegal logging, and the absence of strong and stable institutions (see Lebel, this volume). Combined, these social and institutional factors cause an expansion of settlements, arable agriculture and livestock husbandry into unsuitable, fragile mountain areas and accelerate the disappearance of mountain forests. Their effect can be exacerbated by the simultaneous excessive development of infrastructure (e.g. for tourism and recreation) or by logging for firewood. As soil erosion increases, it turns into a driver in its own right. In the tropics, farmers will abandon low productivity plots a few years after the first clearings, only to shift to and clear new mountain forest plots in the region.

Environmental degradation in mountain areas is especially pronounced, because of their extreme fragility. Mountain ecosystems are characterized by high geomorphic energy and low temperatures, which cause vegetation growth and soil formation to occur very slowly. Soils are usually shallow, young, and highly erodible. Farming in such marginal mountain areas can easily cause environmental imbalance. Once eroded, mountain soils may need hundreds of years to recover.

2. Payments for environmental services

2.1 The concept

One promising instrument in downstream-upstream cooperation for the protection of mountain watersheds involves payments for environmental services (PES). Downstream water users compensate forest owners and landholders in upstream watersheds for, e.g. forest conservation, reforestation or other services that maintain or improve water quantity and quality downstream. By giving an economic value to environmental services, ecosystem protection can become an attractive alternative to other land uses pursued by the forest owners.

So far, there are only a few cases worldwide that involve PES. This paper introduces cases that involve compensation schemes between downstream beneficiaries and upstream suppliers of environmental services in mountain regions around the world. Successful schemes involving PES seem to follow some of the same principles:

- Valuation of environmental services provided by mountain regions, from the vantage point of one or several stakeholder groups downstream;
- Social organization among the respective upstream and downstream negotiating parties that is effective enough to allow for tangible payment agreements;
- Agreement on clear and verifiable targets, and related implementation and monitoring arrangements;
- A solid legal and institutional framework; and

- Provisions for conflict resolution.

The regional dimension. Relatively few effective transboundary river basin management schemes and formal commissions exist. Examples of regional schemes are found in the Mekong, Danube or Rhine river basins. There is a need to compile and synthesize the lessons learned from such transboundary environmental management agreements, including PES, and to apply them to the large number of cases-in-waiting: some 120 of the world's major rivers traverse two or more countries. Also, within countries, major rivers tend to traverse several state-level boundaries. Salient examples of transboundary management needs and shortcomings can be found in Brazil (e.g. the Sao Francisco River, the Rio Doce, or the Parnaiba do Sul), or in India (e.g. the Ganges). Ample documentation and analyses are available on these and other cases.

The global dimension. In the future, there may also be additional opportunities for compensating mountain regions for some global environmental services. For example, mountain forests play an important role for carbon sequestration (Schimel and Braswell, this volume), and this environmental service could be compensated for through the global carbon offset schemes that are part of the Clean Development Mechanism (CDM) under the framework of the Kyoto Protocol on Climate Change. In addition, these CDM compensation schemes could bring great indirect benefits in terms of watershed protection, poverty alleviation and disaster mitigation, especially in the poorer mountain regions of the world.

2.2 Synopsis of existing cases

While the overall number of compensation schemes for environmental services in mountain areas remains rather low worldwide, one can take courage from the fact that the existing schemes have been introduced and run successfully in diverse cultural settings. Indications are that environmental service payments are a promising tool to foster sustainable development in mountainous regions worldwide. The cases described in Table 1 have *some common features*:

- The environmental service underlying the different agreements is almost always water. Siltation of irrigation channels is a close second, in cases where soil erosion is a major issue. The FONAFIFO program in Costa Rica also compensates upstream landowners for the mitigation of greenhouse gas emissions as well as the protection of biodiversity and scenic beauty.
- Problems experienced in the lower reaches of watersheds served as incentives for setting up schemes that compensate upstream landowners for the environmental services of their forests, i.e. the agreements are problem-driven.
- There is usually little interaction between upstream communities and downstream water users, prior to the implementation of compensation schemes.
- In most cases, the expected benefits have not been evaluated in economic terms. The price paid for ecological services has rather been set for pragmatic reasons that reflect political or budgetary considerations.

A first typology emerges: Schemes for PES can be grouped into self-organized private deals, trading schemes and public payment schemes:

Table 1: Summary of different environmental compensation schemes. For details see Koch-Weser and Kahlenborn (2002).

	Downstream problems	Upstream environ. services	Who pays?	Who receives?	Involvement of public institutions	Compensation scheme	Legal set-up
Case1: Australia (State Forests of NSW 2001)	Soil salinisation	Reforestation	Downstream farmer association	Government agency, private upstream landowners	Major involvement; public agency reforests and sells salinity reduction credits	Yearly payments per ha of reforested land for 10 years	Trading scheme
Case 2: Columbia (Echavarria 2002)	Water scarcity, floods, siltation of irrigation channels	Reforestation, erosion control, spring & stream protection	Downstream farmer associations	Government agency, private upstream landowners	Minimal; Agency only designs management plans and distributes the money	Individual contracts	Private deal
Case 3: Costa Rica (Rojas & Aylward 2001)	Siltation of hydro-electric reservoirs, irregular stream flow	Reforestation, sustainable forestry, forest preservation	Hydroelectric companies, government fund (FONAFIFO)	Private upstream landowners	Minimal; provides frame-work for payments, serves as mediator, supplements payments	Yearly payments per ha of enrolled land for 5 years	Private deal
Case 4: Ecuador (Troya & Curtis 1998)	Decreasing water quality & quantity (Quito water supply)	Patrolling the reserve, change in land use practices, reforestation	Water users	Fund, private upstream landowners	Major involvement; agency collects fee and undertakes compensation measures	Individual contracts	Public payment scheme, fee
Case 5: France (Perrot-Maitre & Davis 2001)	Decreasing quality of spring water for mineral water production	Reduction of nutrient runoff and pesticide use	Private bottler of mineral water (Perrier-Vittel)	Upstream farmers	Non-existent	Yearly payments per ha for 18-30 years, pays for new equipment	Private deal
Case 6: Philippines (Francisco et al. 1999)	Decreasing water quality & quantity	Restriction of forest use, watershed protection	Users of recreation facilities, water users	Fund	University plays a major role		
Case 7: USA (Walter & Walter 1999)	Decreasing quality of drinking water (NYC water supply)	Implementation of Whole Farm Plans and best management practices	City and water users (water tax)	Upstream farmers	Major involvement; NYC completely finances the program	Coverage of additional management costs, reduced property tax	Public payment scheme, tax
Case 8: USA (Perrot-Maitre & Davis 2001)	Soil erosion, decreasing water quality	Reforestation, land retirement, implementation of conservation practices	Government	Farmers	Major involvement; the government completely finances the program	Yearly payments per ha for 10-15 years	Public payment scheme

In the case of *self-organized private deals*, government involvement is minimal (e.g. incentive schemes) or non-existent, and payments are made voluntarily by the downstream partner, which is either a private company or a farmers association. These cases can be found on the (sub-)watershed level, where an agreement provides private downstream entities with water services at a lower cost than would be involved, e.g. with the construction of additional water treatment plants or reservoirs. Examples are compensation schemes in France and the Colombian Cauca River Valley, and the FONAFIFO deals in Costa Rica.

Trading schemes occur where governments set either a very strict water quality standard or a cap on total pollution emissions. In Australia, the government aimed at addressing the national problem of increasing soil salination by replanting forests and trading salinity reduction credits to downstream farmers.

Public payment schemes are the most common mechanism. A government entity finances upstream conservation activities or reforestation from general tax revenues or water user fees. The money usually goes into a fund, which is managed by a public-private council. Examples are cases in Ecuador and New York City (NYC).

3. First lessons and recommendations

"Overall, there is no blueprint mechanism that fits all situations – innovative mechanisms will be site-specific, will often involve elements of different approaches, and will vary depending on the nature of the ecosystem services, the number and diversity of stakeholders, and the legal and regulatory framework in place" (Johnson et al. 2001). Some first lessons and recommendations emerge, nevertheless:

- Although Payment for Environmental Services Schemes in mountain regions are not very different from similar schemes dealing with water resources in the plains, they are much less common. One approach for quickly raising their numbers would be to *integrate mountain areas into existing comprehensive environmental payment programs*, such as the Conservation Reserve Program in the USA. Similar programs exist in many countries.
- Downstream water users, such as farmers and hydroelectric companies, have an interest in watershed protection. An existing *strong legal and regulatory framework*, such as the FONAFIFO Fund in Costa Rica, helps with the development of local schemes because it reduces the transaction costs of establishing and maintaining the compensation mechanism.
- Economic instruments seem to work better in an environment of well-established links between nature management actions and products and with well-defined rights and responsibilities of the different parties involved. However, those conditions are the exception rather than the rule in river basin management. Therefore, *stakeholder participation, negotiation and institution building are critically important*, as can be seen in the Cauca River Valley and NYC cases (Tognetti 2001). To be successful, most of the water users should be integrated in a constituency for watershed protection. Institutional engagement appears to be more important in complex cases and in situations where legislation concerning a

certain environmental service is weak.

- Self-organised private deals are likely to develop when there is a strong link between land use actions and impacts on upstream watershed services. The water services provided have to be related to private goods (e.g. bottled water or agricultural products). In these cases, private companies, farmers or households have a strong self-interest in paying upstream landholders for environmental services. For example, hydroelectric companies try to avoid irregular stream flow and siltation of reservoirs. Incentives for downstream farmers are manifold and range from a reduction in soil salinity to a decrease in the siltation of irrigation channels. Another precondition of voluntary contractual arrangements are low transaction costs. Thus, private deals are more likely to occur in smaller watersheds with fewer participants. A problem with voluntary payments is their decline in years of economic crises (e.g. Colombian Cauca River Valley).
- Compensation schemes where private entities, such as companies or farmers associations, fully finance environmental services in mountain regions are *restricted to profitable industries* (Perrier Vittel) or agricultural regions where farmers get good prices for their products (Cauca River Valley). In cases where affected stakeholders are financially weaker, the public sector has to provide some funds to establish a compensation scheme.
- *In contrast to self-organised private deals, public payment schemes usually develop in larger landscape systems* where environmental services are more complex and biophysical relationships are less predictable. Other preconditions are a high number of stakeholders and a scheme in which payments have to be collected from a large number of participants. In other cases, government institutions might be needed to organise upstream interests. Governments who pay for upstream environmental services are interested in protecting the environment, respond to public pressure or try to avoid the even higher costs of, for example, new water filtration plants, as in the case of NYC. In contrast to private entities, the public sector is able to reduce high transaction costs, caused by a large number of stakeholders, by using existing administrative structures. A problem that is mainly associated with public payment schemes is the free rider phenomenon, i.e. some enjoy the service without paying for it.
- Public-private partnerships seem to work particularly well for environmental service agreements. The cooperation between the farmers associations in the Cauca River Valley and the public Cauca Valley Corporation represents a promising example for the efficient operation of a PES scheme, in which farmers take responsibility for financing the project and the public authority carries out the watershed protection measures according to management plans. Thus, *public-private partnerships are most likely to occur when private stakeholders are well organised and the public institution involved has an interest in watershed protection.* In most cases, institutions such as the Ministry for the Environment or an associated regional or local authority are involved.
- Regarding the possibility of organising upstream forest owners and downstream water users, *it appears much easier to form water user associations downstream than to organise upstream landowners.* This is due to the higher financial

resources downstream and their common interest in the environmental service. On the other hand, although much more complicated, it is more urgent to initiate upstream cooperation. Cooperative efforts enable upstream people to formulate their interests and to communicate the environmental services they have to offer, contributing considerably to mountain forest protection. When planning to encourage upstream organisation it is crucial to consider that, compared to lowland regions, the area concerned can be much bigger. An example are large dams that affect people in numerous sub-watersheds. Public entities or NGOs play an important role as initiators of upstream organisation (e.g. Costa Rica where private NGOs carry out technical studies of upstream forest owners' properties and help them with the paperwork necessary to enrol in the FONAFIFO program).

- *PES must be granted for many years* to guarantee a long-term change to sustainable land uses and agricultural practices. Ideally, the contract states that the upstream partner has to sustainably manage resources for an agreed time, even after the payments will have ended (http://www.rainforest-alliance.org/programs/cmc/newsletter/mar01-1.html). Otherwise, farmers might be tempted to clearcut their forests after they stopped receiving compensation.

4. Pointers for starting new PES initiatives

4.1 Parameters

- It will be crucial to the success of any initiative that the resource to be protected is scarce and declining, and that its decline directly affects downstream investments or beneficiaries.
- Compensation must also be high enough to serve as an incentive to upstream forest landowners to change their land-use practices. This is a complex process, in which not only individual farmers, but also communities must collectively change their way of life. Compensation levels should be based on the estimated value or the economic importance of the service (http://www.rainforest-alliance.org/programs/cmc/newsletter/mar01-1.html).
- In many cases, education and assistance is required to enable upland farmers and communities to change their land use patterns. Existing laws and customs have to be taken into account and key stakeholders need to be involved in the planning process early on.
- While implementing long-term payment schemes for environmental services, major assumptions should be monitored and tested and, if necessary, adjusted or revised altogether.
- The financial mechanisms chosen should fit existing institutional parameters and local customs. Great care should be taken not to introduce divisive financing schemes, which could harm equity and peace among involved mountain communities.
- The choice of financial mechanisms will also mirror regional institutional particularities: In areas with weak public institutions, self-organised private deals

are probably the most effective. On the other hand, in areas with strong public institutions, trading or public payment schemes are more likely to be successful.

4.2 Initial questions

- *What ecosystem services are provided?* It is important to identify those services that provide direct benefits to people and to determine which management strategies will result in, e.g. less soil erosion or higher water quality.
- *Can these services be measured and monitored?* In most mountain regions, there is little data on the ecosystem services provided by upstream forests. Thus, scientific investigations and comparisons with similar regions can provide important arguments in negotiations between financers and service providers.
- *What are the rights and responsibilities for resource use and management?* Knowledge about the legal/formal and the customary/informal rights and responsibilities in a watershed is critical to the successful introduction of market mechanisms.
- *Who supplies and who receives the ecosystem service?* A precondition for establishing a PES scheme is to learn who owns or manages the mountain areas that provide the service. On the other hand, there must be people who directly profit from an enhanced ecosystem service in order to use market tools successfully.
- *Are potential participants of the scheme aware of the environmental problem?* If this is not the case, measures have to be taken to put the problem on the local political agenda, or, if only a few parties are involved, to raise their awareness.
- *How can downstream and upstream interests be organised?* The organisation of downstream interests is relatively easy. If user organisations do not already exist, downstream beneficiaries need to be supported with knowledge, and possibly money, to organise themselves. Numerous organisations from other regions can serve as examples. The organisation of upstream interests is much more difficult. Here, it would be important to initiate communication between isolated mountain communities and to provide assistance in formulating their interests and offering environmental services
- *What is the value of the ecosystem service?* The real economic value of ecosystem services is very difficult to determine and, in most cases, can only be roughly estimated. Methods to do this are 1) an evaluation of the costs involved in counteracting negative environmental effects (e.g. NYC); 2) an evaluation of the economic activities that directly depend on the environmental service (e.g. Energía Global de Costa Rica); or 3) a willingness-to-pay survey (e.g. Philippines). Finding the right price will be the result of negotiations between the parties involved.
- *Are beneficiaries willing and able to pay for the ecosystem service? Are suppliers willing and able to provide it?* These are the most important questions. Although one never knows if beneficiaries are willing to pay until someone makes an offer, chances are good when the ecosystem service is scarce or declining, the economic activity linked to it is relatively important and substitutes are expensive or unavailable. In most cases, potential suppliers will only provide the ecosystem service if they are paid as much, or more, than they could obtain from alternative

uses.

- *Is the government or an environmental NGO interested in implementing PES schemes?* The debate over whether or not to introduce market tools to maintain or enhance environmental services is often initiated by governments or NGOs who bring users and service providers together. Thus, it might be crucial for a successful scheme to seek their support.
- *What transaction costs are involved?* Assessing the potential for a PES scheme must recognise the transaction costs arising from stakeholder participation, negotiations, research, monitoring and enforcement. Negotiating with associations of water users or forest owners rather than with individual water users or upstream landowners can reduce transaction costs considerably. Governments or donor agencies might also be willing to pay for PES schemes if the overall concept is cost-effective (http://www.rainforest-alliance.org/programs/cmc/newsletter/mar01-1.html).
- *Which PES scheme is the most suitable in the given situation?* In most cases, the decision will have to be made between purely private deals and public payment schemes because trading schemes only seem to be an option for industrialised countries with a strong legislation. Public schemes are more appropriate in large watersheds.

5. References

Echavarría, M. (2002). Water user associations in the Cauca Valley, Colombia: A voluntary mechanism to promote upstream-downstream cooperation in the protection of rural watersheds. *In* "Land-water linkages in rural watersheds." FAO Case Study Series.

Francisco, H. et al. (1999). "Economic instruments for the sustainable management of natural resources: A case study on the Philippines' forestry sector." UNEP Report, New York.

Johnson, N., White, A., and Perrot-Maître, D. (2001). Developing markets for water services from forests: Issues and lessons for innovators. Forest Trends, World Resources Institute, Katoomba Group, Washington.

Koch-Weser, M., and Kahlenborn (2002). Background Conference Paper for Bishkek Mountain Summit, October 2002.

Perrot-Maître, D., and Davis, P. (2001). Case studies of markets and innovative financial mechanisms for water services from forests. *Water Resources Impact* **1**.

Rojas, M., and Aylward, B. (2001). Cooperation between a small private hydropower producer and a conservation NGO for forest protection: The case of La Esperanza, Costa Rica. *In* "Land-water linkages in rural watersheds." FAO Case Study Series, Rome.

State Forests of NSW (2001). Forest facts: Developing markets for salinity control. Background info. *In* "Water resources impact." (D. Perrot-Maître, and P. Davis, Eds.), Vol. 1 (5). Case Studies of Markets and Innovative Financial Mechanisms for Water Services from Forests.

Tognetti, S. (2001). "Creating incentives for river basin management as a conservation strategy: A survey of the literature and existing initiatives." US World Wildlife Fund, Ecoregion Conservation Strategies Unit, Innovative Landscapes Track.

Troya, R., and Curtis, R. (1998). "Water: Together we can care for it!" Case Study of a Watershed Conservation Fund for Quito, Ecuador.

Walter, M. T., and Walter, M. F. (1999). The New York City Watershed Agricultural Program (WAP): A model for comprehensive planning for water quality and agricultural economic viability. *In* "Water resources impact. Case studies of markets and innovative financial mechanisms for water services from forests." (D. Perrot-Maître, and P. Davis, Eds.), pp. 5-8. Vol. 1 (5).

Synthesis: Future Research Directions

Astrid Björnsen[1*], Ulli Huber[2], Mel Reasoner[1], Bruno Messerli[3], and Harald Bugmann[4]
[1]Mountain Research Initiative, Schwarztorstr. 9/11, CH-3007 Bern, Switzerland
[2]Institute of Plant Sciences (IPS), University of Bern, Switzerland
[3]Brunnweid, 3086 Zimmerwald, Switzerland
[4]Mountain Forest Ecology, Dept. of Environmental Sciences, Swiss Federal Institute of Technology, Zurich, Switzerland
**phone +41 31 328 23 30, fax +41 31 328 23 20, email bjoernsen@sanw.unibe.ch*

1. Introduction

This book provides a snapshot of Global Change research in the world's mountain regions. It results from the increased awareness of mountain issues that was generated during the International Year of Mountains 2002. Similar to a photograph, it represents only one part of reality and does not aim to include all aspects of Global Change research in mountains in a single compilation. To provide a panorama view, the book compiles examples of prominent recent mountain research in both natural and social sciences. An important and recurring theme of this book is the need for an integrated approach to mountain research involving stakeholders and policy-makers besides natural and social scientists. Such an approach pays attention to the complexity of mountain regions not only in terms of the strong environmental gradients along mountain slopes, but also in terms of socio-economic gradients associated with varied access to limited resources and rapidly changing land-use, economic and political circumstances.

The book reflects the fact that mountain regions have become the focus of a growing body of Global Change research in recent years. This trend has been fuelled by the recognition of the critical role of mountains in the functioning of the global geosphere-biosphere system. The inclusion of mountains in Chapter 13 of Agenda 21 during the

U. M. Huber et al. (eds.), Global Change and Mountain Regions, 637-650.

1992 United Nations Conference on Environment and Development underscored the significance of mountains as follows: "As a major ecosystem, representing the complex and interrelated ecology of our planet, mountain environments are essential to the survival of the global ecosystem." Subsequently, the Second Assessment Report of the Intergovernmental Panel on Climate Change included a chapter on the impacts of climate change on mountain regions. And most recently, mountain issues have received recognition in the Convention on Biodiversity, in a special chapter of the Millennium Ecosystem Assessment, and in 4 resolutions of the UN General Assembly between 1998 and 2004.

Several factors contribute to the high importance attached to mountain regions. Apart from covering about one fourth of the Earth's terrestrial surface, these regions provide goods and services to more than half of humanity (Messerli and Ives 1997) and, together with adjacent foothills and valleys, are home to one fourth of the global population (Meybeck et al. 2001). Important goods and services include, for instance, (1) the supply of 30-40% of all freshwater of the Earth (in arid and semi-arid regions, this fraction ranges between 70-100%); (2) hot spots of biodiversity; this is partly due to the compression of "life zones" that is characteristic of mountains, but also due to the low management intensity; (3) major terrestrial carbon pools such as mountain forests and soils; (4) a primary source of forest products; (5) protected areas and refugia for threatened species; and (6) unspoiled recreation areas for a rapidly growing and urbanized world population. In many mountain regions, the ability to provide these goods and services to mountain people and lowland communities is jeopardized by the concerted effects of environmental and anthropogenic driving forces (e.g. natural and human-induced climatic changes, pollution, increasing CO_2 levels, land-use and land cover changes). In addition, globalization, i.e. the growing global integration of social, political, and economic relationships, affects mountain environments. Demographic changes, the incorporation of mountain economies into extra-regional economies, the increasing influence of urban processes, and changes in the location of decision-making and institutional arrangements are only a few examples of the effects of globalization. Although natural drivers of change might be less perceptible at the moment, their importance will reach or even exceed the importance of globalization as a driver of change in the near future.

From a scientific point of view, mountain regions provide excellent opportunities to detect and analyze global change processes and phenomena, mainly because of their high sensitivity to environmental and socio-economic drivers. However, this sensitivity is often coupled with a low adaptive capacity and thus renders mountain regions particularly vulnerable to future environmental changes. Thus, mountain ecosystems and the societies that live in mountains or are dependent upon mountain resources are potentially at risk. Predominantly marginalized people in developing countries with lower adaptive capacities will be most vulnerable to the socio-economic impacts of global change.

In many parts of the world, mainly in the arid tropical and subtropical regions, mountains are the primary and sometimes the only source of freshwater. As these regions are home to more than half of humanity, there is great concern over potential changes in mountain watershed hydrology and how these will affect freshwater supply, particularly in the face of growing per capita demand. The world's population tripled during the 20th century, while at the same time water withdrawals increased by more than sixfold. Moreover, freshwater availability is shrinking due to pollution of many water resources, also in lowlands, overexploitation of aquifers, and adverse effects of climate change. Currently, about 70% of the world's available freshwater is used for agriculture, and the United Nations has forecasted that water scarcity will be the chief constraint to increased food production in the coming decades. It is clear from these facts that water represents a crucial link between many of the world's problems. As a key resource in food production and mobile resource crossing borders of ecosystems and nations, water bears a high potential for conflicts between countries, and between highlands and lowlands.

2. Paleoenvironmental changes

Scientists need to compare contemporary short-term changes with past long-term natural variability to assess the extent of human influence on mountain environments. The view from the past helps to understand the natural and human processes that drive environmental change and to develop future scenarios. For quantitative approaches to the reconstruction of past climatic variability, continuous and well-dated time series are required to evaluate the magnitude and rate of past changes in mountains.

The paleo-environmental articles in Part 2 of this book cover topics ranging from high-resolution reconstructions of environmental proxies (e.g. pollen, diatoms, and ice core chemistry) to the dating of glacial deposits and modeling of past glacier and ecosystem dynamics. Recurring themes in these articles are that: 1) excellent dating control is essential for the reconstruction of the change rate and the global patterns of past environmental variability; 2) the insights gained from multi-proxy approaches by far exceed those gained from the sum of individual records; and 3) the flow of paleo-environmental information from the scientific community to stakeholders and policy makers must be encouraged and facilitated. Two recent recoveries of fossils from mountain settings illustrate the value of a paleo perspective for placing modern environmental change in a long-term context.

During the 2002 field season, Lonnie Thompson and his research team, studying the ice core records in the Cordillera Blanca of Peru, discovered remarkably well-preserved wetland plants that had recently become exposed by retreating permanent snow banks along the edge of the Quelccaya ice cap. Even more remarkably, the plants yielded radiocarbon ages of ca. 5200 years BP. The exceptionally well-preserved condition of the plants makes clear that they were covered by ice during this entire period, i.e. the snow and ice cover in the Cordillera Blanca are currently less extensive than at any time during the last 5200 years.

Another witness of climate change is the human corpse that was discovered melting out of the ice at a 3120 m pass along the Austrian-Italian border in the Ötztal Alps in 1991. At first, the very well preserved condition of the corpse led to the presumption that the unfortunate person had died only recently, but it soon became apparent that the frozen corpse was the body of a Bronze Age man. Subsequent radiocarbon dating confirmed that "Ötzi" was in fact ca. 5300 years old. The climatic significance of Ötzi's emergence is that the current climatic conditions in the central Alps were unprecedented over the last five millennia.

These two examples are among many additional pieces of evidence suggesting that the late 20th century climatic conditions in mountain regions are extraordinary compared with the last millennia. Recent high-resolution reconstructions of Northern Hemisphere mean temperature indicate that the late 20th century warming is unprecedented over the last two millennia. It should therefore not come as a surprise that the signal is turning up at sensitive high-elevation sites where one of the most rapid responses can be expected. In fact, the Northern Hemisphere "hockey stick"-shaped record of the last 2000 years has much in common with long-term continuous records of climate change derived from high-elevation ice cores from tropical glaciers.

Paleoenvironmental information from mountain areas thus provides the benchmark to gauge current and anticipated climate change. The records indicate that global warming is significant and likely to accelerate in the coming decades, with profound effects on mountain regions. Globalization processes that are not related to climate change will exacerbate the consequences even more. If humanity were now taking concrete steps to minimize the potential damage of these combined impacts, there would be grounds for some optimism for the future state of the world's mountain regions.

3. Cryospheric changes

The changes that are currently impacting the cryosphere around the globe are the clearest indication of the dramatic environmental changes in mountain regions. The dynamics of Alpine glaciers, for example, are directly linked to climate, and the reaction time of Alpine glaciers is fast. According to the World Glacier Monitoring Service and the Global Terrestrial Network on Glaciers of GTOS/GCOS, first field results indicate that the extremely warm and dry weather in summer 2003 caused an average thickness loss of glaciers in the European Alps reaching about 3 m water equivalent, which is nearly twice as much as during the previous record year of glacier retreat 1998 (1.6 m). This is roughly five times more than the average loss of 0.65 m per year recorded during the exceptionally warm period 1980-2000, and about an order of magnitude more than the reconstructed average of the 20th century. During the summer of 2003, the glaciers of the Alps lost a staggering 5-10% of their mass to ablation (W. Haeberli, pers. communication). In other words, assuming that this retreat rate was sustained, and it is likely that the hot summer of 2003 becomes the norm rather than the exception (cf. Schär et al. 2004), the glaciers of the Alps would essentially vanish in just a few decades.

Numerous other glaciers of the Globe match the trend of Alpine glaciers that have already lost more than 25% of their volume in the 25 years before 2003 and roughly two thirds of their original volume since 1850. The extrapolation of current recession rates translates into drastically reduced mountain glaciers and permafrost by the end of this century. Joint efforts of long-term monitoring, process studies and modeling are a prerequisite for understanding cryospheric changes occurring in such a pervasive manner.

In its Third Assessment Report, the Intergovernmental Panel on Climate Change (IPCC) has produced a series of future climate scenarios based on several General Circulation Models. A common feature of these scenarios is that the anticipated warming of the next several decades is more pronounced in northern high latitudes. A less well-known feature of these IPCC scenarios is that the expected pattern of warming in the atmosphere is more pronounced at progressively higher elevations in the troposphere along a latitudinal gradient from the Arctic to approximately 30° south of the equator. This has profound implications for mountain regions, many of which are located in the high-altitude zone of anticipated enhanced warming (Fig. 1). A number of independent lines of evidence have indicated that the Arctic is already warming at a higher rate than other parts of the globe (e.g. Overpeck et al. 1997). The rapidly thinning Artic ice pack and increasing thaw penetration into permafrost are two prominent examples of climate change impacts in the western Arctic, which is warming at a rate three to five times faster than the rest of the world. The rapidly receding tropical mountains glaciers, such as Kilimanjaro (Kenya), Quelccaya (Peru) and Dasuopu glacier (Tibet), suggest that enhanced warming may also be occurring at high-elevation sites in low latitudes, which would be consistent with the IPCC scenarios. In stark contrast, only few meteorological stations are situated at high altitudes. The transect along the crest of the North and South American Cordillera (Fig. 1), for example, shows a large observational "data gap" in mountain regions between approximately 40° N and 30° S. Although this gap is filled (at least in part) by climatic information obtained from radiosonde measurements, there remain significant discrepancies between radiosonde data and mountain surface data. Seidel and Free (2003) documented greater warming at tropical mountain locations than at similar altitudes above low elevation stations, which were based on radiosonde information. Figure 1 contains a particularly important message for new research

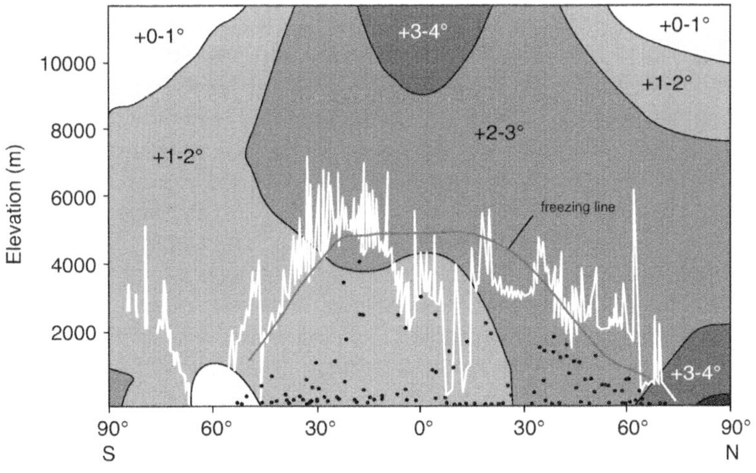

Figure 1. Projected changes in mean annual temperature with 2 x CO_2 along the Cordillera transect. The white line show the highest peaks and the black line the mean annual freezing line. The black dots indicate the existing GCOS stations (Büchler et al. 2004).

strategies and challenges: Assuming that the zone of maximum temperature change, based on 8 General Circulation Model simulations with a doubling of CO_2 (IPCC 2001; R. Bradley, pers. communication), extends from low elevation in the Arctic and higher latitudes to high elevation in the tropics and southern hemisphere-subtropics, then we have to create a network of observatories in the high mountains of all climatic zones, but especially in the tropics and subtropics, to reach this belt of stronger warming. Such a network of observatories could play the important role of an early warning system for climatic and hydrological changes concerning natural processes and human activities in the 21st century. In other words, and as reflected in many vision statements of this book, the detailed information required to assess climatic impacts in mountain environments requires direct measurement at high-elevation sites and thus, additional high-elevation climate monitoring stations.

4. Climatic and hydrological changes

Climate change means not only global warming, but also changes in the pluvial and hydrological regime. This is most important in the tropics and subtropics that host more than two thirds of the world's population. In these regions, where monsoon or mediterranean, seasonal and highly variable climates dominate, mountains play a fundamental role from a hydrological point of view. For future research strategies, the statement of Wallström et al. (2004) must be of highest interest: "Poor access to fresh water means that more than two billion people currently live under what experts call 'severe water stress'. With population growth and economic expansion, this figure is expected to nearly double by 2025. Climate Change would further exacerbate this situation [...] but [...] the impacts of global change are equally complex, as they often combine with local and regional environmental stresses in unexpected ways."

In this vein, we need to pay more attention to the mountains as "water towers" for food production and water supply (Fig. 2 and 3). Sound measurements are scarce and cover only short time periods. Often, the understanding of the spatial and temporal heterogeneity of runoff and discharge conditions in mountain areas is unsatisfactory. Global change research would be quite incomplete if the scientific community should be unable to focus its programs and projects much more strongly on this endangered and vulnerable part of our planet, especially on the most sensitive hydrological systems of the mountains. Viviroli et al. (2003) characterized 19 river basins in various parts of the world with discharge characteristics, specific runoff data and other variables (Fig. 2). The particular hydrological processes of mountain areas are manifested by disproportionally large discharge, typically about twice the amount that could be expected from the areal extent of the mountain section. Thus, mountains account for 20-50% of total discharge in humid areas, while in semi-arid and arid regions, the contribution of mountains to total discharge is as high as 50-90%, with extremes of more than 95%. Moreover, discharge from mountain regions is highly reliable and causes significant reduction of the coefficient of variation of total discharge. Storage of winter precipitation in the form of snow and glaciers, thus transforming winter precipitation into spring and summer runoff, is often essential for land use in the lowlands. Figure 3 shows an overall assessment of the hydrological

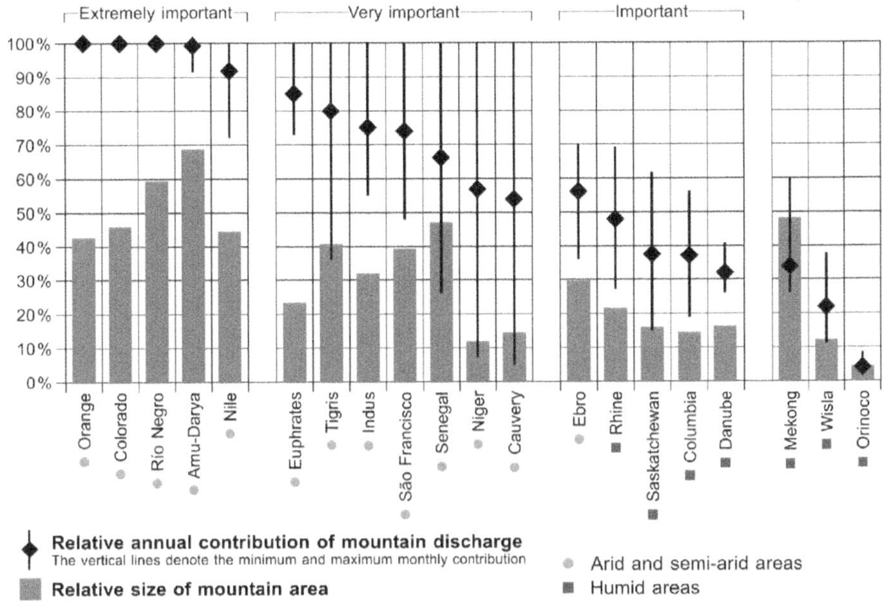

Figure 2. Mean annual mountain contribution to total discharge of freshwater (with an indication of the minimum and maximum monthly contribution) and proportion of mountain area (represented by a gauging station in the vicinity of 1000 m asl on the condition, that there is a mountain relief) relative to the entire catchment for the selected river basins (Viviroli et al. 2003).

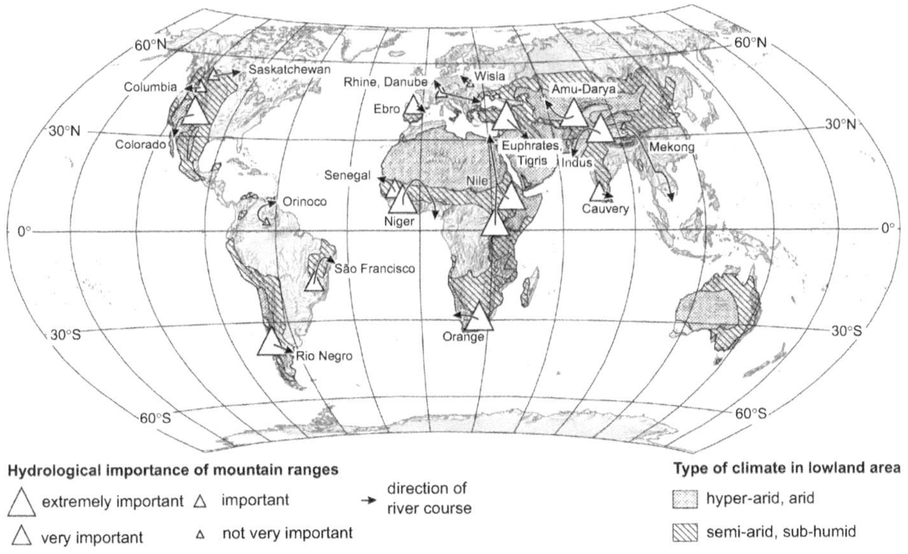

Figure 3. Hydrological significance of mountain ranges for the selected river basins in Figure 2 (Viviroli et al. 2003).

significance for large lowland areas revealing that the world's major "water towers" are found in semi-arid and arid zones, i.e. the drier the lowlands, the more important are the more humid mountain areas.

In the mountains of the middle to northern latitudes, varying snow cover and changing hydrological regimes are evident problems apart from increasing warming and shrinking glaciers. From observations in the Canadian Rocky Mountains we know, for example, that snowpack and glaciers trap volatile organic contaminants from as far away as Eurasia. The deposition of such substances generally increases with elevation. Persistent organochlorine compounds accumulated by 10- to 100-fold between less than 1000 m to more than 3000 m of altitude. The volatilization in relatively warm locations and the condensation in cooler environments lead to enhanced concentrations at high altitudes and high latitudes (Blais et al. 1998). Climate warming may add to the release of many chemicals stored in glaciers, including DDT and other pesticides. The deposited substances thus may enter freshwater food chains where they bio-accumulate to high concentrations. For example, in the Rocky Mountains glaciers and snowpack form the basis for human populations and most ecosystems in the western prairies of Canada. In the case of Canada, at present neither federal nor provincial governments ensure adequate protection, and the scientific basis for the management of cumulative human effects is inadequate (Schindler 2002).

The Canadian example shows that climatic and hydrological changes are fundamental problems for future sustainable mountain development in both the developing and in the developed, the more arid and the more humid regions of our planet. As the chapters in this book show, disciplinary research in mountain regions played a prominent role in the past and will continue to do so in future. Thus, developing hydrological models for the integrated analysis of climatic and biogeophysical conditions at various spatial and temporal scales is still a big challenge. Moreover, the close relationship between natural processes and human activities, especially in mountain areas, has created an increasing demand for integrated natural-human science projects. Inter- or transdisciplinary projects are urgently needed to interlink natural variability and human interferences in highland-lowland systems, combining natural, social-cultural as well as health and engineering sciences with the participation of the local population and policy-makers.

5. Ecological changes

Most of the goods and services from mountain regions are mediated, moderated or produced by mountain ecosystems (cf. Becker and Bugmann 2001). In landscapes characterized by high gravitational energy, ecosystems and their components have a major role for counter-acting this energy, e.g. by maintaining slope stability (Fig. 4) and by reducing the risk of natural hazards such as landslides, mud and debris flows, snow avalanches and other associated damages.

Global changes have multiple effects on mountain ecosystems on a wide range of temporal and spatial scales. Therefore, no single approach or methodology would be able to provide a comprehensive assessment of these effects on mountain ecosystems and societies. Rather, a diversity of complementary tools, methods, and approaches is required. Thus, this diversity of approaches is not evidence of a research community that cannot agree on common protocols and methods, but it is necessary due to the nature of the complex systems we are trying to understand, and to manage in a sustainable manner.

Concerning the linkages between the structure and the function of terrestrial ecosystems (Fig. 4), we need to better understand the associated processes, including the role of species-specific ecological and evolutionary differences. Thus, research on species-specific patterns and processes is highly welcome, and has made considerable progress over the past years (e.g. studies that use "natural experiments" such as elevational gradients, slope/aspect differences at a given elevation, or manipulative experiments). We are convinced that comparative studies using different methods will be the most fruitful approach to elucidate the potential and limitations of each method with respect to our ability to extrapolate the results in time and space.

As a link between different ecosystems, water plays a particular role in mountain regions. Water integrates processes across a mountain catchment, at multiple spatial and temporal scales, and from the headwater down to the oceans. Thereby, it connects the dynamics and fate of terrestrial and aquatic ecosystems in mountain regions. The present volume confirms, for instance, that aquatic biota in mountain streams and lakes can serve as powerful indicators of environmental changes. These aquatic systems are quite sensitive to environmental changes – thus, we may experience losses of biodiversity in mountain lakes that are largely unnoticed by the casual observer and the general public. Simultaneously, evidence is accumulating of the strong impacts of human activities – through N fertilization from urban sprawl and industrial development – on terrestrial and aquatic ecosystems in mountain catchments. What may be particularly worrying is that several simultaneous environmental changes

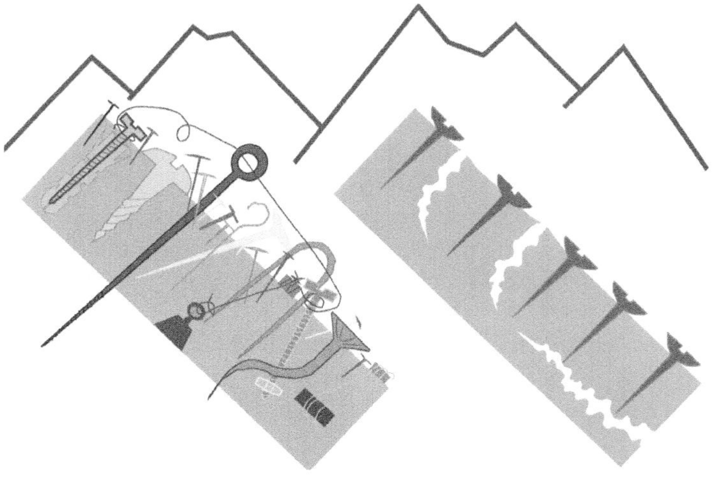

Figure 4. The role of mountain plants for ensuring slope stability. Changes in ecological processes induced by Global Change are likely to have repercussions for slope stability and thus for the integrity of ecosystems and human societies in mountain regions as well as in downstream areas (from Körner et al. 2001).

may have unprecedented negative results. Consequently, this calls for caution against extrapolations of ecosystem behavior from studies that examine single-factor changes at a time only. Consequently, integration across environmental factors and processes should be a high priority in future research.

Instead of taking a "bottom-up" approach that tries to link the findings from a large number of small-scale studies, one can adopt a landscape-ecological view considering an entire mountian region. Several contributions in this book document the beginning of a new research era that can and should integrate the findings from catchment-scale observations and experiments with remote sensing data and modeling activities. These rapidly evolving tools and methods allow us to tackle the problem of the impacts of Global Changes on mountain regions based on a "top-down" approach that complements rather than competes with the more traditional "bottom-up" approaches. This allows us to assess issues such as biodiversity changes and natural hazards (cf. Fig. 4) from a more integrative point of view. For example, from such landscape-scale studies it is likely that the role of mountain regions for the global carbon balance will emerge more clearly.

Ecological changes have often been assessed from a pure natural science point of view. However, land-use practices have become globally pervasive over the past few decades, and they typically have strong impacts on ecological dynamics in mountain catchments. As a matter of fact, land use changes will probably be the most important agents of ecological change in mountain ecosystems (and beyond mountain regions as well) for the coming few decades. It is evident that a lot is known on this topic, but this knowledge needs to be merged with the insights gained on the impacts of the other drivers of change.

Further, long-term data collection needs to complement short-term process studies and modeling efforts, even if many peers and particularly funding agencies may not view this as a high-profile scientific activity at the moment. We need to know the range of historical variability of ecological processes, and need data for testing the models used for providing scenarios of the future environmental, ecological, and societal conditions in mountain regions. Several papers in the present volume provide insights and guidelines on how to achieve efficient, modern monitoring programmes that will have multiple benefits to user groups. The upward shift of alpine vegetation is becoming globally pervasive (e.g. Walther et al. 2002; Root et al. 2003); even though it is not yet possible to prove that these shifts are a first consequence of anthropogenic warming, they are consistent with scientific expectations regarding the impacts of Global Change, and thus are a serious source of concern.

Over the past years, many sub-disciplines addressing ecological changes in mountain regions contributed valuable insights. The research results obtained to date document that the potential impacts of the drivers of change on mountain ecosystems are relatively well known, and that mountain ecosystems are often sensitive to these changes. However, an integrative picture of the joint effects of various drivers of Global Change on landscapes and the relationship to sustainable land use practices still needs attention. Thus, future research efforts should be geared towards integration, to achieve a comprehensive picture of the ecological impacts of Global Changes on mountain ecosystems.

6. The human dimension of Global Change

The first part of this book describing paleoenvironmental, cryospheric, climatic, hydrological and ecological changes shows how global change scientists take stock of the high sensitivity of mountain regions. The importance of high-elevation areas for global change research is also reflected in the widely established observation, monitoring and reporting networks for natural systems in mountain areas. In strong contrast, we lack an analogue on the social systems side, especially in its linkages with natural systems. Despite the wide recognition that the human dimensions of Global Change are at least as important as the biophysical realm, the translation of this into active, interdisciplinary research is much harder to achieve. Such translation, however, is crucial for two reasons. First, because the phenomenon of Global Change is a manifestation of the fact that the current trajectory of human activities is not sustainable at the global scale. Second, because the increasing impact of human actions is no longer local or regional, but has reached a global dimension. In other words, in contrast to the disconnect between the natural and social research communities, global environmental change has tremendous implications for the people living in these highly sensitive regions. *Vice versa*, the land use and land cover change triggered by these people can be fundamental for the hydrological regime of mountain regions and will also affect downstream water resources. Since the middle of the 20th century, such human activities are clearly dominant over other aspects of Global Change.

In this context, it becomes clear that global change projects must consider mountain regions in the first place as the home for several hundred millions of people, most of them living in difficult conditions. About 90% (663 million) of the 720 million mountain people in the world live in developing and transition countries (Huddleston et al. 2003). Many of them have to cope with strong internal and external pressures that often result in conflicts over control of increasingly scarce resources. Against this background, it is obvious that the sensitivity of mountain regions must be translated with fragility or vulnerability when dealing with the human dimension of Global Change.

Subsistence agriculture in mountains may best illustrate the vulnerability of farmers having to cope with both the unprecedented impacts of global change and high fragility. In lowland production areas, where sufficient irrigation and nutrients can be provided, farmers may benefit from increased CO_2 levels and a temperature increase that boost crop yield. In mountain regions, however, the effects of climate change may vary from place to place, and a higher incidence of extreme weather events may affect production on rain-fed slopes without access to irrigation and fertilizers. A failure of timely and sufficient rainfall directly results in crop loss and food insufficiency. Moreover, as temperature and CO_2 availability influence plant phenology and nutrient contents, the change in agroecosystems might trigger an increased level of field and storage pests threatening the household food security of mountain farmers. Hence, the direct and indirect impacts of a changing climate may further marginalize mountain farmers that already operate in a high-risk system separated from mainstream economies.

The performance of agriculture is a crucial factor in determining the degree of vulnerability to food insecurity. High vulnerability and declining resilience have become typical features of many mountain livelihood systems. More than half of the mountain population in developing and transition countries (250-370 million people) is vulnerable to food insecurity (FAO 2002; Huddleston 2003). Of those, 87% live below 2500 m asl where population pressure and increasing demand for grazing land are creating serious sustainability problems for mountain environments and the livelihood systems of the inhabitants. In high mountain areas, the total number of vulnerable rural people is small, yet they represent almost 70% of the population living above 2500 m. The higher prevalence of vulnerability at higher elevations attracts attention to these areas although the total number of vulnerable people is seven times greater at lower elevations (Fig. 5; Huddleston 2003). All together, food insecurity can lead to migration and overuse of natural resources, which again affect local climate, biodiversity, soil erosion, and runoff processes.

These numbers make clear that, when dealing with the human dimension of global change in mountain regions, it is crucial to take vulnerability issues into account. Effective human dimension research in global change must come to grips with the special difficulties of understanding interactions between complex dynamic systems. The close relationship between natural processes and human activities in mountain areas has created a high demand for integrated natural-social science projects to furnish the knowledge necessary for addressing sustainable development issues. Interdisciplinarity is the classical approach for scientific cooperation, but very often we praise interdisciplinarity while we still promote disciplinarity. Transdisciplinarity is much more; it is a joint problem-solving operation that involves science, technology, and society.

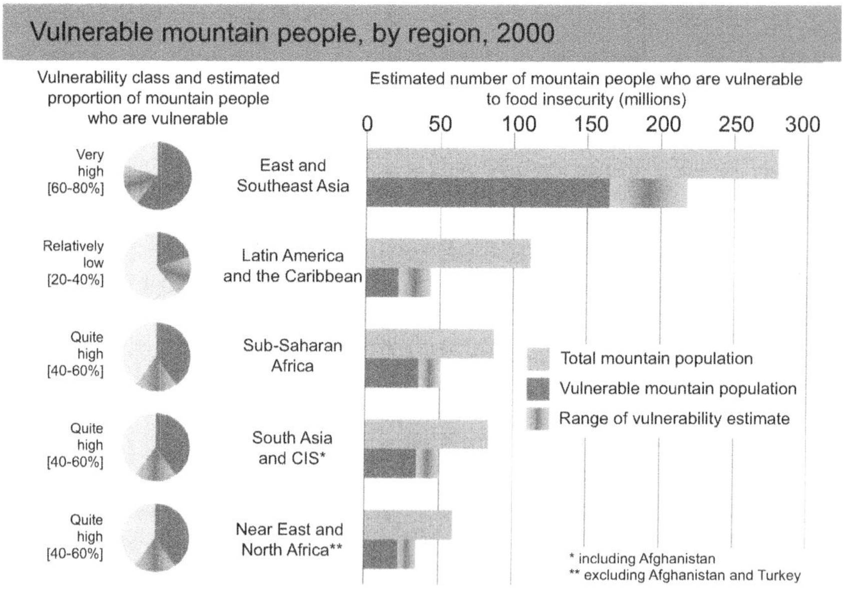

Figure 5. Vulnerable mountain people by region (FAO 2002).

7. Concluding remarks

Each contribution to this State of Knowledge Overview on "Global Change and Mountain Regions" witnesses the change of our planet from a different perspective. Despite their various approaches and views, many contributions show that in recent decades, many environmental indicators have moved outside the range in which they have varied in the past hundred thousand years or even longer; be it the increase in atmospheric carbon dioxide or temperature, glacier retreat, population growth rates or the clearance of tropical forests.

The attempt to allocate the articles to Parts 2-6 of this book made clear that disciplinary thinking has become obsolete in view of the complexity of the Earth system. Even if basic research will remain very important, we must keep in mind that the impacts of Global Change in mountain regions are complex, as they combine global, regional and local natural- and human-induced processes in unexpected ways. This is particularly true as humans became a significant and sometimes dominating environmental force. Never before has the Earth experienced the current suite of simultaneous changes challenging the political decision-makers by its uncertainty, its complexity and its magnitudes and rates of change.

The first step towards meeting the challenge presented by Global Change is to focus on a specific subsystem of the Earth system, such as mountain areas, while acknowledging the complex nature, the ways in which humans are affecting the system, and the economic and societal consequences. The latter is particularly crucial when dealing with often marginalized mountain people that are most vulnerable and exposed to the impacts of Global Change. The second step is the establishment and further development of a dialogue between global change scientists, policy-makers and stakeholders to communicate current knowledge and guide future research. International initiatives such as the MRI may make an essential contribution towards meeting these goals.

8. Acknowledgements

Putting together this book has been like taking a journey across a full range of climates, continents and altitudes. During the ascent, the summit has not always been in sight. Thus, without the spirit and support of the MRI team and the patience of the contributing authors we still would be stuck very near the base camp.

We are grateful that more than sixty lead authors were convinced that this book is worthwhile a contribution. A large number of peer-reviewers provided crucial comments to make this adventure a success. The synthesis chapter gained weight through the input of Ray Bradley who repeatedly elaborated on his ideas of filling data gaps in high altitude regions. Last but not least, we wish to thank Thomas Hofer for placing the broad variety of contributions in a political and institutional framework of Global Change in mountain regions.

An adventure treck such as this State of Knowledge Overview would not have been feasible without the financial support of the MRI and the Swiss Academy of Natural Sciences (SANW), Bern, whose grants are gratefully acknowledged.

9. References

Becker, A. and Bugmann, H., eds. (2001). "Global Change and mountain regions: The Mountain research Initiative." IGBP Report 49, Stockholm.

Blais, J. M., Schindler, D. W., Muir, D. C. G., Donald, D. B., and Rosenberg, B. (1998). Accumulation of persistent organochlorine compounds in mountains of western Canada. *Nature* **395**, 585-588.

Büchler, B., Bradley, R. S., Messerli, B., and Reasoner, M. A. (2004). MRI Newsletter 3: Understanding climate change in mountains. *Mountain Research and Development* **24**, 176-177.

FAO (2002). "Environment, poverty and food insecurity: The vulnerability of mountain environments and people." Special Feature, FAO, Rome.

Huddleston, B., Ataman, E., and d'Ostiani, L. F. (2003). "Towards a GIS-based analysis of mountain environments and populations." Environment and Natural Resources, Working Paper 10. FAO, Rome.

IPCC (2001). Intergovernmental Panel on Climate Change. Vol.: Climate Change, the Scientific Basis. WMO / UNEP, IPCC Secretariat, Paris.

Körner, C., Spehn, E., and Messerli, B. (2001). Mountain Biodiversity matters: Executive summary of the Global Mountain Biodiversity Assessment Conference 2000, Rigi-Kaltbad, Switzerland.

Messerli, B., and Ives, J. D., Eds. (1997). "Mountains of the World: A global priority." Parthenon, Lancaster.

Meybeck, M., Green, P., and Vörösmarty, Ch. (2001). A new typology for mountains and other relief classes: An application to global continental water resources and population distribution. *Mountain Research and Development* **22**, 34-45.

Overpeck, J., Hughen, K., Hardy, D., Bradley, R., Case, R., Douglas, M., Finney, B., Gajewski, K., Facoby, G., Jennings, A., Lamoureaux, S., Lasca, A., MacDonald, G., Moore, J., Retelle, M., Smith, S., Wolfe, A., and Zielinski, G. (1997). Arctic environmental change of the last four centuries. *Science* 278, 1251-1256.

Root, T. L., Price, J. T., Hall, K. R., Schneider, S. H., Rosenzweig, C., and Pounds, J. A. (2003). Fingerprints of global warming on wild animals and plants. *Nature* **421**, 57-60.

Schär, C., Vidale, P. L., Lüthi, D., Frei, C., Häberli, C., Liniger, M. A., and Appenzeller, C. (2004). The role of increasing temperature variability in European summer heatwaves. *Nature* **427**, 332.

Schindler, D. (2002). Freshwater in the Rocky Mountains of Alberta as indicators of the cumulative impacts of climate change and human activities. *In* "Mountain Science Highlights" Nr. 6. The Banff Centre, Canada.

Seidel, D. J., and Free, M. (2003). Comparison of lower-tropospheric temperature climatologies and trends at low and high elevation radio-sonde sites. *Climatic Change* **59**, 53-74.

Viviroli, D., Weingartner, R., and Messerli, B. (2003). Assessing the hydrological significance of the world's mountains. *Mountain Research and Development* **23**, 32-40.

Wallström, M., Bolin, B., Crutzen, P., and Steffen, W. (2004). "The Earth's life-support system is in peril." International Herald Tribune, Editorial page. January 20.

Walther, G.-R., Post, E., Convey, P., Menzel, A., Parmesan, C., Beebee, T. J., Fromentin, J. M., Hoegh-Guldberg, O., and Bairlein, F. (2002). Ecological responses to recent climate change. *Nature* **416**, 389-395.

ADVANCES IN GLOBAL CHANGE RESEARCH

VOLUME 23

The titles published in this series are listed at the end of this volume.

GPSR Compliance

The European Union's (EU) General Product Safety Regulation (GPSR) is a set of rules that requires consumer products to be safe and our obligations to ensure this.

If you have any concerns about our products, you can contact us on ProductSafety@springernature.com

In case Publisher is established outside the EU, the EU authorized representative is:

Springer Nature Customer Service Center GmbH
Europaplatz 3
69115 Heidelberg, Germany

Batch number: 10755458

Printed by Printforce, the Netherlands